Large Sample Inference *for* Long Memory Processes

Large Sample Inference *for* Long Memory Processes

Liudas Giraitis

Queen Mary, University of London, UK

Hira L Koul

Michigan State University, USA

Donatas Surgailis

Vilnius University, Lithuania

Imperial College Press

ICP

Published by

Imperial College Press
57 Shelton Street
Covent Garden
London WC2H 9HE

Distributed by

World Scientific Publishing Co. Pte. Ltd.
5 Toh Tuck Link, Singapore 596224
USA office: 27 Warren Street, Suite 401-402, Hackensack, NJ 07601
UK office: 57 Shelton Street, Covent Garden, London WC2H 9HE

British Library Cataloguing-in-Publication Data
A catalogue record for this book is available from the British Library.

LARGE SAMPLE INFERENCE FOR LONG MEMORY PROCESSES

ISBN-13 978-1-84816-278-5
ISBN-10 1-84816-278-2

Printed in Singapore by World Scientific Printers.

Preface

A discrete-time stationary stochastic process is said to have long memory if its autocovariances tend to zero hyperbolically slowly as the lag tends to infinity, but their sum diverges. Such processes have unbounded spectral densities at the origin. This is unlike the so-called short memory processes where autocovariances are summable, often tending to zero exponentially fast and whose spectral densities are bounded at the origin.

Ever since the work of Hurst (1951, 1956), the proponent of the Aswan High Dam, more and more scientists have found the presence of long memory in their data. Hurst noticed that measurements on the Nile River were not consistent with the assumption of independence. Mandelbrot and van Ness (1968) and Mandelbrot and Wallis (1969a,b) were the first to provide a theoretical justification for this by advancing the idea that the data observed by Hurst follow a long memory process. They proposed the use of fractional Brownian motion, as opposed to classical Brownian motion, to model and analyze various phenomena in hydrology, and more generally, for modeling a long memory time series. The review paper of Lawrance and Kottegoda (1977) nicely summarizes this fact and some of the other stochastic models used in the modeling of river-flow time series.

Box and Jenkins (1970) popularized the idea of obtaining a stationary time series by differencing the given, possibly non-stationary, time series. Numerous time series in economics are found to have this property, i.e., even though the initial time series is not stationary, its dth-order difference, for some positive integer d, is stationary. Subsequently, Granger and Joyeux (1980) and Hosking (1981) found examples of time series whose fractional difference is a short memory process, in particular, white-noise, while the initial series has unbounded spectral density at the origin. Beran (1992a) gives examples from numerous other sciences where data follow long memory. Baillie (1996) cites several references that demonstrate the dramatic empirical success of long memory processes in modeling the volatility of the asset prices and power transforms of stock market returns while Ding and Granger (1996) point out the long memory property of the absolute values and the squares of S&P 500 daily stock market index returns. Willinger, Paxson, Riedi and Taqqu (2003) discuss the importance of long memory processes in network traffic data.

For classical time-series analysis the text of Brockwell and Davis (1991) is an excellent source. The monograph of Beran (1994) provides a nice introduction to some basic notions and applications of long memory processes. However, there have been significant advances in theoretical aspects of long memory processes since the mid-1990s that need to be made available in a unified fashion in one place. The monographs by Doukhan, Oppenheim and Taqqu (2003), Dehling, Mikosch and Sørensen (2002), and Robinson (2003) consist of collections of papers that discuss and review various theoretical results of long memory processes and their applications, while that of Teyssière and Kirman (2007) contains a collection of papers that emphasize the presence of long memory in economics and finance. The text of Palma (2007) summarizes some statistical theory and applications of these processes. Some theoretical aspects of asymptotic theory for long memory processes are discussed in Chapter 5 of the monograph by Taniguchi and Kakizawa (2000). Various connections of long memory with non-stationary and regime switching processes, self-similar processes, and the Hurst phenomenon are outlined in Samorodnitsky (2007).

At present there is a need for a text where an interested reader can methodically learn some basic asymptotic theory techniques found useful in the analysis of statistical inference procedures for long memory processes. This text makes an attempt in this direction. Our goal here is to provide in a concise style a text at the graduate level summarizing theoretical developments both for short and long memory processes and their applications to statistics. It also contains some real-data applications and mentions some unsolved inference problems for interested researchers in the field at the time of writing this monograph.

This book can be used by doctoral students needing to familiarize themselves with the detailed proofs and derivations. It can also be used as a source of theoretical tools for further investigations in econometrics and statistics and as a theoretical background for practical applications and modeling. Parts of the text can also be used by students in Masters' programs in statistics or econometrics.

The literature on long memory processes is vast. We were influenced by numerous works and the ideas of researchers working in the field. We would like to thank our teachers, colleagues, co-authors, and students. In particular, we are grateful to Rainer Dahlhaus, Ronald L. Dobrushin, Peter C.B. Phillips, Peter M. Robinson, and Murad S. Taqqu. We thank Violetta Dalla for providing some of the graphs and simulations included in this text, and Natalia Bailey and Remigijus Leipus for careful reading of the manuscript. We are also grateful to Shama Koul and Rūta Surgailienė for their patience and support during this long quest. We thank Shama and Ajeet Koul for hosting numerous dinners and facilitating a conducive atmosphere for the completion of this project.

The Department of Statistics and Probability at Michigan State University, the School of Economics and Finance of Queen Mary, University of London, and the Institute of Mathematics and Informatics, Vilnius University, have been particularly generous in providing excellent support and working environments.

During the preparation of this monograph the authors were supported in part by the NSF DMS Grants 00-71619, 07-04130, ESRC Grant RES062230790, and a grant from the Lithuanian State Science and Studies Foundation and the Research Council of Lithuania.

Liudas Giraitis
School of Economics and Finance
Queen Mary, University of London

Hira L. Koul
Department of Statistics and Probability
Michigan State University

Donatas Surgailis
Institute of Mathematics and Informatics
Vilnius University

February 10, 2012

Contents

Preface   v

Notation and Conventions   xv

1. Introduction   1

2. Some Preliminaries   7

    2.1   Preliminaries from trigonometry   7
    2.2   Stationarity, spectral density, and autocovariance   9
    2.3   Slowly and regularly varying functions   18
    2.4   Hermite and Appell polynomials   22
    2.5   Miscellaneous preliminaries   29

3. Long Memory Processes   33

    3.1   Characterization of short and long memory processes . . .   33
    3.2   Wold decomposition and linear processes   38
    3.3   Asymptotics of the variance   42
    3.4   Self-similarity, long memory, and fractional Brownian
        motion   45
    3.5   Other classes of long memory processes   52

4. Limit Theory for Sums   55

    4.1   Introduction   55
    4.2   Cumulant criterion for CLT   56
    4.3   CLT for sums of a linear process   62

4.4   Partial-sum process   73

4.5   More results for weighted sums   80

4.6   Sums of a transformed Gaussian process   88

4.7   Non-central limit theorems   101

4.8   Limit theorems for polynomial forms   107

5.   Properties of the DFT and the Periodogram   111

5.1   Introduction   111

5.2   Covariances of the DFT   112

5.3   Some other properties of the DFT and the
      periodogram   123

6.   Asymptotic Theory for Quadratic Forms   132

6.1   Introduction   132

6.2   Asymptotic normality for sums of weighted
      periodograms   133

6.3   Asymptotic normality of quadratic forms   157

      6.3.1   Stronger sufficient conditions for asymptotic
              normality   168

7.   Parametric Models   174

7.1   Introduction   174

7.2   ARFIMA models   174

7.3   GARMA models   185

7.4   Simulation of ARFIMA processes   191

7.5   Aggregation, disaggregation, and long memory   197

8.   Estimation   202

8.1   Introduction   202

8.2   Parametric model: consistency   204

8.3   Parametric model: asymptotic normality   213

      8.3.1   Parametric models: asymptotic normality of
              Whittle estimators   213

      8.3.2   Whittle estimators in ARFIMA models   219

8.4   Semiparametric models   221

       8.4.1 R/S estimation method 222

       8.4.2   Log-periodogram estimator 225

  8.5 Local Whittle estimators 226

       8.5.1   Semiparametric models: consistency of local
                 Whittle estimators 229

       8.5.2   Local Whittle estimators for linear processes . . . 240

  8.6 Signal plus noise processes 245

       8.6.1   Estimation in signal+noise 246

       8.6.2   Application to a stochastic volatility model 255

  8.7 Monte Carlo experiment 258

9. Elementary Inference Problems   265

  9.1 Introduction 265

  9.2 Linear-trend model 266

  9.3 Estimation of the mean μ 272

  9.4 Estimation of the long-run variance 275

  9.5 Testing for long memory and breaks 285

10. Empirical Processes   305

  10.1 Introduction 305

  10.2 Uniform reduction principle 307

      10.2.1 URP: Gaussian case 308

      10.2.2 URP: moving-average case 314

  10.3 Expansion of the empiricals 328

      10.3.1 Expansion of empiricals: Gaussian case 329

      10.3.2 Expansion of empiricals: moving-average case . . 336

11. Regression Models   352

  11.1 Introduction 352

  11.2 First-order asymptotics of M and R estimators 354

      11.2.1 CLT for the LS estimator 359

      11.2.2 First-order asymptotics of M estimators 361

      11.2.3 First-order asymptotics of R estimators 369

  11.3 Higher-order asymptotics of M estimators 376

  11.4 Whittle estimator in linear regression 382

11.5   Non-linear regression with random design   394
11.6   A simulation study   412

12. Non-parametric Regression   415

12.1   Introduction   415
12.2   Uniform design   415
       12.2.1   Some preliminaries for $\hat{\mu}$ 417
       12.2.2   Asymptotic properties of $\hat{\mu}$ 424
       12.2.3   Uniform rate of convergence   425
12.3   Asymptotics of $\hat{\sigma}^2$ 429
       12.3.1   Some preliminaries for $\hat{\sigma}^2$ 430
       12.3.2   MSE and asymptotic distribution of $\hat{\sigma}^2$ 437
12.4   Estimation of $\mathrm{Var}(\hat{\mu}(x))$ 439

13. Model Diagnostics   446

13.1   Introduction   446
13.2   Lack-of-fit tests   447
       13.2.1   Testing for H_0: martingale-difference errors   449
       13.2.2   Lack-of-fit tests: LM moving-average errors   455
13.3   Long memory design and heteroscedastic errors   464
       13.3.1   Lack-of-fit tests: heteroscedastic LMMA errors . .   476
       13.3.2   A Monte Carlo simulation   479
       13.3.3   Application to the forward premium anomaly . .   482
       13.3.4   Application to a foreign exchange data set   484
13.4   Testing a subhypothesis   488
       13.4.1   Random design   490
       13.4.2   Non-random design   501
13.5   Goodness-of-fit of a marginal distribution function   505

14. Appendix   519

14.1   Appell polynomials, Wick products, and diagram
       formulas   519
14.2   Sums of Appell polynomials   528
14.3   Multiple Wiener–Itô integrals   531
14.4   Limit of traces of Toeplitz matrices   545

15. Bibliography 554

Author Index 571

Subject Index 575

Notation and Conventions

The following notation and conventions are used throughout the book. The symbol " := " stands for "by definition". All limits are taken as $n \to \infty$, unless specified otherwise.

$$\mathbb{R} := (-\infty, \infty), \quad \bar{\mathbb{R}} := [-\infty, \infty], \quad \mathbb{R}^+ := [0, \infty).$$

$$\mathbb{Z} := \{0, \pm 1, \pm 2, \cdots\}, \quad \Pi := [-\pi, \pi], \quad \mathbf{i} := (-1)^{1/2}.$$

$a.e. :=$ almost everywhere.

$a.s. :=$ almost surely.

$a \wedge b := \min(a, b), \quad a \vee b := \max(a, b).$

$$B(a, b) := \int_0^1 u^{a-1}(1-u)^{b-1} du, \ a \wedge b > 0.$$

$\mathrm{CLT} :=$ the central limit theorem.

$\mathcal{C}(A) :=$ class of continuous functions defined on a set A.

$\mathrm{C\text{–}S} :=$ the Cauchy–Schwarz inequality.

$\mathrm{DCT} :=$ the dominated convergence theorem.

$\mathrm{ET} :=$ the ergodic theorem.

$\mathrm{d.f.} :=$ distribution function.

$\|g\|_\infty :=$ the supremum norm over the domain of a function g.

$I(A) :=$ indicator of the set A.

$\mathrm{i.i.d.} :=$ independent identically distributed.

$\mathrm{l.h.s.} :=$ left-hand side.

$L_p(A) :=$ class of p-integrable functions defined on a set A, $p \in \mathbb{R}$.

MSE := mean-squared error.

$\mathcal{N}_q(\mathbf{0}, \mathbf{C})$:= a q-variate normal distribution with the mean
vector $\mathbf{0}$ and the covariance matrix $\mathbf{C}$.

$\mathcal{N}(0, 1) := \mathcal{N}_1(0, 1)$.

$\quad \Phi$:= the d.f. of $\mathcal{N}(0, 1)$ r.v.

r.h.s. := right-hand side.

$\quad$ r.v. := random variable or vector.

$O_p(1)$:= a sequence of r.v.'s that is bounded, in probability.

$o_p(1)$:= a sequence of r.v.'s converging to zero, in probability.

sgn($\cdot$) := sign function.

$\quad \lfloor x \rfloor$:= the largest integer not greater than x.

$\quad [x] := \lfloor x \rfloor, \ x \geq 0; \ \lfloor x \rfloor + 1, \ x < 0$.

$\quad z_\alpha := (1 - \alpha)$th percentile of $\mathcal{N}(0, 1)$ distribution, $0 \leq \alpha \leq 1$.

$\quad Z$:= a $\mathcal{N}(0, 1)$ r.v., unless mentioned otherwise.

w.r.t. := with respect to.

$u_p(1)$:= a sequence of stochastic processes converging to zero uniformly
over the time domain, in probability.

Equality of distribution is denoted by $=_D$, while $\to_p$ and $\to_D$, respectively, denote the convergence in probability and in distribution.

For a sequence of stochastic processes $\{Y, Y_n, \ n \geq 1\}$, $Y_n \to_{fdd} Y$ denotes the weak convergence of finite-dimensional distributions of Y_n to the corresponding finite-dimensional distributions of Y, and $Y_n \Rightarrow Y$ means that Y_n converges weakly to Y in the given topology.

For any two real sequences $a_n, b_n, n \geq 1$, $a_n \sim b_n$ denotes convergence $a_n/b_n \to 1$, and $a_n \propto b_n$ means that $C_1 \leq a_n/b_n \leq C_2$, for some $C_1, C_2 > 0$, as $n \to \infty$.

The kth derivative of a smooth function g is denoted by $g^{(k)}$, $k = 1, 2, \cdots$. Often we write $\dot{g} = g^{(1)}$, $\ddot{g} = g^{(2)}$.

Chapter 1

Introduction

In this text the dependence structure of a stationary process is described by its autocovariance and spectral density functions. In addition, it is assumed that the given stationary process has a linear structure with either i.i.d. or white-noise innovations. The rate of decay of the coefficients of a linear process determines the type of dependence, which may be weak or strong.

A discrete-time stationary stochastic process X_j, $j \in \mathbb{Z}$, with finite variance is said to have long memory if its autocovariances $\gamma(k) := \text{Cov}(X_0, X_k) \sim c_\gamma |k|^{-2d}$ tend to zero hyperbolically slowly in the lag k, for some $0 < d < 1/2$ and finite $c_\gamma \neq 0$. The sum of autocovariances of a long memory process diverges, and such processes have an unbounded spectral density at the origin. In contrast, a weakly dependent or short memory process has absolutely summable autocovariances that often tend to zero exponentially fast, and a continuous and bounded spectral density. Numerous classical methods that are useful in analyzing weakly dependent stationary time series are inapplicable to long memory processes. The ultimate statistical inference theory must be broad enough to accommodate both short and long memory processes.

The need for new statistical inference methods arises, for instance, from the fact that under long memory the sample mean estimate $\bar{X}_n := n^{-1} \sum_{j=1}^{n} X_j$ of the mean $\mu = EX_1$ of a linear process is consistent at the rate $n^{d-1/2}$, which is slower than the classical rate $n^{-1/2}$, while $n^{1/2-d}(\bar{X}_n - \mu)$ is still asymptotically normally distributed. Thus the use of the classical confidence intervals for μ based on $n^{1/2}$ scaling is unjustified.

In long memory processes, the dependence between the current observation and the one at a distant future is persistent, i.e., observations that are distant from each other continue to have a linear relationship. This fact alone leads to some surprising results. For example, under long memory, any other location invariant estimator $\hat{\mu}_n$ of μ is a first-order equivalent to $\bar{X}_n$, i.e., $n^{1/2-d}(\hat{\mu}_n - \bar{X}_n)$ tends to zero, in probability. In comparison, for weakly dependent processes, observations distant from each other are approximately uncorrelated, which in turn yields that the weak limit of $n^{1/2}(\hat{\mu}_n - \bar{X}_n)$ is non-degenerate Gaussian, for a large class of estimators $\hat{\mu}_n$. Similar first-order degeneracy is observed between the least-squares estimator and the robust estimators in linear and non-linear regression models when errors follow a long memory moving-average process. This is discussed in Chapters 10 and 11. A resounding conclusion from these results is that in the presence of long memory moving-average errors, there is no gain in using robust estimators of regression parameters over the least-squares estimator, for the purpose of the first-order large sample inference.

In order to develop a rigorous asymptotic theory we need various preliminaries. Chapter 2 reviews some facts from trigonometric series, analysis, and probability. It includes some examples of time-series models and definitions of slowly varying functions and Hermite and Appell polynomials.

Chapter 3 defines short and long memory processes in the time and spectral domains, and provides some characteristics of these processes. This chapter also discusses self-similar processes and fractional Brownian motion and their connection with long memory processes.

A large class of tests and estimators in many time-series models are based on sums and weighted sums of underlying observations. Chapter 4 introduces the basic theoretical tools and provides minimal sufficient conditions for asymptotic normality of the weighted sums of a linear process. The methods of proof include the cumulant method and the method of approximation by m-dependent variables. The methodology presented also provides techniques for analyzing asymptotic distributions of sums of functions of long and short memory Gaussian process. The results of this chapter are useful in deriving asymptotic distributions of various inference procedures in regression models with dependent errors in Chapters 9 and 12.

Given that the sample $X_1, \cdots, X_n$ is fully represented by discrete Fourier transforms (DFTs) $w(u_1), \cdots, w(u_n)$, computed at Fourier frequencies $u_1, \cdots, u_n$, and that the periodogram at frequency u_j equals $|w(u_j)|^2$, the asymptotic theory of Chapter 6 for sums of weighted periodograms of dependent linear variables largely reduces to the asymptotic analysis of rescaled DFTs. Chapter 5 focuses on the asymptotic properties of the covariances of DFTs. It is observed that even under long memory, the standardized DFTs at different discrete frequencies tend to behave asymptotically as uncorrelated identically distributed complex Gaussian random variables.

Under fairly general conditions, specified in Chapter 6, the weighted sums of the periodogram of a stationary linear process have the asymptotic normality property. The Bartlett approximation of the periodogram of a linear process allows it to be approximated by a rescaled periodogram of the underlying noise, which results in asymptotic normality of the sum under Lindeberg–Feller-type condition on the weights, ensuring that the contribution of each term to the sum is small. Section 6.3.1 provides two sets of relatively easy to verify sufficient conditions for asymptotic normality of a weighted integrated periodogram. The results derived in Chapter 6 are useful in proving asymptotic normality of various estimators and tests in parametric and semiparametric time series models in subsequent chapters.

Process X_j, $j \in \mathbb{Z}$, or a statistical model based on this process, is said to be *parametric* if its spectral density (autocovariance) is fully specified by an unknown Euclidean parameter. It is said to be *semiparametric* if its spectral density (autocovariance) is known to depend on an Euclidean parameter and an infinite-dimensional parameter.

Chapter 7 presents some common parametric time-series models that are often used for modeling long, short, and negative memory data. They include parametric ARFIMA (autoregressive fractionally integrated moving-average) and GARMA (Gegenbauer ARMA, or seasonal fractionally differenced) models. Their main properties and a mechanism for generating these processes are described in this chapter. The chapter also includes a description of how an aggregation generates long memory phenomenon in time series.

Chapter 8 is devoted to estimation in parametric and semiparametric time-series models. Consistency and asymptotic normality of Whittle estimators of the underlying parameters in parametric spectral density are established in this chapter. Often in semiparametric models, only the behavior of the spectral density near the origin is specified. For this reason, only Fourier frequencies near the origin are used for inference in these models. Whittle estimators based on Fourier frequencies near the origin are called local Whittle estimators. Consistency and asymptotic normality of these estimators in a large class of semiparametric models are also discussed in Chapter 8. The class of models considered include ARFIMA, signal plus noise, and volatility models like EGARCH.

Chapter 9 discusses some basic tests and applications. Section 9.2 derives confidence intervals for the trend and memory parameters in a simple linear trend regression model with a wide class of linear and possibly nonlinear and non-Gaussian errors. Construction of these intervals requires consistent estimation of the long-run variance that is used to studentize various statistics. Section 9.4 establishes the consistency of time-domain (HAC) and spectral-domain (MAC) estimators of the long-run variance. Section 9.5 focuses on testing for long memory and structural breaks using Lobato–Robinson, R/S, $KPSS$, and V/S tests. An application using real currency exchange data is also included. The results of several simulations are given in the chapter.

The empirical distribution function $F_n(x) := \sum_{j=1}^{n} I(X_j \leq x)/n$, $x \in \mathbb{R}$, is an important process for developing inference about the marginal distribution function F of a stationary process. For a large class of long memory moving-average processes,

$$\sup_{x \in \mathbb{R}} \left| n^{1/2-d}(F_n(x) - F(x)) + f(x)n^{-1/2-d} \sum_{j=1}^{n} X_j \right| = o_p(1),$$

where f is the density of F. Dehling and Taqqu (1989) established this first-order degeneracy result for an empirical process when $\{X_j\}$ is a Gaussian process and called it the *uniform reduction principle*. This finding is in *complete contrast* to the analogous results available under weak dependence where $n^{1/2}(F_n - F)$ converges weakly in Skorokhod space, to a time-transformed Brownian bridge. A similar result holds for the class of more general weighted residual empirical processes. In view of the first-order degeneracy of these empirical processes it is natural to look for their higher-order asymptotic behavior. Chapter 10 is devoted to the study of the first- and higher-order asymptotic analyses of empirical processes.

A classical problem in statistics is to use a vector X of p variables to explain a one-dimensional response Y, $p \geq 1$. This is often done in terms of the regression function $\mu(x) := E(Y|X = x)$, $x \in \mathbb{R}^p$, assuming it exists. A model that formulates inference problems in terms of the regression function is called a regression model. From the 1980s examples from numerous social and physical sciences indicate the need to understand the behavior of various inference procedures in regression models when design process X or error variables have long memory. Chapter 11 uses the results of Chapter 10 to discuss the asymptotic distribution theory of numerous inference procedures in linear and non-linear parametric regression models in the presence of long memory in errors, and when the design vector is either a vector of known constants or a vector of random variables that possibly have long memory. This includes the first-order equivalence of the least-squares and M estimators.

Chapter 12 is devoted to the large sample inference in non-parametric heteroscedastic regression models with short or long memory moving-average errors and uniform non-random design on the unit interval. The consistency and asymptotic normality of Nadaraya–Watson kernel-type estimators of regression and heteroscedasticity functions $\mu(x)$ and $\sigma^2(x)$ are established. The uniform consistency rate for kernel-type estimators of $\mu(x)$ is also discussed in this chapter. To studentize the kernel estimator $\hat{\mu}(x)$, a consistent estimator of $\text{Var}(\hat{\mu}(x))$ is provided that avoids the estimation of the limit variance and the long memory parameter. The results thus obtained are useful in constructing confidence bands for $\mu(x)$.

Chapter 13 deals with testing for the lack-of-fit of a parametric regression model, testing of a subhypothesis in a linear regression model, and testing for the goodness-of-fit of an error d.f. in parametric regression models in the presence of heteroscedasticity and long memory in design or errors. The consistency of classes of kernel- and cross-validation-type estimators of the conditional variance function is established for some linear parametric regression models. The latter result is especially useful for testing the lack-of-fit of a parametric regression model as illustrated in Section 13.3. Regression models with long memory design and errors are known to arise in finance as illustrated by a real-data example of currency exchange rates in Section 13.3.4.

A problem of interest in statistical inference is to see if a subset of variables is significant or not for predicting a response variable in linear regression models. Section 13.4 discusses testing a subhypothesis for the absence of a subset of predictors in linear regression models when errors form a long memory moving average, and when covariates are either random, having short or long memory, or are non-random. Section 13.5 contains some tests of the goodness-of-fit of the marginal d.f. of a stationary long memory process based on an empirical process. The implementation of these tests uses the uniform reduction principle and its refinements, referred to above, such as the higher order asymptotic expansions of Chapter 10. In particular, tests for fitting a d.f. up to unknown mean or variance and an error d.f. in the linear regression model are presented.

Chapter 14 contains some results from probability theory that are of general interest. It gives a definition of *Wick products* that are multivariate generalizations of Appell and Hermite polynomials, and formulas for their moments and cumulants. As an application of these techniques, the limit distribution of the sums of the Appell polynomials of a linear process is derived. It also contains the definition and basic properties of Wiener–Itô integrals and their convergence criteria, and a lemma describing the limit of the trace of the product of matrices. Several of these results are used in the rest of the text.

Chapter 2

Some Preliminaries

In this chapter we will review some preliminary results from analysis. These are needed to proceed further with the rigorous treatment of the relevant subject matter.

2.1 Preliminaries from trigonometry

The following facts about the trigonometric functions are from the well-known monograph of Zygmund I and II (2002), many of which are stated without proof here. Recall $\mathbf{i} := (-1)^{1/2}$ and $\mathbb{Z} = \{0, \pm 1, \cdots\}$. The first fact we recall is that the functions $\{e^{\mathbf{i}nu}; n \in \mathbb{Z}\}$ are orthogonal for u in any interval of length 2π: For every $u \in \mathbb{R}$,

$$\int_u^{u+2\pi} e^{\mathbf{i}(m-n)v} dv = \begin{cases} 0, & m \neq n, \\ 2\pi, & m = n. \end{cases} \tag{2.1.1}$$

The Fourier series expansion of any square integrable complex-valued function f defined on $\Pi := [-\pi, \pi]$ is given by

$$f(u) := \sum_{k=-\infty}^{\infty} c_k e^{\mathbf{i}ku}, \qquad u \in \Pi, \tag{2.1.2}$$

$$c_k := \frac{1}{2\pi} \int_\Pi f(u) e^{-\mathbf{i}ku}\, du, \qquad k \in \mathbb{Z}.$$

Fourier Coefficient of $f(u)$

Note that if we let

$$a_k := \frac{1}{2\pi} \int_\Pi f(u)\cos(ku)du, \qquad b_k := \frac{1}{2\pi} \int_\Pi f(u)\sin(ku)du,$$

then $c_k = a_k - \mathbf{i}b_k$, $c_{-k} = a_k + \mathbf{i}b_k$, $k \in \mathbb{Z}$. In particular, if f is symmetric around zero, i.e., $f(u) \equiv f(-u)$, for all $u \in \Pi$, then $b_k = 0$, $\forall k \in \mathbb{Z}$.

7

For any complex-valued functions $f \in L_p(\Pi)$, $g \in L_q(\Pi)$, with $p, q > 1$, $1/p + 1/q = 1$, and with Fourier coefficients $c_{k,f}$ and $c_{k,g}$,

$$\frac{1}{2\pi} \int_\Pi f(u)\overline{g(u)}du = \sum_{k=-\infty}^{\infty} c_{k,f}\overline{c_{k,g}}. \qquad (2.1.3)$$

This result implies Parseval's identity

$$\frac{1}{2\pi} \int_\Pi |f(u)|^2 du = \sum_{k=-\infty}^{\infty} |c_{k,f}|^2, \quad f \in L_2(\Pi). \qquad (2.1.4)$$

This states that a function $f \in L_2(\Pi)$ has square summable Fourier coefficients, and the equality (2.1.2) is to be understood as the convergence of the sums to f in $L_2(\Pi)$:

$$\lim_{n\to\infty} \int_\Pi \left| \sum_{k=-n}^{n} c_k e^{iku} - f(u) \right|^2 du = 0.$$

A continuous function f is said to belong to the Lipschitz class Λ_α, $0 < \alpha \le 1$, if for some $0 < C < \infty$, $|f(v) - f(u)| \le C|v - u|^\alpha$, for all $u, v \in \Pi$. If $f \in \Lambda_\alpha$, $0 < \alpha < 1$, then $c_j = O(|j|^{-\alpha})$. In addition, if $\alpha > 1/2$, then $\sum_{j\in\mathbb{Z}} |c_j| < \infty$. The property of absolute summability of the Fourier coefficients remains valid for an absolutely continuous f with its first a.e. derivative $\dot{f}$ in $L_p(\Pi)$ for some $p > 1$.

Using the properties of sine and cosine functions, basic definitions and the equality $e^{iu} = \cos(u) + i\sin(u)$, we obtain

$$\sum_{j=1}^{n} e^{iuj} = e^{iu}\frac{1 - e^{inu}}{1 - e^{iu}},$$

$$\sum_{j=1}^{n} e^{i(\frac{2\pi k}{n})j} = 0, \qquad 1 \le k \le n-1, \qquad (2.1.5)$$

$$= n, \qquad k = 0, n.$$

An important function often used in the proofs in this book is the so-called *Dirichlet kernel* given by

$$D_n(u) = \frac{\sin(nu/2)}{\sin(u/2)} = \sum_{j=1}^{n} e^{iju}\, e^{-i(n+1)u/2}, \quad u \in \Pi, n \ge 1. \quad (2.1.6)$$

Some of its useful properties are as follows:

$$|D_n(u)| \le \pi \min(|u|^{-1}, n) \le \pi n^\gamma |u|^{\gamma-1}, \quad \forall 0 \le \gamma \le 1, \quad (2.1.7)$$

$$\le 2\pi n\,(1 + n|u|)^{-1}, \quad \forall u \in \Pi, \qquad (2.1.8)$$

$$\int_\Pi D_n^2(u)du = 2\pi n. \qquad (2.1.9)$$

We also have

$$\left| \sum_{j=0}^{n-1} e^{iju} \right|^2 = \frac{\sin^2(nu/2)}{\sin^2(u/2)} = D_n^2(u), \quad u \in \Pi, \quad n \geq 1. \quad (2.1.10)$$

Note: The above discussion is based on Theorems II.4.7, VI.3.1, VI.3.8, and VII.6.11 (Part I) of Zygmund (2002).

2.2 Stationarity, spectral density, and autocovariance

In this subsection we discuss strict and weak stationarity, spectral density, and autocovariance.

Definition 2.2.1. A real-valued stochastic process $\{X_j\} = \{X_j, j \in \mathbb{Z}\}$ is said to be

(a) *strictly stationary* if the joint distribution of $(X_{j_1+k}, \cdots, X_{j_n+k})$ does not depend on $k \in \mathbb{Z}$ for any $j_1, \cdots, j_n \in \mathbb{Z}$ for all $n \geq 1$;

(b) *covariance stationary* if $EX_j^2 < \infty$ and $\mu = EX_j$ are constant for all $j \in \mathbb{Z}$, and the covariance function $\mathrm{Cov}(X_j, X_{j+k}) = \mathrm{Cov}(X_0, X_k)$ is constant in j for all $j, k \in \mathbb{Z}$.

Covariance stationary processes are also known as *second-order stationary* or *weakly stationary* processes. The *autocovariance function* of such a process $\{X_j\}$ is defined to be

$$\gamma(k) := \mathrm{Cov}(X_j, X_{j+k}) = \mathrm{Cov}(X_0, X_k), \quad j, k \in \mathbb{Z}. \quad (2.2.1)$$

The integer k in (2.2.1) is called the *lag*.

The finite-dimensional distributions of the process $\{X_j\}$ are distributions of the vectors $(X_{j_1}, \cdots, X_{j_n})$ for all $j_1, \cdots, j_n \in \mathbb{Z}$ and $n \geq 1$.

Definition 2.2.2. (*Gaussian process*). A continuous- or discrete-time stochastic process X is said to be Gaussian if all its finite-dimensional distributions are Gaussian.

As is well known, see Loève (1968), a Gaussian process is fully characterized by its mean and covariance functions. Thus the distribution of a Gaussian vector $(X_{j_1}, \cdots, X_{j_n})$ having the mean vector $\mu = (EX_{j_1}, \cdots, EX_{j_n})$ and the covariance matrix $\Sigma := \big(\mathrm{Cov}(X_{j_s}, X_{j_t})\big)_{s,t=1,\cdots,n}$ is denoted by $\mathcal{N}_n(\mu, \Sigma)$.

In particular, the distribution of a covariance stationary Gaussian process $\{X_j\}$ is fully determined by its mean $\mu = EX_j$ and its autocovariance function $\gamma(k)$, $k \in \mathbb{Z}$. While a strictly stationary process with $EX_0^2 < \infty$ is also covariance stationary, the converse holds for Gaussian processes.

Some of the properties of the autocovariance function $\gamma(k)$ are

$$\gamma(0) \geq 0, \quad |\gamma(k)| \leq \gamma(0), \quad \gamma(k) = \gamma(-k), \quad \forall\, k \in \mathbb{Z}.$$

The function $r(k) := \gamma(k)/\gamma(0)$, $k \in \mathbb{Z}$ is called the *autocorrelation* function.

A real function $\gamma(k)$ on the set of integers $\mathbb{Z}$ is said to be *non-negative definite* if for every $n \geq 1$, for every set of n real numbers $a_1, \cdots, a_n$,

$$\sum_{j,k=1}^{n} a_j \gamma(j-k) a_k \geq 0. \tag{2.2.2}$$

A real-valued function γ on $\mathbb{Z}$ is the autocovariance function of a second-order stationary process if, and only if, $\gamma(-k) \equiv \gamma(k)$ and γ satisfies (2.2.2).

A result that characterizes the autocovariance function is Bochner's theorem: A real-valued function γ on $\mathbb{Z}$ satisfies (2.2.2) if, and only if, there exists a right-continuous non-decreasing function F on Π such that $F(-\pi) = 0$ and

$$\gamma(k) = \int_{\Pi} e^{ivk}\, dF(v), \qquad \forall\, k \in \mathbb{Z}.$$

The function F is called the *spectral distribution function*. It is defined on the compact interval Π and has all the properties of a distribution function except that $F(\pi) - F(-\pi)$, although finite, may be not normalized to one, whereas the standardized function $F/\gamma(0)$ is a distribution function. The function F defines the spectral measure on Π.

If F is absolutely continuous w.r.t. the Lebesgue measure λ, then $f := dF/d\lambda$ is called the *spectral density*.

Every zero-mean stationary process $\{X_j\}$ has *spectral representation*

$$X_j = \int_{\Pi} e^{ijv} dZ_X(v), \quad \forall\, j \in \mathbb{Z}, \tag{2.2.3}$$

where Z_X is an orthogonal-increment process on Π with $Z_X(-\pi) = 0$. The above stochastic integral is defined in the mean-square sense, and

$$E\big[dZ_X(v)\big] = 0, \quad E|dZ_X(v)|^2 = dF_X(v),$$
$$E\big[dZ_X(u)d\overline{Z_X(v)}\big] = 0, \qquad v \neq u.$$

Definition 2.2.3. (*Spectral density*). A real-valued function f on Π is called the spectral density of a second-order stationary process $\{X_j, j \in \mathbb{Z}\}$ with the autocovariance function γ if

$$(a) \qquad f \geq 0, \qquad \text{on } \Pi,$$

$$(b) \qquad \gamma(k) = \int_{\Pi} e^{ivk} f(v) dv, \quad \forall k \in \mathbb{Z}.$$

A real-valued function f on Π is the spectral density of a second-order stationary process $\{X_j, j \in \mathbb{Z}\}$ if

$$(i) \qquad f(v) = f(-v), \qquad f(v) \geq 0, \quad \forall v \in \Pi,$$

$$(ii) \qquad \int_{\Pi} f(v) dv < \infty.$$

Condition $f \in L_1(\Pi)$ ensures that the corresponding autocovariance $\gamma(k)$, defined by (b) has the property: $\gamma(k) \to 0$, as $k \to \infty$, i.e. correlation between X_t and X_s vanishes as the distance in time $|t - s|$ increases.

In particular, for any f with properties (i) and (ii) there exists a (strictly) stationary Gaussian process $\{X_j\}$ with spectral density f and autocovariance function as in (b). *In this text, the term "stationary" means both strictly stationary and covariance stationary, unless specified otherwise.*

Suppose that an autocovariance function γ satisfies

$$\sum_{k \in \mathbb{Z}} |\gamma(k)| < \infty. \tag{2.2.4}$$

Then it has a spectral density, which can be written as

$$f(v) = \frac{1}{2\pi} \sum_{k \in \mathbb{Z}} e^{-ikv} \gamma(k), \quad v \in \Pi. \tag{2.2.5}$$

This series converges uniformly in $v \in \Pi$ and f is continuous on Π.

A more general criterion for the existence of spectral density is given in Proposition 2.2.1.

Proposition 2.2.1. *If an autocovariance function γ is square summable, $\sum_{k \in \mathbb{Z}} \gamma^2(k) < \infty$, then it has spectral density $f \in L_2(\Pi)$ given by the Fourier series (2.2.5).*

Proof. It suffices to show that f as defined in (2.2.5) is non-negative, $f(v) \geq 0$ a.e. in Π.

Let $a_1, \cdots, a_n \in \mathbb{R}$ be arbitrary real numbers. Then

$$\int_\Pi f(v) \Big| \sum_{j=1}^n e^{\mathrm{i}jv} a_j \Big|^2 dv = \sum_{j,k=1}^n a_j a_k \int_\Pi f(v) e^{\mathrm{i}(j-k)v} dv$$

$$= \sum_{j,k=1}^n a_j a_k \gamma(j-k) \geq 0,$$

according to (2.2.2). Since the class of all trigonometric polynomials $P(v) = \sum_{j=1}^n e^{\mathrm{i}jv} a_j$ is dense in the space $\mathcal{C}(\Pi)$ of all continuous functions on Π, it follows that for any $g \in \mathcal{C}(\Pi)$,

$$\int_\Pi f(v) |g(v)|^2 dv \geq 0.$$

The last fact implies $f(v) \geq 0$ a.e. in Π, thereby completing the proof.

Sometimes we will write γ_X and f_X for the autocovariance function and the spectral density, respectively, of the given process $\{X_j\}$.

Definition 2.2.4. (*White-noise*).   A covariance stationary sequence of r.v.'s $\{\zeta_j, \, j \in \mathbb{Z}\}$, is said to form *white-noise* with mean 0 and variance σ_ζ^2, and we write $\{\zeta_j\} \sim \mathrm{WN}(0, \sigma_\zeta^2)$, if $E\zeta_0 = 0$, $\gamma(0) := \sigma_\zeta^2 := E\zeta_0^2 < \infty$, and $\gamma(k) \equiv 0$, for all $k \in \mathbb{Z}$, $k \neq 0$.

Note that the autocovariance function γ of white-noise satisfies (2.2.4) and by (2.2.5) the corresponding spectral density is $f(v) = \sigma_\zeta^2 / 2\pi$, $v \in \Pi$.

The simplest example of white-noise is a sequence of independent identically distributed (i.i.d.) r.v.'s ζ_j, $j \in \mathbb{Z}$, with zero mean and variance σ_ζ^2. From now onwards, such a sequence will be denoted by $\{\zeta_j\} \sim \mathrm{IID}(0, \sigma_\zeta^2)$. A white-noise Gaussian sequence is *a priori* an i.i.d. sequence. Since, in general, having zero correlation does not imply independence, the class of white-noise sequences is wider than that of i.i.d. sequences.

Besides including i.i.d. sequences, the class of white-noise processes also includes martingale differences. Since the term "martingale differences" is used in a slightly different context and allows time-varying variance, we give a formal definition.

A filtration $\{\mathcal{F}_j, j \in \mathbb{Z}\}$ is a sequence of σ-fields such that $\mathcal{F}_j \subset \mathcal{F}_{j+1}$, $\forall j \in \mathbb{Z}$. Typically, a filtration is generated by a process $\{Y_i\}$, namely $\mathcal{F}_j = \sigma$-field$(Y_i, i \leq j)$, reflecting the available "information set" up to time j.

Definition 2.2.5. A sequence of r.v.'s $\{u_j, j \in \mathbb{Z}\}$ is said to be a *martingale-difference sequence* with respect to filtration $\{\mathcal{F}_j\}$ if u_j is $\mathcal{F}_j$-measurable, $E|u_j| < \infty$, and $E[u_j|\mathcal{F}_{j-1}] = 0$, for all $j \in \mathbb{Z}$.

The above definition assumes neither stationarity nor the existence of the variances of the u_j's. On the other hand, if a martingale-difference sequence $\{u_j\}$ is covariance stationary, then it is a white-noise sequence since $Eu_j = 0$ and $Eu_j u_k = E[u_j E[u_k|\mathcal{F}_{k-1}]] = 0$ for $j < k$, by the property of conditional expectation. An important distinction of martingale differences from i.i.d. sequences is the fact that the conditional variance $\mathrm{Var}(u_j|\mathcal{F}_{j-1}) = E[u_j^2|\mathcal{F}_{j-1}]$ of the former sequence may be random and depend on the past history $\mathcal{F}_{j-1}$.

Example 2.2.1. Let $\{Y_j, j \geq 1\}$ be a martingale, i.e., $E|Y_j| < \infty$, and
$$E[Y_j|Y_1, \cdots, Y_{j-1}] = Y_{j-1}, \quad j \geq 2, \quad EY_1 = 0.$$
Set $u_1 := Y_1$, $u_j := Y_j - Y_{j-1}$ for $j \geq 2$, and $u_j := 0$ for $j \leq 0$. Then $\{u_j\}$ is a martingale-difference sequence with respect to filtration $\mathcal{F}_j := \sigma$-field$\{Y_i, 1 \leq i \leq j\}$.

Example 2.2.2. Let $\{\zeta_j\} \sim \mathrm{IID}(0,1)$ and $u_j := \zeta_j \zeta_{j-1}$. Then $\{u_j\}$ is a martingale-difference sequence with respect to $\mathcal{F}_j := \sigma$-field$\{\zeta_i, i \leq j\}$, with zero conditional mean $E[u_j|\mathcal{F}_{j-1}] = 0$ and conditional variance $E[u_j^2|\mathcal{F}_{j-1}] = \zeta_{j-1}^2$.

Example 2.2.3. (*Linear filter and transfer function*). Suppose that $\{X_j\}$ is an output of a *linear filter* $\{a_j\}$, $\sum_{j\in\mathbb{Z}} a_j^2 < \infty$ applied to a stationary input process $\{Y_k\}$,
$$X_j = \sum_{k=-\infty}^{\infty} a_{j-k} Y_k, \quad j \in \mathbb{Z}. \tag{2.2.6}$$
The filter is called *causal* if $a_j = 0$ for $j < 0$, and in this case
$$X_j = a_0 Y_j + a_1 Y_{j-1} + a_2 Y_{j-2} + \cdots = \sum_{k=0}^{\infty} a_k Y_{j-k}.$$
The Fourier transform $A(v) := \sum_{k\in\mathbb{Z}} e^{-ivk} a_k$, $v \in \Pi$, is called the *transfer function*. In the case of a causal filter, $A(v) = \sum_{k=0}^{\infty} e^{-ivk} a_k$. The following proposition describes the relationship between the spectral densities f_Y and f_X of the input $\{Y_j\}$ and the output $\{X_j\}$.

Proposition 2.2.2. *Let $\{Y_j\}$ be a stationary zero-mean process with absolutely summable autocovariance $\sum_{k\in\mathbb{Z}}|\gamma_Y(k)| < \infty$, and $\{a_j\}$ be a square summable filter: $\sum_{j\in\mathbb{Z}} a_j^2 < \infty$. Then the filtered process $\{X_j\}$ of (2.2.6) is stationary with zero mean and has the spectral density*

$$f_X(v) = \Big| \sum_{k=-\infty}^{\infty} a_k e^{-ikv}\Big|^2 f_Y(v), \quad v \in \Pi. \tag{2.2.7}$$

Proof. We shall show that the series in (2.2.6) converges in mean square. By the fact $|ab| \le (a^2 + b^2)/2$, for all $a, b \in \mathbb{R}$, for all $i \in \mathbb{Z}$,

$$EX_i^2 = \sum_{k,l\in\mathbb{Z}} a_{i-k}a_{i-l}\gamma_Y(k-l)$$

$$\le \sum_{k,l\in\mathbb{Z}} a_{i-k}^2 |\gamma_Y(k-l)| = \Big(\sum_{k\in\mathbb{Z}} a_k^2\Big)\sum_{l\in\mathbb{Z}}|\gamma_Y(l)| < \infty.$$

Then, for any $i \in \mathbb{Z}$,

$$\gamma_X(k) = EX_iX_{i+k} = \sum_{j,s=-\infty}^{\infty} a_{i-j}a_{i+k-s}\gamma_Y(j-s) \tag{2.2.8}$$

$$= \sum_{j,s=-\infty}^{\infty} a_j a_s \gamma_Y(k+j-s)$$

depends only on k so that $\{X_j\}$ is covariance stationary. Strict stationarity of $\{X_j\}$ follows from that of $\{Y_j\}$ and the moving-average form in (2.2.6).

It remains to show (2.2.7). Since $\{a_j\}$ is square summable, the transfer function $A \in L_2(\Pi)$. By definition,

$$a_j = (2\pi)^{-1} \int_\Pi e^{ijv} A(v)dv \quad \text{and} \quad \gamma_Y(j) = \int_\Pi e^{ijv} f_Y(v)dv$$

are real numbers. An application of the equality (2.1.3) with $f \equiv A$ and $g(v) \equiv 2\pi e^{i(k-s)v} f_Y(v)$, yields

$$\sum_{j=-\infty}^{\infty} a_j\gamma_Y(j+k-s) = (2\pi)^{-1} \int_\Pi f(v)\overline{g(v)}dv$$

$$= \int_\Pi e^{-iv(k-s)} A(v)f_Y(v)dv.$$

Next, apply (2.1.3) to the sum over s in (2.2.8) with $f \equiv A$ and $g(v) := (2\pi)e^{-ivk} A(v)f_Y(v)$ to obtain

$$\gamma_X(k) = \int_\Pi e^{ivk}|A(v)|^2 f_Y(v)dv, \quad k \in \mathbb{Z},$$

which proves (2.2.7). Note that f_Y bounded implies that this $g \in L_2(\Pi)$.

Proposition 2.2.2 remains valid for an absolutely summable filter $\sum_{j \in \mathbb{Z}} |a_j| < \infty$ and any stationary zero-mean process $\{Y_j\}$ with spectral density f_Y.

The following theorem is essentially Theorem 4.10.1 of Brockwell and Davis (1991). It is more general than Proposition 2.2.2. It relates spectral representations and measures of input and output processes.

Theorem 2.2.1. *Let $\{Y_j\}$ be a zero-mean stationary process with spectral representation*

$$Y_j = \int_\Pi e^{ijv} dZ_Y(v), \quad j \in \mathbb{Z},$$

and the spectral distribution function $F_Y(\cdot)$. Suppose a_k, $k \in \mathbb{Z}$ is a sequence of real numbers such that the series $\sum_{j=-n}^{n} a_j e^{-iju}$ converges in $L_2(F_Y)$ norm to $h(e^{-iu})$, as $n \to \infty$.

Then the process $X_j = \sum_{k \in \mathbb{Z}} a_k Y_{j-k}$, $j \in \mathbb{Z}$, is stationary with zero mean, spectral distribution function $dF_X(v) = |h(e^{-iv})|^2 dF_Y(v)$, and spectral representation

$$X_j = \int_\Pi e^{ijv} h(e^{-iv}) dZ_Y(v), \quad j \in \mathbb{Z}.$$

This theorem leads to the relationships between the autocovariances and spectral densities of the input $\{Y_j\}$ and output $\{X_j\}$ processes:

$$\gamma_X(j) = \int_\Pi e^{ijv} |h(e^{-iv})|^2 dF_Y(v), \quad j \in \mathbb{Z},$$
$$f_X(v) = |h(e^{-iv})|^2 f_Y(v), \quad v \in \Pi.$$

Definition 2.2.6. (*Periodogram*). The periodogram $I_X(u)$ at frequency $u \in \Pi$ based on observations $X_1, \cdots, X_n$ is defined by

$$I_X(u) = (2\pi n)^{-1} \Big| \sum_{k=1}^{n} e^{iku} X_k \Big|^2, \quad u \in \Pi. \tag{2.2.9}$$

Given a sample $X_1, \cdots, X_n$ of a stationary process $\{X_j\}$, the periodogram is an empirical version of the spectral density f_X of $\{X_j\}$. Although it is not a consistent estimator of $f_X(u)$, it satisfies $EI_X(u) \to f_X(u)$ at every continuity point u of $f_X(u)$ and has a number of other useful properties, as discussed in Chapter 5.

Example 2.2.4. (*AR*(1) *model*). Consider the first-order autoregressive (AR(1)) process $\{X_j\}$, defined as a solution to the equations

$$X_j - \rho X_{j-1} = \zeta_j, \quad j \in \mathbb{Z}, \tag{2.2.10}$$

where $\{\zeta_j\} \sim \text{WN}(0, \sigma_\zeta^2)$ is a white-noise sequence. Let B denote the backward shift operator, i.e., $BX_j := X_{j-1}$, for all $j \in \mathbb{Z}$.

For $|\rho| < 1$, the AR(1) equations have a stationary solution

$$X_j = (1 - \rho B)^{-1} \zeta_j = \sum_{k=0}^{\infty} \rho^k B^k \zeta_j = \sum_{k=0}^{\infty} \rho^k \zeta_{j-k}, \quad j \in \mathbb{Z}.$$

Hence, $EX_j = 0$, for all $j \in \mathbb{Z}$, and

$$\gamma(0) = \frac{\sigma_\zeta^2}{1 - \rho^2}, \quad \gamma(k) = \rho^{|k|} \gamma(0), \quad k \in \mathbb{Z}.$$

Clearly, the above $\{X_j\}$ is obtained by applying a linear causal filter with exponentially fast decaying weights ρ^k to a white-noise sequence $\{\zeta_j\}$, and the autocovariance $\gamma(k)$ decays to 0 exponentially fast and satisfies (2.2.4). Hence, by (2.2.5), the spectral density of the process $\{X_j\}$ is

$$f_X(v) = \frac{1}{2\pi} \sum_{k \in \mathbb{Z}} e^{-ikv} \gamma(k) = \frac{1}{2\pi} \gamma(0) \sum_{k \in \mathbb{Z}} e^{-ikv} \rho^{|k|}$$

$$= \frac{\sigma_\zeta^2}{2\pi} \frac{1}{1 - 2\rho \cos v + \rho^2} = \frac{\sigma_\zeta^2}{2\pi} \frac{1}{|1 - \rho e^{iv}|^2}, \quad v \in \Pi,$$

since

$$\sum_{k \in \mathbb{Z}} e^{-ikv} \rho^{|k|} = 1 + \sum_{k=1}^{\infty} \rho^k [e^{ikv} + e^{-ikv}]$$

$$= 1 + \frac{\rho e^{iv}}{1 - \rho e^{iv}} + \frac{\rho e^{-iv}}{1 - \rho e^{-iv}} = \frac{1 - \rho^2}{|1 - \rho e^{iv}|^2}.$$

Alternatively, since $\{X_j\}$ is a linear filter of $\{\zeta_j\}$, by Proposition 2.2.2,

$$f_X(v) = \Big| \sum_{k=0}^{\infty} e^{-ikv} \rho^k \Big|^2 f_\zeta(v) = \frac{\sigma_\zeta^2}{2\pi} \frac{1}{|1 - \rho e^{iv}|^2}.$$

Example 2.2.5. (*ARMA*(*p, q*) *model*). Let $p, q \geq 0$ be known integers and $\{\zeta_j\} \sim \text{WN}(0, \sigma_\zeta^2)$ be a white-noise sequence. An *autoregressive moving-average* process $\{X_j, j \in \mathbb{Z}\}$ of order p, q, written as ARMA(p, q), is defined as a stationary solution of the equations

$$X_j - \sum_{k=1}^{p} \phi_k X_{j-k} = \sum_{k=1}^{q} \theta_k \zeta_{j-k} + \zeta_j, \quad j \in \mathbb{Z},$$

where $\phi_1, \cdots, \phi_p$ and $\theta_1, \cdots, \theta_q$ are some real parameters.

If we define

$$\phi(z) := 1 - \sum_{k=1}^{p} \phi_k z^k, \quad \theta(z) := 1 + \sum_{k=1}^{q} \theta_k z^k,$$

the above ARMA(p,q) equations can be rewritten in a compact form $\phi(B)X_j = \theta(B)\zeta_j$, $j \in \mathbb{Z}$, where B is the backward shift operator.

If the polynomials $\phi(\cdot)$ and $\theta(\cdot)$ have no common zeros and all their roots are outside the unit complex disk $\{|z| \leq 1\}$, then the ARMA(p,q) process is causal, i.e., it can be written as a moving average

$$X_j = \sum_{k=0}^{\infty} \psi_k \zeta_{j-k}, \quad j \in \mathbb{Z}, \tag{2.2.11}$$

of the noise $\{\zeta_j, j \in \mathbb{Z}\}$, where the coefficients ψ_k, $k \geq 0$ are determined by comparing the coefficients of z^k of the power series

$$\psi(z) := \sum_{k=0}^{\infty} \psi_k z^k = \frac{\theta(z)}{\phi(z)}, \quad |z| \leq 1.$$

The linear representation (2.2.11) of X_j allows us to compute its autocovariance function in terms of the weights $\{\psi_j\}$,

$$\gamma(k) = \sigma^2 \sum_{j=0}^{\infty} \psi_j \psi_{j+|k|}, \quad k \in \mathbb{Z},$$

whereas Proposition 2.2.2 yields the formula for the spectral density

$$f_X(v) = \frac{|\theta(e^{-iv})|^2}{|\phi(e^{-iv})|^2} f_\zeta(v) = \frac{\sigma_\zeta^2}{2\pi} \frac{|\theta(e^{-iv})|^2}{|\phi(e^{-iv})|^2}, \quad v \in \Pi.$$

A common feature of stationary ARMA models is that they can be written as a linear filter of a white-noise sequence with exponentially decaying coefficients ψ_k, see Subsection 8.3.2. In addition, they have an exponentially decaying autocovariance function $\gamma_X(k)$ and a continuous spectral density $f_X(v)$ bounded away from ∞ and 0.

Finally, if $X_j = Y_j + Z_j$, $j \in \mathbb{Z}$, where $\{Y_j\}$ and $\{Z_j\}$ are two uncorrelated covariance stationary processes with autocovariance functions $\gamma_Y(k)$, $\gamma_Z(k)$, and spectral densities $f_Y(v)$, $f_Z(v)$, respectively, then

$$\gamma_X(k) = \gamma_Y(k) + \gamma_Z(k), \quad k \in \mathbb{Z}, \tag{2.2.12}$$

$$f_X(v) = f_Y(v) + f_Z(v), \quad v \in \Pi.$$

Note: A good source for the proofs of many of the above facts is the book by Brockwell and Davis (1991, p. 85-93, 123).

2.3   Slowly and regularly varying functions

To describe the slow decay of autocovariances, we need to define *regularly* and *slowly* varying functions. The following facts about these functions are taken from Bingham, Goldie and Teugels (1987).

Definition 2.3.1. A function L on $[0, \infty)$ is said to be slowly varying at infinity (in Karamata's sense) if L is positive on $[A, \infty)$ for some $A > 0$, and

$$\lim_{x \to \infty} \frac{L(tx)}{L(x)} = 1, \qquad \forall \, t > 0. \tag{2.3.1}$$

From now on, SV will denote the class of functions varying slowly at infinity. An $L \in SV$ can be negative or positive on $[0, A)$.

   If $L \in SV$, then (2.3.1) holds uniformly in t on each interval $[a, b] \subset (0, \infty)$. The basic representation of SV functions is provided by

Theorem 2.3.1. (Representation Theorem). *A function L is slowly varying if, and only if, it can be written in the form*

$$L(x) = c(x) \exp \left\{ \int_a^x \frac{\zeta(u)}{u} du \right\}, \qquad \forall \, x \geq a, \tag{2.3.2}$$

for some $a > 0$, where $c(x)$, $\zeta(x)$ are bounded functions and $c(x) \to c > 0$, $\zeta(x) \to 0$, as $x \to \infty$.

   Any function L, which has a finite limit $L(x) \to c > 0$, as $x \to \infty$, is slowly varying at infinity. Other examples of slowly varying functions are $\log(x)$, $\log(\log x)$, $(\log x)^\delta$ for any $\delta \in \mathbb{R}$, $\exp((\log x)^\beta)$ for $0 < \beta < 1$, and $x > 1$.

   Some additional facts that will be often used are: Let $L, L_1, L_2 \in SV$ and $a > 0$. Then

$$x^a |L(x)| \to \infty, \quad x^{-a} L(x) \to 0, \quad \frac{\log |L(x)|}{\log x} \to 0, \quad \text{as } x \to \infty, \tag{2.3.3}$$

$$L_1 L_2 \in SV, \qquad L_1 + L_2 \in SV, \qquad |L|^a \in SV.$$

Definition 2.3.2. A function $f(x)$, $x \geq 0$ is a *regularly varying* function with index $\delta \in \mathbb{R}$, if f is positive on $[A, \infty)$ for some $A > 0$ and $\forall t > 0$,

$$\lim_{x \to \infty} \frac{f(tx)}{f(x)} = t^\delta.$$

A regularly varying function can be written as $f(x) = x^\delta L(x)$, for some $L \in SV$.

Observe that $f(tx)/f(x) = t^\delta L(tx)/L(x)$. Regularly varying functions have the following useful properties. As $x \to \infty$,

$$\left| t^\delta \frac{L(tx)}{L(x)} - t^\delta \right| \to 0, \quad \text{uniformly in } t, \tag{2.3.4}$$

on intervals $[a, b] \subset (0, \infty)$, if $\delta = 0$;

on intervals $(0, b]$, $0 < b < \infty$, if $\delta > 0$;

on intervals $[a, \infty)$, $0 < a < \infty$, if $\delta < 0$.

If $\delta > 0$, in (2.3.4) it is additionally assumed that L is bounded on finite intervals $(0, b]$.

Relation (2.3.4) implies that for any $a > 0$, there exist a $C < \infty$ and $x_0 > 0$ such that

$$\left| \frac{L(tx)}{L(x)} - 1 \right| \leq C(t^a + t^{-a}), \quad \forall t > 0, \, x > x_0. \tag{2.3.5}$$

Result (2.3.5) (setting $y = tx$) implies that for $x \geq x_0$, $y > 0$,

$$\left| \frac{L(y)}{L(x)} \right| \leq C(|y/x|^a + |x/y|^a), \tag{2.3.6}$$

$$|L(x)| + 1/|L(x)| \leq C(|x|^a + |x|^{-a}).$$

A large class of autocovariance functions $\gamma(k)$ and spectral densities $f(v)$ in this book are regularly varying functions of the form $\gamma(k) = |k|^{-1+2d}L(|k|)$ or $f(v) = |v|^{-2d}L(1/|v|)$, $|d| < 1/2$.

Zygmund slowly varying functions form a subclass of SV functions, which sometimes appear in applications. A function $L \in SV$ is said to be *Zygmund slowly varying at* ∞, if for any $a > 0$ and some $x_0 > 0$, $x^a L(x)$ is increasing in x and $x^{-a}L(x)$ decreasing in x, $x \geq x_0$. This class of functions, denoted by ZSV, coincides with the class of SV functions that can be represented as (2.3.2) with $c(x) \equiv c$.

If an $L \in ZSV$, then for all $0 < \beta < 1$, as $x \to 0+$,

$$\sum_{j=1}^{\infty} j^{-\beta} L(j) \sin(jx) \sim \Gamma(1 - \beta) \cos(\frac{\pi\beta}{2}) x^{\beta-1} L(1/x), \tag{2.3.7}$$

$$\sum_{j=1}^{\infty} j^{-\beta} L(j) \cos(jx) \sim \Gamma(1 - \beta) \sin(\frac{\pi\beta}{2}) x^{\beta-1} L(1/x). \tag{2.3.8}$$

This result implies that the transfer function $A(v) = \sum_{k=0}^{\infty} a_k e^{-ikv}$ of a causal linear filter $a_k = k^{-1+d}L(k)$, $k = 1, 2, \cdots$, with $0 < d < 1/2$ and $L \in ZSV$, is unbounded, and for some positive and finite constant c_d,

$$|A(v)|^2 \sim c_d |v|^{-2d} L^2(1/|v|), \quad v \to 0. \qquad (2.3.9)$$

The next proposition describes the asymptotic behavior of sums and integrals of regularly varying functions.

Proposition 2.3.1. *Let $L \in SV$ be bounded on finite intervals $(0, a]$, $a > 0$. Then*

$$\sum_{j=1}^{n} j^{-\beta} L(j) \sim \frac{n^{1-\beta} L(n)}{1 - \beta}, \qquad 0 < \beta < 1, \qquad (2.3.10)$$

$$\int_{0}^{n} u^{-\beta} L(u) du \sim \frac{n^{1-\beta} L(n)}{1 - \beta}, \qquad 0 < \beta < 1, \qquad (2.3.11)$$

$$\sum_{j=n}^{\infty} j^{-\beta} L(j) \sim \frac{n^{1-\beta} L(n)}{\beta - 1}, \qquad \beta > 1. \qquad (2.3.12)$$

Proof. To prove (2.3.10), write

$$\frac{1}{L(n)n^{1-\beta}} \sum_{j=1}^{n} j^{-\beta} L(j) = n^{-1} \sum_{j=1}^{n} \left(\frac{j}{n}\right)^{-\beta} + n^{-1} \sum_{j=1}^{n} \left(\frac{j}{n}\right)^{-\beta} \left(\frac{L(j)}{L(n)} - 1\right)$$

$$=: s_{n,1} + s_{n,2},$$

where $s_{n,1} \sim \int_{0}^{1} u^{-\beta} du = (1 - \beta)^{-1}$. We show that

$$s_{n,2} \to 0, \qquad (2.3.13)$$

which yields (2.3.10). Apply (2.3.4) and (2.3.5) to the r.h.s. of

$$\left|\frac{L(j)}{L(n)} - 1\right| = \left|\frac{L((\frac{j}{n})n)}{L(n)} - 1\right|,$$

to conclude that for any $0 < \epsilon < 1$ and $0 < \delta < 1 - \beta$,

$$\max_{\epsilon n \leq j \leq n} \left|\frac{L(j)}{L(n)} - 1\right| \to 0,$$

$$\left|\frac{L(j)}{L(n)} - 1\right| \leq C \left(\frac{n}{j}\right)^{\delta}, \qquad j = 1, \cdots, [\epsilon n].$$

Thus

$$|s_{n,2}| \leq o(1) n^{-1} \sum_{j=[\epsilon n]}^{n} \left(\frac{j}{n}\right)^{-\beta} + C n^{-1} \sum_{j=1}^{[\epsilon n]-1} \left(\frac{j}{n}\right)^{-\beta-\delta}$$

$$\leq o(1) \int_{\epsilon}^{1} u^{-\beta} du + C \int_{0}^{\epsilon} u^{-\beta-\delta} du \to 0, \quad n \to 0, \quad \epsilon \to 0,$$

which proves (2.3.13).

The proofs of (2.3.11) and (2.3.12) are similar to that of (2.3.10). This completes the proof of Proposition 2.3.1.

Until now we considered functions L slowly varying at infinity. Sometimes we also need to consider functions that vary slowly at zero. A realvalued function b on $[0, \pi]$ is said to be *slowly varying at* 0, if b is positive on $(0, a]$ for some $a > 0$, and $\lim_{v \to 0+} b(vt)/b(v) = 1$, for all $t > 0$. It is said to be *Zygmund slowly varying at* 0 if in addition for all $\delta > 0$ and some $v_0 > 0$, $v^\delta b(v)$ is increasing and $v^{-\delta} b(v)$ decreasing in v, for $0 < v \leq v_0$.

Relations (2.3.7) and (2.3.8) can be reversed in the following sense. Let $0 < \beta < 1$, and let a function b be Zygmund slowly varying at 0, and of bounded variation in every interval $(\epsilon, \pi]$, $\epsilon > 0$. Then, as $k \to \infty$,

$$\int_0^\pi \cos(kv)|v|^{-\beta} b(v) dv \sim \Gamma(1 - \beta) \sin\left(\frac{\pi\beta}{2}\right) k^{-1+\beta} b\left(\frac{1}{k}\right), \quad (2.3.14)$$

$$\int_0^\pi \sin(kv)|v|^{-\beta} b(v) dv \sim \Gamma(1 - \beta) \cos\left(\frac{\pi\beta}{2}\right) k^{-1+\beta} b\left(\frac{1}{k}\right). \quad (2.3.15)$$

The next lemma provides a sufficient condition for a function b to be Zygmund slowly varying at 0.

Lemma 2.3.1. *Let b be slowly varying at 0 and differentiable in $(0, a]$ for some $a > 0$ with derivative $\dot{b}$ such that*

$$b(v) > 0, \quad v\dot{b}(v)/b(v) \to 0, \quad v \to 0+. \quad (2.3.16)$$

Then b is Zygmund slowly varying at 0.

If in addition b is piecewise differentiable and $\dot{b}$ is integrable on $(0, \pi)$, then b has a bounded variation on every interval $(\epsilon, \pi]$, $\epsilon > 0$.

Proof. Let $\delta \neq 0$. By assumption (2.3.16), the sign of derivative $d(v^\delta b(v))/dv$, as $v \to 0+$, is the same as that of δ, because

$$\frac{d}{dv}(v^\delta b(v)) = \delta v^{\delta-1} b(v) + v^\delta \dot{b}(v)$$

$$= \delta v^{\delta-1} b(v)\left(1 + \frac{v\dot{b}(v)}{\delta b(v)}\right) \sim \delta v^{\delta-1} b(v), \quad v \to 0+.$$

Hence, $v^\delta b(v) \nearrow$, if $\delta > 0$, and $v^\delta b(v) \searrow$, if $\delta < 0$, for all $v \leq v_0$, for some $v_0 > 0$, i.e., b is Zygmund slowly varying at 0. The second claim of the lemma is a well-known fact.

Note: Various other useful facts about slowly and regularly varying functions can be found in Bingham, Goldie and Teugels (1987). Zygmund I and II (2002) contains a variety of results on the properties of trigonometric series involving hyperbolically decaying functions, see e.g., Theorem V.2.6 and Theorem V.2.24.

2.4   Hermite and Appell polynomials

This section summarizes some basic facts about Hermite and Appell polynomials that play an important role in the following study of properties of non-linear transforms of stationary Gaussian and linear processes. Usually, Hermite polynomials are defined by

$$H_k(x) := (-1)^k e^{x^2/2} \frac{d^k e^{-x^2/2}}{dx^k}, \quad x \in \mathbb{R}, \ k = 0, 1, \cdots . \qquad (2.4.1)$$

They can also be defined by the power series (generating function)

$$\psi(x,t) := e^{tx - t^2/2} = \sum_{k=0}^{\infty} \frac{t^k}{k!} H_k(x), \quad x, t \in \mathbb{R}, \qquad (2.4.2)$$

and arise as the coefficients of the powers of t. This sum converges for all values of x and t. To see the equivalence of definitions in (2.4.1) and (2.4.2), note

$$H_k(x) = \frac{d^k}{dt^k} \psi(x,t)\Big|_{t=0} = e^{x^2/2} \frac{d^k}{dt^k} e^{-(t-x)^2/2}\Big|_{t=0}$$

$$= (-1)^k e^{x^2/2} \frac{d^k}{dx^k} e^{-(t-x)^2/2}\Big|_{t=0}$$

$$= (-1)^k e^{x^2/2} \frac{d^k}{dx^k} e^{-x^2/2}.$$

Furthermore, since $t \to e^{tx - t^2/2}$ is analytic on the whole complex plane, the power series (2.4.2) can be extended to imaginary $t = \mathbf{i}u, u \in \mathbb{R}$, yielding

$$e^{\mathbf{i}xu + u^2/2} = \sum_{k=0}^{\infty} \mathbf{i}^k \frac{u^k}{k!} H_k(x), \quad x, t \in \mathbb{R}, \qquad (2.4.3)$$

$$H_k(x) := (-\mathbf{i})^k \frac{d^k e^{\mathbf{i}xu + u^2/2}}{du^k}\Big|_{u=0}, \quad k = 0, 1, \cdots .$$

Let $Z \sim \mathcal{N}(0,1)$ and Φ, $\phi(x)$ and $\widehat{\phi}(u)$ denote the distribution function, probability density, and characteristic function of Z, respectively, i.e.,

$$\phi(x) = (2\pi)^{-1/2} e^{-x^2/2}, \quad x \in \mathbb{R},$$

$$\widehat{\phi}(u) := E e^{\mathbf{i}uZ} = e^{-u^2/2}, \quad u \in \mathbb{R}.$$

Then the generating function (2.4.3) can be rewritten as

$$e^{\mathbf{i}xu+u^2/2} = \frac{e^{\mathbf{i}xu}}{E\exp\{\mathbf{i}uZ\}} = \sum_{k=0}^{\infty} \mathbf{i}^k \frac{u^k}{k!} H_k(x).$$

This form of generating function shows the relation between the Hermite polynomials and the standard normal distribution:

$$H_k(x) := (-\mathbf{i})^k \frac{d^k}{du^k}\left(\frac{e^{\mathbf{i}xu}}{E\exp\{\mathbf{i}uZ\}}\right)\Big|_{u=0}, \quad k = 0, 1, \cdots. \qquad (2.4.4)$$

Direct calculations show that

$$H_k(-x) = (-1)^k H_k(x), \qquad \forall\, k \geq 0,$$
$$H_0(x) = 1, \quad H_1(x) = x, \quad H_2(x) = x^2 - 1,$$
$$H_3(x) = x^3 - 3x, \quad H_4(x) = x^4 - 6x^2 + 3.$$

We summarize some of the main properties of Hermite polynomials:

$$\dot{H}_k(x) = kH_{k-1}(x), \qquad (2.4.5)$$
$$H_{k+1}(x) - xH_k(x) + kH_{k-1}(x) = 0, \qquad (2.4.6)$$
$$\ddot{H}_k(x) - x\dot{H}_k(x) + kH_k(x) = 0, \qquad (2.4.7)$$
$$\phi(x)H_k(x) = (-1)^k \phi^{(k)}(x), \qquad k \geq 1, x \in \mathbb{R}, \qquad (2.4.8)$$

where $\phi^{(k)}(x)$ denotes the kth derivative of $\phi(x)$.

To prove (2.4.5), differentiate (2.4.2) w.r.t. x and observe that

$$\sum_{k=1}^{\infty} \frac{t^k}{(k-1)!} H_{k-1}(x) = t\psi(x,t) = \frac{\partial\psi(x,t)}{\partial x} = \sum_{k=1}^{\infty} \frac{t^k}{k!} \dot{H}_k(x).$$

Equate the coefficients of these power series to obtain (2.4.5). Relation (2.4.6) follows similarly, by differentiating both sides of (2.4.2) w.r.t. t.

Next, differentiate (2.4.5) and (2.4.6) w.r.t. x to obtain

$$k\dot{H}_{k-1}(x) - x\dot{H}_k(x) + kH_k(x) = 0, \quad \ddot{H}_k(x) = k\dot{H}_{k-1}(x),$$

which proves (2.4.7). Finally, (2.4.8) is equivalent to (2.4.1).

The r.v.'s $H_k(Z)$, $k = 1, 2, \cdots$ are orthogonal, i.e.,

$$EH_k(Z) = 0, \quad EH_k^2(Z) = k!, \qquad k = 1, 2, \cdots, \qquad (2.4.9)$$
$$EH_k(Z)H_m(Z) = 0, \qquad k \neq m = 0, 1, 2, \cdots.$$

The formulas in (2.4.9) easily follow from multiplying the series (2.4.3) at the points u and v and taking the expectation of the product:

$$\sum_{k,j=0}^{\infty} \frac{(iu)^k (iv)^j}{k!j!} E[H_k(Z)H_j(Z)] \tag{2.4.10}$$

$$= E[e^{i(u+v)Z}]e^{u^2/2+v^2/2} = e^{-uv} = \sum_{k=0}^{\infty} \frac{(-1)^k u^k v^k}{k!}$$

and equating the coefficients of $u^k v^j$, $k, j = 0, 1, \cdots$, of the power series on both sides.

Alternately, (2.4.9) is equivalent to saying that Hermite polynomials are orthogonal with respect to the Gaussian measure $\Phi(dx) = \phi(x)dx$: for $k, m = 0, 1, \cdots$,

$$\int_{\mathbb{R}} H_k(x)\phi(x)dx = 0, \quad \int_{\mathbb{R}} H_k^2(x)\phi(x)dx = k!, \tag{2.4.11}$$

$$\int_{\mathbb{R}} H_k(x)H_m(x)\phi(x)dx = 0, \qquad k \neq m.$$

Hermite expansion and rank. It is known that Hermite polynomials H_k, $k = 0, 1, \cdots$, form a complete orthogonal system in the space $L_2(\mathbb{R}, \Phi)$ of Φ-square integrable functions $h : \mathbb{R} \to \mathbb{R}$. Any function $h \in L_2(\mathbb{R}, \Phi)$ can be written as

$$h(x) = \sum_{k=0}^{\infty} \frac{J_k}{k!} H_k(x), \quad \text{a.s. and in } L_2(\mathbb{R}, \Phi), \tag{2.4.12}$$

$$J_k := Eh(Z)H_k(Z) = \int_{\mathbb{R}} h(x)H_k(x)\phi(x)dx, \ k \geq 0.$$

The numbers J_k are called Hermite expansion coefficients. By (2.4.9),

$$Eh(Z) = J_0, \quad \text{Var}(h(Z)) = \sum_{k=1}^{\infty} \frac{J_k^2}{k!}. \tag{2.4.13}$$

If h has kth derivative $h^{(k)} \in L_2(\mathbb{R}, \Phi)$, then in view of the equality (2.4.8), J_k can be written as

$$J_k = (-1)^k \int_{\mathbb{R}} h(x)\phi^{(k)}(x)dx = \int_{\mathbb{R}} h^{(k)}(x)\phi(x)dx = Eh^{(k)}(Z).$$

The *Hermite rank* of an $h \in L_2(\mathbb{R}, \Phi)$ is the index of the first non-zero Hermite coefficient J_k:

$$\text{H-rank}(h) := \min\{k \geq 0; J_k \neq 0\}. \tag{2.4.14}$$

For example, the H-rank of odd powers $x, x^3, x^5, \cdots$ equals 1, and the H-rank of even powers $x^2, x^4, x^6, \cdots$ equals 2. Equality (2.4.3) is the Hermite expansion of the function $h(x) = e^{iux + u^2/2}$.

As another example, consider the family of indicator functions $h_y(x) := I(x \leq y) - \Phi(y)$, $y \in \mathbb{R}$. The H-rank of the function h_y for a fixed y is also 1 because

$$J_0 = E[I(Z \leq y) - \Phi(y)] = 0,$$

$$J_1 = EZ[I(Z \leq y) - \Phi(y)] = \int_{-\infty}^{y} z\phi(z)dz = -\phi(y) \neq 0.$$

The Hermite expansion of the bivariate normal density is also a useful tool. Let X, Y be bivariate normal r.v.'s with means zero, unit variances, and correlation coefficient ρ, $|\rho| < 1$, having the joint density

$$\phi(x, y) := (2\pi\sqrt{(1-\rho^2)})^{-1} \exp\left\{-\frac{1}{2(1-\rho^2)}(x^2 + y^2 - 2\rho xy)\right\}, \quad x, y \in \mathbb{R}.$$

The corresponding characteristic function is

$$\widehat{\phi}(u, v) = \exp\{-(u^2 + v^2)/2 - \rho uv\}, \quad u, v \in \mathbb{R}.$$

The next proposition shows that the orthogonality property of Hermite polynomials extends to bivariate products $EH_k(X)H_j(Y)$, and that the bivariate density $\phi(x, y)$ can be expressed in terms of marginal densities, the correlation coefficient ρ, and Hermite polynomials.

Proposition 2.4.1. (i) (Orthogonality property):

$$EH_k(X)H_j(Y) = 0, \qquad k \neq j, \qquad (2.4.15)$$
$$= \rho^k k!, \qquad k = j.$$

(ii) (Factorization property):

$$\phi(x, y) = \phi(x)\phi(y) \sum_{k=0}^{\infty} \frac{\rho^k}{k!} H_k(x)H_k(y) \qquad (2.4.16)$$

$$= \sum_{k=0}^{\infty} \frac{\rho^k}{k!} \phi^{(k)}(x)\phi^{(k)}(y), \quad x, y \in \mathbb{R}.$$

Proof. (i) As for (2.4.10), multiply the generating functions $e^{iXu + u^2/2}$ and $e^{iYv + v^2/2}$, use (2.4.3), and take the expectation to obtain

$$\sum_{k,j=0}^{\infty} \frac{(iu)^k (iv)^j}{k!j!} EH_k(X)H_j(Y)$$

$$= Ee^{i(uX + vY)}e^{(u^2+v^2)/2} = e^{-\rho uv} = \sum_{k=0}^{\infty} \frac{(-1)^k \rho^k u^k v^k}{k!}.$$

Claim (2.4.15) follows from this equality by comparing coefficients of the powers $u^k v^j$ on both sides.

(ii) By definition,

$$\int_{\mathbb{R}^2} e^{\mathbf{i}(ux+vy)} \phi(x,y)dxdy = \widehat{\phi}(u,v), \qquad u,v \in \mathbb{R}.$$

It suffices to show that the last equality remains valid with $\phi(x,y)$ replaced by the r.h.s. of (2.4.16), denoted by $\phi^*(x,y)$. Thus

$$I := \int_{\mathbb{R}^2} e^{\mathbf{i}(ux+vy)} \phi^*(x,y)dxdy$$

$$= \sum_{k=0}^{\infty} \frac{\rho^k}{k!} \int_{\mathbb{R}} e^{\mathbf{i}ux} H_k(x)\phi(x)dx \int_{\mathbb{R}} e^{\mathbf{i}vy} H_k(y)\phi(y)dy.$$

Property (2.4.8) and integration by parts yield

$$\int_{\mathbb{R}} e^{\mathbf{i}ux} H_k(x)\phi(x)dx = (-1)^k \int_{\mathbb{R}} e^{\mathbf{i}ux}\phi^{(k)}(x)dx$$

$$= (\mathbf{i}u)^k \int_{\mathbb{R}} e^{\mathbf{i}ux}\phi(x)dx = (\mathbf{i}u)^k e^{-u^2/2}.$$

Hence,

$$I = e^{-(u^2+v^2)/2} \sum_{k=0}^{\infty} \frac{\rho^k}{k!} (\mathbf{i}u)^k (\mathbf{i}v)^k = e^{-(u^2+v^2)/2-\rho uv} = \widehat{\phi}(u,v),$$

which completes the proof of the first equality in (2.4.16). The second equality follows from (2.4.8), which also completes the proof.

Furthermore, let h_1, h_2 be two real functions in $L^2(\mathbb{R}, \Phi)$ and

$$J_{jk} := \int_{\mathbb{R}} h_j(x) H_k(x)\phi(x)dx, \qquad j = 1,2, \ k \geq 0,$$

denote Hermite coefficients of their corresponding Hermite expansions. Then, by (2.4.12) and (2.4.15),

$$Eh_1(X)h_2(Y) = \sum_{k,j=0}^{\infty} \frac{J_{1k}J_{2j}}{k!\,j!} EH_k(X)H_j(Y) = \sum_{k=0}^{\infty} \frac{J_{1k}J_{2k}}{k!} \rho^k, \quad (2.4.17)$$

where (X,Y) are the bivariate normal r.v.'s as above.

Let $\kappa = \max(\text{H-rank}(h_1), \text{H-rank}(h_2))$. Then, from (2.4.17) and the C–S inequality, one obtains

$$|\text{Cov}(h_1(X), h_2(Y))| \leq |\rho|^{\kappa} \sum_{k=\kappa}^{\infty} \frac{|J_{1k}J_{2k}|}{k!} \qquad (2.4.18)$$

$$\leq |\rho|^{\kappa} \Big(\sum_{k=\kappa}^{\infty} \frac{J_{1k}^2}{k!} \Big)^{1/2} \Big(\sum_{k=\kappa}^{\infty} \frac{J_{2k}^2}{k!} \Big)^{1/2}$$

$$\leq |\rho|^{\kappa} \sqrt{\text{Var}(h_1(X))\text{Var}(h_2(Y))}.$$

This inequality implies, in particular, that the supremum of $|\text{Cov}(h_1(X),$ $h_2(Y))|$, over all those functions h_1 and h_2 that have $\text{Var}(h_1(X)) = 1 = \text{Var}(h_2(Y))$, is $|\rho|$. This supremum is attained by the identity functions $h_1(x) \equiv h_2(x) \equiv x$.

Appell polynomials. Let X be an arbitrary r.v. having all moments finite. Appell polynomials $A_k(x)$, $x \in \mathbb{R}$, $k = 0, 1, \cdots$, of the r.v. X are defined by the power series (generating function) that involves the characteristic function of X:

$$\frac{e^{iux}}{E \exp\{iuX\}} = \sum_{k=0}^{\infty} i^k \frac{u^k}{k!} A_k(x). \qquad (2.4.19)$$

They can be computed using the rule

$$A_k(x) = (-i)^k \frac{\partial^k}{\partial u^k} \left(\frac{\exp\{iux\}}{E \exp\{iuX\}} \right) \Big|_{u=0}. \qquad (2.4.20)$$

For any $k \geq 0$, $A_k(x)$ is a polynomial of degree k in x whose coefficients are expressed in terms of moments $\mu_j = EX^j$, $j \leq k$. In particular,

$$A_0(x) = 1, \quad A_1(x) = x - \mu_1, \quad A_2(x) = x^2 - 2\mu_1 x + 2\mu_1^2 - \mu_2,$$
$$A_3(x) = x^3 - 3\mu_1 x^2 + 3x(2\mu_1^2 - \mu_2) + 6\mu_1\mu_2 - \mu_3 - 6\mu_1^3.$$

These polynomials satisfy

$$\dot{A}_k(x) = kA_{k-1}(x), \quad EA_k(X) = 0, \quad k = 1, 2, \cdots. \qquad (2.4.21)$$

These facts are proved as follows. Take the derivative on both sides of (2.4.19) with respect to x to obtain

$$iu \sum_{k=0}^{\infty} i^k \frac{u^k}{k!} A_k(x) = \sum_{k=1}^{\infty} i^k \frac{u^k}{k!} \dot{A}_k(x) = iu \sum_{k=0}^{\infty} i^k \frac{u^k}{k!} \frac{\dot{A}_{k+1}(x)}{k+1}.$$

The first claim in (2.4.21) follows by equating the coefficients in this power series. The second claim in (2.4.21) follows by taking expectations on both sides in (2.4.19) after setting $x = X$. This leads to the identity $1 = \sum_{k=0}^{\infty} i^k (u^k/k!) EA_k(X)$, or $EA_k(X) = 0$, $k \geq 1$. Iterating the differentiation rule in (2.4.21) we obtain

$$A_k^{(j)}(x) = k(k-1)\cdots(k-j+1)A_{k-j}(x), \quad 1 \leq j < k, \qquad (2.4.22)$$
$$= k!, \qquad j = k,$$
$$= 0, \qquad j > k.$$

An important property of these polynomials is location invariance in the sense that if we write $A_{k,X}(x)$ for Appell polynomials of X, then

$$A_{k,X-c}(x-c) = A_{k,X}(x), \qquad x, \ c \in \mathbb{R}, \ k \geq 0. \tag{2.4.23}$$

If $X \sim \mathcal{N}(0,1)$, the Appell polynomials coincide with the Hermite polynomials but, in general, unlike these polynomials, Appell polynomials are generally not orthogonal, i.e., for some $k, m, k \neq m$, $E[A_k(X)A_m(X)] \neq 0$.

Any polynomial $h(x), x \in \mathbb{R}$ of degree $m \geq 0$ can be expanded using Appell polynomials:

$$h(x) = \sum_{j=0}^{m} \frac{h_j}{j!} A_j(x). \tag{2.4.24}$$

The coefficients h_j of this expansion can be determined from properties (2.4.21) and (2.4.22). Indeed, for any $0 \leq j \leq m$ these properties yield $Eh^{(j)}(X) = \sum_{k=j}^{m} (h_k/k!)k(k-1)\cdots(k-j+1)EA_{k-j}(X) = h_j$, or

$$h_j = Eh^{(j)}(X), \qquad 0 \leq j \leq m. \tag{2.4.25}$$

The *finite* Appell expansion of order $m \geq 1$ of a non-linear function $h(X)$ is given by

$$h(X) - Eh(X) = \sum_{k=1}^{m} \frac{h_k}{k!} A_k(X) + R_m(X), \tag{2.4.26}$$

provided the coefficients h_k are well defined. When h is smooth, $h_k := Eh^{(k)}(X)$ as in (2.4.25). If h is not smooth but the function $Eh(X+y)$ is smooth at $y=0$, then $h_k := [Eh(X+y)]_{y=0}^{(k)}, k=0,1,\cdots$. In particular, if X has a smooth density f whose derivatives $f^{(k)}, k \leq m$ decay sufficiently fast at $\pm\infty$, the coefficients h_k can be defined as

$$h_k := (-1)^k \int h(x)f^{(k)}(x)dx.$$

When not all of the h_k's in (2.4.26) are equal to zero, the Appell rank of h is defined to be

$$\text{A-rank}(h) := \min\{1 \leq k \leq m; h_k \neq 0\}. \tag{2.4.27}$$

Because of the lack of the orthogonality property, infinite-order Appell expansions require strong analytic assumptions on h and are not particularly useful, at least for the purposes of the present book. However, a finite-order Appell expansion of (2.4.26) involving a remainder term $R_m(X)$ exists under fairly general conditions on h and X. As will be seen in Chapters 10 and 11, such an expansion is found to be useful for approximating the sums of non-linear functions of stationary linear process $X_j, j \in \mathbb{Z}$ and finding the dominating term in the expansion of $h(X_j)$, determined by the Appell rank. A particularly useful case of (2.4.26) corresponds to the first-order Appell expansion of h with A-rank$(h) = 1$:

$$h(X_j) - Eh(X_j) = h_1 X_j + R_1(X_j), \quad h_1 = -\int h(x)\dot{f}(x)dx \neq 0.$$

For linear long memory processes, the linear term $h_1 X_j$ of the above decomposition dominates the remainder term $R_1(X_j)$ in the sense that the autocovariances of the latter process decay faster with the lag as compared to the autocovariances of the linear term. Therefore, when analyzing the limit distribution of sums $\sum_{j=1}^{n} h(X_j)$, one can often replace the non-liner function $h(X_j)$ by a linear term $h_1 X_j$, thus reducing the problem to a much simpler question for linear sums.

Multivariate extensions of Appell and Hermite polynomials are discussed in Chapter 14.

Note: The inequality (2.4.18) is also derived in Rozanov (1967). For more details on Hermite and Appell polynomials see Sansone (1959), Titchmarsh (1986), Giraitis and Surgailis (1986), and Avram and Taqqu (1987).

2.5 Miscellaneous preliminaries

This section contains a central limit theorem for sums of martingale arrays, some moment inequalities, and the statement of Lusin's theorem. Most of the results are presented for the sake of easy reference and without proofs. They are drawn from Natanson (1964) and Hall and Heyde (1980).

CLT for martingale-difference arrays. Corollary 3.1 of Hall and Heyde (1980) provides a useful criterion for the asymptotic normality of sums of martingale differences. For easy reference, we state it as a lemma.

Let $\{\mathcal{F}_{n,i}, 1 \leq i \leq n\}$ be an array of non-decreasing σ-fields $\mathcal{F}_{n,i} \subset \mathcal{F}_{n,i+1}$; $\mathcal{F}_{n,0}$ a trivial σ-field, X_{ni} be $\mathcal{F}_{n,i}$ measurable with finite variance $EX_{ni}^2 < \infty$, $E(X_{ni}|\mathcal{F}_{n,i-1}) = 0$, $1 \leq i \leq n$, and $S_{nj} := \sum_{i=1}^{j} X_{ni}, 1 \leq j \leq n$, $S_{n0} := 0$. Then $\{S_{ni}, \mathcal{F}_{n,i}; 0 \leq i \leq n, n \geq 1\}$ is called a zero-mean square integrable martingale array with differences $\{X_{ni}; 1 \leq i \leq n, n \geq 1\}$.

Lemma 2.5.1. *Let* $\{S_{ni}, \mathcal{F}_{n,i}; 0 \le i \le n, n \ge 1\}$ *be a zero-mean square integrable martingale array with differences* $\{X_{ni}\}$. *Suppose additionally* $\mathcal{F}_{n,i} \subset \mathcal{F}_{n+1,i}$, *for all* $1 \le i \le n$ *and the following conditions hold:*

$$\forall \epsilon > 0, \ \sum_{i=1}^{n} E[X_{ni}^2 I(|X_{ni}| > \epsilon)|\mathcal{F}_{n,i-1}] \to_p 0, \tag{2.5.1}$$

$$\sum_{i=1}^{n} E[X_{ni}^2|\mathcal{F}_{n,i-1}] \to_p \eta^2, \tag{2.5.2}$$

where η *is a positive r.v. Then* $S_{nn} \to_D Y$, *where the r.v.* Y *has characteristic function* $E \exp(-\eta^2 t^2/2)$, $t \in \mathbb{R}$.

If (2.5.2) holds with a constant limit $\eta^2 = a^2$, *then* $S_{nn} \to_D \mathcal{N}(0, a^2)$, *without requiring the condition* $\mathcal{F}_{n,i} \subset \mathcal{F}_{n+1,i}$, *for all* $1 \le i \le n$.

Moment inequalities. We shall use the following two inequalities.

Lemma 2.5.2. *Let* $p \ge 1$ *and* $\{Y_j, \mathcal{F}_j, 1 \le j \le n\}$ *be a martingale-difference sequence with* $E|Y_j|^p < \infty$. *Then, for every* $n \ge 1$,

$$E\Big|\sum_{j=1}^{n} Y_j\Big|^p \le 2 \sum_{j=1}^{n} E|Y_j|^p, \qquad\qquad 1 \le p \le 2, \tag{2.5.3}$$

$$\le C_p \Big(\sum_{j=1}^{n} (E|Y_j|^p)^{2/p} \Big)^{p/2}, \qquad p > 2,$$

with a constant $C_p > 0$ *depending only on* p.

The inequalities (2.5.3) remain valid for $n = \infty$ *as long as the series on the r.h.s. of (2.5.3) is convergent.*

Before proceeding to prove this lemma, we note that the inequality (2.5.3) implies the following bound.

Corollary 2.5.1. *Let* $S = \sum_{j \in \mathbb{Z}} b_j \zeta_j$, *with* $\{\zeta_j\} \sim \text{IID}(0, \sigma_\zeta^2)$, $E|\zeta_0|^p < \infty$, *and* $p \ge 2$. *Then, for any real numbers* $b_j, j \in \mathbb{Z}$, *with* $\sum_{j \in \mathbb{Z}} b_j^2 < \infty$,

$$E|S|^p \le C_p E|\zeta_0|^p \sigma_\zeta^{-p} (ES^2)^{p/2}, \tag{2.5.4}$$

where C_p *depends on* p *but not on the* b_j's *and* $\{\zeta_j\}$.

Proof. Let $1 \le n < \infty$. For $1 \le p \le 2$, (2.5.3) is the well-known von Bahr and Esséen (1965) inequality. For $p \ge 2$, it follows from Rosenthal's inequality (see Hall and Heyde (1980), p. 24):

$$E\Big|\sum_{j=1}^{n} Y_j\Big|^p \le C_p\Big[\sum_{j=1}^{n} E|Y_j|^p + E\Big(\sum_{j=1}^{n} E[Y_j^2|\mathcal{F}_{j-1}]\Big)^{p/2}\Big]. \tag{2.5.5}$$

To show that this inequality implies (2.5.3), recall that for any real numbers a, b, and for $0 < \alpha < 1$, $|a + b|^\alpha \leq |a|^\alpha + |b|^\alpha$. Since $2/p < 1$, apply this fact with $\alpha = 2/p$ to obtain

$$\left(\sum_{j=1}^{n} E|Y_j|^p \right)^{2/p} \leq \sum_{j=1}^{n} \left(E|Y_j|^p \right)^{2/p}.$$

By the C–S inequality applied to the conditional expectation with $p/2 \geq 1$, $E[Y_j^2|\mathcal{F}_{j-1}] \leq (E[|Y_j|^p|\mathcal{F}_{j-1}])^{2/p}$, $1 \leq j \leq n$. The Minkowski inequality says that for any r.v.'s X and Y,

$$(E|X + Y|^r)^{1/r} \leq (E|X|^r)^{1/r} + (E|Y|^r)^{1/r}, \quad r \geq 1.$$

These facts in turn imply that the second term in the upper bound of (2.5.5) is bounded above by

$$E\left(\sum_{j=1}^{n} E[Y_j^2|\mathcal{F}_{j-1}] \right)^{p/2} \leq E\left(\sum_{j=1}^{n} (E[|Y_j|^p|\mathcal{F}_{j-1}])^{2/p} \right)^{p/2}$$

$$\leq \left(\sum_{j=1}^{n} \left(E|Y_j|^p \right)^{2/p} \right)^{p/2}.$$

Consider now the case of an infinite sum. Let $p \geq 2$. We claim

$$E\left| \sum_{j=1}^{\infty} Y_j \right|^p \leq C_p \left(\sum_{j=1}^{\infty} (E|Y_j|^p)^{2/p} \right)^{p/2}. \tag{2.5.6}$$

This inequality is trivially true if the r.h.s. of (2.5.6) is infinite. Assume it is finite. By (2.5.3), for any $1 < m \leq n < \infty$,

$$E\left| \sum_{j=m}^{n} Y_j \right|^p \leq C_p \left(\sum_{j=m}^{n} (E|Y_j|^p)^{2/p} \right)^{p/2}.$$

But this upper bound tends to zero, as $m, n \to \infty$. Hence, by the Cauchy convergence criterion, $\sum_{j=1}^{\infty} Y_j$ converges in L_p norm, and $E|\sum_{j=n+1}^{\infty} Y_j|^p \to 0$, as $n \to \infty$. Next, by the Minkowski inequality,

$$E^{1/p}\left| \sum_{j=1}^{\infty} Y_j \right|^p \leq E^{1/p}\left| \sum_{j=1}^{n} Y_j \right|^p + E^{1/p}\left| \sum_{j=n+1}^{\infty} Y_j \right|^p.$$

Passing to the limit $n \to \infty$, one obtains

$$E\left| \sum_{j=1}^{\infty} Y_j \right|^p \leq \lim_{n \to \infty} E\left| \sum_{j=1}^{n} Y_j \right|^p. \tag{2.5.7}$$

Apply inequality (2.5.3) to the r.h.s. of (2.5.7) to obtain (2.5.6).

For $1 \leq p \leq 2$, extension of the inequality (2.5.3) to the case $n = \infty$ follows using a similar argument as for $p > 2$. This completes the proof.

Lusin's theorem. Occasionally we also need the following result which appears as Theorem 4, Chapter 5 in Natanson (1964).

Theorem 2.5.1. *Let $-\infty < a < b < \infty$. Let f be a measurable function defined on $[a, b]$ which is finite almost everywhere w.r.t. the Lebesgue measure λ. Then, $\forall \delta > 0$, there exists a continuous function g such that $\lambda\{x \in [a, b]; f(x) \neq g(x)\} < \delta$. If f is bounded, then $\sup_{a \leq u \leq b} |g(u)| \leq \sup_{a \leq u \leq b} |f(u)|$.*

Ergodic theorem. Consider a stationary sequence $\{X_j, j \geq 1\}$. Any event A from the σ-field$\{X_j, j \geq 1\}$ can be written as $A = [(X_1, X_2, \cdots) \in C]$, for some Borel set C of $\mathbb{R}^\infty$.

The event A is said to be *invariant* if $A = [(X_k, X_{k+1}, \cdots) \in C]$, for all $k \geq 1$. A stationary sequence $\{X_j, j \geq 1\}$ is said to be *ergodic* if every invariant event has probability zero or one.

The following two results are often used in this text. They are, respectively, Theorems 3.5.7 and 3.5.8 in Stout (1974).

Theorem 2.5.2. *Let $\{X_j, j \geq 1\}$ be a stationary ergodic sequence. Then the following hold:*

(i) *If $E|X_1| < \infty$, then $n^{-1} \sum_{j=1}^{n} X_j \to EX_1$, a.s.*

(ii) *If $\phi : \mathbb{R}^\infty \to \mathbb{R}$ is a measurable function and $Y_j := \phi(X_j, X_{j+1}, \cdots)$, $j \geq 1$, then the sequence $\{Y_j, j \geq 1\}$ is stationary ergodic.*

The result (i) is known as the ergodic theorem (ET). It is well known that any i.i.d. sequence is ergodic. A stationary Gaussian process is ergodic if and only if its spectral distribution function is continuous, and in particular, if spectral density exists; see Rozanov (1967). The above facts combined with Theorem 2.5.2(ii) imply ergodicity of a linear process $X_j = \sum_{k=0}^{\infty} a_k \zeta_{j-k}$ in i.i.d. innovations $\{\zeta_j\}$, and non-linear functions $\{h(X_j)\}$ of an ergodic linear or Gaussian process discussed in the subsequent chapters.

Summation by parts formula. The following formula is used later in this book. For any two sequences $\{y_j, j = 1, 2, \cdots\}$ and $\{z_j, j = 1, 2, \cdots\}$,

$$\sum_{j=1}^{n} y_j z_j = z_n \sum_{j=1}^{n} y_j + \sum_{j=1}^{n-1} (z_j - z_{j+1}) \sum_{k=1}^{j} y_k. \qquad (2.5.8)$$

Chapter 3

Long Memory Processes

In this chapter we shall review short memory processes, define negative memory and long memory processes, and show how they arise in theory.

3.1 Characterization of short and long memory processes

A stationary process is a sequence of random variables that are dependent in time. There exist various characteristics describing the dependence structure of a stationary process, which can be placed between the two extreme scenarios of dependence. The first extreme scenario corresponds to an i.i.d. sequence $\{\zeta_j\} = \{\zeta_j, j \in \mathbb{Z}\}$ with mean μ and variance σ^2, i.e., $\{\zeta_j\} \sim \text{IID}(\mu, \sigma^2)$. The other extreme is a completely dependent sequence $\{X_j \equiv X, j \in \mathbb{Z}\}$, where X is a given r.v., which allows for only trivial inference. In between these two extremes are many other stationary processes, in particular, α-mixing and m-dependent processes defined as follows. Let

$$\alpha(k) := \sup\left\{|P(AB) - P(A)P(B)| : A \in \mathcal{F}_0^-(X), \ B \in \mathcal{F}_k^+(X)\right\},$$

where $\mathcal{F}_k^-(X)$ and $\mathcal{F}_k^+(X)$ are the σ-fields generated by the "past information" $X_s, s \le k$ and the "future information" $X_s, s > k$, respectively.

Definition 3.1.1. A stationary process $\{X_j\}$ is said to be α-mixing if $\alpha(k) \to 0$, as $k \to \infty$.
Given a positive integer m, a stationary process $\{X_j\}$ is called m-dependent if $\alpha(k) = 0, \forall k > m$.

In many cases, α-mixing processes $\{X_j\}$ have asymptotically similar properties as ARMA, including the fast decay of dependence and the correlation between observations X_j and X_k, as the distance $|j - k|$ in time increases.

Under m-dependence, the collections of variables $\{X_s, s \leq k\}$ and $\{X_s, s > k + m\}$ are independent for any $k \in \mathbb{Z}$. A simple example of an m-dependent stationary process is a moving-average process $X_j = a_0 \zeta_j + \cdots + a_m \zeta_{j-m}$, where $\{\zeta_j\} \sim \text{IID}(0, \sigma^2)$.

The rate of decay of the mixing coefficients $\alpha(k)$, as $k \to \infty$, characterizes the degree of dependence between "past" and "future" and between distant observations in time. However, it does not impose any additional assumptions on the structure of the process $\{X_j\}$. Various other measures of dependence and classes of mixing processes have been introduced in the literature; see Ibragimov and Linnik (1971), Dedecker, and Prieur (2007), and Dedecker, Doukhan, Lang, Léon, Louhichi and Prieur (2007). As a rule, mixing conditions are not easy to verify and for concrete classes of processes they may be too restrictive.

For a large class of processes $\{X_j\}$ with a specific structure, e.g., for Gaussian or linear processes, a simpler and more natural way of describing and measuring dependence can be achieved via basic second-order characteristics such as the autocovariance function or spectral density. Concepts defined in terms of the autocovariance function are referred to as time-domain concepts while those defined in terms of the spectral density are referred to as frequency- or spectral-domain concepts. We shall first focus on time-domain dependence.

Definition 3.1.2. A *covariance stationary* process $\{X_j\}$ with autocovariance function $\gamma(k)$ is said to have:
Short memory (SM) if

$$\sum_{k \in \mathbb{Z}} |\gamma(k)| < \infty \quad \text{and} \quad \sum_{k \in \mathbb{Z}} \gamma(k) > 0; \tag{3.1.1}$$

Long memory (LM) if

$$\sum_{k \in \mathbb{Z}} |\gamma(k)| = \infty; \tag{3.1.2}$$

Negative memory, or *antipersistence*, if

$$\sum_{k \in \mathbb{Z}} |\gamma(k)| < \infty \quad \text{and} \quad \sum_{k \in \mathbb{Z}} \gamma(k) = 0. \tag{3.1.3}$$

Time-domain characterizations of short and long memory are based on the asymptotic behavior of the autocovariance $\gamma(k)$, as $k \to \infty$. Under short memory, the summability condition (3.1.1) of the autocovariance implies that $\{X_j\}$ has a continuous bounded spectral density f given by the absolutely converging series in (2.2.5), which does not vanish at the zero frequency, since $f(0) = (2\pi)^{-1} \sum_{k \in \mathbb{Z}} \gamma(k) > 0$, whereas under negative memory as defined by (3.1.3), the process $\{X_j\}$ has a continuous spectral density f that vanishes at zero, i.e., $f(0) = 0$. Under the long memory condition (3.1.2), the spectral density, provided it exists, is generally unbounded at the origin.

White-noise as defined in Definition 2.2.4 has $\gamma(k) = 0$ for $k \neq 0$, and hence short memory and a flat spectrum. An important class of short memory processes where the autocovariances are not zero is that of the stationary ARMA(p, q) processes of Example 2.2.5.

Conditions (3.1.2) and (3.1.3) defining long memory and negative memory are very general. A useful asymptotic theory and statistical inference is possible if we specify the rate of decay of $\gamma(k)$ at infinity. Consider the autocovariances $\gamma(k)$ that hyperbolically decay to zero:

$$\gamma(k) = |k|^{-1+2d} L_\gamma(|k|), \quad \forall k \geq 1, \exists 0 < d < 1/2, L_\gamma \in SV. \tag{3.1.4}$$

This condition is sometimes further reduced to a simpler condition satisfied by basic parametric long memory time-series models: for some $0 < d < 1/2$, and $c_\gamma > 0$,

$$\gamma(k) \sim c_\gamma |k|^{-1+2d}, \quad k \to \infty. \tag{3.1.5}$$

In view of Proposition 2.3.1, under either of these two conditions, $\gamma(k)$ satisfies (3.1.2).

Observe that the power $1 - 2d$ in (3.1.4) varies in the interval $(0, 1)$. On the other hand, if $\gamma(k) = |k|^{-\alpha} L_\gamma(|k|)$ holds with $\alpha > 1$, then $\gamma(k)$ satisfies SM condition (3.1.1), since $L_\gamma(|k|) = o(|k|^a)$, for any $a > 0$. This hyperbolic decay of $\gamma(k)$ with $\alpha > 1$ is sometimes referred to as *moderate short* memory.

Similarly, the negative memory condition (3.1.3) is satisfied by autocovariances having a hyperbolic decay as follows. Assume that for some parameter $-1/2 < d < 0$, and an $L_\gamma \in SV$, either

$$\gamma(k) = -|k|^{-1+2d} L_\gamma(|k|), \ k \geq 1, \qquad \sum_{k \in \mathbb{Z}} \gamma(k) = 0, \tag{3.1.6}$$

or

$$\gamma(k) \sim c_\gamma |k|^{-1+2d}, \ c_\gamma < 0, \qquad \sum_{k \in \mathbb{Z}} \gamma(k) = 0. \qquad (3.1.7)$$

Under either of these two conditions, the underlying process has negative memory, i.e., satisfies (3.1.3). Observe that (3.1.7) is a particular case of (3.1.6). For the requirement of $\gamma(k) < 0$, $k \to \infty$, see the discussion around (3.3.7) below.

In the following, when referring to long memory in the time-domain, we shall assume that $\gamma(k)$ exhibits slow hyperbolic decay (3.1.5) (or, more generally (3.1.4)), whereas for negative memory, it satisfies (3.1.7) (or, more generally (3.1.6)).

Definition 3.1.3. (*Memory parameter*). The parameter d of (3.1.4) to (3.1.7) is called the *memory parameter* of the underlying stationary process. The memory parameter is positive under long memory: $0 < d < 1/2$, and negative under negative memory: $-1/2 < d < 0$. If the underlying process has short memory then the memory parameter is defined as $d = 0$.

We shall now define memory concepts in the frequency domain.

Definition 3.1.4. Suppose the spectral density f of a stationary process $\{X_j\}$ exists, is bounded on $[\epsilon, \pi]$ for any $\epsilon > 0$, and satisfies

$$f(v) \sim c_f |v|^{-2d}, \ v \to 0, \text{ for some } -\frac{1}{2} < d < \frac{1}{2}, c_f > 0. \qquad (3.1.8)$$

The process $\{X_j\}$ is said to have *negative memory*, *short memory*, or *long memory*, depending on whether $d \in (-1/2, 0)$, $d = 0$, or $d \in (0, 1/2)$.

Frequency-domain conditions characterize the local behavior of the spectrum around the zero frequency. For $d = 0$, condition (3.1.8) means that the spectral density is continuous at the zero frequency. Negative memory means that $f(0) = 0$, and long memory indicates that the spectral density f is unbounded at the origin. Condition (3.1.8) can be generalized by including a slowly varying function:

$$f(v) = L_f(1/|v|)|v|^{-2d}, \qquad v \in \Pi, \qquad (3.1.9)$$

for some non-negative $L_f \in SV$.

Definition 3.1.5. A stationary process $\{X_j\}$ is called an I(d) process, with memory parameter d, $|d| < 1/2$, if its spectral density satisfies (3.1.8).

Under additional restrictions, the time-domain and spectral-domain dependence characteristics are equivalent, that is, one can go from one to the other, as is evident from the following proposition. Let $\mathcal{G}$ denote a class of even functions $g(v)$, $v \in \Pi$, which are piecewise differentiable, $\int_\Pi |\dot{g}(v)| dv < \infty$, and as $v \to 0$, $g(v) \to g(0) > 0$, and $v\dot{g}(v) \to 0$.

Proposition 3.1.1. (a) *Let* $f(v) = |v|^{-2d} g(v)$, $v \in \Pi$, *for some* $0 < d < 1/2$ *and* $g \in \mathcal{G}$. *Then, as* $k \to \infty$,

$$\gamma(k) \sim c_\gamma |k|^{-1+2d}, \qquad c_\gamma := g(0) 2\Gamma(1 - 2d) \sin(\pi d). \quad (3.1.10)$$

More generally, if g *is Zygmund slowly varying at zero, and has bounded variation in every interval* $(\epsilon, \pi]$, $\epsilon > 0$, *then (3.1.10) holds with* $g(0)$ *replaced by* $g(1/k)$.

(b) *Let* $\gamma(k) = c_\gamma |k|^{-1+2d}(1 + O(k^{-1}))$, *for some* $0 < d < 1/2$ *and* $c_\gamma > 0$. *Then* f *exists and, as* $v \to 0$,

$$f(v) \sim c_f |v|^{-2d}, \qquad c_f = c_\gamma (1/\pi) \Gamma(2d) \sin\left(\pi(1 - 2d)/2\right). \quad (3.1.11)$$

More generally, if $\gamma(k) = L(k)|k|^{-1+2d}$, *where* $L \in ZSV$, *then (3.1.11) holds with* c_γ *replaced by* $L(1/|v|)$.

Proof. (a) Since $\gamma(k) = 2 \int_0^\pi \cos(kv) f(v) dv$, (3.1.10) follows from (2.3.14) and Lemma 2.3.1.

(b) Let

$$f_1(v) := \frac{c_\gamma}{2\pi} \sum_{j \in \mathbb{Z}: j \neq 0} e^{ijv} |j|^{-1+2d} = \frac{c_\gamma}{\pi} \sum_{j=1}^{\infty} \cos(jv) |j|^{-1+2d}.$$

Then $f(v) = (2\pi)^{-1} \sum_{j \in \mathbb{Z}} e^{ijv} \gamma(|j|) = f_1(v) + O(1)$. Now apply (2.3.8) to the function f_1 with $L(k) \equiv c_\gamma/\pi$, $\beta = 1 - 2d$, to conclude (3.1.11). When $\gamma(k) = L(k)|k|^{-1+2d}$, (3.1.11) follows from (2.3.8).

Proposition 3.1.1 shows that under long memory the hyperbolic decay of the autocovariance function $\gamma(k)$ at the rate $|k|^{-1+2d}$ is roughly equivalent to the hyperbolic growth at the rate $|v|^{-2d}$ of the spectral density $f(v)$ near the frequency $v = 0$. Under short memory, the summability of $\gamma(k)$ implies the existence of the bounded spectral density, whereas the converse implication is generally not true without imposing additional smoothness restrictions on f, such as $f \in \Lambda_\alpha$, $\alpha > 1/2$.

3.2   Wold decomposition and linear processes

Let $\{\zeta_j, j \in \mathbb{Z}\} \sim \mathrm{WN}(0, \sigma^2)$ be white-noise and a_k, $k = 0, 1, \cdots$, be a sequence of non-random real numbers with $\sum_{k=0}^{\infty} a_k^2 < \infty$.

Definition 3.2.1. A linear or a moving-average process is defined by

$$X_j := \sum_{k=0}^{\infty} a_k \zeta_{j-k}, \qquad j \in \mathbb{Z}. \tag{3.2.1}$$

A linear process can be viewed as the output of a causal linear filter $\{a_k\}$ applied to white-noise $\{\zeta_j\}$, called the innovations of $\{X_j\}$. This process is covariance stationary and also strictly stationary provided $\{\zeta_j\}$ is strictly stationary. It has zero mean $EX_0 = 0$, and the autocovariance function

$$\gamma(k) = EX_0 X_k = \sigma^2 \sum_{j=0}^{\infty} a_j a_{j+k}, \quad k = 0, 1, \cdots . \tag{3.2.2}$$

By Proposition 2.2.2, its spectral density is

$$f(v) = \frac{\sigma^2}{2\pi} \Big| \sum_{j=0}^{\infty} e^{-ijv} a_j \Big|^2, \quad v \in \Pi. \tag{3.2.3}$$

According to a fundamental result in Wold (1938), a very wide class of stationary processes can be represented as linear processes.

Theorem 3.2.1. (Wold Decomposition). *Every stationary process* $\{X_j\}$ *with zero mean whose spectral density* f *satisfies*

$$\int_{\Pi} \log f(u) du > -\infty,$$

can be represented as a linear process (3.2.1) *with respect to some white-noise* $\{\zeta_j\}$.

If $\{X_j\}$ is a Gaussian process, the innovations $\{\zeta_j\}$ can be taken to be Gaussian too, and *vice versa*.

The class of linear processes is very large. In general, in (3.2.1) the innovations $\{\zeta_j\}$ need not be independent. However, to develop useful inferences about these processes, it is often assumed that $\{\zeta_j\} \sim \mathrm{IID}(0, \sigma^2)$. As will be seen later in this text, having i.i.d. innovations in these processes also plays an important role in the development of a useful asymptotic distribution theory for their sums and quadratic forms.

The above three types of dependence structure of the linear process $\{X_j\}$ of (3.2.1) are determined by the rate of decay of the moving-average coefficients. If the a_k's are absolutely summable,

$$\sum_{k=0}^{\infty} |a_k| < \infty, \qquad \sum_{k=0}^{\infty} a_k \neq 0, \tag{3.2.4}$$

then $\{X_j\}$ has short memory. Hyperbolically decaying a_k's

$$a_k \sim c_a k^{-1+d}, \qquad k \to \infty, \qquad c_a \neq 0, \tag{3.2.5}$$

make $\{X_j\}$ a long memory process for $0 < d < 1/2$ and a negative memory process for $-1/2 < d < 0$, under the additional condition $\sum_{k=0}^{\infty} a_k = 0$. Sometimes we use the following stronger assumption:

$$a_k = c_a k^{-1+d}(1 + O(k^{-1})), \qquad k \geq 1, \quad c_a \neq 0. \tag{3.2.6}$$

For the next proposition fact we need the definition of gamma and beta functions. $\Gamma(0) := \infty$, and

$$\Gamma(a) := \int_0^{\infty} x^{a-1} e^{-x} dx, \qquad a > 0, \tag{3.2.7}$$

$$\Gamma(a) := \frac{1}{a} \Gamma(a+1), \qquad -1 < a < 0,$$

$$B(a,b) := \frac{\Gamma(a)\Gamma(b)}{\Gamma(a+b)}, \qquad a > -1, b > -1, a+b > -1.$$

Proposition 3.2.1. *Let $\{X_j\}$ be the linear process (3.2.1).*

(i) *If the a_k satisfy (3.2.4), then $\{X_j\}$ has short memory.*

(ii) *If the a_k satisfy (3.2.5) with $0 < d < 1/2$, then $\{X_j\}$ has long memory:*

$$\gamma(k) \sim c_\gamma |k|^{-1+2d}, \qquad k \to \infty, \qquad c_\gamma = \sigma^2 c_a^2 B(d, 1 - 2d). \tag{3.2.8}$$

(iii) *If the a_k satisfy (3.2.6) with $-1/2 < d < 0$ and $\sum_{k=0}^{\infty} a_k = 0$, then $\{X_j\}$ has negative memory: $\sum_{k \in \mathbb{Z}} \gamma(k) = 0$, and $\gamma(k)$ satisfies (3.2.8).*

Proof. (i) From (3.2.2) and the assumption $\sum |a_j| < \infty$, it follows that

$$\sum_{k=0}^{\infty} |\gamma(k)| \leq \sigma^2 \sum_{k=0}^{\infty} \sum_{j=0}^{\infty} |a_j a_{j+k}| = \sigma^2 \left(\sum_{j=0}^{\infty} |a_j| \right)^2 < \infty,$$

$$\sum_{k=-\infty}^{\infty} \gamma(k) = \gamma(0) + 2 \sum_{k=1}^{\infty} \gamma(k) \tag{3.2.9}$$

$$= \sigma^2 \left(\sum_{j=0}^{\infty} a_j^2 + 2 \sum_{k=1}^{\infty} \sum_{j=0}^{\infty} a_j a_{j+k} \right)$$

$$= \sigma^2 \left(\sum_{j=0}^{\infty} a_j \right)^2 > 0.$$

(ii) For $d > 0$, assumption (3.2.5) implies $a_j a_{j+k}(j(j+k))^{1-d} = c_a^2 + \delta(j, k)$, where $\delta(j, k) \to 0$ as $j \to \infty$, uniformly over $k \geq 1$. Hence, as $k \to \infty$,

$$\sum_{j=0}^{\infty} a_j a_{j+k} = a_0 a_k + \sum_{j=1}^{\infty} j^{-1+d}(j+k)^{-1+d}(c_a^2 + \delta(j, k))$$

$$\sim c_a^2 \int_0^{\infty} x^{-1+d}(x+k)^{-1+d}dx + O(k^{-1+d})$$

$$= c_a^2 k^{-1+2d} B(d, 1-2d) + o(k^{-1+2d}),$$

in view of definition (3.2.2), thereby proving part (ii).

(iii) Relation $\sum_{k \in \mathbb{Z}} \gamma(k) = 0$ follows from (3.2.9) and the assumption $\sum_{j=0}^{\infty} a_j = 0$. To prove (3.2.8), assume without loss of generality that $\sigma^2 = 1$. Then

$$k^{1-2d}\gamma(k) := k^{1-2d} \sum_{j=0}^{\infty} a_j a_{j+k} = k^{1-2d} \sum_{j=0}^{\infty} a_j(a_{j+k} - a_k)$$

$$= k^{1-2d} \sum_{0 \leq j \leq [\epsilon k]} [\cdots] + k^{1-2d} \sum_{j > [\epsilon k]} [\cdots] =: U_{1,\epsilon}(k) + U_{2,\epsilon}(k),$$

where $\epsilon > 0$ is a small number specified below.

We shall first prove that $U_{1,\epsilon}(k) \to 0$, as first $k \to \infty$ and then as $\epsilon \to 0$. Let $K \geq 1$ be a fixed large number. For $0 \leq j \leq [\epsilon k]$ and $k \geq K$, by (3.2.6),

$$|a_{j+k} - a_k| \leq |a_k| \left| \frac{a_{j+k}}{a_k} - 1 \right|$$

$$\leq Ck^{-1+d}\left|(\frac{j+k}{k})^{-1+d}\frac{1+O(k^{-1})}{1+O(k^{-1})} - 1\right|$$

$$= Ck^{-1+d}(O(\frac{j}{k}) + O(k^{-1})) \leq Ck^{-2+d}j,$$

since $((j+k)/k)^{-1+d} = 1 + O(j/k)$. Thus,

$$|U_{1,\epsilon}(k)| \leq Ck^{-d} \sum_{0 \leq j \leq [\epsilon k]} |a_j|(j/k)$$

$$\leq Ck^{-1-d} \sum_{0 \leq j \leq [\epsilon k]} j^d \leq C\epsilon^{1+d} \to 0, \quad \epsilon \to 0.$$

Hence, $U_{1,\epsilon}(k)$ can be made arbitrarily small uniformly in $k \geq K$ by a suitable choice of $\epsilon > 0$.

Next, consider $U_{2,\epsilon}(k)$. As in the proof of part (ii) above we obtain,

$$U_{2,\epsilon}(k) \sim c_a^2 \int_\epsilon^{\infty} x^{d-1}[(1+x)^{d-1} - 1]dx, \quad n \to \infty,$$

$$\to c_a^2 B(d, 1-2d), \quad \epsilon \to 0,$$

because, for $-1/2 < d < 0$,

$$\int_0^\infty x^{d-1}[(1+x)^{d-1} - 1]dx = B(d, 1 - 2d).$$

The above equality can be verified using Gradshteyn and Ryzhik (2000, 3.245). This proves part (iii).

Proposition 3.2.1 shows that regularly decaying weights $\{a_k\}$ can generate short, long, and negative memory in time-domain terms. The "smoother version" (3.2.6) of the a_k's, for $d \neq 0$, implies regular asymptotics of the spectral density at the origin.

Proposition 3.2.2. *Assume that* $|d| < 1/2$, $d \neq 0$, $\{a_k\}$ *satisfy (3.2.6), and* $\sum_{k=0}^\infty a_k = 0$ *for* $-1/2 < d < 0$. *Then, as* $v \to 0$,

$$f(v) \sim c_f|v|^{-2d}, \qquad c_f = \frac{\sigma^2}{2\pi}c_a^2\Gamma^2(d). \qquad (3.2.10)$$

Proof. Write the transfer function

$$A(v) = c_a \sum_{j=1}^\infty e^{-ijv}j^{-1+d} + \left(a_0 + \sum_{j=1}^\infty e^{-ijv}(a_j - c_a j^{-1+d})\right)$$

$$=: c_a A_1(v) + A_2(v).$$

Suppose $0 < d < 1/2$. Under assumption (3.2.6),

$$|A_2(v)| \leq C\sum_{j=1}^\infty j^{-2+d} < \infty,$$

whereas (2.3.7), (2.3.8), and (3.2.3) imply that as $v \to 0+$,

$$A_1(v) \sim K_d v^{-d}, \quad K_d = \Gamma(d)(\cos(\frac{\pi d}{2}) - i\sin(\frac{\pi d}{2})), \qquad (3.2.11)$$

$$f(v) = \frac{\sigma^2}{2\pi}|A(v)|^2 \sim \frac{\sigma^2}{2\pi}|c_a K_d|^2|v|^{-2d} = c_f|v|^{-2d},$$

which proves (3.2.10).

Next, suppose $-1/2 < d < 0$. Because $A(0) = 0$, $A(v) = c_a[A_1(v) - A_1(0)] + (A_2(v) - A_2(0))$. Now, $|A_2(v) - A_2(0)| \leq C|v| = o(v^{-2d})$, because $|\dot{A}_2(v)| \leq C\sum_{j=1}^\infty j^{-1+d} < \infty$. On the other hand, by (3.2.11),

$$\dot{A}_1(v) = -i\sum_{j=1}^\infty e^{-ijv}j^d \sim -i K_{d+1}v^{-1-d}, \quad v \to 0+,$$

which in turn implies

$$A_1(v) - A_1(0) = \int_0^v \dot{A}_1(u)du \sim \mathbf{i}\frac{K_{d+1}}{d}v^{-d},$$

as $v \to 0+$. Note that $|K_{d+1}| = |dK_d|$, because $\Gamma(d+1) = d\Gamma(d)$. Hence, $|A(v)|^2 = |c_aK_d|^2|v|^{-2d} + o(|v|^{-2d}) \sim c_a^2\Gamma^2(d)|v|^{-2d}$, which proves (3.2.10).

Note: Various useful integral formulas can be found in Gradshteyn and Ryzhik (2000).

3.3   Asymptotics of the variance

Let $\{X_j\}$ be a second-order stationary process, $S_n = \sum_{j=1}^n X_j$, and $\bar{X} := S_n/n$. It is well known that the variance of the sample mean $\bar{X}$ of white-noise $\{X_j\} \sim \mathrm{WN}(0,\sigma^2)$ is proportional to $1/n$: $\mathrm{Var}(\bar{X}) = \sigma^2/n$. A similar fact holds asymptotically for any stationary process with short memory. However, under long or negative memory, the behavior of this variance is very different.

Hyperbolic decay in the autocovariance function $\gamma(k)$ of the process $\{X_j\}$ results in the hyperbolic growth of $\mathrm{Var}(S_n)$: under long memory $\mathrm{Var}(S_n)$ grows faster than n, the rate for short memory, while it grows slower than n under negative memory. To see this, consider

$$\mathrm{Var}(S_n) = E\Big(\sum_{j=1}^n (X_j - EX_j)\Big)^2 = \sum_{k,j=1}^n \gamma(k-j) \qquad (3.3.1)$$

$$= n \sum_{j=-(n-1)}^{n-1} \Big(1 - \frac{|j|}{n}\Big)\gamma(j).$$

This relation implies a useful bound

$$\mathrm{Var}(n^{-1/2}S_n) \le \sum_{j\in\mathbb{Z}} |\gamma(j)|.$$

Using (2.1.10), $\mathrm{Var}(S_n)$ can be written as

$$\mathrm{Var}(S_n) = \sum_{k,j=1}^n \int_\Pi e^{\mathbf{i}(k-j)v} f(v)dv = \int_\Pi D_n^2(v)\, f(v)dv. \qquad (3.3.2)$$

The above formulas show that long memory or the slow decay of $\gamma(k)$ and the peak of the spectrum f at the origin may significantly increase $\mathrm{Var}(S_n)$. Define

$$s_{\gamma,d}^2 := \sum_{j\in\mathbb{Z}} \gamma(j) = 2\pi f(0), \qquad\qquad d = 0, \qquad (3.3.3)$$

$$s_{\gamma,d}^2 := \frac{1}{d(1+2d)}\, c_\gamma, \qquad\qquad d \neq 0;$$

$$s_{f,d}^2 := 2\pi\, f(0), \qquad\qquad d = 0, \qquad (3.3.4)$$

$$:= 2\frac{\Gamma(1-2d)\sin(\pi d)}{d(1+2d)}\, c_f, \qquad d \neq 0.$$

Proposition 3.3.1. *Suppose that $\{X_j\}$ is a second-order stationary process with memory parameter $d \in (-1/2, 1/2)$. Then the following hold:*

(i) *If $\{X_j\}$ satisfies the time-domain dependence conditions (3.1.1), (3.1.5), or (3.1.7), then*

$$\mathrm{Var}(S_n) \sim s_{\gamma,d}^2\, n^{1+2d}. \qquad (3.3.5)$$

(ii) *If $\{X_j\}$ satisfies the spectral-domain condition (3.1.8), then*

$$\mathrm{Var}(S_n) \sim s_{f,d}^2\, n^{1+2d}. \qquad (3.3.6)$$

Proof. (i) Let $d = 0$, i.e., $\gamma(k)$ satisfies (3.1.1), $\sum |\gamma(k)| < \infty$. By (3.3.1) and the DCT (dominated convergence theorem),

$$n^{-1}\mathrm{Var}(S_n) = \sum_{j=-n}^{n}\left(1 - \frac{|j|}{n}\right)\gamma(j) \to \sum_{j\in\mathbb{Z}}\gamma(j) = 2\pi f(0).$$

Let $d \neq 0$. Then, under (3.1.5) and (3.1.7), $\gamma(k) \sim c_\gamma |k|^{-1+2d}$. In addition, if $d < 0$, then $\sum_{k\in\mathbb{Z}}\gamma(k) = 0$. Write

$$n^{-1-2d}\mathrm{Var}(S_n) = n^{-2d}\sum_{j=-n}^{n}\gamma(j) - 2n^{-2d-1}\sum_{j=1}^{n}j\gamma(j)$$

$$=: s_{n,1} - s_{n,2}.$$

If $d > 0$, then applying (2.3.10) with $\beta = 1 - 2d$ gives

$$s_{n,1} = n^{-2d}\left(2\sum_{j=1}^{n}\gamma(j) + \gamma(0)\right) \to \frac{c_\gamma}{d},$$

whereas if $d < 0$, then $\sum_{j=-n}^{n}\gamma(j) = -2\sum_{j=n+1}^{\infty}\gamma(j)$, and applying (2.3.12) with $\beta = 1 - 2d$ gives

$$s_{n,1} = -2n^{-2d}\sum_{j=n+1}^{\infty}\gamma(j) \to \frac{c_\gamma}{d}.$$

Since $|j|\gamma(j) \sim |j|^{2d}c_\gamma$, applying (2.3.10) with $\beta = -2d$, gives $s_{n,2} \to 2c_\gamma/(1+2d)$. Thus,

$$s_{n,1} - s_{n,2} \to \frac{1}{d}c_\gamma - \frac{2}{1+2d}c_\gamma = \frac{1}{d(1+2d)}c_\gamma,$$

which completes the proof of (i).

(ii) By assumption (3.1.8), $f(v) \sim c_f |v|^{-2d}$, as $v \to 0$. Let $\epsilon > 0$. Using (3.3.2) one can write

$$n^{-1-2d}\text{Var}(S_n) = n^{-1-2d} \int_\Pi D_n^2(v) \, f(v) dv$$

$$= n^{-1-2d} \int_{|v|\le\epsilon} (\cdots) dv + n^{-1-2d} \int_{\epsilon\le|v|\le\pi} (\cdots) dv = q_{n,1} + q_{n,2}.$$

By (2.1.8), $|D_n(v)| \le C$ for $\epsilon \le |v| \le \pi$. Hence,

$$q_{n,2} \le Cn^{-1-2d} \int_{\epsilon\le|v|\le\pi} f(v) dv \to 0.$$

Setting $g(v) = \big((v/2)/\sin(v/2)\big)^2 f(v) v^{2d}$, write

$$q_{n,1} = n^{-1-2d} \int_{|v|\le\epsilon} \Big(\frac{\sin(nv/2)}{v/2}\Big)^2 |v|^{-2d} g(v) dv$$

$$= \int_{|v|\le\epsilon n} \Big(\frac{\sin(v/2)}{v/2}\Big)^2 |v|^{-2d} g(v/n) dv.$$

Since $g(v) \le C$, $|v| \le \epsilon$, and $g(v) \to c_f$, as $v \to 0$, then $g(v/n) \le C$ for $|v| \le \epsilon n$ and, for each fixed v, $g(v/n) \to c_f$. Hence, by the DCT,

$$q_{n,1} \to c_f \int_{-\infty}^{\infty} \frac{\sin^2(v/2)}{(v/2)^2} |v|^{-2d} dv = s_{f,d}^2,$$

where the limiting integral can be explicitly evaluated and is equal to (3.3.4). This completes the proof of the proposition.

Proposition 3.3.1 implies that the standard deviation of $\bar{X}$ is of the order $n^{-1/2+d}$. Thus, under long memory, the consistency rate of the sample mean as an estimator of the population mean is worse than the usual i.i.d. consistency rate $n^{-1/2}$, i.e., long memory impairs the precision, whereas under negative memory the opposite effect is observed.

Definition 3.3.1. *The long-run variance of $\{X_j\}$ is defined to be the constant $s_{X,d}^2 = \lim_{n\to\infty} n^{-1-2d}\text{Var}(S_n)$, provided it exists.*

Under (3.3.5) and (3.3.6), $s_{X,d}^2$ equals $s_{\gamma,d}^2$ and $s_{f,d}^2$, respectively. The long-run variances $s_{\gamma,d}^2$ and $s_{f,d}^2$ are determined by the memory parameter d and the parameters c_γ, c_f characterizing the behavior of the autocovariance $\gamma(k)$ for large k and of spectral density f at the zero frequency, respectively.

Since the long-run variance is always positive, formula (3.3.3) for $s^2_{\gamma,d}$ indicates that as $k \to \infty$, $\gamma(k) \sim c_\gamma |k|^{-1+2d}$: $c_\gamma > 0$ if $d > 0$, and $c_\gamma < 0$ if $d < 0$. Estimation of the long-run variance is discussed in Chapter 9.

Finally, observe that the long memory condition $\gamma(j) = |j|^{-1+2d}L(|j|)$, $j \geq 1$, $0 < d < 1/2$, $L \in SV$, implies

$$\mathrm{Var}(S_n) = \sum_{k,j=1}^{n} \gamma(k-j) \sim \frac{1}{d(1+2d)}\, n^{1+2d}L(n). \qquad (3.3.7)$$

This claim, with $L(n)$ replaced by $-L(n)$, remains true also under negative memory, where $-1/2 < d < 0$ and $\sum_{j \in \mathbb{Z}} \gamma(j) = 0$, indicating that in this case $\gamma(n) < 0$, as $n \to \infty$. The proof of (3.3.7) is the same as that of (3.3.5).

In the literature on long memory processes the growth rate (3.3.7) of the variance is sometimes used as an alternative characteristic of the dependence structure of a stationary process $\{X_k\}$.

We now state some facts that are often used below. Suppose γ is as in (3.1.7) with $0 < d < 1/2$ and $L \in SV$. Then, for any integer $p = 1, 2, \cdots$,

$$\sum_{j=1}^{n}\sum_{k=1}^{n} \gamma^p(k-j) \sim C_1\, n^{2-p(1-2d)}L^p(n), \qquad p(1-2d) < 1, \qquad (3.3.8)$$
$$\sim C_2\, n, \qquad p(1-2d) > 1.$$
$$\sum_{k,j=1}^{n} |\gamma(k-j)|^p \leq Cn^{2-p(1-2d)}L^p(n), \qquad p(1-2d) < 1,$$
$$\leq C_\epsilon n^{1+\epsilon}, \qquad p(1-2d) = 1, \forall \epsilon > 0,$$
$$\leq Cn, \qquad p(1-2d) > 1,$$

where constants C_1, C_2, C, and C_ϵ do not depend on n.

3.4 Self-similarity, long memory, and fractional Brownian motion

A thorough discussion of self-similar processes is given in Samorodnitsky and Taqqu (1994). These processes arise in a natural way as limits of sums of random variables. In this section we discuss the relationship between self-similar processes with continuous time and stationary I(d) (long memory) processes with discrete time.

Let T be equal to $[0, \infty)$ or $\mathbb{R}$. A stochastic process $\{Z(t), t \in T\}$ is fully described by its finite-dimensional distributions

$$F_{t_1,\cdots,t_k}(x_1,\cdots,x_k) := P\big(Z(t_1) \le x_1, \cdots, Z(t_k) \le x_k\big),$$

for $(t_1,\cdots,t_k) \in T^k$, $(x_1,\cdots,x_k) \in \mathbb{R}^k$, $k = 1,2,\cdots$. Processes $\{Y(t)\}$ and $\{Z(t)\}$ are said to have the same distribution if their finite-dimensional distributions coincide, and we write $\{Y(t)\} =_{fdd} \{Z(t)\}$.

A continuous-time stochastic process $\{Z(t), t \in T\}$ is called stationary if for any $h > 0$, $\{Z(t+h)\} =_{fdd} \{Z(t)\}$.

Definition 3.4.1. (i) A real-valued stochastic process $\{Z(t); t \in T\}$ with $Z(0) = 0$ is said to have stationary increments if for any positive integer k, and for any $t_1 < t_2 < \cdots < t_k$ and h, all in T, the joint distribution of $\{Z(t_j + h) - Z(h); 1 \le j \le k\}$ is the same as that of $\{Z(t_j) - Z(0); 1 \le j \le k\}$. In other words, if for any $h \in T$,

$$\{Z(h+t) - Z(h), t \in T\} =_{fdd} \{Z(t) - Z(0), t \in T\}.$$

(ii) A process $\{Z(t); t \in T\}$ is said to be self-similar of index $H > 0$, if for all $a > 0$, the finite-dimensional distributions of $\{Z(at)\}$ are the same as those of $\{a^H Z(t)\}$:

$$\{Z(at), t \in T\} =_{fdd} \{a^H Z(t), t \in T\}.$$

In other words, for any $a > 0$, $k = 1,2,\cdots$, and for any $t_j \in T$, $1 \le j \le k$,

$$\big(Z(at_1), \cdots, Z(at_k)\big) =_{fdd} a^H \big(Z(t_1), \cdots, Z(t_k)\big).$$

The classical example of an H-self-similar stationary increment (H-sssi) process is Brownian motion, with $H = 1/2$. Proposition 3.4.3 below describes all Gaussian H-sssi processes.

Let

$$r_H(s,t) := \frac{1}{2}\Big\{|s|^{2H} + |t|^{2H} - |s-t|^{2H}\Big\}, \quad t,s \in \mathbb{R}. \qquad (3.4.1)$$

Proposition 3.4.1. *Let $\{Z(t)\}$ be an H-sssi process with index $H > 0$, and $Z(0) = 0$, $EZ^2(t) < \infty$, $t \in T$. Then*

$$\mathrm{Cov}(Z(s), Z(t)) = \sigma^2 r_H(s,t), \quad \sigma^2 = EZ^2(1). \qquad (3.4.2)$$

Moreover, if $EZ^2(t) \to 0$, as $t \to 0$, then $0 < H \le 1$.

Proof. By self-similarity, $Z(t) =_D t^H Z(1)$ and $EZ^2(t) = |t|^{2H}\sigma^2 < \infty$. By the stationary increment property, for any $0 \le t \le s$,

$$\text{Cov}(Z(s), Z(t)) = EZ(s)Z(t)$$
$$= \frac{1}{2}\Big\{EZ^2(s) + EZ^2(t) - E\big(Z(s) - Z(t)\big)^2\Big\}$$
$$= \frac{1}{2}\Big\{EZ^2(s) + EZ^2(t) - EZ^2(s - t)\Big\}$$
$$= \frac{\sigma^2}{2}\Big\{|s|^{2H} + |t|^{2H} - |s - t|^{2H}\Big\} = \sigma^2 r_H(s, t).$$

In addition, $t^{2H}\sigma^2 = EZ^2(t) \to EZ^2(0) = 0$, as $t \to 0$, which implies $H > 0$. On the other hand, by the Minkowski inequality,

$$E^{1/2}Z^2(2t) \le E^{1/2}\big(Z(2t) - Z(t)\big)^2 + E^{1/2}Z^2(t) = 2E^{1/2}Z^2(t).$$

In other words, $|2t|^H \le 2|t|^H$, for all $t \ge 0$. Therefore, we must have $2^H \le 2$, which forces $H \le 1$. This completes the proof.

Note that for $H = 1$ the covariance in (3.4.2) equals $\sigma^2 ts$ and therefore the random variables $Z(s)$ and $Z(t)$ are completely correlated: $\text{Corr}(Z(s), Z(t)) = 1$. The resulting 1-self-similar process with stationary increments and finite variance satisfies $Z(t) = tZ(1)$, a.s., and is of little interest. Note also the fact that a process Z having a covariance function given by (3.4.2) does not necessarily imply that it is an H-sssi process, unless Z is Gaussian.

The next proposition shows that H-sssi processes exist and that the autocovariance function of the increment process obtained from an H-sssi process $\{Z(t)\}$ has hyperbolic decay in the lag. Define

$$Y_k := Z(k + 1) - Z(k), \quad k \in \mathbb{Z},$$
$$\gamma_Y(k) := \frac{\sigma^2}{2}\big(|k + 1|^{2H} + |k - 1|^{2H} - 2|k|^{2H}\big), \quad k \in \mathbb{Z}.$$

Proposition 3.4.2. *For any $0 < H < 1$, the following hold:*

(i) $r_H(s, t)$ *in (3.4.1) is a covariance function.*

(ii) *The process $\{Y_k, k \in \mathbb{Z}\}$ is a covariance stationary process with autocovariance function* $\text{Cov}(Y_j, Y_k) = \gamma_Y(j - k)$, $\forall j, k, \in \mathbb{Z}$ *and*

$$\gamma_Y(k) = \sigma^2 H(2H - 1)|k|^{2H-2} + O(k^{-2}), \quad k \to \infty. \tag{3.4.3}$$

Moreover, for $0 < H < 1/2$,

$$\sum_{k \in \mathbb{Z}} \gamma_Y(k) = 0. \tag{3.4.4}$$

Proof. (i) We need to show that

$$\sum_{j=1}^{n}\sum_{k=1}^{n} a_j r_H(t_j, t_k) a_k \geq 0, \tag{3.4.5}$$

for any real numbers $a_1, \cdots, a_n$ and $t_1, \cdots, t_n$. Define $t_0 := 0$, and set $a_0 := -\sum_{j=1}^{n} a_j$. Then $\sum_{j=0}^{n} a_j = 0$, and the l.h.s. of (3.4.5) equals

$$q := -\frac{1}{2}\sum_{j,k=0}^{n} a_j a_k |t_j - t_k|^{2H}.$$

The function $t \rightarrow e^{-\epsilon|t|^{2H}}$, $\epsilon > 0$ is the characteristic function of a symmetric $2H$-stable r.v. η, $Ee^{it\eta} = e^{-\epsilon|t|^{2H}}$. Therefore,

$$0 \leq E\left|\sum_{j=0}^{n} a_j e^{it_j\eta}\right|^2 = \sum_{j,k=0}^{n} a_j a_k Ee^{-i(t_j - t_k)\eta}$$

$$= \sum_{j,k=0}^{n} e^{-\epsilon|t_j - t_k|^{2H}} a_j a_k$$

$$= \sum_{j,k=0}^{n} \left(e^{-\epsilon|t_j - t_k|^{2H}} - 1\right) a_j a_k.$$

Now divide both sides of this equality by ϵ and let $\epsilon \rightarrow 0+$ to obtain

$$0 \leq \epsilon^{-1}\sum_{j,k=0}^{n} \left(e^{-\epsilon|t_j - t_k|^{2H}} - 1\right) a_j a_k \rightarrow 2q.$$

This completes the proof of (i).

(ii) To prove (3.4.3), let $g(x) = |1+x|^{2H} + |1-x|^{2H} - 2$, $x \in \mathbb{R}$. Rewrite

$$\gamma_Y(k) = \sigma^2(k^{2H}/2)g(1/k).$$

Observe that $g(0) = \dot{g}(0) = g^{(3)}(0) = 0$, and $\ddot{g}(0) = 4H(2H-1)$. Using a Taylor expansion, as $x \rightarrow 0$,

$$g(x) = g(0) + x\dot{g}(0) + (x^2/2)\ddot{g}(0) + O(x^4)$$

$$= 2H(2H-1)x^2 + O(x^4),$$

which proves (3.4.3).

For $H < 1/2$, equality (3.4.4) follows because $\sum_{k\in\mathbb{Z}} |\gamma_Y(k)| < \infty$, and $|\sum_{k=-n}^{n} \gamma_Y(k)| = 2\sigma^2|(n+1)^{2H} - (n-1)^{2H}| = O(n^{2H-2}) = o(1)$.

Proposition 3.4.2 indicates that, although an H-sssi process Z is nonstationary, its increments $\{Y(k), k \in \mathbb{Z}\}$ generate a stationary sequence of dependent variables with covariance property (3.4.3).

For $H = 1/2$, $\{Y_k\}$ is an uncorrelated process, whereas for $1/2 < H < 1$, by (3.4.3), $\sum_{k \in \mathbb{Z}} |\gamma_Y(k)| = \infty$, and hence $\{Y_k, k \in \mathbb{Z}\}$ has long memory with memory parameter $d = H - 1/2 \in (0, 1/2)$. For $0 < H < 1/2$, it has negative memory with memory parameter $d = H - 1/2 \in (-1/2, 0)$.

Fractional Brownian motion. The main example of a Gaussian H-self-similar process is standard Brownian motion B on $\mathbb{R}^+ := [0, \infty)$. Recall, that finite-dimensional distributions of a zero-mean Gaussian process $\{B(t)\}$ are fully defined by its covariance function.

Definition 3.4.2. *Brownian motion* is a Gaussian process $\{B(t), t \in \mathbb{R}^+\}$ with $B(0) = 0$, $EB(t) \equiv 0$, and covariance function $\gamma_B(s, t) := E(B_s B_t)$ $= \min(s, t)$.

Note that because $\text{Var}(B(t)) = t$, the process B is non-stationary, but it has stationary increments: $B(t + h) - B(t) \sim \mathcal{N}(0, h)$ for all t and h, that are also orthogonal: for any $t_1 < t_2 < t_3$ increments $B(t_2) - B(t_1)$ and $B(t_3) - B(t_2)$ are independent. Moreover, B is H-self-similar with parameter $H = 1/2$: for any $a > 0$, $\{B(at)\} =_{fdd} a^{1/2}\{B(t)\}$.

Definition 3.4.3. (*Fractional Brownian motion*). Let $0 < H < 1$ be any number. A Gaussian process $B_H = \{B_H(t), t \in \mathbb{R}\}$, with $B_H(0) = 0$, $EB_H(t) \equiv 0$, and covariance function $r_H(s, t)$ in (3.4.1), is called *fractional Brownian motion (fBm)* with parameter $0 < H < 1$.

Fractional Brownian motion often appears in the limit theory of dependent random variables. Note that for $H = 1/2$, $B_{1/2} = B$, which is Brownian motion. For any $0 < H < 1$, B_H is an H-sssi process. Indeed, by Proposition 3.4.2 (i), r_H is a covariance function. By a routine argument, stationarity of the increments follows from (3.4.2).

Finally, for any $a > 0$, Gaussian processes $\{B_H(at)\}$ and $\{a^H B_H(t)\}$ have equal covariances $r_H(at, as) = a^{2H} r_H(s, t)$, which implies the H-self-similarity property, i.e., $\{B_H(at)\} =_{fdd} a^H\{B_H(t)\}$. We have thus proved the following fact.

Proposition 3.4.3. *Any H-sssi Gaussian process with $0 < H < 1$ can be written as σB_H, with some $\sigma > 0$, where B_H is fBm.*

Similarly, as for a general H-sssi process above, increments of fBm generate a stationary process.

Definition 3.4.4. (*Fractional Gaussian noise*). The discrete-time Gaussian process $\{\xi_j := B_H(j+1) - B_H(j), \; j \in \mathbb{Z}\}$ is called *fractional Gaussian noise* with parameter $0 < H < 1$.

Clearly, the process $\{\xi_j\}$ is stationary, $E\xi_j \equiv 0$, $\mathrm{Var}(\xi_0) = 1$, and

$$\gamma_\xi(k) = \frac{1}{2}\left\{|k+1|^{2H} - 2|k|^{2H} + |k-1|^{2H}\right\}, \quad k \in \mathbb{Z}.$$

Because the function $x \mapsto x^{2H}$, $x \geq 0$, is strictly concave (convex) for $H < 1/2$ ($H > 1/2$), we have that $\gamma_\xi(k) < 0$ for $H < 1/2$, $\gamma_\xi(k) = 0$ for $H = 1/2$, and $\gamma_\xi(k) > 0$ for $H > 1/2$, for all $k = 1, 2, \cdots$.

From Proposition 3.4.2 we have the following corollary.

Corollary 3.4.1. *The covariance of the fractional Gaussian noise $\{\xi_j\}$ with parameter $0 < H < 1$ satisfies the following:*

$$\gamma_\xi(k) = H(2H-1)|k|^{2H-2} + O(|k|^{-2}), \quad \text{as } |k| \to \infty.$$
$$\sum_{j \in \mathbb{Z}} \gamma_\xi(k) = 0, \quad 0 < H < 1/2.$$

Corollary 3.4.2. *The fractional Gaussian noise $\{\xi_j\}$ has long memory for $1/2 < H < 1$ and negative memory for $0 < H < 1/2$, with memory parameter $d = H - (1/2)$ in the sense of time-domain definitions (3.1.5) to (3.1.7). For $H = 1/2$, $\{\xi_j\} \sim \mathrm{WN}(0,1)$ is an i.i.d. Gaussian sequence.*

From Taqqu (2003, equation (9.13)), we see that the spectral density of the fractional Gaussian noise $\{\xi_j\}$ satisfies

$$f(v) = (2/\pi)H\Gamma(2H)\sin(\pi H)(1 - \cos(v))\sum_{k \in \mathbb{Z}}|2\pi k + v|^{-1-2H}$$
$$\sim (1/\pi)H\Gamma(2H)\sin(\pi H)v^{1-2H}, \quad v \to 0.$$

Hence, $\{\xi_j\}$ is an I(d) process with $d = H - (1/2)$, in the sense of Definition 3.1.5, i.e., the spectral density f satisfies (3.1.8), with $c_f = \pi^{-1}H\Gamma(2H)\sin(\pi H)$.

Partial sums and self-similarity. Besides second-order characteristics, dependence in a stationary process $\{X_j\}$ can be also characterized by the limit behavior of a partial-sum process

$$S_n(\tau) = \sum_{j=1}^{[n\tau]} X_j, \quad \tau \geq 0,$$

where $[a]$ denotes the integer part of the real number a. Lamperti (1962) showed that the convergence of finite-dimensional distributions of a normalized partial-sum process has strong structural implications: the limit process is H-self-similar.

Theorem 3.4.1. (Lamperti). *Let $\{X_j\}$ be a strictly stationary process and suppose there exist non-random numbers $A_n \to \infty$ and $b \in \mathbb{R}$ such that*

$$A_n^{-1} \sum_{j=1}^{[n\tau]} (X_j - b) \to_{fdd} Z(\tau), \quad \tau \geq 0, \tag{3.4.6}$$

where the limit process $Z(\tau)$, $\tau \geq 0$, is not identically zero. Then $\{Z(\tau)\}$ is a stochastically continuous H-sssi process with some parameter $H > 0$ and the normalization $A_n = n^H L(n)$, for some $L \in SV$.

Thus, the self-similarity index H characterizes the growth of the normalization of sums and partial sums of $\{X_j\}$.

In particular, if a stationary process $\{X_j\}$ has short, long, or negative memory with memory parameter $d \in (-1/2, 1/2)$ and the limit in Theorem 3.4.1 is Gaussian under normalization $A_n = \sqrt{\text{Var}(S_n(1))} \sim s_{X,d} n^{1/2+d}$ as in Proposition 3.3.1, then the limit process Z in (3.4.6) is fractional Brownian motion B_H with $H = 1/2 + d \in (0, 1)$. In the short memory case, $d = 0$, the limit is standard Brownian motion. If the limit is non-Gaussian, then Z is self-similar with parameter $H = d + 1/2$.

In summary, self-similar processes are non-stationary processes whose increments may generate stationary long memory processes. A common characteristic of self-similar processes is that time and space scale differently. The scaling invariance of self-similar processes is statistical and defined in terms of distributions, while the rescaled trajectory of such a process looks very similar to the original at arbitrary, small or large scales.

The main importance of self-similar processes to probability and statistics comes from Lamperti's theorem. Under long memory, the limit distribution of sums and statistics need not be Gaussian, and one can try to relate it to a self-similar process. Examples of non-Gaussian self-similar processes arising as large sample limits of the statistics of Gaussian and linear processes appear in the following chapters. Finally, we note that the class of H-sssi processes is very large and is by no means limited to processes discussed in this text.

Note: Some of the above discussion was derived from Samorodnitsky and Taqqu (1994). Another good reference on self-similar processes is the book by Embrechts and Maejima (2002).

3.5   Other classes of long memory processes

The linear process in (3.2.1) with hyperbolically decaying coefficients as in (3.2.5) is the main model of long memory processes in this book. The main reason is its mathematical tractability and structural simplicity. Several parametric classes of the linear model in (3.2.1) are discussed in detail in Chapter 7.

However, the linear model in (3.2.1) has its drawbacks and sometimes it is not capable of incorporating empirical features ("stylized facts") of some observed time series. The purpose of this section is to give a brief introduction to the reader of some other non-linear models of long and short memory processes, which are studied in the literature. The choice below is somewhat selective since this literature is quite large, and also reflects the interests of the authors of this book.

Non-linear functions of linear processes. Given a linear long memory process $\{X_j\}$ in (3.2.1), (3.2.5), and a non-linear function $h(x), x \in \mathbb{R}$, the transformed, or subordinated, process $\{Y_j\}$ is defined as $Y_j := h(X_j)$. If $\{X_j\}$ has long memory, the transformed process may have long or short memory, depending on h and the parameter d in (3.2.5). For Gaussian $\{X_j\}$, these properties of $\{Y_j\}$ are discussed in detail in Sections 4.6 and 4.7. For a general moving-average process $\{X_j\}$ with i.i.d. innovations, similar but more technical results about $\{Y_j\}$ are obtained in Chapter 10. The asymptotic analysis of sums of non-linear functions of a moving-average process is very important because numerous inference procedures in models generated by these processes are based on such sums.

ARCH(∞) and related heteroscedastic processes. Suppose in (3.2.1), the innovations $\{\zeta_j\} \sim \text{IID}(0,1)$. Then the conditional variance $\text{Var}(X_j | \zeta_s, s < j) = a_0^2$ is constant (non-random). This means that the linear model in (3.2.1) with i.i.d. innovations is homoscedastic. Recall that a stationary process $\{\xi_j\}$ is called *homoscedastic* or *heteroscedastic* depending on whether the conditional variance (squared volatility)

$\sigma_j^2 = \mathrm{Var}(\xi_j | \mathcal{F}_{j-1})$ is constant or not. Here, $\{\mathcal{F}_j\}$ is a non-decreasing family of σ-fields ("historic information set"), with the property that for each j, ξ_j is $\mathcal{F}_j$-measurable. Heteroscedasticity is especially important in financial time series since it accounts for the widely known volatility clustering "stylized fact" of asset returns.

A random process $\{\xi_j\}$ is said to satisfy the ARCH(∞) equations if there exists a sequence of r.v.'s $\{\zeta_j\} \sim \mathrm{IID}(0,1)$ and a deterministic sequence $b_i \geq 0, i = 0, 1, \cdots$ such that

$$\xi_j = \sigma_i \zeta_i, \qquad \sigma_j^2 = b_0 + \sum_{i=1}^{\infty} b_i \xi_{j-i}^2, \qquad j \in \mathbb{Z}. \qquad (3.5.1)$$

Here, ξ_j is measurable w.r.t. the σ-field $\mathcal{F}_j$ generated by $\zeta_s, s \leq j$. This in turn implies that $E(\xi_j | \mathcal{F}_{j-1}) = 0$, for all $j \in \mathbb{Z}$, and a (typically random) conditional variance σ_j^2 given by a non-negative linear combination of the past values ξ_{j-i}^2 in (3.5.1). The class of ARCH(∞) processes includes the celebrated parametric ARCH(p) and GARCH(p, q) models of Engle (1982) and Bollerslev (1986). While the latter processes have exponentially decaying covariance functions and satisfy other weak dependence conditions (see Lindner (2009)), a (squared) ARCH(∞) process can have autocovariances decaying to zero at the rate $k^{-\gamma}$ with $\gamma > 1$ arbitrarily close to 1. However, the squares of an ARCH(∞) process have absolutely summable autocovariances and therefore $\{\xi_j^2\}$ has short memory in the sense of Definition 3.1.2. For details, see e.g., Giraitis, Leipus and Surgailis (2009) and Giraitis, Kokoszka and Leipus (2000).

The linear ARCH (LARCH) model, introduced by Robinson (1991), is defined by the equations

$$\xi_j = \sigma_j \zeta_j, \qquad \sigma_j = b_0 + \sum_{i=1}^{\infty} b_i \xi_{j-i}, \qquad j \in \mathbb{Z}, \qquad (3.5.2)$$

where $\{\zeta_j\}$ satisfy the same conditions as in (3.5.1) and the coefficients b_j satisfy $\sum_{i=1}^{\infty} b_i^2 < \infty$. Note that $\{\xi_j\}$ is a white-noise process while the covariance of the volatility $\{\sigma_j\}$ in (3.5.2),

$$\mathrm{Cov}(\sigma_k, \sigma_0) = \sigma^2 \sum_{i=1}^{\infty} b_i b_{i+k}, \qquad \sigma^2 := E\sigma_0^2 = E\xi_0^2, \quad k \geq 0,$$

has the same form as if $\{\sigma_j\}$ were a linear process. In contrast to (3.5.1), the squared LARCH process can exhibit long memory in the sense of (3.1.5), under similar decay conditions on the coefficients, as shown in Giraitis, Robinson, Surgailis (2000) and Giraitis, Leipus, Robinson, Surgailis (2004). Surgailis (2008) discussed a modification of (3.5.2) with the property that $\sigma_j > c_0 > 0$ is a.s. strictly positive. Giraitis and Surgailis (2002) studied a class of heteroscedastic *bilinear* models, with non-zero conditional mean, which include the linear, ARCH(∞), and LARCH models, and which can exhibit long memory both in the conditional mean and in the conditional variance with different long memory parameters.

Stochastic volatility processes. A stochastic volatility process has the form

$$\xi_j = \sigma_j \zeta_j, \qquad j \in \mathbb{Z}, \tag{3.5.3}$$

where $\{\zeta_j\} \sim \mathrm{IID}(0,1)$, and $\sigma_j > 0$ is a measurable function of the "past" $\mathcal{F}_{j-1}$. It is often assumed that $\sigma_j = f(\eta_j)$, where f is a (non-linear) function, and $\{\eta_j\}$ is a stationary process of some familiar type, e.g., Gaussian or ARMA. The choice $f(x) = e^x$ and a linear process $\{\eta_j\}$ leads to the Exponential GARCH (EGARCH) model of Nelson (1991). The EGARCH model is further discussed in Section 8.6.2. The review papers of Davis and Mikosch (2009) and Hurvich and Soulier (2009) discuss various probabilistic and long memory properties of stochastic volatility processes.

Network traffic and queueing processes. There is an extensive probabilistic and engineering literature dealing with the phenomenon of self-similarity and long memory in network traffic. Probably, the simplest models of network traffic are the ON/OFF model and the infinite source Poisson model. For both models, long memory in the sense of (3.1.4) can be easily established under the assumption that transmission durations are i.i.d. and have Pareto-like tails with parameter $1 < \alpha < 2$, see e.g., Mikosch, Resnick, Rootzén and Stegeman (2002). See Park and Willinger (2000) and other articles in the same volume for an overview of this research area.

Infinite variance processes. The linear model in (3.2.1) and (3.2.5) can be extended to infinite variance i.i.d. innovations; see e.g., Samorodnitsky and Taqqu (1994). Various definitions of long memory for infinite variance processes are discussed in Samorodnitsky and Taqqu (1994) and Puplinskaitė and Surgailis (2010). Examples of this kind arise in the analysis of volatility in finance and network traffic; see Embrechts, Klüppelberg and Mikosch (1997).

Chapter 4

Limit Theory for Sums

4.1   Introduction

A significant part of large-sample statistical inference relies on limit theorems of probability theory for sums and quadratic forms of stationary observations. Short or long memory of the underlying process $\{X_j\}$ may lead to different limits for suitably normalized sums $S_n = \sum_{j=1}^{n} X_j$, and the proofs of such limits generally use different methods. In this book, "central limit theorem" (CLT) refers to a statement that $\{\mathrm{Var}(S_n)\}^{-1/2}(S_n - ES_n)$ converges in distribution to a standard normal random variable. Limit theorems where a suitably standardized S_n converges in distribution to a non-Gaussian r.v. are called "non-central limit theorems".

Most of the proofs of the asymptotic normality of sums and quadratic forms in this text use the following two methods: (a) the method of moments or cumulants, see Proposition 4.2.1, and (b) the approximation of summands by weakly dependent r.v.'s, such as m-dependent r.v.'s or martingale differences, see Proposition 4.2.3, for which the asymptotic distribution theory is straightforward. The proofs of non-central limit theorems use the theory of Wiener–Itô integrals, the L_2 convergence criterion, and approximations by multiple polynomial forms, in particular the so-called "scheme of discrete multiple integrals" of Section 14.3.

Definition 4.1.1. A sequence of r.v.'s Y_n, $n = 1, 2, \cdots$ is said to converge *in distribution* or *weakly* to a r.v. Y, and we write $Y_n \to_D Y$, if

$$F_n(x) := P(Y_n \leq x) \to F(x) := P(Y \leq x), \qquad (4.1.1)$$

at each continuity point $x \in \mathbb{R}$ of F.

Convergence in distribution is equivalent to point-wise convergence of the corresponding characteristic functions

$$Ee^{iuY_n} \to Ee^{iuY}, \quad \forall u \in \mathbb{R}.$$

Under additional conditions, convergence in distribution may be verified using simpler criteria.

4.2   Cumulant criterion for CLT

This section deals with the so-called moment or cumulant criterion for weak convergence of r.v.'s and the properties of joint cumulants. As an application of this method, we prove a CLT for sums of m-dependent r.v.'s.

For any r.v. Y with all finite moments, its kth order cumulant, for $k = 0, 1, \cdots$, is defined as

$$\mathrm{Cum}_k(Y) = (-\mathbf{i})^k \frac{d^k}{du^k} \log Ee^{iuY}\Big|_{u=0}. \tag{4.2.1}$$

Moreover, the characteristic function and the log-characteristic function of Y can be written as a series of moments and cumulants of Y:

$$Ee^{iuY} = \sum_{k=0}^{\infty} \frac{(iu)^k}{k!} EY^k, \quad \log Ee^{iuY} = \sum_{k=0}^{\infty} \frac{(iu)^k}{k!} \mathrm{Cum}_k(Y). \tag{4.2.2}$$

In particular,

$$\mathrm{Cum}_0(Y) = 0, \quad \mathrm{Cum}_1(Y) = EY, \quad \mathrm{Cum}_2(Y) = EY^2 - (EY)^2,$$

$$\mathrm{Cum}_3(Y) = EY^3 - 3E(Y - EY)^2 EY - (EY)^3.$$

Let Y_n, Y be a sequence of r.v.'s having all moments finite for all n. Because of the one-to-one correspondence between moments and cumulants, see Proposition 4.2.2 below, the convergence of moments, as $n \to \infty$,

$$EY_n^k \to EY^k, \quad \forall k = 1, 2, \cdots \tag{4.2.3}$$

is equivalent to the convergence of cumulants

$$\mathrm{Cum}_k(Y_n) \to \mathrm{Cum}_k(Y), \quad \forall k = 1, 2, \cdots. \tag{4.2.4}$$

Either can be used for verification of the convergence in distribution.

Proposition 4.2.1. (Method of moments). *Suppose r.v.'s Y_n, $n = 1, 2, \cdots$, and Y have all finite moments: $E|Y_n|^k < \infty$, $E|Y|^k < \infty$, $k \geq 1$, $n \geq 1$, and the distribution of Y is uniquely determined by its moments. Then convergence of moments (4.2.3) or cumulants (4.2.4) implies the convergence in distribution (4.1.1).*

Although conditions (4.2.3) and (4.2.4) are equivalent, verification of (4.2.4) can be much easier for Gaussian or Poisson limits.

A Gaussian variable $Y \sim \mathcal{N}(0, \sigma^2)$ has the log-characteristic function $\log E e^{iuY} = -\sigma^2 u^2/2$. Therefore,

$$\mathrm{Cum}_k(Y) = 0, \qquad k = 1, 3, 4, \cdots,$$
$$= \sigma^2, \qquad k = 2.$$

A Poisson r.v. Y with the mean and variance parameter $\lambda > 0$, has the characteristic function $E e^{iuY} = \exp\{\lambda(e^{iu} - 1)\}$, which yields

$$\mathrm{Cum}_k(Y) = \lambda, \qquad k = 1, 2, 3, \cdots.$$

These facts yield the following simple cumulant-based criteria.

Corollary 4.2.1. *Suppose r.v.'s Y_n, $n = 1, 2, \cdots$ have all moments finite.*
(a) *If, in addition,*

$$EY_n \to 0, \quad \mathrm{Var}(Y_n) \to \sigma^2, \tag{4.2.5}$$
$$\mathrm{Cum}_k(Y_n) \to 0, \quad \forall k = 3, 4, \cdots,$$

then $Y_n \to_D \mathcal{N}(0, \sigma^2)$.
(b) *If, in addition,*

$$\mathrm{Cum}_k(Y_n) \to \lambda > 0, \qquad \forall k = 1, 2, 3, \cdots,$$

then $Y_n \to_D \mathrm{Poisson}(\lambda)$.

Properties of cumulants. Applying the cumulant criterion to a sum $Y_n = X_1 + \cdots + X_n$, one needs to express the cumulants $\mathrm{Cum}_k(Y_n)$ in terms of joint cumulants of the variables X_j, $j = 1, \cdots, n$.

Let m be a positive integer, and $\{X_j, 1 \le j \le m\}$ be a set of r.v.'s with $E|X_j|^m < \infty$, for all $j = 1, \cdots, m$.

Definition 4.2.1. The *joint cumulants* of $\{X_j, 1 \le j \le m\}$ are defined by

$$\mathrm{Cum}(X_1, \cdots, X_m) \tag{4.2.6}$$
$$= (-i)^m \partial^m \log E \exp\left\{ i \sum_{j=1}^m u_j X_j \right\} \Big/ \partial u_1 \cdots \partial u_m \Big|_{u_1 = \cdots = u_m = 0}.$$

In particular,

$$\text{Cum}(X_1) = EX_1, \qquad \text{Cum}(X_1, X_2) = \text{Cov}(X_1, X_2),$$

$$\text{Cum}(X_1, \cdots, X_m) = \text{Cum}_m(Y), \quad \text{if } X_1 = \cdots = X_m = Y.$$

The joint cumulant is a measure of the joint dependence between the given r.v.'s. We list some of their additional properties:

(c1) (*symmetry*): $\text{Cum}(X_1, \cdots, X_m)$ is symmetric under permutations of $X_1, \cdots, X_m$.

(*multilinearity*): if $X_i = \sum_{j=1}^n c_{ij}\xi_j$ where ξ_j are r.v.'s and c_{ij}, $i = 1, \cdots, m$, $j = 1, \cdots, n$ are some real constants, then

$$\text{Cum}(X_1, \cdots, X_m) = \sum_{j_1, \cdots, j_m=1}^n c_{1j_1} \cdots c_{mj_m} \text{Cum}(\xi_{j_1}, \cdots, \xi_{j_m}).$$

(c2) $\text{Cum}(X_1, \cdots, X_{m_1}, Y_1, \cdots, Y_{m_2}) = 0$, for any collections of mutually independent r.v.'s $\{X_1, \cdots, X_{m_1}\}$ and $\{Y_1, \cdots, Y_{m_2}\}$, where m_1, m_2 are some given positive integers.

(c3) If the r.v.'s $\{X_1, \cdots, X_m\}$ are jointly Gaussian and $m \geq 3$, then $\text{Cum}(X_1, \cdots, X_m) = 0$.

Properties (c1) to (c3) can be derived from equation (4.2.6). They imply the following useful facts. Suppose T_1, T_2 are disjoint subsets of $\{1, 2, \cdots, m\}$ such that the r.v.'s $\{X_j, j \in T_1\}$ and $\{X_j, j \in T_2\}$ are mutually independent, then by $(c2)$, $\text{Cum}(X_1, X_2, \cdots, X_m) = 0$. In particular, for any r.v.'s $X_1, \cdots, X_m$ and constants $a, a_1, \cdots, a_m$,

$$\text{Cum}(a, X_1, \cdots, X_m) = 0,$$

$$\text{Cum}(X_1 + a_1, \cdots, X_m + a_m) = \text{Cum}(X_1, \cdots, X_m).$$

If r.v.'s $X_1, \cdots, X_m$ are independent then by (c1) to (c3), the cumulants of their sum are equal to the sum of the cumulants, i.e.,

$$\text{Cum}_k\Big(\sum_{j=1}^m X_j\Big) = \sum_{j=1}^m \text{Cum}_k(X_j), \quad k = 1, 2, \cdots.$$

Joint cumulants can also be expressed in terms of joint moments and *vice versa*. Let $T = \{1, \cdots, m\}$. Denote by $(V)_r = (V_1, \cdots, V_r)$ a partition of T into r non-empty disjoint subsets $V_1, \cdots, V_r$, $\cup_{i=1}^r V_i = T$, where $r = 1, \cdots, m$. For example,

$$(V_1) = (\{1, \cdots, m\}), \ r = 1; \quad (V_1, V_2) = (\{1, 2\}, \{3, 4, \cdots, m\}), \ r = 2;$$

$$(V_1, \cdots, V_m) = (\{1\}, \cdots, \{m\}), \ r = m.$$

Denote by $\mathcal{V}_T$ the class of all such partitions $(V)_r$, $r = 1, \cdots, m$ of T. For $V = \{j_1, \cdots, j_\ell\} \subset T$, let $EX^V := E\prod_{j \in V} X_j$ and $\mathrm{Cum}(X^V) := \mathrm{Cum}(X_j, j \in V) = \mathrm{Cum}(X_{j_1}, \cdots, X_{j_\ell})$.

Proposition 4.2.2. (Leonov–Shiryaev formulas). *Let $\{X_j, j \in T\}$ be a system of r.v.'s. Then*

$$\mathrm{Cum}(X^T) = \sum_{(V)_r \in \mathcal{V}_T} (-1)^{r-1}(r-1)!\, EX^{V_1} \cdots EX^{V_r}, \qquad (4.2.7)$$

$$EX^T = \sum_{(V)_r \in \mathcal{V}_T} \mathrm{Cum}(X^{V_1}) \cdots \mathrm{Cum}(X^{V_r}). \qquad (4.2.8)$$

The following properties are often used in this book. For centered r.v.'s, $EX_j = 0$,

$$\mathrm{Cum}(X_1, X_2, X_3) = EX_1 X_2 X_3, \qquad (4.2.9)$$

$$\mathrm{Cum}(X_1, X_2, X_3, X_4) = E[X_1 X_2 X_3 X_4] - E[X_1 X_2]E[X_3 X_4]$$
$$- E[X_1 X_3]E[X_2 X_4] - E[X_1 X_4]E[X_2 X_3].$$

Moreover, for any r.v.'s,

$$|\mathrm{Cum}_k(X)| \le CE|X|^k, \qquad (4.2.10)$$

$$|\mathrm{Cum}(X_1, \cdots, X_k)| \le C \max_{1 \le i \le k} E|X_i|^k,$$

where $C < \infty$ depends on k but not on $X, X_1, \cdots, X_k$. Relation $(4.2.10)$ follows from $(4.2.7)$, noting that by the Hölder inequality,

$$|EX^{|V|}| \le E|X|^{|V|} \le (E|X|^k)^{|V|/k},$$

$$|E\prod_{i \in V} X_i| \le \prod_{i \in V} (E|X_i|^k)^{1/k} \le \max_{i \in V}(E|X_i|^k)^{|V|/k},$$

where $|V|$ denotes the number of elements in the set $V \subset \{1, \cdots, k\}$.

Formula $(4.2.8)$ for the product moment EX^T is much simpler for a centered Gaussian system $\{X_j, j \in T\}$. In this case $\mathrm{Cum}(X^V) = 0$, if $|V| \ge 3$. Setting $r_V = EX_i X_j$ for $V = \{i, j\}$, $|V| = 2$, $(4.2.8)$ yields

$$EX^T = \sum_{(V_1, \cdots, V_{n/2}) \in \mathcal{V}_T : |V_i| = 2} r_{V_1} \cdots r_{V_{n/2}} \qquad (4.2.11)$$

provided m is even, $EX^T = 0$ otherwise. The number of partitions in the last sum equals the number of ways to choose $m/2$ pairs out of m objects, which number is equal to

$$\binom{m}{2!, \cdots, 2!} \frac{1}{(m/2)!} = (m-1)!! := 1 \cdot 3 \cdot 5 \cdots (m-1).$$

For a Gaussian variable $X \sim \mathcal{N}(0, \sigma^2)$ from (4.2.11) we obtain

$$EX^m = \sigma^m (m-1)!!, \quad m \text{ is even,}$$
$$= 0, \quad m \text{ is odd.}$$

The above formulas for joint cumulants are used in the proof of the CLT for transformations of a Gaussian stationary process.

Direct analysis of cumulants from moments via (4.2.7) is cumbersome. A rough bound can be obtained from the Cauchy formula of complex analysis. Suppose the moment generating function Ee^{zX} of a r.v. X is analytic on a complex disk $D_r := \{z \in C : |z| < r\}$, $r > 0$. Then there exists $r_0 > 0$ such that $\log Ee^{zX}$ is analytic on the disk D_{r_0} and continuous on its boundary $\partial D_{r_0} := \{z \in C : |z| = r_0\}$. Therefore,

$$\text{Cum}_k(X) = \frac{k!}{2\pi \mathbf{i}} \int_{\partial D_{r_0}} \frac{\log Ee^{zX}}{z^{k+1}} \, dz$$

implying

$$\left| \text{Cum}_k(X) \right| \leq Ck! \, r_0^{-k}, \quad k = 1, 2, \cdots,$$

with $C := \sup_{z \in \partial D_{r_0}} |\log Ee^{zX}|$.

As a simple application of the cumulant method, we prove the following CLT for m-dependent r.v.'s, which will be often used in this text.

Proposition 4.2.3. *Let* $\{X_j\}$ *be a stationary m-dependent process,* $1 \leq m < \infty$, $EX_0^2 < \infty$. *Then*

$$n^{-1/2} \sum_{j=1}^{n} (X_j - EX_j) \to_D \mathcal{N}(0, \sigma^2), \tag{4.2.12}$$

where $\sigma^2 := \sum_{k \in \mathbb{Z}} \text{Cov}(X_0, X_k)$.

Proof. Denote by T_n the l.h.s. of (4.2.12). Write $X_j = X_{j,1} + X_{j,2}$ as a sum of two m-dependent processes by setting

$$X_{j,1} := X_j I(|X_j| \leq \log n), \quad X_{j,2} := X_j I(|X_j| > \log n),$$
$$T_{n,i} := n^{-1/2} \sum_{j=1}^{n} (X_{j,i} - EX_{j,i}), \quad i = 1, 2,$$

then $T_n = T_{n,1} + T_{n,2}$. The claim (4.2.12) will be proved if we show that

$$T_{n,1} \to_D \mathcal{N}(0, \sigma^2), \quad T_{n,2} \to_p 0. \tag{4.2.13}$$

To prove (4.2.13), let $\gamma_{X,i}(k) := \text{Cov}(X_{j,i}, X_{j+k,i})$, $i = 1, 2$. Since $\{X_{j,1}\}$ is m-dependent,

$$\gamma_{X,1}(k) = 0, \quad |k| > m,$$
$$\to \gamma_X(k) = \text{Cov}(X_j, X_{j+k}), \quad |k| \leq m.$$

Therefore, from (3.3.1), we obtain

$$\text{Var}(T_{n,1}) = \sum_{k=-m}^{m} (1 - \frac{|k|}{n})\gamma_{X,1}(k) \to \sum_{k=-m}^{m} \gamma_X(k) = \sigma^2. \quad (4.2.14)$$

In view of (4.2.5) (the cumulant criterion for CLT), to complete the proof of the first part of the claim (4.2.13), it remains to show that

$$\text{Cum}_k(T_{n,1}) \to 0, \quad \forall k = 3, 4 \cdots . \quad (4.2.15)$$

By m-dependence of $\{X_{j,1}\}$, using properties (c1) and (c2) of joint cumulants and (4.2.10), it follows that for all $k = 3, 4, \cdots$,

$$|\text{Cum}_k(T_{n,1})| = n^{-k/2} \left| \sum_{j_1, \cdots, j_k=1}^{n} \text{Cum}(X_{j_1,1}, \cdots, X_{j_k,1}) \right|$$

$$\leq \quad k! n^{-k/2} \sum_{1 \leq j_1 \leq \cdots \leq j_k \leq n} |\text{Cum}(X_{j_1,1}, \cdots, X_{j_k,1})|$$

$$\leq \quad k! n^{-k/2} \sum_{1 \leq j_1 \leq \cdots \leq j_k \leq n : j_k - j_1 \leq km} |\text{Cum}(X_{j_1,1}, \cdots, X_{j_k,1})|$$

$$\leq C n^{1-k/2} \max_{1 \leq j \leq n} E|X_{j,1}|^k \leq C n^{1-k/2} (\log n)^k \to 0.$$

Similarly as in (4.2.14),

$$\text{Var}(T_{n,2}) = \sum_{k=-m}^{m} (1 - \frac{|k|}{n})\gamma_{X,2}(k) \to 0,$$

since $|\gamma_{X,2}(k)| \leq \text{Var}(X_{k,2}) \to 0$, which proves $T_{n,2} \to_p 0$.

Cramér–Wold device. Let F_n denote the distribution function of the random vector $Y_n = (Y_{n,1}, \cdots, Y_{n,k})$, $k \geq 1$. As for the univariate case, the sequence Y_n is said to converge *in distribution* to a random vector $Y = (Y_1, \cdots, Y_k)$, with the distribution function F, and we write $Y_n \to_D Y$, if $F_n(x) \to F(x)$, for all continuity points $x \in \mathbb{R}^k$ of F.

A useful tool for verification of this convergence is the following criterion:

Proposition 4.2.4. (Cramér–Wold device). *Let Y_n, Y be a sequence of random vectors in $\mathbb{R}^k$. If $\sum_{j=1}^{k} a_j Y_{n,j} \to_D \sum_{j=1}^{k} a_j Y_j$ for any constants $a_1, \cdots, a_k \in \mathbb{R}$, then $Y_n \to_D Y$.*

The following fact follows from Slutsky's theorem. It is used often in this text.

Lemma 4.2.1. *Suppose r.v.'s X_n, $Z_{n,\epsilon}$, and $Y_{n,\epsilon}$, $n \geq 1$, $\epsilon > 0$, are such that $X_n = Z_{n,\epsilon} + Y_{n,\epsilon}$. Assume that there is a family of r.v.'s Z, Z_ϵ, $\epsilon > 0$, such that $Z_\epsilon \to_D Z$ as $\epsilon \to 0$. Moreover, assume that $\forall \epsilon > 0$, $Z_{n,\epsilon} \to_D Z_\epsilon$, as $n \to \infty$, and $\limsup_n \mathrm{Var}(Y_{n,\epsilon}) \to 0$ as $\epsilon \to 0$. Then $X_n \to_D Z$.*

A number of proofs of asymptotic normality in this book use the following truncation-type argument. One decomposes $S_n = S_{n,\epsilon} + R_{n,\epsilon}$, $\epsilon > 0$ into the sum of the main term $S_{n,\epsilon}$, and the remainder $R_{n,\epsilon}$, such that

$$S_{n,\epsilon} \to_D Z_\epsilon \sim \mathcal{N}(0, \sigma_\epsilon^2), \quad \mathrm{Var}(S_{n,\epsilon}) \to \sigma_\epsilon^2, \quad n \to \infty, \quad \forall \epsilon > 0;$$

$$\limsup_n \mathrm{Var}(R_{n,\epsilon}) \to 0, \quad \epsilon \to 0.$$

Moreover, if $\sigma^2 = \lim_n \mathrm{Var}(S_n)$ exists, then

$$
\begin{aligned}
|\sigma_\epsilon^2 - \sigma^2| &= |\lim_n \{\mathrm{Var}(S_n - R_{n,\epsilon}) - \mathrm{Var}(S_n)\}| \\
&\leq 2 \limsup_n |\mathrm{Cov}(R_{n,\epsilon}, S_n)| + \limsup_n \mathrm{Var}(R_{n,\epsilon}) \\
&\to 0, \quad \epsilon \to 0.
\end{aligned}
$$

Hence, by Slutsky's theorem, $S_n \to_D Z \sim \mathcal{N}(0, \sigma^2)$.

Note: For other properties of cumulants see Leonov and Shiryaev (1959), Brillinger (1981), Giraitis and Surgailis (1986), and Peccati and Taqqu (2011).

4.3   CLT for sums of a linear process

In this section we discuss asymptotic distributions of sums and weighted sums of a linear process. Special attention is paid to the central and non-central limit theorems for sums of non-linear functions $\{h(X_j)\}$ of a Gaussian process $\{X_j\}$, that stand at the beginning of the long memory literature in the mid-1970s. They demonstrate how the interplay between the dependence intensity of the underlying process $\{X_j\}$ and the Hermite rank of the function h gives rise to a new class of non-linear processes of $h(X_j)$, whose dependence structure is well understood and described by these theorems. As for the i.i.d. Gaussian case, for such processes the asymptotic distribution theory can be based on second-order characteristics, i.e., the covariance function and spectral density. We also present some results on the asymptotic behavior of polynomial forms in i.i.d. r.v.'s, which appear as the main terms in some asymptotic expansions of empirical processes given later in the book.

Let $S_n = \sum_{j=1}^n X_j$. If $\{X_j\}$ is a zero-mean Gaussian process, then $(\mathrm{Var}(S_n))^{-1/2} S_n =_D \mathcal{N}(0,1)$ and the question about the asymptotic distribution of S_n reduces to finding the asymptotics of $\mathrm{Var}(S_n)$. In this section it is assumed that $\{X_j\}$ is a linear process

$$X_j = \sum_{k=0}^{\infty} a_k \zeta_{j-k} = \sum_{k=-\infty}^{j} a_{j-k} \zeta_k, \quad j \in \mathbb{Z} \qquad (4.3.1)$$

with the innovations $\{\zeta_j\} \sim \mathrm{IID}(0, \sigma_\zeta^2)$ and finite variance $EX_j^2 = \sigma_\zeta^2 \sum_{k=0}^{\infty} a_k^2 < \infty$. Theorem 18.6.5 of Ibragimov and Linnik (1971) gives a CLT for S_n under very general conditions. For the sake of completeness it is reproduced here as Theorem 4.3.1, with a somewhat different proof.

Theorem 4.3.1. *Suppose* $\{X_j\}$ *is a stationary linear process (4.3.1), and*

$$\sigma_n^2 := \mathrm{Var}(S_n) \to \infty. \qquad (4.3.2)$$

Then

$$\sigma_n^{-1} S_n \to_D \mathcal{N}(0,1). \qquad (4.3.3)$$

Proof. The proof is based on the cumulant criterion. Without loss of generality, assume that $\sigma_\zeta^2 = 1$. Set $a_k = 0$, $k = -1, -2, \cdots$ in (4.3.1), and let $c_{nj} = \sigma_n^{-1} \sum_{k=1}^n a_{k-j}$, $j \in \mathbb{Z}$. Then

$$\sigma_n^{-1} S_n = \sum_{j=-\infty}^n c_{nj} \zeta_j, \quad \sigma_n^{-2} \mathrm{Var}(S_n) = \sum_{j=-\infty}^n c_{nj}^2 = 1. \qquad (4.3.4)$$

We shall first show

$$c_n := \sup_{-\infty < j \le n} |c_{nj}| = o(1). \qquad (4.3.5)$$

For any $s \in \mathbb{Z}$, $t = 1, 2, \cdots$,

$$\sum_{j=s-t}^s c_{nj}^2 = \sum_{j=s-t}^s \left(c_{n,j-1} + (c_{nj} - c_{n,j-1}) \right)^2,$$

$$c_{n,s}^2 = c_{n,s-t-1}^2 + 2 \sum_{j=s-t}^s c_{n,j-1}(c_{nj} - c_{n,j-1}) + \sum_{j=s-t}^s (c_{nj} - c_{n,j-1})^2.$$

Because $\sum_{-\infty<j\leq n} c_{n,j}^2 = 1$, $\lim_{t\to\infty} c_{n,s-t-1}^2 = 0$, $\forall\, s \in \mathbb{Z}$, $n \geq 1$. Since t is arbitrary, take the limit $t \to \infty$ and use the C–S inequality to obtain

$$c_{n,s}^2 = 2 \sum_{j=-\infty}^{s} c_{n,j-1}(c_{nj} - c_{n,j-1}) + \sum_{j=-\infty}^{s} (c_{nj} - c_{n,j-1})^2 \qquad (4.3.6)$$

$$\leq 2B_n + B_n^2,$$

where $B_n := \left\{ \sum_{j\in\mathbb{Z}} (c_{nj} - c_{n,j-1})^2 \right\}^{1/2}$ does not depend on s. Since by definition $c_{nj} - c_{n,j-1} = \sigma_n^{-1}(a_{1-j} - a_{n-j+1})$,

$$B_n^2 \leq 4\sigma_n^{-2} \sum_{j\in\mathbb{Z}} a_j^2 = o(1),$$

because $\sigma_n \to \infty$ and $EX_1^2 = \sum_{j=0}^{\infty} a_j^2 < \infty$, proving (4.3.5).

We shall now turn to the proof of asymptotic normality in (4.3.3). This proof uses the cumulant criterion in Proposition 4.2.1. Assume first that $E|\zeta_0|^k < \infty$, for all $k \geq 1$. Accordingly, since $E[\sigma_n^{-1}S_n] = 0$, $\mathrm{Var}(\sigma_n^{-1}S_n) = 1$, it suffices to show that

$$\mathrm{Cum}_k(\sigma_n^{-1}S_n) \to 0, \quad \forall\, k \geq 3.$$

Since $\{\zeta_j\} \sim \mathrm{IID}(0,1)$, by properties (c1) to (c3) of joint cumulants,

$$\mathrm{Cum}_k(\sigma_n^{-1}S_n) = \mathrm{Cum}_k(\zeta_0) \sum_{j=-\infty}^{n} c_{nj}^k.$$

Then by (4.3.5), for $k \geq 3$,

$$|\mathrm{Cum}_k(\sigma_n^{-1}S_n)| \leq |\mathrm{Cum}_k(\zeta_0)| c_n^{k-2} \sum_{j=-\infty}^{n} c_{nj}^2 \leq C c_n^{k-2} \to 0. \qquad (4.3.7)$$

This completes the proof of (4.3.3) when all moments of ζ_0 are finite.

Consider now the general case, when some higher moments of innovations ζ_j do not exist. Let

$$\zeta_{j,1} = I(|\zeta_j| \leq |\log c_n|)\zeta_j, \quad \zeta_{j,2} = I(|\zeta_j| > |\log c_n|)\zeta_j,$$

where $|\log c_n| \to \infty$ by (4.3.5). Then

$$S_n = S_{n,1} + S_{n,2}, \quad S_{n,i} := \sum_{j=-\infty}^{n} c_{nj}(\zeta_{j,i} - E\zeta_{j,i}), \quad i = 1, 2.$$

It suffices to show that

$$\sigma_n^{-1}S_{n,1} \to_D \mathcal{N}(0,1), \quad \sigma_n^{-1}S_{n,2} \to_p 0. \qquad (4.3.8)$$

Let $\sigma_{n,i}^2 = \mathrm{Var}(S_{n,i})$, $i = 1, 2$. Then, by (4.3.4),

$$\frac{\sigma_{n,1}^2}{\sigma_n^2} = \frac{\mathrm{Var}(\zeta_{0,1})}{\mathrm{Var}(\zeta_0)} \to 1,$$

implying $\mathrm{Var}(\sigma_n^{-1} S_{n,1}) \to 1$. From the bound (4.2.10) and the definition of $\zeta_{0,1}$, one can estimate

$$|\mathrm{Cum}_k(\zeta_{0,1})| \le CE|\zeta_{0,1}|^k \le C|\log c_n|^k$$

with a constant C depending on k but not on n. Then as in (4.3.7), for $k \ge 3$,

$$|\mathrm{Cum}_k(\sigma_n^{-1} S_{n,1})| \le C|\mathrm{Cum}_k(\zeta_{0,1})|c_n^{k-2} \le C|\log c_n|^k \, c_n^{k-2} \to 0,$$

which proves the first convergence of (4.3.8). The second convergence holds because $\sigma_{n,2}^2/\sigma_n^2 = \mathrm{Var}(\zeta_{0,2})/\mathrm{Var}(\zeta_0) \to 0$. This completes the proof of Theorem 4.3.1.

Remark 4.3.1. By Proposition 3.3.1, a short, long, or negative memory linear process $\{X_j\}$ with memory parameter $|d| < 1/2$ has the property $\mathrm{Var}(S_n) \sim Cn^{1+2d} \to \infty$, which in turn, by Theorem 4.3.1, implies the asymptotic normality result (4.3.3).

The above proof of Theorem 4.3.1 also yields asymptotic normality for weighted sums of i.i.d. r.v.'s with triangular array weights.

Corollary 4.3.1. *Let* $S_n = \sum_{j \in \mathbb{Z}} d_{nj} \zeta_j$, $n \ge 1$, *be the sequence of weighted sums of* $\mathrm{IID}(0, \sigma_\zeta^2)$ *r.v.'s* $\{\zeta_j\}$ *and* d_{nj}, $j \in \mathbb{Z}$ *be an array of real weights such that* $\sum_{j \in \mathbb{Z}} d_{nj}^2 < \infty$ *for all* $n \ge 1$, *and*

$$d_n := \frac{\max_{j \in \mathbb{Z}} |d_{nj}|}{(\sum_{j \in \mathbb{Z}} d_{nj}^2)^{1/2}} \to 0. \tag{4.3.9}$$

Then

$$\sigma_n^{-1} S_n \to_D \mathcal{N}(0, 1), \quad \sigma_n^2 := \mathrm{Var}(S_n) = \sigma_\zeta^2 \sum_{j \in \mathbb{Z}} d_{nj}^2. \tag{4.3.10}$$

Remark 4.3.2. The proof of Theorem 4.3.1 shows that the convergence (4.3.10) remains valid for the sums $S_n = \sum_{j \in \mathbb{Z}} d_{nj} \zeta_{nj}$, $n \ge 1$, where, for each $n \ge 1$, $\{\zeta_{nj}\} \sim \mathrm{IID}(0, \sigma_{\zeta_n}^2)$, as long as (4.3.9) holds and, in addition,

$$\frac{\mathrm{Var}(\zeta_{n1} I(|\zeta_{n1}| \ge |\log d_n|))}{\mathrm{Var}(\zeta_{n1})} \to 0.$$

Since $|\log d_n| \to \infty$, the above condition is satisfied if

$$\liminf_n E\zeta_{n1}^2 > 0, \quad \limsup_n E|\zeta_{n1}|^{2+\delta} < \infty, \quad \exists \delta > 0.$$

Weighted sums. Theorem 4.3.1 can be generalized to weighted sums of linear processes. Let $z_{nj}, j, n \geq 1$ be an array of real numbers, and consider the weighted sums

$$W_n = \sum_{j=1}^{n} z_{nj} X_j. \tag{4.3.11}$$

The following proposition, where $\sigma_n^2 := \mathrm{Var}(W_n)$, gives the limiting distribution of W_n.

Proposition 4.3.1. *Assume $\{X_j\}$ is a stationary linear process (4.3.1). Suppose the weights z_{nj} in W_n and a_j in (4.3.1) satisfy one of the following three conditions:*

(i) $\max_{1 \leq j \leq n} |z_{nj}| = o(\sigma_n)$ *and* $\sum_{j=1}^{n} z_{nj}^2 \leq C\sigma_n^2$,

(ii) $\max_{1 \leq j \leq n} |z_{nj}| = o(\sigma_n)$ *and* $\sum_{j=0}^{\infty} |a_j| < \infty$,

(iii) $|z_{n1}| + |z_{nn}| + \sum_{j=2}^{n} |z_{nj} - z_{n,j-1}| = o(\sigma_n)$.

Then

$$\sigma_n^{-1} W_n \to_D \mathcal{N}(0,1). \tag{4.3.12}$$

Proof. As for (4.3.4), set $a_k = 0$, $k < 0$, and let $c_{nj} = \sigma_n^{-1} \sum_{k=1}^{n} z_{nk} a_{k-j}$, $j \in \mathbb{Z}$. Then one can write

$$\sigma_n^{-1} W_n = \sum_{j=-\infty}^{n} c_{nj} \zeta_j. \tag{4.3.13}$$

Since $\mathrm{Var}(\sigma_n^{-1} W_n) = \sum_{j=-\infty}^{n} \sigma_\zeta^2 c_{nj}^2 = 1$, by Corollary 4.3.1, it suffices to verify relation (4.3.9) for the coefficients c_{nj}, i.e., to show that $\sup_{-\infty < j \leq n} |c_{nj}| = o(1)$ or (4.3.5).

Consider case (i). Clearly, $K_n := \sigma_n / \max_{1 \leq j \leq n} |z_{nj}| \to \infty$ and

$$|c_{nj}| \leq \sigma_n^{-1} \sum_{k=1}^{n} |z_{nk} a_{k-j}| I(|k-j| \geq K_n)$$

$$+ \sigma_n^{-1} \sum_{k=1}^{n} |z_{nk} a_{k-j}| I(|k-j| < K_n) =: q_{n,1j} + q_{n,2j}.$$

By assumption (i) and the C–S inequality,

$$q_{n,1j} \leq \frac{(\sum_{k=1}^{n} z_{nk}^2)^{1/2}}{\sigma_n} (\sum_{k=1}^{n} a_{k-j}^2 I(|k-j| \geq K_n))^{1/2}$$

$$\leq C \sum_{l \geq K_n} a_l^2 \to 0,$$

$$q_{n,2j} \leq \frac{\max_{1 \leq k \leq n} |z_{nk}|}{\sigma_n} \sum_{k=1}^{n} |a_{k-j}| I(|k-j| < K_n)$$

$$\leq K_n^{-1} K_n^{1/2} \Big(\sum_{l=0}^{\infty} a_l^2 \Big)^{1/2} \leq C K_n^{-1/2} \to 0,$$

uniformly in $j \in \mathbb{Z}$, thereby proving (4.3.5).

In case (ii), (4.3.5) follows because

$$|c_{nj}| \leq \sigma_n^{-1} \max_{1 \leq j \leq n} |z_{nj}| \sum_{l=0}^{\infty} |a_l| \leq C \sigma_n^{-1} \max_{1 \leq j \leq n} |z_{nj}|.$$

To verify (4.3.5) in case (iii), we use the bound (4.3.6), which does not depend on any particular form of c_{nj}. It suffices to show that $B_n \to 0$. Define for simplicity $z_{n0} = z_{n,n+1} = 0$. Then one can write

$$c_{nj} - c_{n,j-1} = \sigma_n^{-1} \sum_{k=1}^{n} z_{nk}(a_{k-j} - a_{k+1-j}) = \sigma_n^{-1} \sum_{k=1}^{n+1} (z_{nk} - z_{n,k-1}) a_{k-j}.$$

Hence,

$$B_n^2 = \sum_{j \in \mathbb{Z}} (c_{nj} - c_{n,j-1})^2$$

$$= \sigma_n^{-2} \sum_{k,s=1}^{n+1} (z_{nk} - z_{n,k-1})(z_{ns} - z_{n,s-1}) \sum_{j \in \mathbb{Z}} a_{k-j} a_{s-j}$$

$$\leq \sigma_n^{-2} \sum_{k,s=1}^{n+1} |z_{nk} - z_{n,k-1}||z_{ns} - z_{n,s-1}| \sum_{j=0}^{\infty} a_j^2$$

$$\leq C \sigma_n^{-2} \Big(|z_{n1} + |z_{nn}| + \sum_{k=2}^{n} |z_{nk} - z_{n,k-1}| \Big)^2 \to 0,$$

by condition (iii) of the proposition. This completes the proof.

Verification of conditions (i) to (iii) of Proposition 4.3.1 requires further analysis of the variance $\sigma_n^2 \equiv \mathrm{Var}(W_n) = \sum_{j,k=1}^{n} z_{nj} \gamma_X(j-k) z_{nk}$. The next proposition provides sufficient conditions for (i) and (iii) to hold in terms of the spectral density f of $\{X_j\}$ and weights z_{nk}. Part (a) only requires f to be continuous at 0. This includes spectral densities that may have infinite peaks or may be vanishing on an interval not including the origin. Part (b) focuses on the case when f is bounded away from 0, which includes the case of long memory. Conditions (4.3.16) and (4.3.14) on weights are mild and satisfied in most applications.

Proposition 4.3.2. *Assume $\{X_j\}$, z_{nj}, and W_n are as in (4.3.1) and (4.3.11), with f denoting the spectral density of $\{X_j\}$.*

(a) *Suppose $f(u) \to f(0)$, $u \to 0$, $0 < f(0) < \infty$, and*

$$|z_{n1}| + |z_{nn}| + \sum_{j=2}^{n} |z_{nj} - z_{n,j-1}| = o\Big((\sum_{j=1}^{n} z_{nj}^2)^{1/2}\Big). \qquad (4.3.14)$$

Then Proposition 4.3.1(iii) is satisfied, and (4.3.12) holds. Moreover,

$$\Big(\sum_{j=1}^{n} z_{nj}^2\Big)^{-1} \mathrm{Var}(W_n) \to 2\pi f(0). \qquad (4.3.15)$$

(b) *If $f(u) \geq c > 0$, $|u| \leq u_0$, for some $c > 0$ and $u_0 > 0$, and (4.3.14) is valid, then Proposition 4.3.1(iii) is satisfied and (4.3.12) holds.*

(c) *Suppose there exists a $c > 0$ such that $f(u) \geq c > 0$, $u \in \Pi$, and*

$$\max_{1 \leq j \leq n} |z_{nj}| = o\Big(\big(\sum_{j=1}^{n} z_{nj}^2\big)^{1/2}\Big). \qquad (4.3.16)$$

Then Proposition 4.3.1(i) is satisfied, and hence (4.3.12) holds.

Proof. (a) Let $G(u) := \sum_{j=1}^{n} e^{-iju} z_{nj}$, $u \in \Pi$. Since $\gamma_X(k) = \int_{\Pi} e^{iku} f(u)\, du$, write

$$\sigma_n^2 = \sum_{j,k=1}^{n} z_{nj} \gamma_X(j-k) z_{nk} = \int_{\Pi} f(u)|G(u)|^2 du. \qquad (4.3.17)$$

Observe that

$$\int_{\Pi} |G(u)|^2 du = \int_{\Pi} \sum_{j,k=1}^{n} e^{i(j-k)u} z_{nj} z_{nk}\, du = 2\pi \sum_{j=1}^{n} z_{nj}^2, \qquad (4.3.18)$$

$$\sigma_n^2 = 2\pi f(0) \sum_{j=1}^{n} z_{nj}^2 + i_n, \quad i_n := \sigma_n^2 - 2\pi f(0) \sum_{j=1}^{n} z_{nj}^2.$$

We shall show that

$$i_n = |\sigma_n^2 - 2\pi f(0) \sum_{j=1}^{n} z_{nj}^2| = o\big(\sum_{j=1}^{n} z_{nj}^2\big), \qquad (4.3.19)$$

which proves (4.3.15) and together with (4.3.14) implies condition (iii) of Proposition 4.3.1 and thus (4.3.12).

Let $\epsilon > 0$. Choose $\delta > 0$ such that $\sup_{0 \leq u \leq \delta} |f(u) - f(0)| \leq \epsilon/2\pi$. Then by (4.3.17) and (4.3.18),

$$i_n \leq \int_{|u| \leq \pi} |f(u) - f(0)||G(u)|^2 du$$

$$= \int_{|u| \leq \delta} [\cdots] du + \int_{\delta < |u| \leq \pi} [\cdots] du =: s_{n,1} + s_{n,2},$$

where

$$s_{n,1} \leq \frac{\epsilon}{2\pi} \int_{|u| \leq \delta} |G(u)|^2 du \leq \frac{\epsilon}{2\pi} \int_{\Pi} |G(u)|^2 du = \epsilon \sum_{j=1}^{n} z_{nj}^2.$$

To prove (4.3.19), it suffices to check that for any $\delta > 0$,

$$s_{n,2} = o\left(\sum_{j=1}^{n} z_{nj}^2 \right). \tag{4.3.20}$$

Using the summation by parts formula (2.5.8), write

$$G(u) = \sum_{j=1}^{n-1} \left(\sum_{l=1}^{j} e^{-ilu} \right)(z_{nj} - z_{n,j+1}) + z_{nn} \sum_{l=1}^{n} e^{-ilu}.$$

By (2.1.7), for $j = 1, \cdots, n$, we can bound $|\sum_{l=1}^{j} e^{ilu}| \leq \pi u^{-1} \leq \pi \delta^{-1}$, $\delta \leq u \leq \pi$. Therefore,

$$|G(u)| \leq \pi \delta^{-1} \left(\sum_{j=1}^{n-1} |z_{nj} - z_{n,j+1}| + |z_{nn}| \right) = o\left(\sum_{j=1}^{n} z_{nj}^2 \right)^{1/2},$$

by (4.3.14). Hence,

$$s_{n,2} = o\left(\sum_{j=1}^{n} z_{nj}^2 \right) \int_{\delta < |u| \leq \pi} (f(u) + f(0)) du = o\left(\sum_{j=1}^{n} z_{nj}^2 \right),$$

which implies (4.3.20).

(b) Define $\tilde{f}(u) = cI(|u| \leq u_0)$, $u \in \Pi$. Notice that $f \geq \tilde{f}$, and $\tilde{f}(u) \rightarrow \tilde{f}(0) = c > 0$, $u \rightarrow 0$. Thus, by (a),

$$\sigma_n^2 = \int_{\Pi} f(u)|G(u)|^2 du \geq \int_{\Pi} \tilde{f}(u)|G(u)|^2 du \sim 2\pi \tilde{f}(0) \sum_{j=1}^{n} z_{nj}^2.$$

This fact and (4.3.14) proves Proposition 4.3.1(iii) and implies (4.3.12).

(c) Assumption $f(u) \geq c > 0$ thus implies

$$\sigma_n^2 \geq c \int_{\Pi} |G(u)|^2 du = 2\pi c \sum_{j=1}^{n} z_{nj}^2, \tag{4.3.21}$$

which together with (4.3.16) yields Proposition 4.3.1(i) and therefore proves (4.3.12) and completes the proof of the proposition.

The following proposition, which is valid for any stationary process $\{X_j\}$ not necessarily linear, provides an upper bound for $\text{Var}(W_n)$ and analyzes its asymptotic behavior. These results are useful later when dealing with kernel-type regression function estimators.

Proposition 4.3.3. *Let $\{X_j\}$ be a covariance-stationary process with zero mean and finite variance such that*

$$\sum_{k\in\mathbb{Z}} |\gamma_X(k)| < \infty. \tag{4.3.22}$$

Let W_n be as in (4.3.11). Assume that z_{nj} satisfy (4.3.14). Then

$$\Big(\sum_{j=1}^n z_{nj}^2\Big)^{-1}\text{Var}(W_n) \leq \sum_{k\in\mathbb{Z}} |\gamma_X(k)|, \tag{4.3.23}$$

$$\Big(\sum_{j=1}^n z_{nj}^2\Big)^{-1}\text{Var}(W_n) \to \sigma^2 := \sum_{k\in\mathbb{Z}} \gamma_X(k).$$

Proof. Under (4.3.22),

$$f(u) = (2\pi)^{-1}\sum_{k\in\mathbb{Z}} e^{-iku}\gamma_X(k) \leq (2\pi)^{-1}\sum_{k\in\mathbb{Z}} |\gamma_X(k)| < \infty, \quad u \in \Pi,$$

$$f(u) \to f(0) = (2\pi)^{-1}\sum_{k\in\mathbb{Z}} \gamma_X(k) = (2\pi)^{-1}\sigma^2, \quad u \to 0.$$

Thus, by (4.3.17) and (4.3.18),

$$\text{Var}(W_n) = \int_\Pi f(u)|G(u)|^2 du \leq \sup_{u\in\Pi} f(u)\int_\Pi |G(u)|^2 du$$

$$\leq \sum_{k\in\mathbb{Z}} |\gamma_X(k)| \sum_{j=1}^n z_{nj}^2,$$

which proves the first bound of (4.3.23). The proof of the second bound is the same as that of (4.3.15). This completes the proof of the proposition.

The following result is a multivariate generalization of Proposition 4.3.2.

Theorem 4.3.2. *Let $z_{n,j}^{(i)}$, $i = 1, \cdots, k$, be k arrays of real weights and $\{X_j\}$ be the same as in Proposition 4.3.1. Assume that $W_n^{(i)} := \sum_{j=1}^n z_{n,j}^{(i)} X_j$ and $(\sigma_n^{(i)})^2 := \text{Var}(W_n^{(i)})$, satisfy one of the conditions (i)* to (iii) *of Proposition 4.3.1, for every* $i = 1, \cdots, k$, *and for some (positive definite) matrix* Σ,

$$\left(\mathrm{Cov}(W_n^{(i)}/\sigma_n^{(i)}, \; W_n^{(j)}/\sigma_n^{(j)})\right)_{i,j=1,\cdots,k} \to \Sigma. \qquad (4.3.24)$$

Then $\left(W_n^{(1)}/\sigma_n^{(1)}, \cdots, W_n^{(k)}/\sigma_n^{(k)}\right) \to_D \mathcal{N}_k(0, \Sigma)$.

Proof. As for (4.3.13), write

$$S_n^{(i)} := W_n^{(i)}/\sigma_n^{(i)} = \sum_{j=-\infty}^{n} c_{nj}^{(i)} \zeta_j, \quad i = 1, \cdots, k.$$

To prove the theorem, in view of the Cramér–Wold device, it suffices to show that for every $a' = (a_1, \cdots, a_k) \in \mathbb{R}^k$,

$$S_n := a_1 S_n^{(1)} + \cdots + a_k S_n^{(k)} \to_D \mathcal{N}(0, a'\Sigma a). \qquad (4.3.25)$$

Write $S_n = \sum_{j=-\infty}^{n} d_{nj} \zeta_j$ where $d_{nj} = a_1 c_{nj}^{(1)} + \cdots + a_k c_{nj}^{(k)}$. Condition (4.3.24) implies that

$$\mathrm{Var}(S_n) = E\zeta_1^2 \sum_{j=-\infty}^{n} d_{nj}^2 \to a'\Sigma a.$$

The proof of Proposition 4.3.1 shows that any one of the conditions (i), (ii), or (iii) ensures that $\sup_{-\infty < j \leq n} |c_{nj}^{(i)}| \to 0$, $\forall i = 1, \cdots, k$, which yields $\sup_{-\infty < j \leq n} |d_{nj}| \to 0$. Hence, the coefficients d_{nj} of S_n satisfy the assumptions of Corollary 4.3.1, which imply (4.3.25) and completes the proof.

Remark 4.3.3. In applications, the verification of the conditions for asymptotic normality of the weighted sum W_n in Proposition 4.3.1 often reduces to analyzing the asymptotic behavior of the variance σ_n^2, since the remaining conditions on the weights z_{nj} are usually easy to verify. For example, see the regression application below. Assumption (ii) can be applied to short and negative memory linear processes $\{X_j\}$ that satisfy $\sum_{j=0}^{\infty} |a_j| < \infty$, whereas condition (i) is useful in the case when $\{X_j\}$ has long or short memory. Condition (iii) of Proposition 4.3.1 regarding the weights z_{nj} is precise in the sense that it includes the result of Theorem 4.3.1 corresponding to the case of equal weights $z_{n1} = \cdots = z_{nn} = 1$.

As can be seen from the proofs of Proposition 4.3.1 and Theorem 4.3.2, their conclusions remain valid if the linear process $\{X_j\}$ in W_n is replaced by triangular arrays of linear processes $\{X_{nj} = \sum_{k=0}^{\infty} a_{nk} \zeta_{j-k}, \; j \in \mathbb{Z}\}$,

$n = 1, 2, \cdots$, with innovations $\{\zeta_j\}$ not depending on n, and such that $\mathrm{Var}(X_{n1}) = \sigma_\zeta^2 \sum_{k=0}^\infty a_{nk}^2 < \infty$, $\forall n \geq 1$, and when conditions (i) to (iii) of Proposition 4.3.1 are replaced by the following assumptions:

(i') $\left(\sum_{j=1}^n z_{nj}^2 \right) \left(\sum_{k=0}^\infty a_{nk}^2 \right) = o(\sigma_n^2)$,

(ii') $\left(\max_{1 \leq j \leq n} |z_{nj}| \right) \sum_{k=0}^\infty |a_{nk}| = o(\sigma_n)$,

(iii') $\left(|z_{n1}| + |z_{nn}| + \sum_{j=2}^n |z_{nj} - z_{n,j-1}| \right) \{ \sum_{k=0}^\infty a_{nk}^2 \}^{1/2} = o(\sigma_n)$.

Regression example. Proposition 4.3.1 is useful in deriving limiting distributions of statistics in non-parametric regression models with the uniform non-random design in $[0, 1]$ where the weights in W_n are given by the regression function. Accordingly, let the non-parametric regression function g be continuous on $[0, 1]$ and set

$$z_{nj} = g(j/n), \quad j = 1, \cdots, n. \qquad (4.3.26)$$

Suppose $\{X_j\}$ is a linear process with memory parameter $0 \leq d < 1/2$, and

$$\text{for } 0 < d < 1/2: \quad \gamma_X(k) \sim c_\gamma |k|^{-1+2d}, \quad k \to \infty, \qquad (4.3.27)$$

$$\text{for } d = 0: \quad \sum_{k=0}^\infty |\gamma_X(k)| < \infty, \quad \sum_{k \in \mathbb{Z}} \gamma_X(k) > 0.$$

Let $\|g\|^2 := \int_0^1 g^2(u)du$, $\|z_n\|^2 := \sum_{k=1}^n z_{nk}^2$,

$$v_{d,g}^2 := \|g\|^2 \sum_{k \in \mathbb{Z}} \gamma_X(k), \qquad\qquad d = 0,$$

$$:= c_\gamma \int_0^1 \int_0^1 g(u)g(v)|u - v|^{-1+2d} du dv, \quad 0 < d < 1/2.$$

Corollary 4.3.2. *Suppose a linear process $\{X_j\}$ of (4.3.1) has the memory parameter $0 \leq d < 1/2$, (4.3.27) holds and the weights z_{nj} of W_n satisfy (4.3.26). Then*

$$n^{-1/2-d} W_n \to_D \mathcal{N}(0, v_{d,g}^2). \qquad (4.3.28)$$

Proof. To prove (4.3.28), we shall verify condition (i) of Proposition 4.3.1. We shall prove that

$$\|z_n\|^2 \sim n\|g\|^2, \qquad (4.3.29)$$

$$n^{-1-2d}\mathrm{Var}(W_n) \to v_{d,g}^2, \quad 0 \leq d < 1/2. \qquad (4.3.30)$$

Then $\sigma_n^2 \sim v_{d,g}^2 n^{1+2d}$, which implies $\|z_n\|^2 = O(\sigma_n^2)$ and $\max_{1 \le k \le n} |z_{nk}| \le \sup_{0 \le u \le 1} |g(u)| = o(\sigma_n)$, and completes the verification of condition (i).

By the DCT,

$$n^{-1}\|z_n\|^2 = n^{-1} \sum_{j=1}^{n} g^2\left(\frac{j}{n}\right) = \int_0^1 g^2\left(\frac{[un]+1}{n}\right) du \rightarrow \int_0^1 g^2(u) du \equiv \|g\|^2,$$

which proves (4.3.29).

Let $0 < d < 1/2$. Then (4.3.29), (4.3.27), a change of variables, and the DCT yield

$$n^{-1-2d}\mathrm{Var}(W_n) = n^{-1-2d} \sum_{k,j=1}^{n} z_{nj} z_{n,k} \gamma_X(j-k)$$

$$= n^{-1-2d} c_\gamma \sum_{k,j=1:k\ne j}^{n} g\left(\frac{j}{n}\right) g\left(\frac{k}{n}\right) |j-k|^{-1+2d} + o(1)$$

$$\rightarrow c_\gamma \int_0^1 \int_0^1 g(u) g(v) |u-v|^{-1+2d} du\,dv = v_{d,g}^2.$$

For $d = 0$, setting $z_{nj} = 0$, for $j > n$ and $j \le 0$, write

$$\mathrm{Var}(W_n) = \sum_{s=-(n-1)}^{n-1} \gamma_X(s) p(s), \qquad p(s) := \sum_{k=1}^{n} z_{nk} z_{n,k+s}.$$

By (4.3.29) and the DCT,

$$n^{-1}|p(s)| \le n^{-1} \sum_{k=1}^{n} z_{nk}^2 \le C, \quad \forall\, s,\, n \ge 1,$$

$$n^{-1} p(s) = n^{-1} \sum_{k=1}^{n} g\left(\frac{k}{n}\right) g\left(\frac{k+s}{n}\right) \sim \int_0^1 g\left(\frac{[nu]}{n}\right) g\left(\frac{[nu]+s}{n}\right) du \rightarrow \|g\|^2,$$

for every $s \in \mathbb{Z}$. Since $\sum_{s \in \mathbb{Z}} |\gamma_X(s)| < \infty$, then by the DCT,

$$n^{-1}\mathrm{Var}(W_n) \rightarrow \|g\|^2 \sum_{s \in \mathbb{Z}} \gamma_X(s) \equiv v_{0,g}^2.$$

This completes the proof of (4.3.30) and the corollary.

4.4 Partial-sum process

In this section we shall discuss the weak convergence of the partial-sum process $S_n(\tau) = \sum_{j=1}^{[n\tau]} X_j$, $0 \le \tau \le 1$. Observe that for each n, the re-normalized partial-sum process $\sigma_n^{-1} S_n(\tau), 0 \le \tau \le 1$, is a step function in

τ, belonging to the Skorokhod functional space $\mathcal{D}[0,1]$. While most of the results in this section pertain to the stationary linear process of (4.3.1), we also include two general tightness criteria, Theorem 4.4.1 and Lemma 4.4.1 below, which do not require the linearity assumption.

Recall, say from Billingsley (1968), that a sequence of stochastic processes $Y_n(\cdot)$, $n \geq 1$, in $\mathcal{D}[0,1]$ is said to *converge weakly to a stochastic process* $Y(\cdot) \in \mathcal{D}[0,1]$ if $Ef(Y_n) \to Ef(Y)$, for every bounded function $f : \mathcal{D}[0,1] \to \mathbb{R}$ that is continuous in the Skorokhod J_1-topology in $\mathcal{D}[0,1]$. Such weak convergence implies the convergence in distribution of continuous functionals of this process. For example, one readily concludes

$$\sup_{0 \leq \tau \leq 1} |Y_n(\tau)| \to_D \sup_{0 \leq \tau \leq 1} |Y(\tau)|; \quad \int_0^1 Y_n^2(\tau)d\tau \to_D \int_0^1 Y^2(\tau)d\tau.$$

These facts in turn are useful for implementing various inference procedures based on these statistics.

A well-known procedure for proving weak convergence of a given sequence of stochastic processes consists of two steps: first, proving the weak convergence of finite-dimensional distributions and second, verification of tightness. Recall that a sequence of stochastic processes $Y_n(\cdot)$, $n \geq 1$, in $\mathcal{D}[0,1]$ is said to be *tight* (in the J_1-topology) if for any $\epsilon > 0$ there exists a compact set $K \subset \mathcal{D}[0,1]$ such that $P(Y_n(\cdot) \in K) > 1 - \epsilon$ for all $n \geq 1$. Compact sets of $\mathcal{D}[0,1]$ in the J_1-topology can be characterized by several continuity moduli. See Billingsley (1968) for definitions of these and other concepts, as well as various tightness criteria in $\mathcal{D}[0,1]$.

In this book we deal exclusively with the situation where the limit process $Y(\cdot)$ of $Y_n(\cdot) \in \mathcal{D}[0,1]$ is *continuous*: $P(Y(\cdot) \in \mathcal{C}[0,1]) = 1$. In this case it is natural to discuss the weak convergence of $Y_n(\cdot), n \geq 1$, in the space $\mathcal{D}[0,1]$ endowed with the uniform topology given by the uniform metric $\sup_{t \in [0,1]} |x(t) - y(t)|$, $x(\cdot), y(\cdot) \in \mathcal{D}[0,1]$, and the σ-algebra $\mathcal{B}_0[0,1]$ of subsets of $\mathcal{D}[0,1]$ generated by all closed balls of the uniform metric; see Pollard (1984). For brevity, we call such convergence the *weak convergence in $\mathcal{D}[0,1]$ with uniform metric*. By definition, this is equivalent to the convergence $Ef(Y_n) \to Ef(Y)$, for every bounded $\mathcal{B}_0[0,1]$-measurable function $f : \mathcal{D}[0,1] \to \mathbb{R}$, that is continuous with respect to the uniform metric. Similarly, we say that a sequence of stochastic processes $Y_n(\cdot)$, $n \geq 1$, is *tight in $\mathcal{D}[0,1]$ with uniform metric*, if for any $\epsilon > 0$ there exists a compact set

$K \subset \mathcal{D}[0,1], K \in \mathcal{B}_0[0,1]$, such that $P(Y_n(\cdot) \in K) > 1 - \epsilon$, for all $n \geq 1$.

The uniform topology is stronger than the J_1-topology and therefore weak convergence in $\mathcal{D}[0,1]$ with the uniform topology allows us to extend the class of functionals f for which the convergence $Ef(Y_n) \to Ef(Y)$ holds. Moreover, verification of continuity of f in the uniform metric is relatively simple.

The following theorem gives a sufficient condition for a sequence of stochastic processes $Y_n(\cdot) \in \mathcal{D}[0,1]$, $n \geq 1$, to converge weakly in $\mathcal{D}[0,1]$ with the uniform metric. It is essentially Theorem 15.5, p. 127 of Billingsley (1968). See also Pollard (1984, ch. V.1, Theorem 3).

Theorem 4.4.1. *Let* $\{Y_n(t), 0 \leq t \leq 1\}$, $n \geq 1$, *be a sequence of stochastic processes in* $\mathcal{D}[0,1]$. *Suppose that* $|Y_n(0)| = O_p(1)$ *and that* $\forall\, \epsilon > 0$,

$$\lim_{\eta \to 0} \limsup_{n} P\left(\sup_{|s-t|<\eta} |Y_n(s) - Y_n(t)| \geq \epsilon \right) = 0. \qquad (4.4.1)$$

Then Y_n, $n \geq 1$, *is tight in* $\mathcal{D}[0,1]$ *with uniform metric. Moreover, if* Y *is the weak limit of a subsequence of* Y_n *then* $P(Y \in \mathcal{C}[0,1]) = 1$.

We shall need the following tightness criterion; see Billingsley (1968). For any real-valued function g on $[0,1]$, let

$$\omega_\delta(g) := \sup_{|t-s| \leq \delta} |g(t) - g(s)|, \quad \delta > 0.$$

Lemma 4.4.1. *Suppose there exists a sequence of non-decreasing right continuous functions* F_n, $n \geq 1$, *on* $[0,1]$ *that are uniformly bounded and satisfy*

$$\lim_{\delta \to 0} \limsup_{n \to \infty} \omega_\delta(F_n) = 0. \qquad (4.4.2)$$

Moreover, assume that $\{F_n\}$ *is such that for some* $\alpha > 1/2, \beta > 0, \gamma > 0$, *and* $C > 0$, *and for all* $n \geq 1$ *and* $0 \leq s < t < u \leq 1$,

$$E|Y_n(t) - Y_n(s)|^\gamma |Y_n(u) - Y_n(t)|^\gamma \leq C|F_n(u) - F_n(s)|^{2\alpha}, \qquad (4.4.3)$$

$$E|Y_n(u) - Y_n(s)|^{2\gamma} \leq C\big\{|F_n(u) - F_n(s)|^{2\alpha} \qquad (4.4.4)$$
$$+ n^{-\beta}|F_n(u) - F_n(s)|\big\}.$$

In particular, (4.4.1) is satisfied if

$$E|Y_n(u) - Y_n(s)|^{2\gamma} \leq C|F_n(u) - F_n(s)|^{2\alpha}, \quad 0 \leq s < u \leq 1. \qquad (4.4.5)$$

Proof. The details below are similar to those in Koul (2002), for the proof of Theorem 2.2.1, which in turn have roots in Billingsley (1968). First, we check the following bound: for any $0 \leq s < u \leq 1$,

$$P\left(\sup_{s \leq t \leq u} |Y_n(t) - Y_n(s)| \geq \epsilon \right) \tag{4.4.6}$$

$$\leq K\epsilon^{-2\gamma} |F_n(u) - F_n(s)|^{2\alpha} + P\left(|Y_n(u) - Y_n(s)| \geq \epsilon/2\right).$$

Indeed, let $\delta := u - s$, and $m \geq 1$ be an integer. For $1 \leq j \leq m$, let

$$\xi_j := Y_n\left(\frac{j}{m}\delta + s\right) - Y_n\left(\frac{j-1}{m}\delta + s\right), \quad S_k := \sum_{j=1}^{k} \xi_j, \quad M_m := \max_{1 \leq k \leq m} |S_k|.$$

We will use the following from Billingsley (1968, Theorem 12.1, see also (12.5) and (12.10)). *Suppose there exist non-negative numbers $u_1, \cdots, u_m$, $\gamma > 0$, and $\alpha > 1/2$ such that for $0 \leq i \leq j \leq k \leq m$,*

$$E\left[|S_k - S_j|^{\gamma}|S_j - S_i|^{\gamma}\right] \leq \left(\sum_{r=i+1}^{k} u_r\right)^{2\alpha}. \tag{4.4.7}$$

Then $\forall \epsilon > 0$,

$$P(M_m \geq \epsilon) \leq K\epsilon^{-2\gamma}\left(\sum_{j=1}^{m} u_j\right)^{2\alpha} + P(|S_m| > \epsilon/2), \tag{4.4.8}$$

where $K \equiv K_{\gamma,\alpha}$ depends only on γ, α.

Note that (4.4.7) holds due to (4.4.3), with $u_j = F_n((j/m)\delta + s) - F_n((j-1)\delta/m + s) \geq 0$, satisfying

$$\sum_{j=1}^{m} u_j = \sum_{j=1}^{m} \left\{ F_n\left(\frac{j}{m}\delta + s\right) - F_n\left(\frac{j-1}{m}\delta + s\right) \right\} = F_n(u) - F_n(s).$$

Hence, (4.4.6) follows from (4.4.8) because $S_m = Y_n(u) - Y_n(s)$, and because, by the right continuity of Y_n for each n, $M_m \to \sup_{s \leq t \leq u} |Y_n(t) - Y_n(s)|$ a.s., as $m \to \infty$, for every $n \geq 1$.

Next, for a $\delta > 0$, let $r := [1/\delta] - 1$, $t_j := j\delta$, $j = 1, \cdots, r$, $t_0 := 0$, and $t_r := 1$. Then

$$P\left(\sup_{|s-t|<\delta} |Y_n(s) - Y_n(t)| \geq \epsilon \right) \leq \sum_{i=1}^{r} P\left(\sup_{t_{i-1} \leq s \leq t_i} |Y_n(s) - Y_n(t)| \geq \epsilon/3 \right)$$

$$\leq C\epsilon^{-2\gamma} \sum_{i=1}^{r} |F_n(t_i) - F_n(t_{i-1})|^{2\alpha} + \sum_{i=1}^{r} P\left(|Y_n(t_i) - Y_n(t_{i-1})| \geq \epsilon/6\right),$$

where the first inequality is from Billingsley (1968, (8.9)) and the second follows from (4.4.6). By monotonicity and the uniform boundedness of F_n,

$$\sum_{i=1}^{r} |F_n(t_i) - F_n(t_{i-1})|^{2\alpha} \le \omega_\delta^{2\alpha-1}(F_n) \sum_{i=1}^{r} |F_n(t_i) - F_n(t_{i-1})| \le C\omega_\delta^{2\alpha-1}(F_n).$$

Finally, by Chebyshev's inequality, (4.4.2), and (4.4.4),

$$\sum_{i=1}^{r} P\Big(|Y_n(t_i) - Y_n(t_{i-1})| \ge \epsilon/6\Big) \le C\epsilon^{-2\gamma} \sum_{i=1}^{r} E|Y_n(t_i) - Y_n(t_{i-1})|^{2\gamma}$$

$$\le C\epsilon^{-2\gamma} \sum_{i=1}^{r} \Big(|F_n(t_i) - F_n(t_{i-1})|^{2\alpha} + |F_n(t_i) - F_n(t_{i-1})|n^{-\beta}\Big)$$

$$\le C\epsilon^{-2\gamma}\big(\omega_\delta^{2\alpha-1}(F_n) + n^{-\beta}\big) \to 0,$$

by letting, first $n \to \infty$ and then $\delta \to 0$, as above. This completes the proof of Lemma 4.4.1.

If $F_n \to F$ weakly and F is continuous, then $\sup_{0 \le t \le 1} |F_n(t) - F(t)| \to 0$, and (4.4.2) is *a priori* satisfied.

Returning to the weak convergence of the partial-sum process $S_n(\tau)$, $0 \le \tau \le 1$, of the moving averages X_j as in (4.3.1), we first need to establish the finite-dimensional weak convergence. Assume that $\sigma_n^2 = \text{Var}(S_n) \to \infty$. By the Lamperti theorem, Theorem 3.4.1, if $\sigma_n^{-1} S_n(\tau)$ has a non-degenerate weak limit, then for some $H \in (0, 1]$ and some positive function $L \in SV$,

$$\sigma_n^2 = \text{Var}(S_n) = n^{2H} L(n). \tag{4.4.9}$$

A remarkable feature of $\{X_j\}$ being a linear process is that the Lamperti theorem can be "reversed" in the sense that (4.4.9) alone implies the weak convergence of finite-dimensional distributions of the normalized partial-sum process as in the following proposition.

Proposition 4.4.1. *Suppose a linear process $\{X_j\}$ of (4.3.1) satisfies the conditions of Theorem 4.3.1 and (4.4.9) holds with $0 < H < 1$. Then*

$$\{\sigma_n^{-1} S_n(\tau)\}_{\tau>0} \to_{fdd} \{B_H(\tau)\}_{\tau>0}, \tag{4.4.10}$$

where B_H is a fBm with parameter H.

Proof. Let $T_n(\tau) := \sigma_n^{-1} S_n(\tau)$. Assumption (4.4.9) and $L \in SV$ imply

$$\text{Var}(T_n(\tau)) = \frac{[n\tau]^{2H} L([n\tau])}{n^{2H} L(n)} \to \tau^{2H}.$$

Hence, with r_H as in (3.4.1), for all $0 < s < t$,

$$\text{Cov}(T_n(t), T_n(s)) \qquad\qquad (4.4.11)$$
$$= (1/2)\{\text{Var}(T_n(t)) + \text{Var}(T_n(s)) - \text{Var}(T_n(t) - T_n(s))\}$$
$$\to r_H(t, s).$$

To prove the weak convergence of finite-dimensional distributions of $\{T_n(\tau)\}$ to those of B_H, in view of the Cramér-Wold device, it suffices to show that for every $b_1, \cdots, b_k \in \mathbb{R}$, $0 \le t_1, \cdots, t_k \le 1$, and $k \ge 1$, $S_n := b_1 T_n(t_1) + \cdots + b_k T_n(t_k)$ satisfies

$$S_n \to_D S := b_1 B_H(t_1) + \cdots + b_k B_H(t_k).$$

The proof uses Corollary 4.3.1 as follows. By (4.4.11),

$$\sigma_n^{-2} \text{Var}(S_n) = \sum_{i,j=1}^{k} b_i b_j \text{Cov}(T_n(t_i), T_n(t_j)) \to \sum_{i,j=1}^{k} b_i b_j r_H(t_i, t_j) =: R > 0,$$

because r_H is positive definite. Similarly as in (4.3.4), rewrite $T_n(t) = \sum_{j=-\infty}^{n} c_{nj}(t)\zeta_j$, where $c_{nj}(t) := \sigma_n^{-1} \sum_{k=1}^{[nt]} a_{k-j}$, $j \in \mathbb{Z}$. Then

$$S_n = \sum_{j=-\infty}^{n} \{c_{nj}(t_1) + \cdots + c_{nj}(t_k)\}\zeta_j =: \sum_{j=-\infty}^{n} d_{nj}\zeta_j.$$

Note $\max_{j \le n} |d_{nj}| = o(\sigma_n)$ follows from (4.3.5). Therefore, by Corollary 4.3.1, $S_n \to \mathcal{N}(0, R)$, which completes the proof of Proposition 4.4.1.

Next, we need to show that the process $\sigma_n^{-1} S_n(\cdot)$ is tight in the uniform metric. The following proposition shows that long memory of the summands $\{X_j\}$ alone *a priori* guarantees this.

Proposition 4.4.2. *Let $\{X_j\}$ be a second-order stationary process satisfying (4.4.9) with $1/2 < H < 1$. Then the sequence of partial-sum processes $\sigma_n^{-1} S_n(\cdot)$ is tight with respect to the uniform norm. In addition, if $\sigma_n^{-1} S_n(u) \to_{fdd} S(u)$, where $S(u)$, $0 \le u \le 1$, is a stochastic process, then*

$$\sigma_n^{-1} S_n(\cdot) \Rightarrow S(\cdot), \quad \text{in } \mathcal{D}[0, 1] \text{ with the uniform metric.} \qquad (4.4.12)$$

Proof. To check tightness, we shall verify (4.4.5) for the process $T_n(t) := \sigma_n^{-1} S_n(t)$. Let $G_n(t) := [nt]/n$, $G(t) := t$, $0 \le t \le 1$. Observe that $\sup_t |G_n(t) - G(t)| \to 0$ and G is continuous on $[0, 1]$. By stationarity of increments,

$$E|T_n(t) - T_n(s)|^2 = E\Big(\sigma_n^{-1} \sum_{j=1}^{[nt]-[ns]} X_j\Big)^2 \qquad (4.4.13)$$

$$= \Big(\frac{[nt] - [ns]}{n}\Big)^{2H} \frac{L([nt] - [ns])}{L(n)} \le C\Big(\frac{[nt] - [ns]}{n}\Big)^{2H-\delta}$$

$$= C\,[G_n(t) - G_n(s)]^{2H-\delta},$$

for some $\delta > 0$ by property (2.3.6) of slowly varying functions. Choosing δ such that $\beta := 2H - \delta > 1$, this verifies (4.4.5) for the T_n process with $\gamma = 2$ and completes the proof.

The following proposition where $S_n \equiv S_n(1)$ provides a useful bound for moments of the sums of a linear process. In Proposition 4.4.4 these bounds are used to verify the weak convergence of the partial-sum process when $\{X_j\}$ is a linear process.

Proposition 4.4.3. *Let $\{X_j\}$ be a linear process (4.3.1) such that $E|\zeta_0|^p < \infty$, for some $p \ge 2$. Then*

$$E|S_n|^p \le c(ES_n^2)^{p/2}, \quad \forall n \ge 1, \qquad (4.4.14)$$

where $c > 0$ depends only on p and $E|\zeta_0|^p$.

Proof. Setting $b_{nj} = \sum_{k=1}^{n} a_{k-j} I(j \le k)$, write $S_n = \sum_{j=-\infty}^{n} b_{nj}\zeta_j$. Then $ES_n^2 = \sigma_\zeta^2 \sum_{j=-\infty}^{n} b_{nj}^2 < \infty$ and by inequality (2.5.4)

$$E|S_n|^p = E\Big| \sum_{j=-\infty}^{n} b_{nj}\zeta_j \Big|^p \le CE|\zeta_0|^p \Big(\sum_{j=-\infty}^{n} b_{nj}^2 \Big)^{p/2} = c(ES_n^2)^{p/2},$$

where $c = CE|\zeta_0|^p$ does not depend on n.

Proposition 4.4.4. *Assume that the linear process $\{X_j\}$ of (4.3.1) satisfies (4.4.9) with $0 < H < 1$. For $0 < H \le 1/2$, assume in addition that $E|\zeta_0|^p < \infty$, for some $p > 1/H$. Then*

$$\sigma_n^{-1} S_n(\cdot) \Rightarrow B_H(\cdot), \quad \text{in } \mathcal{D}[0,1] \text{ with the uniform metric}, \qquad (4.4.15)$$

where B_H is a fBm, see Definition 3.4.3.

Proof. Proposition 4.4.1 implies finite-dimensional convergence. For $H \le 1/2$, by (4.4.14) and (4.4.13),

$$E|T_n(t) - T_n(s)|^p \le C\big(E(T_n(t) - T_n(s))^2\big)^{p/2} \le C|G_n(t) - G_n(s)|^{(2H-\delta)p/2}.$$

This verifies (4.4.5) for the T_n process with $\gamma = p$ and $2\beta = (2H - \delta)p > 2$, where $pH > 1$ and $\delta > 0$ is chosen sufficiently small. For $H > 1/2$, (4.4.15) follows from Proposition 4.4.2. This completes the proof.

For ease of reference later we give the following corollary, which pertains to the partial-sum process of certain types of moving averages. It follows from Propositions 3.2.1(ii), 4.4.4, and from (3.3.5).

Corollary 4.4.1. *Suppose* $X_j = \sum_{k=0}^{\infty} a_k \zeta_{j-k}$, $j \in \mathbb{Z}$, *and* $\{\zeta_k\} \sim$ IID$(0, \sigma_\zeta^2)$. *If* $a_k \sim c_0 k^{-1+d}$, *as* $k \to \infty$, $0 < d < 1/2$, *then*
$$\sigma_n^2 \sim s_{\gamma,d}^2 n^{1+2d}, \quad n^{-1/2-d} S_n(\cdot) \Rightarrow s_{\gamma,d} B_{1/2+d}(\cdot), \qquad (4.4.16)$$
in $\mathcal{D}[0,1]$ *with the uniform metric, where* $s_{\gamma,d}^2 := \sigma_\zeta^2 c_0^2 B(d, 1-2d)/d(1+2d)$.

Weak convergence (4.4.16) also holds, if $a_k \sim c_0 k^{-1+d}$, $-1/2 < d < 0$, $\sum_{k=0}^{\infty} a_k = 0$, *and* $E|\zeta_0|^p < \infty$, *for some* $p > 1/(1/2 + d)$.

If $\sum_{k=0}^{\infty} |a_k| < \infty$, $\sum_{k=0}^{\infty} a_k > 0$, *and* $E|\zeta_0|^p < \infty$, *for some* $p > 2$, *then* (4.4.16) *is valid with* $d = 0$ *and* $s_{\gamma,d}^2 = \sigma_\zeta^2 \left(\sum_{k=0}^{\infty} a_k \right)^2$.

4.5   More results for weighted sums

In this section we extend the CLT of Proposition 4.3.2 to weighted sums of a non-linear stationary process $\{X_j\}$ that can be approximated by m-dependent processes and provide some upper bounds for sums and quadratic forms. These results in turn are used to derive CLTs for sums of products of linear processes and weighted sums of squared linear processes.

Weighted sums of m-dependent random variables. Let
$$b_n := \max_{1 \le j \le n} |z_{nj}|, \quad B_n^2 = \sum_{j=1}^{n} z_{nj}^2.$$
The conditions imposed on the weights will include the Lindeberg–Feller-type condition $b_n = o(B_n)$ as in (4.3.16) and (4.3.14). We start with the m-dependent process $\{X_j\}$.

Theorem 4.5.1. *Let* $\{X_j\}$ *be a stationary* m-*dependent process,* $1 \le m < \infty$, $EX_0^2 < \infty$, *and* $\sigma^2 := \sum_{k \in \mathbb{Z}} \text{Cov}(X_0, X_k) > 0$. *Assume that the* z_{nj}'s *satisfy (4.3.14) and (4.3.16). Then*
$$T_n := B_n^{-1} \sum_{j=1}^{n} z_{nj}(X_j - EX_j) \to_D \mathcal{N}(0, \sigma^2). \qquad (4.5.1)$$

Proof. Consider first the case when the X_j's have all finite moments. Then the m-dependent process $\{X_j\}$ has a continuous bounded spectral density,

$$f_X(u) = (2\pi)^{-1} \sum_{k \in \mathbb{Z}} e^{-iku} \gamma_X(k), \quad u \in \Pi, \qquad 2\pi f_X(0) = \sigma^2 > 0.$$

Thus, by (4.3.23), $\mathrm{Var}(T_n) \to \sigma^2$. Because of the cumulant criterion for a CLT, (4.2.5), to prove (4.5.1), it remains to show that

$$\mathrm{Cum}_k(T_n) \to 0, \qquad \forall\, k = 3, 4 \cdots. \tag{4.5.2}$$

By m-dependence of $\{X_j\}$, the properties (c1) to (c2) of joint cumulants, (4.2.10), and by (4.3.16), it follows that for all $k = 3, 4, \cdots$,

$$
\begin{aligned}
|\mathrm{Cum}_k(T_n)| &= B_n^{-k} \Big| \sum_{j_1, \cdots, j_k = 1}^{n} z_{nj_1} \cdots z_{nj_k} \mathrm{Cum}(X_{j_1}, \cdots, X_{j_k}) \Big| \\
&\leq k! b_n^{k-2} B_n^{-k} \sum_{1 \leq j_1 \leq \cdots \leq j_k \leq n} |z_{nj_1} z_{nj_2}| |\mathrm{Cum}(X_{j_1}, \cdots, X_{j_k})| \\
&\leq k! b_n^{k-2} B_n^{-k} \sum_{1 \leq j_1 \leq \cdots \leq j_k \leq n : j_k - j_1 \leq km} (z_{nj_1}^2 + z_{nj_2}^2) \\
&\qquad\qquad\qquad\qquad\qquad\qquad\qquad \times |\mathrm{Cum}(X_{j_1}, \cdots, X_{j_k})| \\
&\leq C(b_n/B_n)^{k-2} B_n^{-2} \max_{1 \leq j \leq n} E|X_j|^k \sum_{j=1}^{n} z_{nj}^2 \leq C(b_n/B_n)^{k-2} \to 0.
\end{aligned}
$$

Consider now the case when some higher moments of X_j's are not finite. Let $L > 0$. Write $X_j = X_{j,1} + X_{j,2}$, where

$$X_{j,1} := X_j I(|X_j| \leq L), \quad X_{j,2} := X_j I(|X_j| > L),$$

$$T_{n,i} := B_n^{-1} \sum_{j=1}^{n} z_{nj}(X_{j,i} - EX_{j,i}), \quad i = 1, 2.$$

Then $T_n = T_{n,1} + T_{n,2}$, and

$$\mathrm{Cov}(X_{k,1}, X_{0,1}) \to \mathrm{Cov}(X_k, X_0), \quad L \to \infty, \quad k = 1, 2, \cdots, m,$$

$$|\mathrm{Cov}(X_{k,2}, X_{0,2})| \leq \mathrm{Var}(X_{0,2}) \to 0, \quad L \to \infty.$$

Since $X_{k,1}, k \in \mathbb{Z}$, has all moments finite, from the above result it follows that for every $L < \infty$, $T_{n,1} \to_D \mathcal{N}(0, \sigma_{1L}^2)$, as $n \to \infty$, where

$$\sigma_{1L}^2 = \sum_{k \in \mathbb{Z}} \mathrm{Cov}(X_{k,1}, X_{0,1}) \to \sum_{k \in \mathbb{Z}} \mathrm{Cov}(X_k, X_0) = \sigma^2, \quad L \to \infty.$$

On the other hand, by (4.3.23),

$$\text{Var}(T_{n,2}) \leq \sum_{|k| \leq m} \text{Cov}(X_{k,2}, X_{0,2}) \leq (2m+1)\text{Var}(X_{0,2}) \to 0,$$

as $L \to \infty$. Thus, (4.5.1) follows from Lemma 4.2.1. This completes the proof of the theorem.

Theorem 4.5.1 allows us to obtain the following general result for the weighted sums of a short memory process $\{X_j\}$ that is not necessarily a moving average of i.i.d. innovations.

Proposition 4.5.1. *Let $\{X_j\}$ be a covariance stationary process satisfying*

$$\sum_{k \in \mathbb{Z}} |\gamma_X(k)| < \infty, \quad \sigma^2 := \sum_{k \in \mathbb{Z}} \gamma_X(k) > 0.$$

In addition, suppose $X_j = Y_{\epsilon,j} + Z_{\epsilon,j}$, where for each $\epsilon > 0$, $\{Y_{\epsilon,j}\}$ is a stationary m-dependent process, and $\{Z_{\epsilon,j}\}$ is a covariance stationary process such that

$$\sum_{k \in \mathbb{Z}} |\gamma_{Z_\epsilon}(k)| \to 0, \quad \epsilon \to 0. \tag{4.5.3}$$

Assume that the z_{nj}'s satisfy (4.3.14) and (4.3.16). Then (4.5.1) still holds.

Proof. Write

$$T_{n,1} := B_n^{-1} \sum_{j=1}^{n} z_{nj}(Y_{\epsilon,j} - EY_{\epsilon,j}), \quad T_{n,2} := B_n^{-1} \sum_{j=1}^{n} z_{nj}(Z_{\epsilon,j} - EZ_{\epsilon,j}).$$

By Theorem 4.5.1 and (4.3.23), for any $\epsilon > 0$,

$$T_{n,1} \to_D \mathcal{N}(0, \sigma_{Y_\epsilon}^2), \quad \sigma_{Y_\epsilon}^2 = \sum_{k \in \mathbb{Z}} \gamma_{Y_\epsilon}(k),$$

$$\text{Var}(T_{n,2}) \leq \sum_{k \in \mathbb{Z}} |\gamma_{Z_\epsilon}(k)|.$$

These facts together with (4.5.3) and Lemma 4.2.1 establish (4.5.1), thereby completing the proof of the proposition.

Upper bounds for variances of sums and quadratic forms. Before proceeding further, we state and prove the following two useful lemmas.

Lemma 4.5.1. *Let $\{\zeta_j\}$ be i.i.d. with $E\zeta_0 = 0$ and $E\zeta_0^4 < \infty$. Let $b(t,s), s, t \in \mathbb{Z}$, be a set of possibly complex numbers with $\sum_{t,s \in \mathbb{Z}} |b(t,s)|^2 < \infty$, and let*

$$Q := \sum_{t,s \in \mathbb{Z}} b(t,s)\zeta_t \zeta_s.$$

Then

$$E\big|Q - EQ\big|^2 \le 4E(\zeta_0^4) \sum_{t,s\in\mathbb{Z}} |b(t,s)|^2. \tag{4.5.4}$$

Proof. Indeed,

$$E\big|Q - EQ\big|^2$$

$$\le 4\Big(E\Big|\sum_{t<s} b(t,s)\zeta_t\zeta_s\Big|^2 + E\Big|\sum_{t>s} b(t,s)\zeta_t\zeta_s\Big|^2$$

$$+ E\Big|\sum_{t=s} b(t,t)(\zeta_t^2 - E\zeta_t^2)\Big|^2\Big)$$

$$= 4(E\zeta_0^2)^2\Big(\sum_{t<s} |b(t,s)|^2 + \sum_{t>s} |b(t,s)|^2\Big) + 4\mathrm{Var}(\zeta_0^2)\sum_{t=s} |b(t,t)|^2$$

$$\le 4E\zeta_0^4 \sum_{t,s\in\mathbb{Z}} |b(t,s)|^2.$$

Lemma 4.5.2. *Let $\{\zeta_j\}$ be i.i.d. r.v.'s with $E\zeta_0 = 0$, $E\zeta_0^4 < \infty$, $\{d_{k,s}^{(i)}\}$, $k \in \mathbb{Z}$, $i = 1,2$, be real numbers and square summable in $s \in \mathbb{Z}$, and set*

$$X_k^{(i)} := \sum_{s\in\mathbb{Z}} d_{k,s}^{(i)}\zeta_{k-s}, \quad k \in \mathbb{Z}, \quad i = 1,2. \tag{4.5.5}$$

Furthermore, let $c_{n,jk}$, $c_{n,j}$, $j,k = 1,\cdots,n$, be possibly complex numbers, and define

$$Q_n := \sum_{j,k=1}^n c_{n,jk} X_j^{(1)} X_k^{(2)}, \qquad S_n := \sum_{j=1}^n c_{n,j} X_j^{(1)} X_j^{(2)}.$$

Then

$$E\big|Q_n - EQ_n\big|^2 \le C \sum_{j,k,l,m=1}^n |c_{n,jk} c_{n,lm}| \, |r_{jl}^{(1)} r_{km}^{(2)}|, \tag{4.5.6}$$

$$E\big|S_n - ES_n\big|^2 \le C \sum_{j,k=1}^n |c_{n,j} c_{n,k}| \, |r_{jk}^{(1)} r_{jk}^{(2)}|,$$

where $r_{jk}^{(i)} = \mathrm{Cov}(X_j^{(i)}, X_k^{(i)})$, $j,k \in \mathbb{Z}$, $i = 1,2$, and $0 < C < \infty$ does not depend on n and the weights $c_{n,jk}$ and $d_{j,k}^{(i)}$.

Proof. To prove (4.5.6), write

$$Q_n = \sum_{t,s\in\mathbb{Z}} b_n(t,s)\zeta_t\zeta_s, \quad b_n(t,s) = \sum_{j,k=1}^n c_{n,jk}\, d_{j,t}^{(1)} d_{k,s}^{(2)}.$$

$$_2|^2 \le C \sum_{t,s\in\mathbb{Z}} |b_n(t,s)|^2 = C \sum_{t,s\in\mathbb{Z}} \left| \sum_{j,k=1}^{n} c_{n,jk} \, d_{j,t}^{(1)} d_{k,s}^{(2)} \right|^2,$$

which implies (4.5.6) and completes the proof of lemma.

Sums of products of linear processes. The next lemma analyzes the properties of the autocovariance function and asymptotic distribution of the sum of a product of two stationary linear processes. This result is useful later in the text.

Let $\{\zeta_j\}$ be i.i.d. with $E\zeta_0 = 0$, $E\zeta_0^4 < \infty$,

$$X_k := Y_{1,k} Y_{2,k}, \quad Y_{i,k} := \sum_{s\in\mathbb{Z}} a_{i,s} \zeta_{k-s}, \quad k \in \mathbb{Z}, \ i = 1, 2, \tag{4.5.7}$$

where $\{a_{i,s}\}$, $i = 1, 2$, $s \in \mathbb{Z}$, are two real sequences of square summable weights. Let $S_n := \sum_{k=1}^{n} X_k$, $\gamma_{i,k} := \mathrm{Cov}(Y_{i,0}, Y_{i,k})$, $i = 1, 2$, $\gamma_{12;k} := E(Y_{1,k} Y_{2,0})$, and $\chi_4 := \mathrm{Cum}_4(\zeta_0)$.

Theorem 4.5.2. *Let X_k, $Y_{1,k}$, $Y_{2,k}$, $k \in \mathbb{Z}$ be as in (4.5.7). With the notation as above, assume*

$$\sum_{k\in\mathbb{Z}} \gamma_{i,k}^2 < \infty, \quad i = 1, 2. \tag{4.5.8}$$

Then

$$\sum_{k\in\mathbb{Z}} |\gamma_X(k)| \le C \prod_{i=1}^{2} \left(\sum_{k\in\mathbb{Z}} \gamma_{i,k}^2 \right)^{1/2}. \tag{4.5.9}$$

Moreover, if $\sigma_\zeta^2 = 1$, then with $\sigma_S^2 := 2\sum_{k\in\mathbb{Z}} \gamma_{1,k} \gamma_{2,k} + \chi_4 \gamma_{12,0}^2$,

$$n^{-1/2}(S_n - ES_n) \to_D \mathcal{N}(0, \sigma_S^2). \tag{4.5.10}$$

If $\sigma_\zeta^2 \neq 1$, then (4.5.10) holds with χ_4 replaced by χ_4/σ_ζ^4.

Proof. Without loss of generality, assume $\sigma_\zeta^2 = 1$. By (4.2.7),

$$\mathrm{Cov}(Y_{1,k} Y_{2,k}, Y_{1,j} Y_{2,j}) = \gamma_{1,k-j} \gamma_{2,k-j} + \gamma_{12,k-j}^2$$
$$+ \mathrm{Cum}(Y_{1,k}, Y_{2,k}, Y_{1,j}, Y_{2,j}).$$

Let $f_i = (\sigma_\zeta^2/2\pi)|A_i|^2$, $u \in \Pi$, be the spectral density of $\{Y_{i,j}\}$, where $A_i(u) := \sum_{s\in\mathbb{Z}} e^{-isu} a_{i,s}$, $i = 1, 2$. Let $f_{12} = (\sigma_\zeta^2/2\pi) A_1 \overline{A_2}$ be the cross-spectral density.

By Parseval's identity (2.1.4),

$$\sum_{k\in\mathbb{Z}} r_{12,k}^2 = 2\pi \int_\Pi |f_{12}(u)|^2 du = 2\pi \int_\Pi f_1(u)f_2(u)du = \sum_{k\in\mathbb{Z}} r_{1,k}r_{2,k}.$$

By the multilinearity property (c1) of the cumulant of Subsection 4.2,

$$\mathrm{Cum}(Y_{1,k}, Y_{2,k}, Y_{1,j}, Y_{2,j}) = \chi_4 \sum_{s\in\mathbb{Z}} a_{1,k-s}\, a_{2,k-s}\, a_{1,j-s}\, a_{2,j-s}.$$

Thus

$$\sum_{k\in\mathbb{Z}} |\mathrm{Cum}(Y_{1,k}, Y_{2,k}, Y_{1,0}, Y_{2,0})| \le |\chi_4| \sum_{k\in\mathbb{Z}} a_{1,k}^2 \sum_{s\in\mathbb{Z}} a_{2,s}^2$$

$$= |\chi_4|\gamma_{1,0}\gamma_{2,0},$$

$$\sum_{k\in\mathbb{Z}} \mathrm{Cum}(Y_{1,k}, Y_{2,k}, Y_{1,0}, Y_{2,0}) = \chi_4 \gamma_{12,0}^2.$$

Hence, by the C–S inequality,

$$\sum_{k\in\mathbb{Z}} |\gamma_X(k)| \le 2 \sum_{k\in\mathbb{Z}} |\gamma_{1,k}\gamma_{2,k}| + |\chi_4|\gamma_{12,0}^2$$

$$\le 2\Big(\sum_{k\in\mathbb{Z}} \gamma_{1,k}^2\Big)^{1/2}\Big(\sum_{k\in\mathbb{Z}} \gamma_{2,k}^2\Big)^{1/2} + |\chi_4|\gamma_{1,0}\gamma_{2,0},$$

which proves (4.5.9) and establishes (4.5.10) for σ_S^2.

Next, we prove the claim of asymptotic normality in (4.5.10). The proof uses a truncation idea. By (3.3.1), (4.5.9), and Parseval's identity (2.1.4),

$$\mathrm{Var}(n^{-1/2}S_n) \le \sum_{k\in\mathbb{Z}} |\mathrm{Cov}(Y_{1,k}Y_{2,k}, Y_{1,0}Y_{2,0})| \tag{4.5.11}$$

$$\le C \prod_{i=1}^2 \Big(\sum_{k\in\mathbb{Z}} \gamma_{i,k}^2\Big)^{1/2} \le C \prod_{i=1}^2 \|f_i\|_{L_2(\Pi)} \le C \prod_{i=1}^2 \|A_i\|_{L_4(\Pi)}^2.$$

Because trigonometric polynomials are dense in $L_4(\Pi)$ then $\forall\, \epsilon > 0$, there exists an integer $m < \infty$ and real sequences $\{q_{i,s},\ s\in\mathbb{Z}\}$, $i = 1,2$, such that $q_{i,s} = 0$, for $|s| > m$, and such that $A_{\epsilon,i}(u) := \sum_{s\in\mathbb{Z}} e^{-isu} q_{i,s}$, $i = 1,2$, and which satisfy

$$\|A_i - A_{\epsilon,i}\|_{L_4(\Pi)} < \epsilon, \quad i = 1,2. \tag{4.5.12}$$

Let

$$Y'_{i,k} := \sum_{s\in\mathbb{Z}} q_{i,k-s}\zeta_s,\ \ k\in\mathbb{Z},\ \ i=1,2;\quad S'_n := \sum_{k=1}^n Y'_{1,k}\, Y'_{2,k}.$$

Note that the processes $\{Y'_{i,k}\}$, $i = 1, 2$, and their product $Y'_{1,k}Y'_{2,k}$, $k \in \mathbb{Z}$, are $2m$-dependent. Therefore, by Proposition 4.2.3,

$$n^{-1/2}(S'_n - E[S'_n]) \to_D \mathcal{N}(0, \sigma^2_{S'}). \tag{4.5.13}$$

Moreover,

$$X_j = Y'_{1,j}Y'_{2,j} + \{(Y_{1,j} - Y'_{1,j})Y_{2,j}\} + \{Y'_{1,t}(Y_{2,t} - Y'_{2,t})\}. \tag{4.5.14}$$

Thus, by (4.5.11) and (4.5.12),

$$n^{-1}\mathrm{Var}(S_n - S'_n) \le 2n^{-1}\mathrm{Var}\Big(\sum_{t=1}^{n}(Y_{1,t} - Y'_{1,t})Y_{2,t}\Big) \tag{4.5.15}$$

$$+ 2n^{-1}\mathrm{Var}\Big(\sum_{t=1}^{n}Y'_{1,t}(Y_{2,t} - Y'_{2,t})\Big)$$

$$\le C \max_{i=1,2}\|A_i - A_{\epsilon,i}\|^2_{L^4(\Pi)} \le C\epsilon^2.$$

This bound together with (4.5.9), (4.5.13), and Lemma 4.2.1 imply (4.5.10), thereby completing the proof of the theorem.

The next lemma gives an approximation to the autocovariance function of a squared stationary linear process along with a rate.

Lemma 4.5.3. *Suppose that $X_k := Y_k^2$, $k \in \mathbb{Z}$, where $Y_k := \sum_{s=0}^{\infty} a_s \zeta_{k-s}$, with $\{\zeta_j\} \sim \mathrm{IID}(0, \sigma_\zeta^2)$. If $a_s \sim c_a s^{-1+d}$, $s \to \infty$, $0 < d < 1/2$, and $c_a > 0$, then as $k \to \infty$,*

$$\gamma_X(k) = 2\gamma_Y^2(k)(1 + o(1)) \tag{4.5.16}$$

$$= 2c_\gamma^2 k^{-2+4d} + o(k^{-2+4d}), \quad c_\gamma = \sigma_\zeta^2 c_a^2 B(d, 1 - 2d).$$

Proof. As in the previous proof,

$$\mathrm{Cov}(Y_k^2, Y_0^2) = 2\gamma_Y^2(k) + \mathrm{Cum}(Y_k, Y_k, Y_0, Y_0).$$

By Proposition 3.2.1(ii) $\gamma_Y(k) \sim c_\gamma|k|^{-1+2d}$ as $k \to \infty$. By assumption, for $s \ge 0$, as $k \to \infty$, $|a_{k+s}| \le C|k+s|^{-1+d} \le Ck^{-1+d}$. Therefore,

$$|\mathrm{Cum}(Y_k, Y_k, Y_0, Y_0)| = |\chi_4|\sum_{s=0}^{\infty}|a_{k+s}a_{k+s}a_s a_s|$$

$$\le Ck^{-2+2d}\sum_{s=0}^{\infty}a_s^2 \le Ck^{-2+2d} = o(k^{-2+4d}),$$

which completes the proof of (4.5.16) and of the lemma.

Remark 4.5.1. If a linear process $\{X_j\}$ with i.i.d. innovations has a square summable autocovariance function, then by Theorem 4.5.2, $\{X_j^2\}$ has short memory, whereas if $\sum_{k\in\mathbb{Z}}\gamma_X^2(k) = \infty$ and the a_j's decay regularly as in Lemma 4.5.3, then $\{X_j^2\}$ has long memory.

Weighted sums of $\{\varepsilon_j^2\}$. We now discuss the CLT for weighted sums of ε_j^2, $j \in \mathbb{Z}$, where $\{\varepsilon_j\}$ is a linear process. This result is useful in deriving the asymptotic distribution of a kernel-type estimator of non-parametric conditional variance function in a heteroscedastic regression model.

Proposition 4.5.2. *Let $\{\zeta_j\}$ be i.i.d. r.v.'s with $E\zeta_0 = 0$, $E\zeta_0^2 = \sigma_\zeta^2$, $E\zeta_0^4 < \infty$, and $\varepsilon_j = \sum_{s=0}^{\infty} a_s\zeta_{j-s}$, $j \in \mathbb{Z}$, be a linear process with auto-covariance γ_ε satisfying*

$$\sum_{k\in\mathbb{Z}}\gamma_\varepsilon^2(k) < \infty. \tag{4.5.17}$$

Assume that the z_{nj}'s satisfy (4.3.14) and (4.3.16). Then

$$\left(\sum_{j=1}^{n} z_{nj}^2\right)^{-1/2} \sum_{j=1}^{n} z_{nj}(\varepsilon_j^2 - E\varepsilon_j^2) \to_D \mathcal{N}(0, \sigma_S^2),$$

$$\sigma_S^2 = \sum_{k\in\mathbb{Z}}\gamma_{\varepsilon^2}(k) = 2\sum_{k\in\mathbb{Z}}\gamma_\varepsilon^2(k) + \chi_4\gamma_\varepsilon^2(0).$$

Proof. The proof uses Proposition 4.5.1 applied to $X_j = \varepsilon_j^2$. We decompose $X_j = \xi_{\epsilon,j} + Z_{\epsilon,j}$ into a sum of an m-dependent process $\{\xi_{\epsilon,j}\}$ and a covariance stationary process $\{Z_{\epsilon,j}\}$ that satisfies (4.5.3). Apply (4.5.14) with $Y_{1,j} = Y_{2,j} = \varepsilon_j$, to obtain

$$X_j = Y_{1,j}'^{\,2} + (Y_{1,j}^2 - Y_{1,j}'^{\,2}) =: \xi_{\epsilon,j} + Z_{\epsilon,j}.$$

By definition the $Y_{1,j}'$'s are m-dependent variables. Therefore, $\{\xi_{\epsilon,j}\}$ is an m-dependent process. Next,

$$Z_{\epsilon,j} = (Y_{1,j} - Y_{1,j}')(Y_{1,j} + Y_{1,j}') =: U_j V_j$$

is a product of two linear processes $\{U_j\}$ and $\{V_j\}$ with the same i.i.d. innovations $\{\zeta_j\}$. By (4.5.12) and (4.5.17), the spectral densities of these processes satisfy

$$\|f_U\| \le C\|A_1 - A_{\epsilon,1}\|_{L_4(\Pi)}^2 \le C\epsilon,$$

$$\|f_V\| \le C\|A_1 + A_{\epsilon,1}\|_{L_4(\Pi)}^2 \le C(\|A_1 - A_{\epsilon,1}\|_{L_4(\Pi)}^2 + \|A_1\|_{L_4(\Pi)}^2)$$

$$\le C(\epsilon + \|f_\varepsilon\|) < \infty.$$

Thus, by (4.5.9),

$$\sum_{k\in\mathbb{Z}}|\gamma_{Z_\varepsilon}(k)| \le C\Big(\sum_{k\in\mathbb{Z}}\gamma_U^2(k)\Big)^{1/2}\Big(\sum_{k\in\mathbb{Z}}\gamma_V^2(k)\Big)^{1/2} \le C\|f_U\|\|f_V\| \le C\epsilon,$$

where C does not depend on ϵ and n, which implies (4.5.3) and completes the proof of the proposition.

Note: The book by Ibragimov and Linnik (1971) contains a number of additional useful results on the classical asymptotic theory of weakly dependent r.v.'s. Proposition 4.4.2 was proved in Taqqu (1975). Corollary 4.4.1 is a special case of the weak convergence result of Davydov (1970) for the partial-sum process of moving-average r.v.'s. Phillips and Solo (1992) developed the CLT and invariance principles for sums of linear processes based on the Beveridge–Nelson decomposition. Peligrad and Utev (2006) extend Ibragimov–Linnik's Theorem 18.6.5 for linear processes with dependent innovations. Wu and Woodroofe (2004) obtain a CLT for the sums of stationary and ergodic sequences using the martingale approximation method. Merlevède, Peligrad and Utev (2006) provide a survey of some recent results on the CLT and its weak invariance principle for stationary processes.

4.6   Sums of a transformed Gaussian process

In this section we shall examine the limit distributions of the sums of non-linear functions of a stationary Gaussian process. Accordingly, suppose $\{X_j\}$ is a stationary Gaussian process and let $h : \mathbb{R} \to \mathbb{R}$ be a non-linear function with $Eh^2(X_0) < \infty$. Then

$$\{Y_j := h(X_j),\, j \in \mathbb{Z}\}$$

is also a stationary process, with covariance $\gamma_Y(j) = \mathrm{Cov}(Y_0, Y_j)$. The process $\{h(X_j)\}$ is sometimes referred to as *subordinated* to $\{X_j\}$, and the latter process is called the *underlying process*. If we additionally know that the Gaussian process $\{X_j\}$ has long memory, with the memory parameter $d_X \in (0, 1/2)$, then some natural questions to ask are whether the process $\{Y_j\}$ also has long memory, what is its memory parameter d_Y, and what is the limit distribution of the sums $\sum_{j=1}^n (Y_j - EY_j)$?

Answering these questions is important for the statistical inference of long memory processes since the Gaussian model is often used as a first approximation to real data, while most statistics are expressed as a sum $\sum_{j=1}^{n} h(X_j)$ of some non-linear function h of observations. Examples of such statistics discussed in this book include the empirical distribution function and a large class of M estimators of the unknown mean EX_0.

Because of the non-linearity of the subordinated process, studying these questions is a non-trivial task. Gaussian processes are probably unique among classes of stationary processes (with linear processes as a possible exception) in the sense that the above questions admit a complete solution. The method which allows us to achieve these goals is the Hermite expansion discussed in Section 2.4. Assume that

$$EX_0 = 0, \quad EX_0^2 = 1; \quad \gamma_X(j) = E(X_0 X_j) \to 0, \quad j \to \infty, \quad (4.6.1)$$
$$Eh(X_0) = 0, \quad Eh^2(X_0) < \infty.$$

Then

$$|\gamma_X(j)| < \gamma_X(0) = 1, \qquad \forall j \geq 1. \quad (4.6.2)$$

Let k_0 be the Hermite rank of h; see (2.4.14). By (2.4.12), the process $Y_j = h(X_j)$ can be written as

$$Y_j = h(X_j) = \sum_{k=k_0}^{\infty} \frac{J_k}{k!} H_k(X_j), \quad (4.6.3)$$

which is a sum of stationary processes $\{H_k(X_j), j \in \mathbb{Z}\}$, with weights $J_k/k!$, $J_k := Eh(X_0)H_k(X_0)$. By Proposition 2.4.1, the *elementary processes* $\{H_k(X_j), j \in \mathbb{Z}\}$ satisfy two important properties, namely,

$$EH_k(X_i)H_m(X_j) = 0, \qquad k \neq m, \quad i, j \in \mathbb{Z}, \quad (4.6.4)$$
$$EH_k(X_0)H_k(X_j) = k! \gamma_X^k(j), \qquad k \geq 1, \quad j \in \mathbb{Z}. \quad (4.6.5)$$

Property (4.6.4) means that $\{H_k(X_j)\}$ and $\{H_m(X_j)\}$ are completely uncorrelated (orthogonal) for $k \neq m$, and therefore the covariance of $\{Y_j\}$ in (4.6.3) can be expressed as a sum of the autocovariances of $\{H_k(X_j)\}$, $k \geq k_0$, as given in (4.6.5). Property (4.6.5) together with (4.6.2) implies that the *dependence in* $\{H_k(X_j)\}$ *decreases with increasing* k. In other words, the first expansion term, $\{H_{k_0}(X_j)\}$ has the largest covariances or the strongest memory among all the other terms in (4.6.3), and therefore the long memory parameter d_Y of $\{Y_j\}$ is the same as the memory parameter d_{k_0} of the lowest expansion term, $\{H_{k_0}(X_j)\}$. Moreover, if $d_{k_0} \in (0, 1/2)$,

$\mathrm{Var}(\sum_{j=1}^{n} H_{k_0}(X_j))$ grows much faster than the variance of sums of the remaining terms in (4.6.3), and therefore $(J_{k_0}/k_0!) \sum_{j=1}^{n} H_{k_0}(X_j)$ completely dominates the limit distribution of $\sum_{j=1}^{n} Y_j$.

We shall now summarize these observations mathematically. From equations (4.6.3) to (4.6.5) we obtain

$$\gamma_Y(j) = \sum_{k,m=k_0}^{\infty} \frac{J_k J_m}{k! m!} E H_k(X_0) H_m(X_j) = \sum_{k=k_0}^{\infty} \frac{J_k^2}{k!} \gamma_X^k(j). \quad (4.6.6)$$

Rewrite (4.6.3) as

$$Y_j = \frac{J_{k_0}}{k_0!} H_{k_0}(X_j) + \mathcal{R}_j, \quad \mathcal{R}_j := \sum_{k=k_0+1}^{\infty} \frac{J_k}{k!} H_k(X_j). \quad (4.6.7)$$

Proposition 4.6.1. *Assume condition (4.6.1) holds. Then*

$$|\gamma_{\mathcal{R}}(j)| = |\mathrm{Cov}(\mathcal{R}_0, \mathcal{R}_j)| \leq \delta_{k_0} |\gamma_X(j)|^{k_0+1}, \quad (4.6.8)$$

$$\gamma_Y(j) = \gamma_X^{k_0}(j) \frac{J_{k_0}^2}{k_0!} + \gamma_{\mathcal{R}}(j) = \gamma_X^{k_0}(j)\Big(\frac{J_{k_0}^2}{k_0!} + \rho_j\Big), \quad (4.6.9)$$

where $\delta_{k_0} := \sum_{k=k_0+1}^{\infty} J_k^2/k!$ *and* $\rho_j \to 0$ *as* $j \to \infty$.

Proof. Claim (4.6.8) follows from (2.4.18), or directly from equations (4.6.3) to (4.6.5):

$$|\gamma_{\mathcal{R}}(j)| = \Big| \sum_{k=k_0+1}^{\infty} \frac{J_k^2}{k!} \gamma_X^k(j) \Big| \leq |\gamma_X(j)|^{k_0+1} \delta_{k_0},$$

since, according to (4.6.6),

$$E h^2(X_0) = \gamma_Y(0) = \sum_{k=k_0}^{\infty} J_k^2/k! < \infty.$$

Claim (4.6.9) follows from (4.6.6) and (4.6.8), with $|\rho_j| = |\gamma_{\mathcal{R}}(j)|/|\gamma_X(j)|^{k_0} \leq |\gamma_X(j)|\delta_{k_0} \to 0$, $j \to \infty$, thereby completing the proof.

From the above proposition it immediately follows that if $\{X_j\}$ has short memory then so does $\{Y_j\}$:

$$\sum_{j\in\mathbb{Z}} |\gamma_Y(j)| \leq C \sum_{j\in\mathbb{Z}} |\gamma_X(j)|^{k_0} \leq C \sum_{j\in\mathbb{Z}} |\gamma_X(j)| < \infty,$$

provided $\sum_{j\in\mathbb{Z}} \gamma_Y(j) \neq 0$. Hence, taking a non-linear function of a short memory Gaussian process cannot result in a long memory process. However, if $\{X_j\}$ has long memory, the subordinated process $\{Y_j\}$ can have either long memory or short memory, depending on the values of d_X and k_0.

Assume that

$$\gamma_X(j) = j^{-(1-2d_X)}L_X(j), \quad j \geq 1, \quad 0 < d_X < 1/2, \quad (4.6.10)$$

for some $L_X \in SV$. Then by (4.6.9), as $j \to \infty$,

$$\gamma_Y(j) = j^{-(1-2d_Y)}L_Y(j) = j^{-(1-2d_X)k_0}L_X^{k_0}(j)(1 + o(1)),$$

for some $L_Y \in SV$, with $L_Y(j)/L_X^{k_0}(j) \to 1$ as $j \to \infty$. Moreover,

$$d_Y := \frac{1}{2} - k_0\left(\frac{1}{2} - d_X\right) \quad (4.6.11)$$
$$= d_X, \quad \text{if } k_0 = 1,$$
$$< d_X, \quad \text{if } k_0 \geq 2.$$

In other words, the transformation $\{h(X_j)\}$ preserves the memory intensity of $\{X_j\}$ if the Hermite rank $k_0 = 1$, and reduces the memory intensity of $\{X_j\}$ when $k_0 \geq 2$. If $(1 - 2d_X)k_0 < 1$, then the process $\{Y_j\}$ has long memory with memory parameter $d_Y \in (0, 1/2)$. The opposite inequality $(1 - 2d_X)k_0 > 1$ implies $\sum_{j \in \mathbb{Z}} |\gamma_Y(j)| < \infty$ and hence short memory of $\{Y_j\}$ provided $\sum_{j \in \mathbb{Z}} \gamma_Y(j) \neq 0$. The boundary case $(1 - 2d_X)k_0 = 1$ is more delicate and will be excluded from the subsequent discussion.

In the long memory case $0 < d_Y < 1/2$, the representation (4.6.7), together with Proposition 4.6.1, reduces the study of the long memory properties of $\{Y_j\}$ to that of $\{H_{k_0}(X_j)\}$. This fact is known as the *reduction principle*. The limit distribution of sums of Hermite polynomials in this case (the non-central limit theorem) is discussed in Section 4.7 below. In the case when the Hermite rank $k_0 = 1$ and $0 < d_X < 1/2$, the large sample properties of $\{Y_j\}$ are similar to those of the underlying Gaussian process $\{X_j\}$.

Next, we turn our attention to the case when the subordinated process $\{Y_j\}$ has short memory. In this case, the decomposition (4.6.7) is not useful since both $\{H_{k_0}(X_j)\}$ and $\{\mathcal{R}_j\}$ have short memory and contribute to the limit distribution of the sums of $\{Y_j\}$. However, the Hermite expansion (4.6.3) is also helpful in the short memory case since it allows us to reduce the proof of asymptotic normality of $\sum_{j=1}^{n} Y_j$ to that of sums of a finite number of Hermite polynomials. Let

$$S_n := \sum_{j=1}^{n} Y_j, \quad Y_j = h(X_j).$$

Theorem 4.6.1. *In addition to (4.6.1), assume*

$$\sum_{j \in \mathbb{Z}} |\gamma_Y(j)| < \infty. \tag{4.6.12}$$

Then

$$n^{-1/2} S_n \to_D \mathcal{N}(0, \sigma^2), \quad \sigma^2 := \sum_{j \in \mathbb{Z}} \gamma_Y(j). \tag{4.6.13}$$

Condition (4.6.12) is equivalent to

$$\sum_{j \in \mathbb{Z}} |\gamma_X(j)|^{k_0} < \infty, \tag{4.6.14}$$

where $k_0 \geq 1$ is the Hermite rank of h.

Proof. The equivalence of conditions (4.6.12) and (4.6.14) follows from Proposition 4.6.1, (4.6.9). By Proposition 3.3.1, (4.6.12) implies $n^{-1} E S_n^2 \to \sigma^2$. In particular, if $\sigma^2 = 0$ then $n^{-1/2} S_n = o_p(1)$ and (4.6.13) trivially holds. Hence, we shall assume $\sigma^2 > 0$ in the following.

To prove (4.6.13), let K be a large positive integer and

$$Y_{j,K} := \sum_{k=k_0}^{K} \frac{J_k}{k!} H_k(X_j), \quad \mathcal{R}_{j,K} := \sum_{k=K+1}^{\infty} \frac{J_k}{k!} H_k(X_j).$$

Then we can rewrite S_n as

$$S_n = \sum_{j=1}^{n} Y_{j,K} + \sum_{j=1}^{n} \mathcal{R}_{j,K} =: S_{n,K} + R_{n,K}.$$

We shall first show that $R_{n,K}$ is negligible if K is chosen large enough,

$$\limsup_{n \to \infty} \mathrm{Var}(n^{-1/2} R_{n,K}) \to 0, \quad K \to \infty.$$

Indeed, by the definitions of $\mathcal{R}_{j,K}$ and $\mathcal{R}_j$ in (4.6.7), these quantities coincide for $k_0 = K$. Therefore, Proposition 4.6.1, (4.6.8), applies, yielding

$$|\gamma_{\mathcal{R},K}(j)| = |\mathrm{Cov}(\mathcal{R}_{0,K}, \mathcal{R}_{j,K})| \leq \delta_K |\gamma_X(j)|^{K+1},$$

where $\delta_K = \sum_{k=K+1}^{\infty} J_k^2/k! \to 0$, $K \to \infty$. Hence, by Proposition 3.3.1,

$$\mathrm{Var}(n^{-1/2} R_{n,K}) \leq \sum_{j \in \mathbb{Z}} |\gamma_{\mathcal{R},K}(j)| \leq \delta_K \sum_{j \in \mathbb{Z}} |\gamma_X(j)|^{k_0} \to 0, \quad (4.6.15)$$

as $K \to \infty$. Thus, by Lemma 4.2.1, to prove (4.6.13), it suffices to prove

$$n^{-1/2} S_{n,K} \to_D \mathcal{N}(0, \sigma_K^2), \quad \forall K > k_0, \tag{4.6.16}$$

$$\sigma_K^2 := \sum_{j \in \mathbb{Z}} \gamma_{Y_K}(j) \to \sigma^2, \quad K \to \infty. \tag{4.6.17}$$

But, (4.6.17) follows from (4.6.12) and (4.6.15), since $\gamma_{Y_K}(j) = \gamma_Y(j) - \gamma_{R,K}(j)$, see (4.6.6).

It remains to prove (4.6.16), which is a difficult part of the proof of this theorem. We use the method of cumulants (see Section 4.2). Let $K \geq k_0$. Because of Corollary 4.2.1, it suffices to show

$$\mathrm{Cum}_p(S_{n,K}) = o(n^{p/2}), \qquad \forall\, p = 3, 4, \cdots. \tag{4.6.18}$$

Let $S_n^{(k)} := \sum_{j=1}^n H_k(X_j)$. By linearity of cumulants, $\mathrm{Cum}_p(S_{n,K}) = \sum_{k_1,\cdots,k_p=k_0}^K (J_{k_1} \cdots J_{k_p}/k_1! \cdots k_p!)\mathrm{Cum}(S_n^{(k_1)}, \cdots, S_n^{(k_p)})$. Since the number of terms in the last sum is finite ($K < \infty$ is fixed), it suffices to show (4.6.18) for each joint cumulant in this sum, in other words, to show that

$$Q_n := \mathrm{Cum}(S_n^{(k_1)}, \cdots, S_n^{(k_p)}) \tag{4.6.19}$$

$$= \sum_{t_1,\cdots,t_p=1}^n \mathrm{Cum}(H_{k_1}(X_{t_1}), \cdots, H_{k_p}(X_{t_p})) = o(n^{p/2})$$

holds for any integers $p \geq 3$ and $k_1 \geq k_0, \cdots, k_p \geq k_0$.

To proceed, we shall need the formula from Section 14.1, which expresses the joint cumulants of Hermite polynomials of a Gaussian process $\{X_j\}$ through the autocovariance function. To do so, we need to introduce some new definitions. Consider the table

$$T = \begin{pmatrix} (1,1) & \cdots & (1,k_1) \\ (2,1) & \cdots & (2,k_2) \\ \cdots & & \\ (p,1) & \cdots & (p,k_p) \end{pmatrix},$$

containing p rows $T_1, \cdots, T_p$ having respective lengths $k_1, \cdots, k_p$. The total number of elements of T is

$$k := \mathrm{Card}(T) = k_1 + \cdots + k_p.$$

A *diagram* is a partition $\mathcal{V} = (V_1, \cdots, V_r)$ of table T by subsets $V_i, i = 1, \cdots, r$, each of which contains exactly *two* elements of T:

$$T = \cup_{i=1}^r V_i, \quad \mathrm{Card}(V_i) = 2, \quad V_i \cap V_j = \emptyset, \quad i \neq j.$$

Elements $V_i \subset T, i = 1, \cdots, r$, of this partition will be called *edges* of the diagram $\mathcal{V}$. Let Γ_T denote the class of all diagrams over table T. It is clear that if k is odd, then a partition of T by pairs does not exist and $\Gamma_T = \emptyset$, whereas if k is even, then necessarily $r = k/2$ and the number of all diagrams equals $(k/2-1)!!$, which is the number of ways to pair elements of T together; see Section 4.2. In the rest of the proof, we shall assume that k is even.

(a)   (b)

Fig. 4.1: Diagrams: (a) connected; (b) non connected.

An edge V_i is said to be *flat* if it is completely contained in a single row of T, i.e., if $V_i \subset T_j$ for some $j = 1, \cdots, p$. Rows T_i and T_j are said to be connected by diagram $\mathcal{V} = (V_1, \cdots, V_{k/2})$ if there is an edge $V_s, s = 1, \cdots, k/2$, and such that $V_s \cap T_i \neq \emptyset$, and $V_s \cap T_j \neq \emptyset$. A diagram $\mathcal{V} = (V_1, \cdots, V_{k/2})$ is called *connected* if T cannot be split into parts T' and T'', each of which consists of some rows of T such that T' and T'' are connected, in other words, if there is no edge V_i intersecting both T' and T''. For example, in Figure 4.1, the table consisting of $p = 4$ rows, each of length 2, has been partitioned in two ways. Note that diagram (a) is connected while (b) is not connected.

Denote by Γ_T^c the class of all connected diagrams over T without flat edges. For a given partition $\mathcal{V} = (V_1, \cdots, V_{k/2}) \in \Gamma_T^c$, denote by $\ell_{ij} = \ell_{ij}(\mathcal{V})$ the number of edges in $\mathcal{V}$ between the ith and the jth rows of T. In the above example, for the left diagram, we have $\ell_{12} = \ell_{23} = \ell_{34} = \ell_{41} = 1$ and $\ell_{ij} = 0$ for the remaining pairs (i, j) of rows, whereas for the right diagram, $\ell_{12} = \ell_{34} = 2$ and $\ell_{ij} = 0$ for the remaining pairs of rows.

By formula (14.1.20) in Section 14.1,

$$Q_n = \sum_{\mathcal{V} \in \Gamma_T^c} Q_n(\mathcal{V}), \quad Q_n(\mathcal{V}) := \sum_{t_1, \cdots, t_p = 1}^{n} \prod_{1 \leq i < j \leq p} \gamma_X^{\ell_{ij}}(t_i - t_j)$$

if k is even, otherwise, $Q_n = 0$. Thus, for (4.6.19), it suffices to show that for any diagram $\mathcal{V} \in \Gamma_T^c$,

$$Q_n(\mathcal{V}) = o(n^{p/2}). \tag{4.6.20}$$

For this purpose, we split $Q_n(\mathcal{V})$ as follows:

$$Q_n(\mathcal{V}) = Q_{n1}^{(L)}(\mathcal{V}) + Q_{n2}^{(L)}(\mathcal{V}),$$

where $L \geq 1$ is a large integer and where

$$Q_{n1}^{(L)}(\mathcal{V}) := \sum_{t_1,\cdots,t_p=1}^{n} \prod_{1 \leq i < j \leq p} \gamma_X^{\ell_{ij}}(t_i - t_j) I(|t_i - t_j| \leq L \text{ if } \ell_{ij} \geq 1)$$

is a sum over $t_1, \cdots, t_p = 1, \cdots, n$ which are "close" to each other, in the sense that $|t_i - t_j| \leq L$ provided the ith and the jth rows of T are connected by the diagram, and $Q_{n2}^{(L)}(\mathcal{V})$ is the sum over the remaining indices $t_1, \cdots, t_p = 1, \cdots, n$. An easy consequence of the definition of $Q_{n1}^{(L)}(\mathcal{V})$ is the fact that for any L fixed,

$$|Q_{n1}^{(L)}(\mathcal{V})| \leq Cn \tag{4.6.21}$$

for some C depending on L but independent of n. Indeed, consider the case when all neighboring pairs of rows are connected: $\ell_{12} \geq 1, \cdots, \ell_{p-1,p} \geq 1$. (The general case can always be reduced to this particular case by a permutation of rows $T_1, \cdots, T_p$.) Then $|\gamma_X(t_i - t_j)| \leq 1$ implies

$$|Q_{n1}^{(L)}(\mathcal{V})| \leq \sum_{t_1,\cdots,t_p=\overline{1,n}:|t_i-t_{i+1}|\leq L, i=1,\cdots,p-1} 1 \leq (2L+1)^{p-1}n.$$

Consider the second term, $Q_{n2}^{(L)}(\mathcal{V})$, in the above decomposition of $Q_n(\mathcal{V})$. By definition, $Q_{n2}^{(L)}(\mathcal{V})$ corresponds to the sum over all indices $t_1, \cdots, t_p = 1, \cdots, n$ such that for some pair (i,j) of rows $|t_i - t_j| \geq L$ and $\ell_{ij} \geq 1$. Hence, the factor $\gamma_X^{\ell_{ij}}(t_i - t_j)$ in the sum for $Q_{n2}^{(L)}(\mathcal{V})$ should be "small" if L is large, since $\gamma_X(t_i - t_j) \to 0$ as $|t_i - t_j| \to \infty$. We shall prove that there exists $\delta(L) \to 0$, $L \to \infty$, such that for any $n \geq 1$

$$|Q_{n2}^{(L)}(\mathcal{V})| \leq \delta(L)n^{p/2}. \tag{4.6.22}$$

Obviously, relations (4.6.21) and (4.6.22) imply (4.6.20), (4.6.18), and the statement of the theorem.

It remains to prove (4.6.22) which is the most difficult step of the proof of the theorem. Note, by definition, $\ell_{ij} = \ell_{ji}$ and

$$\sum_{j=1:j\neq i}^{n} \ell_{ij} = k_i, \quad i = 1, \cdots, p. \tag{4.6.23}$$

Let

$$\nu_1 := \sum_{1 < i < j \leq p} \frac{\ell_{ij}}{k_j}, \qquad \nu_2 := \sum_{1 < i < j \leq p} \frac{\ell_{ji}}{k_i}. \qquad (4.6.24)$$

Then

$$\nu_1 + \nu_2 = \sum_{i=1}^{p} \left(\sum_{j=1: j \neq i}^{p} \frac{\ell_{ij}}{k_i} \right) = p,$$

implying

$$\min(\nu_1, \nu_2) \leq p/2. \qquad (4.6.25)$$

We shall need the auxiliary Lemma 4.6.1, below. To formulate it, let some $\rho_{ij}(t) \geq 0$, $t \in \mathbb{Z}, 1 \leq i, j \leq p, i \neq j$, be given. Define

$$T_n := \sum_{t_1, \cdots, t_p = 1}^{n} \prod_{1 \leq i < j \leq p} \rho_{ij}^{\ell_{ij}}(t_i - t_j), \qquad v_{ij} := \sum_{t \in \mathbb{Z}} \rho_{ij}^{k_i}(t),$$

$$q_1 := \prod_{1 \leq i < j \leq p} v_{ij}^{\ell_{ij}/k_i}, \qquad q_2 := \prod_{1 \leq i < j \leq p} v_{ji}^{\ell_{ji}/k_j}.$$

Lemma 4.6.1. *For any integers $\ell_{ij} = \ell_{ji} \geq 0$, $k_i \geq 1$, satisfying (4.6.23), and any numbers $\rho_{ij}(t) \geq 0$,*

$$T_n \leq n^{p/2} \max(q_1, q_2). \qquad (4.6.26)$$

Proof. Recall the generalized Hölder inequality. For any positive integer k, for any $1 < \beta_1, \cdots, \beta_k < \infty$ such that $\sum_{i=1}^{k} 1/\beta_i = 1$, and for any sequences $\{f_s^{(i)}, s \in \mathbb{Z}\}, i = 1, \cdots, k$, such that $\sum_{s \in \mathbb{Z}} |f_s^{(i)}|^{\beta_i} < \infty, 1 \leq i \leq k$,

$$\left| \sum_{s \in \mathbb{Z}} f_s^{(1)} \cdots f_s^{(k)} \right| \leq \prod_{i=1}^{k} \left(\sum_{s \in \mathbb{Z}} |f_s^{(i)}|^{\beta_i} \right)^{1/\beta_i}.$$

Applying this inequality to the summation in T_n over t_1 and the product $\prod_{j=2}^{p} \rho_{12}^{\ell_{1j}}(t_1 - t_j)$, with $\beta_j = k_1/\ell_{1j}, j = 2, \cdots, p$, since $1/\beta_1 + \cdots + 1/\beta_p = 1$, we obtain

$$\sum_{t_1=1}^{n} \prod_{j=2}^{p} \rho_{1j}^{\ell_{1j}}(t_1 - t_j) \leq \prod_{j=2}^{p} \left(\sum_{t_1=1}^{n} \rho_{1j}^{k_1}(t_1 - t_j) \right)^{\ell_{1j}/k_1} \leq \prod_{j=2}^{p} v_{1j}^{\ell_{1j}/k_1}.$$

The last inequality holds when some of the ℓ_{1j}'s are zero since in that case the corresponding factors on the r.h.s. are equal to 1 by definition. In a similar way, the sum over t_2 can be estimated as

$$\sum_{t_2=1}^{n} \prod_{j=3}^{p} \rho_{2j}^{\ell_{2j}}(t_2 - t_j) \le \Big(\sum_{t_2=1}^{n} 1 \Big)^{\ell_{12}/k_2} \prod_{j=3}^{p} \Big(\sum_{t_2=1}^{n} \rho_{2j}^{k_2}(t_2 - t_j) \Big)^{\ell_{2j}/k_2}$$

$$\le n^{\ell_{12}/k_2} \prod_{j=3}^{p} v_{2j}^{\ell_{2j}/k_2}.$$

In a similar way, estimating the sums over the remaining indices $t_3, \cdots, t_p$, we obtain

$$T_n \le n^{\ell_{12}/k_2} \prod_{j_1=2}^{p} v_{1j_1}^{\ell_{1j_1}/k_1} \prod_{j_2=3}^{p} v_{2j_2}^{\ell_{2j_2}/k_2} \sum_{t_3,\cdots,t_p=1}^{n} \prod_{3\le i<j\le p} \rho_{ij}^{\ell_{ij}}(t_i - t_j)$$

$$\le \prod_{1\le i<j\le 3} n^{\ell_{ij}/k_j} \prod_{i=1}^{3} \prod_{j=i+1}^{p} v_{ij}^{\ell_{ij}/k_j} \sum_{t_4,\cdots,t_p=1}^{n} \prod_{4\le i<j\le p} \rho_{ij}^{\ell_{ij}}(t_i - t_j)$$

$$\le \cdots$$

$$\le \prod_{1\le i<j\le p} n^{\ell_{ij}/k_j} \prod_{1\le i<j\le p} v_{ij}^{\ell_{ij}/k_j}$$

$$= n^{\nu_1} q_1,$$

where ν_1 is defined by (4.6.24). Exactly similarly, by applying Hölder's inequality in the reverse direction from t_p to t_1 we obtain

$$T_n \le n^{\nu_2} q_2.$$

These two inequalities, together with (4.6.25), prove Lemma 4.6.1.

Now, we return to the proof of (4.6.22). Assume that $\ell_{12} \ge 1$ and consider the summands in $Q_{n2}^{(L)}(\mathcal{V})$ with $|t_1 - t_2| > L$:

$$\sum_{t_1,\cdots,t_p=1}^{n} \prod_{1\le i<j\le p} \gamma_X^{\ell_{ij}}(t_i - t_j) I(|t_1 - t_2| > L) =: U_{12}, \qquad \text{say.}$$

Use Lemma 4.6.1 with $\rho_{12}(t) := |\gamma_X(t)| I(|t| > L)$, and $\rho_{ij}(t) := |\gamma_X(t)|$ for the remaining indices i, j in $\{1, \cdots, p\}$. Then, because of (4.6.14),

$$v_{ij} = \sum_{t\in\mathbb{Z}} |\rho_X(t)|^{k_i} < \infty, \qquad (i,j) \ne (1,2), (2,1),$$

$$v_{12} = \sum_{|t|>L} |\rho_X(t)|^{k_1}, \qquad v_{21} = \sum_{|t|>L} |\rho_X(t)|^{k_2},$$

and $\max(v_{12}, v_{21}) \to 0$, as $L \to \infty$. Because the v_{ij}'s do not depend on n and $\max(q_1, q_2) \le C \max(v_{12}, v_{21}) \to 0$, $L \to \infty$, this establishes the bound

$$|U_{12}| \le \delta_{12}(L) n^{p/2},$$

with $\delta_{12}(L)$ independent of n and $\delta_{12}(L) \to 0$ as $L \to \infty$. Since the remaining terms in $Q_{n2}^{(L)}(\mathcal{V})$ can be bounded similarly, this concludes the proof of (4.6.22) and completes the proof of Theorem 4.6.1.

Theorem 4.6.1 is very general and provides a complete answer to the question about the CLT for Gaussian subordinated processes. However, the proof of this theorem is rather involved and uses the diagram technique. One might wonder if there exist simpler ways to prove asymptotic normality for subordinated processes, which do not use cumulants and diagrams. Below, we show that under some additional mild assumptions on $\{X_j\}$, the CLT for $S_n = \sum_{j=1}^{n} h(X_j)$, the sum of Gaussian subordinated sequences $\{h(X_j)\}$, can be proved by approximating S_n by a sum of m-dependent r.v.'s, in the mean square.

As seen above, proving a CLT for the sums S_n is rather easily reduced to that of sums of a finite linear combination of Hermite polynomials; see (4.6.16). Below, we consider the CLT for a sum of a single kth-order Hermite polynomial $H_k(X_j)$ where $\{X_j\}$ is a Gaussian linear process

$$X_j = \sum_{s=0}^{\infty} a_s \zeta_{j-s}, \quad j \in \mathbb{Z}, \tag{4.6.27}$$

in i.i.d. standardized Gaussian innovations $\{\zeta_j\}$ with $\mathrm{Var}(X_0) = 1$.

Proposition 4.6.2. *Let $S_n = \sum_{j=1}^{n} H_k(X_j)$ with $k \geq 1$ and where $\{X_j\}$ is a Gaussian process as in (4.6.27) such that $EX_0^2 = \sum_{s=0}^{\infty} a_s^2 = 1$ and*

$$\sum_{j=0}^{\infty} \left(\sum_{s=0}^{\infty} |a_s a_{s+j}| \right)^k < \infty. \tag{4.6.28}$$

Then

$$n^{-1/2} S_n \to_D \mathcal{N}(0, \sigma^2), \quad \sigma^2 := \sum_{j \in \mathbb{Z}} \gamma_X^k(j) < \infty. \tag{4.6.29}$$

Proof. Let $m \geq 1$ be an integer and introduce the truncated process

$$X_{j,m} := \sum_{s=0}^{m} v_m a_s \zeta_{j-s}, \ j \in \mathbb{Z}, \quad v_m = \left(\sum_{s=0}^{m} a_s^2 \right)^{-1/2}.$$

Note that $v_m^{-2} \to \sum_{s=0}^{\infty} a_s^2 = \mathrm{Var}(X_0) = 1$ as $m \to \infty$. Then $\{X_{j,m}\}$ is a stationary Gaussian process with zero mean, unit variance, and covariance

$$\gamma_{X,m}(j) := \mathrm{Cov}(X_{0,m}, X_{j,m}) = v_m^2 \sum_{s=0}^{m} a_s a_{s+j}. \tag{4.6.30}$$

The process $\{H_k(X_{j,m})\}$ is an m-dependent stationary process with zero mean and covariance

$$\gamma_{H,m}(j) := \text{Cov}(H_k(X_{0,m}), H_k(X_{j,m})) = k!\gamma_{X,m}^k(j).$$

Let $S_{n,m} := \sum_{t=1}^n H_k(X_{t,m})$. By Proposition 4.2.3, for any $m < \infty$ fixed, the sum $S_{n,m}$ satisfies the CLT:

$$n^{-1/2} S_{n,m} \to_D \mathcal{N}(0, \sigma_m^2),$$

with $\sigma_m^2 := \sum_{j \in \mathbb{Z}} \gamma_{H,m}(j) = k! \sum_{j \in \mathbb{Z}} \gamma_{X,m}^k(j)$. By (4.6.30), under assumption (4.6.28),

$$\sigma_m^2 \to k! \sum_{j \in \mathbb{Z}} \gamma_X^k(j) = \sigma^2, \quad m \to \infty.$$

Therefore, to prove (4.6.29), it remains to show that

$$\limsup_{n \to \infty} n^{-1} \text{Var}(S_n - S_{n,m}) \to 0, \quad m \to \infty. \tag{4.6.31}$$

But,

$$\text{Var}(S_n - S_{n,m}) = \sum_{i,j=1}^n \rho_m(j - i),$$

where $\rho_m(j) := \text{Cov}(H_k(X_0) - H_k(X_{0,m}), H_k(X_j) - H_k(X_{j,m}))$. By property (2.4.15) of Hermite polynomials in Proposition 2.4.1, for $j \geq 0$,

$$\text{Cov}(H_k(X_{0,m}), H_k(X_j)) = k! \text{Cov}^k(X_{0,m}, X_j) = k!(v_m \sum_{s=0}^m a_s a_{s+j})^k,$$
$$\text{Cov}(H_k(X_0), H_k(X_{j,m})) = k! \text{Cov}^k(X_0, X_{j,m}) = k!(v_m \sum_{s=0}^{m-j} a_s a_{s+j})^k,$$

whereas $\rho_m(-j) = \rho_m(j)$. Hence, for any $j \geq 0$,

$$\rho_m(j) = k! \left\{ \left(\sum_{s=0}^\infty a_s a_{j+s} \right)^k - \left(v_m \sum_{s=0}^m a_s a_{j+s} \right)^k \right.$$
$$\left. - \left(v_m \sum_{s=0}^{m-j} a_s a_{j+s} \right)^k + \left(v_m^2 \sum_{s=0}^m a_s a_{j+s} \right)^k \right\}.$$

From condition (4.6.28) it is easy to see that $\rho_m(j)$ is dominated in absolute value by a summable sequence independent of m, namely

$$|\rho_m(j)| \leq C \left(\sum_{s=0}^\infty |a_s a_{j+s}| \right)^k.$$

Therefore, by the DCT, since, for each $j \in \mathbb{Z}$, $\rho_m(j) \to 0$, as $m \to \infty$,

$$n^{-1}\mathrm{Var}(S_n - S_{n,m}) \leq \sum_{j \in \mathbb{Z}} |\rho_m(j)| \to 0, \quad m \to \infty,$$

proving (4.6.31). This completes the proof of Proposition 4.6.2.

Because $\gamma_X(j) = \sum_{s=0}^{\infty} a_s a_{s+j}$, when $a_s \geq 0$, condition (4.6.28) becomes $\sum_{j=0}^{\infty} |\gamma_X(j)|^k < \infty$, or condition (4.6.14) of Theorem 4.6.1. Condition (4.6.28) is also satisfied if the coefficients are not necessarily positive but satisfy the short memory restriction

$$\sum_{s=0}^{\infty} |a_s| < \infty, \tag{4.6.32}$$

or have the form

$$a_s = s^{d_X - 1} L(s), \quad 0 < d_X < 1/2, \ L \in SV, \tag{4.6.33}$$

or satisfy the inequality

$$|a_s| \leq C s^{d_X - 1}, \quad \forall s > 0, \quad 0 < d_X < 1/2$$

and $k > 1/(1 - 2d_X)$. The last fact can be verified by estimating the inner sum in (4.6.28) as follows:

$$\sum_{s=0}^{\infty} |a_s a_{s+j}| \leq C \left(j^{d_X - 1} + \sum_{s=1}^{\infty} s^{d_X - 1}(s+j)^{d_X - 1} \right) \tag{4.6.34}$$

$$\leq C \left(j^{d_X - 1} + j^{d_X - 1} \sum_{s=1}^{j} s^{d_X - 1} + \sum_{s>j} s^{2(d_X - 1)} \right)$$

$$= O(j^{-(1 - 2d_X)}).$$

In particular, under the SM condition (4.6.32) or the LM condition (4.6.33), the CLT for subordinated processes $\{h(X_j)\}$ can be proved by simple short memory approximations as in Proposition 4.6.2.

Note: The CLT for functions $h(X_j)$ under similar conditions as in Theorem 4.6.1 was derived by Breuer and Major (1983) and Giraitis and Surgailis (1985). The above proof of Theorem 4.6.1 is an adaptation of Giraitis and Surgailis (1985). For polynomials of a linear process the CLT was derived in Giraitis (1985). Arcones (1994) extended Theorem 4.6.1 to vector-valued Gaussian sequences and provided useful covariance inequalities for Gaussian subordinated functions. Nourdin, Peccati and Podolskij (2011) present a different approach based on Malliavin's calculus to prove the CLT (4.6.13) along with convergence rates.

4.7 Non-central limit theorems

The focus of this subsection is non-central limit theorems for Gaussian sub-ordinated processes $\{Y_j = h(X_j), j \in \mathbb{Z}\}$. The limit distributions in these theorems are defined in terms of multiple Wiener–Itô integrals. Below, we give some basic properties of such integrals, which are used in the subsequent discussion. The proofs, as well as a rigorous definition of multiple Wiener–Itô integrals can be found in the Appendix, Section 14.3.

Let $W = \{W(u), u \in \mathbb{R}\}$ be Brownian motion, also called a Wiener process, on the real line. In other words, $\{W(u), u \geq 0\}$ and $\{W(-u), u \geq 0\}$ are independent Gaussian processes with zero mean and respective covariances $EW(u)W(v) = EW(-u)W(-v) = \min(u, v)$, $u, v \geq 0$. Let $L_2(\mathbb{R}^k)$ denote the Hilbert space of the square integrable function $f : \mathbb{R}^k \to \mathbb{R}$, with finite norm $\|f\|$, where

$$\|f\|^2 := \int_{\mathbb{R}^k} |f(u_1, \cdots, u_k)|^2 du_1 \cdots du_k.$$

A k-tuple Wiener–Itô integral

$$I_k(f) = \int_{\mathbb{R}^k} f(u_1, \cdots, u_k) W(du_1) \cdots W(du_k) \qquad (4.7.1)$$

is defined for any $k = 1, 2, \cdots$ and any $f \in L_2(\mathbb{R}^k)$, as a limit in mean square of k-tuple polynomial forms in Gaussian i.i.d. variables (increments of the Wiener process). In particular, if f is a "simple" function taking a finite number of non-zero constant values $f(j_1, \cdots, j_k)$ on "cubes" $(j_1/n, (j_1 + 1)/n] \times \cdots \times (j_k/n, (j_k + 1)/n] \subset \mathbb{R}^k$, $(j_1, \cdots, j_k) \in \mathbb{Z}^k$, and f "vanishes on the diagonals", in the sense that

$$f(j_1, \cdots, j_k) = 0 \quad \text{unless all } j_1, \cdots, j_k \text{ are different,}$$

then the corresponding integral $I_k(f)$ coincides with the polynomial form

$$I_k(f) = \sum_{j_1, \cdots, j_k \in \mathbb{Z}} f(j_1, \cdots, j_k) \Delta_n W(j_1) \cdots \Delta_n W(j_k),$$

where $\Delta_n W(j) := W((j+1)/n) - W(j/n)$, $j \in \mathbb{Z}$, is an increment of the Wiener process.

Some basic properties of such integrals are as follow. For any $k \geq 1, f \in L_2(\mathbb{R}^k)$, the integral $I_k(f)$ is a r.v. with finite variance, which is linear in f: for any $f, g \in L_2(\mathbb{R}^k)$ and any constants $a, b \in \mathbb{R}$,

$$I_k(af + bg) = aI_k(f) + bI_k(g). \qquad (4.7.2)$$

Moreover, for any $k, m \geq 1$ and any $f \in L_2(\mathbb{R}^k)$, $g \in L_2(\mathbb{R}^m)$

$$EI_k(f) = 0, \quad EI_k^2(f) \leq k!\|f\|^2, \quad EI_k(f)I_m(g) = 0, \quad k \neq m. \quad (4.7.3)$$

The last equation is known as the *orthogonality property of multiple Wiener–Itô integrals*. If $f(u_1, \cdots, u_k)$ is symmetric under permutations of its arguments $u_1, \cdots, u_k$, then

$$EI_k^2(f) = k!\|f\|^2. \quad (4.7.4)$$

From (4.7.3) it immediately follows that the convergence $f_n \to f$ in $L_2(\mathbb{R}^k)$ implies the convergence $E|I_k(f_n) - I_k(f)|^2 \to 0$ and hence the convergence in distribution: for $f_n, f \in L_2(\mathbb{R}^k)$

$$\|f_n - f\| \to 0 \quad \text{implies} \quad I_k(f_n) \to_D I_k(f). \quad (4.7.5)$$

For $k = 1$, $I_1(f)$ is a Gaussian r.v. with zero mean and variance $\|f\|^2 = \int_{\mathbb{R}} |f(u)|^2 du$. This property can be generalized as follows: Let $f \in L_2(\mathbb{R}), \|f\| = 1$. Then for any $k \geq 1$,

$$\int_{\mathbb{R}^k} [f(u_1) \times \cdots \times f(u_k)] W(du_1) \cdots W(du_k) = H_k(Z), \quad (4.7.6)$$

where $Z := I_1(f) \sim \mathcal{N}(0,1)$ and H_k is the kth Hermite polynomial. In particular, from (4.7.6) and (4.7.2), for any $f \in L_2(\mathbb{R})$,

$$\int_{\mathbb{R}^2} f(u_1)f(u_2)W(du_1)W(du_2) = \left(\int_{\mathbb{R}} f(u)W(du) \right)^2 - \int_{\mathbb{R}} |f(u)|^2 du.$$

We shall also need the *change of variables* formula: for any integers $n, k \geq 1$ and any $f \in L_2(\mathbb{R}^k)$,

$$I_k(f) =_D \int_{\mathbb{R}^k} n^{k/2} f(nu_1, \cdots, nu_k)W(du_1) \cdots W(du_k). \quad (4.7.7)$$

The change of variables property is an easy consequence of the linearity property (4.7.2) and the self-similarity property of the Wiener process (see Section 3.4), which can be rewritten in differential form as

$$W(d(nu)) =_D n^{1/2}W(du).$$

We shall now formulate a useful criterion for the convergence in distribution of polynomial forms in i.i.d. variables to a multiple Wiener–Itô integral known as the *scheme of discrete multiple integrals*. Consider an "off-diagonal" polynomial form

$$Q_k(g) := \sum{}' g(j_1, \cdots, j_k)\zeta_{j_1} \cdots \zeta_{j_k} \quad (4.7.8)$$

in i.i.d. r.v.'s $\{\zeta_j\} \sim \mathrm{IID}(0,1)$, with real coefficients $g(j_1, \cdots, j_k)$. Here and below, $\sum'$ denotes the sum over all *mutually distinct* integers $j_1, \cdots, j_k \in \mathbb{Z}$. Since $E(\zeta_{j_1} \cdots \zeta_{j_k} \zeta_{i_1} \cdots \zeta_{i_k}) = 0$, unless $(i_1, \cdots, i_k)$ is a permutation of $(j_1, \cdots, j_k)$, we readily obtain

$$EQ_k^2(g) \le k! {\sum}' g^2(j_1, \cdots, j_k). \tag{4.7.9}$$

Therefore, $Q_k(g)$ converges in mean square if g is square summable. Note that equality in (4.7.9) holds if and only if g is symmetric, or invariant w.r.t. all permutations of $j_1, \cdots, j_k$. With any such sequences of numbers $g_n \in L_2(\mathbb{Z}^k)$ and any $n = 1, 2, \cdots$, we can associate a sequence of functions $\widetilde{g}_n \in L_2(\mathbb{R}^k)$ (the "extension" of g_n to $\mathbb{R}^k$), as defined by

$$\widetilde{g}_n(u_1, \cdots, u_k) := n^{k/2} g_n([u_1 n], \cdots, [u_k n]), \tag{4.7.10}$$

for $(u_1, \cdots, u_k) \in \mathbb{R}^k$, where $[x]$ denotes the integer part of a real number $x \in \mathbb{R}$. The following corollary follows from Proposition 14.3.2.

Corollary 4.7.1. *Let $Q_k(g_n)$, $n = 1, 2, \cdots$, be a sequence of polynomial forms in i.i.d. r.v.'s $\{\zeta_j\} \sim \mathrm{IID}(0,1)$ as in (4.7.8), with coefficients $g_n \in L_2(\mathbb{Z}^k)$, $k \ge 1$. Assume that for some $f \in L_2(\mathbb{R}^k)$*

$$\|\widetilde{g}_n - f\| \to 0, \quad n \to \infty. \tag{4.7.11}$$

Then

$$Q_k(g_n) \to_D I_k(f), \quad n \to \infty. \tag{4.7.12}$$

We now turn to non-central limit theorems. Consider first the limit distribution of simple sums $S_n = \sum_{j=1}^{n} Y_j$ of a Gaussian subordinated process $\{Y_j = h(X_j)\}$ when $\{Y_j\}$ and $\{X_j\}$ both exhibit long memory, with possibly different memory parameters d_X and d_Y. Recall that, by Wold decomposition, under mild assumptions on spectral density, a stationary zero-mean Gaussian process $\{X_j\}$ can be written as a linear process in (4.6.27), with square summable weights a_j, and an i.i.d. Gaussian noise $\{\zeta_k\} \sim \mathcal{N}(0,1)$. Note $\mathrm{Var}(X_0) = 1$ is equivalent to $\sum_{k=0}^{\infty} a_k^2 = 1$.

We shall now assume that the process $\{X_j\}$ has long memory with memory parameter d_X. For simplicity, we shall further suppose that the weights a_j's decay regularly as

$$a_j \sim c_a j^{-1+d_X}, \quad j \to \infty, \quad 0 < d_X < 1/2, \quad c_a > 0. \tag{4.7.13}$$

By Proposition 3.2.1, this implies the time-domain long memory property:

$$\gamma_X(j) \sim c_X |j|^{-1+2d_X}, \quad c_X := c_a^2 B(d_X, 1 - 2d_X). \quad (4.7.14)$$

With k_0 denoting the Hermite rank of h, define

$$g_\tau(u_1, \cdots, u_{k_0}) = \int_0^\tau \prod_{i=1}^{k_0} (v - u_i)_+^{-1+d_X} \, dv, \quad 0 \le \tau \le 1, \quad g := g_1, \quad (4.7.15)$$

where $u_+ = \max(u, 0)$. In Proposition 14.3.7 of the Appendix, it is shown that $g_\tau(\cdot) \in L_2(\mathbb{R}^{k_0})$, $\tau \in [0, 1]$, if $k_0(1 - 2d_X) < 1$, $d_X \in (0, 1/2)$.

The next theorem derives the non-CLT for the sums $S_n = \sum_{j=1}^n Y_j$ when the subordinated process $\{Y_j\}$ has long memory. It uses the definition of d_Y from (4.6.11).

Theorem 4.7.1. *Assume that $\{X_j\}$ and h satisfy assumptions (4.6.1) and (4.7.13), and that $k_0(1 - 2d_X) < 1$. Then $\{Y_j = h(X_j), j \in \mathbb{Z}\}$ has long memory with memory parameter $d_Y = \frac{1}{2} - k_0(\frac{1}{2} - d_X)$, and*

$$n^{-1/2-d_Y} S_n \to_D \frac{J_{k_0} c_a^{k_0}}{k_0!} I_{k_0}(g), \quad (4.7.16)$$

where g is as in (4.7.15).

Proof. Proposition 4.6.1 and (4.7.14) imply

$$\gamma_Y(j) \sim c_Y |j|^{-(1-2d_Y)} \equiv c_Y |j|^{-(1-2d_X)k_0} \quad (4.7.17)$$

with parameters $2d_Y = 1 - k_0(1 - 2d_X)$ and $c_Y = (J_{k_0}^2/k_0!)c_X^{k_0}$. Hence, condition $0 < d_Y < 1/2$ is equivalent to $k_0(1 - 2d_X) < 1$.

To prove (4.7.16), use approximation (4.6.7),

$$h(X_j) = Y_{k_0}(j) + \mathcal{R}_j, \quad Y_{k_0}(j) := \frac{J_{k_0}}{k_0!} H_{k_0}(X_j),$$

$$S_n = \sum_{j=1}^n Y_{k_0}(j) + \sum_{j=1}^n \mathcal{R}_j =: S_{n,k_0} + S_{n,\mathcal{R}}.$$

It suffices to show that

$$n^{-1/2-d_Y} S_{n,k_0} \to_D c_a^{k_0} I_{k_0}(g), \quad (4.7.18)$$

$$\text{Var}(n^{-1/2-d_Y} S_{n,\mathcal{R}}) \to 0. \quad (4.7.19)$$

By (4.6.8), $|\gamma_\mathcal{R}(j)| \le C|\gamma_X(j)|^{k_0+1} = o(|j|^{-1+2d_Y})$. Then

$$n^{-1}\text{Var}(S_{n,\mathcal{R}}) \le \sum_{j=-n}^n |\gamma_\mathcal{R}(j)| = \sum_{j=-n}^n o(|j|^{-1+2d_Y}) = o(n^{2d_Y}),$$

which implies (4.7.19).

To show (4.7.18), set $a(j) = a_j$, $j = 0, 1, \cdots$ or $a(j) = 0$ if $j < 0$, and write X_j as a stochastic integral

$$X_j = \int_{\mathbb{R}} a(j - [u])W(du), \quad j \in \mathbb{Z}.$$

Since $EX_j = 0$ and $EX_j^2 = 1$, by property (4.7.6) $H_{k_0}(X_j)$ can be written as a k-tuple Wiener–Itô integral

$$H_{k_0}(X_j) = \int_{\mathbb{R}^{k_0}} a(j - [u_1]) \cdots a(j - [u_{k_0}])W(du_1) \cdots W(du_{k_0}).$$

Then by (4.7.7) and changing variables $u_i \to u_i n$ in the Wiener–Itô integral,

$$T_n := \sum_{j=1}^{n} H_{k_0}(X_j) = \int_{\mathbb{R}^{k_0}} \sum_{j=1}^{n} \prod_{i=1}^{k_0} a(j - [u_i])W(du_1) \cdots W(du_{k_0})$$

$$= \int_{\mathbb{R}^{k_0}} n^{k_0/2} \sum_{j=1}^{n} \prod_{i=1}^{k_0} a(j - [nu_i])W(du_1) \cdots W(du_{k_0}).$$

This relation and because $\sum_{j=1}^{n} \prod_{i=1}^{k_0} a(j - [nu_i]) = n \int_0^1 \prod_{i=1}^{k_0} a([vn] + 1 - [nu_i])dv$ and $2d_Y = 1 - k_0(1 - 2d_X)$, yields

$$n^{-1/2 - d_Y} T_n = \int_{\mathbb{R}^{k_0}} g_n(u_1, \cdots, u_{k_0})W(du_1) \cdots W(du_{k_0}) \equiv I_{k_0}(g_n),$$

with

$$g_n(u_1, \cdots, u_{k_0}) = n^{k_0(1 - d_X)} \int_0^1 \prod_{i=1}^{k_0} a([nv] + 1 - [n u_i])dv \qquad (4.7.20)$$

$$= \int_0^1 \prod_{i=1}^{k_0} p_n(v, u_i)(v - u_i)_+^{-1 + d_X} dv,$$

where $p_n(v, u) := a([nv] + 1 - [nu])(n(v - u)_+)^{1 - d_X}$. By assumption (4.7.13),

$$\sup_{n \geq 1} \sup_{u,v} |p_n(v, u)| \leq C, \quad p_n(v, u) \to c_a, \quad \forall v > u.$$

This implies, with g as in (4.7.15),

$$\sup_{u_1, \cdots, u_{k_0} \in \mathbb{R}} |g_n(u_1, \cdots, u_{k_0})| \leq Cg(u_1, \cdots, u_{k_0}),$$

$$g_n(u_1, \cdots, u_{k_0}) \to c_a^{k_0} g(u_1, \cdots, u_{k_0}), \quad \forall u_1, \cdots, u_{k_0}.$$

Thus, by the DCT,

$$\|g_n - c_a^{k_0} g\|_{k_0}^2 \to 0, \qquad (4.7.21)$$

and, by property (4.7.5), $I_{k_0}(g_n) \to c_a^{k_0} I_{k_0}(g)$, which completes the proof of (4.7.18) and of Theorem 4.7.1.

The next corollary gives the weak limit of the partial-sum process $S_n(\tau) = \sum_{j=1}^{[n\tau]} Y_j$, $0 \le \tau \le 1$. With g_τ given by (4.7.15), define

$$\mathcal{H}_{k_0}(\tau) = \left(\text{Var}(I_{k_0}(g_1))\right)^{-1/2} I_{k_0}(g_\tau).$$

This process is called a *Hermite process*. Proposition 14.3.7 in the Appendix gives some of its properties.

Corollary 4.7.2. *Under the assumptions of Theorem 4.7.1,*

$$n^{-1/2-d_Y} S_n(\tau) \Rightarrow \frac{J_{k_0} c_a^{k_0}}{k_0!} I_{k_0}(g_\tau), \text{ in } \mathcal{D}[0,1], \text{ for the uniform metric,}$$

$$= \left(\frac{c_Y}{d_Y(1+2d_Y)}\right)^{1/2} \mathcal{H}_{k_0}(\tau), \tag{4.7.22}$$

with c_Y as in (4.7.17).

Proof. Using the Cramér–Wold device, the finite-dimensional weak convergence follows as in the proof of Theorem 4.7.1. Since $d_Y > 0$, tightness follows from Proposition 4.4.2. This completes the proof for the weak convergence.

It remains to prove the equality in (4.7.22). Since under the assumed conditions, the a_j's are as in (4.7.13) then $\gamma_X(j) \sim c_X |j|^{-1+2d_X}$, with $c_X = c_a^2 B(d_X, 1 - 2d_X)$. Then by (4.7.17) and (4.7.14), $\gamma_Y(j) \sim c_Y |j|^{-1+2d_Y}$, with $2d_Y = 1 - k_0(1 - 2d_X)$ and

$$c_Y = \frac{J_{k_0}^2}{k_0!} c_X^{k_0} = \frac{J_{k_0}^2 c_a^{2k_0}}{k_0!} B^{k_0}(d_X, 1 - 2d_X)$$

$$= \frac{J_{k_0}^2 c_a^{2k_0}}{k_0!^2} d_Y(1 + 2d_Y)\text{Var}(I_{k_0}(g_1)),$$

because of property (14.3.39) of a Hermite process as given in the Appendix, which proves (4.7.22).

Note: The non-CLT for functions $h(X_j)$ of a long memory Gaussian process $\{X_j\}$ was first considered by Rosenblatt (1961) and later developed in full generality by Dobrushin and Major (1979) and Taqqu (1979). Surgailis (1982) was the first to study the non-CLT for polynomials of a linear (non-Gaussian) process. Later, Giraitis and Surgailis (1986, 1989) and Avram and Taqqu (1987) related this problem to expansions in Appell polynomials. For non-smooth functions the non-CLT was obtained by Ho and Hsing (1996), Koul and Surgailis (1997), and Surgailis (2000).

4.8 Limit theorems for polynomial forms

In this section we consider polynomial forms
$$Y_j^{(k)} = \sum_{-\infty < s_k < \cdots < s_1 \leq j} a_{j-s_1} \cdots a_{j-s_k} \zeta_{s_1} \cdots \zeta_{s_k}, \quad j \in \mathbb{Z}, \ k \geq 1, \quad (4.8.1)$$
in standardized r.v.'s $\{\zeta_s\} \sim \mathrm{IID}(0,1)$, not necessarily Gaussian, with square summable coefficients $\sum_{s=0}^{\infty} a_s^2 < \infty$. Note that $\{Y_j^{(1)}\}$ is a linear process as in (4.3.1).

For every $k \geq 1$, the series (4.8.1) converges in mean square, and $\{Y_j^{(k)}\}$ is a well-defined zero-mean stationary process with autocovariances
$$\gamma_{Y^{(k)}}(j) := EY_j^{(k)}Y_0^{(k)} = \sum_{0 \leq s_1 < \cdots < s_k < \infty} a_{j+s_1}a_{s_1} \cdots a_{j+s_k}a_{s_k}. \quad (4.8.2)$$
Moreover, polynomial forms $Y_j^{(k)}$ are orthogonal, i.e., for any $j, s \in \mathbb{Z}$,
$$EY_j^{(k)}Y_s^{(m)} = 0, \qquad k \neq m. \quad (4.8.3)$$
These two properties are similar to those of Hermite polynomials of a stationary Gaussian process. Polynomial forms $Y_j^{(k)}$ play a similar role to Hermite polynomials in the asymptotic expansion of transformations and empirical functionals of a linear process, as discussed in this book. We show that the sum $S_{n,k} = \sum_{j=1}^n Y_j^{(k)}$ and the partial-sum process
$$S_{n,k}(\tau) = \sum_{j=1}^{[n\tau]} Y_j^{(k)}, \quad 0 \leq \tau \leq 1, \quad (4.8.4)$$
of such variables have asymptotically the same distribution up to a scale factor as the corresponding sums of Hermite polynomials $H_k(X_j)$ when $\{X_j\}$ is a standardized stationary Gaussian process.

Central limit theorem. As in Proposition 4.6.2, condition (4.6.28) is crucial for the CLT here also. The proof of the central limit theorem below has the advantage that it does not use diagrams but a simple approximation of the summands by finitely dependent r.v.'s. The same method was used for proving the CLT for non-linear functions $H_k(X_j)$ of a linear Gaussian process in Proposition 4.6.2, but the CLT for polynomial forms $\{Y_j^{(k)}\}$ does not require the innovations ζ_j to be Gaussian.

Theorem 4.8.1. *Suppose $Y_j^{(k)}$, $j \in \mathbb{Z}$, $k \geq 1$, are as in (4.8.1) and the a_j's satisfy (4.6.28). Then, for every $k \geq 1$, $\sum_{j \in \mathbb{Z}} |\gamma_{Y^{(k)}}(j)| < \infty$, and*
$$n^{-1/2}S_{n,k} \to_D \mathcal{N}(0, \sigma^2), \qquad \sigma^2 := \sum_{j \in \mathbb{Z}} \gamma_Y(j). \quad (4.8.5)$$

Proof. Let $k \geq 1$. Claim $\sum_{j \in \mathbb{Z}} |\gamma_{Y^{(k)}}(j)| < \infty$ follows from (4.8.2) and (4.6.28) in a routine fashion. If $\sigma^2 = 0$, then (4.8.5) holds trivially. Suppose $\sigma^2 > 0$. Put

$$Y_{j,m} := \sum_{s_k < \cdots < s_1 \leq t} a_{j-s_1,m} \cdots a_{j-s_k,m} \zeta_{s_1} \cdots \zeta_{s_k}, \qquad (4.8.6)$$

where $a_{j,m} := a_j I(0 \leq j \leq m)$ and $1 < m < \infty$. Then $\{Y_{j,m}, j \in \mathbb{Z}\}$ is an mk-dependent stationary process with zero mean and finite variance. Its autocovariance $\rho_{j,m} := \mathrm{Cov}(Y_{0,m}, Y_{j,m})$ has the following properties:

$$\rho_{j,m} = \sum_{0 \leq i_1 < \cdots < i_k < \infty} a_{j+i_1,m} a_{i_1,m} \cdots a_{j+i_k,m} a_{i_k,m}, \quad j \geq 0,$$

$$|\rho_{j,m}| \leq \Big(\sum_{s=0}^{\infty} |a_{j+s} a_s|\Big)^k, \qquad \rho_{j,m} \to \gamma_{Y^{(k)}}(j), \quad \forall j \geq 0, \quad m \to \infty.$$

Write $S_{n,k} = S_{n,k}^{(m)} + (S_{n,k} - S_{n,k}^{(m)})$ with $S_{n,k}^{(m)} := \sum_{j=1}^{n} Y_{j,m}$. Under condition (4.6.28), by Proposition 4.2.3,

$$n^{-1/2} S_{n,k}^{(m)} \to_D \mathcal{N}(0, \sigma_m^2), \quad n \to \infty,$$

$$\sigma_m^2 := \sum_{j \in \mathbb{Z}} \mathrm{Cov}(Y_{0,m}, Y_{j,m}) \to \sigma^2, \quad m \to \infty.$$

To complete the proof, we show that uniformly for $n \geq 1$,

$$n^{-1} \mathrm{Var}(S_{n,k} - S_{n,k}^{(m)}) = \sum_{|j| < n} \rho'_{j,m} \Big(1 - \frac{|j|}{m}\Big) \to 0, \quad m \to \infty, \qquad (4.8.7)$$

where $\rho'_{j,m} := E(Y_j^{(k)} - Y_{j,m})(Y_0^{(k)} - Y_{0,m})$. As for (4.8.2),

$$\rho'_{j,m} = \sum_{0 \leq s_1 < \cdots < s_k} \Big(\prod_{i=1}^{k} a_{j+s_i} a_{s_i} - \prod_{i=1}^{k} a_{j+s_i} a_{s_i,m}$$

$$- \prod_{i=1}^{k} a_{j+s_i,m} a_{s_i} + \prod_{i=1}^{k} a_{j+s_i,m} a_{s_i,m} \Big).$$

Therefore, by (4.6.28) and the definition of $a_{j,m}$,

$$|\rho'_{j,m}| \leq 4 \sum_{s=0}^{\infty} |a_s a_{s+j}|^k, \qquad \sum_{j=0}^{\infty} |\rho'_{j,m}| \leq \sum_{j=0}^{\infty} \Big(4 \sum_{s=0}^{\infty} |a_s a_{s+j}|^k\Big) < \infty,$$

$$\rho'_{j,m} \to 0, \quad m \to \infty, \forall j \in \mathbb{Z}.$$

By the DCT, this in turn implies (4.8.7), and hence completes the proof.

Non-central limit theorem. The non-central limit theorem for the partial-sum process of polynomial forms $\{Y_j^{(k)}\}$ uses the scheme of discrete multiple integrals and the convergence criterion of Corollary 4.7.1. We impose conditions

$$a_j \sim c_a j^{-1+d_X}, \quad j \to \infty, \quad 0 < d_X < 1/2, \ c_a > 0, \qquad (4.8.8)$$

on a_j, which are the same as for the non-CLT Theorem 4.7.1 for functions $h(X_j)$ of a Gaussian process $\{X_j\}$ with moving-average weights a_j. Set $2d_Y = 1 - k(1 - 2d_X)$.

Theorem 4.8.2. *Let $\{Y_j^{(k)}\}$ satisfy (4.8.1), (4.8.8), and $k(1 - 2d_X) < 1$. Then*

$$\gamma_Y(j) \sim \frac{c_a^{2k}}{k!} B^k(d_X, 1 - 2d_X) j^{-1+2d_Y}, \quad j \to \infty, \qquad (4.8.9)$$

$$n^{-1/2-d_Y} S_{n,k}(\tau) \Rightarrow \frac{c_a^k}{k!} I_k(g_\tau), \qquad (4.8.10)$$

in $\mathcal{D}[0,1]$ with the uniform metric, where $I_k(g_\tau)$ and g_τ are as in (4.7.1) and (4.7.15) with $k_0 = k$.

Proof. To show (4.8.9), write the covariance (4.8.2) as

$$\gamma_Y(j) = \frac{1}{k!} \sum{}'' \prod_{i=1}^{k} a_{j+s_i} a_{s_i} = \frac{1}{k!} \int{}'' \prod_{i=1}^{k} a(j + [u_i]) a([u_i]) d^k u,$$

where $\sum''$ denotes the sum over all *mutually distinct* nonnegative integers $s_1, \cdots, s_k$, and $\int''$ the integral over $u_1, \cdots, u_k \geq 0$ such that $[u_l] \neq [u_p]$ $(l \neq p)$. Define the function

$$h_j(u) := a(j + [uj]) a([uj]) (j^2(1+u)u)^{1-d_X}, \quad u \geq 0.$$

Under assumption (4.8.8),

$$\sup_{u \geq 0} |h_j(u)| < \infty, \quad h_j(u) \to c_a^2, \ j \to \infty, \ \forall u > 0.$$

A change of variables and the DCT yield (4.8.9) as follows.

$$j^{k(1-2d_X)} \gamma_Y(j) = \frac{1 + o(1)}{k!} \int_{[0,\infty)^k} \prod_{i=1}^{k} \{h_j(u_i)(u_i(1+u_i))^{d_X-1}\} d^k u$$

$$\to \frac{c_a^{2k}}{k!} \left(\int_0^\infty u^{d_X-1}(1+u)^{d_X-1} du \right)^k$$

$$= \frac{c_a^{2k}}{k!} B^k(d_X, 1 - 2d_X), \quad j \to \infty.$$

To prove (4.8.10), we first show

$$n^{-1/2-d_Y} S_{n,k} \to_D \frac{c_a^k}{k!} I_k(g), \quad g = g_1. \tag{4.8.11}$$

The proof of (4.8.10) will then follow from this result by arguing as in the proof of Corollary 4.7.2.

Set $a_j = 0$ for $j < 0$. Using (4.8.1), we can write

$$n^{-1/2-d_Y} S_{n,k} = \frac{1}{k!} \sum{}' g_n(s_1, \cdots, s_k) \zeta_{s_1} \cdots \zeta_{s_k},$$

$$g_n(s_1, \cdots, s_k) := n^{-1/2-d_Y} \sum_{j=1}^n \prod_{i=1}^k a_{j-s_i},$$

where $\sum'$ denotes the sum over all *mutually distinct* integers $s_1, \cdots, s_k \in \mathbb{Z}$. The extension $\widetilde{g}_n$ of g_n, defined by (4.7.10), equals

$$\widetilde{g}_n(u_1, \cdots, u_k) = n^{k/2} g_n([nu_1], \cdots, [nu_k])$$

$$= n^{k(1-d_X)} \int_0^1 \prod_{i=1}^k a([nv] + 1 - [nu_i]) dv$$

and is the same as (4.7.20) when $k_0 = k$. Hence, by (4.7.21), $\|\widetilde{g}_n - c_a^k g\| \to 0$, which, in view of Corollary 4.7.1, yields the convergence (4.8.11).

Note: The convergence in distribution for off-diagonal polynomial forms to non-Gaussian limits was first considered by Surgailis (1982). Avram and Taqqu (1987) provided a multinomial formula that decomposes an Appell polynomial of a linear process into a sum of polynomial forms and showed that in the long memory case the off-diagonal polynomial is the dominating term. For more details see Surgailis (2003).

Chapter 5

Properties of the DFT and the Periodogram

5.1 Introduction

Large sample inference for a times series $\{X_j\}$ is usually developed either in the time domain or in the frequency domain. Time-domain analysis is carried out using the autocovariance and the sample autocovariance functions $\gamma_X(k)$ and $\hat{\gamma}_X(k)$, while frequency-domain analysis is based on the spectral density f_X of $\{X_j\}$, the discrete Fourier transform (DFT)

$$w_X(u) = (2\pi n)^{-1/2} \sum_{j=1}^{n} e^{-iuj} X_j, \qquad u \in \Pi,$$

and the periodogram

$$I_X(u) = |w_X(u)|^2 = (2\pi n)^{-1} \left| \sum_{j=1}^{n} e^{-iuj} X_j \right|^2, \qquad u \in \Pi,$$

of the sample $X_1, \cdots, X_n$. This chapter analyzes the asymptotic distributional properties of the DFTs $w_X(u_j)$ and the periodogram $I_X(u_j)$, where u_j are the Fourier frequencies given by $u_j := 2\pi j/n$, $j = 0, \cdots, [n/2]$. A consequence of Theorem 5.2.1 is that the standardized DFTs $w_X(u)/f_X^{1/2}(u)$ at a finite number of different Fourier frequencies behave as a set of asymptotically uncorrelated r.v.'s and imply that the periodogram is a mean consistent estimator of spectral density, i.e., $E I_X(u) \sim f_X(u)$. Although $I_X(u)$ is not a consistent estimate of $f_X(u)$, its smoothed version written as a weighted sum allows a fine asymptotic theory, as given in Chapter 6.

Some preliminary facts about DFTs are established in the next section while the asymptotic distribution of the standardized DFTs $w_X(u)/f_X^{1/2}(u)$ and periodograms $I_X(u)/f_X(u)$ is derived in Section 5.3.

111

5.2   Covariances of the DFT

In this section we establish some general properties of DFTs. Given n observations $X_1, \cdots, X_n$, the vector $w_{X,1}, \cdots, w_{X,n}$ of the DFTs is

$$w_{X,j} = (2\pi n)^{-1/2} \sum_{t=1}^{n} X_t e^{-itu_j}, \quad j = 1, \cdots, n.$$

The Fourier frequencies $u_j = 2\pi j / n$, $j = 1, \cdots, n$, are used to construct an orthonormal basis $e_1, \cdots, e_n$ in $\mathbb{C}^n$, and to give the *harmonic representation* of the vector $X = (X_1, \cdots, X_n) \in \mathbb{R}^n$, namely

$$X = a_1 e_1 + \cdots + a_n e_n,$$

where the "coordinates" $a_j := (2\pi)^{1/2} w_{X,j}$s are rescaled DFTs of X at frequency u_j. Note that $|a_j|^2 = 2\pi I_{X,j}$ is the rescaled periodogram. This representation and (2.1.5) readily yield the equality

$$2\pi \sum_{j=1}^{n} I_{X,j} = \frac{1}{n} \sum_{j=1}^{n} \left| \sum_{t=1}^{n} X_t e^{-itu_j} \right|^2 = \frac{1}{n} \sum_{t,s=1}^{n} X_t X_s \left(\sum_{j=1}^{n} e^{iju_{t-s}} \right)$$

$$= \sum_{j=1}^{n} X_j^2 = \|X\|^2.$$

If $(\zeta_1, \cdots, \zeta_n)$ is a $\mathrm{WN}(0,1)$ process, then the DFTs $w_{\zeta,1}, \cdots, w_{\zeta,n}$ are uncorrelated r.v.'s, i.e.,

$$\begin{aligned} E[w_{\zeta,j}\overline{w_{\zeta,k}}] &= \frac{1}{2\pi}, \quad 1 \le k = j \le n, \\ &= 0, \quad 1 \le k < j \le n. \end{aligned} \tag{5.2.1}$$

To prove this, by (2.1.5),

$$E[w_{\zeta,j}\overline{w_{\zeta,k}}] = (2\pi n)^{-1} \sum_{t,s=1}^{n} e^{-i(tu_j - su_k)} E[\zeta_t \zeta_s]$$

$$= (2\pi n)^{-1} \sum_{t=1}^{n} e^{-it(u_j - u_k)} = 0, \quad 1 \le k < j \le n;$$

$$= (2\pi)^{-1}, \quad 1 \le k = j \le n.$$

In general, unlike in the white-noise case, the covariances $E[w_{X,j}\overline{w_{X,k}}] \neq 0$, for $k \neq j$, i.e., the DFTs of a stationary process $\{X_j\}$ with a spectral density f_X, are correlated. However, they become uniformly small when f_X is smooth. Theorem 5.2.1 below shows that if f_X is continuous, their magnitude tends uniformly to 0 over all $j \neq k$, at a rate that depends on the degree of smoothness of f_X. For example, if f_X is a Lipschitz function of order $0 < \beta \leq 1$, then this rate is $n^{-\beta}$ if $\beta < 1$ and $n^{-1} \log n$ if $\beta = 1$. In addition, they also become small when the distance between frequencies $j - k$ increases to infinity.

If the spectral density f_X is discontinuous at 0, as happens to be true for long memory processes, then these covariances are not uniformly small as n increases, but they continue to tend to zero when the frequencies u_j and u_k become more distant from 0, i.e., with the increase in j and k.

Parts (i) and (ii) of Theorem 5.2.1 consider the case where f_X is continuous and bounded, whereas part (iv) covers the case where $f_X(u) = |u|^{-2d} g(u)$, $|d| < 1/2$, where g is a continuous function on Π. The case where g has also a bounded derivative is covered in part (iii). Obtaining the upper bounds in (i) to (iv) of Theorem 5.2.1 requires the innovations ζ_j's in the linear process $\{X_j\}$ to be white-noise only.

For convenience, this theorem is formulated for a cross-spectral density of two stationary linear processes X and Y with the same underlying white-noise innovation process. This allows us to express the cross-spectral density f_{XY} via transfer functions A_X and A_Y as indicated in (5.2.3). In general, the results of Theorem 5.2.1 are valid for any spectral density or cross-spectral density that satisfies the smoothness condition of this theorem and does not require $\{X_j\}$ to be a linear process with i.i.d. innovations.

Assumptions. Consider the two linear processes

$$X_j = \sum_{k=0}^{\infty} a_k \zeta_{j-k}, \qquad Y_j = \sum_{k=0}^{\infty} b_k \zeta_{j-k}, \qquad j \in \mathbb{Z}, \qquad (5.2.2)$$

with $\sum_{k=0}^{\infty} a_k^2 < \infty$, $\sum_{k=0}^{\infty} b_k^2 < \infty$, and the same white-noise innovations $\{\zeta_j\} \sim \mathrm{WN}(0, \sigma_\zeta^2)$. Let

$$A_X(v) := \sum_{k=0}^{\infty} e^{-ikv} a_k, \quad A_Y(v) := \sum_{k=0}^{\infty} e^{-ikv} b_k, \quad v \in \Pi,$$

denote their respective transfer functions and

$$f_X(v) = \frac{\sigma_\zeta^2}{2\pi} |A_X(v)|^2, \quad f_Y(v) = \frac{\sigma_\zeta^2}{2\pi} |A_Y(v)|^2, \quad v \in \Pi,$$

their respective spectral densities.

Denote by $f_{XY}(v)$ a (complex-valued) cross-spectral density, i.e.,

$$f_{XY}(v) := \frac{\sigma_\zeta^2}{2\pi} A_X(v)\overline{A_Y(v)}, \quad v \in \Pi, \qquad (5.2.3)$$

$$E[X_j Y_{j-k}] = \int_\Pi e^{ikv} f_{XY}(v)dv$$

$$= \frac{\sigma_\zeta^2}{2\pi} \sum_{l=0}^\infty a_{l+k}b_l, \quad k \geq 0, \ j \in \mathbb{Z}.$$

Observe that in the case of white-noise $Y_j = \zeta_j$, $j \in \mathbb{Z}$,

$$f_{X\zeta}(v) := \frac{\sigma_\zeta^2}{2\pi} A_X(v), \quad v \in \Pi, \qquad (5.2.4)$$

$$E[X_j Y_{j-k}] = \frac{\sigma_\zeta^2}{2\pi} \int_\Pi e^{ikv} A_X(v)dv = \sigma_\zeta^2 a_k, \quad k \geq 0.$$

To discuss the asymptotic properties of the cross-covariances $E[w_{X,j}\overline{w_{Y,k}}]$, we need to distinguish between the two cases. Case 1 analyzes the covariances at Fourier frequencies from an interval $(-\Delta, \Delta)$, $\Delta < \pi$, e.g., a neighborhood of 0, while case 2 deals with the entire spectrum over Π.

In the first case, the smoothness conditions on f_X, f_Y (A_X, A_Y) are local in the sense that they need to be imposed on an interval $[0, a]$, $a > \Delta$, covering $[0, \Delta]$. In the second case, the smoothness conditions have to be imposed on the whole spectrum Π.

Let $\mathcal{C}[0, a]$ denote the class of complex-valued functions that are continuous on $[0, a]$, and $\Lambda_\beta[0, a]$, $0 < \beta \leq 1$ denote the class of Lipschitz continuous functions with parameter β. We write $h \in \mathcal{C}_{1,\alpha}[0, a]$, $|\alpha| < 1$, if

$$|h(u)| \leq C|u|^{-\alpha}, \quad |\dot{h}(u)| \leq C|u|^{-1-\alpha}, \quad \forall u \in [0, a].$$

The class $\mathcal{C}_{1,\alpha}[0, a]$ allows for an infinite peak and non-differentiability at 0, whereas $\Lambda_\beta[0, a]$ covers continuous piecewise differentiable functions, for example, $h(u) := |u|$ belongs to $\Lambda_1[0, \pi]$.

To measure the smoothness of $h \in \mathcal{C}[0, a]$ we use the continuity modulus

$$\omega_h(\eta) := \sup_{u,v \in [0,a]:|u-v|\leq\eta} |h(u) - h(v)|, \quad \eta \geq 0.$$

For a continuous function h on $[0, a]$, $\omega_h(\eta) \to 0$, $\eta \to 0$, because h is uniformly continuous on $[0, a]$. Moreover, for any $0 < \epsilon < 1$,

$$|h(u) - h(v)| \qquad (5.2.5)$$

$$\leq |h(u) - h(v)|\{I(|u - v| \leq \eta) + I(|u - v| > \eta)\}$$

$$\leq \omega_h(\eta) + C \left(\frac{|u - v|}{\eta}\right)^\epsilon.$$

Define

$$\delta_{n,\epsilon}(h) := \omega_h(n^{-1}\log n) + (\log n)^{-\epsilon}, \quad 0 < \epsilon < 1. \qquad (5.2.6)$$

For the sake of brevity we also need to introduce the functions

$$\ell_n(\epsilon; k) := \frac{\log(2+k)}{(2+k)^{1-\epsilon}} + \frac{\log(2+n-k)}{(2+n-k)^{1-\epsilon}}, \qquad 0 \le k \le j \le n,$$

$$r_{n,jk}(g) := 0, \qquad\qquad g \in \Lambda_1[0, a], \qquad \beta = 1,$$
$$:= n^{-\beta} \ell_n(\beta; j - k), \qquad g \in \Lambda_\beta[0, a], \qquad 0 < \beta < 1,$$
$$:= \delta_{n,\epsilon}(g) \ell_n(\epsilon; j - k), \quad g \in \mathcal{C}[0, a], \qquad \epsilon \in (0, 1).$$

As will be seen below, they appear in several bounds. We are now ready to state

Theorem 5.2.1. *Let either* $\Delta < a < \pi$ *or* $\Delta = a = \pi$. *Then* (i) *to* (v) *hold for all* $0 < |u_k| \le u_j < \Delta$.

(i)   *If* $f_{XY} \in \Lambda_\beta[0, a]$, $0 < \beta \le 1$, *then*

$$\left| E[w_{X,j}\overline{w_{Y,j}}] - f_{XY}(u_j) \right| \le Cn^{-1}\log n, \qquad \beta = 1,$$
$$\le Cn^{-\beta}, \qquad 0 < \beta < 1,$$
$$\left| E[w_{X,j}\overline{w_{Y,k}}] \right| \le Cn^{-1}\log n, \qquad \beta = 1,$$
$$\le Cn^{-\beta}\ell_n(\beta; j-k), \quad 0 < \beta < 1, \; k < j.$$

(ii) *If* $f_{XY} \in \mathcal{C}[0, a]$, *then,* $\forall \epsilon \in (0, 1)$,

$$\left| E[w_{X,j}\overline{w_{Y,j}}] - f_{XY}(u_j) \right| \le C\delta_{n,\epsilon}(f_{XY}),$$
$$\left| E[w_{X,j}\overline{w_{Y,k}}] \right| \le C\delta_{n,\epsilon}(f_{XY})\,\ell_n(\epsilon; j-k), \qquad k < j.$$

(iii) *If* $f_{XY} \in \mathcal{C}_{1,\alpha}[0, a]$, $|\alpha| < 1$, *then*

$$\left| E[w_{X,j}\overline{w_{Y,j}}] - f_{XY}(u_j) \right| \le Cu_j^{-\alpha}j^{-1}\log j,$$
$$\left| E[w_{X,j}\overline{w_{Y,k}}] \right| \le C\left(|u_k|^{-\alpha} + u_j^{-\alpha}\right)j^{-1}\log j, \qquad k < j.$$

(iv) *Suppose* $f_{XY} = hg$, *where* $h \in \mathcal{C}_{1,\alpha}[0, a]$, $|\alpha| < 1$, *and* g *is bounded on* Π. *If, in addition,* $g \in \Lambda_\beta[0, a] \cup \mathcal{C}[0, a]$, $0 < \beta \le 1$, *then*

$$\left| E[w_{X,j}\overline{w_{Y,k}}] - f_{XY}(u_j)I(j = k) \right|$$
$$\le C\left((|u_k|^{-\alpha} + u_j^{-\alpha})j^{-1}\log j + (|u_k|^{-\alpha} \wedge u_j^{-\alpha})r_{n,jk}(g) \right).$$

(v)   *If* $|f_{XY}(u)| \le C|u|^{-\alpha}$, $|\alpha| < 1$, *and* $u \in (0, a]$, *then*

$$\left| E[w_{X,j}\overline{w_{Y,j}}] \right| \le Cu_j^{-\alpha}, \qquad 0 < u_j \le a.$$

The constant C *in* (i) *to* (v) *does not depend on* k, j, *and* n.

Technical lemmas. The proof of this theorem is facilitated by the following two auxiliary lemmas. Recall the definition of the Dirichlet kernel $D_n(u)$ from (2.1.6) and let, with $0 \leq \theta < 1$ and $1 \leq |k| \leq j < n/2$,

$$I_\theta(x) := \int_{\mathbb{R}} (1 + |u|)^{-1}(1 + |u - x|)^{-1+\theta} du, \quad x \in \mathbb{R}, \tag{5.2.7}$$

$$d_{n,jk}(\theta) := n^{-1} \int_\Pi |D_n(u - u_k)||D_n(u - u_j)|^{1-\theta} du,$$

$$R_{n,jk} := n^{-1} \int_\Pi |f(u) - f(u_j)| \, |D_n(u - u_k)||D_n(u - u_j)| du.$$

The next lemma provides upper bounds for the first two entities.

Lemma 5.2.1. *Let $0 \leq \theta < 1$. Then the following holds:*
(i) *For all $x \in \mathbb{R}$,*

$$I_\theta(x) \leq C(2 + |x|)^{-1+\theta}\log(2 + |x|). \tag{5.2.8}$$

(ii) *For $1 \leq |k| \leq j < n/2$, $n \geq 1$,*

$$d_{n,jk}(\theta) \leq Cn^{-\theta}\left(\frac{\log(2 + j - k)}{(2 + j - k)^{1-\theta}} + \frac{\log(2 + n - j + k)}{(2 + n - j + k)^{1-\theta}}\right) \tag{5.2.9}$$

$$\equiv Cn^{-\theta}\ell_n(\theta, j - k), \qquad 0 \leq \theta < 1,$$

$$\leq Cn^{-1}\log n, \qquad \theta = 1.$$

Proof. Let $0 \leq \theta < 1$. Note that $I_{n,\theta}(x) \leq I_{n,\theta}(|x|)$ for all $x \in \mathbb{R}$. Therefore, it suffices to prove (5.2.8) for $x \geq 0$. Let $[\cdots]$ denote the integrand in $I_{n,\theta}$. Then

$$\int_{|u| \leq x/2} [\cdots] \leq C \int_{|u| \leq x/2} (1 + |u|)^{-1}(1 + x)^{-1+\theta} du$$

$$\leq C(2 + x)^{-1+\theta}\log(2 + x),$$

$$\int_{x/2 \leq |u| \leq 2x} [\cdots] \leq C \int_{x/2 \leq |u| \leq 2x} (1 + x)^{-1}(1 + |u - x|)^{-1+\theta} du$$

$$\leq C(2 + x)^{-1+\theta}\log(2 + x),$$

$$\int_{|u| > 2x} [\cdots] \leq C \int_{|u| > 2x} (1 + |u|)^{-2+\theta} du \leq C(2 + x)^{-1+\theta},$$

which proves (5.2.8).

To prove (5.2.9), recall first that by (2.1.8), $|D_n(u)| \leq 2\pi n(1 + n|u|)^{-1}$, $|u| \leq \pi$. Notice that for any $\pi \leq a < 2\pi$, there exists a sufficiently large $C = C(a)$ such that the last inequality extends to

$$|D_n(u)| \leq Cn(1 + n|u|)^{-1}, \quad |u| \leq a. \tag{5.2.10}$$

Recall that $D_n(u)$ is 2π-periodic function. Hence,

$$d_{n,jk}(\theta) = n^{-1}\int_\Pi |D_n(u)||D_n(u-u_{j-k})|^{1-\theta}du$$

$$\leq Cn^{1-\theta}\int_\Pi(1+n|u|)^{-1}(1+n|(u-u_{j-k})(\mathrm{mod}\,2\pi)|)^{-1+\theta}du.$$

Because $u-u_{j-k}\in[-3\pi,\pi]$, for all $u\in\Pi$, $|k|\leq j$, and $2\pi=u_n$,

$$(u-u_{j-k})(\mathrm{mod}\,2\pi)=u-u_{j-k}, \qquad u-u_{j-k}\in\Pi,$$

$$=u-u_{j-k-n}, \qquad u-u_{j-k}+2\pi\in\Pi.$$

After a change of variables $un\to v$, we arrive at

$$d_{n,jk}(\theta)\leq Cn^{-\theta}\{I_\theta(nu_{j-k})+I_\theta(nu_{j-k-n})\},$$

which together with (5.2.8) and $nu_{j-k}=2\pi(j-k)$ yields (5.2.9).

For $\theta=1$, by (5.2.10), $d_{n,jk}(1)\leq C\int_\Pi(1+n|u|)^{-1}du\leq Cn^{-1}\log n$, which completes the proof of (5.2.9).

The next lemma provides upper bounds for $R_{n,jk}$.

Lemma 5.2.2. *Let either* $0<\Delta<a\leq\pi$ *or* $\Delta=a=\pi$. *For all* $1\leq|k|\leq j$, *such that* $u_j\leq\Delta$, *the following hold:*
(i) *If* $f_{XY}\in\Lambda_\beta[0,a]$, *then*

$$R_{n,jk}\leq Cn^{-1}\log n, \qquad\qquad \beta=1,$$

$$\leq Cn^{-\beta}\ell_n(\beta;j-k), \quad 0<\beta<1.$$

(ii) *If* $f_{XY}\in\mathcal{C}[0,a]$, *then for any* $\epsilon\in(0,1)$, *with* $\delta_{n,\epsilon}(f_{XY})$ *as in (5.2.6)*,

$$R_{n,jk}\leq C\delta_{n,\epsilon}(f_{XY})\ell_n(\epsilon;j-k).$$

(iii) *If* $f_{XY}\in\mathcal{C}_{1,\alpha}[0,a]$, *then*

$$R_{n,jk}\leq C(|u_k|^{-\alpha}+u_j^{-\alpha})j^{-1}\log j. \tag{5.2.11}$$

In (i) *to* (iii), $C>0$ *does not depend on* k,j, *and* n.

Proof. Write f for f_{XY} in this proof and

$$R_{n,jk}=\int_{a\leq|u|\leq\pi}[\cdots]+\int_{|u|\leq a}[\cdots]=:t_{1,jk}+t_{2,jk}.$$

First, we shall evaluate $t_{1,jk}$. If $a=\pi$ then $t_{1,jk}=0$. If $a<\pi$ then by an assumption of the lemma, $|u_k|\leq u_j\leq\Delta<a$, and thus $0<a-\Delta\leq$

$|u - u_i| \leq a + \Delta < 2\pi$, for $i = j, k$, and for $a \leq |u| \leq \pi$. Therefore, by (5.2.10), $|D_n(u - u_k)D_n(u - u_j)| \leq C$, for some $C > 0$. Hence,

$$t_{1,jk} \leq Cn^{-1} \int_{a \leq |u| \leq \pi} (|f(u)| + |f(u_j)|) du \leq C(1 + |f(u_j)|)n^{-1},$$

which implies that $t_{1,jk}$ satisfies the above bounds (i) to (iii) of the lemma.

It remains to show that the bounds (i) to (iii) are valid for $t_{2,jk}$. Observe that $u - u_j \in [-2\pi, \pi]$ for $u \in \Pi$. By (5.2.10),

$$|D_n(u - u_j)| = |D_n(u - u_j + 2\pi)| \tag{5.2.12}$$
$$\leq 2\pi\, n(1 + n|u - u_j|)^{-1}, \qquad u - u_j \in [-\pi, \pi],$$
$$\leq 2\pi\, n(1 + n|u - u_j + 2\pi|)^{-1}, \quad u - u_j \in [-2\pi, -\pi].$$

Furthermore, $|f(u) - f(u_j)| = |f(u + 2\pi) - f(u_j)|$, because of the 2π periodicity of f.

(i) Suppose that $f \in \Lambda_\beta(\Pi)$. Then

$$|f(u) - f(u_j)| \leq C|u - u_j|^\beta, \qquad u - u_j \in [-\pi, \pi],$$
$$|f(u) - f(u_j)| = |f(u + 2\pi) - f(u_j)|$$
$$\leq C|u + 2\pi - u_j|^\beta, \qquad u - u_j \in [-2\pi, -\pi],$$

which, upon combining with (5.2.12), yields

$$|f(u) - f(u_j)||D_n(u - u_j)|^\beta \leq C, \quad |u| \leq a, \quad u_j \leq \Delta.$$

Thus, $t_{2,jk} \leq Cd_{n,jk}(\beta)$, which, together with (5.2.9), proves (i) for $t_{2,jk}$.

(ii) Let $f \in \mathcal{C}[0, a]$. Use (5.2.5) with $\eta = n^{-1} \log n$ and $\delta_{n,\epsilon}(h)$ of (5.2.6), to obtain

$$|f(u) - f(u_j)| \leq \omega_f(n^{-1} \log n) + (\log n)^{-\epsilon}(n|u - u_j|)^\epsilon$$
$$\leq C\delta_{n,\epsilon}(f)(1 + n|u - u_j|)^\epsilon.$$

Together with (5.2.12), this yields $|f(u) - f(u_j)||D_n(u - u_j)|^\epsilon \leq C\delta_{n,\epsilon}(f)$ for $|u| \leq a$ and $u_j \leq \Delta$. Hence, $t_{2,jk} \leq C\delta_{n,\epsilon}(f)d_{n,jk}(\epsilon)$, which, together with (5.2.9), proves (ii).

(iii) Let $f \in \mathcal{C}_{1,\alpha}[0, a]$. Case (1): $a < \pi$. Then $|u - u_j|, |u - u_k| \leq a + \Delta < 2\pi$, and by (5.2.10),

$$t_{2,jk} \leq C \int_{|u| \leq a} \frac{n|f(u) - f(u_j)|}{(1 + n|u - u_k|)(1 + n|u - u_j|)} du.$$

Write

$$t_{2,jk} = \int_{3u_j/2 \le |u| \le a} [\cdots] + \int_{|u| \le |u_k|/2} [\cdots]$$
$$+ \int_{-3u_j/2}^{-|u_k|/2} [\cdots] + \int_{|u_k|/2}^{u_j/2} [\cdots] + \int_{u_j/2}^{3u_j/2} [\cdots].$$

It suffices to show that each integral satisfies the bound (iii) of the lemma. The definition of $\mathcal{C}_{1,\alpha}[0,a]$ implies that $|f(u)| \le C|u|^{-\alpha}$ and $|\dot{f}(u)| \le C|u|^{-1-\alpha}$ for $|u| \le a$. Observe that

$$\int_{|u| \le 3u_j} n(1 + n|u|)^{-1} du \le C \log j.$$

Hence,

$$\int_{3u_j/2 \le |u| \le a} [\cdots] \le C \int_{3u_j/2}^{a} \frac{u^{-\alpha} + u_j^{-\alpha}}{nu^2} du$$
$$\le C(nu_j)^{-1} u_j^{-\alpha} \le Cj^{-1} u_j^{-\alpha};$$

$$\int_{|u| \le |u_k|/2} [\cdots] \le C \int_{0}^{|u_k|/2} \frac{|u|^{-\alpha} + u_j^{-\alpha}}{n|u_k|u_j} du$$
$$\le C \frac{|u_k|^{-\alpha} + u_j^{-\alpha}}{nu_j} \le Cj^{-1}(|u_k|^{-\alpha} + u_j^{-\alpha});$$

$$\int_{-3u_j/2}^{-|u_k|/2} [\cdots] \le C \frac{|u_k|^{-\alpha} + u_j^{-\alpha}}{nu_j} \int_{-3u_j/2}^{-|u_k|/2} \frac{n}{1 + n|u - u_k|} du$$
$$\le C(|u_k|^{-\alpha} + u_j^{-\alpha})j^{-1} \log j;$$

$$\int_{|u_k|/2}^{u_j/2} [\cdots] \le C \frac{|u_k|^{-\alpha} + u_j^{-\alpha}}{nu_j} \int_{|u_k|/2}^{u_j/2} \frac{n}{1 + n|u - u_k|} du$$
$$\le C(|u_k|^{-\alpha} + u_j^{-\alpha})j^{-1} \log j.$$

To bound the last integral, by the mean-value theorem, $|f(u) - f(u_j)| \le Cu_j^{-1-\alpha}|u - u_j|$, for $u_j/2 \le u \le 3u_j/2$. Thus,

$$\int_{u_j/2}^{3u_j/2} [\cdots] \le Cu_j^{-1-\alpha} \int_{u_j/2}^{3u_j/2} \frac{1}{1 + n|u - u_k|} du$$
$$\le Cu_j^{-\alpha}j^{-1} \log j,$$

which completes the proof of (5.2.11) for $t_{2,jk}$ and hence for $R_{n,jk}$.

Case (2): $a = \pi$. Suppose that $u_j \in (0, \pi/2]$. Then $|u - u_j|, |u - u_k| \leq 3\pi/2$ for $u \in \Pi$. Thus, by (5.2.10),

$$|D(u - u_j)D(u - u_k)| \leq Cn(1 + n|u - u_k|)^{-1}(1 + n|u - u_j|)^{-1},$$

which implies

$$R_{n,jk} \leq R'_{n,jk} := \int_\Pi \frac{n|f(u) - f(u_j)|}{(1 + n|u - u_k|)(1 + n|u - u_j|)} du.$$

Here the bound (iii) for $R'_{n,jk}$ follows from the bounds derived in case (1) above, setting $a = \pi$.

Suppose that $u_j \in (\pi/2, \pi]$. Write

$$R_{n,jk} = \int_{|u| \leq \pi/2} [\cdots] + \int_{\pi/2 \leq |u| \leq \pi} [\cdots].$$

In the first integral $|u - u_j|, |u - u_k| \leq 3\pi/2$, for $|u| \leq \pi/2$. Therefore, by (5.2.10), $\int_{|u| \leq \pi/2} [\cdots] \leq CR'_{n,jk}$ is upper bounded by (iii).

Next, since $f \in \mathcal{C}_{1,\alpha}(\Pi)$, for $|u| \in [\pi/2, \pi]$, the bound

$$|f(u) - f(u_j)| = |f(|u|) - f(|u_j|)|$$

$$\leq \sup_{\pi/2 \leq x \leq \pi} |\dot{f}(x)| |u - u_j| \leq C|u - u_j|,$$

which together with (5.2.12), implies

$$|f(u) - f(u_j)||D_n(u - u_j)| \leq C, \quad |u| \in [\pi/2, \pi].$$

Hence, by (5.2.9), $\int_{\pi/2 \leq |u| \leq \pi} [\cdots] \leq Cd_{n,jk}(1) \leq Cn^{-1}\log n$ satisfies (iii), which completes the proof of the lemma.

Proof of Theorem 5.2.1. As in the above proof, write f for f_{XY}. Let

$$E_{j,k}(u) := (2\pi n)^{-1} \sum_{t=1}^n e^{-it(u_j - u)} \sum_{s=1}^n e^{is(u_k - u)}.$$

Observe that

$$E_{j,j}(u) = (2\pi n)^{-1}|\sum_{t=1}^n e^{-it(u_j - u)}|^2 = (2\pi n)^{-1}D_n^2(u_j - u),$$

$$|E_{j,k}(u)| = (2\pi n)^{-1}|D_n(u_j - u)D_n(u_k - u)|.$$

Then, for $1 \leq |k| \leq j \leq n/2$, we can write

$$Ew_{X,j}\overline{w_{Y,k}} = (2\pi n)^{-1} \sum_{t,s=1}^n e^{-i(u_j t - u_k s)} E[X_t Y_s]$$

$$= \int_\Pi f(u)E_{j,k}(u)du,$$

$$Ew_{X,j}\overline{w_{Y,j}} = \int_\Pi f(u)E_{j,j}(u)du.$$

Observe that (2.1.5) implies the orthogonality property, i.e., for $1 \leq j, k < n/2$,

$$\int_{\Pi} E_{j,k}(u)du = \sum_{t=1}^{n} e^{-it(u_j - u_k)} = 0, \quad j \neq k, \qquad (5.2.13)$$

$$\int_{\Pi} E_{j,-k}(u)du = \sum_{t=1}^{n} e^{-it(u_j + u_k)} = 0. \qquad \times\times$$

Therefore, for $1 \leq |k| < j < n/2$,

$$E[w_{X,j}\overline{w_{Y,j}}] - f(u_j) = \int_{\Pi} (f(u) - f(u_j))E_{j,j}(u)du,$$

$$E[w_{X,j}\overline{w_{Y,k}}] = \int_{\Pi} (f(u) - f(u_j))E_{j,k}(u)du.$$

Then

$$\left| E[w_{X,j}\overline{w_{Y,k}}] - f(u_j)I(j = k) \right| \qquad (5.2.14)$$

$$\leq \int_{\Pi} |f(u) - f(u_j)||E_{j,k}(u)|du = R_{n,jk},$$

where $R_{n,jk}$ is as in (5.2.7).

Therefore, to prove the theorem, it suffices to show that $R_{n,jk}(f) := R_{n,jk}$, $1 \leq |k| \leq j$, for all j such that $u_j \leq \Delta$, satisfies the corresponding bounds of (i) to (iv) of the theorem.

Proof of (i) to (iii). The corresponding bounds of $R_{n,jk}$ are derived in Lemma 5.2.2.

Proof of (iv). Because now $f = hg$, where g is bounded from above, we obtain

$$|f(u) - f(u_j)| = |h(u)g(u) - h(u_j)g(u_j)|$$

$$\leq C\left(|h(u) - h(u_j)| + u_j^{-\alpha}|g(u) - g(u_j)|\right).$$

Hence,

$$R_{n,jk}(f) = \int_{\Pi} |f(u) - f(u_j)||E_{j,k}(u)|du$$

$$\leq C\left(\int_{\Pi} |h(u) - h(u_j)||E_{j,k}(u)|du \right.$$

$$\left. + u_j^{-\alpha}\int_{\Pi} |g(u) - g(u_j)||E_{j,k}(u)|du\right)$$

$$= C(R_{n,jk}(h) + u_j^{-\alpha}R_{n,jk}(g)).$$

Because $h \in C_{1,\alpha}[0, a]$, $R_{n,jk}(h)$ satisfies the bound of (iii) of Lemma 5.2.2,

$$R_{n,jk}(h) \leq C(|u_k|^{-\alpha} + u_j^{-\alpha})j^{-1} \log j. \tag{5.2.15}$$

It remains to evaluate $u_j^{-\alpha} R_{n,jk}(g)$. Similarly, if $g \in \Lambda_\beta[0, a]$ then $R_{n,jk}(g)$ satisfies the bounds in (i) of Lemma 5.2.2. In particular, if $\beta = 1$ then

$$u_j^{-\alpha} R_{n,jk}(g) \leq C u_j^{-\alpha} n^{-1} \log n \leq u_j^{-\alpha} j^{-1} \log j,$$

which along with (5.2.15) yields the upper bound of (iv) of the theorem for $R_{n,jk}(f)$.

If $0 < \beta < 1$, then by (i),

$$u_j^{-\alpha} R_{n,jk}(g) \leq C u_j^{-\alpha} n^{-\beta} \ell_n(\beta, j - k). \tag{5.2.16}$$

Observe that if $j/2 \leq |k| \leq j$ then $u_j^{-\alpha} \leq C|u_k|^{-\alpha} \wedge u_j^{-\alpha}$, and (5.2.15) and (5.2.16) imply (iv) for $R_{n,jk}(f)$.

For $1 \leq |k| \leq j/2$, $j \leq n/2$, use $n - j + k \geq j/2$, to bound the r.h.s. of (5.2.16) by

$$C u_j^{-\alpha} n^{-\beta} \frac{\log(2 + j)}{(2 + j)^{1-\beta}} \leq C u_j^{-\alpha} j^{-1} \log j,$$

which yields (iv) for $R_{n,jk}(f)$.

Proof of (v). Let

$$\tilde{f}(u) := C|u|^{-\alpha}, \quad |u| \leq a,$$
$$:= |f(u)|, \quad a < |u| \leq \pi,$$

where C is chosen such that $|f(u)| \leq \tilde{f}(u)$, $0 < u \leq a$. Note that $\tilde{f} \in C_{1,\alpha}[0, a]$. Then

$$|E[w_{X,j}\overline{w_{Y,j}}]| \leq \int_\Pi |f(u)|E_{j,j}(u)du \leq \int_\Pi \tilde{f}(u)E_{j,j}(u)du$$
$$\leq \int_\Pi |\tilde{f}(u) - \tilde{f}(u_j)|E_{j,j}(u)du + \tilde{f}(u_j)\int_\Pi E_{j,j}(u)du$$
$$\leq C u_j^{-\alpha}, \quad 0 < u_j \leq a,$$

which follows from (iii), and the fact that $\int_\Pi E_{j,j}(u)du = 1$, by (2.1.9). This completes the proof of the theorem.

Note: For long and negative memory processes, a variant of the result (iii) of Theorem 5.2.1 was derived by Robinson (1995a). In the statement of his Theorem 2, the bound $j^{-1} \log j$ was replaced by the weaker bound $k^{-1} \log j$, $k < j$.

5.3   Some other properties of the DFT and the periodogram

In this section we establish some asymptotic distributional properties of the DFT and the periodogram of a linear process $\{X_j\}$ that allows for short, long, and negative memory. We show that, in general, under weak assumptions the standardized DFTs

$$v_{X,j} := \frac{w_X(u_j)}{f_X^{1/2}(u_j)}, \quad 1 \le j \le n/2,$$

at different frequencies u_j's asymptotically behave as a sequence of independent complex Gaussian variables. The restriction $1 \le j \le n/2$ is equivalent to $0 < u_j \le \pi$.

To begin with, we first establish some properties of the covariances of the standardized DFTs.

Proposition 5.3.1. *Suppose that the linear process $\{X_j\}$ is as in (5.2.2) and has the spectral density*

$$f_X(u) = |u|^{-2d} g(u), \quad u \in \Pi, \quad |d| < 1/2, \tag{5.3.1}$$

where $g(u)$, $u \in \Pi$, is a continuous function bounded away from 0. Then the following hold:
(i) *If $d = 0$ then*

$$\max_{0 < u_j, u_k < \pi} \left| \mathrm{Cov}\left(v_{X,j}, \overline{v_{X,k}}\right) - I(j = k) \right| \to 0, \tag{5.3.2}$$

$$\max_{0 < u_j, u_k \le \pi} \left| \mathrm{Cov}\left(v_{X,j}, v_{X,k}\right) \right| \to 0.$$

(ii) *If $d \ne 0$ then for any $K = K_n < n/2$ and $K_n \to \infty$,*

$$\max_{u_K \le u_j, u_k < \pi} \left| \mathrm{Cov}\left(v_{X,j}, \overline{v_{X,k}}\right) - I(j = k) \right| \to 0, \tag{5.3.3}$$

$$\max_{u_K \le u_j, u_k < \pi} \left| \mathrm{Cov}\left(v_{X,j}, v_{X,k}\right) \right| \to 0.$$

Proof. When $d = 0$, (5.3.2) follows from Theorem 5.2.1(ii). If $d \ne 0$ then (5.3.3) follows from Theorem 5.2.1(iv), noting that the r.h.s. of (iv), divided by $f^{1/2}(u_j) f^{1/2}(u_k)$, for $1 \le k \le j \le n/2$, can be bounded above as

$$\frac{(u_k^{-2d} + u_j^{-2d}) j^{-1} \log(j) + (u_k^{-2d} \wedge u_j^{-2d}) r_{n,jk}(g)}{f^{1/2}(u_j) f^{1/2}(u_k)}$$

$$\le C\{(\tfrac{j}{k})^{|d|} j^{-1} \log(j) + r_{n,jk}(g)\} \le C(j^{-1+|d|} \log(j) + o(1)),$$

uniformly over j, k, which implies (5.3.3), and completes the proof of the proposition.

The global restriction $g(u) \geq \delta$ used in (5.3.1) can be replaced by the local restriction $g(u_k) \geq \delta$, $g(u_j) \geq \delta$, for some $\delta > 0$, when obtaining a bound for $|\text{Cov}(v_{X,k}, \overline{v_{X,j}}) - I(j = k)|$.

The next theorem derives the asymptotic distribution of a vector of standardized DFTs $v_{X,j}$ and periodograms $I_X(u_j)/f_X(u_j)$ computed at $p \geq 1$ different frequencies $0 < u_{j_1} < \cdots < u_{j_p} \leq \pi$, where $j_i \equiv j_i(n)$ may vary with n, showing that asymptotically they behave as a vector of independent complex normal and exponential r.v.'s with mean 1, respectively. Recall that if $Y \sim \chi_2^2$ has a chi-square distribution with two degrees of freedom then $Y/2$ has an exponential distribution with parameter 1.

Theorem 5.3.1. *Let $\{X_j\}$ be a linear process with weights a_j and innovations $\{\zeta_j\}$ as in (5.2.2), and with the spectral density f_X as in (5.3.1). Assume that the ζ_j's are i.i.d. r.v's. In addition, if $d \in (-1/2, 0)$, assume $\sum_{k=0}^{\infty} |a_k| < \infty$.*

Then, for any $K \leq j_1 < \cdots < j_p < n/2$,

$$\left(v_{X,j_1} \cdots, v_{X,j_p}\right) \to_D \frac{1}{\sqrt{2}}(Z_1 + iZ_2, \cdots, Z_{2p-1} + iZ_{2p}), \qquad (5.3.4)$$

$$\left(\frac{I_X(u_{j_1})}{f_X(u_{j_1})}, \cdots, \frac{I_X(u_{j_p})}{f_X(u_{j_p})}\right) \to_D \frac{1}{2}(Z_1^2 + Z_2^2, \cdots, Z_{2p-1}^2 + Z_{2p}^2),$$

where $Z_1, \cdots, Z_{2p}$ is a vector of independent standardized normal r.v.'s, and $K = 1$ if $d = 0$ and $K = K_n < n/2$, $K_n \to \infty$, if $d \neq 0$.

Proof. Write $w_X(u) = (2\pi n)^{-1/2} \sum_{k=1}^{n} e^{-iku} X_k = \alpha(u) + i\beta(u)$, where

$$\alpha(u) = \frac{1}{\sqrt{2\pi n}} \sum_{k=1}^{n} \cos(uk) X_k, \quad \beta(u) = -\frac{1}{\sqrt{2\pi n}} \sum_{k=1}^{n} \sin(uk) X_k,$$

denote the ordinates (real and imaginary parts) of the DFT $w_X(u)$. Let $S_{n,2k-1} := \alpha(u_{j_k})$ and $S_{n,2k} := \beta(u_{j_k})$, $k = 1, \cdots, p$. We show that

$$\left(S_{n,1}, S_{n,2}, \cdots, S_{n,2p-1}, S_{2p}\right) \to_D \frac{1}{\sqrt{2}}(Z_1, Z_2, \cdots, Z_{2p-1}, Z_{2p}), \quad (5.3.5)$$

which, in turn, implies (5.3.4).

Note that the sums $S_{n,i}$, $i = 1, \cdots, 2p$, can be written as weighted sums of the linear process $\{X_k\}$, $S_{n,i} = \sum_{k=1}^{n} z_{nk;i} X_k$, with weights

$$z_{nk; 2i-1} = \frac{\cos(u_{j_i} k)}{\sqrt{2\pi n f_X(u_{j_i})}}, \quad z_{nk; 2i} = \frac{-\sin(u_{j_i} k)}{\sqrt{2\pi n f_X(u_{j_i})}}, \quad i = 1, \cdots, p.$$

We show below that

$$\mathrm{Cov}(S_{n,1}, \cdots, S_{n,2p}) \to \frac{1}{2}\mathrm{diag}(1, \cdots, 1)_{2p \times 2p}, \tag{5.3.6}$$

$$\max_{1 \le k \le n} |z_{nk;\,i}| = o(1), \quad i = 1, \cdots, 2p, \quad \text{for } |d| < 1/2, \tag{5.3.7}$$

$$\sum_{k=1}^{n} z_{nk;\,i}^2 = O(1), \quad i = 1, \cdots, 2p, \quad \text{for } 0 \le d < 1/2. \tag{5.3.8}$$

By (5.3.6), $\sigma_{n,i}^2 := \mathrm{Var}(S_{n,i}) \to 1/2$. Hence, (5.3.7) and (5.3.8) imply that for $d \ge 0$, the weights $z_{nk;\,i}$ satisfy assumption (i), and, for $d < 0$, assumption (ii) of Proposition 4.3.1, respectively. This, together with Theorem 4.3.2 and (5.3.6), implies (5.3.5).

Proof of (5.3.6). By Proposition 5.3.1, for $i = 1, \cdots, p$,

$$E|v_{X,j_i}|^2 = ES_{n,2i-1}^2 + ES_{n,2i}^2 \to 1,$$

$$Ev_{X,j_i}^2 = ES_{n,2i-1}^2 - ES_{n,2i}^2 + 2iES_{n,2i-1}S_{n,2i} \to 0,$$

which yields

$$ES_{n,2i-1}^2 \to \frac{1}{2}, \quad ES_{n,2i}^2 \to \frac{1}{2}, \quad ES_{n,2i-1}S_{n,2i} \to 0. \tag{5.3.9}$$

Next, by Proposition 5.3.1, for $1 \le k < j \le p$,

$$\begin{aligned} Ev_{X,j_i}\overline{v_{X,j_k}} &\equiv E(S_{n,2i-1} + iS_{n,2i})(S_{n,2k-1} - iS_{n,2k}) \\ &= ES_{n,2i-1}S_{n,2k-1} + ES_{n,2i}S_{n,2k} \\ &\quad + i\big(ES_{n,2i}S_{n,2k-1} - ES_{n,2i-1}S_{n,2k}\big) \to 0, \\ Ev_{X,j_i}v_{X,j_k} &\equiv E(S_{n,2i-1} + iS_{n,2i})(S_{n,2k-1} + iS_{n,2k}) \\ &= ES_{n,2i-1}S_{n,2k-1} - ES_{n,2i}S_{n,2k} \\ &\quad + i\big(ES_{n,2i}S_{n,2k-1} + ES_{n,2i-1}S_{n,2k}\big) \to 0, \end{aligned}$$

which implies that

$$ES_{n,2i-1}S_{n,2k-1} \to 0, \quad ES_{n,2i}S_{n,2k} \to 0,$$

$$ES_{n,2i}S_{n,2k-1} \to 0, \quad ES_{n,2i-1}S_{n,2k} \to 0,$$

and therefore, together with (5.3.9), completes the proof of (5.3.6).

Proof of (5.3.7). By assumption, $f_X(u) \le C|u|^{-2d}$. Therefore, for $1 \le i \le p$, $1/f^{1/2}(u_i) \le Cu_i^{|d|} \le Cn^{|d|}$, and

$$|z_{nk;\,i}| \le Cn^{-1/2+|d|} \to 0,$$

uniformly in k for $1 \le i \le 2p$, which yields (5.3.7).

Proof of (5.3.8). For $d \geq 0$, $f^{-1}(u) \leq C$. Hence, $|z_{nk;i}| \leq Cn^{-1/2}$, for $1 \leq k \leq n$, which implies (5.3.8) and so completes the proof of the theorem.

Bartlett approximation. As an example of the application of Theorem 5.2.1, we consider the Bartlett approximation for a standardized periodogram $I_X(u_j)/f_X(u_j)$ of a linear process.

The process $\{X_j\}$, (5.2.2), is the output of a linear filter applied to a white-noise process $\{\zeta_j\}$. By Proposition 2.2.2, the spectral density f_X of $\{X_j\}$ is related to the spectral density $f_\zeta(u) = \sigma_\zeta^2/2\pi$, $u \in \Pi$, of $\{\zeta_j\}$ by

$$f_X(u) = |A_X(u)|^2 f_\zeta(u), \quad u \in \Pi, \tag{5.3.10}$$

where $A_X(u) = \sum_{k=0}^{\infty} e^{-iku} a_k$ is the transfer function of $\{X_j\}$. Theorem 5.3.2 below states

$$I_X(u) = |A_X(u)|^2 I_\zeta(u) + o_p(1), \quad u \in \Pi.$$

Because of (5.3.10), we can rewrite this as

$$\frac{I_X(u)}{f_X(u)} = \frac{I_\zeta(u)}{f_\zeta(u)} + o_p(1), \quad u \in \Pi.$$

This relation is known as the Bartlett approximation. To justify it, we shall use the bounds for the covariances $\mathrm{Cov}(w_X(u_j), \overline{w_X(u_k)})$ and the cross-covariance $\mathrm{Cov}(w_X(u_j), \overline{w_\zeta(u_k)})$ obtained in Theorem 5.2.1. The derivation of the bound for the cross-covariance involves conditions on the transfer function $A_X(u)$ of $\{X_j\}$. To bring some transparency into these conditions and to make them operational, we introduce the following class of linear processes.

Example 5.3.1. Consider a class of linear processes $\{X_j\}$ with memory parameter $|d| < 1/2$ defined as

$$X_j = (1 - B)^{-d} Y_j = Y_j + \theta_1 Y_{j-1} + \theta_2 Y_{j-2} + \cdots, \quad j \in \mathbb{Z}, \tag{5.3.11}$$

where $\{Y_j\}$ is a short memory linear process,

$$Y_j = \sum_{k=0}^{\infty} b_k \zeta_{j-k}, \ j \in \mathbb{Z}, \quad \sum_{k=0}^{\infty} |b_k| < \infty, \quad \sum_{k=0}^{\infty} b_k \neq 0, \tag{5.3.12}$$

and $\{\zeta_j\} \sim \mathrm{WN}(0, \sigma_\zeta^2)$. The process $\{Y_j\}$ has the transfer function $A_Y(u) = \sum_{k=0}^{\infty} b_k e^{-iku}$ and the spectral density

$$f_Y(u) = |A_Y(u)|^2 (\sigma_\zeta^2/2\pi).$$

Both f_Y and A_Y are continuous because the b_k's are absolutely summable. Moreover, $f_Y(0) > 0$.

The weights θ_k are defined by the expansion $(1 - x)^{-d} = \sum_{k=0}^{\infty} \theta_k x^k$, $|x| < 1$, and have the property $\sum_{k=0}^{\infty} \theta_k^2 < \infty$; see Section 7.2. The operator $(1 - B)^{-d}$ is understood as $\sum_{k=0}^{\infty} \theta_k B^k$.

By Proposition 2.2.2, $\{X_j\}$ is a stationary sequence with the following spectral density and transfer function:

$$f_X(u) = |h(u)|^2 f_Y(u), \quad h(u) := (1 - e^{-iu})^{-d}, \qquad (5.3.13)$$
$$A_X(u) = h(u) A_Y(u), \quad u \in (0, \pi].$$

The square summability of the θ_k's and the absolute summability of the b_k's ensure that $\{X_j\}$ can be written as a linear process $X_j = \sum_{k=0}^{\infty} a_k \zeta_{j-k}$, $j \in \mathbb{Z}$, with coefficients a_k defined by the expansion $(1-B)^{-d} \sum_{k=0}^{\infty} b_k B^k = \sum_{k=0}^{\infty} a_k B^k$; see e.g., Section 7.2.

The transfer function A_X is factored as in (iv) of Theorem 5.2.1, since $A_Y(u)$ is a continuous function on $[0, \pi]$ and h is differentiable and satisfies $h \in C_{1,2d}(0, \pi)$, i.e.,

$$|h(u)| \le C|u|^{-2d}, \quad |\dot{h}(u)| \le C|u|^{-1-2d}, \quad \forall u \in [0, \pi], \qquad (5.3.14)$$
$$|h(u)| \sim |u|^{-2d}, \quad u \to 0.$$

To prove (5.3.14), note that $\dot{h}(u) = -\mathbf{i}\, d(1 - e^{-iu})^{-1-d}$,

$$|1 - e^{-iu}| = |e^{iu/2} - e^{-iu/2}| = 2|\sin(u/2)| \le |u|,$$
$$2/\pi \le \frac{\sin(u/2)}{u/2} \le 1, \quad u \in (0, \pi).$$

Hence, for all $|d| < 1/2$, $d \ne 0$, and all $u \in \Pi$,

$$C_1|u|^{-d} \le |h(u)| \le C_2|u|^{-d}, \qquad C_1, C_2 > 0, \qquad (5.3.15)$$
$$|\dot{h}(u)| \le C|u|^{-1-d},$$

which in turn yields (5.3.14).

The above imply that f_X satisfies

$$f_X(u) \sim |u|^{-2d} f_Y(0), \quad u \to 0, \qquad (5.3.16)$$

with memory parameter $|d| < 1/2$, and is continuous and bounded on any interval $[\epsilon, \pi]$, for all $0 < \epsilon < \pi$. Thus, for $d = 0$, $d \in (0, 1/2)$, and $d \in (-1/2, 0)$ the process $\{X_j\}$ has short, long, and negative memory, respectively.

The application of Theorem 5.2.1 to the short memory process $\{Y_j\}$ of (5.3.12) involves the assumption that $A_Y \in \Lambda_\beta[0, \pi]$, $0 < \beta \leq 1$, which, in turn, is satisfied if the coefficients b_j in (5.3.12) are such that

$$\sum_{k=0}^{n} |kb_k| = O(n^{1-\beta}). \tag{5.3.17}$$

This is proved in Theorem 1 of Móricz (2006). It implies that $A_Y, f_Y \in \Lambda_\beta$, $0 < \beta \leq 1$, if

$$|b_k| = O(k^{-1-\beta-\delta}), \quad \exists \delta > 0. \tag{5.3.18}$$

In summary, the transfer function A_X of the process $\{X_j\}$ of (5.3.11) can be factored as $A_X = hA_Y$, where A_Y is a continuous bounded function and $h \in C_{1,2d}(0, \pi)$, satisfying (5.3.15). If $d = 0$ then $A_X = A_Y$ and $h \equiv 1$. Moreover, if $|b_k| = O(k^{-1-\beta})$, $0 < \beta < 1$, then $A_Y \in \Lambda_\beta[0, \pi]$. If $|b_k| = O(k^{-2-\delta})$, $\delta > 0$, then $A_Y \in \Lambda_1[0, \pi]$ and $A_X \in C_{1,2d}(0, \pi)$.

Theorem 5.3.2. *Let $\{X_j\}$ be a linear process as in (5.2.2), with white-noise innovations $\{\zeta_j\}$ and spectral density f_X satisfying (5.3.1) with $|d| < 1/2$. Assume that f_X is bounded away from zero on $[\epsilon, \pi]$ for any $\epsilon \in (0, \pi)$ and that the transfer function A_X can be factored as $A_X = hA_Y$, where A_Y is a continuous bounded function and h satisfies (5.3.15). Then the following hold:*

(i) For $1 \leq j < n/2$,

$$\frac{I_X(u_j)}{f_X(u_j)} = \frac{I_\zeta(u_j)}{f_\zeta(u_j)} + r_{n,j}, \quad \max_{u_K \leq u_j < \pi} E|r_{n,j}| = o(1), \tag{5.3.19}$$

where $K = 1$, if $d = 0$ and $K = K_n < n/2$, $K_n \to \infty$, if $d \neq 0$.

(ii) In addition, if the ζ_j's are i.i.d. r.v.'s and $E\zeta_0^4 < \infty$, then

$$\max_{u_K \leq u_j \leq \pi} E|r_{n,j}|^2 = o(1), \tag{5.3.20}$$

$$\max_{u_K \leq u_j, u_k < \pi} \left| \mathrm{Cov}\left(\frac{I_X(u_j)}{f_X(u_j)}, \frac{I_X(u_k)}{f_X(u_k)}\right) - I(j = k) \right| = o(1). \tag{5.3.21}$$

(iii) Moreover, if $d = 0$ and $A_X \in \Lambda_1[0, \pi]$, then (5.3.19) and (5.3.21) hold with $o(1)$ replaced by $O(n^{-1/2}\sqrt{\log n})$. If $d \neq 0$ and $h := A_X$ satisfies (5.3.15), then

$$E|r_{n,j}| \leq C\left(\frac{\log j}{j}\right)^{1/2}, \quad E|r_{n,j}|^2 \leq C\frac{\log j}{j}, \quad 1 \leq j \leq n/2.$$

Before proceeding to prove the above theorem, we state the following remark about Lemma 4.5.1 used in the proof.

Remark 5.3.1. Lemma 4.5.1 and part (ii) of Theorem 5.3.2 remain valid if $\{\zeta_j\}$ is a martingale-difference sequence such that

$$E[\zeta_j|\mathcal{F}_{j-1}] = 0, \quad E[\zeta_j^2|\mathcal{F}_{j-1}] = \sigma_\zeta^2, \quad E\zeta_j^4 \le C < \infty, \quad j \in \mathbb{Z},$$

where $\mathcal{F}_j$ is the σ-field generated by ζ_s, $s \le j$. Then (4.5.4) follows because now $\{\zeta_j\}$ is a WN$(0, \sigma_\zeta^2)$ sequence, and because, for any $t_1 > s_1$, $t_1 \ge t_2$, and $t_2 > s_2$,

$$E[\zeta_{t_1}\zeta_{s_1}\zeta_{t_2}\zeta_{s_2}] = 0, \qquad t_1 \ne t_2 \text{ or } s_1 \ne s_2,$$
$$= \sigma_\zeta^4, \qquad t_1 = t_2 \text{ and } s_1 = s_2.$$

Indeed,

$$E[\zeta_{t_1}\zeta_{s_1}\zeta_{t_2}\zeta_{s_2}] = E[E(\zeta_{t_1}|\mathcal{F}_{t_1-1})\zeta_{s_1}\zeta_{t_2}\zeta_{s_2}] = 0, \qquad t_1 > t_2,$$
$$E[\zeta_{t_1}^2\zeta_{s_1}\zeta_{s_2}] = E[E(\zeta_{t_1}^2|\mathcal{F}_{t_1-1})\zeta_{s_1}\zeta_{s_2}]$$
$$= \sigma_\zeta^2 E[\zeta_{s_1}\zeta_{s_2}] = 0, \qquad t_1 = t_2, s_1 \ne s_2,$$
$$= \sigma_\zeta^4, \qquad t_1 = t_2, s_1 = s_2.$$

Proof of Theorem 5.3.2. (i) We begin with the bound

$$|r_{n,j}| = \big||v_{X,j}|^2 - |v_{\zeta,j}|^2\big|$$
$$\le 2|v_{X,j} - v_{\zeta,j}||v_{\zeta,j}| + |v_{X,j} - v_{\zeta,j}|^2, \qquad 1 \le j \le n/2.$$

Then

$$E|r_{n,j}| \le 2\big(E|v_{X,j} - v_{\zeta,j}|^2\big)^{1/2}(E|v_{\zeta,j}|^2)^{1/2} + E|v_{X,j} - v_{\zeta,j}|^2,$$
$$E|v_{X,j} - v_{\zeta,j}|^2 = E|v_{X,j}|^2 + E|v_{\zeta,j}|^2 - E[v_{X,j}\overline{v_{\zeta,j}}] - E[\overline{v_{X,j}}v_{\zeta,j}].$$

By assumption, $A_X = hA_Y$ satisfies condition (iv) of Theorem 5.2.1, which implies that uniformly over $K \le j < n/2$,

$$E|v_{X,j}|^2 - 1 = o(1), \qquad E[v_{X,j}\overline{v_{\zeta,j}}] - 1 = o(1). \tag{5.3.22}$$

Thus, $E|v_{X,j} - v_{\zeta,j}|^2 = o(1)$, $E|v_{\zeta,j}|^2 = 1$, and $E|r_{n,j}| = o(1)$, thereby completing the proof of (5.3.19).

(ii) As seen in the proof of Theorem 5.3.1, $v_{X,j}$ and $v_{\zeta,j}$ can be written as weighted sums of the linear process $\{X_j\}$, which in turn, can be rewritten as moving averages of ζ_t's, e.g., as in (4.3.4),

$$v_{X,j} = \sum_{t \in \mathbb{Z}} b_{X,t}\zeta_t, \qquad v_{\zeta,j} = \sum_{t \in \mathbb{Z}} b_{\zeta,t}\zeta_t,$$

with complex weights $b_{X,t}$, $b_{\zeta,t}$ such that $\sum_{t\in\mathbb{Z}}|b_{X,t}|^2 < \infty$ and $\sum_{t\in\mathbb{Z}}|b_{\zeta,t}|^2 < \infty$. Then

$$|v_{X,j}|^2 - |v_{\zeta,j}|^2 = \sum_{s,t\in\mathbb{Z}}(b_{X,s}\overline{b_{X,t}} - b_{\zeta,s}\overline{b_{\zeta,t}})\zeta_s\zeta_t.$$

Observe that $Ev_{X,j}\overline{v_{\zeta,j}} = \sigma_\zeta^2\sum_{t\in\mathbb{Z}}b_{X,t}\overline{b_{\zeta,t}}$, and

$$E|v_{X,j}|^2 = \sigma_\zeta^2\sum_{t\in\mathbb{Z}}|b_{X,t}|^2, \qquad E|v_{\zeta,j}|^2 = \sigma_\zeta^2\sum_{t\in\mathbb{Z}}|b_{\zeta,t}|^2.$$

Thus, by (4.5.4), uniformly in j,

$$E|r_{n,j}|^2 = E(|v_{X,j}|^2 - |v_{\zeta,j}|^2)^2 \leq C\sum_{t,s\in\mathbb{Z}}|b_{X,t}\overline{b_{X,s}} - b_{\zeta,t}\overline{b_{\zeta,s}}|^2 \qquad (5.3.23)$$

$$= C\sum_{t,s\in\mathbb{Z}}(b_{X,t}\overline{b_{X,s}} - b_{\zeta,t}\overline{b_{\zeta,s}})(\overline{b_{X,t}}b_{X,s} - \overline{b_{\zeta,t}}b_{\zeta,s})$$

$$\leq C\big|(E|v_{X,j}|^2)^2 - 2|E[v_{X,j}\overline{v_{\zeta,j}}]|^2 + (E|v_{\zeta,j}|^2)^2\big|.$$

Bounds (5.3.23) and (5.3.22) prove (5.3.20).

Next we prove (5.3.21). Using (5.3.19), write

$$\mathrm{Cov}(|v_{X,j}|^2, |v_{X,k}|^2) = \mathrm{Cov}(|v_{\zeta,j}|^2 + r_{n,j}, |v_{\zeta,k}|^2 + r_{n,k})$$

$$= \mathrm{Cov}(|v_{\zeta,j}|^2, |v_{\zeta,k}|^2) + q_{n,jk},$$

$$q_{n,jk} := \mathrm{Cov}(|v_{\zeta,j}|^2, r_{n,k}) + \mathrm{Cov}(r_{n,j}, |v_{\zeta,k}|^2) + \mathrm{Cov}(r_{n,j}, r_{n,k}).$$

By the C–S inequality, one obtains

$$|q_{n,jk}| = (E|v_{\zeta,j}|^4)^{1/2}(Er_{n,k}^2)^{1/2} + (Er_{n,j}^2)^{1/2}(E|v_{\zeta,k}|^4)^{1/2}$$

$$+ (Er_{n,j}^2)^{1/2}(Er_{n,k}^2)^{1/2}. \qquad (5.3.24)$$

We show below that uniformly over $1 \leq j, k < n/2$,

$$\mathrm{Cov}(|v_{\zeta,j}|^2, |v_{\zeta,k}|^2) = I(j = k) + O(n^{-1}), \qquad (5.3.25)$$

$$E|v_{\zeta,j}|^4 \leq C,$$

which, together with (5.3.20) and (5.3.24), implies $\max_{K\leq j,k<n/2} q_{n,jk} = o(1)$, and proves (5.3.21).

For any zero-mean r.v.'s Y_1, Y_2, Z_1, and Z_2,

$$\mathrm{Cov}(Y_1Y_2, Z_1Z_2) = EY_1Y_2Z_1Z_2 - E[Y_1Y_2]E[Z_1Z_2]$$

$$= E[Y_1Z_1]E[Y_2Z_2] + E[Y_1Z_2]E[Y_2Z_1] - \mathrm{Cum}(Y_1, Y_2, Z_1, Z_2).$$

Therefore,

$$\text{Cov}\big(|v_{\zeta,j}|^2, |v_{\zeta,k}|^2\big) = \text{Cov}(v_{\zeta,j}\overline{v_{\zeta,j}}, v_{\zeta,k}\overline{v_{\zeta,k}})$$
$$= |E[v_{\zeta,j}\overline{v_{\zeta,k}}]|^2 + |E[v_{\zeta,j}v_{\zeta,k}]|^2 + \text{Cum}(v_{\zeta,j}, \overline{v_{\zeta,j}}, v_{\zeta,k}, \overline{v_{\zeta,k}}).$$

By (5.2.1), $E[v_{\zeta,j}\overline{v_{\zeta,k}}] = I(j = k)$, and as in (5.2.1), it follows that $E[v_{\zeta,k}v_{\zeta,k}] = 0$ if $1 \le j + k < n$. Next, by the cumulant properties (c1) and (c2) of Section 4.2,

$$\text{Cum}(v_{\zeta,j}, \overline{v_{\zeta,j}}, v_{\zeta,k}, \overline{v_{\zeta,k}})$$
$$= \frac{1}{(2\pi n)^2 f_\zeta(u_j) f_\zeta(u_k)} \sum_{t=1}^{n} \text{Cum}(\zeta_t, \zeta_t, \zeta_t, \zeta_t)$$
$$= \frac{1}{n\sigma_\zeta^4} \text{Cum}_4(\zeta_0) = O(n^{-1}),$$

which proves the first claim of (5.3.25). In turn, it yields

$$E|v_{\zeta,j}|^4 = \text{Cov}(|v_{\zeta,j}|^2, |v_{\zeta,j}|^2) + (E|v_{\zeta,j}|^2)^2 \le C,$$

which proves the second bound of (5.3.25).

Claim (iii) follows by the same argument as above where we can use the following bounds instead of (5.3.22). For $d = 0$,

$$E|v_{X,j}|^2 - 1 = O(n^{-1}\log n), \qquad E[v_{X,j}\overline{v_{\zeta,j}}] - 1 = O(n^{-1}\log n),$$

and for $d \ne 0$,

$$E|v_{X,j}|^2 - 1 = O(j^{-1}\log j), \qquad E[v_{X,j}\overline{v_{\zeta,j}}] - 1 = O(j^{-1}\log j).$$

These bounds hold by (i) and (iii) of Theorem 5.2.1.

Note: For short memory linear processes, various useful results on the statistical properties of the DFT and the periodogram can be found in Brockwell and Davis (1991). Their Theorems 10.3.1 and 10.3.2 contain similar results to those obtained in Theorems 5.3.1 and 5.3.2 above. Robinson (1995a) discusses some properties of the DFT of a linear process that allows for long memory while Lahiri (2003) studied the first-order asymptotics of the DFT with a data taper. Davis and Mikosch (1999) investigate some properties of the maximum of the periodogram for linear processes with short memory while Phillips (2007) discusses the joint asymptotic normality of an increasing collection of DFTs. The behavior of averaged periodograms of nonstationary processes was studied in Robinson and Marinucci (2001).

Chapter 6

Asymptotic Theory for Quadratic Forms

6.1 Introduction

In this chapter we discuss the asymptotic properties of quadratic forms and the sums of weighted periodograms. We will develop the asymptotic distribution theory for estimates of integrals

$$K(b_n) := \int_0^\pi b_n(u) f_X(u) du, \tag{6.1.1}$$

where $b_n, n \geq 1$, is a sequence of real-valued functions on Π and f_X is the spectral density of a stationary process $\{X_j\}$. These integrals arise naturally in many situations, e.g., the spectral d.f. F can be written as

$$F(y) := \int_{-\pi}^y f_X(u) du = \int_0^y f_X(u) du + \int_0^\pi f_X(u) du, \quad 0 \leq y \leq \pi,$$

and the autocovariance function of $\{X_j\}$ is

$$\text{Cov}(X_k, X_0) = 2 \int_0^\pi \cos(ku) f_X(u) du, \qquad k = 0, 1, 2, \cdots.$$

In these examples b does not depend on n. But sometimes one is interested in estimating a smoothed version of f_X. In such cases b will typically depend on the sample size n. For example, consider the problem of approximating $f_X(u)$ at a point $u = u_0 \in (0, \pi)$ by the Fejér kernel density smoothing method. In this case $b(u) \equiv b_n(u) := (2\pi n)^{-1} |D_n(u - u_0)|^2$; see (2.1.6). If f_X is continuous, and n is large,

$$\int_0^\pi b_n(u) f_X(u) du = \frac{1}{2\pi n} \int_0^\pi |D_n(u - u_0)|^2 f_X(u) du$$

$$\sim \frac{1}{2\pi n} \int_0^\pi |D_n(u - u_0)|^2 du f_X(u_0) \sim f_X(u_0).$$

132

In order to carry out large sample inference for the parameter sequence $K(b_n)$, it is necessary to develop the asymptotic distribution theory for its estimates. Since f_X is unknown, a natural estimator of $K(b_n)$ is obtained by the plug-in method where f_X is replaced by its empirical counterpart, the periodogram I_X, based on the data $X_1, \cdots, X_n$, arriving at the estimator

$$\hat{K}(b_n) := \int_0^\pi b_n(u) I_X(u) du.$$

Deriving asymptotic distributions of $\hat{K}(b_n) - K(b_n)$ for a general sequence of weight functions b_n may be prohibitively complicated. In Section 6.3 we discuss the asymptotic distribution of $\hat{K}(b)$ where b does not depend on n and the underlying process is a linear process. The two sets of relatively easy to verify sufficient conditions for this are given in Subsection 6.3.1. These results prove the CLT for a large class of quadratic forms.

Sometimes it is also useful to replace the integrals $\hat{K}(b_n)$ and $K(b_n)$ by their respective discretized versions,

$$\hat{Q}(b_n) := \frac{2\pi}{n} \sum_{j=1}^{[n/2]} b_n(u_j) I_X(u_j), \quad Q(b_n) := \frac{2\pi}{n} \sum_{j=1}^{[n/2]} b_n(u_j) f_X(u_j),$$

where $u_j = 2\pi j/n$, $j = 1, \cdots, [n/2]$. The estimator $\hat{Q}(b_n)$ is a weighted sum of periodograms and has some theoretical advantages.

In Section 6.2 below we provide the theoretical tools to establish the asymptotic normality of $\hat{Q}(b_n) - Q(b_n)$ and to evaluate the mean-squared error $E(\hat{Q}(b_n) - Q(b_n))^2$, when $\{X_j\}$ is a linear process with i.i.d. innovations. Our conditions for asymptotic normality of these weighted sums are simple and resemble the Lindeberg–Feller condition for the weighted sums of i.i.d. r.v.'s, regardless of the dependence structure of the process $\{X_j\}$. Because of the translation invariance property of the periodogram, $I_{X+c}(u_j) = I_X(u_j)$, for all $c \in \mathbb{R}$, and for $j = 1, \cdots, n-1$, $\hat{Q}(b_n)$ is also translation invariant. This means that we do not need to de-mean the data to compute $\hat{Q}(b_n)$.

6.2 Asymptotic normality for sums of weighted periodograms

In this section we discuss various asymptotic properties of weighted sums of normalized periodograms generated by stationary moving-average processes. Such sums can be treated as a discretized version of the quadratic forms discussed in Section 6.3, which are written as integrals of a weighted periodogram. Accordingly, let

$$X_j = \sum_{k=0}^{\infty} a_k \zeta_{j-k}, \quad j \in \mathbb{Z}, \quad \sum_{k=0}^{\infty} a_k^2 < \infty, \tag{6.2.1}$$

where $\{\zeta_j, \, j \in \mathbb{Z}\}$, are $\mathrm{IID}(0,1)$ standardized random variables.

We assume that the spectral density f_X of the process $X_j, j \in \mathbb{Z}$, satisfies

$$f_X(u) = |u|^{-2d} g(u), \quad |u| \le \pi, \tag{6.2.2}$$

for some $|d| < 1/2$, where $g(u)$ is a continuous function, such that

$$0 < C_1 \le g(u) \le C_2 < \infty, \quad u \in \Pi, \quad (\exists \, 0 < C_1, C_2 < \infty).$$

As shown in Section 7.2, the $\mathrm{ARFIMA}(p, d, q)$ model satisfies this assumption for all $d \in (-1/2, 1/2)$.

Condition (6.2.2) allows us to derive the mean-squared error bounds of the estimates discussed in Theorem 6.2.5 below. To derive their asymptotic normality and some delicate Bartlett-type approximations, we shall additionally need to assume that the transfer function $A_X(u) := \sum_{k=0}^{\infty} e^{-iku} a_k$, $u \in \Pi$, is differentiable in $(0, \pi)$ and its derivative $\dot{A}_X$ has the property

$$|\dot{A}_X(u)| \le C|u|^{-1-d}, \quad |u| \le \pi. \tag{6.2.3}$$

Since $f_X(u) = |A_X(u)|^2/2\pi$, $u \in \Pi$, by assumptions (6.2.2) and (6.2.3),

$$|\dot{f}_X(u)| \le C |u|^{-1-2d}, \quad u \in \Pi. \tag{6.2.4}$$

This result for f_X is slightly weaker than assumption (6.2.3) on A_X.

Conditions (6.2.2) and (6.2.3) are formulated this way to cover long and negative memory models, with $|d| < 1/2$, $d \ne 0$. The case when f_X and A_X are bounded continuous or Lipschitz continuous functions will be discussed later. Assumption (6.2.3) will also be relaxed.

To proceed further we need some more notation. Divide the interval $[0, \pi]$ into a set of *discrete Fourier frequencies* $u_j = 2\pi j/n \in [0, \pi]$, $j = 0, 1, \cdots, [n/2]$. Let, for $j = 0, \cdots, [n/2]$,

$$w_{X,j} = \frac{1}{\sqrt{2\pi n}} \sum_{k=1}^{n} e^{-iu_j k} X_k, \quad w_{\zeta,j} = \frac{1}{\sqrt{2\pi n}} \sum_{k=1}^{n} e^{-iu_j k} \zeta_k, \tag{6.2.5}$$

$$I_{x,\zeta} = \frac{1}{2\pi m}\sum_{k=1}^{m}\sum_{\ell=1}^{m}e^{-i\lambda k}X_k X_\ell e^{i\lambda \ell} =$$

$$\frac{1}{2\pi m}\sum\sum X_k X_\ell e^{i(\ell-k)\lambda}$$

denote the DFTs of $\{X_j\}$ and $\{\zeta_j\}$, computed at frequencies u_j. The corresponding periodograms, transfer functions, and spectral densities of $\{X_j\}$ and the noise $\{\zeta_j\}$ at frequency u_j are, respectively, denoted by

$$I_{X,j} := |w_{X,j}|^2, \quad I_{\zeta,j} := |w_{\zeta,j}|^2, \quad A_{X,j} := A_X(u_j), \quad A_{\zeta,j} := 1,$$

$$f_{X,j} := f_X(u_j), \quad f_{\zeta,j} := f_\zeta(u_j) \equiv \frac{1}{2\pi}, \qquad j = 0,1,\cdots,[n/2].$$

Our focus here is to establish the asymptotic normality of the quadratic forms $Q_{n,X} = \sum_{j=1}^{\tilde{n}} b_{n,j} I_{X,j}$, where $\{b_{n,j}, \ j = 1,\cdots,\tilde{n}\}$ is an array of real numbers, which may depend on n, and $\tilde{n} = [n/2] - 1$. This in turn is facilitated by first developing the asymptotic distribution theory for the sums

$$S_{n,X} = \sum_{j=1}^{\tilde{n}} b_{n,j} \frac{I_{X,j}}{f_{X,j}}.$$

Moreover, asymptotic analysis of these sums is more illustrative of the methodology used. Note that since $u_{n/2} \sim \pi$, the sum $\sum_{j=1}^{\tilde{n}}$ is a discrete analog of the integral $\int_0^\pi$.

Asymptotic normality of $S_{n,X}$. An important role in the asymptotic analysis of $S_{n,X}$ will be played by a Bartlett-type approximation

$$\frac{I_{X,j}}{f_{X,j}} \sim \frac{I_{\zeta,j}}{f_{\zeta,j}} = 2\pi I_{\zeta,j}, \quad j = 1,\cdots,\tilde{n}.$$

Our first goal is to approximate $S_{n,X}$ by the sum

$$S_{n,\zeta} = \sum_{j=1}^{\tilde{n}} b_{n,j} \frac{I_{\zeta,j}}{f_{\zeta,j}} \equiv \sum_{j=1}^{\tilde{n}} b_{n,j} 2\pi I_{\zeta,j}. \tag{6.2.6}$$

Let

$$R_n := S_{n,X} - S_{n,\zeta}, \quad b_n := \max_{j=1,\cdots,\tilde{n}} |b_{n,j}|, \quad B_n^2 := \sum_{j=1}^{\tilde{n}} b_{n,j}^2, \tag{6.2.7}$$

$$q_n^2 := B_n^2 + \text{Cum}_4(\zeta_0)\frac{1}{n}\Big(\sum_{j=1}^{\tilde{n}} b_{n,j}\Big)^2.$$

We shall show later that $\text{Var}(S_{n,\zeta}) = q_n^2$; see (6.2.25).

By definition $S_{n,X} = S_{n,\zeta} + R_n$. Lemma 6.2.1 below provides an upper bound of order $b_n \log^2 n$ for ER_n^2. The main term $S_{n,\zeta}$ is a quadratic form in i.i.d. r.v.'s. Its asymptotic normality is established under minimal conditions on the weights $b_{n,j}$ in Lemma 6.2.2.

The following theorem proves the asymptotic normality for $S_{n,X}$ under the Lindeberg–Feller-type condition (6.2.8) of asymptotic uniform negligibility of the weights $b_{n,j}$, which is analogous to the asymptotic normality condition for the sums of i.i.d. r.v.'s $\sum_{j=1}^{\tilde{n}} b_{n,j}\zeta_j$, $\{\zeta_j\} \sim \text{IID}(0,1)$. Thus, the asymptotic properties of the sums of the normalized periodograms $I_{X,j}/f_{X,j}$ resemble those of the sums of i.i.d. r.v.'s. For other properties of $I_{X,j}/f_{X,j}$, see Section 5.3.

Because of the property of a periodogram, $I_{X+\mu}(u_j) = I_X(u_j)$, $\mu \in \mathbb{R}$, $j = 1, \cdots, n-1$, the process $\{X_j\}$ is automatically de-meaned. All the results obtained below also remain valid for a process $\{X_j\}$ of (6.2.1) that has non-zero mean.

Theorem 6.2.1. *Suppose the linear process $\{X_j, j \in \mathbb{Z}\}$ of (6.2.1) is such that $E\zeta_0^4 < \infty$ and satisfies assumptions (6.2.2) and (6.2.3). For the weights $b_{n,j}$ assume*

$$\frac{\max_{j=1,\cdots,\tilde{n}} |b_{n,j}|}{\left(\sum_{j=1}^{\tilde{n}} b_{n,j}^2\right)^{1/2}} = \frac{b_n}{B_n} \to 0. \tag{6.2.8}$$

Then the following hold:

$$ES_{n,X} = \sum_{j=1}^{\tilde{n}} b_{n,j} + o(q_n), \quad \text{Var}(S_{n,X}) = q_n^2 + o(q_n^2), \tag{6.2.9}$$

$$q_n^{-1}\left(S_{n,X} - \sum_{j=1}^{\tilde{n}} b_{n,j}\right) \to_D \mathcal{N}(0,1).$$

Moreover,

$$\min\left(1, \text{Var}(\zeta_0^2)/2\right) B_n^2 \le q_n^2 \le (1 + |\text{Cum}_4(\zeta_0)|) B_n^2. \tag{6.2.10}$$

Proof. To prove (6.2.10), use the definition of q_n and the C–S inequality to obtain the upper bound. The lower bound is derived in (6.2.27) of Lemma 6.2.2 below.

To estimate $ES_{n,X}$, recall that $R_n = S_{n,X} - S_{n,\zeta}$. Then write $ES_{n,X} = E[S_{n,\zeta}] + E[R_n]$. By (6.2.24) of Lemma 6.2.2 below, $ES_{n,\zeta} = \sum_{j=1}^{\tilde{n}} b_{n,j}$. Use (6.2.15) of Lemma 6.2.1, (6.2.8), and (6.2.10), to obtain

$$E|R_n| \le (ER_n^2)^{1/2} = o(B_n) = o(q_n). \tag{6.2.11}$$

This in turn completes the proof of the first claim in (6.2.9).

To estimate $\mathrm{Var}(S_{n,X})$, use

$$\mathrm{Var}(S_{n,X}) = \mathrm{Var}(S_{n,\zeta}) + \mathrm{Var}(R_n) + 2\mathrm{Cov}(S_{n,\zeta}, R_n).$$

By (6.2.25), $\mathrm{Var}(S_{n,\zeta}) = q_n^2$, which together with (6.2.11) yields

$$\mathrm{Var}(R_n) \le ER_n^2 = o(q_n^2),$$

$$|\mathrm{Cov}(S_{n,\zeta}, R_n)| \le (\mathrm{Var}(S_{n,\zeta}))^{1/2}(\mathrm{Var}(R_n))^{1/2} = o(q_n^2),$$

and proves the second claim in (6.2.9).

Finally, since $ES_{n,\zeta} = \sum_{j=1}^{\tilde{n}} b_{n,j}$ and $ER_n^2 = o(q_n^2)$,

$$S_{n,X} - \sum_{j=1}^{\tilde{n}} b_{n,j} = S_{n,X} - ES_{n,\zeta}$$

$$= S_{n,\zeta} - E[S_{n,\zeta}] + R_n = S_{n,\zeta} - ES_{n,\zeta} + o_p(q_n).$$

This together with (6.2.26) of Lemma 6.2.2 implies the asymptotic normality result in (6.2.9).

The following lemma provides two types of upper bounds for the mean-squared error ER_n^2, which are useful in approximating $S_{n,X}$ by $S_{n,\zeta}$.

Lemma 6.2.1. *Assume that $\{X_j\}$ of (6.2.1) satisfies (6.2.2) and (6.2.3), and $E\zeta_0^4 < \infty$. Then the following hold:*

$$E(R_n - ER_n)^2 \le Cb_n^2 \log^3 n, \quad and \tag{6.2.12}$$

$$\le Cb_n B_n,$$

$$|ER_n| \le Cb_n \log^2 n, \quad and \tag{6.2.13}$$

$$= o(B_n), \quad if \quad b_n = o(B_n).$$

Consequently,

$$E(S_{n,X} - S_{n,\zeta})^2 \le Cb_n^2 \log^4 n, \tag{6.2.14}$$

$$E(S_{n,X} - S_{n,\zeta})^2 = o(B_n^2), \quad if \quad b_n = o(B_n). \tag{6.2.15}$$

To prove Lemma 6.2.1, we need two auxiliary results. The following proposition provides a general approximation bound.

Proposition 6.2.1. *Let $\{Y_{nj}^{(i)}, j = 1, \cdots, n\}, i = 1, 2, n \ge 1$, be the two sets of linear processes*

$$Y_{nj}^{(i)} = \sum_{k\in\mathbb{Z}} b_{nj}^{(i)}(k)\zeta_k, \qquad \sum_{k\in\mathbb{Z}} |b_{nj}^{(i)}(k)|^2 < \infty, \quad i = 1, 2,$$

where $\{b_{nj}^{(i)}(k)\}$ *are possibly complex-valued weights. Assume* $\{\zeta_k\} \sim$ IID$(0,1)$ *and* $E\zeta_0^4 < \infty$. *Then, for any real weights* $c_{n,j}$, $j = 1, \cdots n$,

$$\text{Var}\Big(\sum_{j=1}^{n} c_{n,j}\{|Y_j^{(1)}|^2 - |Y_j^{(2)}|^2\}\Big) \tag{6.2.16}$$

$$\leq C \sum_{j,k=1}^{n} |c_{n,j}c_{n,k}|\big||r_{n,jk}^{11}|^2 + |r_{n,jk}^{22}|^2 - 2|r_{n,jk}^{12}|^2\big|,$$

where

$$r_{n,jk}^{il} := E[Y_{nj}^{(i)}\,\overline{Y_{nk}^{(l)}}] = \sum_{t\in\mathbb{Z}} b_{nj}^{(i)}(t)\,\overline{b_{nk}^{(l)}(t)}, \quad i,l = 1,2.$$

Proof. Write

$$Q_n := \sum_{j=1}^{n} c_{n,j}\{|Y_j^{(1)}|^2 - |Y_j^{(2)}|^2\}$$

$$= \sum_{t,s\in\mathbb{Z}} \Big(\sum_{j=1}^{n} c_{n,j}\{b_{nj}^{(1)}(t)\overline{b_{nj}^{(1)}(s)} - b_{nj}^{(2)}(t)\overline{b_{nj}^{(2)}(s)}\}\Big)\zeta_t\zeta_s$$

$$=: \sum_{t,s\in\mathbb{Z}} B_n(t,s)\zeta_t\zeta_s.$$

But, by Lemma 4.5.1,

$$E\big|Q_n - EQ_n\big|^2 \leq 4E\zeta_0^4 \sum_{t,s\in\mathbb{Z}} |B_n(t,s)|^2.$$

Therefore,

$$\sum_{t,s\in\mathbb{Z}} |B_n(t,s)|^2 = \sum_{j,k=1}^{n} c_{n,j}c_{n,k} \sum_{t,s\in\mathbb{Z}} \{b_{nj}^{(1)}(t)\overline{b_{nj}^{(1)}(s)} - b_{nj}^{(2)}(t)\overline{b_{nj}^{(2)}(s)}\}$$

$$\times \{\overline{b_{nk}^{(1)}(t)}b_{nk}^{(1)}(s) - \overline{b_{nk}^{(2)}(t)}b_{nk}^{(2)}(s)\}$$

$$= \sum_{j,k=1}^{n} c_{n,j}c_{n,k}(|r_{n,jk}^{11}|^2 + |r_{n,jk}^{22}|^2 - |r_{n,jk}^{12}|^2 - |r_{n,kj}^{12}|^2).$$

This completes the proof of the proposition.

Now note that

$$S_{n,X} - S_{n,\zeta} = \sum_{j=1}^{\tilde{n}} b_{n,j}\Big(\frac{I_{X,j}}{f_{X,j}} - \frac{I_{\zeta,j}}{f_{\zeta,j}}\Big) \tag{6.2.17}$$

$$= \sum_{j=1}^{\tilde{n}} \frac{b_{n,j}}{f_{X,j}}\{I_{X,j} - f_{X,j}\frac{I_{\zeta,j}}{f_{\zeta,j}}\}.$$

Also, recall that $I_{X,j} = |w_{X,j}|^2$ and $I_{\zeta,j} = |w_{\zeta,j}|^2$, and that the DFTs $w_{X,j}$ and $w_{\zeta,j}$ are linear processes with complex-valued coefficients. The corollary below, which follows from Proposition 6.2.1, is useful in analyzing sums of the type appearing in (6.2.17). Let $f_{X\zeta}(u) = (2\pi)^{-1}A_X(u)$ denote the cross-spectral density of $\{X_j\}$ and $\{\zeta_j\}$; see the definition in (5.2.3).

Corollary 6.2.1. *Suppose that $\{X_j\}$ is a linear process as in (6.2.1) and $E\zeta_0^4 < \infty$. Then for any real weights $c_{n,j}$, $j = 1, \cdots, n$, the following holds:*

$$\mathrm{Var}\Big(\sum_{j=1}^{\tilde{n}} c_{n,j}\{I_{X,j} - f_{X,j}\frac{I_{\zeta,j}}{f_{\zeta,j}}\}\Big) \leq C(s_{n,1} + s_{n,2}), \qquad (6.2.18)$$

where

$$s_{n,1} := C\sum_{j=1}^{\tilde{n}} c_{n,j}^2 \Big((E|w_{X,j}|^2 - f_{X,j})^2 + f_{X,j}\Big|E|w_{X,j}|^2 - f_{X,j}\Big|$$
$$+ f_{X,j}\Big|E[w_{X,j}\overline{w_{\zeta,j}}] - f_{X\zeta,j}\Big|^2 + f_{X,j}^{3/2}\Big|E[w_{X,j}\overline{w_{\zeta,j}}] - f_{X\zeta,j}\Big|\Big),$$
$$s_{n,2} := \sum_{1 \leq k < j \leq \tilde{n}} |c_{n,j}c_{n,k}|\Big(|E[w_{X,j}\overline{w_{X,k}}]|^2 + f_{X,k}|E[w_{X,j}\overline{w_{\zeta,k}}]|^2\Big).$$

Proof. Observe that
$$\frac{f_{X,j}I_{\zeta,j}}{f_{\zeta,j}} = |A_{X,j}|^2 I_{\zeta,j} = |A_{X,j}w_{\zeta,j}|^2.$$
Define r.v.'s $Y_{nj}^{(1)} = w_{X,j}$ and $Y_{nj}^{(2)} = A_{X,j}w_{\zeta,j}$. They can be written as moving averages of the ζ_j's with complex weights. Therefore, by Proposition 6.2.1, the l.h.s. of (6.2.18) can be bounded above by

$$C\sum_{j,k=1}^{\tilde{n}} |c_{n,j}c_{n,k}|\Big|E[w_{X,j}\overline{w_{X,k}}]|^2$$

$$+ |A_{X,j}|^2|A_{X,k}|^2|E[w_{\zeta,j}\overline{w_{\zeta,k}}]|^2 - 2|A_{X,k}|^2|E[w_{X,j}\overline{w_{\zeta,k}}]|^2\Big|$$

$$= C\Big(\sum_{j=k=1}^{\tilde{n}}[\cdots] + \sum_{k \neq j}[\cdots]\Big) =: C(s'_{n,1} + s'_{n,2}).$$

By (5.2.1) of Section 5.2, $E|w_{\zeta,j}|^2 = 1/2\pi$ and $E[w_{\zeta,j}\overline{w_{\zeta,k}}] = 0$ for $1 \leq k < j \leq \tilde{n}$. Recall also that $f_{X,j} = |A_{X,j}|^2/(2\pi)$. Therefore,

$$s'_{n,1} = \sum_{j=1}^{\tilde{n}} c_{n,j}^2\Big|(E|w_{X,j}|^2)^2 + f_{X,j}^2 - 4\pi f_{X,j}|E[w_{X,j}\overline{w_{\zeta,j}}]|^2\Big|,$$

$$s'_{n,2} = \sum_{1 \leq k < j \leq \tilde{n}} |c_{n,j}c_{n,k}|\Big(|E[w_{X,j}\overline{w_{X,k}}]|^2 + f_{X,k}|E[w_{X,j}\overline{w_{\zeta,k}}]|^2\Big).$$

Observe that $s'_{n,2} = s_{n,2}$. To estimate $s'_{n,1}$, let

$$A := (E|w_{X,j}|^2)^2 - f^2_{X,j}, \quad B := |E[w_{X,j}\overline{w_{\zeta,j}}]|^2 - |f_{X\zeta,j}|^2.$$

The term in $|\cdots|$ in $s'_{n,1}$ can be written as

$$(E|w_{X,j}|^2)^2 + f^2_{X,j} - 4\pi f_{X,j}|E[w_{X,j}\overline{w_{\zeta,j}}]|^2$$
$$= (A - 4\pi f_{X,j}B) + (2f^2_{X,j} - 4\pi f_{X,j}|f_{X\zeta,j}|^2)$$
$$= A - 4\pi f_{X,j}B,$$

because $4\pi f_{X,j}|f_{X\zeta,j}|^2 = 4\pi f_{X,j}|A_{X,j}|^2/(2\pi)^2 = 2f^2_{X,j}$.

Next, note that $||z_1|^2 - |z_2|^2| \le |z_1 - z_2|^2 + 2|z_1 - z_2||z_2|$, for any complex numbers z_1, z_2, and that $|f_{X\zeta,j}| = |A_{X,j}|/(2\pi) \le f^{1/2}_{X,j}$. Therefore, applying the above bound to $|A|$ and $|B|$, one obtains

$$|A - 4\pi f_{X,j}B| \le |A| + 4\pi f_{X,j}|B|$$
$$\le (E|w_{X,j}|^2 - f_{X,j})^2 + 2f_{X,j}\Big|E|w_{X,j}|^2 - f_{X,j}\Big|$$
$$+ 4\pi f_{X,j}\Big|E[w_{X,j}\overline{w_{\zeta,j}}] - f_{X\zeta,j}\Big|^2$$
$$+ 8\pi f^{3/2}_{X,j}\Big|E[w_{X,j}\overline{w_{\zeta,j}}] - f_{X\zeta,j}\Big|,$$

which shows that $s'_{n,1} \le Cs_{n,1}$ and completes the proof of the corollary.

Proof of Lemma 6.2.1. Recall that $R_n = S_{n,X} - S_{n,\zeta}$. We shall prove (6.2.12) and (6.2.13). Since $ER^2_n \le 2E(R_n - ER_n)^2 + 2(ER_n)^2$, then these two facts together imply (6.2.14) and (6.2.15).

To prove (6.2.12), use Corollary 6.2.1, to obtain

$$\text{Var}(R_n) = \text{Var}\Big(\sum_{j=1}^{\tilde{n}} \frac{b_{n,j}}{f_{X,j}}\{I_{X,j} - \frac{f_{X,j}I_{\zeta,j}}{f_{\zeta,j}}\}\Big) \le C(s_{n,1} + s_{n,2}),$$

where

$$s_{n,1} := C\sum_{j=1}^{\tilde{n}} \frac{b^2_{n,j}}{f^2_{X,j}}\Big[(E|w_{X,j}|^2 - f_{X,j})^2 + f_{X,j}\Big|E|w_{X,j}|^2 - f_{X,j}\Big|$$
$$+ f_{X,j}\Big|E[w_{X,j}\overline{w_{\zeta,j}}] - f_{X\zeta,j}\Big|^2 + f^{3/2}_{X,j}\Big|E[w_{X,j}\overline{w_{\zeta,j}}] - f_{X\zeta,j}\Big|\Big],$$
$$s_{n,2} := \sum_{1\le k<j\le\tilde{n}} \frac{|b_{n,j}b_{n,k}|}{f_{X,j}f_{X,k}}\Big(\Big|E[w_{X,j}\overline{w_{X,k}}]\Big|^2 + f_{X,k}\Big|E[w_{X,j}\overline{w_{\zeta,k}}]\Big|^2\Big).$$

It remains to show that $s_{n,1}$ and $s_{n,2}$ satisfy (6.2.12) and (6.2.13).

Part (iii) of Theorem 5.2.1 provides bounds for $E[w_{X,j}\overline{w_{X,k}}]$ and $E[w_{X,j}\overline{w_{\zeta,k}}]$. Recall that the spectral density f_X satisfies (6.2.2), whereas, by (6.2.3), the cross-spectral density $f_{X\zeta}(u) = (2\pi)^{-1}A_X(u)$ has the property $|f_{X\zeta}(u)| \leq C|u|^{-d}$ and $|\dot{f}_{X\zeta}(u)| \leq C|u|^{-1-d}$, $u \in \Pi$. Therefore, they satisfy the conditions of part (iii) of Theorem 5.2.1, which implies

$$|E|w_{X,j}|^2 - f_{X,j}| \leq C|u_j|^{-2d}j^{-1}\log j, \qquad (6.2.19)$$

$$|E[w_{X,j}\overline{w_{\zeta,j}}] - f_{X\zeta,j}| \leq C|u_j|^{-d}j^{-1}\log j,$$

where C does not depend on j and n. By (6.2.2), $f_{X,j}^{-1} \leq Cu_j^{2d}$, and hence

$$s_{n,1} \leq C\sum_{j=1}^{\tilde{n}} b_{n,j}^2(j^{-1}\log j).$$

From this we readily obtain

$$s_{n,1} \leq Cb_n^2\log n\sum_{j=1}^{\tilde{n}}j^{-1} \leq Cb_n^2\log^2 n, \quad \text{and}$$

$$s_{n,1} \leq Cb_n\sum_{j=1}^{\tilde{n}}|b_{n,j}|(j^{-1}\log j)$$

$$\leq Cb_n(\sum_{j=1}^{\tilde{n}}b_{n,j}^2)^{1/2}(\sum_{j=1}^{\tilde{n}}j^{-2}\log^2 j)^{1/2} \leq Cb_nB_n,$$

which proves that $s_{n,1}$ satisfies both bounds of (6.2.12).

Next, by Theorem 5.2.1(iii), for all $1 \leq k < j \leq \tilde{n}$,

$$\left|E[w_{X,j}\overline{w_{X,k}}]\right| \leq C(u_j^{-2d} + u_k^{-2d})j^{-1}\log j,$$

$$\left|E[w_{X,j}\overline{w_{\zeta,k}}]\right| \leq C(u_j^{-d} + u_k^{-d})j^{-1}\log j.$$

Since, by (6.2.2),

$$(f_jf_k)^{-1}(u_j^{-2d} + u_k^{-2d})^2 \leq 2C(u_ju_k)^{2d}(u_j^{-4d} + u_k^{-4d}) \leq C(j/k)^{2|d|},$$

$$f_j^{-1}(u_j^{-d} + u_k^{-d})^2 \leq 2Cu_j^{2d}(u_j^{-2d} + u_k^{-2d}) \leq C(j/k)^{2|d|},$$

we can estimate

$$s_{n,2} \leq C\sum_{1 \leq k < j \leq \tilde{n}}|b_{n,j}b_{n,k}|\left(\frac{j}{k}\right)^{2|d|}\frac{\log^2 j}{j^2}. \qquad (6.2.20)$$

Bound $|b_{n,j}b_{n,k}|$ by b_n^2, to obtain

$$s_{n,2} \leq Cb_n^2\log^2 n\sum_{1 \leq k < j \leq \tilde{n}}\frac{1}{k^{2|d|}j^{2-2|d|}} \leq Cb_n^2\log^2 n\sum_{1 \leq k < \tilde{n}}\frac{1}{k}$$

$$\leq Cb_n^2\log^3 n,$$

which, for $|d| < 1/2$, implies the first estimate of (6.2.12). Next, bound $|b_{n,j}|$ by b_n in (6.2.20), to obtain

$$s_{n,2} \le Cb_n \sum_{1 \le k < j \le \tilde{n}} |b_{n,k}| \frac{\log^2 j}{k^{2|d|} j^{2-2|d|}} \le Cb_n \sum_{1 \le k \le \tilde{n}} |b_{n,k}| \frac{\log^2 k}{k}$$

$$\le Cb_n \Big(\sum_{1 \le k \le \tilde{n}} b_{n,k}^2 \Big)^{1/2} \Big(\sum_{1 \le k \le \tilde{n}} \frac{\log^4 k}{k^2} \Big)^{1/2} \le Cb_n B_n,$$

thereby establishing the second bound of (6.2.12).

To show (6.2.13), recall that $(f_{X,j}/f_{\zeta,j})E|w_{\zeta,j}|^2 = f_{X,j}$. Therefore,

$$ER_n = \sum_{j=1}^{\tilde{n}} \frac{b_{n,j}}{f_{X,j}} \Big(E|w_{X,j}|^2 - \frac{f_{X,j}}{f_{\zeta,j}} E|w_{\zeta,j}|^2 \Big)$$

$$= \sum_{j=1}^{\tilde{n}} \frac{b_{n,j}}{f_{X,j}} \Big(E|w_{X,j}|^2 - f_{X,j} \Big).$$

Then, by (6.2.19) and (6.2.2),

$$|ER_n| \le C \sum_{j=1}^{\tilde{n}} |b_{n,j}| j^{-1} \log j \le Cb_n \sum_{j=1}^{\tilde{n}} j^{-1} \log j \le Cb_n \log^2 n,$$

which implies the first bound in (6.2.13).

To derive the second bound, let $K = (B_n/b_n)^{1/2}$. Notice that if $b_n = o(B_n)$, then as $K \to \infty$, $b_n K = (b_n/B_n)^{1/2} B_n = o(B_n)$. Thus,

$$|ER_n| \le C \Big(\sum_{j=1}^{K-1} |b_{n,j}| j^{-1} \log j + \sum_{j=K}^{\tilde{n}} |b_{n,j}| j^{-1} \log j \Big) \qquad (6.2.21)$$

$$\le C \Big\{ b_n K + \Big(\sum_{j=K}^{\tilde{n}} b_{n,j}^2 \Big)^{1/2} \Big(\sum_{j=K}^{\infty} j^{-2} \log^2 j \Big)^{1/2} \Big\}$$

$$= o(B_n).$$

This completes the proof of the second estimate in (6.2.13), and also of the Lemma 6.2.1.

Now we return to establishing the asymptotic normality of a *quadratic form in i.i.d. r.v.'s* ζ_j's. The CLT for quadratic forms in i.i.d. r.v.'s has been thoroughly investigated; see Guttorp and Lockhart (1988). The following theorem summarizes useful criteria for asymptotic normality, as given in Theorem 2.1 in Bhansali, Giraitis and Kokoszka (2007a). Let

$$Q_n = \sum_{i,j=1}^{n} c_{n,ij} \zeta_i \zeta_j$$

be a quadratic form where $c_{n,ts}$ are entries of a real symmetric $n \times n$ matrix $C_n = \{c_{n,ts},\ t,s = 1, \cdots, n\}$. Let $\|C_n\| = (\sum_{t,s=1}^n c_{n,ts}^2)^{1/2}$ and $\|C_n\|_{sp} = \max_{\|x\|=1} \|C_n x\|$ denote its Euclidean and spectral norms, respectively.

Theorem 6.2.2. *Suppose that $\{\zeta_j\} \sim \mathrm{IID}(0,1)$ and $E\zeta_0^4 < \infty$. Then*

$$\frac{\|C_n\|_{sp}}{\|C_n\|} \to 0, \tag{6.2.22}$$

implies

$$(\mathrm{Var}(Q_n))^{-1/2}(Q_n - EQ_n) \to_D \mathcal{N}(0,1).$$

In addition, if $\sum_{t=1}^n c_{n,tt}^2 = o(\|C_n\|^2)$, then $\mathrm{Var}(Q_n) \sim 2\|C_n\|^2$. Furthermore, in this case, if $E\zeta_0^4 < \infty$ is replaced by $E|\zeta_0|^{2+\delta} < \infty$, for some $\delta > 0$, then $(2\|C_n\|^2)^{-1/2}(Q_n - EQ_n) \to_D \mathcal{N}(0,1)$.

The following lemma derives the asymptotic distribution of the sum $S_{n,\zeta}$ of (6.2.6).

Lemma 6.2.2. *Let $\{\zeta_j\} \sim \mathrm{IID}(0,1)$ and $E\zeta_0^4 < \infty$. If*

$$b_n = o(B_n), \tag{6.2.23}$$

then

$$ES_{n,\zeta} = \sum_{j=1}^{\tilde{n}} b_{n,j}, \tag{6.2.24}$$

$$\mathrm{Var}(S_{n,\zeta}) = q_n^2, \tag{6.2.25}$$

$$q_n^{-1}(S_{n,\zeta} - ES_{n,\zeta}) \to_D \mathcal{N}(0,1), \tag{6.2.26}$$

with B_n and q_n^2 as in (6.2.7). Moreover,

$$q_n^2 \geq \min\left(1, \mathrm{Var}(\zeta_0^2)/2\right) B_n^2. \tag{6.2.27}$$

Proof. Write

$$S_{n,\zeta} = \frac{1}{n} \sum_{t,s=1}^n \sum_{j=1}^{\tilde{n}} e^{i(t-s)u_j} b_{n,j} \zeta_s \zeta_t = \sum_{t,s=1}^n c_n(t-s) \zeta_s \zeta_t,$$

where $c_n(t) := n^{-1} \sum_{j=1}^{\tilde{n}} b_{n,j} \cos(tu_j)$, $t = 1, 2, \cdots$. The matrix $C_n = (c_n(t-s))_{t,s=1,\cdots,n}$ is a symmetric $n \times n$ matrix with real entries. Hence,

(6.2.24) follows because the ζ_j's are IID$(0,1)$. For the same reason,

$$\text{Var}(S_{n,\zeta}) = 2 \sum_{s,t=1:\,t\neq s}^{n} c_n^2(t-s) + \text{Var}(\zeta_0^2) \sum_{t=1}^{n} c_n^2(t-t) \quad (6.2.28)$$

$$= 2\|C_n\|^2 + \text{Cum}_4(\zeta_0)n^{-1}\left(\sum_{j=1}^{\tilde{n}} b_{n,j}\right)^2$$

$$\geq \min(2, \text{Var}(\zeta_0^2))\|C_n\|^2,$$

since $\text{Var}(\zeta_0^2) - 2 = E\zeta_0^4 - 3 = \text{Cum}_4(\zeta_0)$ and $c_n(0) = n^{-1}\sum_{j=1}^{\tilde{n}} b_{n,j}$.

Next, we show that the weights $c_n(t-s)$ satisfy

$$\|C_n\|^2 = 2^{-1}B_n^2, \quad (6.2.29)$$

$$\|C_n\|_{sp} = o(\|C_n\|). \quad (6.2.30)$$

By Theorem 6.2.2, (6.2.30) implies

$$(\text{Var}(S_{n,\zeta}))^{-1/2}(S_{n,\zeta} - E[S_{n,\zeta}]) \to_D \mathcal{N}(0,1),$$

$$\text{Var}(S_{n,\zeta}) = B_n^2 + \text{Cum}_4(\zeta_0)n^{-1}\left(\sum_{j=1}^{\tilde{n}} b_{n,j}\right)^2,$$

which proves (6.2.26), whereas (6.2.29) with (6.2.28) proves (6.2.27).

To prove (6.2.29), use the definition of $c_n(t)$ to write

$$\|C_n\|^2 = \sum_{t,s=1}^{n} c_n^2(t-s) \quad (6.2.31)$$

$$= n^{-2} \sum_{j,k=1}^{\tilde{n}} b_{n,j}b_{n,k} \sum_{s,t=1}^{n} \cos((t-s)u_j)\cos((t-s)u_k).$$

Since $\tilde{n} = [n/2] - 1$, then $j + k < n$ in the above sums. We will use the following equality: for $1 \leq j, k \leq m$, $j + k < n$, and $a, b \in \mathbb{R}$,

$$\sum_{t=1}^{n} \cos(tu_j + a)\cos(tu_k + b) = \frac{n}{2}\cos(a-b)I(j=k). \quad (6.2.32)$$

To prove this equality, use the fact that $\cos(x) = (e^{ix} + e^{-ix})/2$ to write the l.h.s. of (6.2.32) as

$$\sum_{t=1}^{n} \frac{1}{4}\Big(e^{it(u_j+u_k)}e^{i(a+b)} + e^{-it(u_j+u_k)}e^{-i(a+b)}$$

$$+ e^{it(u_j-u_k)}e^{i(a-b)} + e^{-it(u_j-u_k)}e^{-i(a-b)}\Big).$$

Since by (2.1.5), $\sum_{t=1}^{n} e^{itu_l} = n(I(l=0)+I(l=n))$, this expression reduces to $(n/4)(e^{i(a-b)} + e^{-i(a-b)})I(j=k) = (n/2)\cos(a-b)I(j=k)$.

Hence, applying (6.2.32) in (6.2.31), yields (6.2.29):

$$\|C_n\|^2 = 2^{-1}\sum_{j=1}^{\tilde{n}} b_{n,j}^2 = 2^{-1}B_n^2.$$

To prove (6.2.30), let $x \in \mathbb{R}^n$ be such that $\|x\|^2 = 1$. Then

$$\|C_n x\|^2 = \sum_{t=1}^{n}(\sum_{s=1}^{n} c_n(t-s)x_s)^2 \qquad (6.2.33)$$

$$= \sum_{s,v=1}^{n} x_s x_v \left(\sum_{t=1}^{n} c_n(t-s)c_n(t-v)\right),$$

where, by (6.2.32),

$$\sum_{t=1}^{n} c_n(t-s)c_n(t-v)$$

$$= n^{-2}\sum_{j,k=1}^{\tilde{n}} b_{n,j}b_{n,k}\sum_{t=1}^{n}\cos((t-s)u_j)\cos((t-v)u_k)$$

$$= \frac{1}{2n}\sum_{j=1}^{\tilde{n}} b_{n,j}^2 \cos((s-v)u_j).$$

Hence,

$$\|C_n x\|^2 = \frac{1}{2n}\sum_{j=1}^{\tilde{n}} b_{n,j}^2 \sum_{s,v=1}^{n}\cos((s-v)u_j)x_s x_v.$$

Thus, by the equality,

$$\sum_{s,v=1}^{n}\cos((s-v)u_j)x_s x_v = |\sum_{s=1}^{n} e^{isu_j}x_s|^2,$$

$$\|C_n x\|^2 = \frac{1}{2n}\sum_{j=1}^{\tilde{n}} b_{n,j}^2 |\sum_{s=1}^{n} e^{isu_j}x_s|^2 \leq \frac{1}{2n}b_n^2\sum_{j=1}^{n} |\sum_{s=1}^{n} e^{isu_j}x_s|^2$$

$$= \frac{1}{2n}b_n^2\sum_{t,s=1}^{n}\sum_{j=1}^{n} e^{i(t-s)u_j}x_t x_s.$$

By (2.1.5), $\sum_{j=1}^{n} e^{i(t-s)u_j} = nI(t=s)$. Therefore,

$$\|C_n x\|^2 \leq \frac{1}{2}b_n^2\sum_{t=s=1}^{n} x_t^2 = \frac{1}{2}b_n^2\|x\|^2, \qquad \|C_n\|_{sp} \leq (1/\sqrt{2})b_n.$$

Since $b_n = o(B_n)$, and $B_n = \sqrt{2}\|C_n\|$ by (6.2.29), this proves (6.2.30), and also completes the proof of the lemma.

The next corollary is an application of Theorem 6.2.2 and Lemma 6.2.2.

Corollary 6.2.2. *Let $\{\zeta_j\} \sim \mathrm{IID}(0, \sigma_\zeta^2)$ and $1 \le m = m_n \le \tilde{n}$ be a sequence of positive integers such that $m = m_n \to \infty$. Then*

$$m^{-1} \sum_{j=1}^{m} \frac{I_{\zeta,j}}{f_{\zeta,j}} \to_p 1. \qquad (6.2.34)$$

Furthermore, if $E\zeta_0^4 < \infty$, then

$$\sqrt{m}\Big(\frac{1}{m} \sum_{j=1}^{m} \frac{I_{\zeta,j}}{f_{\zeta,j}} - 1\Big) \to_D \mathcal{N}(0,1), \qquad \text{if } m = o(n), \qquad (6.2.35)$$

$$\to_D \mathcal{N}\Big(0, 1 + \frac{\mathrm{Cum}_4(\zeta_0)}{2\sigma_\zeta^4}\Big), \qquad \text{if } m = \tilde{n}.$$

Proof. Let $b_{n,j} = 1$, $j = 1, \cdots, m$ and $b_{n,j} = 0$, $j = m+1, \cdots, \tilde{n}$. Then

$$S_{n,\zeta} \equiv \sum_{j=1}^{\tilde{n}} b_{n,j} \frac{I_{\zeta,j}}{f_{\zeta,j}} = \sum_{j=1}^{m} \frac{I_{\zeta,j}}{f_{\zeta,j}}, \qquad B_n^2 \equiv \sum_{j=1}^{\tilde{n}} b_{n,j}^2 = m,$$

$$v_n^2 \equiv B_n^2 + \mathrm{Cum}_4(\zeta_0) \frac{1}{n} \Big(\sum_{j=1}^{\tilde{n}} b_{n,j}\Big)^2 = m + \frac{m^2}{n} \mathrm{Cum}_4(\zeta_0).$$

Since $E[I_{\zeta,n}/f_{\zeta,j}] = 1$ then $E[m^{-1}S_{n,\zeta}] = 1$. By (6.2.28) and (6.2.29), $\mathrm{Var}(S_{n,\zeta}) \le CB_n^2 = Cm$. Thus, $\mathrm{Var}(m^{-1}S_{n,\zeta}) \to 0$, which in turn completes the proof of (6.2.34). Equation (6.2.35) follows by a straightforward application of Lemma 6.2.2.

A general case: Asymptotic normality of $Q_{n,X}$. We now focus on

$$Q_{n,X} := \sum_{j=1}^{\tilde{n}} b_{n,j} I_{X,j}.$$

The Bartlett approximation $I_{X,j} \sim f_{X,j}(I_{\zeta,j}/f_{\zeta,j})$ suggests approximating $Q_{n,X}$ by the sum

$$Q_{n,\zeta} := \sum_{j=1}^{\tilde{n}} (b_{n,j} f_{X,j})\Big(\frac{I_{\zeta,j}}{f_{\zeta,j}}\Big) = \sum_{j=1}^{\tilde{n}} b_{n,j} f_{X,j}(2\pi) I_{\zeta,j}.$$

Corollary 6.2.1 provides tools for establishing the variance and the mean-squared error of $Q_{n,X} - Q_{n,\zeta}$.

In Theorem 6.2.1 above, the spectral density f_X can be unbounded at 0, but is differentiable on $(0, \pi)$. Then the asymptotic normality of the sums $S_{n,X} = \sum_{j=1}^{\tilde{n}} b_{n,j} (I_{X,j}/f_{X,j})$ holds under the Lindeberg–Feller-type condition (6.2.8) on the weights $b_{n,j}$.

Now we turn to the case when f_X is bounded and continuous on Π, with no assumptions about its differentiability, i.e., $d = 0$ in (6.2.2). In addition, we assume that f_X is bounded away from 0 and ∞:

$$0 < C_1 \leq f_X(u) \leq C_2 < \infty, \ u \in \Pi, \ \exists \, 0 < C_1, C_2 < \infty. \qquad (6.2.36)$$

The restriction $f_X(u) \geq C_1 > 0$ simplifies the conditions.

Theorem 6.2.3 below shows that under the Lindeberg–Feller-type condition (6.2.8) on the weights $b_{n,j}$, the continuity of f_X, or more precisely, the continuity of the transfer function A_X, suffices for the asymptotic normality of the centered sums $Q_{n,X} - EQ_{n,X}$. To obtain an upper bound on the variance $\mathrm{Var}(Q_{n,X})$ it suffices to assume f_X to be continuous, whereas the satisfactory asymptotics of $EQ_{n,X}$ requires f_X to be Lipschitz(β), $\beta > 1/2$; see Theorem 6.2.5.

By Lemma 6.2.2, $Q_{n,\zeta}$ has the following mean and variance:

$$v_n^2 := \sum_{j=1}^{\tilde{n}} (b_{n,j} f_{X,j})^2 + \mathrm{Cum}_4(\zeta_0) \frac{1}{n} \left(\sum_{j=1}^{\tilde{n}} b_{n,j} f_{X,j} \right)^2,$$

$$\mathrm{Var}(Q_{n,\zeta}) = v_n^2, \qquad EQ_{n,\zeta} = \sum_{j=1}^{\tilde{n}} b_{n,j} f_{X,j}.$$

Observe that $Q_{n,X}$, $Q_{n,\zeta}$, and v_n^2, respectively, are like $S_{n,X}$, $S_{n,\zeta}$, and q_n^2 of (6.2.6) and (6.2.7) with $b_{n,j}$ replaced by $b_{n,j} f_{X,j}$. Let

$$b_{f,n} := \max_{j=1,\cdots,\tilde{n}} |b_{n,j}| f_{X,j}, \qquad B_{f,n}^2 := \sum_{j=1}^{\tilde{n}} (b_{n,j} f_{X,j})^2.$$

Similarly as in (6.2.10), we can show that the variance v_n^2 has the same order as $B_{f,n}^2$, i.e., for some $C_1, C_2 > 0$,

$$C_1 B_{f,n}^2 \leq v_n^2 \leq C_2 B_{f,n}^2, \qquad \text{and} \qquad (6.2.37)$$

$$C_1 B_n^2 \leq v_n^2 \leq C_2 B_n^2, \qquad \text{under (6.2.36)}.$$

Let $\mathcal{C}(\Pi)$ denote the class of bounded (complex-valued) continuous functions on Π, and $\Lambda_\beta(\Pi)$, $0 < \beta \leq 1$ denote Lipschitz continuous functions of order β. The next theorem establishes the asymptotic normality of $Q_{n,X}$.

Theorem 6.2.3. *Suppose the linear process $\{X_j, j \in \mathbb{Z}\}$ of (6.2.1) is such that $E\zeta_0^4 < \infty$ and the real weights $b_{n,j}$ satisfy (6.2.8).*

If f_X satisfies (6.2.36) and the transfer function A_X of $\{X_j\}$ is continuous, then

$$\text{Var}(Q_{n,X}) = v_n^2 + o(v_n^2), \tag{6.2.38}$$

$$v_n^{-1}(Q_{n,X} - EQ_{n,X}) \to_D \mathcal{N}(0,1).$$

In addition, if $f_X \in \Lambda_\beta(\Pi)$, with $1/2 < \beta \leq 1$, then

$$EQ_{n,X} = \sum_{j=1}^{\tilde{n}} b_{n,j} f_{X,j} + o(v_n), \tag{6.2.39}$$

$$v_n^{-1}\left(Q_{n,X} - \sum_{j=1}^{\tilde{n}} b_{n,j} f_{X,j}\right) \to_D \mathcal{N}(0,1).$$

In the next theorem we extend the result of asymptotic normality of $Q_{n,X}$ to the case where the spectral density f_X is not bounded in the neighborhood of 0, i.e., $d > 0$, or is not bounded away from 0, i.e., $d < 0$. In these cases the second bound of (6.2.37) does not hold. The Lindeberg–Feller-type condition (6.2.8) now has to be formulated using the weights $b_{n,j} f_{X,j}$ and we need to make some additional smoothness assumptions on A_X in a small neighborhood of 0. We assume that A_X can be factored into a product $A_X = hG$ of a differentiable function h, which may have a pole at 0, and a continuous bounded function G. In particular, if A_X satisfies (6.2.3), we take $G \equiv 1$.

Theorem 6.2.4. *Suppose $\{X_j, j \in \mathbb{Z}\}$ is the linear process of (6.2.1) with $E\zeta_0^4 < \infty$. Assume that f_X satisfies (6.2.2) with parameter $|d| < 1/2$, and the transfer function A_X can be factored as $A_X = hG$, where G is continuous and bounded away from 0 and ∞, h is differentiable having derivative $\dot{h}$, and for some $0 < C, C_1, C_2 < \infty$,*

$$C_1|u|^{-d} \leq |h(u)| \leq C_2|u|^{-d}, \tag{6.2.40}$$

$$|\dot{h}(u)| \leq C|u|^{-1-d}, \quad 0 < |u| \leq \pi.$$

If, in addition, the real weights $b_{n,j}$ are such that

$$\frac{\max_{j=1,\cdots,\tilde{n}} |b_{n,j} f_{X,j}|}{(\sum_{j=1}^{\tilde{n}} (b_{n,j} f_{X,j})^2)^{1/2}} \equiv \frac{b_{f,n}}{B_{f,n}} \to 0, \tag{6.2.41}$$

then (6.2.38) still holds.

If, in addition, $g \in \Lambda_\beta(\Pi)$, with $\beta > 1/2$, then (6.2.39) also holds.

Proof of Theorems 6.2.3 and 6.2.4. Write

$$Q_{n,X} - EQ_{n,X} = Q_{n,\zeta} - EQ_{n,\zeta} + r_n,$$

$$r_n := Q_{n,X} - Q_{n,\zeta} - E[Q_{n,X} - EQ_{n,\zeta}].$$

The proof of both theorems follows from this decomposition and Lemmas 6.2.2 and 6.2.3. The latter lemma will be proved shortly. In (i) and (ii) of this lemma it is shown that $Er_n^2 = o(v_n^2)$ under the assumptions of Theorems 6.2.3 and 6.2.4. Therefore, the result (6.2.38) of Theorem 6.2.3 follows, noticing that, by Lemma 6.2.2, under assumption (6.2.41), $\mathrm{Var}(Q_{n,\zeta}) = v_n^2 + o(v_n^2)$ and $v_n^{-1}(Q_{n,\zeta} - EQ_{n,\zeta}) \to_D \mathcal{N}(0,1)$.

The second result (6.2.39) of these theorems follows from (6.2.38) and (6.2.53) of Theorem 6.2.5 below. This completes the proof of the theorems.

Lemma 6.2.3 below shows that the order of approximation of $Q_{n,X} - EQ_{n,X}$ by $Q_{n,\zeta} - EQ_{n,\zeta}$ is determined by the smoothness of the transfer function A_X. For example, by Lemma 6.2.3(i), if A_X is bounded and continuous, then

$$Q_{n,X} - EQ_{n,X} = Q_{n,\zeta} - EQ_{n,\zeta} + o_p(v_n), \qquad (6.2.42)$$

where $v_n^2 = \mathrm{Var}(Q_{n,\zeta})$. If, in addition, A_X has a bounded derivative, then the order improves to $o_p(n^{-1/2}(\log n) v_n)$ without requiring any additional assumptions on $b_{n,j}$. Lemma 6.2.3(ii) shows that if A_X is discontinuous at 0, then (6.2.42) is valid under an additional regularity behavior on A_X in a neighborhood of 0, as long as the weights $b_{n,j}$ satisfy (6.2.41).

To state the lemma, we need the following notation. For a complex-valued function $h(u)$, $u \in \Pi$, define

$$\epsilon_{n,h} := n^{-1} \log^2 n, \qquad h \in \Lambda_1(\Pi),$$
$$:= n^{-\beta}, \qquad h \in \Lambda_\beta(\Pi), \, 0 < \beta < 1,$$
$$:= \delta_n, \qquad \delta_n \to 0, \quad h \in \mathcal{C}(\Pi).$$

Lemma 6.2.3. *Assume that $\{X_j\}$ is as in (6.2.1), $E\zeta_0^4 < \infty$, and f_X satisfies (6.2.2) with parameter $|d| < 1/2$. Then*

$$Q_{n,X} - EQ_{n,X} = Q_{n,\zeta} - EQ_{n,\zeta} + r_n, \qquad (6.2.43)$$

where r_n can be bounded as follows:
(i) *If $A_X \in \Lambda_\beta(\Pi)$, $0 < \beta \leq 1$, or $A_X \in \mathcal{C}(\Pi)$, then*

$$Er_n^2 \leq C\epsilon_{n,A_X} B_n^2 = o(v_n^2). \qquad (6.2.44)$$

(ii) *If* $A_X = hG$, *where* h *satisfies (6.2.40) and either* $G \in \mathcal{C}(\Pi)$ *or* $G \in \Lambda_\beta(\Pi)$, $0 < \beta \leq 1$, *then*

$$Er_n^2 \leq C\big(\min(b_{f,n}^2 \log^3 n, \, b_{f,n} B_{f,n}) + \epsilon_{n,G} B_{f,n}^2 \big), \qquad (6.2.45)$$

$$\leq C \min(b_{f,n}^2 \log^3 n, \, b_{f,n} B_{f,n}), \quad G \in \Lambda_1(\Pi).$$

If, in addition, (6.2.41) holds, then

$$Er_n^2 = o(v_n^2). \qquad (6.2.46)$$

Proof. Let

$$\widetilde{R}_n = Q_{n,X} - Q_{n,\zeta} = \sum_{j=1}^{\tilde{n}} b_{n,j} \{ I_{X,j} - \frac{f_{X,j} I_{\zeta,j}}{f_{\zeta,j}} \}.$$

By Corollary 6.2.1,

$$\mathrm{Var}(\widetilde{R}_n) \leq C(s_{n,1} + s_{n,2}), \qquad (6.2.47)$$

where

$$s_{n,1} := C \sum_{j=1}^{\tilde{n}} b_{n,j}^2 \Big[\big(E|w_{X,j}|^2 - f_{X,j}\big)^2 + f_{X,j}\big|E|w_{X,j}|^2 - f_{X,j}\big|$$

$$+ f_{X,j}\big|E[w_{X,j}\overline{w_{\zeta,j}}] - f_{X\zeta,j}\big|^2 + f_{X,j}^{3/2}\big|E[w_{X,j}\overline{w_{\zeta,j}}] - f_{X\zeta,j}\big| \Big],$$

$$s_{n,2} := \sum_{1 \leq k < j \leq \tilde{n}} |b_{n,j} b_{n,k}| \Big(\big|E[w_{X,j}\overline{w_{X,k}}]\big|^2 + f_{X,k}\big|E[w_{X,j}\overline{w_{\zeta,k}}]\big|^2 \Big).$$

Proof of (i). We shall show that

$$E(\widetilde{R}_n - E\widetilde{R}_n)^2 \leq C\epsilon_{n,A_X} B_n^2, \qquad (6.2.48)$$

which, in view of (6.2.37), proves (6.2.44). The proof of (6.2.48) is similar to that of Lemma 6.2.1. For the sake of completeness, we provide the details.

We shall consider three cases separately.

Case 1. $A_X \in \Lambda_1(\Pi)$. Then, by Theorem 5.2.1(i),

$$\big|E|w_{X,j}|^2 - f_{X,j}\big| \vee \big|E[w_{X,j}\overline{w_{\zeta,j}}] - f_{X\zeta,j}\big| \leq Cn^{-1}\log n,$$

$$\big|E[w_{X,j}\overline{w_{X,k}}]\big| \vee \big|E[w_{X,j}\overline{w_{\zeta,k}}]\big| \leq Cn^{-1}\log n, \quad 1 \leq k < j \leq \tilde{n}.$$

Therefore,

$$s_{n,1} \leq Cn^{-1}\log n \sum_{j=1}^{\tilde{n}} b_{n,j}^2 = Cn^{-1}\log n \, B_n^2,$$

$$s_{n,2} := Cn^{-2}\log^2 n \sum_{1 \leq k < j \leq \tilde{n}} |b_{n,j} b_{n,k}|$$

$$\leq Cn^{-1}\log^2 n \sum_{j=1}^{\tilde{n}} b_{n,j}^2 = Cn^{-1}\log^2 n \, B_n^2,$$

which proves (6.2.48).

Case 2. $A_X \in \Lambda_\beta(\Pi)$, $0 < \beta < 1$. Then by Theorem 5.2.1(i),

$$|E|w_{X,j}|^2 - f_{X,j}| \vee |E[w_{X,j}\overline{w_{\zeta,j}}] - f_{X\zeta,j}| \leq Cn^{-\beta},$$
$$|E[w_{X,j}\overline{w_{X,k}}]| \vee |E[w_{X,j}\overline{w_{\zeta,k}}]| \leq n^{-\beta}\ell_n(\beta; n - k), \qquad k < j.$$

Note that for $1 \leq k < j \leq \tilde{n} < n/2$, $j - k \leq n - j + k$, and hence

$$\ell_n(\beta; j - k) \leq C\frac{\log(2 + j - k)}{(2 + j - k)^{1-\beta}}, \tag{6.2.49}$$
$$(n^{-\beta}\ell_n(\beta; j - k))^2 \leq C\frac{\log^2(2 + j - k)}{n^\beta(2 + j - k)^{2-\beta}}.$$

Apply this fact to obtain that for $0 < \beta < 1$,

$$s_{n,1} \leq Cn^{-\beta}\sum_{j=1}^{\tilde{n}} b_{n,j}^2 = Cn^{-\beta}B_n^2,$$

$$s_{n,2} \leq C\sum_{1 \leq k < j \leq \tilde{n}} |b_{n,j}b_{n,k}| (n^{-\beta}\ell_n(\beta; j - k))^2$$

$$\leq Cn^{-\beta}\sum_{1 \leq k < j \leq \tilde{n}} |b_{n,j}b_{n,k}| \frac{\log^2(2 + j - k)}{(2 + j - k)^{2-\beta}}$$

$$\leq Cn^{-\beta}\sum_{j=1}^{\tilde{n}} b_{n,j}^2 \sum_{u=0}^{\infty} \frac{\log^2(2 + u)}{(2 + u)^{2-\beta}} \leq Cn^{-\beta}B_n^2,$$

which proves (6.2.48).

Case 3. $A_X \in \mathcal{C}(\Pi)$. By Theorem 5.2.1(ii),

$$|E|w_{X,j}|^2 - f_{X,j}| \vee |E[w_{X,j}\overline{w_{\zeta,j}}] - f_{X\zeta,j}| \leq C\delta_n,$$
$$|E[w_{X,j}\overline{w_{X,k}}]| \vee |E[w_{X,j}\overline{w_{\zeta,k}}]| \leq C\delta_n\ell(\epsilon, j - k), \qquad 1 \leq k < j \leq \tilde{n},$$

for any $0 < \epsilon < 1/2$, with some $\delta_n \to 0$ that does not depend on k, j, and n. Next, observe that (6.2.48) follows by the same argument as in case 2 above. This completes the proof of (i) of the lemma.

Proof of (ii). First, we prove (6.2.45). As above, we need to bound $s_{n,1}$ and $s_{n,2}$ of (6.2.47).

Recall that $f_X = |A_X|^2/(2\pi)$, $f_{X\zeta} = A_X/(2\pi)$, $A_X = h(u)G(u)$, where h satisfies (6.2.40), which together with (6.2.2) implies that G is bounded away from infinity and zero. For $1 \leq k \leq j \leq \tilde{n}$, define

$$\tilde{r}_{n,jk} = 0, \qquad\qquad\qquad G \in \Lambda_1(\Pi),$$

$$= n^{-\beta} \frac{\log(2+j-k)}{(2+j-k)^{1-\beta}}, \qquad G \in \Lambda_\beta(\Pi), \ 0 < \beta < 1,$$

$$= \delta_n \frac{\log(2+j-k)}{(2+j-k)^{1-\epsilon}}, \qquad G \in \mathcal{C}(\Pi), \ 0 < \epsilon < 1/2,$$

where $\delta_n \to 0$.

By Theorem 5.2.1(iv) and (6.2.49), for $1 \leq k \leq j \leq \tilde{n}$,

$$\left| E[w_{X,j}\overline{w_{X,k}}] - f_{X,j}I(j=k) \right|$$
$$\leq C\{(u_k^{-2d} + u_j^{-2d})j^{-1}\log j + (u_k^{-2d} \wedge u_j^{-2d})\tilde{r}_{n,jk}\}$$
$$\left| E[w_{X,j}\overline{w_{\zeta,k}}] - f_{X\zeta,j}I(j=k) \right|$$
$$\leq C\{(u_k^{-d} + u_j^{-d})j^{-1}\log j + (u_k^{-d} \wedge u_j^{-d})\tilde{r}_{n,jk}\}.$$

Since $f_X = |A_X|^2/(2\pi) = |hG|^2/(2\pi)$, the assumptions on h and G used in part (ii) of the lemma imply that for all $u \in \Pi$,

$$f_X(u) \leq C|u|^{-2d}, \qquad f_X^{-1}(u) \leq C|u|^{2d},$$

$$|f_{X\zeta}(u)| \leq C|u|^{-d}, \qquad |f_{X\zeta}^{-1}(u)| \leq C|u|^d.$$

Therefore, for $1 \leq k \leq j$,

$$(f_{x,j}f_{x,k})^{-1}(u_k^{-2d} + u_j^{-2d})^2 \leq C|j/k|^{2|d|},$$

$$(f_{x,j}f_{x,k})^{-1}(u_k^{-2d} \wedge u_j^{-2d})^2 \leq C,$$

$$(f_{X,j})^{-1}(u_k^{-d} + u_j^{-d})^2 \leq C|j/k|^{2|d|},$$

$$(f_{X,j})^{-1}(u_k^{-d} \wedge u_j^{-d})^2 \leq C.$$

To prove (6.2.45) of (ii), we shall use the bound (6.2.47). It suffices to show that $s_{n,1} + s_{n,2}$ can be bounded by the r.h.s. of (6.2.45). The above bounds readily yield that

$$s_{n,1} \leq C\sum_{j=1}^{\tilde{n}}(b_{n,j}f_{X,j})^2(j^{-1}\log j + \tilde{r}_{n,jj}),$$

$$s_{n,2} \leq C\sum_{1 \leq k < j \leq \tilde{n}} |b_{n,j}f_{X,j}| |b_{n,k}f_{X,k}| \left((\frac{j}{k})^{2|d|}\frac{\log^2 j}{j^2} + \tilde{r}_{n,jk}^2 \right).$$

The argument used in evaluating $s_{n,1}$ and $s_{n,2}$ in Lemma 6.2.1 yields

$$\sum_{j=1}^{\tilde{n}}(b_{n,j}f_{X,j})^2\frac{\log j}{j} + \sum_{1 \leq k < j \leq \tilde{n}} |b_{n,j}f_{X,j}||b_{n,k}f_{X,k}|(\frac{j}{k})^{2|d|}\frac{\log^2 j}{j^2}$$
$$\leq C\min\left(b_{f,n}^2\log^3(n), \ b_{f,n}B_{f,n}\right),$$

whereas estimation in cases 2 and 3 above yields

$$\sum_{j=1}^{\tilde{n}} (b_{n,j} f_{X,j})^2 \tilde{r}_{n,jk} + \sum_{1 \le k < j \le \tilde{n}} |b_{n,j} f_{X,j}| |b_{n,k} f_{X,k}| \tilde{r}_{n,jk}^2 \le C \epsilon_{n,G} B_{f,n}^2.$$

Therefore,

$$s_{n,1} + s_{n,2} \le C \{ \min (b_{f,n}^2 \log^3(n), \ b_{f,n} B_{f,n}) + \epsilon_{n,G} B_{f,n}^2 \}, \qquad (6.2.50)$$

which proves (6.2.45).

Observe that $\epsilon_{n,G} \to 0$. Therefore, (6.2.45), (6.2.41), and (6.2.37) imply (6.2.46). This completes the proof of the lemma.

As above, proving the CLT for $v_n^{-1}(Q_{n,X} - \sum_{j=1}^{\tilde{n}} b_{n,j} f_{X,j})$ requires some smoothness of the spectral density f_X and the transfer function A_X. Conditions on A_X can be relaxed if we wish to establish only an upper bound for the mean-squared error of the estimator $Q_{n,X}$ of $\sum_{j=1}^{\tilde{n}} b_{n,j} f_{X,j}$ as is shown in the next theorem.

Theorem 6.2.5. *Let $\{X_j\}$ be as in (6.2.1) and $E\zeta_0^4 < \infty$. Assume that $f_X(u) = |u|^{-2d} g(u)$, $|d| < 1/2$, where g is a continuous function, bounded away from 0 and ∞.*
(i) *Then*

$$E(Q_{n,X} - EQ_{n,X})^2 \le C B_{f,n}^2. \qquad (6.2.51)$$

(ii) *In addition,*

$$E\left(Q_{n,X} - \sum_{j=1}^{\tilde{n}} b_{n,j} f_{X,j}\right)^2 \le C B_{f,n}^2, \qquad (6.2.52)$$

in each of the following three cases.
    (c1) $d = 0$, $g \in \Lambda_\beta(\Pi)$, $1/2 < \beta \le 1$,
    (c2) $d \ne 0$, $g \in \Lambda_\beta(\Pi)$, $1/2 < \beta \le 1$,
    (c3) $|\dot{f}_X(u)| \le C u^{-1-2d}$, $0 < u \le \pi$.
Moreover, in case (c1),

$$EQ_{n,X} - \sum_{j=1}^{\tilde{n}} b_{n,j} f_j = o(B_{f,n}). \qquad (6.2.53)$$

If the $b_{n,j}$ satisfy (6.2.41), then (6.2.53) holds also in cases (c2) and (c3).

Proof. (i) Recall that $I_{X,j} = |w_{X,j}|^2$. By Proposition 6.2.1,

$$E(Q_{n,X} - EQ_{n,X})^2 = \text{Var}\left(\sum_{j=1}^{\tilde{n}} b_{n,j} I_{X,j}\right)$$

$$\leq C \sum_{j,k=1}^{\tilde{n}} |b_{n,j} b_{n,k}| \, |E[w_{X,j} \overline{w_{X,k}}]|^2.$$

For $j = k$, bounding $(E|w_{X,j}|^2)^2 \leq 2(E|w_{X,j}|^2 - f_{X,j})^2 + 2f_{X,j}^2$, and letting

$$s'_{n,1} := \sum_{j=1}^{\tilde{n}} b_{n,j}^2 (E|w_{X,j}|^2 - f_{X,j})^2,$$

$$s'_{n,2} := \sum_{1 \leq k < j \leq \tilde{n}} |b_{n,j} b_{n,k}| \, |E[w_{X,j} \overline{w_{X,k}}]|^2,$$

we obtain

$$E(Q_{n,X} - EQ_{n,X})^2 \leq C(s'_{n,1} + s'_{n,2} + B_{f,n}^2). \tag{6.2.54}$$

Under the assumptions of this theorem, by Theorem 5.2.1(iv), for $1 \leq k < j \leq \tilde{n}$ and $0 < \epsilon < 1/2$,

$$|E|w_{X,j}|^2 - f_{X,j}| \leq C u_j^{-2d}(j^{-1} \log j + \delta_n),$$

$$|E[w_{X,j} \overline{w_{X,k}}]| \leq C(u_k^{-2d} + u_j^{-2d})j^{-1} \log j + (u_k^{-2d} \wedge u_j^{-2d})\delta_n \ell(\epsilon, j - k),$$

where $\delta_n \to 0$. Observe that $s'_{n,1} \leq s_{n,1}$ and $s'_{n,2} \leq s_{n,2}$, where $s_{n,1}$ and $s_{n,2}$ are as in (6.2.18). Therefore, the same argument as used in proving (6.2.50) implies that $s_{n,1} + s_{n,2}$ satisfies the bound (6.2.50), which in turn yields

$$s_{n,1} + s_{n,2} \leq C\left(b_{f,n} B_{f,n} + \epsilon_{n,G} B_{f,n}^2\right) \leq C B_{f,n}^2,$$

since $b_{f,n} \leq B_{f,n}$. This completes the proof of (6.2.51).

(ii) Observe that

$$
\begin{aligned}
|E|w_{X,j}|^2 - f_{X,j}| &\leq C u_j^{-2d} n^{-\beta}, &&\text{in case (c1)}\\
&\leq C u_j^{-2d}(j^{-1} \log j + n^{-\beta}), &&\text{in case (c2)}\\
&\leq C u_j^{-2d}(j^{-1} \log j), &&\text{in case (c3)}
\end{aligned}
$$

by parts (i), (iv), and (iii) of Theorem 5.2.1, respectively. Let

$$q_n := \left|EQ_{n,X} - \sum_{j=1}^{\tilde{n}} b_{n,j} f_{X,j}\right| = \left|\sum_{j=1}^{\tilde{n}} b_{n,j}(E|w_{X,j}|^2 - f_{X,j})\right|.$$

Under the assumptions of the theorem, $f_{X,j}^{-1} \leq Cu^{2d}$, $0 < u \leq \pi$. Thus, in case (c1),

$$q_n \leq C \sum_{j=1}^{\tilde{n}} |b_{n,j} f_{X,j}| n^{-\beta} \leq Cn^{1/2-\beta} \Big(\sum_{j=1}^{\tilde{n}} (b_{n,j} f_{X,j})^2 \Big)^{1/2} \qquad (6.2.55)$$

$$= o(B_{f,n}),$$

which proves (6.2.52) and (6.2.53).

In case (c2),

$$q_n \leq C \sum_{j=1}^{\tilde{n}} |b_{n,j} f_{X,j}| (j^{-1} \log n + n^{-\beta}).$$

By the same argument as in the proof of (6.2.21), it follows that

$$\sum_{j=1}^{\tilde{n}} |b_{n,j} f_{X,j}| j^{-1} \log j = O(B_{f,n}),$$

$$= o(B_{f,n}), \quad \text{if (6.2.41) holds,}$$

which together with (6.2.55) yields (6.2.52) and (6.2.53).

In case (c3), the proof of (6.2.52) and (6.2.53) is the same as in case (c2). This completes the proof of the theorem.

Remark 6.2.1. Consider now the sum

$$Q_{n,X} = \sum_{j=1}^{\theta n} b_{n,j} I_{X,j}, \qquad 0 < \theta < 1/2, \qquad (6.2.56)$$

where summation is taken over a fraction $\{1, \cdots, \theta n\}$ of the set $\{1, \cdots, \tilde{n}\}$. Then the periodograms $I_{X,j}$ used in $Q_{n,X}$ are based on the frequencies u_j from the zero neighborhood $[0, \Delta]$, $\Delta = 2\pi\theta$, which is a subinterval of $[0, \pi]$. Observe that smoothness conditions on f_X and A_X are only required to obtain upper bounds on the covariances $E[w_{X,j} \overline{w_{X,k}}]$ and $E[w_{X,j} \overline{w_{\zeta,k}}]$ in Theorem 5.2.1. Therefore, it follows that in order for these bounds to be valid at frequencies $u_j \in [0, \Delta]$ it suffices to impose smoothness conditions on f_X and A_X on a slightly larger interval $[0, a]$, $a > \Delta$, covering $[0, \Delta]$.

Hence, for the sum $Q_{n,X}$ of (6.2.56), all the results derived in this section above hold if the conditions on f_X and A_X are satisfied on some interval $[0, a]$, with $a > \Delta$, instead of $[0, \pi]$. For example, Theorem 6.2.3 holds if A_X is continuous on $[0, a]$. Theorem 6.2.4 holds if on $[0, a]$, $A_X = hG$

and h, f satisfy (6.2.40) and (6.2.2). Theorem 6.2.5 is valid, if on $[0, a]$, $f_X(u) = |u|^{-2d}g(u)$ and g is Lipschitz continuous with order $\beta > 1/2$.

No restrictions are required on f_X on the interval $[a, \pi]$, except the integrability condition $\int_a^\pi f_X(u)du < \infty$.

Remark 6.2.2. (i) Consider the stationary ARFIMA(p, d, q) model

$$\phi(B)X_j = \theta(B)(1 - B)^{-d}\zeta_j, \quad j \in \mathbb{Z}, \qquad \{\zeta_j\} \sim \text{IID}(0, \sigma_\zeta^2).$$

We shall show that this model satisfies the smoothness and differentiability conditions (6.2.2) and (6.2.3) pertaining to the spectral density f_X and the transfer function A_X. Indeed, when the complex roots of the polynomials $\phi(z)$ and $\theta(z)$ lie outside the unit circle $\{|z| \leq 1\}$, the operator $\phi(B)^{-1}\theta(B) = \sum_{k=0}^\infty b_k B^k$ can be written as a series of powers of B^k and the above equations can be rewritten in the form (5.3.11) as

$$X_j = (1 - B)^{-d}Y_j, \quad Y_j = \phi(B)^{-1}\theta(B)\zeta_j = \sum_{k=0}^\infty b_k\zeta_{j-k}, \quad j \in \mathbb{Z},$$

where $\{Y_j\}$ is a short memory process with absolutely summable weights b_k; see Section 7.2. Its transfer and spectral density functions

$$A_Y(u) = \frac{\theta(e^{-iu})}{\phi(e^{-iu})} = \sum_{k=0}^\infty b_k e^{-iuk}, \quad f_Y(u) = \frac{\sigma_\zeta^2}{2\pi}|A_Y(u)|^2, \quad u \in \Pi,$$

are continuous and bounded away from 0. Moreover, A_Y and f_Y have bounded derivatives. Then, by (5.3.13), with $h(u) = (1 - e^{-iu})^{-d}$,

$$f_X(u) = |h(u)|^2 f_Y(u) \equiv |u|^{-2d}g(u), \tag{6.2.57}$$
$$A_X(u) = h(u)A_Y(u),$$

where $g(u) = (|h(u)|^2/|u|^{-2d})f_Y(u) = (2|\sin(u/2)|/|u|)^{-2d}f_Y(u)$ is a continuous function on Π, bounded away from 0 and ∞. Hence, f_X satisfies assumption (6.2.2).

Moreover, by (5.3.14),

$$|\dot{A}_X(u)| \leq C(|\dot{h}(u)\|A_Y(u)| + |h(u)\|\dot{A}_Y(u)|)$$
$$\leq C|1 - e^{-iu}|^{-d-1} \leq C|u|^{-d-1}, \qquad 0 < |u| < \pi,$$

and hence A_X satisfies (6.2.3).

(ii) Now consider a more general process $\{X_j\}$,

$$X_j = (1 - B)^{-d}Y_j, \quad j \in \mathbb{Z}, \qquad |d| < 1/2,$$

defined as in (5.3.11). Using (5.3.13), it has spectral density and transfer function as in (6.2.57). Hence, the same argument as used in (6.2.57) shows that f_X satisfies (6.2.2) with parameter $|d| < 1/2$. Although A_X may not satisfy (6.2.3), because only A_Y is continuous, A_X can be factored as required for Theorems 6.2.4 and 6.2.5. Hence, these theorems are applicable.

Note: Asymptotic normality for the quadratic forms $Q_{n,X}$ with weights $b_{n,j} \equiv b_j$ that do not vary with n was investigated by Hannan (1973); see also Proposition 10.8.6 of Brockwell and Davis (1991). Their proof required the summability condition $\sum_{k=0}^{\infty} k^{1/2} |a_k| < \infty$ on the coefficients a_k of the linear process $\{X_j\}$ and was based on the Bartlett approximation of the periodogram $I_{X,j}/f_{X,j}$ by the periodogram $I_{\zeta,j}/f_{\zeta,j}$ of the noise. Robinson (1995b) established asymptotic normality of the sum $S_{n,X}$ for a particular case of weights $b_{n,j} = \log(j/m) - m^{-1} \sum_{k=1}^{m} \log(k/m)$, $j = 1, \cdots, m$, where $m = m_n \to \infty$, $m = o(n)$.

6.3   Asymptotic normality of quadratic forms

In this section we discuss the asymptotic normality of a class of quadratic forms of stationary and ergodic linear processes under general sufficient conditions that allow for long memory. Two sets of additional sufficient conditions are given in Subsection 6.3.1.

Let a_k, $k = 0, 1, \cdots$, be a square summable sequence of real numbers, not necessarily depending on any parameter. Consider the linear process

$$X_j = \sum_{k=0}^{\infty} a_k \zeta_{j-k} \equiv \sum_{k=-\infty}^{j} a_{j-k} \zeta_k, \quad j \in \mathbb{Z}, \tag{6.3.1}$$

where ζ_j, $j \in \mathbb{Z}$, is now a sequence of i.i.d. r.v.'s with zero mean and variance σ^2. The weights a_k can be written as

$$a_k = \int_{\Pi} e^{ikv} \hat{a}(u) du, \quad k \in \mathbb{Z}, \quad \text{with} \quad \hat{a}(u) := \frac{1}{2\pi} \sum_{k=0}^{\infty} e^{-iku} a_k,$$

where $\hat{a}(u)$ is a square integrable $L_2(\Pi)$ function. The spectral density f of the process X_j is

$$f(u) = \frac{\sigma^2}{2\pi} |\sum_{k=0}^{\infty} e^{-iku} a_k|^2 = 2\pi\sigma^2 |\hat{a}(u)|^2, \quad u \in \Pi. \tag{6.3.2}$$

Clearly, the process $\{X_j\}$ of (6.3.1) is strictly stationary.

et h be a real-valued even function in $L_1(\Pi)$, and

$$\mathcal{Q}_n := 2\pi n \int_\Pi h(u)I(u)du = \sum_{s=1}^n \sum_{t=1}^n b_{t-s} X_s X_t, \qquad (6.3.3)$$

be the integrated weighted periodogram corresponding to the variables X_j with real weights $b_j = \int_\Pi e^{ij\tilde{u}} h(u)du$, $j \in \mathbb{Z}$. Note $b_j = b_{-j}$, $j = 1, 2, \cdots$.

Our goal here is first to give some general sufficient conditions under which a suitably standardized $\mathcal{Q}_n$ converges in distribution to a normal r.v. These weak conditions are expressed in terms of the asymptotic behavior of the product of the two $n \times n$ Toeplitz matrices

$$R_n := (\gamma_{k-j})_{k,j=1,\cdots,n}, \quad B_n := (b_{k-j})_{k,j=1,\cdots,n},$$

where $\gamma_j \equiv \gamma_X(j)$. Later we shall provide simpler sufficient conditions for the asymptotic normality of $\mathcal{Q}_n$ in terms of f and h or γ_k and b_k.

Let $\chi_4 = E\zeta_0^4 - 3(E\zeta_0^2)$ denote the fourth-order cumulant of the r.v. ζ_0 and let $\mathrm{tr}(A)$ stand for the trace of a matrix A. Let

$$\sigma_Q^2 := 16\pi^3 \int_\Pi (f(u)h(u))^2 \, du + \left(2\pi \int_\Pi f(u)h(u)\, du\right)^2 \chi_4. \quad (6.3.4)$$

We are now ready to state the following central limit theorem.

Theorem 6.3.1. *Suppose* $\{X_j, j \in \mathbb{Z}\}$ *is a stationary linear process (6.3.1) and* $E\zeta_0^4 < \infty$. *In addition, assume that*

$$\frac{\mathrm{tr}((R_n B_n)^2)}{n} \longrightarrow (2\pi)^3 \int_\Pi (f(u)h(u))^2 \, du < \infty. \qquad (6.3.5)$$

If $\sigma^2 = 1$ *then*

$$n^{-1/2}(\mathcal{Q}_n - E\mathcal{Q}_n) \to_D \mathcal{N}(0, \sigma_Q^2). \qquad (6.3.6)$$

If $\sigma^2 \neq 1$ *then the same conclusion holds with* χ_4 *replaced by* χ_4/σ^4 *in* σ_Q^2.

The meaning of the asymptotic normality condition (6.3.5) can be easily seen for a Gaussian process $\{X_j\}$ for which

$$\mathrm{Var}(\mathcal{Q}_n) = 2\mathrm{tr}((R_n B_n)^2).$$

Hence, for $\mathcal{Q}_n$ to satisfy the CLT, it suffices to require that the variance $\mathrm{Var}(\mathcal{Q}_n) \sim nC$ increases at the rate n with the correct constant

$$C = 16\pi^3 \int_\Pi (f(u)h(u))^2 \, du.$$

Condition (6.3.5) might be difficult to check. To make this theorem operational, Subsection 6.3.1 provides a number of examples where this condition is satisfied.

The proof of Theorem 6.3.1 is based on approximating $\mathcal{Q}_n$ by the sum

$$S_n := \sum_{k=1}^{n} Y_{1,k} Y_{2,k}, \qquad Y_{i,k} := \sum_{s \in \mathbb{Z}} g_{i,s} \zeta_{k-s}, \quad k \in \mathbb{Z}, \ i = 1, 2, \qquad (6.3.7)$$

of the product of two linear processes $Y_{1,k}$ and $Y_{2,k}$, where $\{g_{i,k}\}$, $i = 1, 2$, are two sequences of real numbers that are square summable in $k \in \mathbb{Z}$ as defined below.

Let

$$\hat{g}_1(u) := (2\pi)^{1/2} \hat{a}(u) |h(u)|^{1/2}, \qquad (6.3.8)$$

$$\hat{g}_2(u) := (2\pi)^{1/2} \hat{a}(u) |h(u)|^{1/2} \operatorname{sgn}(h(u)), \quad u \in \Pi,$$

$$g_{i,k} := \int_{\Pi} e^{iku} \hat{g}_i(u) du, \quad i = 1, 2, \quad k \in \mathbb{Z}.$$

By (6.3.8) and (6.3.2), the spectral density f_i and autocovariance $\gamma_{i,k}$ of the process $\{Y_{i,k}\}$ is

$$f_i(u) = 2\pi |\hat{g}_i(u)|^2 = 2\pi f(u) |h(u)|, \quad i = 1, 2, \qquad (6.3.9)$$

$$\gamma_{1,k} = \gamma_{2,k} = \int_{\Pi} e^{iku} f_1(u) du, \ k \in \mathbb{Z}, \ i = 1, 2.$$

The usefulness of this approximation of Q_n by S_n is that the processes $\{Y_{i,k}\}, i = 1, 2$, can have "much shorter memory" (are less dependent) than the original process $\{X_k\}$ and so the problem of asymptotic normality for Q_n is reduced to the much simpler problem for S_n. In particular, it follows from the form of the transfer functions $|\hat{g}_i(u)| = (2\pi)^{1/2} |\hat{a}(u)| |h(u)|$, that a singularity $|\hat{a}(u)| \to \infty$, $u \to 0$, can be compensated for by letting $|h(u)| \to 0$ as $u \to 0$. One may also say that a slow decay of a_k is compensated for by a fast oscillatory decay of b_k.

In the Whittle estimation application, Theorem 8.3.1 below, $|\hat{g}_i(u)| = (2\pi)^{1/2} |\hat{a}(u)| |h(u)|^{1/2} = 1/(2\pi)$, $i = 1, 2$. In this case, the corresponding processes $\{Y_{i,k}, k \in \mathbb{Z}\}$, $i = 1, 2$, in (6.3.7) have constant spectral densities $f_i = 2\pi |\hat{g}_i|^2 \equiv 1/2\pi$, i.e., $\{Y_{i,k}\}$ is white-noise.

The next lemma gives the desired approximation of Q_n by S_n.

Lemma 6.3.1. *Assume that (6.3.1) and (6.3.5) hold. Let the $\hat{g}_i$'s, $i = 1, 2$, in S_n be as in (6.3.8). Then*

$$\operatorname{Var}(\mathcal{Q}_n - S_n) = o(n). \qquad (6.3.10)$$

The proof of Lemma 6.3.1 will appear later in this section. Here we now show how it together with Theorem 4.5.2 yields Theorem 6.3.1.

Proof of Theorem 6.3.1. Clearly, the weights $g_{i,k}$, $k \in \mathbb{Z}$, of (6.3.8) are real. According to (6.3.5), fh belongs to $L_2(\Pi)$. Therefore,

$$|\hat{g}_i(u)|^2 = 2\pi |\hat{a}(u)|^2 |h(u)| = f(u)h(u)$$

also belongs to $L_2(\Pi)$ and by (2.1.4),

$$\sum_{k \in \mathbb{Z}} g_{i,k}^2 < \infty, \qquad i = 1, 2.$$

Hence, according to (6.3.2) and (6.3.5), the corresponding $\{Y_{i,k}\}$, $i = 1, 2$, are well-defined stationary linear processes in $k \in \mathbb{Z}$, with spectral densities $f_i = 2\pi |\hat{g}_i|^2 \in L_2(\Pi)$, and covariances $\gamma_{i,k}$ satisfying $\sum_{k \in \mathbb{Z}} \gamma_{i,k}^2 < \infty$, $i = 1, 2$. Therefore, by Theorem 4.5.2,

$$n^{-1/2}(S_n - ES_n) \to_D \mathcal{N}(0, \sigma_S^2), \qquad (6.3.11)$$

$$\sigma_S^2 := 2 \sum_{k \in \mathbb{Z}} \gamma_{1,k} \gamma_{2,k} + \chi_4 \gamma_{12;0}^2.$$

Theorem 6.3.1 now follows from (6.3.11) and Lemma 6.3.1, as the equality $\sigma_Q^2 = \sigma_S^2$ of the limit variances on the r.h.s. of (6.3.4) and (6.3.11) can be identified by using the definition of $\hat{g}_j$ and Parseval's identity (2.1.4). Indeed, by (6.3.9),

$$2 \sum_{k \in \mathbb{Z}} \gamma_{1,k} \gamma_{2,k} = 2 \sum_{k \in \mathbb{Z}} \gamma_{1,k}^2 = 4\pi \int_\Pi f_1^2(u)\, du$$

$$= 16\pi^3 \int_\Pi |\hat{g}_1(u)|^4\, du = 16\pi^3 \int_\Pi (f(u)h(u))^2\, du.$$

Also,

$$\gamma_{12;0} = \sum_{k \in \mathbb{Z}} g_{1,k} g_{2,k} = 2\pi \int_\Pi \hat{g}_1(u)\overline{\hat{g}_2(u)}\, du$$

$$= 4\pi^2 \int_\Pi |\hat{a}(u)|^2 h(u)\, du = 2\pi \int_\Pi f(u)h(u)\, du.$$

This implies $\sigma_Q^2 = \sigma_S^2$ and completes the proof of Theorem 6.3.1.

We shall now focus on the proof of Lemma 6.3.1. This in turn is facilitated by the following auxiliary lemma. Let

$$I_{a,b}(z) := \int_{-\infty}^{\infty} (1 + |v|)^{-a}(1 + |v + z|)^{-b} dv, \qquad z \in \mathbb{R},$$

$$I'_{a,b}(z) := \int_{-\infty}^{\infty} |v|^{-a}|v + z|^{-b} dv, \qquad z \in \mathbb{R}.$$

seux ℓ_u *[* *6.3.1*

Lemma 6.3.2. *For all* $0 < a, b < 1$, $a + b > 1$,

$$I_{a,b}(z) \leq C(1 + |z|)^{-a-b+1}, \tag{6.3.12}$$

$$I'_{a,b}(z) = C|z|^{-a-b+1}, \quad z \in \mathbb{R}, \tag{6.3.13}$$

where C *does not depend on* z.

Proof. Changing variables $v = zx$, one obtains

$$I'_{a,b}(z) = |z|^{-a-b+1} \int_{-\infty}^{\infty} |x|^{-a} |x + 1|^{-b} dx = C|z|^{-a-b+1}, \quad z \in \mathbb{R},$$

which proves (6.3.13).

To prove (6.3.12), consider first the case $|z| \geq 2$. Because $|z|^{-1} \leq 2(1 + |z|)^{-1}$ for $|z| \geq 2$,

$$I_{a,b}(z) \leq I'_{a,b}(z) = C|z|^{-a-b+1}, \quad |z| \geq 2.$$

Next, suppose $|z| \leq 2$. For $|v| \geq 4$, $1 + |v + z| \geq |v|/2$, and for $|v| < 4$, $1 + |v + z| \geq 1$, and hence,

$$I_{a,b}(z) \leq \int_{|v| \leq 4} |v|^{-a} dv + \int_{|v| > 4} 2^b |v|^{-a-b} dv = C < \infty,$$

which implies (6.3.12), since $1 \leq 4(1 + |z|)^{-1}$ for $|z| \leq 2$. This completes the proof of the lemma.

Next, let λ denote the Lebesgue measure on Π and set

$$d^k u := du_1 du_2 \cdots du_k, \quad u := (u_1, \cdots, u_k)' \in \Pi^k, \quad k = 1, \cdots, 4.$$

Recall the definition of the Dirichlet kernel D_n from (2.1.6) and let for $u := (u_1, u_2, u_3, u_4)' \in \Pi^4$,

$$\ell_n(u) := D_n(u_1 - u_3) D_n(u_2 - u_3) D_n(u_1 - u_4) D_n(u_2 - u_4), \tag{6.3.14}$$

$$\mu_n(A) := \frac{1}{n} \int_A \ell_n(u) d^4 u. \qquad D_m(M) = \sum_{s=1}^{\infty} e^{js} e^{-\tilde{\imath}\, (m+\|u\|)}$$

Note that μ_n is a signed measure on the Borel sets of Π^4, and $|\mu_n|(du) := n^{-1} |\ell_n(u)| d^4 u$ is its variation.

We are interested in the analysis of integrals with respect to μ_n. Since $n^{-1} D_n(u) \to 2\pi I(u = 0)$, Fox and Taqqu (1987) showed in their Lemma 7.1 that for any real-valued continuous function h on Π^4,

$$\int_{\Pi^4} h \, d\mu_n \to (2\pi)^3 \int_{\Pi} h(u, \cdots, u) \, du. \tag{6.3.15}$$

We need an extension of this to specific discontinuous functions given in the following lemma.

Lemma 6.3.3. (i) *Let h_j, $j = 1, \cdots, 4$, be bounded measurable functions on Π, and let $h(u_1, \cdots, u_4) = \prod_{j=1}^{4} h_j(u_j)$. Then (6.3.15) holds.*

(ii) *For each $0 \leq g \in L_1(\Pi)$,*

$$\int_{\Pi^4} g(u_1)\, d|\mu_n| \leq C \int_{\Pi} g(u)\, du, \quad n \geq 1, \tag{6.3.16}$$

where $C > 0$ does not depend on n.

Proof. First we prove part (ii). Let $\psi_n(u) := 1/(1 + n|u|)$, $u \in \Pi$. Extend $\psi_n(u)$ to $\mathbb{R}$, setting $\psi_n(u + 2\pi k) = \psi_n(u)$, $\forall\, k \in \mathbb{Z}$. Then from (2.1.8),

$$|D_n(u)| \leq 2\pi n\, \psi_n(u), \quad u \in \mathbb{R}. \tag{6.3.17}$$

Moreover, by a change of variable $n(u + y_1) = v$ and by (6.3.12) we see that for all $0 < \alpha, \beta < 1$, $\alpha + \beta > 1$, $y_1, y_2 \in \Pi$,

$$n \int_{\Pi} \psi_n^{\alpha}(u + y_1)\, \psi_n^{\beta}(u + y_2)\, du \tag{6.3.18}$$

$$= n \int_{\Pi} \psi_n^{\alpha}(u)\, \psi_n^{\beta}(u + y_2 - y_1)\, du$$

$$\leq I_{\alpha,\beta}((y_2 - y_1)(\operatorname{mod} 2\pi)) \leq C_{\alpha,\beta}\psi_n^{\alpha+\beta-1}(y_2 - y_1).$$

Applying (6.3.17) and (6.3.18), we obtain, with $u_5 = u_1$,

$$\int_{\Pi^4} g(u_1)\, d|\mu_n| = n^{-1} \int_{\Pi^4} g(u_1) \prod_{j=1}^{4} |D_n(u_j - u_{j+1})|\, d^4u \tag{6.3.19}$$

$$\leq n^3 \int_{\Pi^4} g(u_1) \prod_{j=1}^{4} \psi_n(u_j - u_{j+1})\, d^4u$$

$$\leq C n \int_{\Pi^2} g(u_1)\psi_n^{2-\eta}(u_1 - u_2)\, du_1 du_2 \leq C \int_{\Pi} g(u)\, du,$$

for an arbitrary $0 < \eta < 1$, which proves (6.3.16).

For part (i), by Theorem 2.5.1, for any $\epsilon > 0$ there exist continuous functions $\tilde{h}_j$, $j = 1, \cdots, 4$, such that, with $A_j := \{u \in \Pi : \tilde{h}_j(u) \neq h_j(u)\}$,

$$\sup_{u \in \Pi} |\tilde{h}_j(u)| \leq \sup_{u \in \Pi} |h_j(u)|, \quad \lambda(A_j) \leq \epsilon, \quad j = 1, \cdots, 4. \tag{6.3.20}$$

Hence,

$$\int_{\Pi^4} h \, d\mu_n = \int_{\Pi^4} \prod_{j=1}^4 \tilde{h}_j(u_j) \, d\mu_n + \int_{\Pi^4} r \, d\mu_n,$$

where r is a function on Π^4 such that

$$\left| r(u_1, \cdots, u_4) \right| \le C \sum_{j=1}^4 I(u_j \in A_j).$$

By (6.3.15) and (6.3.20),

$$\int_{\Pi^4} \prod_{j=1}^4 \tilde{h}_j(u_j) \, d\mu_n \to (2\pi)^3 \int_{\Pi} \prod_{j=1}^4 \tilde{h}_j(u) \, du,$$

$$\left| \int_{\Pi} \prod_{j=1}^4 \tilde{h}_j(u) \, du - \int_{\Pi} \prod_{j=1}^4 h_j(u) \, du \right| \le C\lambda\left(\bigcup_{j=1}^4 A_j \right) \le C\epsilon.$$

By (6.3.16),

$$\left| \int_{\Pi^4} r \, d\mu_n \right| \le C \sum_{j=1}^4 \int_{\Pi^4} I(u_j \in A_j) \, d|\mu_n| \le C\lambda\left(\bigcup_{j=1}^4 A_j \right) \le C\epsilon.$$

Hence, $\left| \int_{\Pi^4} r \, d\mu_n \right|$ can be made arbitrarily small by an appropriate choice of ϵ, thereby completing the proof of (6.3.15) and the lemma.

Next, for any functions p, p' in $L_2(\Pi^4)$, define

$$\langle p, p' \rangle_n := \int_{\Pi^4} p(u_1, u_2, u_3) \overline{p'(u_1, u_2, u_4)} \, d\mu_n(u), \tag{6.3.21}$$

$$\|p\|_n^2 := \langle p, p \rangle_n$$

$$= n^{-1} \int_{\Pi^2} \left| \int_\Pi p(u_1, u_2, v) D_n(u_1 - v) D_n(u_2 - v) dv \right|^2 du_1 du_2.$$

The bilinear form (6.3.21) obviously satisfies the C–S and triangle inequalities:

$$\left| \langle p, p' \rangle_n \right| \le \|p\|_n \|p'\|_n, \quad \|p + p'\|_n \le \|p\|_n + \|p'\|_n.$$

On the other hand, $\|p\|_n$ is not a norm, since $\|p\|_n = 0$ does not imply $p(u_1, u_2, y) = 0$ in Π^3 (take, for example, $p(u_1, u_2, y) = \exp(\mathbf{i}\, 3ny)$).

Proof of Lemma 6.3.1. In this proof, C and $C(\cdot)$ are generic constants, which may depend on the quantities in brackets, and the indices k, k_1, and k_2 in various sums are understood to vary over $\mathbb{Z}$, unless mentioned otherwise.

$$\| \hat{a}_{u_1} \overline{\hat{a}(u_2)} h_{(\gamma)} - \hat{a}(u_1) \hat{a}(u_2)(|h(u_1)|\| $$

Next, let $\delta_k = I(k = 0)$, $a_k := 0$, $k < 0$ and set

$$d_{1,n}(k_1, k_2) := \sum_{t,s=1}^{n} b_{t-s}\, a_{t-k_1}\, a_{s-k_2},$$

$$d_{2,n}(k_1, k_2) := \sum_{t,s=1}^{n} \delta_{t-s}\, g_{1,t-k_1}\, g_{2,s-k_2} = \sum_{t=1}^{n} g_{1,t-k_1} g_{2,t-k_2},$$

$$\Delta_n := \sum_{k_1,k_2} \Big[d_{1,n}(k_1, k_2) - d_{2,n}(k_1, k_2) \Big]^2,$$

where $g_{i,k}$, $i = 1, 2$, are as in (6.3.8). Direct calculations show that

$$\mathcal{Q}_n - S_n = \sum_{k_1,k_2 \in \mathbb{Z}} \big(d_{1,n}(k_1, k_2) - d_{2,n}(k_1, k_2) \big) \zeta_{k_1} \zeta_{k_2}. \qquad (6.3.22)$$

Hence, by (4.5.4),

$$\operatorname{Var}(\mathcal{Q}_n - S_n) \le C\, \Delta_n. \qquad (6.3.23)$$

Now set

$$p_1(u_1, u_2, y) := \hat{a}(u_1)\, \overline{\hat{a}(u_2)}\, h(y), \qquad (6.3.24)$$

$$p_2(u_1, u_2, y) := \hat{g}_1(u_1)\, \overline{\hat{g}_2(u_2)} \hat{\delta}(y) \qquad (6.3.25)$$

$$= \hat{a}(u_1)\overline{\hat{a}(u_2)}(|h(u_1)|\,|h(u_2)|)^{1/2}\mathrm{sgn}(h(u_2)).$$

We shall now show that

$$\Delta_n = (2\pi)^{-2} n \|p_1 - p_2\|_n^2, \qquad (6.3.26)$$

$$\|p_1 - p_2\|_n^2 \to 0, \quad n \to \infty, \qquad (6.3.27)$$

which, in turn, will complete the proof of the lemma.

Let

$$H(u_1, u_2, y) := \hat{a}(u_1)\hat{a}(u_2)h(y) - (2\pi)^{-1}\hat{g}(u_1)\hat{g}(u_2), \quad D_n^*(u) := \sum_{j=1}^{n} e^{iuj}.$$

Recall that the Fourier transform of the δ_k's is $\hat{\delta}(y) = (2\pi)^{-1}\sum_{k \in \mathbb{Z}} e^{-iky}\delta_k$ $= 1/2\pi$. Replacing $a_j, b_j, \delta_j, g_{1,j}$, and $g_{2,j}$ by the corresponding Fourier integrals, we can write

$$d_{1,n}(k_1, k_2) - d_{2,n}(k_1, k_2)$$

$$= \sum_{t,s=1}^{n} (b_{t-s}\, a_{t-k_1}\, a_{s-k_2} - \delta_{t-s}\, g_{1,t-k_1}\, g_{2,s-k_2})$$

$$= \sum_{t,s=1}^{n} \int_{\Pi^3} e^{iy(t-s)} e^{iu_1(t-k_1)} e^{iu_2(s-k_2)} H(u_1, u_2, y)\, dy\, du_1\, du_2$$

ser ($l_1(u_2)$) $\Delta_u = 2$

$$= \int_{\Pi^2} e^{-i(u_1 k_1 + u_2 k_2)}$$
$$\times \left[\int_\Pi D_n^*(y + u_1) D_n^*(-y + u_2) H(u_1, u_2, y)\, dy \right] du_1 du_2.$$

Thus, by Parseval's identity in $L_2(\Pi^2)$,

$$(2\pi)^2 \Delta_n = \int_{\Pi^2} \left| \int_\Pi D_n^*(y + u_1) D_n^*(-y + u_2) H(u_1, u_2, y)\, dy \right|^2 du_1 du_2.$$

Change variable u_2 to $-u_2$, observe that $D_n^*(u) = e^{i(n+1)u/2} D_n(u)$ by (2.1.6), and use the equality

$$H(u_1, -u_2, y) = p_1(u_1, u_2, y) - p_2(u_1, u_2, y),$$

to obtain

$$(2\pi)^2 \Delta_n = \int_{\Pi^2} \left| \int_\Pi D_n(y + u_1) D_n(-y - u_2) H(u_1, -u_2, y)\, dy \right|^2 du_1 du_2$$
$$= \int_{\Pi^2} \left| \int_\Pi D_n(y - u_1) D_n(y - u_2)(p_1(u_1, u_2, y) \right.$$
$$\left. - p_2(u_1, u_2, y))\, dy \right|^2 du_1 du_2 = n\|p_1 - p_2\|_n^2 \quad \text{meara}$$

which proves (6.3.26).

To prove (6.3.27), we use a truncation argument. Let $K > 0$ and $B_K := \{u, v \in \Pi : |\hat{a}(u)| < K^{1/2}, |h(v)| < K\}$. Introduce the truncated functions

$$p_1^K(u_1, u_2, y) := p_1(u_1, u_2, y) I((u_1, y) \in B_K) I((u_2, y) \in B_K), \quad (6.3.28)$$
$$p_2^K(u_1, u_2, y) := p_2(u_1, u_2, y) I((u_1, u_1) \in B_K) I((u_2, u_2) \in B_K).$$

Then from (6.3.26), by the triangular inequality,

$$n^{-1}\Delta_n \leq C\|p_1 - p_2\|_n^2$$
$$\leq C\left(\|p_1^K - p_2^K\|_n^2 + \|p_1 - p_1^K\|_n^2 + \|p_2 - p_2^K\|_n^2 \right).$$

Hence, (6.3.27) will be proved if we show that

$$\lim_n \|p_1^K - p_2^K\|_n = 0, \quad \forall\, K, \tag{6.3.29}$$
$$\lim_{K \to \infty} \limsup_n \|p_1 - p_1^K\|_n = 0, \tag{6.3.30}$$
$$\lim_{K \to \infty} \limsup_n \|p_2 - p_2^K\|_n = 0. \tag{6.3.31}$$

Consider (6.3.29). Let

$$G(u_1, \cdots, u_4) = \sum_{j,k=1}^2 p_j^K(u_1, u_2, u_3) \overline{p_k^K(u_1, u_2, u_4)} (-1)^{j+k}.$$

Then $\|p_1^K - p_2^K\|_n^2 = \int_{\Pi^4} G \, d\mu_n \to 0$ by Lemma 6.3.3, because G is a linear combination of a finite number of functions satisfying the conditions of this lemma, and $G(u, \cdots, u) = 0$, $u \in \Pi$ (note that $p_1(u, u, u) = p_2(u, u, u) = |\hat{a}(u)|^2 h(u)$ and $p_1^K(u, u, u) = p_2^K(u, u, u)$, $u \in \Pi$, by definition).

Next, we prove (6.3.30). We have

$$\|p_1\|_n^2 = \|p_1^K + (p_1 - p_1^K)\|_n^2 \qquad (6.3.32)$$
$$= \|p_1^K\|_n^2 + 2\langle p_1^K, p_1 - p_1^K \rangle_n + \|p_1 - p_1^K\|_n^2.$$

Consequently,

$$\|p_1 - p_1^K\|_n^2 \leq \left| \|p_1\|_n^2 - \|p_1^K\|_n^2 \right| + 2\left| \langle p_1^K, p_1 - p_1^K \rangle_n \right| \qquad (6.3.33)$$
$$=: r_n'(K) + 2r_n''(K).$$

Let $A_K = \{u \in \Pi : |\hat{a}(u)| \geq K^{1/2} \text{ or } |h(u)| \geq K\}$ and for any set A, let A^c denote its complement. By the definitions of p_1, $\mu_n(u)$, and because $f(u) = 2\pi |\hat{a}(u)|^2$ and by condition (6.3.5),

$$\|p_1\|_n^2 = \int_{\Pi^4} p_1(u_1, u_2, u_3) \overline{p_1(u_1, u_2, u_4)} \, d\mu_n(u) \qquad (6.3.34)$$

$$= (2\pi)^{-2} n^{-1} \int_{\Pi^4} f(u_1) f(u_2) h(u_3) h(u_4) \ell(u) d^4 u$$

$$= (2\pi)^{-2} n^{-1} \operatorname{tr}((R_n B_n)^2) \to 2\pi \int_{\Pi} \left(f(u) h(u) \right)^2 du,$$

while, according to Lemma 6.3.3,

$$\lim_n \|p_1^K\|_n^2 = 2\pi \int_{\Pi} \left(f(u) h(u) \right)^2 I(A_K^c) \, du. \qquad (6.3.35)$$

Hence,

$$\lim_n r_n'(K) = 2\pi \int_{\Pi} \left(f(u) h(u) \right)^2 I(A_K) du \to 0, \ K \to \infty.$$

Consider the second term in (6.3.33), $r_n''(K) = \left| \int_{\Pi^4} G'' \, d\mu_n \right|$. We have

$G''(u_1, \cdots, u_4)$
$$:= p_1^K(u_1, u_2, u_3) \overline{(p_1(u_1, u_2, u_4) - p_1^K(u_1, u_2, u_4))}$$
$$= |\hat{a}(u_1)|^2 |\hat{a}(u_2)|^2 h(u_3) h(u_4) I \left(|\hat{a}(u_i)|^2 \leq K, i = 1, 2, \ |h(u_3)| < K \right)$$
$$\times \left[1 - I(|\hat{a}(u_i)|^2 < K, i = 1, 2, \ |h(u_4)| < K) \right].$$

By the definition of p_1 and p_1^K and by (6.3.8), we see that $G''(u_1, \cdots, u_4) \neq 0$ implies that $|\hat{a}(u_1)|^2 < K$, $|\hat{a}(u_2)|^2 < K$, $|h(u_3)| < K$, and $|h(u_4)| \geq K$, a.e. in $d^4 u$. Therefore,

$$G'' = G'' \prod_{i=1}^{2} I(|\hat{a}(u_i)|^2 \leq K)I(|h(u_3)| < K, |h(u_4)| \geq K).$$

For $L > K$, write $G'' = G_1'' + G_2''$, where

$$G_1'' := G'' I(|h(u_4)| < L), \quad G_2'' := G'' I(|h(u_4)| \geq L).$$

Note that the function G_1'' satisfies the assumptions of Lemma 6.3.3. Hence,

$$\int_{\Pi^4} G_1'' \, d\mu_n \to (2\pi)^3 \int_{\Pi} G_1''(u, u, u, u) \, du = 0, \quad n \to \infty,$$

since $G_1''(u, u, u, u) = 0$. On the other hand,

$$|G_2''(u_1, \cdots, u_4)| \leq K^3 |h(u_4)| I(|h(u_4)| \geq L),$$

and by Lemma 6.3.3(ii),

$$\left| \int_{\Pi^4} G_2'' \, d\mu_n \right| \leq CK^3 \int_{\Pi^4} |h(u_4)| I(|h(u_4)| \geq L) d|\mu_n|$$

$$\leq CK^3 \int_{\Pi} |h(u_4)| I(|h(u_4)| \geq L) du_4, \quad n \geq 1.$$

Therefore, for any $\epsilon > 0$ and $K > 0$, we can choose L such that $|r_n''(K)| \leq \epsilon$, as $n \to \infty$, which proves (6.3.30).

It remains to prove (6.3.31). Let

$$k(u_1, u_2, y) := p_2(u_1, u_2, y) - p_2^K(u_1, u_2, y).$$

Then

$$|k(u_1, u_2, y)| \leq C \sum_{i=1}^{2} I(u_i \in A_K) |f(u_1)f(u_2)h(u_1)h(u_2)|^{1/2}.$$

By the C–S inequality,

$$\|p_2 - p_2^K\|_n^2 \leq \int_{\Pi^4} |k(u_1, u_2, u_3)k(u_1, u_2, u_4)| \, d|\mu_n(u)|$$

$$\leq \sum_{i=1}^{2} \int_{\Pi^4} I(u_i \in A_K)f(u_1)f(u_2)|h(u_1)h(u_2)| \, d|\mu_n(u)|$$

$$\leq 2 \left(\int_{\Pi^4} I(u_1 \in A_K)f^2(u_1)h^2(u_1) \, d|\mu_n(u)| \right)^{1/2}$$

$$\times \left(\int_{\Pi^4} f^2(u_2)h^2(u_2) \, d|\mu_n(u)| \right)^{1/2}.$$

Since $(fh)^2 \in L_1(\Pi)$, applying Lemma 6.3.3, it follows that, for all $n \geq 1$,

$$\|p_2 - p_2^K\|_n^2 \leq C \int_{\Pi} I(u \in A_K)f^2(u)h^2(u) \, du \to 0, \quad K \to \infty.$$

Note that this bound does not depend on n and tends to zero as $K \to \infty$. This proves (6.3.31) and, consequently, Lemma 6.3.1 as well.

6.3.1   *Stronger sufficient conditions for asymptotic normality*

In this subsection we discuss some easy to verify but stronger sufficient conditions for the CLT of quadratic forms Q_n. They are formulated in terms of the spectral density f and function h and imply the asymptotic behavior of $\text{tr}((R_n B_n)^2)$ given by (6.3.5).

Consider the following two conditions.

$$f \in L_{2p}, \; h \in L_{2q}, \; \text{for some } p \geq 1, \, q \geq 1, \, 1/p + 1/q \leq 1. \qquad (6.3.36)$$

$$f(u) \leq C|u|^{-\alpha}, \; |h(u)| \leq C|u|^{-\beta}, \quad \forall u \in \Pi, \qquad (6.3.37)$$

for some $-1 < \alpha, \, \beta < 1, \, \alpha + \beta < 1/2$.

We shall prove the following:

Theorem 6.3.2. *Let $\{X_j\}$, f, h, and Q_n be as above. Then either (6.3.36) or (6.3.37) implies condition (6.3.5), and Q_n satisfies the normal approximation (6.3.6).*

Proof. Suppose (6.3.36) holds. Recall the definitions of p_j, p_j^K, $j = 1, 2$, from (6.3.24) and (6.3.28). According to (6.3.32) and (6.3.34),

$$(2\pi)^{-2} n^{-1} \text{tr}((R_n B_n)^2) = \|p_1\|_n^2 = \|p_1^K\|_n^2 + \rho_{n,K},$$

where

$$|\rho_{n,K}| \leq 2|\langle p_1^K, p_1 - p_1^K \rangle_n| + \|p_1 - p_1^K\|_n^2$$
$$\leq 2\|p_1^K\|_n \, \|p_1 - p_1^K\|_n + \|p_1 - p_1^K\|_n^2.$$

By (6.3.35), as $K \to \infty$,

$$\lim_n \|p_1^K\|_n^2 = 2\pi \int_\Pi \big(f(u)h(u)\big)^2 I(A_K^c) du \to 2\pi \int_\Pi \big(f(u)h(u)\big)^2 du.$$

Hence, Theorem 6.3.2 will follow from

$$\lim_{K \to \infty} \limsup_{n \to \infty} \|p_1 - p_1^K\|_n^2 = 0. \qquad (6.3.38)$$

Let

$$F(u_1, \cdots, u_4) := f(u_1)f(u_2)|h(u_3)| \, |h(u_4)|,$$
$$B_{K,j} := \{f(u_j) \geq 2\pi K\}, \; j = 1, 2, \; B_{K,j} := \{|h(u_j)| \geq K\}, \; j = 3, 4.$$

Then according to the definition of p_1 and p_1^K,

$$\left\| p_1 - p_1^K \right\|_n^2 \leq \sum_{j=1}^{4} \int_{\Pi^4} F\,I(B_{K,j})\,d|$$

Thus, (6.3.38) will follow if we show that for

$$\lim_{K \to \infty} \limsup_{n \to \infty} I_{j,K} = 0.$$

By symmetry, it suffices to prove (6.3.40) for $j = 1$. According to Höln... inequality, with $p^{-1} + q^{-1} = 1$,

$$I_{1,K} \leq \left(\int_{\Pi^4} f^{2p}(u_1) I(B_{K,1})\,d|\mu_n| \right)^{1/2p}$$

$$\times \left(\int_{\Pi^4} f^{2p}(u_2)\,d|\mu_n| \right)^{1/2p} \left(\int_{\Pi^4} |h(u_3)|^{2q}\,d|\mu_n| \right)^{1/q}.$$

By Lemma 6.3.3(ii),

$$I_{1,K} \leq C \left(\int_{\Pi} f^{2p}(u_1) I(B_{K,1})\,du_1 \right)^{1/2p}$$

$$\times \left(\int_{\Pi} f^{2p}(u_2)\,du_2 \right)^{1/2p} \left(\int_{\Pi} |h(u_3)|^{2q}\,du_3 \right)^{1/q}$$

$$\to 0, \quad K \to \infty.$$

This completes the proof of Theorem 6.3.2 under (6.3.36).

Now suppose (6.3.37) holds. The case $\alpha\beta \geq 0$ follows from the previous case, since in this case f and h satisfy (6.3.36). Now let $\alpha > 0$ and $\beta < 0$. Assume without loss of generality that $\alpha + \beta > 0$. Again, it suffices to prove (6.3.40) for $j = 1$. As f is bounded outside a neighborhood of the origin $u = 0$, there is a sequence $\epsilon = \epsilon(K) \to 0$, $K \to \infty$, such that

$$\{f(u) \geq K\} \subset \{|u| < \epsilon\}.$$

Set

$$W_1 = \{|u_1| \leq \epsilon,\ |u_2| > 2\epsilon\}, \quad W_2 = \{|u_1| \leq \epsilon,\ |u_2| \leq 2\epsilon\}.$$

Then, as in (6.3.39),

$$\left\| p_1 - p_1^K \right\|_n^2 \leq \sum_{j=1}^{2} \int_{\Pi^4} F I(W_j)\,d|\mu_n| =: I'_{\epsilon,1} + I'_{\epsilon,2}.$$

It remains to show that

$$\lim_{\epsilon \to 0} \limsup_{n \to \infty} I'_{\epsilon,i} = 0, \quad i = 1, 2. \tag{6.3.41}$$

we estimate $I'_{\epsilon,1}$. Since $\beta < 0$ and $\alpha > 0$ then $F(u_1, u_2, u_3, u_4)$
$_1) \le C|u_1|^{-\alpha}\epsilon^{-\alpha}$ and

$$I'_{\epsilon,1} \le C\epsilon^{-\alpha} \int_{\Pi^4} |u_1|^{-\alpha} I(W_1) \, d|\mu_n|.$$

Let $\psi_n(u)$ be same as in (6.3.18). Then, for $|u_1| \le \epsilon$ and $2\epsilon \le |u_2| \le \pi$, $|\psi_n(u_1 - u_2)| \le C/(n\epsilon)$, where $C > 0$ does not depend on n and ϵ. Using the same argument as in (6.3.19), for any $\eta \in (0, 1)$ and $\epsilon > 0$, we obtain

$$I'_{\epsilon,1} \le C\epsilon^{-\alpha} n \int_{\Pi^2} |u_1|^{-\alpha} I(|u_1| \le \epsilon, |u_2| \ge 2\epsilon) \, \psi_n^{2-\eta}(u_1 - u_2) du_1 du_2$$

$$\le C\epsilon^{-\alpha} n(n\epsilon)^{-2+\eta} \int_{\Pi} |u_1|^{-\alpha} du_1 \to 0, \quad n \to \infty,$$

which yields (6.3.41) for $i = 1$.

To estimate $I'_{\epsilon,2}$, let $v(u, u_3, u_4) := n^{-1/2}|D_n(u - u_3)D_n(u - u_4)|$. Using the definition (6.3.14) of μ_n and the bound

$$F(u_1, u_2, u_3, u_4) \le C|u_1 u_2|^{-\alpha}|u_3 u_4|^{-\beta},$$

we find that

$$I'_{\epsilon,2} \le C \int_{\Pi^4} I(W_2) F \, d|\mu_n|$$

$$\le C \int_{\Pi^2} |u_3 u_4|^{-\beta} \Big| \int_{|u| \le 2\epsilon} |u|^{-\alpha} v(u, u_3, u_4) du \Big|^2 du_3 du_4$$

$$\le 2C \int_{\Pi^2 : |u_3| \le |u_4|} [\cdots] du_3 du_4,$$

because $v(u, u_3, u_4)$ is symmetric with respect to u_3 and u_4. Denote

$$B_1 = \{u : |u| > |u_3 u_4|^{1/2}/2\}, \quad B_2 = \{u : |u| \le |u_3 u_4|^{1/2}/2\}.$$

Then $I'_{\epsilon,2} \le C(r_{\epsilon,1} + r_{\epsilon,2})$, where, for $i = 1, 2,$

$$r_{\epsilon,i} := \int_{\Pi^2 : |u_3| \le |u_4|} |u_3 u_4|^{-\beta} \Big| \int_{|u| \le 2\epsilon} I(B_i)|u|^{-\alpha} v(u, u_3, u_4) du \Big|^2 du_3 du_4.$$

To analyze $r_{\epsilon,1}$, let $|u_3| \le |u_4|$, and $u \in B_1 \cap [-2\epsilon, 2\epsilon]$, and recall that $0 < \alpha + \beta < 1/2$. Then

$$|u|^{-\alpha} \le 2|u_3 u_4|^{-\alpha/2}, \quad |u_3| \le 4\epsilon, \quad |u_3 u_4|^{-(\alpha+\beta)} \le |u_3|^{-2(\alpha+\beta)},$$

and

$$r_{\epsilon,1} \le C \int_{\Pi^2 : |u_3| \le 4\epsilon} |u_3|^{-2(\alpha+\beta)} \Big| \int_{\Pi} v(u, u_3, u_4) du \Big|^2 du_3 du_4$$

$$\le C \int_{\Pi^4} |u_3|^{-2(\alpha+\beta)} I(|u_3| \le 4\epsilon) d|\mu_n|.$$

Apply Lemma 6.3.3(ii) to the r.h.s. to obtain

$$r_{\epsilon,1} \leq C \int_{|u_3| \leq 4\epsilon} |u_3|^{-2(\alpha+\beta)} du_3 \leq C\epsilon^{1-2(\alpha+\beta)}, \quad n \geq 1, \qquad (6.3.42)$$

where $\epsilon^{1-2(\alpha+\beta)} \to 0$, $\epsilon \to 0$.

To bound $r_{\epsilon,2}$, recall from (2.1.7), $|D_n(u)| \leq \pi n^\delta |u|^{\delta-1}$, $0 < \delta < 1$, $u \in \Pi$. Since $0 < \alpha < 1$ and $0 < \alpha + \beta < 1/2$, choose $0 < \delta' < \delta < \alpha \wedge 1/2$, such that $2\delta + \beta > 0$ and $2(\alpha + \beta + \delta - \delta') < 1$. Assume that $|u_3| \leq |u_4| \leq \pi$ and $u \in B_2 \cap [-2\epsilon, 2\epsilon]$. Then $|u| \leq |u_4|/2$, $|u - u_4| \geq |u_4|/2$, and

$$|D_n(u - u_3)| \leq Cn^{\delta'} |u - u_3|^{-1+\delta'}, \quad |D_n(u - u_4)| \leq Cn^{1/2-\delta} |u_4|^{-1/2-\delta}.$$

Hence, by (6.3.13),

$$\left| \int_{|u| \leq 2\epsilon} I(B_2) |u|^{-\alpha} v(u, u_3, u_4) du \right|^2$$

$$\leq Cn^{2(\delta'-\delta)} |u_4|^{-1-2\delta} \left(\int_{\mathbb{R}} |u|^{-\alpha} |u - u_3|^{-1+\delta'} du \right)^2$$

$$\leq Cn^{2(\delta'-\delta)} |u_4|^{-1-2\delta} |u_3|^{-2\alpha+2\delta'},$$

and

$$r_{\epsilon,2} \leq Cn^{2(\delta'-\delta)} \int_{\Pi^2 : |u_3| \leq |u_4|} |u_3|^{-\beta-2\alpha+2\delta'} |u_4|^{-1-(\beta+2\delta)} du_4 du_3$$

$$\leq Cn^{2(\delta'-\delta)} \int_\Pi |u_3|^{-2(\alpha+\beta+\delta-\delta')} du_3 \leq Cn^{2(\delta'-\delta)} \to 0.$$

This, together with (6.3.42), proves (6.3.41) for $j = 2$, thereby completing the proof of Theorem 6.3.2.

Other sufficient conditions. We conclude this section by presenting two more sufficient conditions each of which implies (6.3.5).

$$\gamma \in L_p(\mathbb{Z}), \ b \in L_q(\mathbb{Z}), \ \text{for } 1/p + 1/q \geq 3/2, \ p \wedge q \geq 1. \qquad (6.3.43)$$

$$f \in L_2(\Pi), \ h \in L_2(\Pi), \ fh \in L_2(\Pi), \ \text{and} \qquad (6.3.44)$$

$$\int_\Pi f^2(u) h^2(u - y) \, du \to \int_\Pi f^2(u) h^2(u) \, du, \quad y \to 0.$$

Lemma 6.3.4. *Let $\{X_j\}$, f, h, and $\mathcal{Q}_n$, be as above. Then either (6.3.43) or (6.3.44) implies (6.3.5), and the normal approximation (6.3.6) holds.*

Proof. Note that the condition (6.3.43) is stronger than (6.3.36). By the Hausdorff–Young theorem, (see Zygmund (2002), Theorem XII.2.3, p. 101),

$$\left(\int_\Pi |f(u)|^{p'} du \right)^{1/p'} \leq C_p \left(\sum_{j \in \mathbb{Z}} |\gamma(j)|^p \right)^{1/p}, \quad 1 < p \leq 2, \ p' = p/(p-1).$$

This and condition (6.3.43) imply that $f \in L_{p'}(\Pi)$, $h \in L_{q'}(\Pi)$, with $p' = p/(p-1)$ and $q' = q/(q-1)$. Note that $1/p' + 1/q' = 2 - (1/p + 1/q) < 1/2$. Hence, (6.3.36) is satisfied, which implies (6.3.5) by Theorem 6.3.2.

To show that (6.3.44) implies (6.3.5), as in the proof of Theorem 6.3.2, it suffices to show (6.3.40) for $j = 1$, i.e.,

$$I_{1,K} \equiv \int_{\Pi^4} FI(B_{K,1}) d|\mu_n| \to 0, \quad n \to \infty, \ K \to \infty. \qquad (6.3.45)$$

Denote

$$\ell_K := \int_{\Pi^4} f^2(u_1) I(f(u_1) \geq K) h^2(u_3) \, d|\mu_n|, \quad K \geq 0,$$

$$\theta_K(v) := \int_{\Pi} f^2(u) I(f(u) \geq K) h^2(u - v) \, du, \quad v \in \mathbb{R}.$$

Condition (6.3.44) ensures that

$$\theta_K(v) \to \theta_K(0), \quad v \to 0, \quad \forall K \geq 0. \qquad (6.3.46)$$

By the C–S inequality and symmetry, $I_{1,K} \leq (\ell_K)^{1/2}(\ell_0)^{1/2}$. By (6.3.17) and (6.3.18), $\ell_K \leq Cn \int_{\Pi} \theta_K(v) \psi_n^\nu(v) \, dv$ for some $\nu \in (1, 2)$. A standard argument and (6.3.46) imply

$$n \int_{\Pi} \theta_K(v) \psi_n^\nu(v) \, dv \to C\theta_K(0), \quad n \to \infty, \quad \forall K \geq 0,$$

where $C = \int_{\mathbb{R}} (1 + |v|)^{-\nu} dv < \infty$ does not depend on K. Hence,

$$\limsup_n I_{1,K} \leq C(\theta_K(0))^{1/2}(\theta_0(0))^{1/2} \to 0, \quad K \to \infty,$$

since $\theta_K(0) = \int_{\Pi} I(f(u) \geq K) f^2(u) h^2(u) \, du \to 0$, $K \to \infty$, because $fh \in L_2(\Pi)$. This proves (6.3.45).

The next lemma contains an upper bound for the variance of

$$Q_n := 2\pi n \int_{\Pi} h_n(u) I(u) du,$$

where h_n is a real-valued even function that may depend on n. Hence, Q_n is a more general quadratic form than $\mathcal{Q}_n$ of (6.3.3).

Lemma 6.3.5. *Let $\{X_j\}$ be a stationary linear process (6.3.1) and $E\zeta_0^4 < \infty$. Suppose that for some $-1 < \alpha, \beta < 1$ and $k_n \geq 0$,*

$$f(u) \leq C|u|^{-\alpha}, \quad |h_n(u)| \leq k_n|u|^{-\beta}, \quad \forall u \in \Pi, \ n \geq 1.$$

If $\alpha + \beta < 1/2$, then $\mathrm{Var}(Q_n) \leq Ck_n^2 n$, $\forall n \geq 1$.

Moreover, if $1/2 < \alpha < 1$, $\beta = 0$, then $\mathrm{Var}(Q_n) \leq Ck_n^2 n^{2\alpha}$, $\forall n \geq 1$.

Proof. By (6.3.22), (6.3.23), and (6.3.34),

$$\text{Var}(Q_n) \le Cn \int_{\Pi^4} f(u_1)f(u_2)h_n(u_3)h_n(u_4)\,d\mu_n(u)$$

$$= C \int_{\Pi^2} h_n(u_3)h_n(u_4)|\int_\Pi f(u)D_n(u-u_3)D_n(u-u_4)du|^2 du_3 du_4$$

$$\le Ck_n^2 \int_{\Pi^2} |u_3|^{-\beta}|u_4|^{-\beta}|\int_\Pi f(u)D_n(u-u_3)D_n(u-u_4)du|^2 du_3 du_4$$

$$=: Ck_n^2 n\|p_1\|_n^2,$$

where p_1 is defined by (6.3.24) with $h(u) = |u|^{-\beta}$. In the proof of Theorem 6.3.2 it was shown that when $\beta + \alpha < 1/2$, $\|p_1\|_n^2 \sim C$. This, in turn, implies the first claim of the lemma.

Let $1/2 < \alpha < 1$ and $\beta = 0$. Then the above bound, (2.1.8), and (2.1.5) together yield

$$\text{Var}(Q_n) \le Ck_n^2 \int_{\Pi^2} |\int_\Pi f(u)D_n(u-u_3)D_n(u-u_4)du|^2 du_3 du_4$$

$$\le Ck_n^2 \int_{\Pi^2} f(u_1)f(u_2)|D_n(u_1-u_2)|^2 du_1 du_2$$

$$\le Ck_n^2 \int_{\Pi^2} (|u_1||u_2|)^{-\alpha}n^2(1+n|u_1-u_2|)^{-2} du_1 du_2$$

$$\le Ck_n^2 n^{2\alpha} \int_{\mathbb{R}^2} (|u_1||u_2|)^{-\alpha}(1+|u_1-u_2|)^{-2} du_1 du_2 \le Ck_n^2 n^{2\alpha},$$

which implies the second claim of the lemma.

Note: This section is based on the work of Giraitis and Surgailis (1990). When the observable process is Gaussian, Fox and Taqqu (1987) prove an analog of Theorem 6.3.1. Bhansali, Giraitis and Kokoszka (2007b) extend the CLT of Theorem 6.3.1 to the case when the weight function h in Q_n depends on n. Simple sufficient conditions for the CLT of quadratic forms that can be written as a sequence of multiple stochastic integrals can be found in Nualart and Peccati (2005). Wu and Shao (2007) established a similar result using a method of martingale approximation in a more general set-up than considered here. For non-Gaussian weak limits, see Terrin and Taqqu (1990).

Chapter 7

Parametric Models

7.1 Introduction

This chapter describes some parametric models that are often used for modeling long, short, and negative memory time series. We provide definitions and the main properties of the parametric ARFIMA (autoregressive fractionally integrated moving average) and GARMA (Gegenbauer ARMA, or seasonal fractionally differenced) models. The spectrum of an ARFIMA model has an infinite peak or zero at the frequency zero. In comparison, a GARMA model generates long memory time series with long-range periodic behavior at a fixed frequency (pole) ω, different from the origin. We discuss the causality and invertibility of these processes and provide some formulas and asymptotics for their autocovariances and spectral densities.

In Section 7.5 we study the contemporaneous aggregation procedure and show that long memory processes can be obtained by aggregation of independent random coefficient AR(1) models.

7.2 ARFIMA models

ARFIMA$(0, d, 0)$ **model.** Most long memory time series, whose autocovariances are not absolutely summable, have spectral densities exploding to infinity at the origin in a hyperbolic fashion, i.e., $f(u) \sim c|u|^{-2d}$, $0 < d < 1/2$, $c > 0$, as $u \to 0$. This means that their energy spectrum is concentrated at low frequencies. A class of such time series, called fractional differenced noise (or ARFIMA$(0, d, 0)$), was introduced by Granger and Joyeux (1980) and Hosking (1981) to model phenomena in economics.

For an integer $d = 1, 2, \cdots$, the operator $\nabla^d := (1 - B)^d$ denotes the differencing operator, $\nabla X_j = X_j - X_{j-1}$, $\nabla^2 X_j = (1 - 2B + B^2)X_j = X_j - 2X_{j-1} + X_{j-2}$, and so on. Box and Jenkins (1970) observed that often a time series may not be stationary but its first- or second-order differences form a stationary process. Such models are known as ARIMA (autoregressive integrated moving average) models. ARFIMA models generalize ARIMA models to non-integer values of d.

For a non-integer real number $d > -1$, the difference operator $\nabla^d = (1-B)^d$, where $BX_j := X_{j-1}$, is defined by means of the binomial expansion

$$\nabla^d = \sum_{k=0}^{\infty} \binom{d}{k}(-B)^k = \sum_{k=0}^{\infty} \pi_{k,d} B^k,$$

where

$$\binom{d}{k} := \frac{(d)!}{(k)!(d-k)!} = \frac{d(d-1)\cdots(d-k+1)}{1 \times 2 \times \cdots \times k}, \qquad (7.2.1)$$

$$\pi_{k,d} := (-1)^k \binom{d}{k} = \prod_{s=1}^{k} \frac{s-1-d}{s} = \frac{\Gamma(k-d)}{\Gamma(k+1)\Gamma(-d)}$$

and $\Gamma(\cdot)$ is the gamma function (3.2.7). Hence, $\pi_{k,d} = \binom{k-d-1}{k}$ and $\pi_{0,d} = 1$.

Recall that $(x)! := \Gamma(x+1)$, $x = 0, 1, \cdots$, and for $k = 1, 2, \cdots, 0 < x < 1$, $\Gamma(x-k)$ is defined as $\Gamma(x) = (x-1)\Gamma(x-1) = \cdots = (x-1)\cdots(x-k)\Gamma(x-k)$.

Definition 7.2.1. (*ARFIMA*(0, d, 0) *process*). The process $\{X_j, j \in \mathbb{Z}\}$ is said to be an ARFIMA(0, d, 0) process with $d \in (-1/2, 1/2)$ if $\{X_j\}$ is a stationary solution with zero mean of the difference equations

$$\nabla^d X_j = \zeta_j, \quad j \in \mathbb{Z}, \qquad (7.2.2)$$

where $\zeta_j \sim \text{WN}(0, \sigma_\zeta^2)$ is a white-noise sequence.

All series in this section are understood to be converging in mean square. In particular, definition (7.2.2) imposes the requirement on solution X_j that the series $\nabla^d X_j \equiv \sum_{k=0}^{\infty} \pi_{k,d} X_{j-k}$ should converge in mean square.

Inverting the operator ∇^d yields a linear process with innovations ζ_j:

$$X_j = \nabla^{-d} \zeta_j = \Big(\sum_{k=0}^{\infty} \pi_{k,-d} B^k\Big)\zeta_j = \sum_{k=0}^{\infty} \pi_{k,-d} \zeta_{j-k}, \quad j \in \mathbb{Z}. \qquad (7.2.3)$$

The coefficients $\pi_{k,-d}$ are sometimes called the impulse response weights. We note that $\pi_{0,-d} = 1$. We shall show below that this process is a unique stationary solution to equations (7.2.2). First we list some of its properties.

Assume that $d \in (-1/2, 1/2)$, $d \neq 0$. Stirling's formula

$$\Gamma(x) \sim (2\pi)^{1/2} e^{-x+1} (x-1)^{x-1/2}, \quad x \to \infty, \qquad (7.2.4)$$

yields that for any $a \in \mathbb{R}$,

$$\frac{\Gamma(x+a)}{\Gamma(x)} \sim x^a, \quad x \to \infty, \qquad (7.2.5)$$

(see Gradshteyn and Ryzhik (2000), 8.328.2). This implies $\Gamma(k+d)/\Gamma(k+1) \sim k^{d-1}$ as $k \to \infty$ and leads to the asymptotics

$$\pi_{k,-d} \sim (\Gamma(d))^{-1} k^{d-1}, \quad k \to \infty. \qquad (7.2.6)$$

Hence, the weights $\pi_{k,-d}$ are square summable, $\{X_j\}$ is a stationary process with zero mean, and $\mathrm{Var}(X_j) = \sigma_\zeta^2 \sum_{k=0}^\infty \pi_{k,-d}^2$.

The autocovariance and autocorrelation functions $\gamma_X(j)$, $r_X(j) = \gamma_X(j)/\gamma_X(0)$ and the spectral density f_X have the following properties.

Theorem 7.2.1. *Let $\{X_j\}$ be as in (7.2.3) with $d \in (-1/2, 1/2)$, $d \neq 0$. Then*

(i) $\pi_{0,-d} = 1$ *and*

$$\pi_{k,-d} = \frac{\Gamma(k+d)}{\Gamma(k+1)\Gamma(d)} = \prod_{s=1}^k \frac{s-1+d}{s}, \qquad k = 1, 2, \cdots,$$

$$= \frac{1}{\Gamma(d)} k^{-1+d} \big(1 + O(k^{-1})\big), \qquad k \to \infty.$$

Moreover, for $-1/2 < d < 0$,

$$\sum_{k=0}^\infty \pi_{k,-d} = 0, \quad \sum_{j \in \mathbb{Z}} \gamma_X(j) = 0. \qquad (7.2.7)$$

(ii) *The spectral density is*

$$f_X(v) = \frac{\sigma_\zeta^2}{2\pi} |1 - e^{-iv}|^{-2d} = \frac{\sigma_\zeta^2}{2\pi} |2\sin(v/2)|^{-2d}, \quad v \in \Pi, \qquad (7.2.8)$$

$$\sim \frac{\sigma_\zeta^2}{2\pi} |v|^{-2d}, \quad v \to 0.$$

(iii) *For $j = 1, 2, \cdots$,*

$$\gamma_X(0) = \sigma_\zeta^2 \frac{\Gamma(1-2d)}{\Gamma^2(1-d)}, \tag{7.2.9}$$

$$\gamma_X(j) = \sigma_\zeta^2 \frac{\Gamma(j+d)}{\Gamma(j-d+1)} \frac{\Gamma(1-2d)}{\Gamma(1-d)\Gamma(d)},$$

$$r_X(j) = \frac{\Gamma(j+d)}{\Gamma(j-d+1)} \frac{\Gamma(1-d)}{\Gamma(d)} = \prod_{k=1}^{j} \frac{k-1+d}{k-d}.$$

In addition, as $j \to \infty$,

$$\gamma_X(j) \sim c_\gamma j^{-1+2d}, \quad c_\gamma := \frac{\Gamma(1-2d)}{\Gamma(d)\Gamma(1-d)} \sigma_\zeta^2. \tag{7.2.10}$$

Proof. (i) Because of Stirling's formula (7.2.4), for a $q > 0$,

$$\log \Gamma(x) = \frac{1}{2} \log(2\pi) + (x - \frac{1}{2}) \log x - x + O(x^{-1}), \quad x \to \infty,$$

$$\log(x+q) = \log x + \log(1 + q/x) = \log x + (q/x) + O(x^{-2}).$$

Hence, as $x \to \infty$,

$$\log\left(\frac{\Gamma(x)}{\Gamma(x+q)}\right) = (x - \frac{1}{2}) \log x - (x + q - \frac{1}{2}) \log(x+q) - q + O(x^{-1})$$

$$= -q \log x + O(x^{-1}),$$

which in turn implies

$$\frac{\Gamma(x)}{\Gamma(x+q)} = x^{-q}(1 + O(x^{-1})), \quad x \to \infty, \tag{7.2.11}$$

$$\frac{\Gamma(k+d)}{\Gamma(k+1)} = \frac{\Gamma(k+d)}{\Gamma(k+d+1-d)} = k^{-1+d}(1 + O(k^{-1})), \quad k \to \infty.$$

This together with (7.2.1) proves the first part of (i). The equalities in (7.2.7) follow by noting that for $d < 0$, the $\pi_{k,-d}$ are absolutely summable, so that $\sum_{k=0}^{\infty} \pi_{k,-d} = (1-1)^{-d} = 0$ and $\sum_{j\in\mathbb{Z}} \gamma_X(j) = \sigma_\zeta^2 \left(\sum_{k=0}^{\infty} \pi_{k,-d} \right)^2 = 0$. This completes the proof of part (i).

(ii) We use Proposition 2.2.2 to prove (7.2.8). Because $\{X_j\}$ is a linear process with square summable weights $\pi_{k,-d}$, and $\{\zeta_j\}$ is a WN sequence with spectral density $f_\zeta(v) = \sigma_\zeta^2/2\pi$, we obtain by (2.2.7),

$$f_X(v) = \frac{\sigma_\zeta^2}{2\pi} \left| \sum_{k=0}^{\infty} \pi_{k,-d} e^{-ikv} \right|^2.$$

Since, by the definition of $\pi_{k,-d}$, series $\sum_{k=0}^{\infty} \pi_{k,-d} e^{-ikv}$ converges to $(1 - e^{-iv})^{-d}$ in $L_2(\Pi)$, the r.h.s. of the above expression converges, in $L_2(\Pi)$, to $\frac{\sigma_\zeta^2}{2\pi}|1 - e^{-iv}|^{-2d}$, for all $v \in \Pi$. Claim (7.2.8) can also be proved by using Theorem 2.2.1.

(iii) By definition of $\gamma_X(j)$ and (7.2.8),

$$\gamma_X(j) = \int_{-\pi}^{\pi} \cos(jv) f_X(v) dv = \int_0^{2\pi} \cos(jv) f_X(v) dv, \quad j \geq 0, \quad (7.2.12)$$

$$= \frac{\sigma_\zeta^2}{2\pi} \int_0^{2\pi} \cos(jv) |2\sin(v/2)|^{-2d} dv$$

$$= \sigma_\zeta^2 \frac{(-1)^j \Gamma(1 - 2d)}{\Gamma(j - d + 1)\Gamma(-j - d + 1)}.$$

The last equality above follows from the identity

$$\int_0^{\pi} \cos(kv) \sin^{\nu-1}(v) dv = \frac{\pi \cos(k\pi/2)\Gamma(\nu + 1)}{2^{\nu-1}\nu\Gamma((k + \nu + 1)/2)\Gamma((-k + \nu + 1)/2)}$$

of Gradshteyn and Ryzhik (2000, 3.631.8). The equality

$$\frac{\Gamma(-d + 1)}{\Gamma(-j - d + 1)} = (-d)(-1 - d) \cdots (-j - d + 1)$$

$$= (-1)^j (d)(1 + d) \cdots (j + d - 1) = (-1)^j \frac{\Gamma(j + d)}{\Gamma(d)},$$

together with equation (7.2.12) yields (7.2.9).

Claim (7.2.10) follows from (7.2.9), applying (7.2.5). Alternatively, by Proposition 3.2.1, (7.2.10) follows from (7.2.3), (7.2.6), and (7.2.7).

The next theorem establishes the existence and uniqueness of a solution to ARFIMA equations (7.2.2). This solution has the infinite moving-average representation (7.2.3).

Theorem 7.2.2. *Suppose $d \in (-1/2, 1/2)$, $d \neq 0$, and $\{\zeta_j\}$ is an ergodic white-noise process. Then $\{X_j\}$ of (7.2.3) is a unique stationary ergodic zero-mean solution of equations (7.2.2).*

Moreover, $\{X_j\}$ is invertible:

$$\zeta_j = \nabla^d X_j = \sum_{k=0}^{\infty} \pi_{k,d} X_{j-k}, \quad j \in \mathbb{Z}, \quad \pi_{0,d} = 1, \quad (7.2.13)$$

$$\pi_{k,d} = \frac{\Gamma(k - d)}{\Gamma(k + 1)\Gamma(-d)} = \prod_{s=1}^{k} \frac{s - 1 - d}{s}, \quad k = 1, 2, \cdots,$$

$$= \frac{1}{\Gamma(-d)} k^{-1-d}\big(1 + O(k^{-1})\big), \quad k \to \infty.$$

Proof. Let X_j be as in (7.2.3). Because $\zeta_j \sim \mathrm{WN}(0, \sigma_\zeta^2)$, due to (2.2.3), it has spectral representation $\zeta_j = \int_\Pi e^{\mathrm{i}jv} dZ_\zeta(v)$. Since $\sum_{k=0}^\infty \pi_{k,-d} e^{-\mathrm{i}kv}$ converges to $(1 - e^{-\mathrm{i}v})^{-d}$ in $L_2(\Pi)$, then by Theorem 2.2.1,

$$X_j = \nabla^{-d} \zeta_j \equiv \sum_{k=0}^\infty \pi_{k,-d} \zeta_{j-k} = \int_\Pi e^{\mathrm{i}jv}(1 - e^{-\mathrm{i}v})^{-d} dZ_\zeta(v), \quad (7.2.14)$$

and $f_X(v) = |1 - e^{-\mathrm{i}v}|^{-2d} f_\zeta(v) = (\sigma_\zeta^2/2\pi)|1 - e^{-\mathrm{i}v}|^{-2d} f_\zeta(v)$.

Similarly, because $\sum_{k=0}^\infty \pi_{k,d} e^{-\mathrm{i}kv}$ converges to $(1 - e^{-\mathrm{i}v})^d$ in $L_2(F_X)$,

$$\nabla^d X_j = \sum_{k=0}^\infty \pi_{k,d} X_{j-k}$$

$$= \int_\Pi e^{\mathrm{i}jv} \left(\sum_{k=0}^\infty \pi_{k,d} e^{-\mathrm{i}kv} \right)(1 - e^{-\mathrm{i}v})^{-d} dZ_\zeta(v)$$

$$= \int_\Pi e^{\mathrm{i}jv} dZ_\zeta(v) = \zeta_j, \quad \forall j \in \mathbb{Z}.$$

This, together with (7.2.1) and Theorem 7.2.1(i), proves that the series $\nabla^d X_j$ is mean-square convergent. Hence, X_j of (7.2.3) satisfies (7.2.13). Moreover, by Theorem 3.5.8 of Stout (1974) the moving average (7.2.3) of ergodic noise is also ergodic, so $n^{-1} \sum_{j=1}^n X_j \to EX_1 = 0$, a.s.

To establish uniqueness, let $\{Y_j\}$ be another solution to (7.2.2), i.e., $\nabla^d Y_j \equiv \sum_{k=0}^\infty \pi_{k,d} Y_{j-k} = \zeta_j$, and let $Y_j = \int_\Pi e^{\mathrm{i}jv} dZ_Y(v)$ be its spectral representation. By the definition of a solution, series $\nabla^d Y_j$ is mean-square convergent. Hence, by Theorem 2.2.1, the process

$$\zeta_j = \nabla^d Y_j = \int_\Pi e^{\mathrm{i}jv}(1 - e^{-\mathrm{i}v})^d dZ_Y(v)$$

is stationary with spectral measure $dF_\zeta(v) = |1 - e^{-\mathrm{i}v}|^{2d} dF_Y(v)$. Since $\{\zeta_j\} \sim \mathrm{WN}(0, \sigma_\zeta^2)$, then $dF_\zeta(v) = (\sigma_\zeta^2/2\pi) dv$, implying $|1 - e^{-\mathrm{i}v}|^{2d} dF_Y(v) = (\sigma_\zeta^2/2\pi) dv$. The last fact implies that $dF_Y(v) = dF_X(v) = f_X(v) dv$ for $v \neq 0$. Since F_Y is non-decreasing, the only difference between F_Y and F_X is a possible jump at $v = 0$. In other words, it follows that

$$dF_Y(v) = (\sigma_\zeta^2/2\pi)|1 - e^{-\mathrm{i}v}|^{-2d} dv + (\sigma_c^2/2\pi)\delta_0(v) \qquad (7.2.15)$$

$$= dF_X(v) + (\sigma_c^2/2\pi)\delta_0(v), \quad v \in \Pi,$$

where δ_0 is the degenerate measure at $v = 0$ and $\sigma_c^2 \geq 0$ is some nonnegative constant. From the spectral representation, the process $\{Y_j\}$ can be written as $Y_j = \tilde{X}_j + c$, where

$$\tilde{X}_j := \int_\Pi e^{\mathrm{i}jv} I(|v| > 0) dZ_Y(v), \qquad c := Z_Y(0+) - Z_Y(0-).$$

From this and (7.2.15) it follows that $\{\tilde{X}_j\}$ is a stationary process with spectral density $f_{\tilde{X}} = f_X$ and c is possibly a r.v. with zero mean and variance $Ec^2 = \sigma_c^2$. By the assumption of ergodicity of $\{Y_j\}$, $0 = EY_1 = \lim_n n^{-1} \sum_{j=1}^n Y_j = \lim_n n^{-1} \sum_{j=1}^n \tilde{X}_j + c = c$, a.s. Hence, $c = 0$, $Y_j = \tilde{X}_j$, and $\zeta_j = \nabla^d \tilde{X}_j = \int_\Pi e^{ijv}(1 - e^{-iv})^d I(|v| > 0) dZ_Y(v)$, a.s. Finally, applying Theorem 2.2.1 again, we obtain

$$X_j = \sum_{k=0}^\infty \pi_{k,-d}\zeta_{j-k}$$
$$= \int_\Pi e^{ijv}(1 - e^{-iv})^{-d}(1 - e^{-iv})^d I(|v| > 0)dZ_Y(v)$$
$$= \int_\Pi e^{ijv} I(|v| > 0)dZ_Y(v) = Y_j,$$

which completes the proof of the theorem.

Relations (7.2.8) and (7.2.10) imply that $\{X_j\}$ has long memory, with memory parameter d, if $0 < d < 1/2$, and negative memory if $-1/2 < d < 0$. In the latter case, $\sum_{j \in \mathbb{Z}} \gamma_X(j) = 0$.

ARFIMA(p, d, q) **model.** The class of ARFIMA(p, d, q) (fractionally integrated ARMA(p, q)) processes, introduced by Granger and Joyeux (1980) and Hosking (1981), allows us to model a larger variety of autocovariances and spectral densities than ARFIMA$(0, d, 0)$ models, and includes the ARMA(p, q) class of short memory models.

Definition 7.2.2. (*ARFIMA*(p, d, q) *process*). The stationary zero-mean process $\{X_j, j \in \mathbb{Z}\}$ is said to be an ARFIMA(p, d, q) process with parameter $d \in (-1/2, 1/2)$ if it satisfies the difference equations

$$\phi(B)\nabla^d X_j = \theta(B)\zeta_j, \quad j \in \mathbb{Z}, \tag{7.2.16}$$

where $\{\zeta_j\}$ is a WN$(0, \sigma_\zeta^2)$ sequence, and $\phi(z) = 1 - \phi_1 z - \cdots - \phi_p z^p$ and $\theta(z) = 1 + \theta_1 z + \cdots + \theta_q z^q$ are polynomials of degrees $p, q \geq 0$, respectively, that have no common zeros.

Writing

$$\phi(B)X_j = \theta(B)Y_j, \quad Y_j := \nabla^{-d}\zeta_j, \quad j \in \mathbb{Z},$$

the process $\{X_j\}$ can be regarded as an ARMA(p, q) process driven by ARFIMA$(0, d, 0)$ process $\{Y_j\}$. Note also that an ARFIMA$(p, 0, q)$ model is the same as an ARMA(p, q) model.

A solution $\{X_j\}$ of equations (7.2.16) is said to be *causal* if it can be represented as a one-side moving average:

$$X_j = \sum_{k=0}^{\infty} a_k \zeta_{j-k}, \quad j \in \mathbb{Z}, \quad \sum_{k=0}^{\infty} a_k^2 < \infty,$$

and *invertible* if the corresponding representation

$$\zeta_j = \sum_{k=0}^{\infty} \nu_k X_{j-k}, \quad j \in \mathbb{Z}, \quad \sum_{k=0}^{\infty} \nu_k^2 < \infty, \tag{7.2.17}$$

holds for some real numbers $\{\nu_k\}$, and the series in (7.2.17) converges in mean square.

To define a solution, assume

$$|\phi(z)| > 0, \quad |\theta(z)| > 0, \quad \forall |z| \le 1. \tag{7.2.18}$$

Set

$$A(B) := (1 - B)^{-d} \phi(B)^{-1} \theta(B).$$

Under (7.2.18), for $|d| < 1/2$, the function $A(z) \equiv (1 - z)^{-d} \theta(z) / \phi(z) = \sum_{k=0}^{\infty} a_{k,-d} z^k$, $|z| \le 1$, can be expanded into a converging series. Consider the process

$$X_j = A(B)\zeta_j = \sum_{k=0}^{\infty} a_{k,-d} \zeta_{j-k}, \quad a_{0,-d} = 1, \quad j \in \mathbb{Z}. \tag{7.2.19}$$

Before formally showing that this is an infinite moving-average representation of a stationary solution to ARFIMA equations, we investigate the properties of the weights $a_{k,-d}$. The following theorem establishes their square summability, and hence implies the existence of $EX_j = 0$, $\mathrm{Var}(X_j) = \sigma_\zeta^2 \sum_{k=0}^{\infty} a_{k,-d}^2$, and

$$\gamma_X(j) = \sigma_\zeta^2 \sum_{k=0}^{\infty} a_{k,-d} a_{k+j,-d}, \quad j \ge 0. \tag{7.2.20}$$

It also gives some asymptotic properties of the autocovariances $\gamma_X(j)$ and the spectral density f_X.

Theorem 7.2.3. *Suppose that* $|d| < 1/2$ *and (7.2.18) holds. Then the process* $\{X_j\}$ *of (7.2.19) has the following properties:*
(i) (ARMA(p, q) *model). If* $d = 0$, *then, for some* $\alpha > 0$ *and* $C > 0$,

$$|a_{k,0}| \le Ce^{-\alpha k}, \quad k \ge 1, \tag{7.2.21}$$

$$|\gamma_X(j)| \le Ce^{-\alpha j}, \quad j \ge 1.$$

(ii) (ARFIMA(p, d, q) model). *If $d \neq 0$ then with $\nu := \frac{\theta(1)}{\phi(1)}$ and $c_a := \frac{\nu}{\Gamma(d)}$,*

$$a_{k,-d} = \nu \pi_{k,-d} + O(k^{-2+d}) = c_a k^{d-1} + O(k^{-2+d}), \qquad (7.2.22)$$

$$\gamma_X(j) \sim \nu^2 \gamma_Y(j) \sim c_\gamma j^{-1+2d}, \quad c_\gamma = c_a^2 B(d, 1 - 2d)\sigma_\zeta^2.$$

In addition, if $d \in (-1/2, 0)$, then

$$\sum_{k=0}^{\infty} a_{k,-d} = 0, \quad \sum_{j \in \mathbb{Z}} \gamma_X(j) = 0.$$

(iii) *The spectral density of $\{X_j\}$ is*

$$f_X(v) = \frac{\sigma_\zeta^2}{2\pi} |2\sin(v/2)|^{-2d} \left| \frac{\theta(e^{-iv})}{\phi(e^{-iv})} \right|^2, \quad v \in \Pi, \qquad (7.2.23)$$

$$\sim \frac{\sigma_\zeta^2}{2\pi} \left| \frac{\theta(1)}{\phi(1)} \right|^2 |v|^{-2d}, \quad v \to 0.$$

Proof. (i) Let $d = 0$. Write

$$\frac{\theta(z)}{\phi(z)} = \sum_{k=0}^{\infty} b_k z^k, \quad |z| \leq 1. \qquad (7.2.24)$$

Because of (7.2.18), $\theta(z)/\phi(z)$ is analytic for $|z| \leq 1 + \epsilon$, for some $\epsilon > 0$. Hence, $|\theta(z)/\phi(z)| < \infty$, $\forall |z| \leq 1 + \epsilon$, and the convergence of the series $\theta(z)/\phi(z)$ implies that $|b_k|(1+\epsilon)^k \to 0$ as $k \to \infty$. In turn, this yields, with $\alpha = \log(1 + \epsilon)$,

$$|b_k| \leq C(1+\epsilon)^{-k} = C\exp(-k\alpha), \quad k \geq 1. \qquad (7.2.25)$$

This proves (7.2.21) with $a_{k,0} \equiv b_k$. The bound (7.2.25) readily yields the upper bound (7.2.21) for $\gamma_X(j)$.

(ii) Next, suppose $d \neq 0$. Then

$$A(z) = (1 - z)^{-d}\frac{\theta(z)}{\phi(z)} = \left(\sum_{i=0}^{\infty} \pi_{i,-d} z^i \right)\left(\sum_{s=0}^{\infty} b_s z^s \right) \equiv \sum_{k=0}^{\infty} a_{k,-d} z^k,$$

with $a_{k,-d} = \sum_{s=0}^{k} \pi_{s,-d} b_{k-s}$, $k \geq 0$. Write,

$$a_{k,-d} = \pi_{k,-d}\sum_{s \leq k} b_{k-s} - \pi_{k,-d}\sum_{s < 0} b_{k-s} + \sum_{s \leq k}(\pi_{s,-d} - \pi_{k,-d})b_{k-s}$$

$$:= r_{1k} + r_{2k} + r_{3k}.$$

By (i) of Theorem 7.2.1, $\pi_{s,-d} = \frac{1}{\Gamma(d)}k^{-1+d} + O(k^{-2+d})$. Then

$$r_{1k} = \nu \pi_{s,-d} = c_a k^{-1+d} + O(k^{-2+d}),$$

because $\sum_{s=0}^{\infty} b_s = \theta(1)/\phi(1) = \nu$. By (7.2.25), $|r_{2k}| \le Ck^{-1+d}e^{-\alpha k}$ $\times \sum_{s<0} e^{\alpha s} \le Ck^{-2+d}$. Next, we show that

$$|r_{3k}| \le Ck^{-2+d}. \qquad (7.2.26)$$

By (7.2.25),

$$|r_{3k}| \le C\sum_{s=0}^{k} |\pi_{s,-d} - \pi_{k,-d}|e^{-\alpha(k-s)} =: C\Big(\sum_{s=0}^{k/2}[\cdots] + \sum_{s=k/2}^{k}[\cdots]\Big).$$

Clearly, $\sum_{s=0}^{k/2}[\cdots] \le Ce^{-\alpha k/2}\sum_{s=0}^{k/2} 1 \le Ck^{-2+d}$. For $k/2 \le s \le k$, by the mean-value theorem,

$$|\pi_{s,-d} - \pi_{k,-d}| \le C|s^{-1+d} - k^{-1+d}| + C|s^{-2+d} + k^{-2+d}| \le C|k-s|k^{-2+d}.$$

Thus

$$\sum_{s=k/2}^{k}[\cdots] \le Ck^{-2+d}\sum_{s=k/2}^{k} |k-s|e^{-\alpha(k-s)} \le Ck^{-2+d}\sum_{v=0}^{k}|v|e^{-\alpha v} \le Ck^{-2+d},$$

which proves (7.2.26). The above facts imply (7.2.22) for $a_{k,-d}$.

For $d \in (-1/2, 0)$, by the first equality in (7.2.7),

$$\sum_{k=0}^{\infty} a_{k,-d} = \sum_{s=0}^{\infty} \pi_{s,-d} \sum_{k=0}^{\infty} b_k = 0,$$

and therefore,

$$\sum_{j\in\mathbb{Z}} \gamma_X(j) = \sigma_\zeta^2\Big(\sum_{k=0}^{\infty} a_{k,-d}\Big)^2 = 0.$$

Finally, claim (7.2.22) about $\gamma_X(j)$ follows by Proposition 3.2.1.

(iii) By the same argument as in (7.2.8), (7.2.22) implies (7.2.23) .

Remark 7.2.1. Theorem 7.2.3 shows that the coefficients $a_{j,-d}$ of an infinite moving-average representation of an ARFIMA(p, d, q) model and its autocovariances $\gamma_X(j)$ are asymptotically proportional to their analogs in an ARFIMA$(0, d, 0)$ model, i.e.,

$$a_{j,-d} = (\nu + o(1))\pi_{j,-d}, \quad \gamma_X(j) = (\nu^2 + o(1))\gamma_Y(j),$$

where the scaling constant $\nu = \sum_{k=0}^{\infty} b_k$ is the sum of the impulse response coefficients of the ARMA(p, q) model of (7.2.24).

The following theorem establishes the existence and uniqueness of an invertible moving-average (MA) solution to ARFIMA equations and demonstrates the existence of its infinite-order autoregressive representation:

$$\sum_{k=0}^{\infty} b_{k,d} X_{j-k} = \zeta_j, \quad j \in \mathbb{Z}, \tag{7.2.27}$$

where the autoregressive (AR) weights $b_{k,d}$ are defined by the expansion $(1-z)^d \phi(z)/\theta(z) = \sum_{k=0}^{\infty} b_{k,d} z^k$, $|z| \leq 1$, $b_{0,d} = 1$, and for $d \neq 0$ they have the property

$$b_{k,d} = (\nu' + o(1))\pi_{k,d} \sim \frac{\nu'}{\Gamma(-d)} k^{-1-d}, \quad k \to \infty, \quad \nu' = \frac{\phi(1)}{\theta(1)}.$$

Its proof is similar to that of Theorem 7.2.2.

Theorem 7.2.4. *Suppose* $d \in (-1/2, 1/2)$, $\{\zeta_j\}$ *is an ergodic process, and (7.2.18) holds. Then the process* $\{X_j\}$ *of (7.2.19) is a unique invertible stationary ergodic zero-mean solution of the ARFIMA equations (7.2.16).*

Remark 7.2.2. Hosking (1981) derived the autocovariance function $\gamma_Y(j)$ for the two-parameter ARFIMA$(1, d, 0)$ process given by

$$(1 - \phi B)(1 - B)^d Y_j = \zeta_j, \quad |d| < 1/2, \quad |\phi| < 1, \quad \zeta_j \sim \text{WN}(0, \sigma_\zeta^2).$$

This process can be written as a first-order autoregressive model $(1 - \phi B)Y_j = X_j$ with ARFIMA$(0, d, 0)$ noise $X_j = (1 - B)^{-d}\zeta_j$. Its long-term behavior will be similar to that of $\{X_j\}$ controlled by the memory parameter d, whereas the short-term behavior will also depend on the parameter ϕ.

The hypergeometric function is

$$F(a, b; c; z) = 1 + \frac{ab}{c \cdot 1} z + \frac{a(a+1)b(b+1)}{c(c+1) \cdot 1 \cdot 2} z^2 + \cdots \tag{7.2.28}$$

$$= \sum_{k=0}^{\infty} \frac{\Gamma(c)\Gamma(a+k)\Gamma(b+k)}{\Gamma(k+1)\Gamma(c+k)\Gamma(a)\Gamma(b)} z^k.$$

The weights $a_{k,-d}$ and $b_{k,d}$ of the infinite moving-average and autoregressive representations (7.2.19) and (7.2.27) of this model are given by

$$a_{k,-d} = \frac{\Gamma(k+d)}{\Gamma(k+1)\Gamma(d)} F(1, -k; 1-k-d; \phi) \sim \frac{1}{(1-\phi)\Gamma(d)} k^{-1+d},$$

$$b_{k,d} = \frac{\Gamma(k-d-1)}{\Gamma(k)\Gamma(-d)}\left(1 - \phi - \frac{1+d}{k}\right) \sim \frac{1-\phi}{\Gamma(-d)} k^{-1-d}, \quad k \to \infty.$$

The autocovariance function $\gamma_Y(j)$ of $\{Y_j\}$ can be expressed in terms of the autocovariance $\gamma_X(j)$ of $\{X_j\}$:

$$\gamma_Y(j) = \gamma_X(j)\frac{F(1, k+d; k+1-d; \phi) + F(1, -k+d; -k+1-d; \phi)}{(1-\phi)^2},$$

$$\gamma_Y(0) = \frac{\Gamma(-2d+1)}{\Gamma^2(-d+1)}\frac{F(1, d; 1-d; \phi)}{1+\phi}.$$

Explicit formulas for the autocovariance function of a general ARFIMA$(0, d, 1)$ model can be found in Hosking (1981) and for a general ARFIMA(p, d, q) model in Sowell (1992).

7.3 GARMA models

GARMA(p, d, q) model. The analysis of dependent data exhibiting seasonal behavior is important for many applications. Hosking (1981) was perhaps the first author to mention the possibility of using a generalization of a fractionally differenced model to allow for seasonality. Andel (1986) and Gray, Zhang and Woodward (1989, 1994) independently developed this idea further. They analyzed the model

$$\phi(B)(1 - 2\cos(\omega) \cdot B + B^2)^d X_j = \theta(B)\zeta_j, \quad j \in \mathbb{Z}, \qquad (7.3.1)$$

for $0 \leq \omega \leq \pi, |d| < 1/2$, where $\{\zeta_j\}$ is a $\mathrm{WN}(0, \sigma_\zeta^2)$ sequence. This model may exhibit long memory periodic behavior at any frequency $0 \leq \omega \leq \pi$ of the spectrum. In (7.3.1), $\phi(z) = 1 - \phi_1 z - \cdots - \phi_p z^p$ and $\theta(z) = 1 + \theta_1 z + \cdots + \theta_q z^q$ are polynomials of degrees $p, q \geq 0$, respectively, that have no common zeroes.

The GARMA models that we discuss have zeros or singularities of order d ($|d| < 1/2$) on the unit circle that allow us to model long and short memory data containing seasonal periodicities. Such time series exhibit both *long-term persistence and quasiperiodic behavior*, and their correlation function resembles a superposition of hyperbolically damped sine waves. The spectral density of such time series may have several peaks and zeros on $[-\pi, \pi]$.

Consider the operator

$$\nabla_\omega^d := (1 - e^{i\omega}B)^d(1 - e^{-i\omega}B)^d = (1 - 2\cos\omega \cdot B + B^2)^d$$

and the expansion

$$(1 - 2\cos\omega \cdot B + B^2)^d := \sum_{k=0}^{\infty} \psi_{k,d} B^k \qquad (7.3.2)$$

in powers of the back shift operator, B. Here, $0 \le \omega \le \pi$ is a fixed frequency of the spectrum, $d \neq 0$ is a fractional differencing degree, and the coefficients $\psi_{k,d}$ are given by

$$\psi_{k,d} = C_k^{(-d)}(\cos\omega), \qquad (7.3.3)$$

where $C_k^{(d)}(x)$ are Gegenbauer (or ultraspherical) polynomials; see Szegö (1975). These polynomials are orthogonal on $[-1, 1]$ with the weight function $(1 - x^2)^{d-1/2}$, and are usually defined by the generating function

$$(1 - 2xz + z^2)^{-d} = \sum_{k=0}^{\infty} C_k^{(d)}(x) z^k, \quad |z| < 1, \quad d \neq 0.$$

Note that if $\cos\omega = \pm 1$, $\omega = 0, \pi$, then

$$C_k^{(d)}(\pm 1) = (\pm 1)^k \binom{2d + k - 1}{k}.$$

Notice that $C_0^{(d)}(x) \equiv 1$ and the operator ∇_0^d is equal to the operator ∇^{2d} of the ARFIMA$(0, 2d, 0)$ model.

Definition 7.3.1. (*GARMA*$(0, d, 0)$ *process*). The process $\{X_j, j \in \mathbb{Z}\}$ is said to be a GARMA$(0, d, 0)$ process with parameters $d \in (-1/2, 1/2)$, $d \neq 0$, and $\omega \in (0, \pi)$ if $\{X_j\}$ is a stationary solution with zero mean of the difference equations

$$\nabla_\omega^d X_j = \zeta_j, \quad j \in \mathbb{Z}, \qquad (7.3.4)$$

where $\zeta_j \sim \mathrm{WN}(0, \sigma_\zeta^2)$ is a white-noise sequence.

A solution of equations (7.3.4) can be written as a linear process

$$X_j = \nabla_\omega^{-d} \zeta_j = \sum_{k=0}^{\infty} \psi_{k,-d} \zeta_{j-k}, \quad \psi_{0,-d} = 1, \quad j \in \mathbb{Z}, \qquad (7.3.5)$$

with $\psi_{k,-d} \equiv C_k^{(d)}(\cos\omega)$, $k = 0, 1, \cdots$.

An approximation formula (8.21.14) from Szegö (1975), applied with $p = 1$ and $0 < \omega < \pi$ gives

$$\psi_{k,-d} = C_k^{(d)}(\cos\omega) \qquad (7.3.6)$$

$$= \frac{2}{(2\sin\omega)^d} \pi_{k,-d} \cos\{(k + d)\omega - d\pi/2\} + O(k^{d-2})$$

$$= \frac{2}{\Gamma(d)(2\sin\omega)^d} k^{d-1} \cos\{(k + d)\omega - d\pi/2\} + o(k^{d-1}),$$

as $k \to \infty$, because $\pi_{k,-d} \equiv \binom{j+d-1}{j} \sim k^{d-1}/\Gamma(d)$, by (7.2.6).

Hence, unlike the weights of linear solutions of ARFIMA models, which decay at the rate k^{d-1}, the weights $\psi_{k,-d}$ of the above GARMA process tend to zero in an oscillating fashion as $k^{d-1}\cos(\alpha k+\beta)$, for some $\alpha,\ \beta \in \mathbb{R}$.

Relation (7.3.6) implies that the weights $\psi_{k,-d}$ are square summable for all $d \in (-1/2, 1/2)$, $d \neq 0$. Therefore, the linear process $\{X_j\}$ of (7.3.5) has zero mean and finite variance. The following theorem gives the main properties of a solution of a GARMA model.

Theorem 7.3.1. *The following statements hold for every* $|d| < 1/2$, $d \neq 0$, *and* $0 < \omega < \pi$:

(i) *The process* $\{X_j\}$ *of (7.3.5) is a unique stationary solution of the* $GARMA(0,d,0)$ *equations (7.3.4).*

(ii) *The solution* $\{X_j\}$ *is invertible.*

(iii) *Its spectral density is*

$$f_X(v) = \frac{\sigma_\zeta^2}{2\pi}|1 - e^{-i(v-\omega)}|^{-2d}|1 - e^{-i(v+\omega)}|^{-2d} \qquad (7.3.7)$$

$$= \frac{\sigma_\zeta^2}{2\pi}|2\sin(\frac{v-\omega}{2})|^{-2d}|2\sin(\frac{v+\omega}{2})|^{-2d}, \quad v \in \Pi,$$

$$\sim \frac{\sigma_\zeta^2}{2\pi}|2\sin\omega|^{-2d}|v-\omega|^{-2d}, \quad v \to \omega.$$

(iv) *If* $d > 0$, *then*

$$\gamma_X(j) = c_\gamma j^{-1+2d}(\cos(j\omega) + o(1)), \quad j \to \infty, \qquad (7.3.8)$$

with $c_\gamma := 2\sigma_\zeta^2\pi^{-1}(2\sin\omega)^{-2d}\Gamma(1-2d)\sin(\pi d)$.

Proof. The proof of (i) to (iii) uses the square summability of $\{\psi_{k,-d}\}$ in a similar fashion as in the case of the ARFIMA$(0,d,0)$ model discussed in Section 7.2, and therefore is omitted.

To prove (iv), write the spectral density as

$$f_X(v) = |v-\omega|^{-2d}h(v), \quad 0 \le v \le \pi,$$

$$h(v) := \frac{\sigma_\zeta^2}{2\pi}|\frac{2\sin(\frac{v-\omega}{2})}{v-\omega}|^{-2d}|2\sin(\frac{v+\omega}{2})|^{-2d}.$$

Observe that h has a continuous first derivative $\dot{h}$ in $v \in [0,\pi]$, $\forall\, 0 < \omega < \pi$, and $h(v) \to \sigma_\zeta^2|2\sin\omega|^{-2d}/2\pi$, $v \to \omega$. To derive (7.3.8), apply Lemma 7.3.1 below to $\gamma_X(j) = 2\int_0^\pi \cos(jv)f(v)dv$.

Lemma 7.3.1. *Let* $f(v) = |v - \omega|^{-2d} h(v)$, $v \in [0, \pi]$, *where* $0 < d < 1/2$, $0 < \omega < \pi$, *and let* $h(v)$, $v \in [0, \pi]$, *be a piecewise differentiable function, such that*

$$\int_0^\pi |\dot{h}(v)| dv < \infty, \quad h(v) \to h(\omega) > 0, \quad (v - \omega)\dot{h}(v) \to 0, \quad v \to \omega. \quad (7.3.9)$$

Then, as $j \to \infty$,

$$\int_0^\pi f(v) \cos(jv) dv = c_{d,\omega} j^{-1+2d} (\cos(j\omega) + o(1)), \quad (7.3.10)$$

$$c_{d,\omega} := 2h(\omega)\Gamma(1 - 2d)\sin(\pi d).$$

Proof. Let $h_\omega(v) := h(v+\omega)$. Assumptions (7.3.9) and Lemma 2.3.1 imply that function $h_\omega(\pm v)$ has bounded variation, is a Zygmund slowly varying function as $v \to 0$, and $\lim_{v \to 0} h_\omega(v) = h(\omega)$.

Denote by $R(j)$ the l.h.s. of (7.3.10). Then

$$R(j) = \int_{-\omega}^{\pi-\omega} |v|^{-2d} h_\omega(v) \cos(j(v + \omega)) dv. \quad (7.3.11)$$

Since $\cos(j(v + \omega)) = \cos(jv)\cos(j\omega) - \sin(jv)\sin(j\omega)$, letting

$$r_{j,1} := \int_{-\omega}^{\pi-\omega} |v|^{-2d} h_\omega(v) \cos(jv) dv, \quad r_{j,2} := \int_{-\omega}^{\pi-\omega} |v|^{-2d} h_\omega(v) \sin(jv) dv,$$

we can write $R(j) = r_{j,1} \cos(j\omega) - r_{j,2} \sin(j\omega)$. By (2.3.14) and (2.3.15),

$$r_{j,1} = \int_0^{\pi-\omega} |v|^{-2d} h_\omega(v) \cos(jv) dv + \int_0^\omega |v|^{-2d} h_\omega(-v) \cos(jv) dv$$
$$= 2j^{-1+2d} h(\omega)\Gamma(1 - 2d)\sin(\pi d) + o(j^{-1+2d}),$$

$$r_{j,2} = \int_0^{\pi-\omega} |v|^{-2d} h_\omega(v) \sin(jv) dv - \int_0^\omega |v|^{-2d} h_\omega(-v) \sin(jv) dv$$
$$= o(j^{-1+2d}),$$

which proves (7.3.10).

Theorem 7.3.1 shows that, when $d > 0$ ($d < 0$), the spectral density f_X of the GARMA$(0, d, 0)$ process $\{X_j\}$ has two peaks (zeros) at frequencies ω and $-\omega$, and otherwise is bounded away from 0 and ∞. If $d > 0$, the auto-covariance function $\gamma_X(j)$ decays as a dampened sine wave $j^{-1+2d} \cos(j\omega)$, and ω defines the length of a cycle. If $d > 0$, the covariance function $\gamma_X(j)$ is absolutely not summable, $\sum_{j \in \mathbb{Z}} |\gamma_X(j)| = \infty$, although the series $\sum_{j \in \mathbb{Z}} \gamma_X(j) = 2\pi f_X(0) < \infty$ converges.

Remark 7.3.1. Gegenbauer polynomials $c_k \equiv c_k(x) := C_k^{(d)}(x)$, $k \geq 0$, $|x| \leq 1$, can be computed in several ways (see Chung (1996), p. 239). The general expression for Gegenbauer polynomials c_k is

$$(1 - xz + z^2)^{-d} = \sum_{k=0}^{\infty} c_k z^k, \quad |z| \leq 1,$$

$$c_k = \sum_{s=0}^{[k/2]} \frac{(-1)^s \Gamma(d + k - s)}{\Gamma(d)\Gamma(s+1)\Gamma(k - 2s + 1)} (2x)^{k-2s}, \quad k \geq 0,$$

where $[j/2]$ is the integer part of $j/2$. The simplest way to compute c_k is to use a recursion formula:

$$c_k = 2x\left(\frac{d-1}{k} + 1\right)c_{k-1} - \left(2\frac{d-1}{k} + 1\right)c_{k-2}, \quad k \geq 2,$$

where $c_0 = 1$, $c_1 = 2dx$, and $c_2 = 2d(d+1)x^2 - d$.

Hence, the coefficients $\psi_{k,-d} = C_k^{(d)}(\cos(\omega)) = c_k(\cos(\omega))$, $k \geq 0$ of the MA representation (7.3.5) of the GARMA$(0, d, 0)$ process $\{X_j\}$ can be computed by setting $x = \cos(\omega)$.

The autocovariances $\gamma_X(j)$, $j \geq 0$, of the GARMA$(0, d, 0)$ process $\{X_j\}$ as in (7.3.4), setting $x = \cos(\omega)$, for $|x| < 1$, are

$$\frac{\sigma_\zeta^2}{2\sqrt{\pi}} \Gamma(1 - 2d)(2\sin(\omega))^{1/2-2d}\{P_{j-1/2}^{2d-1/2}(x) + (-1)^j P_{j-1/2}^{2d-1/2}(-x)\},$$

where $P_a^b(x)$ are the associated Legendre functions, see Chung (1996). The associated Legendre functions $P_{j-1/2}^b(x)$, $j = 0, 1, \cdots$, can be calculated using the recursion

$$P_a^b(x) = \frac{2a - 1}{a - b} x P_{a-1}^b(x) - \frac{a + b - 1}{a - b} P_{a-2}^b(x).$$

The initial terms required for this recursion are

$$P_{-1/2}^{2d-1/2}(x) = \left(\frac{1+x}{1-x}\right)^{d-1/4} \frac{1}{\Gamma(\frac{3}{2} - 2d)} F\left(\frac{1}{2}, \frac{1}{2}; \frac{3}{2} - 2d; \frac{1-x}{2}\right),$$

$$P_{1/2}^{2d-1/2}(x) = \left(\frac{1+x}{1-x}\right)^{d-1/4} \frac{1}{\Gamma(\frac{3}{2} - 2d)} F\left(-\frac{1}{2}, \frac{3}{2}; \frac{3}{2} - 2d; \frac{1-x}{2}\right),$$

where $F(a, b; c; z)$ is the hypergeometric function (7.2.28).

Now we turn to the existence of a solution of the general GARMA model defined by (7.3.1).

The model (7.3.1) can be written as an ARMA(p, q) model driven by a GARMA$(0, d, 0)$ process $\{Y_j\}$, i.e.,

$$\phi(B)X_j = \theta(B)Y_j, \quad Y_j := \nabla_\omega^{-d}\zeta_j, \quad j \in \mathbb{Z}. \tag{7.3.12}$$

Expand the function

$$g(z) := (1 - 2\cos(\omega) \cdot z + z^2)^{-d}\phi(z)^{-1}\theta(z)$$

$$= \Big(\sum_{i=0}^{\infty} \psi_{i,-d}z^i\Big)\Big(\sum_{s=0}^{\infty} b_s z^s\Big)$$

$$= \sum_{k=0}^{\infty} g_{k,-d}z^k, \quad g_{k,-d} = \sum_{s=0}^{k} \psi_{s,-d}b_{k-s}, \quad g_{0,-d} = 1,$$

in powers of z^k, where $\psi_{k,-d}$ and b_k are as in (7.3.2) and (7.2.24).

To obtain an upper bound on $|g_{k,-d}|$, which is needed to prove its square summability, recall from (7.3.6) that $|\psi_{k,-d}| = O(k^{d-1})$. Therefore, by an argument similar to the one used to evaluate $a_{k,-d}$ in the proof of (7.2.22),

$$|g_{k,-d}| = O(k^{d-1}), \quad k \to \infty. \tag{7.3.13}$$

The derivation of the exact asymptotics of $g_{k,-d}$ is more involved and we omit it in this text.

Since the weights $g_{k,-d}$ are square summable, if $|d| < 1/2$, $d \neq 0$, they define a stationary zero mean and finite variance linear process

$$X_j = \Big(\sum_{k=0}^{\infty} g_{k,-d}B^k\Big)\zeta_j = \sum_{k=0}^{\infty} g_{k,-d}\zeta_{j-k}, \quad g_{0,-d} = 1, \quad j \in \mathbb{Z}. \tag{7.3.14}$$

The next theorem shows that this process $\{X_j\}$ is a unique stationary solution to the GARMA equations (7.3.1). It also describes relationships between the spectral densities and autocovariances of the process $\{X_j\}$ and the GARMA$(0, d, 0)$ process $Y_j = \nabla_\omega^{-d}\zeta_j$. Its proof is similar to that of Theorem 7.3.1.

Theorem 7.3.2. *Suppose $0 < \omega < \pi$ and (7.2.18) holds. Then the following hold:*

(i) Process $\{X_j\}$ of (7.3.14) is a unique invertible stationary solution of the GARMA equations (7.3.1) having the spectral density

$$f_X(v) = \Big|\frac{\theta(e^{-iv})}{\phi(e^{-iv})}\Big|^2 f_Y(v), \quad v \in \Pi, \tag{7.3.15}$$

$$\sim \Big|\frac{\theta(e^{-i\omega})}{\phi(e^{-i\omega})}\Big|^2 \frac{\sigma_\zeta^2}{2\pi}|2\sin\omega|^{-2d}|v - \omega|^{-2d}, \quad v \to \omega, \quad \forall |d| < 1/2, d \neq 0.$$

(ii) *If $d > 0$, then*

$$\gamma_X(j) \sim \left| \frac{\theta(e^{-i\omega})}{\phi(e^{-i\omega})} \right|^2 \gamma_Y(j)$$

$$\sim \left| \frac{\theta(e^{-i\omega})}{\phi(e^{-i\omega})} \right|^2 c_\gamma j^{-1+2d} (\cos(j\omega) + o(1)), \quad j \to \infty,$$

where c_γ is the same as in (7.3.8).

Note: The approximation for $C_k^{(d)}(x)$ given by formula (13) in Gray *et al.* (1989) has a slightly different constant from that in (7.3.6) above. Giraitis and Leipus (1995) and Woodward, Cheng, and Gray (1998) analyzed the GARMA model with a finite number of poles in the spectrum. As a special case, this model includes seasonal fractionally differenced models; see Porter-Hudak (1990) and Hassler (1994). Estimation in GARMA models is discussed in Chung (1996), Giraitis, Hidalgo, and Robinson (2001), Palma and Chan (2005), Hidalgo (2005), and Reisen, Rodrigues and Palma (2006).

7.4   Simulation of ARFIMA processes

Suppose that we wish to generate a realization $X_0, X_1, \cdots, X_n$ of a stationary Gaussian process $\{X_j\}$ with a given autocovariance function $\gamma_X(0)$, $\gamma_X(1), \cdots, \gamma_X(n)$. The Davies–Harte algorithm, first given in Davies and Harte (1987), provides a fast and exact method of generating such time series, which we now describe.

First, form a sequence $\gamma_X(0), \gamma_X(1), \cdots, \gamma_X(n-1), \gamma_X(n), \gamma_X(n-1),$ $\cdots, \gamma_X(1)$, and compute its finite Fourier transforms:

$$g_{k,n} := \sum_{j=0}^{n-1} e^{\frac{i\pi kj}{n}} \gamma_X(j) + \sum_{j=n}^{2n-1} e^{\frac{i\pi kj}{n}} \gamma_X(2n-j), \quad k = 0, \cdots, 2n-1.$$

The series $g_{k,n}$ is real. A sufficient condition for the validation of the Davies–Harte procedure is the set of restrictions

$$g_{k,n} \geq 0, \qquad k = 0, \cdots, 2n-1. \tag{7.4.1}$$

These restrictions are satisfied by the autocovariance function of the ARFIMA$(0, d, 0)$ model, $|d| < 1/2$; see Craigmile (2003).

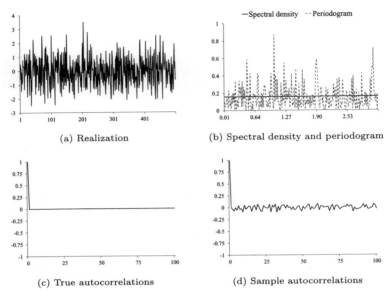

Fig. 7.4.1: IID $N(0,1)$ process: (a) realization; (b) spectral density (solid line) and periodogram (dashed line); (c) true autocorrelations; (d) sample autocorrelations.

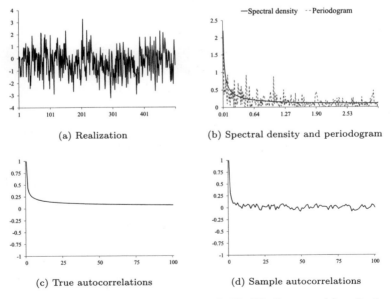

Fig. 7.4.2: ARFIMA$(0,d,0)$, $d = 0.3$ process with IID $N(0,1)$ errors: (a) realization; (b) spectral density (solid line) and periodogram (dashed line); (c) true autocorrelations; (d) sample autocorrelations.

Let $\{\zeta_k\} \sim \text{IID}(0,1)$ be a Gaussian i.i.d. sequence. Define independent complex normal random variables

$$
\begin{aligned}
Z_k &= \sqrt{2}\zeta_0, & k &= 0, \\
&= \zeta_{2k-1} + i\zeta_{2k}, & k &= 1, \cdots, n-1, \\
&= \sqrt{2}\zeta_{2n-1}, & k &= n, \\
&= \bar{Z}_{2n-k}, & k &= n+1, \cdots, 2n-1,
\end{aligned}
$$

where $\bar{Z}_{2n-k}$ is the complex conjugate of Z_{2n-k}. Then the simulated time series

$$
X_j = \frac{1}{2\sqrt{n}} \sum_{k=0}^{2n-1} e^{i\pi kj/n} \sqrt{g_{k,n}}\, Z_k, \quad j = 0, \cdots, n,
$$

has the required Gaussian distribution.

When condition (7.4.1) does not hold, we may need to resort to a computationally more involved Durbin–Levinson algorithm; see Brockwell and Davis (1991), p. 169.

In practice, realizations of a general Gaussian ARFIMA(p,d,q), $|d| < 1/2$, model can be obtained as follows. Rewrite it as

$$
(1 - \phi_1 B - \cdots - \phi_p B^p)Y_j = (1 + \theta_1 B + \cdots + \theta_q B^q)X_j, \quad j \in \mathbb{Z},
$$

where $X_j = (1-B)^{-d}\zeta_j$ is a Gaussian ARFIMA$(0,d,0)$ process. Generate $X_1, \cdots, X_n$ by the Davies–Harte algorithm, set $Y_0 = Y_1 = \cdots = Y_{p-1} = 0$, and use the following recursion to compute

$$
Y_j = \phi_1 Y_{j-1} + \cdots + \phi_p Y_{j-p} + X_j + \theta_1 X_{j-1} + \cdots + \theta_q X_{j-q}, \quad j \geq p.
$$

Then, for $N > 1$ sufficiently large, $\{Y_j, j \geq N\}$ will approximate an ARFIMA(p,d,q) model.

The general linear process $\{X_j\}$ may be specified by the coefficients a_k of its infinite moving-average representation:

$$
X_j = \zeta_j + a_1\zeta_{j-1} + a_2\zeta_{j-2} + \cdots, \quad j \in \mathbb{Z},
$$

where $\{\zeta_k\} \sim \text{IID}(0,\sigma_\zeta^2)$ and $a_0^2 + a_1^2 + \cdots < \infty$, which ensures the finiteness of $\text{Var}(X_j)$ for all j. If the weights a_k are known and the ζ_j's can be simulated, then for sufficiently large q, we may approximate X_j by an MA of order q:

$$
X_j = \zeta_j + a_1\zeta_{j-1} + a_2\zeta_{j-2} + \cdots + a_q\zeta_{j-q}, \quad j \in \mathbb{Z}.
$$

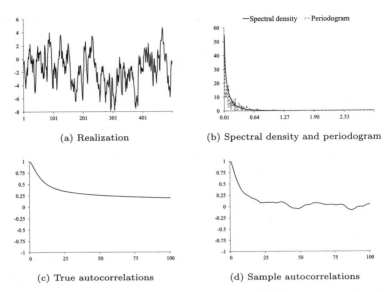

(a) Realization

(b) Spectral density and periodogram

(c) True autocorrelations

(d) Sample autocorrelations

Fig. 7.4.3: ARFIMA$(1, d, 0)$, $\rho = 0.8$, $d = 0.3$ process with IID N$(0, 1)$ errors: (a) realization; (b) spectral density (solid line) and periodogram (dashed line); (c) true autocorrelations; (d) sample autocorrelations.

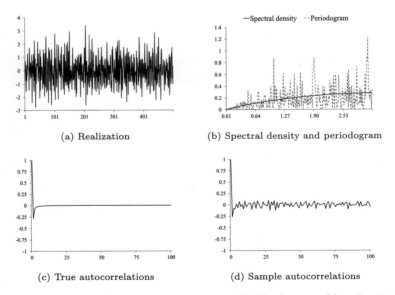

(a) Realization

(b) Spectral density and periodogram

(c) True autocorrelations

(d) Sample autocorrelations

Fig. 7.4.4: ARFIMA$(0, d, 0)$, $d = -0.4$ process with IID N$(0, 1)$ errors: (a) realization; (b) spectral density (solid line) and periodogram (dashed line); (c) true autocorrelations; (d) sample autocorrelations.

Simulated examples. Figures 7.4.1 to 7.4.6 show simulated examples of $IID(0, 1)$, $ARFIMA(0, d, 0)$, $ARFIMA(1, d, 0)$, and $GARMA(0, d, 0)$ processes. In each figure, part (a) shows a simulated realization of length 500, part (b) shows the periodogram and the true spectral density and parts (c) and (d) contain the true and sample autocorrelations.

Figure 7.4.1 confirms that the sample autocorrelations of an $IID(0, 1)$ process vanish at non-zero lags, and the extremes of the periodogram seemingly do not have any dominating peaks.

Figure 7.4.2 shows that the presence of long memory in an ARFIMA process induces non-periodic wandering behavior in its realization. In addition, its sample autocorrelations decay to zero slowly, and the periodogram has a sharp peak at the zero frequency. As seen from Figure 7.4.3, these patterns are further strengthened by the additional presence of a strong autoregressive factor.

Figure 7.4.4 shows that the periodogram of an ARFIMA process with negative memory has a tendency to drop to zero at the origin.

Figures 7.4.5 and 7.4.6 confirm the periodic nature of a GARMA process illustrated by the sinusoidal slowly decaying behavior of the sample autocorrelations and the periodogram, which peaks at a non-zero frequency.

The patterns of the asymptotic behavior of the true spectral density and true autocorrelation function are consistent with the theoretical findings of Sections 7.2 and 7.3.

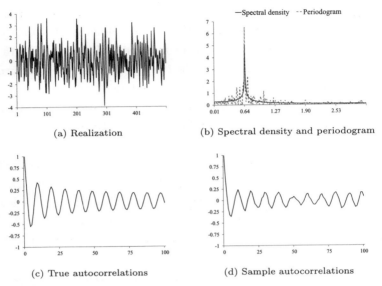

Fig. 7.4.5: GARMA$(0, d, 0)$, $d = 0.3$, $\omega = \arccos(0.8)$ process with IID $N(0, 1)$ errors: (a) realization; (b) spectral density (solid line) and periodogram (dashed line); (c) true autocorrelations; (d) sample autocorrelations.

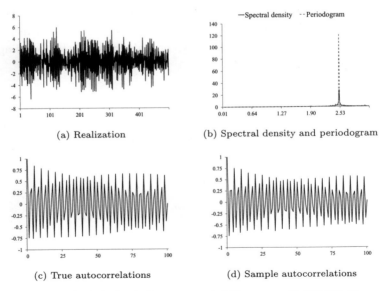

Fig. 7.4.6: GARMA$(0, d, 0)$, $d = 0.45$, $\omega = \arccos(-0.8)$ process with IID $N(0, 1)$ errors: (a) realization; (b) spectral density (solid line) and periodogram (dashed line); (c) true autocorrelations; (d) sample autocorrelations.

7.5 Aggregation, disaggregation, and long memory

The aggregation procedure provides a physical explanation of the long memory phenomenon in numerous physical and social sciences like hydrology, economics, finance, and communication networks. It also gives a practical tool for simulating long memory processes.

The contemporaneous aggregation is a summation procedure of identically distributed *elementary* random processes $\{X_j^{(k)}\}$, $k \geq 1$, over index k. Elementary processes are usually short memory AR models with random coefficients. The *aggregate* $\{Y_j = \lim_{n \to \infty} Y_j^n\}$, if it exists, is defined as the limit of the partial aggregation process $Y_j^n = A_n^{-1} \sum_{k=1}^n X_j^{(k)}$, $j \geq 1$, with some normalization A_n, and is independent of the random coefficients of the elementary process.

Contemporaneous aggregation was introduced by Granger (1980). He showed that aggregating dynamic micro-relationships described by the dynamic equations of an AR(1) model can generate long memory at the macro level and may lead to a class of aggregates that have very different properties and dynamics from the aggregated elementary processes.

More precisely, consider the AR(1) model

$$X_j = \rho X_{j-1} + \zeta_j, \quad j \in \mathbb{Z}, \quad \zeta_j \sim \text{IID}(0, \sigma_\zeta^2) \tag{7.5.1}$$

with a random coefficient $\rho \in (-1, 1)$ that is independent of $\{\zeta_j\}$. Suppose

$$E[\frac{1}{1 - \rho^2}] < \infty. \tag{7.5.2}$$

This AR(1) model has a zero-mean stationary linear solution with autocovariance and spectral density as follows:

$$X_j = \sum_{k=0}^{\infty} \rho^k \zeta_{j-k}, \quad j \in \mathbb{Z}, \tag{7.5.3}$$

$$\gamma_X(k) = \sigma_\zeta^2 E[\frac{\rho^k}{1 - \rho^2}], \quad \forall k \geq 0,$$

$$f_X(v) = \frac{\sigma_\zeta^2}{2\pi} E|1 - \rho e^{-iv}|^{-2}, \quad v \in \Pi.$$

Assumption (7.5.2) ensures that the series (7.5.3) converges in the mean square, and yields the above formula for $\gamma_X(k)$. To derive the spectral density $f_X(v)$, let $\zeta_j = \int_\Pi e^{ijv} dZ_\zeta(v)$ be the spectral representation of $\{\zeta_j\}$.

Under (7.5.2), the series of r.v.'s $\sum_{s=0}^{\infty} e^{-isv} \rho^s$ converges to $(1 - \rho e^{-iv})^{-1}$ in L_2 and is independent of Z_ζ. Hence,

$$X_k = \sum_{s=0}^{\infty} \rho^s \zeta_{k-s} = \int_\Pi e^{ikv} \sum_{s=0}^{\infty} e^{-isv} \rho^s \, dZ_\zeta(v)$$

$$= \int_\Pi e^{ikv} (1 - \rho e^{-iv})^{-1} \, dZ_\zeta(v),$$

$$\gamma_X(k) = E[X_0 \overline{X}_k] = \int_\Pi e^{-ikv} E|1 - \rho e^{-iv}|^{-2} E|dZ_\zeta(v)|^2$$

$$= \int_\Pi e^{ikv} \{ \frac{\sigma_\zeta^2}{2\pi} E|1 - \rho e^{-iv}|^{-2} \} dv, \quad k \geq 1,$$

since $E|dZ_\zeta(v)|^2 = (\sigma_\zeta^2/2\pi)dv$. By definition, this proves the existence of the spectral density f_X as in (7.5.3).

The distribution of the coefficient ρ is called the *mixture* distribution, and its density φ is called the *mixture* density. Assume now that the random coefficient ρ takes values in $(0, 1)$ and has the beta mixture density

$$\varphi(v) = \frac{1}{B(p, q)} v^{p-1}(1 - v)^{q-1}, \quad 0 < v < 1, \tag{7.5.4}$$

with parameters $p > 0$ and $q > 0$.

The next proposition shows that an AR(1) model with a beta mixture density has long memory if $1 < q < 2$, and short memory if $q > 2$. Let $c_\rho := \Gamma(p + q)/\Gamma(p)$.

Proposition 7.5.1. *Suppose that the random coefficient ρ in the AR(1) model (7.5.1) has the beta mixture density (7.5.4). Then the following hold:*
(i) Series $\{X_j\}$ in (7.5.3) converges a.s., $\forall q > 1$, and is a stationary solution of (7.5.1).
(ii) The autocovariance and spectral density of $\{X_j\}$ satisfy:

$$\gamma_X(k) = \frac{\sigma_\zeta^2 c_\rho}{q - 1} k^{-q+1}(1 + O(k^{-1})), \quad k \to \infty, \quad \forall q > 1, \tag{7.5.5}$$

$$f_X(v) \sim c_X |v|^{-2+q}, \quad v \to 0, \quad \forall 1 < q < 2, \tag{7.5.6}$$

$$c_X = \frac{\sigma_\zeta^2 c_\rho}{q - 1} \pi^{-1} \Gamma(2 - q) \sin(\frac{\pi(q - 1)}{2}).$$

Moreover, $\mathrm{Var}(X_0) = \infty$ *if* $0 < q \leq 1$.

Proof. First we show that for all $q > 0$,

$$E[\rho^k] = c_\rho k^{-q}(1 + O(k^{-1})), \quad k \to \infty. \tag{7.5.7}$$

By definition

$$E[\rho^k] = \frac{1}{B(p,q)} \int_0^1 v^{k+p-1}(1-v)^{q-1}dv \tag{7.5.8}$$

$$= \frac{\Gamma(p+q)}{\Gamma(p)} \frac{\Gamma(k+p)}{\Gamma(k+p+q)}.$$

By (7.2.11),

$$\frac{\Gamma(k+p)}{\Gamma(k+p+q)} = k^{-q}(1 + O(k^{-1})), \quad k \to \infty,$$

which together with (7.5.8) proves (7.5.7).

(i) Recall the Radamacher–Menchoff convergence theorem for orthogonal r.v.'s. Suppose a sequence of zero mean and uncorrelated r.v.'s Y_j satisfies $\sum_{k=1}^{\infty}(\log^2 k)EY_k^2 < \infty$. Then $\sum_{k=1}^n Y_k \to \sum_{k=1}^{\infty} Y_k$ a.s.; see Theorem 2.3.2 of Stout (1974).

Write the process X_j in (7.5.3) as $X_j = \sum_{k=0}^{\infty} Y_k$, $Y_k := \rho^k \zeta_{j-k}$. Clearly, the r.v.'s Y_k's have zero mean and are uncorrelated. Moreover, by (7.5.7),

$$\sum_{k=1}^{\infty}(\log^2 k)EY_k^2 \leq C \sum_{k=1}^{\infty}(\log^2 k)k^{-q} < \infty, \quad q > 1.$$

Hence, for $q > 1$, series (7.5.3) converges a.s. The process $\{X_j\}$ of (7.5.3) satisfies $X_j - \rho X_{j-1} = \zeta_j$, so it is a solution of (7.5.1).

(ii) Relation (7.5.7) implies (7.5.5):

$$\gamma_X(k) = \sigma_\zeta^2 E[\frac{\rho^k}{1-\rho^2}] = \sigma_\zeta^2 \sum_{j=0}^{\infty} E[\rho^{k+j}]$$

$$= \sigma_\zeta^2 c_\rho \left(\sum_{j=0}^{\infty}(k+j)^{-q}\right)(1 + O(k^{-1}))$$

$$= (q-1)^{-1}\sigma_\zeta^2 c_\rho k^{-q+1}(1 + O(k^{-1})),$$

since $s_k := \sum_{j=0}^{\infty}(k+j)^{-q} = (q-1)^{-1}k^{-q+1}(1 + O(k^{-1}))$, which follows from the estimates

$$s_k \leq \int_k^{\infty} u^{-q}du = (q-1)^{-1}k^{-q+1},$$

$$s_k \geq \int_{k+1}^{\infty} u^{-q}du = (q-1)^{-1}(k+1)^{-q+1}$$

$$= (q-1)^{-1}k^{-q+1}(1 + O(k^{-1})).$$

Property (7.5.5) of $\gamma_X(j)$ yields (7.5.3) for the spectral density f_X, by part (b) of Proposition 3.1.1.

Finally, for $0 < q \leq 1$, (7.5.7) implies

$$\mathrm{Var}(X_0) \geq \lim_{n \to \infty} \sigma_\zeta^2 \sum_{k=0}^n E[\rho^{2k}] = \infty.$$

Aggregation. For *aggregation*, the mixture density φ is *a priori* given and the objective is to characterize the properties of the aggregated process $\{Y_j\}$, obtained by summing independent copies of X_j, in particular, the behavior of the spectral density and covariance function.

We now provide a precise description of the aggregation procedure of the elementary process AR(1).

Let $\{X_j\}$ be a stationary solution of the AR(1) model (7.5.1) with a random coefficient. Let $\{X_j^{(k)}\}$, $k \geq 1$, be its independent copies given by

$$X_j^{(k)} = \rho^{(k)} X_{j-1}^{(k)} + \zeta_j^{(k)}, \quad j \in \mathbb{Z},$$

where

(a) $\{\zeta_j^{(k)}\}$ are independent copies of $\{\zeta_j\} \sim \mathrm{IID}(0, \sigma_\zeta^2)$;

(b) $\{\rho^{(k)}\}$ are independent copies of the r.v. ρ;

(c) ρ satisfies condition (7.5.2);

(d) $\rho^{(k)}$ are independent of $\{\zeta_j^{(k)}\}$, $k = 1, 2, \cdots$.

For each $n \geq 1$, define the partial aggregation process,

$$Y_j^n = n^{-1/2} \sum_{k=1}^n X_j^{(k)}, \quad j \in \mathbb{Z}.$$

Because Y_j^n is a rescaled sum of r.v.'s $X_j^{(k)}$, $k = 1, \cdots, n$, which are independent copies of X_j, its autocovariance and spectral density coincide with those of $\{X_j\}$:

$$\gamma_{Y^n}(j) \equiv \gamma_X(j), \quad j \in \mathbb{Z}, \quad f_{Y^n} \equiv f_X. \tag{7.5.9}$$

Since the summands $X_j^{(k)}$ are i.i.d. r.v.'s with zero mean and finite variance, the standard CLT implies that for every j, as $n \to \infty$,

$$Y_j^n \to_D \mathcal{N}(0, \mathrm{Var}(X_0)). \tag{7.5.10}$$

Let $\{Y_j\}$ be a Gaussian process with zero mean and autocovariance $\gamma_Y(j) \equiv \gamma_X(j)$, $j \in \mathbb{Z}$. Then (7.5.9) and (7.5.10) and the Cramér–Wold device yield

$$\{Y_j^n, j \in \mathbb{Z}\} \to_D \{Y_j, j \in \mathbb{Z}\}, \tag{7.5.11}$$

$$\gamma_Y(k) = \gamma_X(k), \quad k \in \mathbb{Z}; \quad f_Y = f_X.$$

To conclude, AR(1) equations with a non-random parameter $0 < \rho < 1$ generate a short memory process. If parameter $\rho \in (0,1)$ is random, the AR(1) equations define a stationary model $\{X_j\}$, whose autocovariance $\gamma_X(k) = E[\rho^k/(1-\rho^2)]$ may decay to zero hyperbolically slowly at rate $k^{-\alpha}$, with $\alpha > 0$, e.g., $\alpha = q - 1 > 0$ in (7.5.5). The decay rate is determined by the asymptotic properties of the mixture density φ in the neighborhood of 1. If $0 < \alpha < 1$, process $\{X_j\}$ has long memory.

The aggregation procedure, based on sampling from an AR(1) model with a random coefficient, preserves the autocovariance and spectral density but the limit aggregate is Gaussian and independent of ρ.

Disaggregation. Suppose that we observe the aggregated process $\{Y_j\}$ with spectral density f_Y. Assuming that this data process was accumulated by aggregating a sample from an AR(1) or other short memory dynamics, we may want to recover the mixture density and reconstruct the individual elementary process. If the individual model is AR(1), then $\gamma_Y(k) = \sigma_\zeta^2 E[\rho^k/(1 - \rho^2)]$, $k \geq 0$, and the mixture density φ can be estimated from the sample covariance function $\widehat{\gamma}_Y(k)$. The estimation of mixture φ belonging to a parametric family was considered by Robinson (1978), and using the orthogonal Gegenbauer expansion, by Leipus, Oppenheim, Philippe and Viano (2006).

Note: The aggregation of micro level processes leading to long memory at macro level and some applications were discussed in Haubrich and Lo (2001), Oppenheim and Viano (2004), Zaffaroni (2004) and Beran, Schutzner and Ghosh (2010). Dacunha-Castelle and Fermin (2006) proved that a large set of long memory processes, including classical long memory processes, is obtained by the aggregation of short memory processes having infinitely differentiable spectral densities. Puplinskaitė and Surgailis (2010) discussed the aggregation of independent AR(1) processes with infinite variance and various long memory properties of the limit α-stable aggregated process.

Chapter 8

Estimation

8.1 Introduction

This chapter discusses some methods of estimation of the underlying parameters in moving-average processes including the long memory parameter d. Broadly speaking, the moving-average processes considered here may be classified into parametric and semiparametric processes. A process or an underlying model is said to be *parametric* if its spectral density or autocovariance function is known up to an unknown Euclidean parameter. A process is said to be *semiparametric* if its spectral density or autocovariance function is known to depend on an Euclidean parameter and an unknown infinite-dimensional parameter. Examples of the former models include ARFIMA and GARMA processes of the previous chapter while an example of the latter would be a moving-average model whose spectral density is like $|u|^{-d}g(u)$, $|d| < 1/2$, where g is some unspecified function.

When the only parameter of interest is d, the drawback of parametric methods is that they imply unnecessary assumptions on the spectral density. Since d determines the behavior of the spectral density f at zero, one can construct consistent estimators of d without any prior knowledge of f outside an arbitrary small neighborhood of zero. These methods are referred to as local semiparametric methods.

The two types of estimators that are often used in practice are Whittle and local Whittle estimators. The former is used in parametric models while the latter is used in semiparametric models. Both are relatively easy to compute, compared to the maximum likelihood estimator in the Gaussian case. Whittle estimators are asymptotically efficient when the underlying process is Gaussian. Local Whittle estimators are used especially in semi-parametric models where the behavior of the spectral densities is known only near the origin in a semiparametric fashion. A finite sample simulation study of Taqqu and Teverovsky (1997) shows that this estimator is also less sensitive to heavy tail innovation distributions compared to the Whittle estimator and some other estimators.

Section 8.2 focuses on the consistency of Whittle estimators in the parametric set-up not only under long memory but also when the spectral density of the underlying process vanishes at the origin. This result is stated in Theorem 8.2.1. This section ends by showing the applicability of Theorem 8.2.1 to the ARFIMA and GARMA models of the previous chapter.

The focus of Section 8.3 is the asymptotic normality of these estimators. Using the results of Section 6.3, the asymptotic normality of Whittle estimators is established in Subsection 8.3.1 for parametric models under minimal sufficient conditions on the given spectral density. Subsection 8.3.2 establishes the validity of these conditions for ARFIMA(p, d, q) models for $-1/2 < d < 1/2$.

Sections 8.4 to 8.6 deal with the estimation of the parameters appearing in the underlying spectral density in non- and semiparametric short and long memory models. Section 8.4 discusses the R/S and the log-periodogram method of estimation without any proofs while Section 8.5 focuses on the consistency of local Whittle estimators for general stationary and ergodic processes whose spectral density $f(u) \sim c|u|^{-2d}$, $u \to 0$, $-1/2 < d < 1/2$, $c > 0$. The rates of consistency are also discussed here. These results are applicable to a large class of processes, including some non-linear models. In particular, they are applied in Section 8.6 to consistently estimate the underlying parameters of a signal process when observing a signal plus noise process. The general results on consistency are also used to prove $\log(n)$-consistency of the Whittle estimator of the long memory parameter appearing in long memory moving-average error processes in regression models with non-linear trends. Some further applications for a volatility process are given in Subsection 8.6.2. This chapter ends with a simulation study.

8.2   Parametric model: consistency

This section discusses the consistency of the Whittle estimator when the underlying process is a short or long memory linear process with a known parametric spectral density. Let Θ be a subset of $\mathbb{R}^q$, and $\{\zeta_j\} \sim \mathrm{WN}(0, \sigma^2)$ be a white-noise process. For $\theta \in \Theta$, let $a_k(\theta), k = 0, 1, \cdots$, be sequences of real numbers. Assume that observations $X_1, \cdots, X_n$ are from the *linear (moving-average)* process

$$X_j := \sum_{k=0}^{\infty} a_k(\theta)\zeta_{j-k}, \quad a_0(\theta) = 1, \quad j \in \mathbb{Z}, \tag{8.2.1}$$

$$\sum_{k=0}^{\infty} a_k^2(\theta) < \infty, \quad \theta \in \Theta.$$

By Proposition 2.2.2, the spectral density of the process $\{X_j\}$ has a parametric form

$$f(u) \equiv f_{\theta,\sigma^2}(u) = \frac{\sigma^2}{2\pi} s_\theta(u), \tag{8.2.2}$$

$$s_\theta(u) := \left| \sum_{k=0}^{\infty} a_k(\theta)e^{iku} \right|^2, \quad u \in \Pi, \ \theta \in \Theta.$$

As mentioned in Section 3.2 above, by Wold decomposition, the class of stationary processes having linear representation (8.2.1) is very large.

Denote the periodogram based on $X_1, \cdots, X_n$ by

$$I_X(u) := \frac{1}{2\pi n} \left| \sum_{j=1}^{n} X_j e^{iju} \right|^2. \tag{8.2.3}$$

The quadratic form that is useful for developing inference procedures is the integrated weighted periodogram

$$Q_X(\theta) := \int_\Pi \frac{I_X(u)}{s_\theta(u)} du = \frac{1}{n} \sum_{j=1}^{n} \sum_{k=1}^{n} b_{j-k}(\theta) X_j X_k, \tag{8.2.4}$$

$$b_j(\theta) := \frac{1}{2\pi} \int_\Pi \frac{e^{iju}}{s_\theta(u)} du, \quad j \in \mathbb{Z}, \ \theta \in \Theta.$$

Let $\boldsymbol{X}_n := (X_1, \cdots, X_n)$, and define the objective function

$$\Lambda_n(\boldsymbol{X}_n; \sigma, \theta) := \frac{1}{2\sigma^2} Q_X(\theta) + \log \sigma. \tag{8.2.5}$$

Throughout here, σ_0, θ_0 denote the true parameter values of σ, θ, respectively, and $\Omega = (0, \infty) \times \Theta$.

Definition 8.2.1. *Whittle estimators* of σ_0, θ_0 based on $\boldsymbol{X}_n$ are defined as

$$(\hat{\sigma}_n, \hat{\theta}_n) := \operatorname{argmin}_{(\sigma,\theta)\in\Omega} \Lambda_n(\boldsymbol{X}_n; \sigma, \theta). \qquad (8.2.6)$$

Clearly,

$$\hat{\theta}_n = \operatorname{argmin}_{\theta\in\Theta} Q_X(\theta), \quad \hat{\sigma}_n^2 = Q_X(\hat{\theta}_n). \qquad (8.2.7)$$

These estimators were first discussed by Whittle (1953). Although we are making no Gaussianity assumptions on $\{X_j\}$, the estimator (8.2.6) is suggested by the Gaussian maximum likelihood obtained by minimizing $\mathcal{L}_n/n$, where $\mathcal{L}_n$ is the negative of the log-likelihood:

$$\mathcal{L}_n := \frac{n}{2}\log(2\pi) + n\log\sigma + \frac{1}{2}\log|\Gamma_{n,\theta}| + \frac{1}{2\sigma^2}\boldsymbol{X}_n'\Gamma_{n,\theta}^{-1}\boldsymbol{X}_n.$$

Here, $\Gamma_{n,\theta}$ is the $n\times n$ positive definite covariance matrix with (j,k)th entry $E(X_j X_k)/\sigma^2$, and its determinant is $|\Gamma_{n,\theta}|$. This function is not easy to analyze from the minimization point of view. Because

$$n^{-1}\log(|\Gamma_{n,\theta}|) \to 0, \quad \left|n^{-1}\boldsymbol{X}_n'\Gamma_{n,\theta}^{-1}\boldsymbol{X}_n - Q_X(\theta)\right| \to_p 0,$$

then $\Lambda_n + \log(2\pi)/2$ is an approximation of $\mathcal{L}_n/n$. This gives the reason for defining $\hat{\sigma}_n^2$, $\hat{\theta}_n$ as above.

Our aim here is to prove the consistency of the estimators $(\hat{\sigma}_n^2, \hat{\theta}_n)$ under minimal conditions on the underlying model. For this purpose, in addition to (8.2.1), we shall assume that $X_j, j \in \mathbb{Z}$, is an ergodic sequence. Later, when proving asymptotic normality of these estimators, we shall need the stronger assumption that the $\zeta_k, k \in \mathbb{Z}$, in (8.2.1) are i.i.d. r.v.'s.

To proceed, recall that the Kolmogorov formula for one-step mean-square prediction error τ^2 of a stationary process with spectral density f is given by

$$\tau^2 = 2\pi \exp\left\{\frac{1}{2\pi}\int_{\Pi}\log f(u)du\right\}. \qquad (8.2.8)$$

This formula implies that $\tau^2 = 0$ if $\int_\Pi \log f(u)du = -\infty$, and $\tau^2 > 0$ if and only if $\int_\Pi \log f(u)du > -\infty$. Note that $-\infty \le \log f < \infty$, a.e., because $\log f \le f$ and f is integrable on Π.

Under the condition $a_0(\theta) = 1$ the formula (8.2.8) holds *a priori* with $\tau^2 = \sigma^2$, since the innovations ζ_j are the one-step predictor errors of the process $\{X_j\}$, and the variance $\sigma^2 = \operatorname{Var}(\zeta_0)$ is the mean-squared error of the one-step predictor.

For any stationary AR(p) process with the spectral density

$$\frac{\sigma^2}{2\pi} \prod_{k=1}^{p} |1 - a_k e^{-iku}|^{-2}, \quad |a_k| < 1,$$

the equality $\tau^2 = \sigma^2$ is valid because $\int_\Pi \log |1 - ae^{-iku}|^2 du = 0$, for all non-zero integers k and $|a| < 1$. See Brockwell and Davis (1991, p. 191) for a proof that $\tau^2 = \sigma^2$ in the case when f is continuous, symmetric, and bounded away from zero. For a more general f, see Hannan (1973, p. 137).

By (8.2.8), s_θ of (8.2.2) satisfies the *normalization condition*

$$\int_\Pi \log s_\theta(u) du = 0, \quad \forall\, \theta \in \Theta. \tag{8.2.9}$$

To prove the consistency of the Whittle estimators, we also need the following approximation lemma. For a positive integer M and a function h on Π, let

$$g_k := \frac{1}{2\pi} \int_\Pi e^{iku} h(u) du, \quad k \in \mathbb{Z}, \quad h_M(u) := \frac{1}{M} \sum_{j=0}^{M-1} \sum_{|k| \le j} g_k e^{-iku}.$$

Lemma 8.2.1. *For any continuous function h on Π with $h(-\pi) = h(\pi)$,*

$$\sup_{u \in \Pi} |h_M(u) - h(u)| \to 0, \quad as\ M \to \infty. \tag{8.2.10}$$

Proof. Let

$$V_M(u) := \sum_{|k| \le M} (1 - \frac{|k|}{M}) e^{-iku}, \quad u \in \Pi.$$

Then we can rewrite

$$h_M(u) = \sum_{|k| \le M} (1 - \frac{|k|}{M}) g_k e^{-iku} = \frac{1}{2\pi} \int_\Pi h(v) V_M(u - v) dv.$$

But,

$$V_M(u) = \frac{1}{M} \sum_{j=0}^{M-1} \sum_{k=0}^{M-1} e^{i(j-k)u} = \frac{1}{M} \left| \sum_{j=0}^{M-1} e^{iju} \right|^2$$

$$= \frac{1}{M} \left[\frac{\sin(Mu/2)}{\sin(u/2)} \right]^2 = \frac{1}{M} D_M^2(u), \quad u \in \Pi,$$

where D_M is the Dirichlet kernel of (2.1.6). Thus for any $\eta > 0$,

$$0 < V_M(u) \le \frac{1}{M \sin^2(\eta/2)}, \quad \eta \le |u| \le \pi.$$

Hence, for any $\eta > 0$,

$$\sup_{\eta \leq |u| \leq \pi} V_M(u) \to 0, \quad \text{as } M \to \infty. \tag{8.2.11}$$

Also, recall from (2.1.9) that

$$\int_\Pi V_M(u)du = 2\pi. \tag{8.2.12}$$

Next, extend the definition of h to $\mathbb{R}$ by letting $h(y + 2\pi) = h(y)$. Here use the assumption $h(-\pi) = h(\pi)$. A change of variable yields

$$h_M(u) - h(u) = \frac{1}{2\pi} \int_\Pi \{h(u - v) - h(u)\} V_M(v) dv.$$

Now split the integral over two sets, $|v| \leq \eta$ and $\eta < |v| \leq \pi$, to obtain, using (8.2.11), (8.2.12), and the uniform continuity of h on Π, that

$$\sup_{u \in \Pi} |h_M(u) - h(u)| \leq \sup_{|z-y| \leq \eta} |h(z) - h(y)| + 2\|h\|_\infty \sup_{\eta < z \leq \pi} |V_M(z)|$$
$$\to 0, \quad \text{as } M \to \infty, \ \eta \to 0.$$

This completes the proof of the lemma.

The next lemma deals with ergodic stationary stochastic processes. From now on in this section (σ_0^2, θ_0) will denote the fixed parameter values under which all probability statements are supposed to hold, and we will write $I(u)$ for $I_X(u)$.

Lemma 8.2.2. *Suppose the stochastic process $\{X_j\}$ is ergodic and stationary with the spectral density $f = \sigma_0^2 s_{\theta_0}/2\pi$. Let Θ_0 be a compact subset of $\mathbb{R}^q$ and $h_\theta(u)$ be a family of functions on $\Pi \times \Theta_0$ such that $h_\theta(\pi) \equiv h_\theta(-\pi)$.*
(i) Suppose, in addition, $h_\theta(u)$ are continuous functions on $\Pi \times \Theta_0$. Then, uniformly in $\theta \in \Theta_0$,

$$\int_\Pi h_\theta(u)I(u)du \to \int_\Pi h_\theta(u)f(u)du, \quad a.s. \tag{8.2.13}$$

(ii) Suppose, in addition, for each $0 < K < \infty$, $\max(-K, h_\theta(u) \wedge K)$ is a continuous function on $\Pi \times \Theta_0$, there exists an integrable function h such that $|h_\theta(u)| \leq h(u)$ on $\Pi \times \Theta_0$ and the spectral density f is bounded. Then, uniformly in $\theta \in \Theta_0$, the convergence (8.2.13) holds in probability.

Proof. (i) First, we shall prove (8.2.13). Let

$$g_k(\theta) := \frac{1}{2\pi} \int_\Pi e^{iku} h_\theta(u) du, \quad k \in \mathbb{Z}, \quad T_n(\theta) := \int_\Pi h_\theta(u) I(u) du.$$

For a positive integer M, let

$$q_{M,\theta}(u) := \frac{1}{M} \sum_{j=0}^{M-1} \sum_{|k| \le j} g_k(\theta) e^{-iku} \equiv \sum_{|k| \le M} \left(1 - \frac{|k|}{M}\right) g_k(\theta) e^{-iku}$$

By (8.2.10), we can take M large enough so that

$$\sup_{u \in \Pi, \theta \in \Theta_0} |h_\theta(u) - q_{M,\theta}(u)| \le \eta. \tag{8.2.14}$$

Let $\eta > 0$ and $v^2 := EX_0^2$. By the ergodic Theorem 2.5.2(i), $\int_\Pi I(u) du = n^{-1} \sum_{j=1}^n X_j^2 \to v^2$, a.s. Then, uniformly in $\theta \in \Theta_0$, and for all sufficiently large n,

$$\left| T_n(\theta) - \int_\Pi q_{M,\theta}(u) I(u) du \right| = \left| \int_\Pi [h_\theta(u) - q_{M,\theta}(u)] I(u) du \right|$$
$$\le \eta \int_\Pi I(u) du = \eta \, n^{-1} \sum_{j=1}^n X_j^2 \le 2\eta \, v^2, \quad \text{a.s.}$$

Moreover,

$$\int_\Pi q_{M,\theta}(u) I(u) du = \sum_{|k| \le M} \left(1 - \frac{|k|}{M}\right) g_k(\theta) \int_\Pi e^{-iku} I(u) du$$
$$= \sum_{|k| \le M} \left(1 - \frac{|k|}{M}\right) g_k(\theta) c_k,$$

where c_k is the sample autocovariance

$$c_k = c_{-k} := \frac{1}{n} \sum_{j=1}^{n-k} X_j X_{j+k}, \quad 0 \le k \le n-1.$$

By Theorem 2.5.2, for each $|k| \le M$, as $n \to \infty$, c_k converges almost surely to $\gamma(k) := \int_\Pi e^{-iuk} f(u) du$. Therefore, the above expression tends almost surely to

$$\sum_{|k| \le M} \left(1 - \frac{|k|}{M}\right) g_k(\theta) \gamma(k) = \int_\Pi \sum_{|k| \le j} \left(1 - \frac{|k|}{M}\right) g_k(\theta) e^{-iku} f(u) du$$
$$= \int_\Pi q_{M,\theta}(u) f(u) du \to \int_\Pi h_\theta(u) f(u) du$$

on letting $\eta \to 0$ in (8.2.14). This completes the proof of part (i).

(ii) It remains to prove part (ii) of the lemma. For $0 < K < \infty$, let
$h_\theta^-(u) := \max(-K, h_\theta(u) \wedge K)$ and $h_\theta^+(u) := h_\theta(u)\mathrm{I}(|h_\theta(u)| > K)$. Then
$|h_\theta^-(u)| \leq |h_\theta(u)|$ on $\Pi \times \Theta_0$, and

$$\int_\Pi h_\theta(u)I(u)du = t_n(\theta) + r_n(\theta),$$

where

$$t_n(\theta) = \int_\Pi h_\theta^-(u)I(u)du, \qquad |r_n(\theta)| \leq 2\int_\Pi |h_\theta^+(u)|I(u)du.$$

We shall show that, uniformly in $\theta \in \Theta_0$,

$$t_n(\theta) \to \int_\Pi h_\theta(u)f(u)du, \quad \text{a.s.,} \quad n \to \infty, \; K \to \infty. \quad (8.2.15)$$

$$r_n(\theta) \to_p 0, \quad n \to \infty, \; K \to \infty. \quad (8.2.16)$$

These two facts imply the convergence in (8.2.13), in probability.

To prove (8.2.16), it suffices to verify that

$$E \sup_{\theta \in \Theta_0} |r_n(\theta)| \to 0, \quad n \to \infty, \; K \to \infty. \quad (8.2.17)$$

Since $h_\theta^-(u)$ is continuous in u, θ, and uniformly bounded by $h(u)$, and f is
a bounded function, then by (8.2.13), uniformly in $\theta \in \Theta_0$,

$$\lim_{K \to \infty} \lim_{n \to \infty} t_n(\theta) = \lim_{K \to \infty} \int_\Pi h_\theta^-(u)f(u)du = \int_\Pi h_\theta(u)f(u)du, \quad \text{a.s.}$$

Hence, by (2.1.9),

$$E \sup_{\theta \in \Theta_0} |r_n(\theta)| \leq 2\int_{h(u)>K} h(u)\,E\big(I(u)\big)du$$

$$= \frac{1}{\pi n} \int_\Pi \int_{h(u)>K} |D_n(x+u)|^2 f(x)h(u)dxdu$$

$$\leq Cn^{-1} \int_\Pi \int_{h(u)>K} |D_n(x+u)|^2 h(u)dxdu$$

$$= C \int_{h(u)>K} h(u)du \quad \to 0, \quad K \to \infty.$$

This completes the proof of part (ii), and hence of the lemma.

We shall now address the consistency of $(\hat{\sigma}_n^2, \hat{\theta}_n)$. Consider the following
assumption.

Assumption (a0). The parameter space Θ is compact, parameter $(\sigma, \theta) \in \Omega := (0, \infty) \times \Theta$ determines the spectral density $\sigma^2 s_\theta(u)/2\pi$ uniquely, and the function $1/(s_\theta(u) + a)$ is continuous in $(u, \theta) \in \Pi \times \Theta$, for all $a > 0$. Moreover, the true value (σ_0, θ_0) lies in the interior of Ω and $f = \sigma_0^2 s_{\theta_0}/2\pi$.

Lemma 8.2.3. *Suppose the process $\{X_j\}$ of (8.2.1) with spectral density f as in (8.2.2) is ergodic and satisfies Assumption (a0). Then*

$$\frac{1}{2\pi} \int_\Pi \frac{s_{\theta_0}(u)}{s_\theta(u)} du > 1, \qquad \forall \theta \neq \theta_0. \tag{8.2.18}$$

If, in addition, the function $1/f$ is continuous on Π, then

$$\int_\Pi \frac{I(u)}{f(u)} du \to 2\pi, \quad a.s. \tag{8.2.19}$$

If, in addition, the function f is continuous in $u \in \Pi$ and $1/f$ is integrable, then the convergence in (8.2.19) holds in probability.

Proof. Claim (8.2.18) trivially holds if the integral is infinite. If it is finite, then from (8.2.9), the integrand $s_{\theta_0}(u)/(2\pi s_\theta(u))$ can be taken to be the spectrum of a stationary process, with unit prediction variance. Hence, the l.h.s. of (8.2.18) being the variance of the process, is always larger than the prediction variance unless the spectrum is constant, which is ruled out by the identifiability part of Assumption (a0). This completes the proof of (8.2.18).

If the function $1/f$ is continuous on Π, then (8.2.19) follows from Lemma 8.2.2(i). If the function f is continuous and $1/f$ is integrable, then the convergence in (8.2.19) holds in probability by Lemma 8.2.2(ii), thereby completing the proof of the lemma.

Theorem 8.2.1. (Consistency of Whittle estimators). *Suppose an observable moving-average process $\{X_j, j \in \mathbb{Z}\}$, of (8.2.1) is ergodic and has the spectral density $f(u) = \frac{\sigma_0^2}{2\pi} s_{\theta_0}(u)$, and suppose the functions $s_\theta, \theta \in \Theta$, satisfy Assumption (a0).*

    (i) *If, additionally, $1/f$ is continuous on Π, then*

$$\hat{\theta}_n \to \theta_0, \qquad \hat{\sigma}_n^2 \to \sigma_0^2, \quad a.\,s. \tag{8.2.20}$$

    (ii) *If, in addition, f is continuous and $1/f$ is integrable on Π, then*

$$\hat{\theta}_n \to_p \theta_0, \quad \hat{\sigma}_n^2 \to_p \sigma_0^2. \tag{8.2.21}$$

Proof. The proof is by contradiction. To emphasize the dependence on n write Q_n for Q_X. Suppose $\hat{\theta}_n$ is not consistent for θ_0. Then by the compactness of Θ, there is a subsequence $\tilde{\theta}_m = \hat{\theta}_{n_m}$ of $\hat{\theta}_n$ converging to some $\vartheta \in \Theta$, $\vartheta \neq \theta_0$. For any $a > 0$, $1/(s_{\theta_0}(u) + a)$ is a continuous function in (u, θ) by Assumption (a0). Then, by (8.2.13) of Lemma 8.2.2,

$$\liminf_m Q_m(\tilde{\theta}_m) \geq \liminf_m \int_\Pi \frac{I(u)}{s_{\tilde{\theta}_m}(u) + a} du = \frac{\sigma_0^2}{2\pi} \int_\Pi \frac{s_{\theta_0}(u)}{s_\vartheta(u) + a} du,$$

$$\rightarrow \frac{\sigma_0^2}{2\pi} \int_\Pi \frac{s_{\theta_0}(u)}{s_\vartheta(u)} du, \quad a \rightarrow 0, \quad \text{a.s.}$$

By (8.2.18), $\frac{1}{2\pi} \int_\Pi \frac{s_{\theta_0}(u)}{s_\vartheta(u)} du > 1$, and hence

$$\liminf_m Q_m(\tilde{\theta}_m) > \sigma_0^2, \quad \text{a.s.}$$

But on the other hand, in view of the definition (8.2.7) of the estimator $\hat{\theta}_n$, $Q_m(\tilde{\theta}_m) \leq Q_m(\theta)$, for all $\theta \in \Theta$. Therefore,

$$\limsup_m Q_m(\tilde{\theta}_m) \leq \limsup_m Q_m(\theta_0).$$

If $1/f(u)$ is continuous in u, then by (8.2.19), as $m \rightarrow \infty$,

$$Q_m(\theta_0) \rightarrow \frac{\sigma_0^2}{2\pi} \int_\Pi \frac{s_{\theta_0}(u)}{s_{\theta_0}(u)} du = \sigma_0^2,$$

a.s., whereas if $f(u)$ is continuous and $1/f(u)$ is integrable, then convergence holds in probability.

Hence, the contradiction, $\limsup_m Q_m(\tilde{\theta}_m) \leq \sigma_0^2 < \liminf_m Q_m(\tilde{\theta}_m)$, a.s. or in the second case, in probability, unless $\vartheta = \theta_0$. Therefore, $\hat{\theta}_n \rightarrow \theta_0$ a.s. in case (i), and in probability in case (ii).

The above proof also implies that in case (i),

$$\limsup_n Q_n(\hat{\theta}_n) \leq \sigma_0^2 \leq \liminf_n Q_n(\hat{\theta}_n), \quad \text{a.s.}$$

Hence, $\hat{\sigma}_n^2 = Q_n(\hat{\theta}_n) \rightarrow \sigma_0^2$, a.s. Similarly, $\hat{\sigma}_n^2 \rightarrow_p \sigma_0^2$, in case (ii). This completes the proof of the theorem.

Consistency in ARFIMA and GARMA models. Suppose $\{X_j\}$ is an ARFIMA(p, d, q) process (7.2.16) with ergodic white-noise innovations $\{\zeta_j\}$ and known orders p and q. By (7.2.23), its spectral density is

$$f_{\sigma,\theta}(u) = \frac{\sigma^2}{2\pi} s_\theta(u), \tag{8.2.22}$$

$$s_\theta(u) = |2\sin(u/2)|^{-2d} \left| \frac{\theta(e^{-iu})}{\phi(e^{-iu})} \right|^2, \quad u \in \Pi.$$

The parameters of interest are $\sigma^2 = \text{Var}(\zeta_0)$ and $\theta = (d, \nu)$, $\nu = (\phi_1, \cdots, \phi_p, a_1, \cdots, a_q)$ where $\phi_1, \cdots, \phi_p$ and $a_1, \cdots, a_q$ are coefficients of the polynomials ϕ and θ, respectively. We need to assume the following.

Assumption (b). Parameter ν takes values in a compact set $\Theta_1 \subset \mathbb{R}^{p+q}$ such that

$$\inf_{\nu \in \Theta_1} \inf_{|z| \leq 1} (|\phi(z)| \wedge |\theta(z)|) > 0.$$

The true value θ_0 lies in the interior of $\Theta := [-1/2, 1/2] \times \Theta_1$.

Assumption (b) ensures that the function $|\theta(e^{-iu})|^2 / |\phi(e^{-iu})|^2$, $u \in \Pi$, is continuous and bounded away from 0 and ∞, for every $\nu \in \Theta_1$, and that the ARFIMA equations have a stationary ergodic linear solution (8.2.1). In Section 7.2 it was shown that parameter θ determines the ARFIMA process and spectral density uniquely. Clearly, the functions s_θ, $\theta \in \Theta$, of (8.2.22) satisfy Assumption (a0).

Moreover, with $c_f := (\sigma^2/2\pi)|\theta(1)/\phi(1)|^2 > 0$,

$$f_{\theta, \sigma^2}(u) \sim c_f |u|^{-2d}, \quad u \to 0. \tag{8.2.23}$$

So, an observable process $\{X_j\}$ with spectral density $f = f_{\theta_0, \sigma_0^2}$ has long memory, if $0 < d_0 < 1/2$, short memory if $d_0 = 0$, and negative memory if $-1/2 < d_0 < 0$. If $d_0 \geq 0$, the inverse spectral density $1/f$ is continuous on Π, whereas if $d_0 < 0$ then f is continuous on Π and $(1/f) \in L_1(\Pi)$. Hence, Theorem 8.2.1 is applicable to ARFIMA models yielding the following:

Corollary 8.2.1. *Suppose $\{X_j\}$ is an ARFIMA(p, d, q) process with $p, q \geq 1$ known integers and unknown parameters (d_0, ν_0, σ_0^2). Assume also that Assumption (b) holds. Then the Whittle estimators are consistent:*

$$(\hat{d}_n, \hat{\nu}_n, \hat{\sigma}_n^2) \to (d_0, \nu_0, \sigma_0^2), \quad a.\,s., \qquad d_0 \in [0, 1/2),$$

$$\text{in probability}, \quad d_0 \in (-1/2, 0).$$

Suppose now that the observable process $\{X_j\}$ belongs to a family of GARMA(p, d, q) processes (7.3.1), with spectral density

$$f_{\sigma, \theta}(u) = \frac{\sigma^2}{2\pi} s_\theta(u),$$

$$s_\theta(u) = |2\sin(\frac{u-\omega}{2})|^{-2d} |2\sin(\frac{u+\omega}{2})|^{-2d} \left| \frac{\theta(e^{-iu})}{\phi(e^{-iu})} \right|^2, \quad u \in \Pi,$$

determined by parameters $\theta = (d, \nu, \omega)$, $\nu = (\phi_1, \cdots, \phi_p, a_1, \cdots, a_q)$, $\omega \in [0, \pi]$, and $\sigma^2 = \mathrm{Var}(\zeta_0)$. Assume that parameters $\theta = (d, \nu, \omega)$ take values in a compact set $\Theta := [-1/2, 1/2] \times \Theta_1 \times [0, \pi]$ where $\Theta_1 \subset \mathbb{R}^{p+q}$ satisfies Assumption (b). Suppose that the true value $\theta_0 = (d_0, \nu_0, \omega_0)$ lies in the interior of Θ. Then $|d_0| < 1/2$ and $0 < \omega_0 < \pi$ and the above functions s_θ, $\theta \in \Theta$, satisfy Assumption (a0). Hence, Theorem 8.2.1 is applicable to this model yielding the following:

Corollary 8.2.2. *Suppose $\{X_j\}$ is a GARMA(p, d, q) process with $p, q \geq 1$ known and unknown parameters $(d_0, \nu_0, \omega_0, \sigma_0^2)$. Assume, additionally, that Θ_1 satisfies Assumption (b). Then the Whittle estimators are consistent:*

$$(\hat{d}_n, \hat{\nu}_n, \hat{\omega}_n, \hat{\sigma}_n^2) \to (d_0, \nu_0, \omega_0, \sigma_0^2), \quad a.\,s.\,, \qquad d_0 \in [0, 1/2),$$

$$in\ probability, \quad d_0 \in (-1/2, 0).$$

8.3 Parametric model: asymptotic normality

The aim of this section is to obtain the asymptotic distribution of Whittle estimators of the underlying parameters σ^2, θ, appearing in the parametric spectral density of an observable moving-average process (8.2.1) with i.i.d. innovations. The asymptotic normality of Whittle estimators $\hat{s}_n^2$ and $\hat{\theta}_n$ is discussed in Subsection 8.3.1. It uses the results on the asymptotic normality of quadratic forms of linear processes of Section 6.3.

8.3.1 *Parametric models: asymptotic normality of Whittle estimators*

In Section 8.2, we showed the consistency of the Whittle estimators of (σ^2, θ) based on the observable linear process

$$X_j := \sum_{k=0}^{\infty} a_k(\theta)\zeta_{j-k}, \quad a_0(\theta) = 1, \quad j \in \mathbb{Z}, \tag{8.3.1}$$

where $(\sigma, \theta) \in \Omega := (0, \infty) \times \Theta$, with Θ a compact subset of $\mathbb{R}^q$. In this section we shall now derive the asymptotic normality of these estimators.

Here, the weights $a_k(\theta)$ have properties as in (8.2.1), and the innovations $\{\zeta_j, j \in \mathbb{Z}, \}$ are assumed to be i.i.d. r.v.'s with zero mean and variance σ^2

and finite fourth moment, which also implies that $\{X_j\}$ is ergodic. We additionally suppose that Assumption (a0) is satisfied.

Recall that $\{X_j\}$ has spectral density $f_{\sigma^2,\theta}(u) = \sigma^2 s_\theta(u)/2\pi$ as in (8.2.2), and we let $f(u) = \sigma_0^2 s_{\theta_0}(u)/2\pi$ denote the true spectral density, where σ_0^2 and θ_0 are the true values of the parameters for the observed data. Recall that $\theta = (\theta_1, \cdots, \theta_q)'$ and $s_\theta^{-1}(u)$ stands for $1/s_\theta(u)$. In Section 8.2 it was shown that assumption $a_0(\theta) = 1$ of (8.3.1) is equivalent to the normalization condition (8.2.9).

To prove the asymptotic normality of the Whittle estimators $\hat{\sigma}_n^2$ and $\hat{\theta}_n$ given by (8.2.6), we need the following additional conditions.

Assumptions (a1)–(a4). There exists a small compact ball $\Theta_0 \subset \Theta$ centered at θ_0, such that

(a1) $\int_\Pi \log s_\theta(u)\, du$ is twice differentiable in $\theta \in \Theta_0$ under the integral sign.

(a2)   The functions $s_\theta(u)$, $\frac{\partial}{\partial \theta_j} s_\theta^{-1}(u)$, and $\frac{\partial^2}{\partial \theta_j \partial \theta_k} s_\theta^{-1}(u)$ are continuous at all $(u, \theta) \in \{\Pi \backslash \{0\}\} \times \Theta_0$ for $j, k = 1, \cdots, q$.

(a3)   There exist $-1 < \alpha, \beta < 1$ such that $\alpha + \beta < 1/2$ and

$$s_{\theta_0}(u) \le C|u|^{-\alpha}, \quad |(\partial/\partial u)s_{\theta_0}(u)| \le C|u|^{-\alpha-1},$$
$$|(\partial/\partial \theta_j)\, s_{\theta_0}^{-1}(u)| \le C|u|^{-\beta}, \quad 1 \le j \le q,\ u \in \Pi \backslash \{0\}.$$

(a4)   If $\alpha \le 0$, then there exists $g \in L_1(\Pi)$ such that for $j, k = 1, \cdots, q$,

$$|(\partial^2/\partial \theta_j \partial \theta_k)\, s_\theta^{-1}(u)| \le g(u), \quad (u, \theta) \in \{\Pi \backslash \{0\}\} \times \Theta_0.$$

If $\alpha > 0$, then $\{(\partial^2/\partial \theta_j \partial \theta_k)\, s_\theta^{-1}(u), j, k = 1, \cdots, q\}$ are continuous in $(u, \theta) \in \Pi \times \Theta_0$.

In this section, for any differentiable function $h(\theta)$, we will let $\dot{h}(\theta)$ denote the $q \times 1$ vector of the first partial derivatives of h and $\ddot{h}(\theta)$ will denote the $q \times q$ matrix of the second partial derivatives, w.r.t. θ:

$$\dot{h}(\theta) := (\frac{\partial}{\partial \theta_1}, \cdots, \frac{\partial}{\partial \theta_q})'h(\theta), \quad \ddot{h}(\vartheta) := (\frac{\partial^2}{\partial \theta_j\, \partial \theta_k} h(\theta))_{j,k=1,\cdots,q}.$$

For example,

$$\dot{s}_\theta^{-1}(u)' := (\frac{\partial}{\partial \theta_1}, \cdots, \frac{\partial}{\partial \theta_q})s_\theta^{-1}(u), \quad \ddot{s}_\theta^{-1}(u) := (\frac{\partial^2 s_\theta^{-1}(u)}{\partial \theta_j\, \partial \theta_k})_{j,k=1,\cdots,q}.$$

Let

$$W(\theta) := \int_\Pi s_\theta(u)\ddot{s}_\theta^{-1}(u)\, du. \tag{8.3.2}$$

We are now ready to state and prove:

Theorem 8.3.1. (Asymptotic normality of Whittle estimators). *Let* $\{X_j, j \in \mathbb{Z}\}$ *be an observable moving-average process (8.3.1) having parametric spectral density* $f = \sigma_0^2 s_{\theta_0}/2\pi$, *with* s_θ *satisfying Assumptions (a0) and (a1) to (a4). Then*

$$n^{1/2}(\hat{\theta}_n - \theta_0) \to_D \mathcal{N}_q(0, 4\pi W^{-1}(\theta_0)), \tag{8.3.3}$$

$$n^{1/2}(\hat{\sigma}_n^2 - \sigma_0^2) \to_D \mathcal{N}(0, E\zeta_0^4 - \sigma_0^4). \tag{8.3.4}$$

Proof. The following proof is classical in nature. Recall

$$Q_n(\theta) \equiv Q_X(\theta) = \int_\Pi \frac{I(u)}{s_\theta(u)} du.$$

The consistency of $\hat{\theta}_n$ for θ_0, guaranteed by Theorem 8.2.1, implies that, as $n \to \infty$, the minimizer $\hat{\theta}_n$ of (8.2.6) belongs to Θ_0, with probability tending to 1, and solves the equation $\dot{Q}_n(\hat{\theta}_n) = 0$, where $\dot{Q}_n(\theta) = \int_\Pi \dot{s}_\theta^{-1}(u)I(u)du$. Set $\ddot{Q}_n(\theta) = \int_\Pi \ddot{s}_\theta^{-1}(u)I(u)du$. By the mean-value theorem, for $|\theta_n^* - \theta_0| \le |\hat{\theta}_n - \theta_0|$,

$$0 = \dot{Q}_n(\hat{\theta}_n) = \dot{Q}_n(\theta_0) + \ddot{Q}_n(\theta_n^*)(\hat{\theta}_n - \theta_0). \tag{8.3.5}$$

Consider first $\ddot{Q}_n(\theta_n^*)$. Since the components of $\partial^2/\partial\theta^2 \, s_\theta^{-1}(u)$ satisfy assumption (a4), by Lemma 8.2.2(ii) above,

$$\ddot{Q}_n(\theta) \to_p \frac{\sigma_0^2}{2\pi} \int_\Pi s_{\theta_0}(u)\ddot{s}_\theta^{-1}(u) \, du,$$

uniformly in $\theta \in \Theta_0$. This together with the consistency of $\hat{\theta}_n$ for θ_0 imply

$$\ddot{Q}_n(\theta_n^*) \to_p \frac{\sigma_0^2}{2\pi} \int_\Pi s_{\theta_0}(u)\ddot{s}_{\theta_0}^{-1}(u) \, du = \frac{\sigma_0^2}{2\pi} W(\theta_0). \tag{8.3.6}$$

In view of these observations and (8.3.5), (8.3.3) will follow from the following lemma.

Lemma 8.3.1. *Under the assumptions of Theorem 8.3.1,*

$$n^{1/2}(\dot{Q}_n(\theta_0) - E\dot{Q}_n(\theta_0)) \to_D \mathcal{N}_q(0, \sigma_0^4 W(\theta_0)/\pi), \tag{8.3.7}$$

$$n^{1/2} E\dot{Q}_n(\theta_0) \to 0. \tag{8.3.8}$$

Proof. To prove (8.3.7), it suffices to show that for any $c \in \mathbb{R}^q$,

$$n^{1/2} c' (\dot{Q}_n(\theta_0) - E\dot{Q}_n(\theta_0)) \to_D \mathcal{N}(0, \sigma_0^4 c' W(\theta_0) c / \pi). \qquad (8.3.9)$$

Let $c \in \mathbb{R}^q$ and

$$T_{n,c} := \sum_{t,s=1}^{n} b_{t-s} X_t X_s, \quad h(u) := \frac{1}{2\pi} c' \dot{s}_{\theta_0}^{-1}(u), \quad b_j := \int_{\Pi} e^{iju} h(u) du, \quad j \in \mathbb{Z}.$$

Then the l.h.s. of (8.3.9) is equal to $n^{-1/2} T_{n,c}$.

With conditions (a0) to (a4), $|f(u)| \le C|u|^{-\alpha}$ and $|h(u)| \le C|u|^{-\beta}$ for $u \ne 0$ with $-1 < \alpha, \beta < 1$, $\alpha + \beta < 1/2$, and therefore $h(u)$ and $s_{\theta_0}(u)$ satisfy the asymptotic normality condition (6.3.37) above. Hence, by Theorems 6.3.1 and 6.3.2, $n^{-1/2}(T_{n,c} - ET_{n,c}) \to_D \mathcal{N}(0, \tau_c^2)$, where

$$\tau_c^2 := 16\pi^3 \int_{\Pi} (f(u)h(u))^2 \, du + \frac{\chi_4}{\sigma_0^4} \left(2\pi \int_{\Pi} f(u)h(u) \, du \right)^2.$$

The normalization condition (8.2.9), assumption (a1), and the equality $s_{\theta_0}(u) \dot{s}_{\theta_0}^{-1}(u) = (\partial/\partial\theta) \log s_\theta^{-1}(u)$ yield

$$\int_{\Pi} f(u) h(u) \, du = \frac{\sigma_0^2}{(2\pi)^2} c' \frac{\partial}{\partial\theta} \int_{\Pi} \log s_{\theta_0}^{-1}(u) \, du = 0.$$

Moreover, the first term in the r.h.s. of τ_c^2 equals

$$\frac{\sigma_0^4}{\pi} c' \int_{\Pi} s_{\theta_0}^2(u) \dot{s}_{\theta_0}^{-1}(u) \dot{s}_{\theta_0}^{-1}(u)' du \, c = \frac{\sigma_0^4}{\pi} c' \tilde{W}(\theta_0) c,$$

where $\tilde{W}(\theta_0) := (\tilde{w}_{jk}(\theta))_{j,k=1,\cdots,q}$,

$$\tilde{w}_{jk}(\theta_0) := \int_{\Pi} s_{\theta_0}^2(u) \Big[\frac{\partial}{\partial\theta_j} s_{\theta_0}^{-1}(u) \Big] \Big[\frac{\partial}{\partial\theta_k} s_{\theta_0}^{-1}(u) \Big] du.$$

We shall now prove that $\tilde{W}(\theta_0) = W(\theta_0)$. According to (8.2.9), (a1) to (a3), and the identity $\dot{s}_{\theta_0} \equiv -s_{\theta_0}^2 \dot{s}_{\theta_0}^{-1}$,

$$0 = \int_{\Pi} \frac{\partial^2}{\partial\theta_j \partial\theta_k} \log s_{\theta_0}^{-1}(u) \, du$$

$$= \int_{\Pi} s_\theta(u) \frac{\partial^2}{\partial\theta_j \partial\theta_k} s_{\theta_0}^{-1}(u) \, du - \int_{\Pi} s_\theta^2(u) \Big[\frac{\partial}{\partial\theta_j} s_{\theta_0}^{-1}(u) \Big] \Big[\frac{\partial}{\partial\theta_k} s_{\theta_0}^{-1}(u) \Big] du,$$

thereby completing the proof of (8.3.9) and of (8.3.7).

To prove (8.3.8), let

$$s_c^*(u) := \int_{\Pi} s_{\theta_0}(u - v) c' \dot{s}_{\theta_0}^{-1}(v) \, dv.$$

By the definitions (8.2.4) and (2.1.6) of $Q_n(\theta)$ and D_n,

$$m_n := n^{1/2} E\, T_{n,c} \tag{8.3.10}$$

$$= (2\pi)^{-1} n^{-1/2} \int_{\Pi^2} \sum_{t,s=1}^{n} e^{i(t-s)(u+v)} f(u) c' \dot{s}_{\theta_0}^{-1}(v)\, du\, dv$$

$$= C n^{-1/2} \int_{\Pi^2} D_n^2(u+v) s_{\theta_0}(u) c' \dot{s}_{\theta_0}^{-1}(v)\, du\, dv$$

$$= C n^{-1/2} \int_{\Pi} D_n^2(u) s_c^*(u)\, du.$$

We shall show that

$$|s_c^*(u)| \le C|u|^\gamma, \quad u \in \Pi, \tag{8.3.11}$$

for some $1/2 < \gamma < 1$, which together with the bound (2.1.8) implies

$$n^{-1/2} \int_{\Pi} D_n^2(u) s_c^*(u)\, du \le C n^{-1/2} \int_{\Pi} \frac{n^2}{1+(nu)^2} |u|^\gamma\, du \tag{8.3.12}$$

$$\le C n^{1/2-\gamma} \int_{-\infty}^{\infty} (1+u^2)^{-1} |u|^\gamma\, du$$

$$\le C n^{1/2-\gamma} \to 0,$$

and completes the proof of (8.3.8).

We shall now prove (8.3.11). By (8.2.9) and (a1),

$$\int_{\Pi} s_{\theta_0}(u) c' \dot{s}_{\theta_0}^{-1}(u)\, du = c' \frac{\partial}{\partial \theta} \int_{\Pi} \log s_{\theta_0}^{-1}(u)\, du = 0.$$

Therefore,

$$s_c^*(u) = \int_{\Pi} \left(s_{\theta_0}(u-v) - s_{\theta_0}(v) \right) c' \dot{s}_{\theta_0}^{-1}(v)\, dv.$$

Hence,

$$|s_c^*(u)| \le \int_{|v|\le 3|u|} [\cdots]\, dv + \int_{3|u|\le |v|\le \pi} [\cdots]\, dv =: s_1(u) + s_2(u).$$

Use the bounds in (a3) for $s_{\theta_0}(v)$ and $\dot{s}_{\theta_0}^{-1}(v)$ to obtain

$$|s_1(u)| \le C \int_{|v|\le 3|u|} (|u-v|^{-\alpha} + |v|^{-\alpha}) |v|^{-\beta}\, dv$$

$$\le C|u|^{1-\alpha-\beta} \int_{|v|\le 3} (|1-v|^{-\alpha} + |v|^{-\alpha}) |v|^{-\beta}\, dv$$

$$\le C|u|^\gamma,$$

where $\gamma = 1 - \alpha - \beta > 1/2$. On the other hand, if $|v| > 3|u|$, then $|v|/2 \leq |v - u| \leq 2|v|$. By the mean-value theorem and (a3),

$$|s_{\theta_0}(u - v) - s_{\theta_0}(v)| \leq \sup_{\{|v|/2 \leq y \leq 2|v|\}} |\frac{d}{dy} s_{\theta_0}(y)| \, |u| \leq C|u| \, |v|^{-\alpha-1}.$$

Let $\epsilon > 0$. Use this bound and (a3), to obtain

$$\begin{aligned}
&|s_{\theta_0}(u - v) - s_{\theta_0}(v)| \, |s_{\theta_0}^{-1}(v)| \\
&\leq |s_{\theta_0}(u - v) - s_{\theta_0}(v)|^{1/2+\epsilon} \, |s_{\theta_0}(u - v) - s_{\theta_0}(v)|^{1/2-\epsilon} \, |s_{\theta_0}^{-1}(v)| \\
&\leq C(|u| \, |v|^{-\alpha-1})^{1/2+\epsilon} |v|^{-\alpha(1/2-\epsilon)} |v|^{-\beta} \\
&= C|u|^{1/2+\epsilon} v^{-\phi},
\end{aligned}$$

where $\phi = (\alpha + 1)(1/2 + \epsilon) + \alpha(1/2 - \epsilon) + \beta = \alpha + \beta + 1/2 + \epsilon < 1$, for a sufficiently small $\epsilon > 0$, because $\alpha + \beta < 1/2$. Thus

$$|s_2(u)| \leq C|u|^{1/2+\epsilon} \int_\Pi |v|^{-\phi} dv \leq C|u|^{1/2+\epsilon}.$$

This completes the proof of (8.3.11).

Hence, $|m_n| \to 0$, thereby completing the proof of (8.3.8). This also shows that for every $c \in \mathbb{R}^q$, $n^{-1/2} ET_{n,c} \to 0$, thereby completing the proof of the Lemma 8.3.1.

To complete the proof of Theorem 8.3.1 it remains to show (8.3.4). By definition, $\hat{\sigma}_n^2 = Q_n(\hat{\theta}_n)$. By the Taylor expansion, for $|\theta_n^* - \theta_0| \leq |\hat{\theta}_n - \theta_0|$,

$$\begin{aligned}
\hat{\sigma}_n^2 &\equiv Q_n(\theta_0) + (\hat{\theta}_n - \theta_0)' \dot{Q}_n(\theta_0) \\
&\quad + \frac{1}{2}(\hat{\theta}_n - \theta_0)' \ddot{Q}_n(\theta_n^*)(\hat{\theta}_n - \theta_0) = Q_n(\theta_0) + O_p(n^{-1}).
\end{aligned}$$

The last claim follows because by (8.3.3), (8.3.6), and Lemma 8.3.1, $\dot{Q}_n(\theta_0) = O_p(n^{-1/2})$, $\|\hat{\theta}_n - \theta_0\| = O_p(n^{-1/2})$, and $\|\ddot{Q}_n(\theta_n^*)\| = O_p(1)$.

Now, by the same argument as used in proving (8.3.7), it follows that $n^{-1/2}(Q_n(\theta_0) - EQ_n(\theta_0)) \to_D \mathcal{N}(0, A^2)$, where

$$\begin{aligned}
A^2 &:= 16\pi^3 \int_\Pi \left(f(u)s_{\theta_0}^{-1}(u)/2\pi\right)^2 du + \frac{\chi_4}{\sigma_0^4}\left(2\pi \int_\Pi f(u)s_{\theta_0}^{-1}(u)/2\pi \, du\right)^2 \\
&= 2\sigma_0^4 + \chi_4 = E\zeta_0^4 - \sigma_0^4.
\end{aligned}$$

To complete the proof of (8.3.4) it remains to show that

$$EQ_n(\theta_0) = \sigma_0^2 + o(n^{-1/2}).$$

Let $s^*(u) = \int_\Pi s_{\theta_0}(u-v)s_{\theta_0}^{-1}(v)\,dv$. Then, as in (8.3.10),

$$EQ_n(\theta_0) = \frac{\sigma_0^2}{(2\pi)^2 n} \int_\Pi D_n^2(u)s^*(u)\,du.$$

Write

$$EQ_n(\theta_0) = \frac{\sigma_0^2}{(2\pi)^2 n} \int_\Pi D_n^2(u)(s^*(u) - s^*(0))\,du$$

$$+ s^*(0)\frac{\sigma_0^2}{(2\pi)^2 n} \int_\Pi D_n^2(u)\,du$$

$$=: q_{n,1} + q_{n,2}, \quad \text{say.}$$

The same argument used in proving (8.3.11) shows that $|s^*(u) - s^*(0)| \le C|u|^\gamma$ for all $u \in \Pi$ and some $\gamma > 1/2$, which as in (8.3.12) implies that $q_{n,1} = O(n^{-\gamma}) = o(n^{-1/2})$. This proves (8.3.4), since by (2.1.9), $q_{n,2} = (s^*(0)/2\pi)\sigma_0^2 = \sigma_0^2$. This completes the proof of Theorem 8.3.1.

8.3.2 *Whittle estimators in ARFIMA models*

The above results apply to one of the most popular parametric time series models, the ARFIMA(p,d,q) model of (7.2.16), with i.i.d. innovations $\{\zeta_j\} \sim \text{IID}(0,\sigma^2)$ and $|d| < 1/2$.

The parameters of interest $\theta = (d,\nu) \equiv (d,\phi_1,\cdots\phi_p,a_1,\cdots,a_q)$ include the memory parameter d and the coefficients of the polynomials $\phi(z)$ and $\theta(z)$, as well as the variance σ^2 of the innovations $\{\zeta_j\}$. Consequently, the parameter set $\Theta = [-1/2,1/2] \times \Theta_1$ is the same as in Corollary 8.2.1. Then, for every $\theta = (d,\nu) \in \Theta$, an ARFIMA$(p,d,q)$ model has a unique stationary solution that can be written as a moving-average process (8.3.1) with $a_0(\theta) = 1$ and spectral density $f_{\theta,\sigma^2}(v) = \frac{\sigma^2}{2\pi}s_\theta(v)$ as in (8.2.22). Since the true parameter $\theta_0 = (d_0,\nu_0)$ is the inner point of the compact set Θ, $|d_0| < 1/2$. The next proposition establishes the asymptotic normality of the Whittle estimator for ARFIMA models.

Proposition 8.3.1. *Suppose that $\{X_j\}$ is an ARFIMA(p,d,q) process with parameters (d_0,ν_0,σ_0^2) with $p,q \ge 1$ known and $\Theta = [-1/2,1/2] \times \Theta_1$, where Θ_1 satisfies Assumption (b) of Section 8.2. Then the Whittle estimator $(\hat\theta_n, \hat\sigma_n^2)$ of (θ_0, σ_0^2) is asymptotically normal, i.e. (8.3.3) and (8.3.4) are true.*

Proof. Corollary 8.2.1 established the consistency of Whittle estimators. By Theorem 8.3.1, to show asymptotic normality, it remains to prove that the functions s_θ, $\theta \in \Theta$, satisfy Assumptions (a1) to (a4).

The functions $s_\theta(u)$, $\dot{s}_\theta^{-1}(u)$, and $\ddot{s}_\theta^{-1}(u)$ are continuous at all $(u, \theta) \in \{\Pi \backslash \{0\}\} \cap \Theta$. Hence, by (8.2.22), conditions (a0), (a1), and (a2) are satisfied. To check condition (a3), note that

$$s_{\theta_0}(u) \leq C|u|^{-2d_0}, \quad |(\partial/\partial u)s_{\theta_0}(u)| \leq C|u|^{-2d_0-1},$$

$$|(\partial/\partial\theta_j)s_{\theta_0}(u)| \leq C|u|^{-2d_0}|\log|u||, \quad u \neq 0.$$

Hence, (a3) is satisfied with $\alpha = 2d_0$ and $\beta = -2d_0 - \epsilon$ for any $0 < \epsilon < 1/2$.

To check (a4) we consider two cases. First, if $d_0 \leq 0$, then the spectral density $f = \sigma_0^2 s_{\theta_0}/2\pi$ is bounded, and for a small $\epsilon > 0$,

$$\sup_{\eta \in \Theta_1} \sup_{d:|d-d_0| \leq \epsilon} |(\partial^2/\partial\theta_j\partial\theta_j) s_\theta^{-1}(u)| \leq C|u|^{2d_0-2\epsilon}(2\log|u|)^2 \equiv g(u),$$

where function $g(u)$ is integrable because $2d_0 - 2\epsilon > -1$. Secondly, if $d_0 > 0$, then the spectral density f has a pole at zero and the derivatives $(\partial^2/\partial\theta_j\partial\theta_k) s_\theta^{-1}(u)$ are continuous, i.e.,

$$\frac{\partial^2}{\partial d^2} s_\theta^{-1}(u) = 4|2\sin(\frac{u}{2})|^{2d}(\log|2\sin(\frac{u}{2})|)^2 \frac{|\theta(e^{iu})|^2}{|\phi(e^{iu})|^2}$$

is a continuous function in $d \in (d_0 - \epsilon, d_0 + \epsilon)$ and $\eta \in \Theta_1$ when $0 < \epsilon < d_0$. So, assumptions (a0) to (a4) are satisfied with $\Theta_0 = [d_0 - \epsilon, d_0 + \epsilon] \times \Theta_1$. This completes the proof.

Consider now the parametric family ARFIMA$(0, d, 0)$, $|d| < 1/2$. Then $\theta = d$ and the spectral density has the form

$$f_{d,\sigma^2}(u) = \frac{\sigma^2}{2\pi}|2\sin(\frac{u}{2})|^{-2d}, \quad u \in \Pi.$$

Theorem 8.3.1 applies. To compute the asymptotic variance of $\hat{d}$, note that $W(d_0)$ is a 1×1 matrix, and

$$\frac{\partial^2}{\partial d^2} s_d^{-1}(u) = 4|2\sin(\frac{u}{2})|^{2d}\left(\log|2\sin(\frac{u}{2})|\right)^2.$$

Therefore,

$$W(d_0) := \int_\Pi s_{d_0}(u)\frac{\partial^2}{\partial d^2} s_{d_0}^{-1}(u)\, du = 4\int_\Pi (\log|2\sin\frac{u}{2}|)^2\, du = 2\pi^3/3.$$

Hence, by Theorem 8.3.1, we obtain the following result:

Corollary 8.3.1. Let $\{X_j\}$ be an ARFIMA$(0, d_0, 0)$ process, with parameter $|d_0| < 1/2$. Then

$$n^{1/2}(\hat{d}_n - d_0) \to_D \mathcal{N}(0, 6/\pi^2).$$

Note that the limiting variance does not depend on the parameter d_0.

Note: Many details of Section 8.2 are based on the work of Hannan (1973) where he also established the asymptotic normality of the Whittle estimate under the condition $\sum_{k=0}^{\infty} k\, a_k^2(\theta) < \infty$, which excludes the case of long memory. When the observable process is Gaussian, Fox and Taqqu (1986) proved Theorem 8.3.1 under somewhat more stringent conditions on the parametric spectral density than (a0) to (a4), in the long memory case where $0 < d < 1/2$. The above proofs are somewhat influenced by some proofs in their paper and are based on the work of Giraitis and Surgailis (1990). Giraitis and Taqqu (1999) discussed the asymptotic properties of the Whittle estimator for some non-Gaussian processes. Dahlhaus (1989) proved the asymptotic normality of the maximum likelihood estimator when the observable process is long memory Gaussian. Velasco and Robinson (2000) proved the asymptotic normality of the Whittle estimator for non-stationary processes.

8.4 Semiparametric models

Often even in the context of parametric ARFIMA(p, d, q) models, the main parameter of interest is d. As seen in the previous sections, when the spectral density f of the model is correctly parameterized, d (and other parameters) can be estimated at the $\sqrt{n}$-consistency rate by the parametric Whittle estimators. But, if the model is misspecified, for example if in the ARFIMA(p, d, q) model the orders p, q are under-specified, then the parametric estimates are typically inconsistent. Fortunately, even if f remains unspecified away from the zero frequency, the estimates of d considered below in these types of semiparametric models are found to be useful.

In the semiparametric models considered here it is assumed that for some $c > 0$ and the memory parameter $-1/2 < d < 1/2$,

$$f(u) \sim c|u|^{-2d}, \quad u \to 0. \tag{8.4.1}$$

This model is called semiparametric because no form of the spectral density is assumed. All assumptions relate to the behavior of the spectral density near the origin. The parameters of interest are c and d. Recall from (8.2.23) that the spectral density of an ARFIMA(p, d, q) time series satisfies (8.4.1) with $|d| < 1/2$.

In this section we shall discuss the three common estimation procedures for the parameters c, d in (8.4.1). The first two are the so-called rescaled range and the Geweke–Porter-Hudak estimation methods while the third is the LW estimation method. The common feature of both log-periodogram and LW estimators is that they do not require any restrictions on the spectral density f away from zero, apart from being integrable on Π. We shall discuss the former two methods briefly in the next subsection without many proofs while the other method will be discussed in more detail in the subsequent sections.

8.4.1 *R/S estimation method*

One of the earliest model-free (non-parametric) estimation methods was based on the rescaled range, the so-called R/S method. This method was first used by Hurst (1951, 1956) to study the behavior of the Nile and various reservoirs. In these applications X_j typically denotes the total inflow of water into a reservoir in the time unit j, over n units of time so that $1 \leq j \leq n$. Let $\bar{X}$ denote the average inflow over n units of time. Then $X_j - \bar{X}$ denotes the deviation of the inflow from the mean for the time unit j, and at time k the amount of water in the reservoir exceeds the average $\bar{X}$ by $\sum_{j=1}^{k}(X_j - \bar{X})$. Assuming that the reservoir is never to be empty or overfull, the minimum amount of the water in it at time n will be $-\min_{1 \leq k \leq n} \sum_{j=1}^{k}(X_j - \bar{X})$, so that the minimum volume of the reservoir at the end of n units of time is equal to

$$R_n := \max_{1 \leq k \leq n} \sum_{j=1}^{k}(X_j - \bar{X}) - \min_{1 \leq k \leq n} \sum_{j=1}^{k}(X_j - \bar{X}).$$

The letter R in R_n stands for the range. Ripple (1957) was the first scientist to use this statistic in hydrology. The statistic R_n is location invariant but not scale invariant. That is, R_n based on $\{aX_j + b\}$ remains the same for all real numbers b while it is seriously affected by the scale factor a. To make it scale invariant Hurst proposed to use its scaled version based on the ratio R_n/S_n, where

$$S_n := \left(\frac{1}{n} \sum_{j=1}^{n}(X_j - \bar{X})^2\right)^{1/2}.$$

Besides being invariant under location and scale change, Mandelbrot and Wallis (1969b) pointed out that the ratio R_n/S_n is also robust against extreme deviations from normality, including for the infinite variance case.

In his famous studies, Hurst averaged the values of this ratio over several n values and for numerous rivers and reservoirs. He found that this average fluctuated around n^H with H typically about 0.74. The property that $H > 1/2$ is known as the Hurst effect for this reason. Formally this phenomena is defined as follows.

Definition 8.4.1. (*Hurst exponent*). A positive number H is called a Hurst exponent if, as $n \to \infty$,

$$\frac{1}{n^H} \frac{R_n}{S_n} \to_D \text{ to a non-degenerate r.v. } \mathcal{R}_1.$$

Example 8.4.1. Increments of a self-similar process of index $H \in (0,1)$ show the Hurst effect. To see this let $\{Z(t), t \geq 0\}$ be an H-sssi process of index $H > 0$ with $Z(0) = 0$ and $EZ^2(t) < \infty$, $t \geq 0$ as described in Section 3.4 above. Define the increment process $\{X_j := Z(j) - Z(j - 1), j = 1, \cdots, n\}$, and observe that $\sum_{j=1}^{k}(X_j - \bar{X}) = Z_k - (k/n)Z_n$, $k = 1, \cdots, n$. Definition 3.4.1 for a self-similar process implies that $\{Z(j), j = 1, \cdots, n\} =_D \{n^H Z(j/n), j = 1, \cdots, n\}$, which yields, with $\{\Gamma(t) := Z(t) - tZ(1), 0 \leq t \leq 1\}$,

$$\begin{aligned} R_n &= \max_{0 \leq k \leq n} (Z(k) - (k/n)Z(n)) - \min_{0 \leq k \leq n} (Z(k) - (k/n)Z(n)) \\ &=_D n^H \big[\max_{0 \leq k \leq n} (Z(k/n) - (k/n)Z(1)) - \min_{0 \leq k \leq n} (Z(j/n) - (k/n)Z(1)) \big] \\ &\approx n^H \big[\max_{0 \leq t \leq 1} \Gamma(t) - \min_{0 \leq t \leq 1} \Gamma(t) \big]. \end{aligned}$$

By Proposition 3.4.2, the increments $\{X_j\}$ of the Z process form a covariance stationary process with autocovariance function $\gamma_X(j) \sim c_\gamma j^{2H-2}$, $j \to \infty$. Now assume Z is such that $S_n^2 \to_p \sigma^2 := \text{Var}(X(1)) \equiv \text{Var}(Z(1))$. Hence,

$$\frac{1}{n^H} \frac{R_n}{S_n} \to_D \mathcal{R}_1 = \frac{1}{\sigma} \Big(\max_{0 \leq t \leq 1} \Gamma(t) - \min_{0 \leq t \leq 1} \Gamma(t) \Big).$$

More generally, suppose that $\{X_j, j \in \mathbb{Z}\}$ is a stationary ergodic process satisfying (8.4.1) and such that for some $H \in (0,1)$ and $C > 0$,

$$\text{Var}(\sum_{j=1}^{n} X_j) \sim Cn^{2H}. \tag{8.4.2}$$

In addition, suppose

$$\frac{1}{n^H} \sum_{j=1}^{[nt]} (X_j - \bar{X}) \Rightarrow c_H B_H(t), \ 0 \le t \le 1,$$

in $\mathcal{D}[0,1]$ with the uniform metric, where c_H is a positive constant and B_H is fractional Brownian motion with parameter $H \in (0,1)$ and covariance function given by (3.4.1). This and the identity

$$\sum_{j=1}^{k} (X_j - \bar{X}) = \sum_{j=1}^{k} (X_j - EX_j) - \frac{k}{n} \sum_{j=1}^{n} (X_j - EX_j)$$

together imply

$$\frac{R_n}{n^H} \to_p c_H \Big(\max_{0 \le t \le 1} B_H^0(t) - \min_{0 \le t \le 1} B_H^0(t) \Big),$$

where $B_H^0(t) = B_H(t) - t B_H(1)$ is a *fractional Brownian bridge*.

Combining this with the fact that $S_n^2 \to_p \text{Var}(X_1)$, we see that

$$\frac{1}{n^H} \frac{R_n}{S_n} \to_p \frac{c_H \Big(\max_{0 \le t \le 1} B_H^0(t) - \min_{0 \le t \le 1} B_H^0(t) \Big)}{\left(\text{Var}(X_1) \right)^{1/2}} = \mathcal{R}_1.$$

This result forms a theoretical foundation for the R/S method. Taking logarithms of both sides, we obtain a heuristic identity

$$\log(R_n/S_n) \sim H \log n + \log \mathcal{R}_1,$$

which can also be written as

$$\hat{H} - H = O_p(\log^{-1} n), \text{ with } \hat{H} = \log(R_n/S_n)/\log n,$$

and which shows that H can be interpreted as the slope of a regression line of $\log(R_n/S_n)$ against $\log n$ with random errors $\log \mathcal{R}_1$. The point of the R/S analysis is to consider many subsamples of varying size n from a given sample $X_1, \cdots, X_n$, i.e., to pick several (large) values of n, such as 10, 20, 30, etc., and plot the graph of

$$(\log n, \log(R_n/S_n))$$

for these values of n, to estimate the slope of the regression line; see e.g., Mandelbrot and Taqqu (1979) or Beran (1994).

For the Hurst effect, this graph should look like a straight line especially for large values of n. The slope of this line then provides an R/S estimate of H. In addition, the equality $d = H - 1/2$ yields the estimator $\hat{d} = \hat{H} - 1/2$ for the memory parameter d in (8.4.1).

Note: For some recent results on R/S estimators see Li, Yu, Carriquiry and Kliemann (2011).

8.4.2 *Log-periodogram estimator*

The earliest estimator for a semiparametric model (8.4.1) was that of Geweke and Porter-Hudak (1983), originally developed for the ARFIMA$(0, d, 0)$ model (7.2.2). Recall from (7.2.8), the spectral density of this model is $f(u) = (\sigma^2/2\pi)|2\sin(u/2)|^{-2d}$, so that $\log f(u) = \log(\sigma^2/2\pi) - 2d\log(4\sin^2(u/2))$. This gives a reason to introduce the following regression model.

Let $1 < m < n/2$, $u_j := 2\pi j/n$, and $a_j := -\log(4\sin^2(u_j/2))$, for $1 \le j \le m$. Let $E = 0.577215\cdots$ be the Euler's constant and $\tilde{c} = \log(\sigma^2/2\pi) - E$. Consider the model

$$\log(I(u_j)) = \log(\frac{I(u_j)}{f(u_j)}) + \log(f(u_j)) = \tilde{c} + 2da_j + U_j,$$

$$U_j = \log(\frac{I(u_j)}{f(u_j)}) + E, \quad 1 \le j \le m.$$

Using the regression model analogy, Geweke and Porter-Hudak (GPH) proposed to estimate d by the least-squares estimator in this model. If the pseudo-errors U_j behave like mean-zero i.i.d. r.v.'s then this is a reasonable estimation procedure.

This estimator is a member of a larger class of estimators obtained by minimizing the objective function

$$K_n(c, d) = m^{-1} \sum_{j=1}^{m} k\big(f_{c,d}^*(u_j), I(u_j)\big), \tag{8.4.3}$$

with respect to c and d. Here, $k(.,.)$ is a pseudo-distance between the approximate spectral density $f_{c,d}^*(u) = c|u|^{-2d}$ and the periodogram $I(u)$. The logarithmic objective function $k(f, g) = (\log f - \log g)^2$ leads to the *log-periodogram estimator* of c and d in (8.4.1) proposed by Geweke and Porter-Hudak (1983). Minimization of K_n in this case is equivalent to minimizing the sum of square deviations

$$\sum_{j=1}^{m} \big(\log c - 2d\log u_j - \log I(u_j)\big)^2$$

w.r.t. $c > 0$ and $d \in [-1/2, 1/2]$. The log-periodogram estimators of c, d are thus obtained by fitting the least-squares line $\log c - 2d\log u_j$ to the observations $\log I(u_j)$ at the design points $\log(u_j)$, $1 \le j \le m$. The corresponding estimators of c, d are the so-called Geweke–Porter-Hudak estimators

$$\hat{d}_{GPH} = -\frac{\sum_{j=1}^{m} \nu_j \log I(u_j)}{2\sum_{j=1}^{m} \nu_j^2},$$

$$\hat{c} := m^{-1}\sum_{j=1}^{m} \log I(u_j) - 2\hat{d}_{GPH}m^{-1}\sum_{j=1}^{m} \log u_j,$$

$$\nu_j = \log j - m^{-1}\sum_{k=1}^{m} \log k \equiv \log(u_j) - m^{-1}\sum_{k=1}^{m} \log(u_k).$$

Let $\eta_j^* = I(u_j)/(cu_j^{-2d_0})$, where d_0 is the true value of d for which (8.4.1) holds. Simple algebra yields an alternative form

$$\hat{d}_{GPH} - d_0 = -\frac{\sum_{j=1}^{m} \nu_j \log \eta_j^*}{2\sum_{j=1}^{m} \nu_j^2} = -\frac{1}{2m}\Big\{\sum_{j=1}^{m} \nu_j \log \eta_j^*\Big\}(1 + o(1)),$$

since, uniformly in $1 \le j \le m$,

$$\nu_j = \log(\frac{j}{m}) + 1 + O(\frac{\log m}{m}), \qquad \sum_{j=1}^{m} \nu_j^2 = m(1 + O(\frac{\log m}{m})).$$

These estimators are easy to compute. But until the mid-1990s no rigorous asymptotic distribution theory was available. Robinson (1995b) was the first to provide the rigorous sufficient conditions under which the asymptotic distribution of $\sqrt{m}(\hat{d}_{GPH} - d_0)$ is $\mathcal{N}(0, \pi^2/24)$.

8.5 Local Whittle estimators

Here, we shall focus on the semiparametric local Whittle estimation method of the memory parameter d and scaling constant c of spectral density f of (8.4.1). In this section we discuss some general sufficient conditions under which local Whittle (LW) estimators are consistent. It also contains detailed proofs that show how their rates of convergence and asymptotic distributions can be derived.

Most of the previous results for these estimators are based on the assumption that $\{X_j, j \in \mathbb{Z}\}$ is linear. However, in empirical applications, e.g., financial econometrics, non-linear models are rather common. The goal here is thus to discuss consistency of these estimators under conditions that will allow for such non-linearity. The general conditions discussed below are naturally satisfied by signal plus noise processes and linear moving-average models.

Let X_j, $j = 1, \cdots, n$, be n observations from a stationary process $\{X_j\}$ with an unknown mean μ, variance σ^2, and spectral density f satisfying (6.2.2), which specifies the local behavior of f at the zero frequency. Recall that if $d = 0$, then $\{X_j\}$ has *short memory*; if $0 < d < 1/2$, then the process $\{X_j\}$ has *long memory*, whereas when $-1/2 < d < 0$, it is *antipersistent*.

Unlike in the case of the parametric models of Section 8.3 where the underlying spectral density is known up to an Euclidean parameter, in semiparametric models, the spectral density generally is unknown except for its behavior near the origin as specified above.

The estimation of the memory parameter d in semiparametric models *a priori* uses information about the spectral density near zero only. It relies on the fact that the spectral density $f(u)$ for u close to 0 can be approximated by $c|u|^{-2d}$. While parametric methods take into account the whole band of frequencies, the semiparametric method uses information only from a neighborhood of zero. Besides (6.2.2), it imposes no additional parametric specification or restrictions on f.

To define the LW estimators, first consider the approximate log-likelihood of a Gaussian process with spectral density f given by

$$-\frac{1}{2\pi} \int_\Pi \left(\log f(u) + \frac{I(u)}{f(u)} \right) du.$$

For the parametric spectral density, $f = f_{\sigma^2,\theta} = (\sigma^2/2\pi)s_\theta$, this expression reduces to $-\Lambda_n(X; \sigma, \theta)$ of (8.2.5), and the minimizers of $\Lambda_n(X; \sigma, \theta)$ w.r.t. σ and θ lead to the Whittle estimators given by (8.2.6).

In semiparametric models, the behavior of f is specified only in a neighborhood about the origin. For this reason the above parametric approach is modified as follows. Reduce the range of integration in the approximate log-likelihood from $|u| \leq \pi$ to the low frequencies $|u| \leq u_m$, and replace $f(u)$ by its approximation $c|u|^{-2d}$, where $u_j := 2\pi j/n$, $j = 1, \cdots, n/2$, and m is a positive integer, $m < n/2$. Then an analog of $\Lambda_n(X; \sigma, \theta)$, which is suitable here is

$$\frac{1}{2\pi} \int_0^{u_m} \left\{ \log(c|u|^{-2d}) + \frac{I(u)}{c|u|^{-2d}} \right\} du.$$

This is further approximated by

$$\frac{1}{2\pi}\frac{1}{m}\sum_{j=1}^{m}\Big\{\log(c|u_j|^{-2d}) + \frac{I(u_j)}{c|u_j|^{-2d}}\Big\}.$$

Local Whittle estimators of (c, d) are defined as minimizers of this expression w.r.t. (c, d) or equivalently of

$$\frac{1}{m}\sum_{j=1}^{m}\Big\{\log c - 2d\log(|u_j|) + \frac{1}{c}|u_j|^{2d}I(u_j)\Big\}.$$

The minimum of this expression w.r.t. c is achieved by $c(d) = m^{-1}\sum_{j=1}^{m}|u_j|^{2d}I(u_j)$. Replacing c by $c(d)$ and u_j by $2\pi j/n$ in the above expression yields

$$U_n(d) = \log\Big(\frac{1}{m}\sum_{j=1}^{m}u_j^{2d}I(u_j)\Big) - \frac{2d}{m}\sum_{j=1}^{m}\log u_j \qquad (8.5.1)$$

$$= \log\Big(\frac{1}{m}\sum_{j=1}^{m}j^{2d}I(u_j)\Big) - \frac{2d}{m}\sum_{j=1}^{m}\log j.$$

Then the LW estimator of d is

$$\widehat{d} := \text{argmin}_{[-1/2,1/2]}U_n(d), \qquad (8.5.2)$$

while that of the parameter c is

$$\widehat{c} := c(\widehat{d}) = m^{-1}\sum_{j=1}^{m}u_j^{2\widehat{d}}I(u_j). \qquad (8.5.3)$$

Note that the above estimators are also members of the class of estimators obtained by minimizing the distance $K_n(c, d)$ of (8.4.3) when $k(f, g) = \log f + g/f$.

The consistency of these estimators is proved in the next subsection under fairly general ergodicity-type conditions. As in the parametric case, these conditions neither assume the Gaussianity of the observable process nor i.i.d. innovations in its moving-average representation. These results are thus applicable to a large class of processes, including non-linear models. We also discuss a set of stronger but easy to verify sufficient conditions, which allow us to evaluate the rates of convergence and establish the asymptotic normality of suitably standardized LW estimators. These results are also shown to be applicable to a signal plus noise process where the signal dominates the noise. Several examples of practical applications of these results are given later in the chapter.

8.5.1 Semiparametric models: consistency of local Whittle estimators

Suppose that a stationary process $\{X_j\}$ has a spectral density

$$f(u) = |u|^{-2d} g(u), \qquad u \in \Pi, \ |d| < 1/2, \qquad (8.5.4)$$
$$\sim c|u|^{-2d}, \qquad u \to 0,$$

where g is continuous on $[0, a] \subset [0, \pi]$, $a > 0$, with $c := g(0) > 0$.

Let c_0 and d_0 denote the true parameter values of c and d, which are the objects of estimation, and let $m = m_n$ be a *bandwidth* sequence of positive integers satisfying

$$m \to \infty, \qquad m = o(n). \qquad (8.5.5)$$

As seen in Theorem 8.2.1, the ergodicity of the process $\{X_j\}$ suffices to prove the consistency of Whittle estimators in parametric models. In the current set-up, we shall replace the ergodicity assumption by the weak law of large numbers assumption for the normalized periodogram at the Fourier frequencies. Accordingly, let

$$\eta_j = \frac{I(u_j)}{f(u_j)}, \qquad j = 1, \cdots, m. \qquad (8.5.6)$$

If g is continuous in an interval $[0, a + \delta]$, $a > 0, \delta > 0$, then by Theorem 5.2.1(iv), the standardized periodograms $\{\eta_j\}$ at frequencies from a smaller interval $(0, a)$, have the property

$$\max_{u_K \le u_j \le a} |E\eta_j - 1| \to 0, \qquad K \to \infty, \ n \to \infty. \qquad (8.5.7)$$

In addition,

$$\max_{0 < u_j \le a} E\eta_j \le C, \qquad n \ge 1. \qquad (8.5.8)$$

If $\{X_j\}$ is a linear process with i.i.d. innovations, then the results derived in Theorem 5.3.2 for the covariances of η_j show that they behave like weakly dependent r.v.'s with unit mean, and

$$\max_{u_K \le u_k < u_j \le a} |\mathrm{Cov}(\eta_j, \eta_k)| \to 0, \qquad K \to \infty. \qquad (8.5.9)$$

Equations (8.5.7) and (8.5.9) imply that the η_j's satisfy a weak law of large numbers (WLLN):

$$m^{-1} \sum_{j=1}^{m} \eta_j \to_p 1. \qquad (8.5.10)$$

For linear processes convergence (8.5.10) is established in Theorem 8.5.2(i) below. Some sufficient conditions under which the WLLN holds for a wide class of stationary processes are provided later on in this section.

The variables η_j are invariant with respect to the mean of X_j since the periodogram $I(u_j) \equiv I_X(u_j)$ is location invariant, i.e., $I_X(u_j) = I_{X-c}(u_j)$, for all $c \in \mathbb{R}$ and for all $j = 1, \cdots, n-1$.

The consistency of the LW estimators $\widehat{d}$ and $\widehat{c}$ of (8.5.2) and (8.5.3) is based on the following two assumptions.

Assumption A. $\{X_j\}$ is a stationary process with spectral density f satisfying (8.5.4) with the true parameters c_0 and d_0.

Assumption B. For any $m = m_n$ satisfying (8.5.5), the normalized periodograms η_j, $1 \leq j \leq m$, satisfy the WLLN (8.5.10).

Let

$$T_m := \frac{\sum_{j=1}^{m} \nu_j \, u_j^{2d_0} I(u_j)}{\sum_{j=1}^{m} \nu_j \, \log(j/m) g(u_j)}, \qquad \nu_j := \log j - \frac{1}{m} \sum_{k=1}^{m} \log k. \qquad (8.5.11)$$

To describe the bias, define

$$B_m := \frac{\sum_{j=1}^{m} \nu_j \, g(u_j)}{\sum_{j=1}^{m} \nu_j \, \log(j/m) g(u_j)}. \qquad (8.5.12)$$

Besides establishing the consistency of $\widehat{d}$ and $\widehat{c}$, the next theorem also gives an asymptotic expansion for $\widehat{d}$. It shows, under mild conditions, that $\widehat{d} - d_0$ can be decomposed into a stochastic term $T_m - ET_m$, a bias term B_m, and an error term that depends on the smoothness of g appearing in (8.5.4).

The original literature for LW estimation requires the differentiability of f with its derivative $\dot{f}$ satisfying

$$|\dot{f}(u)| \leq C|u|^{-1-2d}, \quad 0 < u \leq a. \qquad (8.5.13)$$

Define

$$\begin{aligned} r_{n,g} &= o(1), & g \in \mathcal{C}[0,a], & \qquad (8.5.14) \\ &= n^{-\alpha}, & g \in \Lambda_\alpha[0,a], \ 0 < \alpha \leq 1, \\ &= 0, & \text{if } (8.5.13) \text{ holds}. \end{aligned}$$

Theorem 8.5.1. (Consistency of local Whittle estimators). *If $\{X_j\}$ and m satisfy Assumptions A and B and (8.5.5), then the following holds:*

(i) (Consistency of $\widehat{d}$) *The LW estimator $\widehat{d}$ is consistent for d_0 and satisfies the following expansion:*

$$\widehat{d} \to_p d_0, \tag{8.5.15}$$

$$\widehat{d} - d_0 = -\frac{1}{2}T_m(1 + o_p(1)). \tag{8.5.16}$$

(ii) (Bias) *For $g \in \mathcal{C}[0, a] \cup \Lambda_\alpha[0, a]$, $0 < \alpha \leq 1$, or f as in (8.5.13),*

$$\widehat{d} - d_0 = -\frac{1}{2}\{(T_m - ET_m) + B_m\}(1 + o_p(1)) \tag{8.5.17}$$

$$+ O_p(m^{-1}\log^3 m) + O_p(r_{n,g}).$$

(iii) (Consistency of $\widehat{c}$) *In addition, if $\widehat{d} - d_0 = o_p(1/\log n)$, then*

$$\widehat{c} \to_p c_0. \tag{8.5.18}$$

Remark 8.5.1. Part (ii) of Theorem 8.5.1 shows that the approximation (8.5.17) is highly precise if f satisfies (8.5.13) or g is sufficiently smooth.

The bias formula B_m and T_m in Theorem 8.5.1 are adjusted to a sample size n. As n increases, T_m simplifies and is approximated by the well-known entity in the literature for LW estimation:

$$T_m = \left(m^{-1}\sum_{j=1}^{m} \nu_j \frac{I(u_j)}{c_0 u_j^{-2d_0}}\right)(1 + o_p(1)).$$

To see this, use the fact that

$$\nu_j = \log(j/m) + 1 + O(m^{-1}\log m), \quad 1 \leq j \leq m, \tag{8.5.19}$$

$$g(u_j) = g(0) + o(1),$$

uniformly for $0 < u_j \leq u_m \to 0$, to obtain

$$m^{-1}\sum_{j=1}^{m} \nu_j \log(\frac{j}{m})\frac{g(u_j)}{g(0)} \to \int_0^1 (\log(u) + 1)\log(u)du = 1. \tag{8.5.20}$$

Before proceeding to the proof of Theorem 8.5.1, we state the following properties of the bias B_m of $\widehat{d}$. They require the function g to satisfy

$$g(u) = c_0 + b|u|^\beta + o(|u|^\beta), \quad u \to 0, \quad 0 < \beta \leq 2. \tag{8.5.21}$$

Proposition 8.5.1. *The following hold for the B_m of (8.5.12):*

(i) $\qquad B_m = \dfrac{1 + o(1)}{g(0)m}\sum_{j=1}^{m} \nu_j\left(g(u_j) - g(0)\right).$

(ii) *If g satisfies (8.5.21), then*

$$B_m = u_m^\beta \frac{b}{c_0} \frac{\beta}{(1+\beta)^2} + o((\frac{m}{n})^\beta). \qquad (8.5.22)$$

(iii) *If g has a continuous second derivative $\ddot{g}$ at the origin, then*

$$B_m = u_m^2 \frac{\ddot{g}(0)}{g(0)} \frac{1}{9} + o((\frac{m}{n})^2). \qquad (8.5.23)$$

Proof. (i) follows from (8.5.12), applying $\sum_{j=1}^m \nu_j = 0$ and (8.5.20).
(ii) If g satisfies (8.5.21), then by (i),

$$B_m = \frac{b}{c_0 m} \sum_{j=1}^m (\log(j/m) + 1) u_j^\beta + o(u_m^\beta)$$

$$= u_m^\beta \frac{b}{c_0} \int_0^1 (\log(u) + 1) u^\beta du + o(u_m^\beta) \sim u_m^\beta \frac{b}{c_0} \frac{\beta}{(1+\beta)^2}.$$

(iii) Since the spectral density is an even function, $\dot{g}(0) = 0$ and $g(u) = g(0) + (\ddot{g}(0)/2) u^2 + o(u^2)$, as $u \to 0$. Thus, g satisfies (8.5.21) with $\beta = 2$, and (iii) follows from (ii).

The proof of Theorem 8.5.1 is facilitated by the following auxiliary lemma.

Lemma 8.5.1. *Assume that an array of non-negative r.v.'s, $y_{m,j}$, $1 \le j \le m$, $m \ge 1$, and a real function $w(u,d)$, $(u,d) \in [0,1] \times I$, $I = [a,b] \subset \mathbb{R}$, have the following properties:*

For any $0 < \tau \le 1$, as $m \to \infty$,

$$[\tau m]^{-1} \sum_{j=1}^{[\tau m]} y_{m,j} \to_p 1, \qquad (8.5.24)$$

$$E|y_{m,j}| \le C, \qquad 1 \le j \le m, \ m \ge 1. \qquad (8.5.25)$$

There exists $\alpha \in (0,1)$, such that

$$\sup_{d \in I} |w(u,d)| \le C u^{-\alpha}, \quad u \in [0,1], \qquad (8.5.26)$$

$$\sup_{b \le u \le 1} \sup_{d \in I} |\frac{\partial}{\partial u} w(u,d)| \le C, \quad \text{for any fixed } 0 < b < 1.$$

Then, as $m \to \infty$,

$$\sup_{d \in I} |m^{-1} \sum_{j=1}^m w(\frac{j}{m}, d) y_{m,j} - \int_0^1 w(u,d) \, du| \to_p 0, \qquad (8.5.27)$$

$$\sup_{d \in I} |m^{-1} \sum_{j=1}^m w(\frac{j}{m}, d) y_{m,j} - m^{-1} \sum_{j=1}^m w(\frac{j}{m}, d)| \to_p 0.$$

Proof. We prove only the first result in (8.5.27). The steps of the proof are elementary and also yield the second claim in (8.5.27). Let $0 < b \le 1$. Write

$$m^{-1} \sum_{j=1}^{m} w(\frac{j}{m}, d) y_{m,j} = m^{-1} \sum_{j=[bm]}^{m} [\cdots] + m^{-1} \sum_{j=1}^{[bm]-1} [\cdots]$$

$$=: s_{m,1}(d) + s_{m,2}(d).$$

We shall show that

$$\sup_{d \in I} \left| s_{m,1}(d) - \int_{b}^{1} w(u,d) du \right| \to_{p} 0, \quad m \to \infty, \quad \forall b > 0, \quad (8.5.28)$$

$$\sup_{d \in I} |s_{m,2}(d)| \to_{p} 0, \quad m \to \infty, \ b \to 0,$$

$$\sup_{d \in I} \left| \int_{0}^{b} w(u,d) du \right| \to 0, \quad b \to 0,$$

which proves (8.5.27).

To prove the first bound of (8.5.28), write $[b,1] = \Delta_1 \cup \cdots \cup \Delta_K$ as a union of K non-intersecting intervals $\Delta_j = [b_j, b_{j+1})$ of length $|\Delta_j| = (1-b)/K \le 1/K$. Use the step function:

$$w_{\Delta}(u,d) = \sum_{l=1}^{K} w(b_l, d) I(u \in \Delta_l),$$

and set

$$s_{m,1}^{\Delta}(d) := m^{-1} \sum_{j=1}^{m} w_{\Delta}(\frac{j}{m}, d) y_{m,j}.$$

We shall show that for any $0 < b < 1$,

$$\sup_{d \in I} |s_{m,1}(d) - s_{m,1}^{\Delta}(d)| \to 0, \quad m \to \infty, \ K \to \infty, \quad (8.5.29)$$

$$\sup_{d \in I} \left| s_{m,1}^{\Delta}(d) - \int_{b}^{1} w^{\Delta}(u,d) du \right| \to 0, \quad m \to \infty,$$

$$\sup_{d \in I} \left| \int_{b}^{1} w^{\Delta}(u,d) du - \int_{b}^{1} w(u,d) du \right| \to 0, \quad K \to \infty,$$

which implies the first result of (8.5.28). By (8.5.26) and the mean-value theorem,

$$\sup_{u \in [b,1], d \in I} |w_{\Delta}(u,d) - w(u,d)| \le K^{-1} \sup_{u \in [b,1], d \in I} |\frac{\partial}{\partial u} w(u,d)| \quad (8.5.30)$$

$$\le K^{-1} C,$$

where C does not depend on K. Therefore, as $m \to \infty$,

$$|s_{m,1}(d) - s^{\triangle}_{m,1}(d)| \leq \frac{C}{Km} \sum_{j=[bm]}^{m} y_{m,j}$$

$$\leq \frac{C}{Km} \sum_{j=1}^{m} y_{m,j} \leq \frac{C}{K}, \quad \text{in probability,}$$

$$= o_p(1), \quad K \to \infty,$$

by (8.5.24), which proves the first result of (8.5.29).

To obtain the second result of (8.5.29), note that

$$\left| s^{\triangle}_{m,1}(d) - \int_b^1 w^{\triangle}(u, d) du \right|$$

$$\leq \sum_{l=1}^{K} |w(b_l, d)| \left| m^{-1} \sum_{(j/m) \in \Delta_l} y_{m,j} - |\Delta_l| \right|$$

$$\leq C \sum_{l=1}^{K} \left| m^{-1} \sum_{(j/m) \in \Delta_l} y_{m,j} - |\Delta_l| \right| \to_p 0, \quad m \to \infty,$$

by (8.5.26), and because

$$m^{-1} \sum_{(j/m) \in \Delta_l} y_{m,j} = m^{-1} \left(\sum_{j=1}^{[b_{l+1}m]} y_{m,j} - \sum_{j=1}^{[b_l m]} y_{m,j} \right) + o_p(1),$$

$$\to_p b_{l+1} - b_l = |\Delta_l|,$$

by (8.5.24) and (8.5.25). The last result of (8.5.29) holds because of (8.5.30).

To establish the second bound of (8.5.28), by (8.5.26) and (8.5.24),

$$E[\sup_{d \in I} |s_{m,2}(d)|] \leq C m^{-1} \sum_{j=1}^{[bm]} (j/m)^{-\alpha} E[y_{j,m}]$$

$$\leq C m^{-1+\alpha} \sum_{j=1}^{[bm]} j^{-\alpha} \leq C b^{1-\alpha} \to 0, \quad b \to 0.$$

To prove the third bound of (8.5.28), note that by (8.5.26),

$$\sup_{d \in I} \left| \int_0^b w(u, d) du \right| \leq C \int_0^b u^{-\alpha} du \leq C b^{1-\alpha} \to 0, \quad b \to 0,$$

which completes the proof of the lemma.

Proof of Theorem 8.5.1. (i) Let

$$\tilde{\eta}_j = c_0^{-1} u_j^{2d_0} I(u_j), \quad j = 1, \cdots, m,$$

denote the periodogram standardized by approximating the spectral density $c_0 u^{-2d_0}$ in the neighborhood of 0. Note that by (8.5.4), $\tilde{\eta}_j = \eta_j g(u_j)/g(0)$, where

$$\max_{j=1,\cdots,m} \left| \frac{g(u_j)}{g(0)} - 1 \right| \leq \sup_{0 \leq u \leq u_m} \frac{|g(u) - g(0)|}{g(0)} \to 0, \quad u_m \to 0. \quad (8.5.31)$$

Therefore, (8.5.10) and (8.5.8) imply that, under (8.5.5),

$$m^{-1} \sum_{j=1}^{m} \tilde{\eta}_j \to_p 1, \quad E\tilde{\eta}_j \leq C, \quad 1 \leq j \leq m. \quad (8.5.32)$$

To proceed further, we need to introduce

$$V_n(d) := m^{-1} \sum_{j=1}^{m} \left(\frac{j}{m} \right)^{2d} \tilde{\eta}_j, \quad (8.5.33)$$

$$V(d) := \int_0^1 u^{2d} du = (1 + 2d)^{-1}, \quad d > -1.$$

(i) *Proof of (8.5.15).* By definition, a local Whittle estimator $\hat{d}$ minimizes the objective function $U_n(d)$ of (8.5.1). To show consistency, we need to prove that for any $\epsilon > 0$, the probability that the minimizer of $U_n(d)$ lies in the interval $|d - d_0| \leq \epsilon$ tends to 1, as $n \to \infty$. For this purpose, it suffices to show that for any $\epsilon > 0$, there exists $\delta > 0$ such that, with $J = [-1/2, 1/2] \cap \{d : |d - d_0| > \epsilon\}$,

$$P\left\{ \inf_{d \in J} (U_n(d) - U_n(d_0)) \geq \delta \right\} \to 1, \quad \text{as } n \to \infty. \quad (8.5.34)$$

Split $J = I \cup I'$ into two sets:

$$I = J \cap \{d : d - d_0 \geq -1/2 + \xi\}, \quad I' = J \cap \{d : d - d_0 < -1/2 + \xi\},$$

where $\xi > 0$ will be chosen later. Use the fact that $m^{-1} \sum_{j=1}^{m} \log(j/m) = -1 + o(1)$, to write

$$U_n(d) - U_n(d_0) = \log V_n(d - d_0) - \log V_n(0) + 2(d - d_0) + o(1).$$

For $d \in I$, apply Lemma 8.5.1, with $y_{m,j} := \tilde{\eta}_j$, $1 \leq j \leq m$, and $w(u, d) := u^{2(d-d_0)}$. Note that $|w(u, d)| \leq Cu^{-1+2\xi}$, $u \in (0, 1)$, i.e., (8.5.26) holds, and that $y_{m,j}$ satisfy (8.5.24) because of (8.5.32). Hence, by Lemma 8.5.1, $\sup_{d \in I} |V_n(d - d_0) - V(d - d_0)| \to_p 0$. Use this together with the fact that $-\log(1 + x) + x > 0$ for $x > -1$, to obtain

$$\sup_{d \in I} \{U_n(d) - U_n(d_0)\}$$
$$= \sup_{d \in I} \{\log V(d - d_0) - \log V(0) + 2(d - d_0)\} + o_p(1)$$
$$= \sup_{d \in I} \{-\log(1 + 2(d - d_0)) + 2(d - d_0)\} + o_p(1) \geq \delta_\epsilon > 0.$$

Now let $d \in I'$. For $d - d_0 \le -1/2 + \xi$, $V_n(d - d_0) \ge V_n(-1/2 + \xi)$, whereas by Lemma 8.5.1, $V_n(-1/2 + \xi) \to V(-1/2 + \xi) = (2\xi)^{-1}$. Therefore,

$$U_n(d) - U_n(d_0)$$
$$\ge \log V_n(-1/2 + \xi) - \log V_n(0) + 2(d - d_0) + o_p(1)$$
$$\ge -\log(2\xi) + 2(-1/2 - d_0) + o_p(1) \ge 1,$$

with probability tending to 1, as $n \to \infty$, when $\xi > 0$ is chosen small. This completes the proof of the consistency (8.5.15).

Proof of (8.5.16). It suffices to show that for some $\epsilon > 0$,

$$(\widehat{d} - d_0)I(|\widehat{d} - d_0| \le \epsilon) = -\frac{1}{2}T_m(1 + o_p(1)). \tag{8.5.35}$$

Since by (8.5.15),

$$I(|\widehat{d} - d_0| \le \epsilon) = 1 + o_p(1), \quad I(|\widehat{d} - d_0| > \epsilon) = o_p(1),$$

then from (8.5.35) we conclude

$$\widehat{d} - d_0 = -\frac{1}{2}T_m(1 + o_p(1)) + (\widehat{d} - d_0)I(|\widehat{d} - d_0| > \epsilon)$$
$$= -\frac{1}{2}T_m(1 + o_p(1)) + (\widehat{d} - d_0)o_p(1),$$

which yields

$$\widehat{d} - d_0 = -\frac{1}{2}T_m(1 + o_p(1)), \tag{8.5.36}$$

thereby proving (8.5.16).

Choose $\epsilon > 0$ such that $(d_0 - \epsilon, d_0 + \epsilon) \subset (-1/2, 1/2)$. Let $\widehat{d} \in (d_0 - \epsilon, d_0 + \epsilon)$, i.e., $|\widehat{d} - d_0| < \epsilon$. Since $\widehat{d}$ minimizes $U_n(d)$ at an inner point of $(-1/2, 1/2)$, then $\dot{U}_n(\widehat{d}) = 0$. Next, notice that using notation

$$S_n(d) := \frac{1}{m} \sum_{j=1}^{m} (\frac{j}{m})^{2d} \nu_j \tilde{\eta}_j$$

and V_n of (8.5.33), we can write the derivative $\dot{U}_n(d)$ as

$$\dot{U}_n(d) = \frac{2S_n(d - d_0)}{V_n(d - d_0)},$$

so that $\dot{U}_n(\widehat{d}) = 2S_n(\widehat{d} - d_0)/V_n(\widehat{d} - d_0)$. Since we assume $|\widehat{d} - d_0| \le \epsilon$, then

$$V_n(\widehat{d} - d_0) \ge \frac{1}{m} \sum_{j=1}^{m} (\frac{j}{m})^{2\epsilon} \tilde{\eta}_j \to_p \int_0^1 u^{2\epsilon} du > 0, \tag{8.5.37}$$

by Lemma 8.5.1. Therefore, $\dot{U}_n(\hat{d}) = 0$ implies $S_n(\hat{d} - d_0) = 0$.

Hence, by the mean-value theorem, with $|d^*| \le |\hat{d} - d_0|$,

$$S_n(\hat{d} - d_0) - S_n(0) = \dot{S}_n(d^*)(\hat{d} - d_0), \tag{8.5.38}$$

$$\hat{d} - d_0 = -\frac{S_n(0)}{\dot{S}_n(d^*)}.$$

Observe that

$$S_n(0) = \frac{1}{c_0 m} \sum_{j=1}^{m} \nu_j u_j^{2d_0} I(u_j).$$

Set $s_n := (mc_0)^{-1} \sum_{j=1}^{m} \nu_j \log(j/m) g(u_j)$. Note that $s_n \to 1$; see Remark 8.5.1. We will show that

$$\left| \dot{S}_n(d^*) - 2s_n \right| \to_p 0, \tag{8.5.39}$$

which yields $\dot{S}_n(d^*) = 2s_n(1 + o_p(1))$ and together with (8.5.38) implies (8.5.35). Let

$$r_j := \left(\frac{j}{m}\right)^{2d^*} \tilde{\eta}_j - \frac{g(u_j)}{c_0} - (\tilde{\eta}_j - 1),$$

$$q_{n,1} := m^{-1} \sum_{j=1}^{m} \nu_j \log\left(\frac{j}{m}\right)(\tilde{\eta}_j - 1), \quad q_{n,2} := m^{-1} \sum_{j=1}^{m} \nu_j \log\left(\frac{j}{m}\right) r_j.$$

Then

$$\dot{S}_n(d^*) - 2s_n = 2m^{-1} \sum_{j=1}^{m} \nu_j \log\left(\frac{j}{m}\right)\left(\left(\frac{j}{m}\right)^{2d^*} \tilde{\eta}_j - \frac{g(u_j)}{c_0}\right)$$

$$=: 2(q_{n,1} + q_{n,2}).$$

Since the η_j's satisfy the WLLN (8.5.32), then by (8.5.19) and (8.5.27) of Lemma 8.5.1, $q_{n,1} = o_p(1)$. However, since $|d^*| \le |\hat{d} - d_0| \le \epsilon$, then by the mean-value theorem and (8.5.31), uniformly in $1 \le j \le m$,

$$|r_j| \le \left|\left(\frac{j}{m}\right)^{2d^*} - 1\right| \tilde{\eta}_j + \left|\frac{g(u_j)}{c_0} - 1\right|$$

$$\le 2|d^*| \left|\log\left(\frac{j}{m}\right)\right| \sup_{|\eta| \le 2|d^*|} \left(\frac{j}{m}\right)^{\eta} \tilde{\eta}_j + o(1)$$

$$\le 2|\hat{d} - d_0| \left|\log\left(\frac{j}{m}\right)\right| \left(\frac{j}{m}\right)^{-2\epsilon} \tilde{\eta}_j + o(1).$$

Use this bound and $|\nu_j| \le C(|\log(j/m)| + 1)$, to obtain

$$|q_{n,2}| \le C|\hat{d} - d_0| \left\{ m^{-1} \sum_{j=1}^{m} (|\log\left(\frac{j}{m}\right)| + 1)^3 (j/m)^{-2\epsilon} \tilde{\eta}_j \right\} + o(1).$$

By Lemma 8.5.1, the term in the curly brackets of the r.h.s. above is bounded in probability by

$$\int_0^1 (|\log u| + 1)^3 u^{-2\epsilon} du < \infty.$$

By (8.5.15), $|\hat{d} - d_0| = o_p(1)$, which yields $q_{n,2} = o_p(1)$ and completes the proof of (8.5.39) and (8.5.16).

(ii) *Proof of (8.5.17)*. Using (8.5.20), we can bound

$$|ET_m - B_m| \le Cm^{-1} \Big| \sum_{j=1}^m \nu_j \frac{g(u_j)}{g(0)} (E\eta_j - 1) \Big|$$

$$\le Cm^{-1} \sum_{j=1}^m (|\log(j/m)| + 1)|E\eta_j - 1|.$$

Under the assumptions on f and g in (ii), by Theorem 5.2.1(iv),

$$\Big| \frac{E[I(u_j)]}{f(u_j)} - 1 \Big| \le C(j^{-1} \log j + r_{n,g}), \quad 1 \le j \le m,$$

with $r_{n,g}$ as in (8.5.14). Therefore,

$$|ET_m - B_m| \le Cm^{-1} \sum_{j=1}^m (|\log(j/m)| + 1)(j^{-1} \log j + r_{n,g})$$

$$\le C(m^{-1} \log^3 m + r_{n,g}),$$

which together with (8.5.16) implies (8.5.17).

(iii) *Proof of (8.5.18)*. Write

$$\hat{c} = m^{-1} \sum_{j=1}^m u_j^{2\hat{d}} I(u_j) = m^{-1} \sum_{j=1}^m u_j^{2d_0} I(u_j) + m^{-1} \sum_{j=1}^m (u_j^{2\hat{d}} - u_j^{2d_0}) I(u_j)$$

$$=: r_{n,1} + r_{n,2}.$$

Notice that, by Assumption B, which yields (8.5.32),

$$r_{n,1} = c_0 m^{-1} \sum_{j=1}^m \tilde{\eta}_j \to_p c_0.$$

To show $r_{n,2} = o_p(1)$, use $|\log(u_j)| \le C \log n$, $1 \le j \le m$, and the assumption $\hat{d} - d_0 = o_p(1/\log n)$, to obtain

$$\max_{j=1,\cdots,m} |u_j^{2(\hat{d}-d_0)} - 1| = \max_{j=1,\cdots,m} |\exp(2(\hat{d} - d_0) \log(u_j)) - 1|$$

$$= O_p\Big(\max_{j=1,\cdots,m} |(\hat{d} - d_0) \log(u_j)| \Big) = o_p(1).$$

Thus

$$|r_{n,2}| \leq m^{-1} \sum_{j=1}^{m} |u_j^{2(\hat{d}-d_0)} - 1| u_j^{2d_0} I(u_j)$$

$$= o_p(1) m^{-1} \sum_{j=1}^{m} u_j^{2d_0} I(u_j) = o_p(1) r_{n,1} = o_p(1),$$

which completes the proof of (iii) and of the Theorem 8.5.1.

Some sufficient conditions for WLLN property (8.5.10). Suppose $\{X_j\}$ is a linear process (5.2.2) with white-noise innovations $\{\zeta_j\}$, not necessarily i.i.d. Let $\eta_{X,j} := I_{X,j}/f_{X,j}$ and $\eta_{\zeta,j} := I_{\zeta,j}/f_{\zeta,j}$. The following proposition shows that under some mild assumptions on the spectral density f_X and the transfer function A_X, the WLLN condition for the $\eta_{X,j}$'s is implied by the WLLN property of the standardized periodograms $\eta_{\zeta,j}$ of the white-noise process.

Proposition 8.5.2. *Suppose that f_X satisfies (8.5.4) with parameter $|d| < 1/2$, and that A_X can be factored as $A_X = h A_Y$, with a continuous bounded function A_Y and h satisfying (5.3.15). If for any $m = m_n$ as in (8.5.5),*

$$m^{-1} \sum_{j=1}^{m} \eta_{\zeta,j} \to_p 1, \tag{8.5.40}$$

then the $\eta_{X,j}$'s satisfy (8.5.10).

In particular, (8.5.10) holds for the martingale-difference sequence $\{\zeta_j\}$ defined in Remark 5.3.1.

Proof. By Theorem 5.3.2(i),

$$\eta_{X,j} = \eta_{\zeta,j} + r_{n,j}, \qquad \max_{u_K \leq u_j \leq \pi} E|r_{n,j}| = o(1), \tag{8.5.41}$$

where $K = 1$ if $d = 0$ and for any $K = K_n < n/2$, $K_n \to \infty$ if $d \neq 0$. By Theorem 5.2.1(iv), $E\eta_{X,j} \leq C$ for all $1 \leq j \leq m$. Hence,

$$m^{-1} \sum_{j=1}^{m} \eta_{X,j} = m^{-1} \sum_{j=1}^{m} \eta_{\zeta,j} + m^{-1} \sum_{j=1}^{m} r_{n,j} = m^{-1} \sum_{j=1}^{m} \eta_{\zeta,j} + o_p(1),$$

which together with (8.5.40) implies (8.5.10).

Because of Remark 5.3.1, equation (8.5.41) still holds when $\{\zeta_j\}$ is a martingale-difference sequence, which completes the proof.

8.5.2   Local Whittle estimators for linear processes

In this subsection we shall apply the general results of the previous section to derive the consistency and asymptotic normality of the LW estimator $\widehat{d}$ when the observable process is the linear process (6.2.1) with i.i.d. innovations $\{\zeta_j\}$. Many of the derivations below are influenced by the pioneering work of Robinson (1995b).

For consistency of $\widehat{d}$, we need only $E\zeta_0^2 < \infty$, while its asymptotic normality requires that $E\zeta_0^4 < \infty$, and the following condition on the transfer function A_X: for some $0 < a \leq \pi$,

$$A_X(u) = h(u)G(u), \tag{8.5.42}$$

$$C_1|u|^{-d} \leq |h(u)| \sim C_2|u|^{-d}, \ u \to 0, \ \exists C_1, C_2 > 0, \quad |d| < 1/2,$$

$$|\dot{h}(u)| \leq C|u|^{-1-d}, \qquad 0 \leq u \leq a,$$

where G is a continuous function on $[0, a]$, possibly complex-valued, and bounded away from 0. See the examples given in Remark 8.5.4. Recall the definition of B_m from (8.5.12).

Theorem 8.5.2. *Let $\{X_j\}$ be a linear process as in (6.2.1) with the spectral density $f(u) = |u|^{-2d_0}g(u)$, $u \in \Pi$, $|d| < 1/2$, where g is continuous and positive on $[0, a]$ for some $0 < a \leq \pi$. Also assume the bandwidth m satisfies (8.5.5). Then the following holds:*

(i) *$\{X_j\}$ satisfies Assumption B.*

(ii) *$(\widehat{d}, \widehat{c})$ are consistent estimates of (d_0, c_0), satisfying (8.5.15), (8.5.16), and (8.5.18).*

(iii) *Suppose, in addition, $E\zeta_0^4 < \infty$, A_X satisfies (8.5.42), $g \in \Lambda_\alpha$, $\alpha > 1/2$, and satisfies (8.5.21) with $0 < \beta \leq 2$. Then with $Z_m = T_m - ET_m$,*

$$\widehat{d} - d_0 = -\frac{1}{2}Z_m - \frac{1}{2}B_m + o_p(m^{-1/2}) + o_p((m/n)^\beta), \tag{8.5.43}$$

$$\mathrm{Var}(Z_m) \sim \frac{1}{\sum_{j=1}^m \nu_j^2} \sim \frac{1}{m}, \qquad \sqrt{m}\, Z_m \to_D \mathcal{N}(0,1),$$

$$B_m = u_m^\beta \frac{b}{c_0} \frac{\beta}{(1+\beta)^2} + o((\frac{m}{n})^\beta).$$

Moreover,

$$\mathrm{Var}(Z_m) \sim \frac{\sum_{j=1}^m \nu_j^2 g^2(u_j)}{\left(\sum_{j=1}^m \nu_j \log(j/m)g(u_j)\right)^2}. \tag{8.5.44}$$

Proof. (i) To verify Assumption B, we need to show (8.5.10). Decompose η_j as in (8.5.41): $\eta_j = I_{\zeta,j}/f_\zeta(u_j) + r_{n,j}$, $1 \le j \le m$. Then $m^{-1} \sum_{j=1}^m \eta_j = q_{n,1} + q_{n,2}$, where

$$q_{n,1} := m^{-1} \sum_{j=1}^m \frac{I_{\zeta,j}}{f_\zeta(u_j)}, \quad q_{n,2} := m^{-1} \sum_{j=1}^m r_{n,j}.$$

By Corollary 6.2.2, $q_{n,1} \to_p 1$. We show that $E|q_{n,2}| = o(1)$, which implies $q_{n,2} = o_p(1)$ and completes the proof of (8.5.10).

Let $K = \log m$. By (8.5.8) and (8.5.41),

$$\max_{1 \le j \le m} E|r_{n,j}| \le C, \quad \max_{K \le j \le m} E|r_{n,j}| = o(1).$$

Hence,

$$E|q_{n,2}| \le m^{-1} \sum_{j=1}^{K-1} E|r_{n,j}| + m^{-1} \sum_{j=K}^m E|r_{n,j}|$$
$$\le Cm^{-1} \log m + o(1) = o(1).$$

(ii) Since $\{X_j\}$ satisfies Assumptions A and B, then (8.5.15), (8.5.16), and (8.5.18) hold by parts (i) and (ii) of Theorem 8.5.1.

(iii) Under Assumptions A and B, expansion (8.5.17) of Theorem 8.5.1 implies (8.5.43) with $Z_m := T_m - ET_m$, and $r_{n,g} = O_p(n^{-\alpha}) = o_p(m^{-1/2})$, since $g \in \Lambda_\alpha$, $\alpha > 1/2$. To show $Z_m \to_D \mathcal{N}(0,1)$, we use Theorem 6.2.4. Write $Z_m = Q_{n,X} - EQ_{n,X}$, $Q_{n,X} = \sum_{j=1}^m b_{n,j} I_{n,j}$, whereby (8.5.11),

$$b_{n,j} = \nu_j \frac{g(u_j)}{A_n f_{X,j}}, \quad A_n := \sum_{j=1}^m \nu_j \log(\frac{j}{m}) g(u_j).$$

We now show that A_X, f, and $b_{n,j}$ satisfy the conditions of Theorem 6.2.4. Our conditions on A_X and $f = f_X$ stated on a small interval $[0, a]$ are analogous to those of Theorem 6.2.4 formulated on $[0, \pi]$. Since summation in $Q_{n,X}$ is over $j \le m = o(n)$, by Remark 6.2.1, the assumptions on A_X and f can be restricted to an interval $[0, a]$, $a > 0$, thereby establishing the claims for A_X and f.

To verify that the $b_{n,j}$'s satisfy assumption (6.2.41), observe that $g(u_j) \ge c > 0$ for $j = 1, \cdots, m$. Thus,

$$B_{f,n}^2 \equiv \sum_{j=1}^m b_{n,j}^2 f_{X,j}^2 = \frac{1}{A_n^2} \sum_{j=1}^m \nu_j^2 g^2(u_j) \ge \frac{c}{A_n^2} \sum_{j=1}^m \nu_j^2,$$
$$b_{f,n} \equiv \max_{j=1,\cdots,m} |b_{n,j}| f_{X,j} \le CA_n^{-1} \max_{j=1,\cdots,m} |\nu_j|.$$

Therefore,

$$\frac{b_{f,n}}{B_{f,n}} \le C \frac{\max_{j=1,\cdots,m} |\nu_j|}{\left(\sum_{j=1}^m \nu_j^2\right)^{1/2}} \to 0,$$

since $|\nu_j| \le C \log m$, for $1 \le j \le m$, and

$$\sum_{j=1}^m \nu_j^2 = (1 + o(1))m \int_0^1 \left(\log u - \int_0^1 \log v \, dv\right)^2 du \qquad (8.5.45)$$

$$= (1 + o(1))m,$$

which proves (6.2.41).

This completes the verification of the conditions of Theorem 6.2.4. Consequently, by the C–S inequality and because $m = o(n)$,

$$\frac{1}{n}\left(\sum_{j=1}^m b_{n,j} f_{X,j}\right)^2 \le \frac{m}{n} B_{f,n}^2 = o(B_{f,n}^2),$$

and $\mathrm{Var}(Q_{m,X})^{-1/2}(Q_{n,X} - EQ_{n,X}) \to_D \mathcal{N}(0,1)$ and

$$\mathrm{Var}(Q_{m,X}) \sim v_n^2 = \sum_{j=1}^m (b_{n,j} f_{X,j})^2 + \mathrm{Cum}_4(\zeta_0)\frac{1}{n}\left(\sum_{j=1}^m b_{n,j} f_{X,j}\right)^2$$

$$= (1 + o(1))B_{f,n}^2. \qquad (8.5.46)$$

This proves the asymptotic normality of Z_m in (8.5.43) and (8.5.44).

Finally, we evaluate $B_{f,n}^2$. By the continuity of g at 0, $\max_{1\le j\le m} |g(u_j) - g(0)| \to 0$, which in turn implies

$$\sum_{j=1}^m \nu_j^2 g^2(u_j) = (1 + o(1))g^2(0)\sum_{j=1}^m \nu_j^2,$$

$$A_n = (1 + o(1))g(0)\sum_{j=1}^m \nu_j \log(\frac{j}{m}) = (1 + o(1))g(0)\sum_{j=1}^m \nu_j^2,$$

$$B_{f,n}^2 = (1 + o(1))\left(\sum_{j=1}^m \nu_j^2\right)^{-1} = (1 + o(1))m^{-1},$$

which, in view of (8.5.46), proves the second claim in (8.5.43).

The last claim in (8.5.43) follows from (8.5.22). This completes the proof of (iii) and also of the theorem.

Remark 8.5.2. Expansion (8.5.43) of Theorem 8.5.2 can be used to obtain the following conclusions. If the bandwidth parameter m increases not too fast, i.e., if $m = o(n^{2\beta/(1+2\beta)})$, then $B_m = O((m/n)^\beta) = o(m^{-1/2})$, and

$$2\Big(\sum_{j=1}^{m} \nu_j^2\Big)^{1/2}(\widehat{d} - d_0) \to_D \mathcal{N}(0, 1), \qquad (8.5.47)$$

$$2m^{1/2}(\widehat{d} - d_0) \to_D \mathcal{N}(0, 1).$$

On the other hand, if $n^{2\beta/(2\beta+1)}/m \to 0$ and $m = o(n)$, then

$$2(n/m)^\beta(\widehat{d} - d_0) \to_p -B, \quad B := (b/c_0)(2\pi)^\beta \beta/(1 + \beta)^2,$$

whereas, if $m = n^{2\beta/(2\beta+1)}$, we then have

$$2m^{1/2}(\widehat{d} - d_0) \to_D \mathcal{N}(-B, 1).$$

Finite sample simulations suggest that in (8.5.47), the first normalization may be preferable for small to moderate bandwidths m.

Remark 8.5.3. The formulas (8.5.12) and (8.5.44) for the bias B_m and $\mathrm{Var}(Z_m)$ are dependent on the shape of the function g at low frequencies for finite samples. Since g is not known, in applications we often use $B_m = O((m/n)^\beta)$ and $\mathrm{Var}(Z_m) \sim (\sum_{j=1}^m \nu_j^2)^{-1} \sim m^{-1}$.

The asymptotic results of Theorem 8.5.2 require imposing regularity conditions on A_X and f only in a small neighborhood of 0, $[0, a]$. For a large enough n, such an interval will contain all frequencies $u_1, \cdots, u_m$ used in the LW estimation. In finite sample applications, based on frequencies $0 < u_1 < \cdots < u_m$, we have to assure that $a > u_m$.

Remark 8.5.4. The assumptions on the spectral density f and transfer function A_X of the linear process $\{X_j\}$ in Theorem 8.5.2 are fairly general. As seen in Remark 6.2.2, they are satisfied by an ARFIMA(p, d, q) process, and by a more general linear process

$$X_j = (1 - B)^{-d} Y_j, \qquad j \in \mathbb{Z}, \qquad |d| < 1/2,$$

defined in (5.3.11), where $\{Y_j\}$ is a short memory process with absolutely summable coefficients and A_Y is continuous. From (5.3.13) we have $A_X = h A_Y$ where $h(u) = (1 - e^{-iu})^{-d}$ is differentiable. This shows condition (8.5.42) is satisfied by the above process $\{X_j\}$.

If $\{X_j\}$ is an ARFIMA(p, d, q) process, then the corresponding $\{Y_j\}$ is an ARMA process, all derivatives of A_Y are bounded, and thus A_X is also differentiable. Then one can also decompose $A_X = hG$ with $h = A_X$ and $G = 1$. In addition, observe that

$$f_Y(u) = f_Y(0) + \ddot{f}_Y(0)u^2/2 + O(u^4),$$

$$f_X(u) = |h(u)|^2 f_Y(u) = |u|^{-2} g(u),$$

$$g(u) := \left(\frac{|u|}{|2\sin(u/2)|}\right)^{2d} f_Y(u)$$

$$= \left(1 - \frac{du^2}{12} + O(u^4)\right)\left(f_Y(0) + \frac{\ddot{f}_Y(0)}{2}u^2 + O(u^4)\right)$$

$$= c_0 + bu^2 + O(u^4), \quad c_0 = f_Y(0), \quad b = \frac{\ddot{f}_Y(0)}{2} - \frac{f_Y(0)d}{12}.$$

Hence, g satisfies (8.5.21) with $\beta = 2$.

Selection of the optimal bandwidth. For a linear process $\{X_j\}$, by Theorem 8.5.2,

$$\text{MSE}(\widehat{d}) := E(\widehat{d} - d_0)^2 \sim \frac{1}{4}\left(\frac{1}{m} + B_m^2\right).$$

An m that minimizes this MSE is called an optimal m and denoted by m_{opt}. To find m_{opt}, assume that g in (8.5.4) has a non-vanishing continuous second derivative in the neighborhood of zero. Then

$$m_{opt} = n^{4/5}\left(\frac{3}{4\pi}\right)^{4/5}\left|\frac{2g(0)}{\ddot{g}(0)}\right|^{2/5} =: n^{4/5}C_{opt}. \tag{8.5.48}$$

Indeed, by (8.5.23),

$$\frac{1}{m} + B_m^2 \sim \frac{1}{m} + u_m^4\left(\frac{\ddot{g}(0)}{9\,g(0)}\right)^2 = \frac{1}{m} + m^4 u_1^4\left(\frac{\ddot{g}(0)}{9\,g(0)}\right)^2.$$

Minimizing this expression w.r.t. m readily yields (8.5.48).

Clearly, C_{opt} depends on the unknown function g. In order to use this optimal choice we must be able to estimate it in a given situation. To illustrate, consider the case when $\{X_j\}$ is an AR(1) process with the coefficient $|\rho| < 1$, $\rho \neq 0$. Then $d_0 = 0$, and (8.5.4) reads $f = g$, where g is the spectral density of an AR(1) model,

$$g(u) = c(1 + \rho^2 - 2\rho\cos(u))^{-1}, \quad u \in \Pi,$$

$$g(0) = c(1 - \rho)^{-2}, \quad \ddot{g}(0) = -2c\rho(1 - \rho)^{-4},$$

$$2g(0)/\ddot{g}(0) = -(1 - \rho)^2/\rho.$$

Then

$$C_{opt} = \left(\frac{3}{4\pi}\right)^{4/5} \left|\frac{(1-\rho)^2}{\rho}\right|^{2/5}.$$

Since ρ is the correlation at lag one, a plug-in estimator of C_{opt} is obtained by replacing ρ by the sample correlation $\hat{r}_1$, yielding

$$\widehat{m}_{opt} = n^{4/5}\left(\frac{3}{4\pi}\right)^{4/5} \left|\frac{(1-\hat{r}_1)^2}{\hat{r}_1}\right|^{2/5}.$$

To restrict $\widehat{m}_{opt}$ to the set $1, \cdots, n/2$, Lobato and Robinson (1998) suggested using the truncation

$$
\begin{aligned}
\tilde{m}_{opt} &= \widehat{m}_{opt}, & &\text{if } 0.06n^{4/5} \leq \widehat{m}_{opt} \leq 1.2n^{4/5}, & (8.5.49)\\
&= 0.06n^{4/5}, & &\text{if } \widehat{m}_{opt} < 0.06n^{4/5},\\
&= 1.2n^{4/5}, & &\text{if } \widehat{m}_{opt} > 1.2n^{4/5},
\end{aligned}
$$

where the choice of the truncation level is based on simulations.

Computing confidence intervals for d, based on (8.5.47) with $\beta = 2$, we can use the bandwidth $m^* = \min(m_{opt}, 0.8n^{0.8})$, which improved coverage probabilities as seen in Tables 8.7.3 and 8.7.4 of Section 8.7 below.

Bandwidths $\tilde{m}_{opt}$ and m^* are easy to compute. In practice the data-based automatic bandwidth selection rule (8.5.49) is used also for general processes, since in simulations it often outperforms the fixed bandwidth selection.

8.6 Signal plus noise processes

In this section we discuss estimation of the memory parameter of a stationary signal Y_j observed with additive noise. More precisely, let

$$X_j = Y_j + Z_j, \quad j \in \mathbb{Z}, \tag{8.6.1}$$

be observables, where Y_j stands for the signal and Z_j for the noise. These types of models often arise in practice where typically the noise process $\{Z_j\}$ is small relative to the input process $\{Y_j\}$. For example, when one detrends data, the noise Z_j is non-stationary but negligible since its magnitude tends to 0 with n uniformly over $j = 1, \cdots, n$. Another example is obtained when taking the logarithm of the volatility, which leads to a decomposition (8.6.1) with a long memory signal Y_j and white-noise Z_j.

For a long memory process $\{\xi_j, \ j \in \mathbb{Z}\}$, let d_ξ denote its long memory parameter and $\widehat{d_\xi}$ stand for the local Whittle estimator of d_ξ based on $\xi_j, 1 \leq j \leq n$.

In general, if a signal dominates stationary noise, i.e., has a larger memory parameter, the memory parameter d_Y of the signal Y_j can be consistently estimated from the observable process X_j. However, noise can significantly increase the bias in finite samples and larger samples will be needed to achieve the same precision as when estimating d_Y without any noise. We discuss below the consistent estimation of d_Y under fairly general conditions where the noise process is either stationary or non-stationary. They neither require the Y_j's to be a moving-average process, nor the noise Z_j to be i.i.d. or independent of Y_j.

To appreciate the effect of noise on the estimation of the memory parameter of a signal, consider the following example. Let $X_j = Y_j + Z_j$ be the sum of two independent stationary Gaussian processes. Then $\{X_j\}$ is a stationary Gaussian process with the spectral density $f_X = f_Y + f_Z$, and due to Wold decomposition, the Gaussian sequence can be written as a linear sequence (6.2.1) with i.i.d. innovations, so that Theorem 8.5.2 applies to $\hat{d}_X$. Assuming that $\{Z_j\} \sim \text{IID}(0, 1)$ and the spectral density of $\{Y_j\}$ satisfies $f_Y(u) = |u|^{-2d_Y}/2\pi, |d_Y| < 1/2$, we obtain

$$f_X(u) = \frac{1}{2\pi}\left(|u|^{-2d_Y} + 1\right) = \frac{1}{2\pi}|u|^{-2d_Y}\left(1 + |u|^{2d_Y}\right), \quad d_Y > 0,$$
$$= \frac{1}{2\pi}\left(1 + |u|^{-2d_Y}\right), \quad d_Y < 0.$$

Therefore, X_j has memory parameter $d_X = \max(d_Y, d_Z) = \max(d_Y, 0)$. If Y_j dominates, i.e., $d_Y > 0$, then d_Y can be estimated by the local Whittle estimate $\hat{d}_X$. Notice that the smoothness parameter $\beta = 2d_Y$ of f_X is smaller than that of f_Y. This in turn will mean that the estimator $\hat{d}_X$ will have a larger bias than $\hat{d}_Y$; see (8.5.22).

8.6.1 *Estimation in signal+noise*

In this subsection we extend the domain of application of the LW estimation method to the signal plus noise model (8.6.1) where the signal process Y_j may be correlated with the noise process Z_j. The noise process can be stationary short or long memory or stochastically negligible.

First we relax the assumption of the stationarity of observations in Theorem 8.5.1, to make it applicable for a signal plus noise process. Here r.v.'s X_j cannot be treated as a stationary process, the spectral density of X_j may not exist and the memory parameter d_X may not be defined. However, we can still compute the LW estimators $\widehat{c}$ and $\widehat{d}_X$ based on observations $X_1, \cdots, X_n$.

To proceed, we need the following result. For a sample $(X_1, \cdots, X_n)$, and real numbers $c_0 > 0$ and $|d_0| < 1/2$, define

$$\eta_j^* := c_0^{-1} u_j^{2d_0} I_X(u_j), \quad i = 1, \cdots, m.$$

Theorem 8.6.1 below says that the LW estimators $\widehat{d}$, $\widehat{c}$ will be consistent for those values of the parameters d_0 and c_0 for which the standardized periodograms η_j^* satisfy properties (8.6.2) and (8.6.4).

By the WLLN property if m satisfies (8.5.5), then

$$m^{-1} \sum_{j=1}^m \eta_j^* \to_p 1. \tag{8.6.2}$$

In some cases it may be difficult to show that

$$E\eta_j^* \leq C, \quad 1 \leq j \leq m. \tag{8.6.3}$$

This restriction can be replaced by a slightly weaker assumption. We assume that the η_j^*'s can be bounded by a linear combination of some nonnegative r.v.'s $\eta_{j,1}$, $\eta_{j,2}$ and s_n such that the following holds:

$$\eta_j^* \leq C(\eta_{j,1} + s_n \eta_{j,2}), \quad 1 \leq j \leq m, \tag{8.6.4}$$
$$E\eta_{j,1} \leq C, \qquad E\eta_{j,2} \leq C, \quad s_n = O_p(1).$$

Note that here the s_n's do not depend on j and C nor on n and j. Condition (8.6.4) is applicable in linear and non-linear regression models.

In the following theorem, the X_j's are not assumed to have a particular spectral density. Therefore, expansion (8.5.17) of Theorem 8.5.1 is replaced by its large sample analog. The proof of this theorem is similar to that of Theorem 8.5.1. We note that Lemma 8.5.1 remains valid if (8.5.25) is replaced by requiring the $y_{m,j}$'s to satisfy the weaker assumption (8.6.4). In view of this fact, the only difference in the proof is that in (8.5.39) one replaces s_n by its limit value 1.

Theorem 8.6.1. *Let $X_1, X_2, \cdots$, be a sequence of r.v.'s. Suppose that for some $c_0 > 0$ and $|d_0| < 1/2$, this sequence has the property of WLLN (8.6.2) and satisfies either (8.6.3) or (8.6.4). Then*

$$\widehat{d}_X \to_p d_0, \tag{8.6.5}$$

$$\widehat{d}_X - d_0 = -\frac{1}{2}\tilde{T}_m(1 + o_p(1)), \quad \tilde{T}_m = m^{-1}\sum_{j=1}^{m} \nu_j \frac{I_X(u_j)}{c_0 u_j^{-2d_0}}.$$

In addition, if $\widehat{d}_X - d_0 = o_p(1/\log n)$, then

$$\widehat{c} \to_p c_0. \tag{8.6.6}$$

The next theorem shows that for a wide class of signal plus noise processes, where the X_j's are as in (8.6.1) and where we are interested in estimating the memory parameter d_Y of the Y process, the LW estimator $\widehat{d}_X$ can estimate d_Y with the rate n^ϵ, $\epsilon > 0$.

Theorem 8.6.2. *Let $\{Y_j\}$ and $\{Z_j\}$ be covariance stationary processes with respective spectral densities f_Y and f_Z satisfying*

$$f_Y(u) = c_Y|u|^{-2d_Y} + o(|u|^{-2d_Y}), \quad f_Z(u) \le c_Z|u|^{-2d_Z}, \quad \text{as } u \to 0,$$

with $-1/2 < d_Y, d_Z < 1/2$. Also, suppose $\{Y_j\}$ satisfies Assumptions A and B.

(i) Suppose, additionally, $X_j = Y_j + Z_j$, $j \in \mathbb{Z}$, $d_Y > d_Z$. Then

$$\widehat{d}_X \to_p d_Y, \tag{8.6.7}$$

$$\widehat{d}_X - d_Y = (\widehat{d}_Y - d_Y)(1 + o_p(1)) + O_p(s_n), \tag{8.6.8}$$

where $s_n := (m/n)^{d_Y - d_Z}$.

(ii) If additionally, $X_j = Y_j + r_n Z_j$, $1 \le j \le n$, $n \ge 1$, where the r.v.'s r_n are such that

$$s_n := r_n(\frac{m}{n})^{d_Y - d_Z}, \qquad d_Y > d_Z, \tag{8.6.9}$$

$$:= r_n(\frac{1}{n})^{d_Y - d_Z}, \qquad d_Y \le d_Z,$$

satisfies

$$s_n \to_p 0, \tag{8.6.10}$$

then (8.6.7) and (8.6.8) hold with s_n of (8.6.9).

Remark 8.6.1. Case (i) of Theorem 8.6.2 shows that even if we observe Y_j with noise, it is possible to estimate its memory parameter d_Y consistently by $\widehat{d}_X$ as long as assumptions A and B hold for the $\{Y_j\}$ process and the memory parameter of the noise process is dominated by that of the signal process. For example, this result is applicable to a Gaussian or a linear process $\{Y_j\}$ for which Assumption A *a priori* implies the WLLN of Assumption B. Theorem 8.6.2 does not put any restrictions on the dependence between the signal Y_j and the noise Z_j, or the distribution of Z_j, except that $d_Y > d_Z$.

Case (ii) of the above theorem deals with negligible noise in residuals. As illustrated in the example below, it can be used in regression models with random regressors to show that the LW estimator of the memory parameter of the error process based on residuals is $\log n$ consistent.

Example 8.6.1. Consider the model

$$X_j = \beta Z_j + \varepsilon_j, \quad j \in \mathbb{Z}, \quad \beta \in \mathbb{R},$$

where $\{\varepsilon_j\}$ and $\{Z_j\}$ are two linear processes with memory parameters $0 < d_Z$, $d_\varepsilon < 1/2$. To estimate d_ε, use the LW estimator $\widehat{d}_{\hat{\varepsilon}}$ based on the residuals

$$\hat{\varepsilon}_j = X_j - \hat{\beta} Z_j = \varepsilon_j + (\hat{\beta} - \beta) Z_j, \tag{8.6.11}$$

where $\hat{\beta}$ is the ordinary least-squares (OLS) estimator of β. Write $\hat{\varepsilon}_j = Y_j + r_n Z_j$, where $Y_j = \varepsilon_j$ and $r_n = \hat{\beta} - \beta$. By Theorem 8.6.2(ii), $\widehat{d}_{\hat{\varepsilon}} - d_\varepsilon = O_p(\widehat{d}_{\hat{\varepsilon}} - d_\varepsilon) + O_p(s_n)$ where $s_n \le r_n n^{(d_Z - d_\varepsilon) \vee 0}$. Therefore, consistency of $\widehat{d}_{\hat{\varepsilon}}$ requires negligibility of s_n.

Case (1) If $EZ_j = 0$, then $\hat{\beta}$ has the property $\hat{\beta} - \beta = O_p(n^{d_Z + d_\varepsilon - 1})$; see Section 11.5. Therefore,

$$s_n = O_p(n^{d_\varepsilon + d_Z - 1}) = o_p\left(\frac{1}{\log n}\right), \quad d_Z < d_\varepsilon,$$

$$= O_p(n^{2d_Z - 1}) = o_p\left(\frac{1}{\log n}\right), \quad d_Z > d_\varepsilon.$$

Case (2) If $EZ_j \ne 0$, then $\hat{\beta} - \beta = O_p(n^{d_\varepsilon - 1/2})$; see Section 11.5, and

$$s_n = O_p(n^{d_\varepsilon - 1/2}) = o_p\left(\frac{1}{\log n}\right), \quad d_Z < d_\varepsilon,$$

$$= O_p(n^{d_Z - 1/2}) = o_p\left(\frac{1}{\log n}\right), \quad d_Z > d_\varepsilon.$$

Consequently, $\widehat{d}_\varepsilon - d_\varepsilon = o_p(1/\log n)$.

The next theorem deals with stochastically negligible trends.

Theorem 8.6.3. *Suppose that a stationary process $\{Y_j\}$ satisfies Assumption B and has spectral density f_Y,*

$$f_Y(u) = c_Y |u|^{-2d_Y} + o(|u|^{-2d_Y}), \quad u \to 0, \quad |d_Y| < 1/2.$$

Let $r_{n,j}, j = 1, \cdots, n$, be arrays of r.v.'s such that

$$\Delta_n := \sum_{j=1}^{n-1} |r_{n,j} - r_{n,j+1}| = O_p(n^{d_Y - 1/2}). \tag{8.6.12}$$

Consider $X_j = Y_j + r_{n,j}, j = 1, \cdots, n$. Then

$$\widehat{d}_X - d_Y = (\widehat{d}_Y - d_Y)(1 + o_p(1)) + o_p(m^{-1/2}) = o_p(1), \tag{8.6.13}$$

$$\widehat{d}_X \to_p d_Y.$$

In addition, if $\widehat{d}_Y - d_Y = o_p(1/\log n)$, then

$$\widehat{c} \to_p c_Y. \tag{8.6.14}$$

As an application of this theorem, consider the case where $r_{n,j} = j^\alpha$ with $\alpha \le d_Y - 1/2$, then $\Delta_n = n^\alpha$, and (8.6.12) holds true. This theorem allows us to estimate the memory parameter of errors in regression models with deterministic design.

Example 8.6.2. Consider the linear regression

$$X_j = \beta j + \varepsilon_j, \qquad j \in \mathbb{Z},$$

with moving-average errors $\{\varepsilon_j\}$ having memory parameter $0 < d_\varepsilon < 1/2$. Here, the OLS estimator $\hat\beta$ of β has the property $\hat\beta - \beta = O_p(n^{d_\varepsilon - 3/2})$; see Section 9.2 below. To apply Theorem 8.6.3, write the residuals as $\hat\varepsilon_j = X_j - \hat\beta j = \varepsilon_j + r_{n,j}$, $r_{n,j} = (\hat\beta - \beta)j$. Then $\Delta_n = |\hat\beta - \beta|n = O_p(n^{d_\varepsilon - 1/2})$, (8.6.12) holds, and the disturbances $r_{n,j}$'s have negligible impact on the quality of the LW estimation of d_ε.

Similarly, in the non-linear regression model

$$X_j = \beta \mu(j/n) + \varepsilon_j, \qquad j \in \mathbb{Z},$$

with moving-average errors $\{\varepsilon_j\}$ and sufficiently smooth μ, the slope parameter β can be estimated consistently with the rate $\hat\beta - \beta = O_p(n^{d_\varepsilon - 1/2})$; see Section 11.2 below. Write

$$\hat\varepsilon_j = X_j - \hat\beta\mu(j/n) = \varepsilon_j + r_{n,j}, \qquad r_{n,j} = (\beta - \hat\beta)\mu(j/n).$$

Assume, that $\mu(\cdot)$ has a bounded derivative on $[0, 1]$ so that $|\mu(j/n) - \mu((j+1)/n)| \le C/n$. Then

$$\Delta_n = |\hat{\beta} - \beta| \sum_{j=1}^{n-1} |\mu(\frac{j}{n}) - \mu(\frac{j+1}{n})| = O_p(n^{d_\varepsilon - 1/2}),$$

and therefore (8.6.12) holds true.

In both cases above, Theorem 8.6.3 implies that

$$\hat{d}_{\hat{\varepsilon}} - d_\varepsilon = (\hat{d}_\varepsilon - d_\varepsilon)(1 + o_p(1)) + o_p(m^{-1/2}),$$

so that the estimator $\hat{d}_{\hat{\varepsilon}}$ of the memory parameter d_ε of errors ε_j, $j \in \mathbb{Z}$, based on residuals $\hat{\varepsilon}_j$ has the same asymptotic magnitude and distribution as the estimator $\hat{d}_\varepsilon$ based on ε_j's.

Proof of Theorem 8.6.2. We prove the theorem for case (ii), since (i) follows from (ii) setting $r_n = 1$. Let

$$w_{X,j} := (2\pi n)^{-1/2} \sum_{t=1}^{n} X_t e^{-itu_j}, \qquad I_{X,j} = |w_{X,j}|^2,$$

$$\tilde{\eta}_{Y,j} := (1/c_Y) u_j^{2d_Y} I_{Y,j}, \qquad \tilde{\eta}_{Z,j} = (1/c_Z) u_j^{2d_Z} I_{Z,j},$$

$$\eta_j^* := (1/c_Y) u_j^{2d_Y} I_{X,j}, \qquad j = 1, \cdots, m.$$

We shall show that,

$$m^{-1} \sum_{j=1}^{m} \eta_j^* \to_p 1, \tag{8.6.15}$$

$$\eta_j^* \le C(\eta_{j,1} + s_n \eta_{j,2}), \quad 1 \le j \le m, \tag{8.6.16}$$

with s_n as in (8.6.9), $\eta_{j,1} = \tilde{\eta}_{Y,j}$, and $\eta_{j,2} = \tilde{\eta}_{Z,j}$. Under the assumptions for f_X and f_Y in this theorem, by Theorem 5.2.1(v),

$$E\tilde{\eta}_{Y,j} \le C, \quad E\tilde{\eta}_{Z,j} \le C, \quad j = 1, \cdots, m, \tag{8.6.17}$$

whereas $s_n = o_p(1)$, by (8.6.10). Hence, the r.v.'s η_j^*, with parameters c_Y and d_Y, satisfy conditions (8.6.2) and (8.6.4) of Theorem 8.6.1, which implies the consistency claim $\hat{d}_X \to_p d_Y$ of (8.6.7). The proof of the second claim (8.6.8) will be given later.

Now we show (8.6.15) and (8.6.16). Write

$$\frac{1}{m} \sum_{j=1}^{m} \eta_j^* = \frac{1}{m} \sum_{j=1}^{m} \frac{I_{Y,j}}{c_Y u_j^{-2d_Y}} + \frac{1}{m} \sum_{j=1}^{m} \frac{I_{X,j} - I_{Y,j}}{c_Y u_j^{-2d_Y}} \tag{8.6.18}$$

$$:= q_{n,1} + q_{n,2},$$

where $q_{n,1} \to_p 1$, since by assumption the Y_j's satisfy Assumption B; see also (8.5.32). To prove (8.6.15), it remains to show that $q_{n,2} \to_p 0$. We have

$$w_{X,j} = w_{Y,j} + r_n w_{Z,j}, \qquad (8.6.19)$$

$$|I_{X,j} - I_{Y,j}| \le r_n^2 I_{Z,j} + 2|r_n| |w_{Y,j}| |w_{Z,j}|$$

$$\le 2r_n^2 I_{Z,j} + I_{Y,j},$$

estimating $2|r_n| |w_{Y,j}| |w_{Z,j}| \le |w_{Y,j}|^2 + r_n^2 |w_{Z,j}|^2$. Thus

$$|q_{n,2}| \le C\Big(r_n^2 \frac{1}{m} \sum_{j=1}^{m} u_j^{2d_Y} I_{Z,j} + |r_n| \frac{1}{m} \sum_{j=1}^{m} u_j^{2d_Y} |w_{Y,j}| |w_{Z,j}|\Big).$$

Using s_n as in (8.6.9), we can bound

$$r_n^2 u_j^{2d_Y} I_{Z,j} \le C s_n^2 u_j^{2d_Z} I_{Z,j} = C s_n^2 \tilde{\eta}_{Z,j}, \qquad (8.6.20)$$

$$|r_n| u_j^{2d_Y} |w_{Y,j}| |w_{Z,j}| \le C|s_n| (u_j^{d_Y} |w_{Y,j}|) u_j^{d_Z} |w_{Z,j}|$$

$$\le C|s_n| (\tilde{\eta}_{Y,j} \tilde{\eta}_{Z,j})^{1/2}.$$

Hence,

$$u_j^{2d_Y} |I_{X,j} - I_{Y,j}| \le C\big(s_n^2 \tilde{\eta}_{Z,j} + |s_n| (\tilde{\eta}_{Y,j} \tilde{\eta}_{Z,j})^{1/2}\big), \qquad (8.6.21)$$

$$|q_{n,2}| \le C\big(s_n^2 \frac{1}{m} \sum_{j=1}^{m} \tilde{\eta}_{Z,j} + |s_n| \frac{1}{m} \sum_{j=1}^{m} (\tilde{\eta}_{Y,j} \tilde{\eta}_{Z,j})^{1/2}\big).$$

By (8.6.17),

$$E\Big(\frac{1}{m} \sum_{j=1}^{m} \tilde{\eta}_{Z,j}\Big) \le C, \quad E\Big(\frac{1}{m} \sum_{j=1}^{m} (\tilde{\eta}_{Y,j} \tilde{\eta}_{Z,j})^{1/2}\Big) \le C,$$

which implies $|q_{n,2}| = O_p(s_n) = o_p(1)$, since $s_n = o_p(1)$ by assumption (8.6.10).

To obtain (8.6.16), use (8.6.19) and (8.6.20),

$$\eta_j^* \le C\big(u_j^{2d_Y} I_{Y,j} + r_n^2 u_j^{2d_Y} I_{Z,j}\big) \le C(\tilde{\eta}_{Y,j} + s_n^2 \tilde{\eta}_{Z,j}).$$

It remains to establish the expansion (8.6.8). Since the η_j^*'s satisfy the assumptions of Theorem 8.6.1 with $d_0 = d_Y$ and $c_0 = c_Y$, by (8.6.5),

$$\widehat{d}_X - d_Y = -\frac{1}{2}\Big(m^{-1} \sum_{j=1}^{m} \nu_j \frac{I_{X,j}}{c_Y u_j^{-2d_Y}}\Big)(1 + o_p(1)).$$

Moreover, because the Y_j's satisfy Assumptions A and B, by Theorem 8.5.1 and Remark 8.5.1, $\widehat{d}_Y \to_p d_Y$ and

$$\widehat{d}_Y - d_Y = -\frac{1}{2}\Big(m^{-1}\sum_{j=1}^{m} \nu_j \frac{I_{Y,j}}{c_Y u_j^{-2d_Y}}\Big)(1 + o_p(1)).$$

Hence,

$$\widehat{d}_X - d_Y = \big(\widehat{d}_Y - d_Y + Q_n\big)(1 + o_p(1)), \qquad (8.6.22)$$

$$Q_n = m^{-1}\sum_{j=1}^{m} \nu_j u_j^{2d_Y}(I_{X,j} - I_{Y,j}).$$

Using (8.6.21), and

$$|\nu_j| \le C(|\log(j/m)| + 1), \quad j = 1, \cdots, m, \qquad (8.6.23)$$

we obtain

$$|Q_n| \le O_p(1)m^{-1}\sum_{j=1}^{m}(|\log(\tfrac{j}{m})| + 1)\{s_n^2 \tilde{\eta}_{Z,j} + s_n(\tilde{\eta}_{Y,j}\tilde{\eta}_{Z,j})^{1/2}\}.$$

But, by (8.6.17),

$$E\Big(\frac{1}{m}\sum_{j=1}^{m}(|\log(\tfrac{j}{m})| + 1)\tilde{\eta}_{Z,j}\Big) \le C,$$

$$E\Big(\frac{1}{m}\sum_{j=1}^{m}|\log(\tfrac{j}{m})| + 1)(\tilde{\eta}_{Y,j}\tilde{\eta}_{Z,j})^{1/2}\Big) \le C.$$

This in turn implies $|Q_n| = O_p(s_n)$, which proves (8.6.8), thereby yielding $\widehat{d}_X \to_p d_Y$ and completing the proof of Theorem 8.6.2.

Proof of Theorem 8.6.3. This proof is similar to that of Theorem 8.6.2. First, we verify (8.6.15) and (8.6.16). These two equations, as seen in the proof of Theorem 8.6.2, yield the expansion (8.6.22) and that $\widehat{d}_X \to_p d_Y$.

Proof of (8.6.15). Because of (8.6.18), it suffices to prove $q_{n,2} = o_p(1)$. Let

$$w_{r,j} := (2\pi n)^{-1/2}\sum_{k=1}^{n} r_{n,k}e^{-iku_j}, \qquad I_{r,j} = |w_{r,j}|^2.$$

Then, as for (8.6.19),

$$w_{X,j} = w_{Y,j} + w_{r,j}, \quad |I_{X,j} - I_{Y,j}| \le I_{r,j} + 2|w_{Y,j}|\,|w_{r,j}|.$$

We shall need the bound

$$|w_{r,j}| \le C\Delta_n j^{-1}\sqrt{n}, \qquad 1 \le j \le m. \qquad (8.6.24)$$

To derive it, note that by (2.1.5) and (2.1.7), for $0 < u_j < \pi$,

$$\sum_{l=1}^{n} e^{ilu_j} = 0, \quad \left| \sum_{l=1}^{k} e^{ilu_j} \right| \leq \pi u_j^{-1} \leq nj^{-1}, \quad k = 1, \cdots, n-1,$$

and by summation by parts (2.5.8),

$$|w_{r,j}| = \frac{1}{\sqrt{2\pi n}} \left| \sum_{k=1}^{n-1} (r_{n,k} - r_{n,k+1}) \sum_{l=1}^{k} e^{-ilu_j} + r_{n,n} \sum_{l=1}^{n} e^{-ilu_j} \right|$$

$$\leq C n^{1/2} j^{-1} \sum_{k=1}^{n-1} |r_{n,k} - r_{n,k+1}| = C \Delta_n j^{-1} \sqrt{n}.$$

Recall that by (8.6.12), $\Delta_n = O_p(n^{d_Y - 1/2})$. Therefore, by (8.6.24) and $|c_Y^{-1/2} u_j^{d_Y} w_{Y,j}| = (\tilde{\eta}_{Y,j})^{1/2}$,

$$u_j^{d_Y} |w_{r,j}| \leq C(j/n)^{d_Y} \Delta_n j^{-1} \sqrt{n} = O_p(1) j^{d_Y - 1}, \quad (8.6.25)$$

$$u_j^{2d_Y} |I_{X,j} - I_{Y,j}| \leq O_p(1)(j^{2d_Y - 2} + j^{d_Y - 1} (\tilde{\eta}_{Y,j})^{1/2})$$

$$= O_p(1)(1 + \tilde{\eta}_{Y,j}).$$

This, together with (8.6.18), implies

$$|q_{n,2}| \leq \frac{C}{m} \sum_{j=1}^{m} u_j^{2d_Y} |I_{X,j} - I_{Y,j}|$$

$$= O_p(1) m^{-1} \sum_{j=1}^{m} \left(j^{2d_Y - 2} + j^{d_Y - 1} (\tilde{\eta}_{Y,j})^{1/2} \right)$$

$$= O_p(1)(m^{-1} + T_n), \quad T_n := m^{-1} \sum_{j=1}^{m} j^{d_Y - 1} (\tilde{\eta}_{Y,j})^{1/2}.$$

Since by (8.6.17) $E\tilde{\eta}_{Y,j} \leq C$, for all $1 \leq j \leq m$, then $ET_n \leq Cm^{-1} \sum_{j=1}^{m} j^{d_Y - 1} = O(m^{-1/2}/\log^2 m)$ and $T_n = O_p(m^{-1/2}/\log^2 m)$, because $d_Y < 1/2$. Therefore,

$$q_{n,2} = O_p(m^{-1/2}/\log^2 m) = o_p(m^{-1/2}), \quad (8.6.26)$$

which completes the proof of (8.6.15).

Proof of (8.6.16). By (8.6.25),

$$\eta_j^* \leq C(u_j^{2d_Y} I_{Y,j} + u_j^{2d_Y} |I_{X,j} - I_{Y,j}|) \leq C\tilde{\eta}_{Y,j} + O_p(1)(1 + \tilde{\eta}_{Y,j}),$$

which yields (8.6.16) with $\eta_{j,1} = \eta_{j,2} = 1 + \tilde{\eta}_{Y,j}$, $s_n = O_p(1)$.

Proof of (8.6.13). The consistency $\widehat{d}_X \to_p d_Y$ was shown above. As in the proof of Theorem 8.6.2, the first claim in (8.6.13) follows from (8.6.22), if we show that $Q_n = o_p(m^{-1/2})$. By (8.6.23) and (8.6.26),

$$|Q_n| \le C(\log m)m^{-1}\sum_{j=1}^{m} u_j^{2d_Y}|I_{X,j} - I_{Y,j}|$$
$$= O_p(m^{-1/2}/\log m) = o_p(m^{-1/2}).$$

Proof of (8.6.14). By $\widehat{d}_Y - d_Y = o_p(1/\log n)$ and (8.6.13), $\widehat{d}_X - d_Y = o_p(1/\log n)$. Since (8.6.15) holds, then (8.6.14) follows by the same argument as in the proof of (8.5.18). This completes the proof of theorem.

8.6.2 Application to a stochastic volatility model

In this section we shall discuss the estimation of the memory parameter of some stochastic volatility models. We show that these processes can be decomposed into signal plus noise processes, so that the results of Section 8.6.1 apply.

By a stochastic volatility model we usually understand a stationary process $r_j, j \in \mathbb{Z}$, of the form

$$r_j = \varepsilon_j \sigma_j, \quad j \in \mathbb{Z}, \quad \varepsilon_j \sim \mathrm{IID}(0,1),$$

where the (volatility) process $\sigma_j > 0$ is a function of the past information up to time $j - 1$. Let $\mathcal{F}_{j-1}$ be the σ-field generated by past information $r_s, s \le j - 1$. If $\varepsilon_s, s \ge j$ are independent of $\mathcal{F}_{j-1}$, then under suitable moment assumptions on ε_j and σ_j,

$$E[r_j|\mathcal{F}_{j-1}] = 0, \qquad \sigma_j^2 = \mathrm{Var}(r_j|\mathcal{F}_{j-1}),$$

and $\{r_j\}$ is a white-noise process: for any $j > k$,

$$E[r_j r_k] = E\big[r_k E[r_j|\mathcal{F}_{j-1}]\big] = 0.$$

However, the dependence structure of the powers $\{X_j = |r_j|^\delta, \; j \in \mathbb{Z}\}$, is the same as that of $\{|\sigma_j|^\delta\}$, $\delta > 0$. To illustrate this, decompose X_j into

$$X_j = Y_j + Z_j, \tag{8.6.27}$$
$$Y_j := a|\sigma_j|^\delta, \quad Z_j := (|\varepsilon_j|^\delta - a)|\sigma_j|^\delta, \quad a = E|\varepsilon_j|^\delta.$$

Here, $\{Z_j\}$ is a white-noise process and $\{Y_j\}$ is a sequence of dependent r.v.'s. Hence, if $\{Y_j\}$ has long memory, the X_j's also have long memory and $d_X = d_Y$. As was shown in the previous section, d_Y can be estimated by a local Whittle estimate $\widehat{d}_X$ based on observations $X_1, \cdots, X_n$.

It is often assumed that the volatility process $\sigma_j = h(\eta_j)$, $j \in \mathbb{Z}$, is a nonlinear function of a stationary Gaussian or linear process $\{\eta_j\}$. Robinson (2001) showed that a wide class of these types of models with Gaussian $\{\eta_j\}$, allow for long memory in volatility.

EGARCH model. The choice of $h(\eta_j) = \exp(\eta_j)$ is used in the Exponential Generalized ARCH (EGARCH) model proposed by Nelson (1991). A related class of stochastic volatility models with long memory in $\{\sigma_j\}$ was introduced in Breidt, Crato and de Lima (1998) and Harvey (1998):

$$r_j = \sigma_j \varepsilon_j, \qquad \sigma_j = e^{\eta_j}, \qquad \eta_j = a_0 + \sum_{k=1}^{\infty} a_k \zeta_{j-k}, \qquad (8.6.28)$$

where $\{\zeta_k\} \sim \mathrm{IID}(0, \sigma_\zeta^2)$, independent of $\{\varepsilon_k\}$. Under the assumption that the weights a_k are as in the ARFIMA model, e.g., $a_k \sim ck^{d-1}$, $0 < d < 1/2$, and the ζ_k's are Gaussian, Harvey (1998) showed that

$$\gamma_X(k) = \mathrm{Cov}(X_0, X_k) = \mathrm{Cov}(|r_0|^\delta, |r_k|^\delta) = (EX_0)^2(e^{\delta^2 \gamma_\eta(k)} - 1)$$
$$\sim (\delta EX_0)^2 \gamma_\eta(k), \quad k \to \infty.$$

This relation implies that the processes $\{X_j\}$ and $\{Y_j\}$ have the same memory parameters, $d_X = d_\eta$. Surgailis and Viano (2002) obtained a similar result without imposing a Gaussianity assumption on $\{\zeta_k\}$.

The next proposition discusses the consistent estimation of d_η.

Proposition 8.6.1. *Let $\{r_j\}$ be as in (8.6.28). Suppose we observe X_1, $\cdots, X_n$ from the model $X_j = |r_j|^\delta$, $j \in \mathbb{Z}$, $\delta > 0$, where $\{\eta_j\}$ is a stationary standardized Gaussian ARFIMA(p, d_η, q) process, $0 < d_\eta < 1/2$, $d_\eta \neq 1/4$. Then, under (8.5.5) and $E|\varepsilon_0|^{2\delta} < \infty$,*

$$\widehat{d}_X - d_\eta = O_p\big(m^{-1/2} + (m/n)^{d_\eta \wedge (1/2 - d_\eta)}\big). \qquad (8.6.29)$$

Proof. Decompose $X_j = |r_j|^\delta$ into a signal plus noise process as in (8.6.27) with the signal proportional to η_j, and then apply part (i) of Theorem 8.6.2.

First, we verify that the signal $\{\eta_j\}$ satisfies Assumptions A and B of Theorem 8.6.2. In Section 7.2 it was shown that $\{\eta_j\}$ satisfies

$$\gamma_\eta(j) \sim c_\gamma |j|^{-1+2d_\eta}, \qquad f_\eta(v) \sim c_f |v|^{-2d_\eta},$$

and, hence Assumption A, whereas Assumption B holds by Theorem 8.5.2(i). Moreover, (8.5.21) holds with $\beta = 2$ as seen from Remark 8.5.4. By Theorem 8.5.2(iii), we also obtain

$$\widehat{d}_\eta = O_p(m^{-1/2} + (m/n)^2). \tag{8.6.30}$$

This will be used later.

Next, we derive a signal plus noise decomposition for X_j. Let $h(x) = \exp(\delta x)$, $x \in \mathbb{R}$, $\delta > 0$. Since $|\sigma_j|^\delta \equiv h(\eta_j)$ is a square integrable function of a Gaussian r.v. $\{\eta_j\}$, it has Hermite expansion

$$h(\eta_j) - Eh(\eta_j) = J_1 H_1(\eta_j) + \sum_{k=2}^{\infty} \frac{J_k}{k!} H_k(\eta_j) =: J_1\eta_j + \mathcal{R}_j,$$

where $H_1(x) = x$, and by (2.4.12) $J_1 = E\eta_0 h(\eta_0) = E\eta_0 \exp(\delta\eta_0) = \delta E \exp(\delta\eta_0)$. Use this in (8.6.27), to obtain

$$X_j = aJ_1\eta_j + \tilde{Z}_j, \qquad \tilde{Z}_j := a\mathcal{R}_j + Z_j. \tag{8.6.31}$$

Now the "signal" is a Gaussian process $\{\eta_j\}$ and $\{\tilde{Z}_j\}$ is the noise. By Propositions 2.4.1 and 4.6.1,

$$\mathrm{Cov}(\eta_j, \mathcal{R}_k) = 0, \quad j, k \in \mathbb{Z},$$

$$\gamma_\mathcal{R}(j) = (J_2^2/2)\gamma_\eta^2(j)(1 + o(1)), \quad j \to \infty.$$

Hence, $\{\tilde{Z}_j\}$ is uncorrelated with $\{\eta_j\}$, and

$$\gamma_{\tilde{Z}}(j) \sim c\gamma_\eta^2(j) \sim c_1|j|^{-2(1-2d_\eta)}, \quad j \to \infty, \quad c, c_1 > 0. \tag{8.6.32}$$

To apply Theorem 8.6.2(i), we need to show that

$$f_{\tilde{Z}}(u) \leq c|u|^{-2d_{\tilde{Z}}}, \quad u \to 0, \quad \exists d_{\tilde{Z}} < d_\eta. \tag{8.6.33}$$

Case $0 < d_\eta < 1/4$. Here $\gamma_{\tilde{Z}}(j)$ is absolutely summable, $f_{\tilde{Z}}$ is continuous and bounded, and hence (8.6.33) holds with $d_{\tilde{Z}} = 0 < d_\eta$.

Case $1/4 < d_\eta < 1/2$. Because $d_{\tilde{Z}}$ satisfies $\gamma_{\tilde{Z}}(j) \sim c|j|^{-1+2d_{\tilde{Z}}}$ and by (8.6.32), $1 - 2d_{\tilde{Z}} = 2(1 - 2d_\eta)$, and hence $0 < d_{\tilde{Z}} = 2d_\eta - (1/2) < d_\eta$. Thus, $\{\tilde{Z}_j\}$ exhibits the time-domain long memory property with parameter $d_{\tilde{Z}} > 0$. We can also verify that $f_{\tilde{Z}}(v) \sim c|v|^{-2d_{\tilde{Z}}}$, establishing (8.6.33).

Therefore, the signal and the noise in decomposition (8.6.31) satisfy the assumptions of Theorem 8.6.2(i). Hence, $d_\eta - d_{\tilde{Z}} = d_\eta \wedge (1/2 - d_\eta)$ together with (8.6.30), gives

$$\widehat{d}_X - d_\eta = O_p(\widehat{d}_\eta - d_\eta)(1 + o_p(1)) + O_p((m/n)^{d_\eta \wedge (1/2-d_\eta)})$$
$$= O_p(m^{-1/2} + (m/n)^{d_\eta \wedge (1/2-d_\eta)}).$$

This completes the proof of the proposition.

Remark 8.6.2. Proposition 8.6.1 shows that the LW estimator allows us to estimate the long memory parameter of the powers $X_j = |r_j|^\delta$, $r_j = \varepsilon_j \sigma_j = \varepsilon_j \exp(\eta_j)$, of an EGARCH model.

An alternative way of decomposing the EGARCH model $X_j = |r_j|^\delta$ into signal plus noise is to apply a logarithmic transformation, i.e.,

$$U_j := \log X_j = Y_j + Z_j, \quad Y_j := \delta\eta_j, \quad Z_j := \delta \log |\varepsilon_j|. \qquad (8.6.34)$$

These Z_j's are i.i.d. shocks, and thus the memory parameter d_η can be estimated by the local Whittle estimator $\widehat{d}_U$. For a linear process $\{\eta_j\}$, the asymptotic properties of $\widehat{d}_U$ were analyzed by Arteche (2004) and Hurvich, Moulines and Soulier (2005).

If $\{\eta_j\}$ is Gaussian, then the processes $\{\log X_j\}$, $\{\eta_j\}$, and $\{X_j\}$ have the same long memory parameter d_η, which can be consistently estimated by the local Whittle estimator based on X_j or $\log X_j$.

8.7   Monte Carlo experiment

The behavior of LW estimator $\widehat{d}_X$ for finite samples can be analyzed by a Monte Carlo experiment. The following experiment used $10,000$ replications with sample sizes $n = 1024$ and 2048, and bandwidth parameters $m = [n^{0.5}]$, $[n^{0.6}]$, $[n^{0.7}]$, $[n^{0.8}]$, and $\widetilde{m}_{opt}$; see (8.5.49).

Tables 8.7.1 and 8.7.2 show the bias and, in parenthesis, the MSE of $\widehat{d}_X$ when $\{X_j\}$ is a Gaussian ARFIMA$(0, d, 0)$ process with memory parameter $d = -0.4, -0.2, 0, 0.2, 0.4$. The data was generated using a Davies and Harte (1987) algorithm. The estimator $\widehat{d}_X$ seems to have negative bias when the process has short or long memory, whereas for negative memory the bias tends to be positive. For each fixed choice of m and n, the MSE appears to be robust w.r.t. the values of d. It is interesting to note that among the chosen four bandwidths, $m = [n^{0.8}]$ minimizes the finite sample MSE. Recall that the large sample optimal bandwidth m_{opt} of (8.5.48) is of this order.

Simulations based on ARFIMA(p, d, q) models usually show that the bias of the LW estimator significantly increases when the roots of the corresponding ARMA(p, q) polynomials approach the unit circle. Therefore, in ARFIMA(p, d, q) models the quality of $\widehat{d}_X$ may not be as good as for an ARFIMA$(0, d, 0)$ model. The quality of the confidence intervals can be improved using the optimal bandwidth selection rule. To illustrate, consider a stationary Gaussian ARFIMA$(1, d, 0)$ process $\{X_j\}$ with memory parameter $d_X = -0.3, 0, 0.3$ and AR(1) parameter $r = 0, 0.5, 0.8$. Let $t(m) = 2\sqrt{m}(\widehat{d}_X - d_X)$ and $t^*(m) = 2\left(\sum_{j=1}^n \nu_j^2\right)^{1/2}(\widehat{d}_X - d_X)$ with the ν_j's as in (8.5.47). Recall that $t^*(m)$ allows us to obtain a better normal approximation for small samples. By (8.5.47), $t(m) \sim \mathcal{N}(0, 1)$ and $t^*(m) \sim \mathcal{N}(0, 1)$. Tables 8.7.3 to 8.7.5 show the coverage probabilities of 95% confidence intervals for d_X based on studentizations $t(\widehat{m}^*)$ and $t^*(\widehat{m}^*)$ and $10,000$ replications, setting $\widehat{m}^* = \min\{\widetilde{m}_{opt}, 0.8n^{0.8}\}$ where $\widetilde{m}_{opt}$ is as in (8.5.49). The spectral density of the ARFIMA model $\{X_j\}$ satisfies (8.5.21) with $\beta = 2$. Applying the CLT of Remark 8.5.2 requires $m = o(n^{2\beta/(1+2\beta)}) = o(n^{0.8})$. For this reason we bound $\widehat{m}^*$ from above by $0.8n^{0.8}$. Tables 8.7.3 and 8.7.4 show that for $d_X \geq 0$, the coverage probabilities are close to the nominal (0.95); however, for larger r smaller $\widehat{m}^*$ are selected, which results in a longer confidence interval. Table 8.7.5 shows that using the fixed bandwidth $m = 0.8n^{0.8} = 115$, $n = 500$, the coverage probabilities deteriorate very quickly as r increases.

Tables 8.7.6 and 8.7.7 give the bias and MSE of $\widehat{d}_X$ for the signal plus noise process $X_j = Y_j + Z_j$, where $\{Y_j\}$ and $\{Z_j\}$ are Gaussian ARFIMA$(0, d_Y, 0)$ and ARFIMA$(0, d_Z, 0)$ processes, respectively. The two memory parameters are chosen to satisfy $d_Y > d_Z$. In particular, we use the following choices $d_Y = 0, 0.2, 0.4$ and $d_Z = -0.4, -0.2, 0, 0.2$. Additionally, $\{Y_j\}$ and $\{Z_j\}$ are independent and have unit variances. The data in these tables show, as the theory predicts, that the noise significantly increases the bias of the estimator. Bias tends to decrease when the difference $d_Y - d_Z$ increases, and it remains negative when the signal and the noise are independent. For a fixed n and m, the MSE varies with d_X and the bandwidth minimizing MSE depends on d_Y and d_Z. Overall, it appears that the bandwidth parameter $m = [n^{0.6}]$ results in the lowest MSE.

Both theory and simulations suggest that the LW estimator also remains consistent for non-linear time series, e.g., signal plus noise processes. However, the presence of noise worsens the behavior of the estimator for finite samples and a larger sample size is needed to achieve a desirable accuracy. For practical applications the simulation results suggest the use of $m = [n^{0.6}]$ for signal plus noise and other non-linear models. Although the choice of the optimal bandwidth parameter remains an open problem, the results of Tables 8.7.1 to 8.7.7 indicate that the $\widehat{d}_X$ computed using the "optimal" bandwidth $\widetilde{m}_{opt}$ of Subsection 8.5 tends to outperform the fixed bandwidth choice.

d	$m = [n^{0.5}]$	$[n^{0.6}]$	$[n^{0.7}]$	$[n^{0.8}]$	$\widetilde{m}_{opt}$
			$n = 1024$		
−0.4	0.021	0.011	0.008	0.013	0.008
	(0.010)	(0.005)	(0.002)	(0.001)	(0.002)
−0.2	−0.006	−0.003	−0.001	0.004	0.002
	(0.012)	(0.005)	(0.002)	(0.001)	(0.002)
0	−0.010	−0.006	−0.004	−0.002	−0.002
	(0.013)	(0.005)	(0.002)	(0.001)	(0.001)
0.2	−0.008	−0.005	−0.004	−0.007	−0.004
	(0.012)	(0.005)	(0.002)	(0.001)	(0.003)
0.4	−0.012	−0.003	−0.002	−0.009	−0.003
	(0.010)	(0.005)	(0.002)	(0.001)	(0.005)

Table 8.7.1: Bias and MSE of $\hat{d}$ for X ARFIMA$(0, d, 0)$.

d	$m = [n^{0.5}]$	$[n^{0.6}]$	$[n^{0.7}]$	$[n^{0.8}]$	$\widetilde{m}_{opt}$
			$n = 2048$		
−0.4	0.016	0.009	0.007	0.011	0.006
	(0.007)	(0.003)	(0.002)	(0.001)	(0.001)
−0.2	−0.004	−0.002	0.000	0.004	0.002
	(0.008)	(0.003)	(0.001)	(0.001)	(0.001)
0	−0.007	−0.004	−0.002	−0.001	−0.001
	(0.008)	(0.003)	(0.001)	(0.001)	(0.001)
0.2	−0.006	−0.003	−0.002	−0.005	−0.002
	(0.008)	(0.003)	(0.001)	(0.001)	(0.001)
0.4	−0.005	0.001	0.001	−0.006	0.001
	(0.007)	(0.003)	(0.001)	(0.001)	(0.003)

Table 8.7.2: Bias and MSE of $\hat{d}$ for X ARFIMA$(0, d, 0)$.

				$n = 500$		
d	r	CP $t(\widehat{m}^*)$	CP $t^*(\widehat{m}^*)$	AL $t(\widehat{m}^*)$	AL $t^*(\widehat{m}^*)$	$\widehat{m}^*$
-0.3	0	91.51	93.91	0.20	0.22	97.64
-0.3	0.5	37.77	43.81	0.23	0.26	72.02
-0.3	0.8	27.53	37.37	0.35	0.41	32.06
0	0	92.01	94.06	0.18	0.20	114.73
0	0.5	86.49	92.24	0.33	0.39	34.88
0	0.8	80.64	92.07	0.53	0.71	13.88
0.3	0	89.20	93.69	0.29	0.33	46.36
0.3	0.5	91.14	95.71	0.52	0.69	14.63
0.3	0.8	88.60	95.46	0.69	1.05	8

Table 8.7.3: 95% coverage probabilities (CP) and the average lengths (AL) of confidence intervals (CI) based on $t(\widehat{m}^*)$ and $t^*(\widehat{m}^*)$, and average bandwidth $\widehat{m}^* = \min\{\widetilde{m}_{opt}, [0.8n^{0.8}]\}$ for Gaussian ARFIMA$(1, d, 0)$.

				$n = 1000$		
d	r	CP $t(\widehat{m}^*)$	CP $t^*(\widehat{m}^*)$	AL $t(\widehat{m}^*)$	AL $t^*(\widehat{m}^*)$	$\widehat{m}^*$
-0.3	0	92.72	94.21	0.15	0.16	169.71
-0.3	0.5	34.14	38.16	0.18	0.19	124.41
-0.3	0.8	22.29	29.06	0.26	0.30	55.81
0	0	93.18	94.45	0.14	0.15	199.97
0	0.5	89.23	93.08	0.25	0.28	60.56
0	0.8	84.72	92.45	0.40	0.49	24.07
0.3	0	91.52	94.16	0.22	0.24	78.12
0.3	0.5	91.92	95.98	0.40	0.49	24.71
0.3	0.8	92.18	96.39	0.51	0.67	15

Table 8.7.4: 95% coverage probabilities (CP) and the average lengths (AL) of confidence intervals (CI) based on $t(\widehat{m}^*)$ and $t^*(\widehat{m}^*)$, and average bandwidth $\widehat{m}^* = \min\{\widetilde{m}_{opt}, [0.8n^{0.8}]\}$ for Gaussian ARFIMA$(1, d, 0)$.

			$n = 500$		
d	r	CP $t\,(m)$	CP $t^*\,(m)$	AL $t\,(m)$	AL $t^*\,(m)$
-0.3	0	91.35	93.51	0.18	0.2
-0.3	0.5	0.29	0.4	0.18	0.2
-0.3	0.8	0	0	0.18	0.2
0	0	92.01	94.01	0.18	0.2
0	0.5	0.43	0.62	0.18	0.2
0	0.8	0	0	0.18	0.2
0.3	0	91.75	93.77	0.18	0.2
0.3	0.5	0.69	0.95	0.18	0.2
0.3	0.8	0	0	0.18	0.2

Table 8.7.5: 95% coverage probabilities (CP) and the average lengths (AL) of confidence intervals (CI) based on $t\,(m)$ and $t^*\,(m)$, and bandwidth $m = \left[0.8n^{0.8}\right] = 115$ for Gaussian ARFIMA$(1, d, 0)$.

		$n = 1024$				
d_Y	d_Z	$m = [n^{0.5}]$	$[n^{0.6}]$	$[n^{0.7}]$	$[n^{0.8}]$	$\widetilde{m}_{opt}$
0	-0.4	-0.041	-0.053	-0.073	-0.099	-0.089
		(0.014)	(0.008)	(0.008)	(0.011)	(0.009)
0	-0.2	-0.054	-0.057	-0.065	-0.072	-0.069
		(0.015)	(0.009)	(0.007)	(0.006)	(0.006)
0.2	-0.4	-0.028	-0.043	-0.076	-0.111	-0.136
		(0.013)	(0.007)	(0.008)	(0.017)	(0.020)
0.2	-0.2	-0.046	-0.061	-0.086	-0.117	-0.121
		(0.014)	(0.009)	(0.010)	(0.015)	(0.016)
0.2	0	-0.058	-0.062	-0.071	-0.083	-0.077
		(0.016)	(0.009)	(0.007)	(0.008)	(0.008)
0.4	-0.2	-0.046	-0.072	-0.126	-0.195	-0.144
		(0.013)	(0.011)	(0.018)	(0.039)	(0.024)
0.4	0	-0.072	-0.094	-0.130	-0.171	-0.123
		(0.016)	(0.014)	(0.019)	(0.030)	(0.018)
0.4	0.2	-0.076	-0.082	-0.094	-0.110	-0.085
		(0.017)	(0.012)	(0.011)	(0.013)	(0.011)

Table 8.7.6: Bias and MSE of $\hat{d}_X$ for signal+noise process $X = Y + Z$, where Y is Gaussian ARFIMA$(0, d_Y, 0)$, and Z is Gaussian ARFIMA$(0, d_Z, 0)$.

d_Y	d_Z	$m = [n^{0.5}]$	$[n^{0.6}]$	$[n^{0.7}]$	$[n^{0.8}]$	$\widetilde{m}_{opt}$
				$n = 2048$		
0	−0.4	−0.030	−0.042	−0.062	−0.090	−0.080
		(0.009)	(0.005)	(0.005)	(0.009)	(0.007)
0	−0.2	−0.045	−0.051	−0.059	−0.069	−0.067
		(0.010)	(0.006)	(0.005)	(0.005)	(0.005)
0.2	−0.4	−0.018	−0.031	−0.060	−0.111	−0.123
		(0.008)	(0.004)	(0.005)	(0.013)	(0.016)
0.2	−0.2	−0.034	−0.049	−0.073	−0.107	−0.110
		(0.009)	(0.006)	(0.007)	(0.012)	(0.013)
0.2	0	−0.049	−0.056	−0.065	−0.078	−0.071
		(0.010)	(0.006)	(0.006)	(0.007)	(0.006)
0.4	−0.2	−0.028	−0.053	−0.102	−0.174	−0.120
		(0.008)	(0.006)	(0.012)	(0.031)	(0.016)
0.4	0	−0.052	−0.078	−0.113	−0.158	−0.109
		(0.010)	(0.009)	(0.014)	(0.025)	(0.014)
0.4	0.2	−0.065	−0.075	−0.088	−0.104	−0.080
		(0.012)	(0.009)	(0.009)	(0.011)	(0.009)

Table 8.7.7: Bias and MSE of $\hat{d}_X$ for signal+noise process $X = Y + Z$, where Y is Gaussian ARFIMA$(0, d_Y, 0)$ and Z is Gaussian ARFIMA$(0, d_Z, 0)$.

Note: The contents of the above Sections 8.5 to 8.6.2 are based on the works of Robinson (1995b) and Dalla, Giraitis and Hidalgo (2006). A number of semiparametric estimators of d_0 have been developed for Gaussian and linear processes, besides the ones discussed above. Log-periodogram and local Whittle estimators were introduced and explored by Geweke and Porter-Hudak (1983), Künsch (1987), and Robinson (1995a, 1995b). Robinson (1994) and Lobato and Robinson (1996) provide estimators of the long memory parameter using a discrete version of the integrated periodogram. A time-domain estimator is discussed in Hall, Koul and Turlach (1997).

Moulines and Soulier (1999) and Hurvich and Brodsky (2001) proposed broadband estimators while Shimotsu and Phillips (2005) discussed exact local Whittle estimation methods. Abadir, Distaso and Giraitis (2007) extended the definition of local Whittle estimators for non-stationary processes and proved their consistency and asymptotic normality. Moulines and Soulier (2003) provide a review of semiparametric estimation methods. Velasco (1999a, 1999b) and Hurvich and Chen (2000) discuss semiparametric estimation of the memory parameter using tapered estimators. Chen and Deo (2006) study estimation of misspecified long memory models. Giraitis, Robinson and Samarov (1997, 2000) discuss rate optimal and adaptive local Whittle estimators.

Signal plus noise models (with i.i.d. innovations) have drawn much attention as they arise after taking the logarithmic transformation of the stochastic volatility model, and were introduced by Taylor (1994) and explored by Harvey, Ruiz and Shephard (1994). Results for the log-periodogram estimator in these models and signal plus noise models were obtained by Deo and Hurvich (2001) and Sun and Phillips (2003). Hurvich *et al.* (2005) and Arteche (2004) obtained similar results for local Whittle estimators, assuming that signal and noise are independent processes. Some of the results presented here have roots in Dalla *et al.* (2006).

Automatic data-based bandwidth selection rules for local Whittle estimates are discussed in Henry and Robinson (1996) and Lobato and Robinson (1998), and for the log-periodogram estimators in Hurvich and Beltrao (1994), Hurvich and Deo (1998), and Hurvich, Deo and Brodsky (1998). The inference of impulse response weights was investigated by Baillie and Kapetanios (2010). The simulations for Tables 8.7.1 to 8.7.7 were completed by V. Dalla.

Chapter 9

Elementary Inference Problems

9.1 Introduction

In this chapter we discuss large sample inference in a simple linear-trend model with dependent stationary errors. Asymptotic confidence intervals for the trend and memory parameters are derived in Section 9.2 while those for the mean are derived in Section 9.3. These confidence intervals are applicable for a wide class of linear and possibly non-linear, and non-Gaussian error processes. They exhibit good coverage accuracy and are easy to implement.

An important entity to be estimated consistently for making inference feasible is the long-run variance since it is used to studentize various statistics. Section 9.4 establishes the consistency of a time-domain long-run variance HAC estimator and its spectral-domain alternative MAC, under a variety of conditions on the observable process.

Section 9.5 focuses on testing for long memory and structural breaks. Asymptotic null distributions and the consistency of the Lobato–Robinson, R/S, $KPSS$, and V/S tests are discussed under a variety of assumptions on the observable process. An application to real currency exchange data is also given in this section. Several simulation results are reported throughout the chapter.

9.2   Linear-trend model

In this section we shall discuss asymptotic confidence intervals for the trend
and memory parameters in the linear-trend model

$$X_j = \alpha + \beta j + \varepsilon_j, \quad j = 1, 2, \cdots, n, \tag{9.2.1}$$

where the ε_j's are assumed to be stationary, possibly having long memory
with $E\varepsilon_0 = 0$. This is a simple model that provides a nice example for seeing
the effect of long memory in errors on some classical inference procedures.

Recall Definition 3.1.5 for an I(d) process. We suppose $\{\varepsilon_j, j \in \mathbb{Z}\}$, is an
I(d_0) *process* with memory parameter $d_0 \in (-1/2, 1/2)$ and spectral density
f_ε satisfying assumption

$$f_\varepsilon(u) = |u|^{-2d_0} g(u) \sim c_0 |u|^{-2d_0}, \quad u \to 0, \tag{9.2.2}$$

for some even function $g \geq 0$ with $\lim_{u \to 0} g(u) = c_0 > 0$. Observe that this
is equivalent to assuming that L_f in (3.1.9) is g and $c_0 = g(0)$.

The main goal here is to obtain large sample confidence intervals for the
parameters d_0, α, and β in the model (9.2.1) when $\{\varepsilon_j\} \sim$ I(d_0), $|d_0| < 1/2$.
We do this by using ordinary least-squares (LS) estimators. Because errors
are correlated we may base these intervals on generalized LS (GLS) estimators, which in turn requires the specification of the autocovariance structure
explicitly. Even assuming the autocovariance structure is correctly specified and known, the loss of efficiency by using LS over GLS estimators is
not substantial. For example, Table 1 of Yajima (1988) implies that the
maximal loss of asymptotic efficiency by an LS estimator compared to a
GLS estimator is 11% when estimating α and β in model (9.2.1).

We first derive the asymptotic distribution of the LS estimators of α, β
under mild restrictions on $\{\varepsilon_j\}$, which are satisfied by linear and a large
class of non-linear and non-Gaussian processes. These estimators are

$$\widehat{\beta} = \frac{\sum_{j=1}^{n} (j - \frac{n+1}{2})(X_j - \bar{X})}{\sum_{j=1}^{n} (j - \frac{n+1}{2})^2}, \quad \widehat{\alpha} = \bar{X} - \widehat{\beta}(n+1)/2, \tag{9.2.3}$$

where $\bar{X} = n^{-1} \sum_{j=1}^{n} X_j$.

First we obtain an asymptotic expression for the *long-run variance* s_ε^2
of $\{\varepsilon_j\}$. By Proposition 3.3.1,

$$s_\varepsilon^2 := s_{\varepsilon, d_0}^2 = \lim_{n \to \infty} E \left(n^{-1/2 - d_0} \sum_{j=1}^{n} \varepsilon_j \right)^2 \tag{9.2.4}$$

$$= \lim_{n\to\infty} n^{-1-2d} \int_{-\pi}^{\pi} \left(\frac{\sin(nu/2)}{\sin(u/2)}\right)^2 f_\varepsilon(u)du = p(d_0)c_0,$$

where c_0 is defined in (9.2.2) and

$$p(d) := \int_{-\infty}^{\infty} \left(\frac{\sin(u/2)}{u/2}\right)^2 |u|^{-2d}du = \begin{cases} 2\frac{\Gamma(1-2d)\sin(\pi d)}{d(1+2d)}, & \text{if } d \neq 0, \\ 2\pi, & \text{if } d = 0. \end{cases}$$

The rates of convergence of the LS estimators of β and α depend on the memory parameter d_0, and their asymptotic distributions depend on the scaling constant s_ε^2, which needs to be estimated. Note that the long-run variance s_ε^2, is defined by parameters c_0 and d_0.

To derive the asymptotic distributions of the LS estimators $\widehat{\beta}$ and $\widehat{\alpha}$ for a general error process $\{\varepsilon_j\}$, we need the following condition on its partial-sum process. Let

$$Y_n(v) := n^{-1/2-d_0} \sum_{j=1}^{[nv]} \varepsilon_j, \quad 0 \leq v \leq 1. \tag{9.2.5}$$

Assumption FDD. There exists a real-valued stochastic process $Y_\infty(v)$, $0 \leq v \leq 1$, such that the finite-dimensional distributions of Y_n converge weakly to those of Y_∞, i.e., $Y_n \to_{fdd} Y_\infty$.

Under assumption (9.2.2),

$$E[Y_n^2(v)] \to s_\varepsilon^2 v^{1+2d_0}, \quad \forall\, 0 \leq v \leq 1, \tag{9.2.6}$$

$$\sup_{0 \leq v \leq 1, n \geq 1} E[Y_n^2(v)] < \infty, \tag{9.2.7}$$

$$E[Y_n(v)Y_n(\tau)] \to \frac{\sigma_\zeta^2}{2}(v^{1+2d_0} + \tau^{1+2d_0} - |v-\tau|^{1+2d_0})$$
$$\equiv \sigma_\zeta^2 r_{1/2+d_0}(v,\tau), \quad \forall 0 \leq v, \tau \leq 1,$$

where the covariance $r_{1/2+d_0}$ is the same as in (3.4.1). Thus,

$$EY_\infty^2(v) = s_\varepsilon^2 v^{1+2d_0}, \qquad EY_\infty^2(1) = s_\varepsilon^2, \tag{9.2.8}$$

$$E[Y_\infty(v)Y_\infty(\tau)] = \sigma_\zeta^2 r_{1/2+d_0}(v,\tau), \quad \forall 0 \leq v, \tau \leq 1.$$

Let $W_{d_0} := s_\varepsilon^{-1} Y_\infty$. Note that $W_{d_0} \equiv B_H$ is the standard fractional Brownian motion with parameter $H = 1/2 + d_0$ when Y_∞ is a Gaussian process, of Section 3.4.

Moreover, if $\{\varepsilon_j\}$ is a linear process with i.i.d. innovations, then under (9.2.2) it satisfies Assumption FDD. This follows by Proposition 4.4.1, because under (9.2.2), $\text{Var}(\sum_{j=1}^n \varepsilon_j) \sim s_\varepsilon^2 n^{1+2d}$.

Theorem 9.2.1. *Assume that the model (9.2.1) holds with $\{\varepsilon_j\}$ as an $I(d_0)$ process for some $d_0 \in (-1/2, 1/2)$. Then, uniformly in $n \geq 1$,*

$$E(\widehat{\beta} - \beta)^2 \leq C n^{-3+2d_0}, \qquad E(\widehat{\alpha} - \alpha)^2 \leq C n^{-1+2d_0}. \qquad (9.2.9)$$

In addition, if the error process $\{\varepsilon_j\}$ satisfies Assumption FDD with Gaussian limit Y_∞, then

$$\frac{n^{1/2-d_0}}{s_\varepsilon \sigma_\alpha(d_0)}(\widehat{\alpha} - \alpha) \to_D \mathcal{N}(0,1), \qquad \frac{n^{3/2-d_0}}{s_\varepsilon \sigma_\beta(d_0)}(\widehat{\beta} - \beta) \to_D \mathcal{N}(0,1),$$

where, for $|d| < 1/2$,

$$\sigma_\alpha(d) := \left(1 + 36\left(\frac{1}{2d+3} - \frac{1}{4}\right)\right)^{1/2}, \qquad \sigma_\beta(d) := 12\left(\frac{1}{2d+3} - \frac{1}{4}\right)^{1/2}.$$

To make this theorem operational, we need to estimate the unknown parameters d_0 and s_ε.

Corollary 9.2.1. *Suppose that the assumptions of Theorem 9.2.1 are satisfied and there exist estimators $\hat{d}$ and $\hat{s}_\varepsilon^2$ such that*

$$\hat{d} = d + o_p(1/\log n), \qquad \hat{s}_\varepsilon^2 = s_\varepsilon^2 + o_p(1). \qquad (9.2.10)$$

Then

$$\frac{n^{1/2-\hat{d}}}{\hat{s}_\varepsilon \sigma_\alpha(\hat{d})}(\widehat{\alpha} - \alpha) \to_D \mathcal{N}(0,1), \qquad \frac{n^{3/2-\hat{d}}}{\hat{s}_\varepsilon \sigma_\beta(\hat{d})}(\widehat{\beta} - \beta) \to_D \mathcal{N}(0,1).$$

Thus, an asymptotic confidence interval of size $1 - \gamma$ for β is

$$\left[\widehat{\beta} - \frac{\sigma_\beta(\hat{d})\hat{s}_\varepsilon}{n^{3/2-\hat{d}}} z_{\gamma/2}, \ \widehat{\beta} + \frac{\sigma_\beta(\hat{d})\hat{s}_\varepsilon}{n^{3/2-\hat{d}}} z_{\gamma/2}\right],$$

where z_γ is the $(1 - \gamma)$th percentile of the standard normal distribution. Similarly, an asymptotic confidence interval of size $1 - \gamma$ for α is

$$\left[\widehat{\alpha} - \frac{\sigma_\alpha(\hat{d})\hat{s}_\varepsilon}{n^{1/2-\hat{d}}} z_{\gamma/2}, \ \widehat{\alpha} + \frac{\sigma_\alpha(\hat{d})\hat{s}_\varepsilon}{n^{1/2-\hat{d}}} z_{\gamma/2}\right]. \qquad (9.2.11)$$

Note that asymptotically the length of this interval for α increases as d_0 approaches the upper bound $1/2$.

To derive limiting distributions for the LS estimators $\widehat{\beta}$ and $\widehat{\alpha}$ in the case when $\{\varepsilon_j\}$ satisfies Assumption FDD with a non-Gaussian limiting process Y_∞, we need to introduce

$$Z_\beta(d) = -\int_0^1 \left(W_d(v) - vW_d(1)\right)dv,$$

$$Z_\alpha(d) := 6\int_0^1 \left(W_d(v) - vW_d(1)\right)dv + W_d(1), \qquad |d| < 1/2.$$

Set
$$Z_\beta(0,1) := \frac{Z_\beta(d_0)}{\sqrt{EZ_\beta^2(d_0)}}, \quad Z_\alpha(0,1) := \frac{Z_\alpha(d_0)}{\sqrt{EZ_\alpha^2(d_0)}}.$$

Random variables $Z_\beta(0,1)$ and $Z_\alpha(0,1)$ have zero means and unit variances. Therefore, whenever Y_∞ is Gaussian, $Z_\alpha(0,1)$ and $Z_\beta(0,1)$ have standard normal distributions.

Now, from (9.2.8) and direct integration, it follows that
$$EZ_\beta^2(d) = \sigma_\beta^2(d)/(12)^2, \quad EZ_\alpha^2(d) = \sigma_\alpha^2(d).$$

Theorem 9.2.2. *Assume that $\{\varepsilon_j\}$ satisfies Assumption FDD with non-Gaussian limit Y_∞. Then*
$$\frac{n^{1/2-d}}{s_\varepsilon \sigma_\alpha(d)}(\widehat{\alpha} - \alpha) \to_D Z_\alpha(0,1), \quad \frac{n^{3/2-d}}{s_\varepsilon \sigma_\beta(d)}(\widehat{\beta} - \beta) \to_D Z_\beta(0,1). \quad (9.2.12)$$

Proofs of Theorems 9.2.1 and 9.2.2. First we prove (9.2.12), which also implies the asymptotic normality results of Theorem 9.2.1. To describe the limit distributions of the LS estimators $\widehat{\beta}$ and $\widehat{\alpha}$, we shall use variables $Z_\beta(d_0)$ and $Z_\alpha(d_0)$. Rewrite $\widehat{\beta} - \beta$ as
$$n^{3/2-d_0}(\widehat{\beta} - \beta) = V_n/D_n, \quad (9.2.13)$$
where, with $\bar{\varepsilon} = n^{-1}\sum_{j=1}^n \varepsilon_j$,
$$V_n := n^{-3/2-d_0}\sum_{j=1}^n (j - \frac{n+1}{2})(\varepsilon_j - \bar{\varepsilon}),$$
$$D_n := n^{-3}\sum_{j=1}^n (j - \frac{n+1}{2})^2 = \frac{n^2-1}{12n^2} \to \frac{1}{12}.$$

We will show that
$$V_n \to_D s_\varepsilon Z_\beta(d_0). \quad (9.2.14)$$

Then these facts will imply the second claim in (9.2.12).

To prove (9.2.14), rewrite
$$V_n = -n^{-3/2-d_0}\sum_{j=1}^{n-1}\sum_{k=1}^j (\varepsilon_k - \bar{\varepsilon}) \quad (9.2.15)$$
$$= -\sum_{j=1}^{n-1} n^{-3/2-d_0}\left(\sum_{k=1}^j \varepsilon_k - \frac{j}{n}\sum_{k=1}^n \varepsilon_k\right)$$
$$= -\int_0^1 n^{-1/2-d_0}\left(\sum_{k=1}^{[nv]} \varepsilon_k - \frac{[nv]}{n}\sum_{k=1}^n \varepsilon_k\right)dv.$$

Thus,

$$V_n = -\int_0^1 \left(Y_n(v) - \frac{[nv]}{n}Y_n(1)\right)dv = -\int_0^1 \left(Y_n(v) - vY_n(1)\right)dv + o_p(1).$$

Note that $\{Y_n(v), 0 \leq v \leq 1\}$, $n \geq 0$, is a sequence of real-valued stochastic processes with sample paths in the space $L_2[0,1]$, i.e., $\int_0^1 Y_n^2(v)dv < \infty$, $n \geq 1$, a.s. The map $Y :\mapsto \int_0^1 (Y(v) - vY(1))^2 dv$ is a continuous and bounded functional from $L_2([0,1])$ to $\mathbb{R}$. Assumptions FDD, (9.2.6), and (9.2.7) imply that the process Y_n satisfies the conditions of the weak convergence criterion in the space $L_2[0,1]$ by Crámer and Kadelka (1986). Hence,

$$V_n \to_D -\int_0^1 \left(Y_\infty(v) - vY_\infty(1)\right)dv = s_\varepsilon Z_\beta(d_0),$$

thereby completing the proof of (9.2.14).

Since $\widehat{\alpha} - \alpha = (\beta - \widehat{\beta})(n+1)/2 + \bar{\varepsilon}$, using (9.2.13) we can write

$$n^{1/2-d_0}(\widehat{\alpha} - \alpha) = -6V_n(1 + o(1)) + n^{1/2-d_0}\bar{\varepsilon}.$$

Similarly,

$$n^{1/2-d_0}(\widehat{\alpha} - \alpha)$$

$$= 6(1 + o(1))\int_0^1 \left(Y_n(v) - vY_n(1)\right)dv + Y_n(1) + o_p(1)$$

$$\to_D 6\int_0^1 \left(Y_\infty(v) - vY_\infty(1)\right)dv + Y_\infty(1) = s_\varepsilon\sigma_\alpha(d_0)Z_\alpha(0,1).$$

This completes the proof of (9.2.12).

The mean-squared error bounds (9.2.9) for $\widehat{\beta}$ and $\widehat{\alpha}$ follow from (9.2.15), the C–S inequality, and (9.2.7).

Remark 9.2.1. Observe that V_n of (9.2.13) can be written as a weighted sum of $\{\varepsilon_j\}$:

$$V_n = \sum_{j=1}^n z_{nj}\varepsilon_j, \quad z_{nj} = n^{-3/2-d_0}\left(j - \frac{n+1}{2}\right).$$

Therefore, if $\{\varepsilon_j\}$ is a linear process with i.i.d. innovations, then the convergence (9.2.14) can be established by applying Proposition 4.3.1.

Estimation of d_0 and s_ε^2 in (9.2.12). We now discuss some estimators for the unknown parameters d_0, c_0, and s_ε in the linear-trend model (9.2.1). Let $\widehat{d_{\widetilde{\varepsilon}}}$ and $\widehat{c_{\widetilde{\varepsilon}}}$ denote the LW estimators (8.5.2) and (8.5.3) of d_0 and c_0, based on the detrended data

$$\widehat{\varepsilon}_j = X_j - \widehat{\beta}j = \alpha + \varepsilon_j + r_{n,j}, \quad j = 1, \cdots, n,$$

written as a signal plus noise process with $r_{n,j} := (\beta - \widehat{\beta})j$. Note that under (9.2.2), by (9.2.9), $\widehat{\beta} - \beta = O_p(n^{-3/2+d_0})$. Therefore,

$$\sum_{j=1}^{n-1} |r_{n,j} - r_{n,j+1}| = O_p(n^{d_0-1/2}).$$

Hence, the noise $r_{n,j}$ is stochastically negligible, i.e., it satisfies assumption (8.6.12) of Theorem 8.6.3.

If, in addition, $\{\varepsilon_j\}$ satisfies Assumptions A and B of Subsection 8.5.1, and the LW estimator $\widehat{d_\varepsilon}$ of d_0, based on $\varepsilon_j, j = 1, \cdots, n$, has the property $\widehat{d_\varepsilon} - d_0 = o_p(1/\log n)$, then by Theorem 8.6.3,

$$\widehat{d_{\widetilde{\varepsilon}}} - d_0 = (\widehat{d_\varepsilon} - d_0)(1 + o_p(1)) + o_p(m^{-1/2}) = o_p\left(\frac{1}{\log n}\right), \quad (9.2.16)$$

$$\widehat{c_{\widetilde{\varepsilon}}} := \frac{1}{m} \sum_{j=1}^{m} u_j^{2\widehat{d_{\widetilde{\varepsilon}}}} I_{n,\widetilde{\varepsilon}}(u_j) \to_p c_0.$$

Thus, because of (9.2.16) the long-run variance s_ε^2 can be estimated by Robinsons's (2005b) MAC estimator

$$\widehat{s_{\widetilde{\varepsilon}}^2} := p(\widehat{d_{\widetilde{\varepsilon}}})\widehat{c_{\widetilde{\varepsilon}}} \to_p p(d_0)c_0 \equiv s_\varepsilon^2.$$

For more details on MAC estimators see Section 9.4 below.

Example 9.2.1. Assume $\{\varepsilon_j\}$ to be a stationary ARFIMA(p, d_0, q) model, for some $|d_0| < 1/2$, and with i.i.d. innovations having a finite fourth moment. Then $\{\varepsilon_j\}$ satisfies Assumption A, whereas Assumption B holds by Theorem 8.5.2(i). Thus, by (9.2.16) and Remark 8.5.2,

$$2\sqrt{m}(\widehat{d_{\widetilde{\varepsilon}}} - d_0) \to_D \mathcal{N}(0, 1), \qquad \text{if } m = o(n^{4/5}),$$
$$(n/m)^2(\widehat{d_{\widetilde{\varepsilon}}} - d_0) = O_p(1), \qquad \text{if } m \geq Cn^{4/5}.$$

The corresponding estimators of c_0 and s_ε^2 are also consistent.

9.3　Estimation of the mean μ

Here we consider the location model

$$X_j = \mu + \varepsilon_j, \quad j = 1, 2, \cdots, n. \tag{9.3.1}$$

A natural choice for an estimate of the mean parameter μ for this model is the sample mean $\bar{X} = n^{-1} \sum_{j=1}^{n} X_j$. If $\{\varepsilon_j\}$ satisfies Assumption FDD with a Gaussian limit $Y_\infty(v) = s_\varepsilon W_{d_0}(v)$, then

$$\frac{n^{1/2-d_0}}{s_\varepsilon} (\bar{X} - \mu) \to_D \mathcal{N}(0,1), \tag{9.3.2}$$

whereas if the limit $Y_\infty(v)$ is non-Gaussian, then (9.3.2) holds with $\mathcal{N}(0,1)$ replaced by $W_{d_0}(1)$.

In the former case, if estimators $\widehat{d}$ and $\hat{s}_\varepsilon^2$ satisfy (9.2.10), then

$$\left[\bar{X} - \frac{\hat{s}_\varepsilon}{n^{1/2-\widehat{d}}} z_{\gamma/2}, \; \bar{X} + \frac{\hat{s}_\varepsilon}{n^{1/2-\widehat{d}}} z_{\gamma/2} \right]$$

provides an asymptotic confidence interval of size $1 - \gamma$ for μ. The length of this confidence interval is an increasing function of $\widehat{d}$. In particular, this implies that the stronger the long memory, i.e., the closer d_0 is to $1/2$, the larger the confidence interval. This interval is shorter than the corresponding interval (9.2.11) derived for the location parameter α because of the extra term $\sigma_\alpha(d)$ appearing in (9.2.11), with $1 < \sigma_\alpha(d) < \sqrt{10}$.

Remark 9.3.1. Assumption FDD is satisfied by a wide class of error processes $\{\varepsilon_j\}$. In a number of cases the limit process Y_∞ is Gaussian, which in turn results in Gaussian asymptotic distributions for $\widehat{\alpha}$ and $\widehat{\beta}$. For example, Proposition 4.4.1 implies this when $\{\varepsilon_j\}$ is a linear or Gaussian process.

Assumption FDD is also valid if $\varepsilon_j = h(X_j), j \in \mathbb{Z}$, is a non-linear square integrable function of a Gaussian process $\{X_j\}$, with memory parameter $d_X \in [0, 1/2)$. If $\{\varepsilon_j\}$ has short memory, or if the function h has Hermite rank $k_0 = 1$, then by Theorems 4.6.1 and 4.7.1, the limit Y_∞ is Gaussian, whereas for $k_0 \geq 2$, Assumption FDD may also hold with Y_∞ non-Gaussian.

Simulation results. We shall now present some simulation results assessing the finite sample performance of the above point and interval estimators.

We consider the class of linear error processes $\{\varepsilon_j\}$ generated by the Gaussian fractional ARFIMA$(1, d_0, 0)$ model with unit variance, and the autoregressive parameter ρ equal to $-0.5, 0, 0.5$. The number of replications is 10,000, the sample size $n = 500$, and the bandwidth parameter $m = \lfloor n^{0.65} \rfloor = 56$. Under some additional restrictions, the optimal m can be chosen using the data-driven methods of Henry and Robinson (1996) or Andrews and Sun (2004), but there is no known general way to choose an optimal bandwidth for Whittle estimators. For the definition of optimal m, see Subsection 8.5.2.

Table 9.3.1 shows the results for the LW estimator $\widehat{d_{\widehat{\varepsilon}}}$ of the memory parameter d_0 when the series is detrended. The performance of the estimator for $d_0 \in (-1/2, 1/2)$ shows that when $\rho = 0$, d_0 is estimated very accurately over the range considered. The bias and MSE of $\widehat{d_{\widehat{\varepsilon}}}$ are generally slightly higher than those of $\widehat{d_{\varepsilon}}$, reflecting the contribution of the estimation of β. The MSE is reassuringly stable across the interval for d_0.

In the case when $\rho = -0.5$, the quality of the estimator does not seem to be affected by much. When $\rho = 0.5$, the bias and MSE of the LW estimator are higher than those when $\rho = 0$. The estimator becomes positively biased, and the value of the bias seems roughly constant across the different d_0.

Corollary 9.2.1 established the consistency and a CLT for the LS estimators $\widehat{\beta}$ and $\widehat{\alpha}$ when $d_0 \in (-1/2, 1/2)$. The variance and the convergence rates of estimators are functions of d_0, and the rate of convergence of the estimators is inversely proportional to the value of d_0. The rate of $\widehat{\beta}$ approaches n and the rate of $\widehat{\alpha}$ to $n^0 = 1$ when d_0 gets closer to the upper value $1/2$. The results in Table 9.3.2 reveal that the precision of the estimator $\widehat{\beta}$ is very high for low values of d_0, and then decreases (implying wider confidence intervals) as d_0 tends to $1/2$. When d_0 decreases, the rate of convergence to the limiting distribution increases and this translates into narrower confidence intervals (CIs). When using the asymptotic CIs of β, the calculated coverage probabilities (CPs) imply that an increase in d_0 or ρ reduce the coverage.

The results for the location parameter α are shown in Table 9.3.3 and they display qualitatively similar results to those of Table 9.3.2. Note that the results in Tables 9.3.2 and 9.3.3 are invariant to the values of α and β, which are set equal to zero in the tables.

d_0	Bias	MSE	Bias	MSE	Bias	MSE
	$\rho = -0.5$		$\rho = 0$		$\rho = 0.5$	
-0.3	-0.01338	0.00668	-0.00842	0.00670	0.09298	0.01492
0	-0.03021	0.00674	-0.01019	0.00702	0.09135	0.01499
0.3	-0.04197	0.00871	-0.01488	0.00703	0.08664	0.01420

Table 9.3.1: Bias and mean-squared error of $\widehat{d}_{\widehat{\varepsilon}}$ as ρ varies.

d_0	95% CI	CP	95% CI	CP	95% CI	CP
	$\rho = -0.5$		$\rho = 0$		$\rho = 0.5$	
-0.4	0.00021	0.97	0.00031	0.98	0.00060	0.99
-0.2	0.00035	0.92	0.00053	0.94	0.00121	0.97
0	0.00076	0.90	0.00116	0.93	0.00287	0.98
0.2	0.00188	0.90	0.00287	0.92	0.00750	0.96
0.4	0.00507	0.89	0.00778	0.91	0.02084	0.96

Table 9.3.2: Lengths of 95% confidence intervals (CI) and coverage probabilities (CP) of $\widehat{\beta}$ for $\beta = 0$ as ρ varies.

d_0	95% CI	CP	95% CI	CP	95% CI	CP
	$\rho = -0.5$		$\rho = 0$		$\rho = 0.5$	
-0.4	0.05644	0.97	0.08250	0.97	0.16254	0.98
-0.2	0.09500	0.94	0.14478	0.94	0.34104	0.97
0	0.21933	0.91	0.33442	0.92	0.87299	0.97
0.2	0.60364	0.91	0.92695	0.92	2.91031	0.95
0.4	2.22987	0.87	3.43777	0.90	6.12008	0.92

Table 9.3.3: Lengths of 95% confidence intervals (CI) and coverage probabilities (CP) of $\widehat{\alpha}$ when $\alpha = 0$ as ρ varies.

Note: The problem of estimating the regression parameters in linear regression models with non-random designs and long memory stationary errors with the memory parameter $d \in (-1/2, 1/2)$, has been studied extensively in the literature. Yajima (1988, 1991) derived sufficient conditions for the consistency and asymptotic normality of LS estimators in these models. Dahlhaus (1995) suggested an efficient weighted LS estimator and investigated its asymptotic properties for a polynomial regression with stationary errors.

In the long memory case, the theoretical asymptotic properties of sample mean estimators were investigated by Adenstedt (1974). Hall, Jing and Lahiri (1998) used sample window methods to construct confidence intervals for the mean μ of long memory processes that are non-linear transforms of a Gaussian sequence. They studied the asymptotic properties of the studentized sample mean estimator, which allows us to establish confidence intervals for μ for a wide class of short and long memory processes. Estimation of the parameters in the trend models allowing for non-stationary errors was discussed in Deo and Hurvich (1998). The discussion in this section benefited from Abadir, Distaso and Giraitis (2011). Surgailis and Viano (2002) showed that an EGARCH process satisfies Assumption FDD with a Gaussian limit Y_∞. Non-linear regression models with long memory errors were investigated by Ivanov and Leonenko (2004, 2008). Robinson (2005a) suggested an adaptive semiparametric estimation method for the case of a polynomial regression with fractionally integrated errors.

9.4 Estimation of the long-run variance

For a covariance stationary process $\{X_j\}$, let $\varepsilon_j = X_j - EX_j$, $j \in \mathbb{Z}$. In empirical studies, it is important to have consistent estimates of standard errors (SEs). The SE s_ε of the standardized sample mean is the square root of the long-run variance s_ε^2 of (9.2.4). Applying a consistent estimator of s_ε in the studentized sample mean (9.3.2), gives a standard normal distribution, and allows us to construct confidence intervals for the mean. Consistent estimators of the long-run variance s_ε^2 are required in a number of statistical procedures based on partial sums of the observable process, e.g., in the *KPSS* and *V/S* tests used in econometrics to test for unit roots and long memory, which are discussed in the next subsection.

We briefly discuss the two estimators of the long-run variance of a stationary I(d) process that are found useful in the presence of long memory and antipersistence. They are the Bartlett-kernel heteroscedasticity and autocorrelation consistent (HAC) and the memory and autocorrelation consistent (MAC) estimators. We provide two sets of conditions for their consistency, one for a general second-order stationary process and one for a linear process.

HAC estimator. To define the HAC estimator, which is suitable for long memory and antipersistence of an I(d_0) process, let

$$\widehat{\gamma}_j := n^{-1} \sum_{t=1}^{n-j} (X_t - \bar{X})(X_{t+j} - \bar{X}), \quad 0 \le j < n; \quad \bar{X} := n^{-1} \sum_{i=1}^{n} X_i,$$

be the sample autocovariances of X_j, $j \in \mathbb{Z}$, centered around the sample mean $\bar{X}$. For a known value of d_0, the HAC estimator is defined to be

$$\bar{s}_q^2(d_0) := q^{-1-2d_0} \sum_{i,j=1}^{q} \widehat{\gamma}_{|i-j|} = q^{-2d_0} \Big(\widehat{\gamma}_0 + 2 \sum_{k=1}^{q} (1 - k/q) \widehat{\gamma}_k \Big), \qquad (9.4.1)$$

where the bandwidth parameter $q \equiv q_n$ satisfies $q \to \infty$, $q = o(n)$. But since d_0 is not known, we must plug in a $\log(n)$-consistent estimator of d_0 in this entity. Accordingly, let $\widehat{d}$ be an estimator of d_0 such that

$$\widehat{d} - d_0 = o_p(1/\log n). \qquad (9.4.2)$$

Then a HAC estimator of s_ε^2 is defined to be $\bar{s}_q^2(\widehat{d})$. To establish the consistency of this estimator we need the following:

Assumption M. $\{X_j, j \in \mathbb{Z}\}$ is a fourth-order stationary I(d_0) process for some $d_0 \in (-1/2, 1/2)$, with the fourth-order cumulant satisfying

$$\sum_{t_1,t_2,t_3=-\infty}^{\infty} |\mathrm{Cum}(X_0, X_{t_1}, X_{t_2}, X_{t_3})| \leq C, \qquad d_0 < 0,$$

$$\max_{t_3} \sum_{t_1,t_2=-n}^{n} |\mathrm{Cum}(X_0, X_{t_1}, X_{t_2}, X_{t_3})| \leq C n^{2d_0}, \qquad d_0 \geq 0, \ n \geq 1.$$

As stated in the following theorem, Assumption M guarantees consistency but not necessarily the other second-order asymptotic properties of the HAC estimator. For a linear process Assumption M is not needed.

Theorem 9.4.1. *Suppose $\{X_j\}$ is an I(d_0) process, $|d_0| < 1/2$, with spectral density f satisfying $f(u) \leq C|u|^{-2d_0}$, $u \in \Pi$, and where $\widehat{d}$ is an estimator of d_0 satisfying (9.4.2). For the bandwidth q assume that $q \to \infty$, and for some $0 < \gamma < 1$,*

$$q = O(n^\gamma), \qquad 0 \leq d_0 < 1/2, \qquad (9.4.3)$$

$$q = O(n^{1/2}), \qquad -1/2 < d_0 < 0.$$

Then $\bar{s}_q^2(\widehat{d}) \to_p s_\varepsilon^2$ in each of the following two cases:
(i) $\{X_j\}$ is a linear process (9.4.4) and $E\zeta_0^4 < \infty$.
(ii) $\{X_j\}$ satisfies Assumption M.

It has been observed in practical applications that a HAC estimator is sensitive to the bandwidth choice, even when the underlying process has short memory. The following MAC estimators partly overcome this problem.

MAC estimator. In general, the problem of estimating the long-run variance s_ε^2 of an $I(d_0)$ process is closely related to that of d_0 and $c_0 = \lim_{u \to 0} |u|^{2d_0} f(u)$. The equality (9.2.4) implies that s_ε^2 is just a scaling of the parameter c_0 by the continuous function $p(d)$, so in the usual short memory case of $d_0 = 0$, $s_\varepsilon^2 = 2\pi f(0)$ and $c_0 = f(0)$.

Now, let

$$\widehat{c}_m(d) := m^{-1} \sum_{j=1}^{m} u_j^{2d} I(u_j), \quad \widehat{s}_m^2(d) := p(d)\widehat{c}_m(d), \quad |d| < 1/2.$$

For an $I(d_0)$ process $\{\varepsilon_j\}$, the MAC estimator is defined to be

$$\widehat{s}_\varepsilon^2 := \widehat{s}_m^2(\widehat{d}),$$

where $\widehat{d}$ is an estimator of d_0 satisfying (9.4.2).

Recall from Section 8.5, under assumption (9.4.2), that $\widehat{c}_m(\widehat{d}) \to_p c_0$ for a wide class of $I(d_0)$ processes. This fact and the continuity of the function $p(d)$ imply the consistency of $\widehat{s}_\varepsilon^2$ for s_ε^2, i.e., $\widehat{s}_\varepsilon^2 \to_p s_\varepsilon^2$, in a routine fashion.

Example 9.4.1. We shall now discuss HAC and MAC estimators when the given $I(d_0)$ process is a linear process:

$$X_j = \mu + \varepsilon_j, \qquad \varepsilon_j = \sum_{k=0}^{\infty} a_k \zeta_{j-k}, \qquad j \in \mathbb{Z}, \tag{9.4.4}$$

where the ζ_j's are standardized i.i.d. r.v.'s with finite fourth moment $E\zeta_0^4 < \infty$.

Consider first the HAC estimator. Under some additional assumptions on the coefficients of a linear process, for each $|d_0| < 1/2$ the rate of convergence of the HAC estimator is

$$|\bar{s}_q^2(d_0) - s_\varepsilon^2| = O_p\big((q/n)^{(1/2) \wedge (1 - 2d_0)} + q^{-1-2d_0}\big).$$

Moreover, the MSE of the HAC estimator $\bar{s}_q^2(\widehat{d})$ is minimized asymptotically by bandwidth q of the order

$$q \propto \begin{cases} n^{1/(3+4d_0)}, & -1/2 < d_0 < 1/4, \\ n^{1/2-d_0}, & 1/4 < d_0 < 1/2. \end{cases} \tag{9.4.5}$$

If $\{\varepsilon_j\}$ has short or moderate long memory $-1/2 < d_0 < 1/4$, then the HAC estimator has a Gaussian limit distribution, whereas for heavy long memory $1/4 < d_0 < 1/2$, the asymptotic distribution is non-Gaussian. For short memory processes with $d_0 = 0$, (9.4.5) suggests the optimal band-width to be $q = [n^{1/3}]$.

Next, consider the MAC estimator. In addition, assume that $\{\varepsilon_j\}$ has spectral density (9.2.2) where g satisfies (8.5.21) with $\beta = 2$ and the transfer function A_ε satisfies (6.2.3). Computation of the MAC estimator involves two bandwidth parameters. The first one, m, is used in the estimator $\widehat{c}_m$ and the other, m_w, is used in the LW estimator $\widehat{d}$. Suppose $m_w \propto n^{4/5}$ and $m \propto n^{4/5}$. Then

$$(\widehat{s}_m^2(\widehat{d}) - s_X^2)^2 = O_p\left(\frac{1}{m} + \frac{\log n}{m_w}\right). \tag{9.4.6}$$

The MSE-optimal bandwidth m is therefore the one that grows at the maximal rate of $n^{4/5}$.

Unlike the HAC estimator, the asymptotic properties of the MAC estimator do not depend on the memory parameter d, and its asymptotic distribution is always Gaussian. In general, a MAC estimator is less sensitive to the bandwidth choice m compared to a HAC estimator. Proofs of the above statements can be found in Abadir, Distaso, and Giraitis (2009).

Example 9.4.2. To illustrate the range of applicability of the general Assumption M imposed on the fourth-order cumulant, we show its validity for the linear process $\{X_j\}$ of Example 9.4.1. For simplicity, set $a_j = 0$, $j \leq 0$. Then by the cumulant properties (c1) to (c2) of Section 4.2,

$$\text{Cum}(X_0, X_{t_1}, X_{t_2}, X_{t_3}) = \text{Cum}_4(\zeta_0) \sum_{k=-\infty}^{\infty} a_k a_{k+t_1} a_{k+t_2} a_{k+t_3}.$$

Case $d_0 \in (-1/2, 0]$. If, in addition, $\sum_k |a_k| < \infty$, then

$$\sum_{t_1, t_2, t_3 = -n}^{n} |\text{Cum}(X_0, X_{t_1}, X_{t_2}, X_{t_3})| \leq C\left(\sum_{s=-\infty}^{\infty} |a_s|\right)^4 < \infty.$$

Case $d_0 \in (0, 1/2)$. If, in addition, $\exists\, c > 0$, $a_j \sim cj^{d_0-1}$, $j \to \infty$, then

$$\sum_{t_1, t_2 = -n}^{n} |\text{Cum}(X_0, X_{t_1}, X_{t_2}, X_{t_3})| \leq C \sum_{|t_1|, |t_2| \leq n} \sum_{s=-\infty}^{\infty} |a_s a_{s+t_1} a_{s+t_2} a_{s+t_3}|$$

$$\leq C\left(\sum_{|s| \leq 2n} |a_s a_{s+t_3}| \sum_{|t_1|, |t_2| \leq 3n} |a_{t_1}||a_{t_2}|\right.$$

$$\left. + \sum_{|s| > 2n} |a_s a_{s+t_3}| \sum_{|t_1|, |t_2| \leq n} |a_{s+t_1}||a_{s+t_2}|\right).$$

Note that $\sum_{s=0}^{\infty} a_s^2 < \infty$, $\sum_{|s| \leq n} |a_s| \leq C n^{d_0}$, and $|a_{s+t}| \leq C n^{-1+d_0}$ for $|s| > 2n$ and $|t| \leq n$. Thus,

$$\max_{t_3} \sum_{t_1, t_2 = -n}^{n} |\mathrm{Cum}(X_0, X_{t_1}, X_{t_2}, X_{t_3})|$$

$$\leq C \Big(\sum_{s=-\infty}^{\infty} a_s^2 \Big) \Big(\Big(\sum_{|t| \leq 3n} |a_t| \Big)^2 + n^{2d_0} \Big) \leq C n^{2d_0}.$$

Observe that by (7.2.22), the above properties of the a_j's and also Assumption M are satisfied by the fractional ARFIMA(p, d, q) process (7.2.16).

Proof of Theorem 9.4.1. Under assumption (9.4.2) on $\hat{d}$, $q^{2\hat{d}} = q^{2d_0}(1 + o_p(1))$, implying $\bar{s}_q^2(\hat{d}) = \bar{s}_q^2(d_0)(1 + o_p(1))$. It thus suffices to show that $\bar{s}_q^2(d_0) \to_p s_\varepsilon^2$.

Letting $\tilde{\gamma}_j := n^{-1} \sum_{i=1}^{n-|j|}(X_i - EX_1)(X_{i+|j|} - EX_1)$, $|j| < n$, write

$$\bar{s}_q^2(d_0) = v_{n,1} + v_{n,2},$$

$$v_{n,1} := q^{-2d_0} \sum_{|j| \leq q} \Big(1 - \frac{|j|}{q} \Big) \tilde{\gamma}_j, \quad v_{n,2} := q^{-2d_0} \sum_{|j| \leq q} \Big(1 - \frac{|j|}{q} \Big) (\hat{\gamma}_j - \tilde{\gamma}_j).$$

We first prove a general result that is also used in Section 9.5 below: If $\{X_j\}$ is an I(d_0) process, $|d_0| < 1/2$, then

$$E\bar{s}_q^2(d_0) \to s_\varepsilon^2. \tag{9.4.7}$$

This will follow if we prove

$$Ev_{n,1} \to s_\varepsilon^2, \quad E|v_{n,2}| \to 0. \tag{9.4.8}$$

Subsequently we show that under assumptions (i) or (ii) of Theorem 9.4.1,

$$\mathrm{Var}(v_{n,1}) \to 0, \tag{9.4.9}$$

which together with (9.4.8) proves $\bar{s}_q^2(d_0) \to_p s_\varepsilon^2$.

Proof of (9.4.8). First, we verify that $Ev_{n,1} \to s_\varepsilon^2$. Because $E\tilde{\gamma}_j = \gamma_\varepsilon(j)(1 - j/n)$, $j = 1, \cdots, n$,

$$Ev_{n,1} = q^{-2d_0} \sum_{|j| \leq q} \Big(1 - \frac{|j|}{q} \Big) \Big(1 - \frac{|j|}{n} \Big) \gamma_\varepsilon(j) = r_{n,1} - r_{n,2}, \tag{9.4.10}$$

$$r_{n,1} := q^{-2d_0} \sum_{|j| \leq q} \Big(1 - \frac{|j|}{q} \Big) \gamma_\varepsilon(j), \quad r_{n,2} := q^{-2d_0} \sum_{|j| \leq q} \Big(1 - \frac{|j|}{q} \Big) \frac{|j|}{n} \gamma_\varepsilon(j).$$

It suffices to show that $r_{n,1} \to s_\varepsilon^2$ and $r_{n,2} \to 0$. By (9.2.4),

$$r_{n,1} = q^{-1-2d_0} \sum_{j,k=1}^{q} \gamma_\varepsilon(j-k) \to s_\varepsilon^2.$$

Let $\theta_j := (1 - |j|/q)j$, $j = 1, \cdots, q$. Summation by parts and $\theta_q = 0$ yield

$$r_{n,2} = 2q^{-2d_0}n^{-1} \sum_{j=1}^{q} \theta_j \gamma_\varepsilon(j) = 2q^{-2d_0}n^{-1} \sum_{j=1}^{q-1}(\theta_j - \theta_{j+1}) \sum_{i=1}^{j} \gamma_\varepsilon(i).$$

By (9.2.2), $f_\varepsilon(u) \le C|u|^{-2d_0}$, $|u| \le \epsilon$, for a small $\epsilon > 0$, and by (2.1.8), $|D_j(u)| \le 2\pi j(1 + j|u|)^{-1}$, $u \in \Pi$. Thus,

$$\left| \sum_{i=1}^{j} \gamma_\varepsilon(i) \right| = \left| \int_\Pi \sum_{k=1}^{j} e^{iku} f_\varepsilon(u) du \right| \le \int_\Pi |D_j(u)| f_\varepsilon(u) du$$

$$\le C \left\{ \int_{|u| \le \epsilon} \frac{j}{1+j|u|} |u|^{-2d_0} du + \int_{\epsilon \le |u| \le \pi} \epsilon^{-1} f_\varepsilon(u) du \right\}$$

$$\le C(j^{2d_0 \vee 0} + I(d_0 = 0) \log j).$$

Since $|\theta_j - \theta_{j+1}| \le 2$, then under (9.4.3), $r_{n,2} \le Cq^{-2d_0}n^{-1} \sum_{j=1}^{q-1} j^{2d_0 \vee 0}$ $\le Cq^{-(2d_0 \wedge 0)}(q/n) \to 0$, for $0 < |d_0| < 1/2$, whereas for $d_0 = 0$, $r_{n,2} \le Cn^{-1} \sum_{j=1}^{q-1} \log j \le C(q/n) \log q \to 0$.

To prove the second claim in (9.4.8), bound $|v_{n,2}| \le q^{-2d_0} \sum_{|j| \le q} |\hat{\gamma}_j - \tilde{\gamma}_j|$. Observe that with $\mu = EX_0$,

$$\hat{\gamma}_j - \tilde{\gamma}_j = \frac{n-|j|}{n}(\bar{X} - \mu)^2 - \frac{\bar{X} - \mu}{n} \left(\sum_{i=1}^{n-|j|}(X_i - \mu) + \sum_{i=1}^{n-|j|}(X_{i+|j|} - \mu) \right).$$

By (9.2.4), $E(\sum_{i=1}^{n}(X_i - \mu))^2 \le Cn^{1+2d_0}$, which together with the fourth-order stationarity gives

$$E|\hat{\gamma}_j - \tilde{\gamma}_j| \le E(\bar{X} - \mu)^2 + 2n^{-1} E^{1/2}(\bar{X} - \mu)^2 E^{1/2} \left(\sum_{i=1}^{n-|j|}(X_i - \mu) \right)^2$$

$$\le Cn^{-1+2d_0} + Cn^{-1/2+d_0}(n - |j|)^{-1/2+d_0} \le Cn^{-1+2d_0},$$

$$E|v_{n,2}| \le Cq^{-2d_0} \sum_{|j| \le q} n^{-1+2d_0} \le C(q/n)^{1-2d_0} \to 0.$$

Proof of (9.4.9). Use $\tilde{\gamma}_j = \int_\Pi e^{iju} I_\varepsilon(u) du$, $|j| \le n$, where $I_\varepsilon(u) := (2\pi n)^{-1}|\sum_{k=1}^{n} e^{iku}\varepsilon_k|^2$, and $b_n(u) := q^{-1-2d_0}|D_q(u)|^2$ to write

$$v_{n,1} = q^{-1-2d_0} \sum_{j,k=1}^{q} \tilde{\gamma}_{j-k} = \int_\Pi b_n(u) I_\varepsilon(u) du. \qquad (9.4.11)$$

(i) Assume that $\{\varepsilon_j\}$ is a linear process. For $d_0 \leq 1/4$, set $\beta := \min(1, 1/2 - 2d_0) - \delta$, where $\delta > 0$ is small. By $|D_q(u)| \leq \min(q, \pi|u|^{-1})$,

$$b_n(u) \leq q^{1-2d_0-\beta}|D_q(u)|^\beta \leq k_n|u|^{-\beta}, \quad u \in \Pi, \quad k_n := \pi q^{1-\beta-2d_0}.$$

Then $\beta + 2d_0 < 1/2$, and by (9.4.3), $k_n = o(n^{1/2})$. Thus, by Lemma 6.3.5, $\mathrm{Var}(v_{n,1}) \leq Ck_n^2/n \to 0$.

For $1/4 < d_0 < 1/2$, use $b_n(u) \leq k_n := q^{1-2d_0}$ and Lemma 6.3.5 to obtain $\mathrm{Var}(v_{n,1}) \leq Ck_n^2 n^{4d_0-2} = C(q/n)^{2(1-2d_0)} \to 0$.

(ii) Suppose now that $\{X_j\}$ satisfies Assumption M. Then

$$i_{n,\varepsilon} := \mathrm{Var}(v_{n,1}) = q^{-4d_0} \sum_{|s|,|t|\leq q} (1 - |s|/q)(1 - |t|/q)\mathrm{Cov}(\tilde{\gamma}_s, \tilde{\gamma}_t),$$

$$\mathrm{Cov}(\tilde{\gamma}_s, \tilde{\gamma}_t) = n^{-2} \sum_{i=1}^{n-|s|} \sum_{j=1}^{n-|t|} \mathrm{Cov}(\varepsilon_i \varepsilon_{i+|s|}, \varepsilon_j \varepsilon_{j+|t|}),$$

where $\varepsilon_j = X_j - EX_j$. Observe that $\mathrm{Cov}(\varepsilon_i \varepsilon_{i+|s|}, \varepsilon_j \varepsilon_{j+|t|})$ equals

$$\mathrm{Cum}(X_i, X_{i+|s|}, X_j, X_{j+|t|}) + \gamma_\varepsilon(i-j)\gamma_\varepsilon(i-j+|s|-|t|)$$
$$+ \gamma_\varepsilon(i-j-|t|)\gamma_\varepsilon(i+|s|-j).$$

Denote by $\{\eta_j\}$ a Gaussian process with the spectral density $f_\eta = f_\varepsilon$. Then $\gamma_\eta(k) = \gamma_\varepsilon(k)$, $k \geq 0$. Since $\{\eta_j\}$ can be written as a linear process and the corresponding fourth-order cumulant of the η_j's equals 0, then $i_{n,\varepsilon} = i_{n,\eta} + (i_{n,\varepsilon} - i_{n,\eta})$. By (i) above, $i_{n,\eta} \to 0$, whereas

$$J_n := |i_{n,\varepsilon} - i_{n,\eta}| \leq q^{-4d_0} \sum_{|s|,|t|\leq q} \sum_{i=1}^{n-|s|} \sum_{j=1}^{n-|t|} |\mathrm{Cum}(X_i, X_{i+|s|}, X_j, X_{j+|t|})|.$$

By the fourth-order stationarity of $\{X_j\}$, $\mathrm{Cum}(X_i, X_{i+|s|}, X_j, X_{j+|t|}) = \mathrm{Cum}(X_0, X_{|s|}, X_{j-i}, X_{j-i+|t|})$. Thus, by Assumption M, for $d_0 \geq 0$,

$$J_n \leq q^{-4d_0} n^{-1} \max_{|u|\leq 2n} \sum_{|s|,|t|\leq q} \sum_{k=-n}^{n} |\mathrm{Cum}(X_0, X_{|s|}, X_k, X_u)|$$
$$\leq C(q/n)^{1-2d_0} q^{-2d_0} \to 0,$$

whereas for $d_0 \in (-1/2, 0)$,

$$J_n \leq q^{-4d_0} n^{-1} \sum_{|s|,|k|,|u|\leq 2n} |\mathrm{Cum}(X_0, X_s, X_k, X_u)| \leq Cq^{-4d_0} n^{-1} \to 0,$$

by (9.4.3). This completes the proof of the theorem.

Simulation study. The main purpose of the HAC and MAC estimators is to provide consistent estimators of the long-run variance s_ε^2 for studentizing various statistics. To illustrate the finite sample performance of these estimators, we will calculate their MSEs. Moreover, to check the precision of the normal approximation of the t-statistic

$$t := \frac{n^{1/2-\widehat{d}}(\bar{X} - \mu)}{\widehat{s}_\varepsilon} \to_D \mathcal{N}(0,1), \qquad (9.4.12)$$

we will study the closeness of the coverage probabilities (CPs) of the 95% asymptotic confidence intervals for mean μ to the nominal level, considering how the choice of bandwidths affects the closeness of these CPs.

In this simulation, $\{X_j\}$ is taken to be a linear Gaussian ARFIMA $(1, d, 0)$ process with unit standard deviation, and different values of ρ (AR parameter) and d. The number of replications is 5,000, and the sample size $n = 500, 1000$. The memory parameter d is estimated using the LW estimator $\widehat{d}$ with bandwidth $m_w = [n^{0.65}]$. Simulations carried out with $m_w = [n^{0.5}]$, $[n^{0.8}]$ were dominated by results obtained for $m_w = [n^{0.65}]$.

Table 9.4.1 shows the MSEs of the HAC estimator $\bar{s}_q^2(\widehat{d})$ when q is chosen according to the asymptotically-optimal rule (9.4.5). The MSEs are comparable to the optimal MSEs calculated for different values of the bandwidth q, except when d and ρ are simultaneously large. Table 9.4.3 shows the MSEs of the MAC estimator $\widehat{s}_m^2(\widehat{d})$ calculated for different values of the bandwidth m. The minimum MSE value for each d and n reveals the accuracy of the simple bandwidth rule that resulted from (9.4.6): almost all the optima are for $m = n^{4/5}$. Both tables show that the MSEs of the HAC and MAC estimators usually increase when $|d|$ or $|\rho|$ increase.

Tables 9.4.2 and 9.4.4 show CPs for μ using the HAC estimator $\bar{s}_q^2(\widehat{d})$ with q chosen by the rule (9.4.5) and the MAC estimator computed for the three bandwidths $m = [n^{0.6}]$, $[n^{0.7}]$, $[n^{0.8}]$, respectively. Both the HAC and MAC estimators give comparable CPs. CPs approach the nominal 95% level as the sample size increases. They are close to the 95% level except when $d \to 0.5$ or when ρ becomes negative. The bandwidth $m = [n^{0.8}]$ tends to give better CPs for MAC, and this is in line with the findings reported in Table 9.4.3.

These simulation results suggest the use of the straightforward choice (9.4.5) for bandwidth for HAC, and $m = [n^{0.8}]$ for MAC, and $m_w = [n^{0.65}]$

for the LW estimator of d. When the data do not follow an ARFIMA model, procedures based on the LW estimation of c_0 lead to better MSEs with smaller m, $m = [n^{0.65}]$, $[n^{0.7}]$. In a practical application of MAC, it would be advisable to reduce m to $m = [n^{0.65}]$, $[n^{0.7}]$, or use the automatic bandwidth choice m_{opt} of Section 8.5.

Note: Popular references in econometrics for the estimation of covariance matrices include White (1980), Newey and West (1987), and Andrews and Monahan (1992). In statistics, the literature goes further back to Jowett (1955) and Hannan (1957). The procedures for estimating covariance matrices account for heteroscedasticity and autocorrelation of unknown form, for short memory models. Robinson (2005b) introduced the MAC estimator of the covariance matrix and established its consistency, leaving open the issue of its higher-order expansion. Phillips, Sun and Jin (2007) considered HAC estimation with sharp origin kernels. The paper of Abadir *et al.* (2009) discusses second-order asymptotic expansions for HAC and MAC estimators for I(d) models. These results in turn are used to explain the problem of sensitivity to the choice of bandwidth and to provide some practical advice for the selection of the MSE-optimal bandwidth.

d	$n = 500$			$n = 1000$		
	$\rho = -0.5$	$\rho = 0$	$\rho = 0.5$	$\rho = -0.5$	$\rho = 0$	$\rho = 0.5$
$d = -0.4$	4.864	2.516	3.300	4.721	2.046	5.409
$d = -0.2$	0.457	0.306	2.068	0.439	0.258	3.808
$d = 0$	0.219	0.117	2.064	0.242	0.051	2.926
$d = 0.2$	0.224	0.049	3.231	0.241	0.029	4.093
$d = 0.4$	1.296	0.164	11.480	1.218	0.101	12.527

Table 9.4.1: MSEs of HAC estimator $\bar{s}_q^2(\widehat{d})$ when q is chosen using rule (9.4.5).

d	$n = 500$			$n = 1000$		
	$\rho = -0.5$	$\rho = 0$	$\rho = 0.5$	$\rho = -0.5$	$\rho = 0$	$\rho = 0.5$
$d = -0.4$	0.970	0.958	0.942	0.984	0.948	0.950
$d = -0.2$	0.932	0.918	0.958	0.928	0.922	0.956
$d = 0$	0.914	0.896	0.962	0.922	0.930	0.958
$d = 0.2$	0.890	0.878	0.970	0.926	0.910	0.970
$d = 0.4$	0.822	0.830	0.870	0.884	0.870	0.898

Table 9.4.2: CPs for μ using t-ratio (9.4.12) and HAC estimator $\bar{s}_q^2(\widehat{d})$ with q chosen using (9.4.5).

d	$n = 500$			$n = 1000$		
	$\rho = -0.5$	$\rho = 0$	$\rho = 0.5$	$\rho = -0.5$	$\rho = 0$	$\rho = 0.5$
	$m = \lfloor n^{0.6} \rfloor = 41$			$m = \lfloor n^{0.6} \rfloor = 63$		
$d = -0.4$	4.632	2.264	15.770	4.641	1.892	28.134
$d = -0.2$	0.519	0.656	4.186	0.552	0.494	6.278
$d = 0$	0.276	0.134	2.866	0.271	0.112	3.973
$d = 0.2$	0.260	0.059	5.616	0.280	0.034	5.269
$d = 0.4$	1.195	0.231	26.383	1.134	0.155	31.099
	$m = \lfloor n^{0.7} \rfloor = 77$			$m = \lfloor n^{0.7} \rfloor = 125$		
$d = -0.4$	4.460	1.913	9.449	4.365	1.668	19.986
$d = -0.2$	0.507	0.894	2.453	0.510	0.262	3.939
$d = 0$	0.241	0.081	1.644	0.263	0.051	2.605
$d = 0.2$	0.236	0.027	3.575	0.266	0.018	3.827
$d = 0.4$	1.157	0.194	14.549	1.136	0.130	19.078
	$m = \lfloor n^{0.8} \rfloor = 144$			$m = \lfloor n^{0.8} \rfloor = 251$		
$d = -0.4$	3.138	1.475	3.044	3.454	1.344	5.342
$d = -0.2$	0.329	0.211	0.900	0.364	0.111	1.133
$d = 0$	0.150	0.072	0.517	0.186	0.021	0.809
$d = 0.2$	0.140	0.018	1.682	0.185	0.010	1.381
$d = 0.4$	0.769	0.235	4.667	0.840	0.201	6.192

Table 9.4.3: MSEs of MAC estimator $\widehat{s}_m^2(\widehat{d})$.

d	$n = 500$			$n = 1000$		
	$\rho = -0.5$	$\rho = 0$	$\rho = 0.5$	$\rho = -0.5$	$\rho = 0$	$\rho = 0.5$
	$m = \lfloor n^{0.6} \rfloor = 41$			$m = \lfloor n^{0.6} \rfloor = 63$		
$d = -0.4$	0.968	0.984	0.992	0.962	0.986	0.980
$d = -0.2$	0.924	0.948	0.966	0.922	0.926	0.960
$d = 0$	0.894	0.912	0.962	0.912	0.922	0.972
$d = 0.2$	0.886	0.912	0.974	0.914	0.918	0.984
$d = 0.4$	0.838	0.856	0.934	0.844	0.868	0.932
	$m = \lfloor n^{0.7} \rfloor = 77$			$m = \lfloor n^{0.7} \rfloor = 125$		
$d = -0.4$	0.982	0.972	0.982	0.976	0.982	0.972
$d = -0.2$	0.942	0.940	0.948	0.912	0.934	0.970
$d = 0$	0.894	0.930	0.966	0.916	0.924	0.942
$d = 0.2$	0.880	0.896	0.958	0.896	0.940	0.944
$d = 0.4$	0.876	0.870	0.920	0.864	0.866	0.916
	$m = \lfloor n^{0.8} \rfloor = 144$			$m = \lfloor n^{0.8} \rfloor = 251$		
$d = -0.4$	0.990	0.984	0.948	0.996	0.974	0.962
$d = -0.2$	0.936	0.910	0.934	0.942	0.928	0.902
$d = 0$	0.916	0.890	0.922	0.936	0.924	0.936
$d = 0.2$	0.918	0.922	0.920	0.938	0.920	0.942
$d = 0.4$	0.902	0.862	0.828	0.912	0.884	0.852

Table 9.4.4: CPs for μ based on the t-ratio (9.4.12) and the MAC estimator $\widehat{s}_m^2(\widehat{d})$.

9.5 Testing for long memory and breaks

In this section we shall analyze two tests for the presence of long memory in the given time series, applicable to linear processes. These tests are based on a LW estimator of d and on a Lagrange multiplier-type test statistic by Lobato and Robinson (1998). We also discuss several tests, such as R/S, *KPSS*, and V/S tests, that do not require linearity of the observable process and are applicable to a wider class of processes. The re-scaled variance V/S test differs from the *KPSS* test because of correction using a mean and is shown to achieve a slightly better balance of size and power than the R/S and *KPSS* tests.

Assume that the observable process $\{X_j\}$ satisfies (9.3.1) with an unknown mean μ, where $\{\varepsilon_j\}$ is an I(d_0) process with zero mean and memory parameter $|d_0| < 1/2$. Recall that the spectral density f of an I(d_0) process satisfies assumption (9.2.2) and by (9.2.4),

$$\text{Var}(S_n) \sim s_\varepsilon^2 n^{1+2d_0}, \tag{9.5.1}$$

where s_ε^2 is the long-run variance of $\{\varepsilon_j\}$. Note that $\{\varepsilon_j\}$ does not have to be a linear process.

Now recall that $d_0 = 0$ means the process has short memory while for $0 < d_0 < 1/2$, it has long memory. Thus, we are led to test the null and alternative hypotheses

$$H_0 : d_0 = 0, \quad \text{vs.} \quad H_1 : 0 < d_0 < 1/2.$$

Lobato–Robinson test. Let $\{\varepsilon_j\}$ in (9.3.1) be the linear process in (9.4.4) and m be the bandwidth used in $\widehat{d}$ of (8.5.2).

Then a test for H_0 can be based on

$$t(m) = 2\sqrt{m}\,\widehat{d}$$

or the *Lobato–Robinson* statistic

$$t_{LR}(m) = -\frac{\sqrt{m}\sum_{j=1}^{m} \nu_j I_n(u_j)}{\sum_{j=1}^{m} I_n(u_j)},$$

computed using observations $X_1, \cdots, X_n$, where $m = m_n$ is the bandwidth parameter, $m \to \infty$, $m = o(n)$, and the ν_j's are as in (8.5.11). In the next proposition we assume that the spectral density f_ε of $\{\varepsilon_j\}$ has the property

$$f_\varepsilon(u) = |u|^{-2d_0}(c_0 + bu^2 + o(u^2)), \quad u \to 0, \ |d_0| < 1/2, \ c_0 > 0, \tag{9.5.2}$$

and the bandwidth m satisfies $m = o(n^{4/5})$.

Proposition 9.5.1. *Let $\{X_j\}$ be as in (9.3.1). Suppose that $\{\varepsilon_j\}$ is the linear process in (9.4.4) with transfer function A_ε as in Theorem 8.5.2(iii) and spectral density f_ε satisfying (9.5.2). Then the following hold:*
Under H_0 (short memory),

$$t(m) \to_D \mathcal{N}(0,1), \qquad t_{LB}(m) \to_D \mathcal{N}(0,1). \qquad (9.5.3)$$

Under H_1 (long memory),

$$t(m) \to_p +\infty, \quad t_{LB}(m) \to_p +\infty. \qquad (9.5.4)$$

The test that rejects H_0 in favor of H_1 whenever $t(m) > z_\alpha$ or whenever $t_{LB}(m) > z_\alpha$, has an asymptotic size $1 - \alpha$ and is consistent, i.e., $P(t(m) > z_\alpha) \to 1$ and $P(t_{LB}(m) > z_\alpha) \to 1$, if $d_0 > 0$.

The original Lobato–Robinson test, $t_{LR}^2(m)$, is asymptotically χ_1^2 distributed and was used to test H_0 against a two-sided alternative $H_1' : d_0 \neq 0$. The proof below shows that the tests $t_{LR}(m)$ and $t(m)$ are asymptotically equivalent under H_0. However, a simulation study indicates that $t_{LR}(m)$ is preferable because of the better finite sample size and power properties of the chosen models. Moreover, to implement these tests, it is essential to use the automatic bandwidth selection rule, e.g., taking $m = \widehat{m}^* := \min(\widetilde{m}_{opt}, 0.8n^{4/5})$ where $\widetilde{m}_{opt}$ is as in (8.5.49).

Proof of Proposition 9.5.1. By Theorems 8.5.2 and 8.5.1, under (9.2.2),

$$2\sqrt{m}(\hat{d} - d_0) = -\sqrt{m}\,T_m + o_p(1) \to_D \mathcal{N}(0,1), \qquad (9.5.5)$$

$$T_m = m^{-1} \sum_{j=1}^{m} \nu_j\, c_0^{-1} u_j^{2d_0} I_n(u_j),$$

$$m^{-1} \sum_{j=1}^{m} u_j^{2d_0} I_n(u_j) \to_p c_0.$$

Under H_0, (9.5.5) implies (9.5.3). Now suppose H_1 holds. Then

$$2\sqrt{m}\,\hat{d} = 2\sqrt{m}(\hat{d} - d_0) + 2\sqrt{m}\,d_0 = O_p(1) + 2\sqrt{m}\,d_0 \to_p +\infty,$$

by (9.5.5). To show $t_{LR}(m) \to_p +\infty$, let $\widetilde{\eta}_j = c_0^{-1} u_j^{2d_0} I_n(u_j)$. Rewrite

$$m^{-1/2} t_{LR}(m) = -\frac{m^{-1} \sum_{j=1}^{m} \nu_j\, (j/m)^{-2d_0} \widetilde{\eta}_j}{m^{-1} \sum_{j=1}^{m} (j/m)^{-2d_0} \widetilde{\eta}_j}.$$

By Theorem 8.5.2(i), $\{\varepsilon_j\}$ satisfies Assumption B, and thus the $\tilde{\eta}_j$'s have the property of WLLN in (8.5.32). Thus, by Lemma 8.5.1,

$$m^{-1/2}t_{LR}(m) \to_p -\frac{\int_0^1 (\log(u)+1)u^{-2d_0}\,du}{\int_0^1 u^{-2d_0}\,du} = \frac{2d_0}{1-2d_0} > 0,$$

which completes the proof of consistency (9.5.4).

Other tests for long memory. Let $\{X_j\}$ be as in (9.3.1). Testing procedures, based on the *R/S*, *KPSS*, and *V/S* tests do not require $\{\varepsilon_j\}$ in (9.3.1) to be a linear process with i.i.d. innovations.

Denote by $X_1, \cdots, X_n$ the observed sample, and by

$$S'_k = \sum_{j=1}^{k}(X_j - \bar{X}) \equiv \sum_{j=1}^{k}(\varepsilon_j - \bar{\varepsilon}), \ k = 1, \cdots, n,$$

the partial sums centered by the sample mean $\bar{X}$. Recall the partial-sum process $Y_n(v)$, $0 \le v \le 1$ of ε_j's from (9.2.5).

Since the *R/S* statistic is based on a max-type functional, to derive its asymptotic distribution, we shall need the following:

Assumption W. The process of partial sums Y_n converges weakly

$$Y_n \Rightarrow Y_\infty, \tag{9.5.6}$$

to a Gaussian limit Y_∞ in the space $\mathcal{D}[0,1]$, endowed with the Skorokhod topology and uniform metric.

The limit Y_∞ can be written as $Y_\infty = s_\varepsilon W_{d_0}$, $|d_0| < \infty$, where $W_{d_0} = B_{1/2+d_0}$ is the fractional Brownian motion and s_ε^2 is the long-run variance of $\{\varepsilon_j\}$. Assumption W is slightly stronger than the convergence of finite-dimensional distributions, Assumption FDD of Section 9.2.

The *R/S* test requires an estimator $\bar{s}_\varepsilon^2$ of the long-run variance, estimating s_ε^2 consistently in the case of short memory $d_0 = 0$. For that purpose, we shall use either the HAC estimator of (9.4.1) or the MAC estimator with $d_0 = 0$, respectively, given as

$$\bar{s}_\varepsilon^2 := \bar{s}_q^2(0) = \hat{\gamma}_0 + 2\sum_{k=1}^{q}(1 - k/q)\hat{\gamma}_k, \tag{9.5.7}$$

$$\bar{s}_\varepsilon^2 = 2\pi\,\hat{c}_m = 2\pi\,m^{-1}\sum_{j=1}^{m} I_n(u_j). \tag{9.5.8}$$

The HAC estimator is used in the literature when discussing the *R/S*, *KPSS*, and *V/S* tests for long memory.

We shall assume below that $\bar{s}_\varepsilon$ is either a HAC or MAC estimator. Under short memory, $d = 0$, we shall suppose that

$$\bar{s}_\varepsilon^2 \to_p s_\varepsilon^2. \tag{9.5.9}$$

In Section 9.4 it was shown that consistency (9.5.9) is valid for a wide class of I(0) processes; see Theorem 9.4.1.

Modified R/S test. The rescaled range, or R/S analysis discussed in Section 8.4, focuses on estimating the limit of the ratio

$$\log(R_n/S_n)/\log n,$$

called the Hurst coefficient. In Section 8.4 it was shown that the asymptotic distribution of the sequence of statistics $n^{-H}(R_n/S_n)$ depends on the correlation structure of the data and is not parameter free, and therefore it cannot be used to construct a test for H_0. To avoid this difficulty, Lo (1991) proposed the modified R/S statistic

$$R/S(n) := n^{-1/2}\frac{R_n}{\bar{s}_\varepsilon}, \quad R_n = \max_{j=1,\cdots,n} S_j - \min_{j=1,\cdots,n} S_j,$$

where $\bar{s}_\varepsilon^2$ is a consistent estimator of the long-run variance s_ε^2 of the ε_j's, e.g., (9.5.7) or (9.5.8). He applied it in a statistical hypothesis testing procedure to detect long memory of observables $\{X_j\}$.

The modified R/S statistic has been extensively used to detect long memory in speculative assets. By allowing the bandwidth q of the HAC estimator $\bar{s}_\varepsilon$ to increase slowly with the sample size, the asymptotic null distribution of $R/S(n)$ is parameter free and robust to many forms of weak dependence in the data.

The *KPSS* statistic was introduced by Kwiatkowski, Phillips, Schmidt, and Shin (1992) as a tool for testing trend stationarity against a unit root alternative. In the context of testing for long memory in a stationary sequence this statistic is equal to

$$KPSS(n) = \frac{1}{\bar{s}_\varepsilon^2 n^2} \sum_{k=1}^{N} S_k'^2. \tag{9.5.10}$$

The *V/S* statistic or rescaled variance statistic introduced in Giraitis, Kokoszka, Leipus and Teyssière (2003) is of the form

$$V/S(n) = \frac{1}{\bar{s}_\varepsilon^2 n^2} \sum_{k=1}^{n} (S_k' - \bar{S}')^2, \quad \bar{S}' = n^{-1} \sum_{k=1}^{n} S_k'.$$

The range of the partial sums S'_k used in the R/S statistic is replaced by the second moment of S'_k in the *KPSS* statistic and by the centered second moment in the *V/S* statistic. The *KPSS* statistic (9.5.10) may be seen as a rescaled sample second moment of the partial sums. It is expected that the "corrected for a mean" statistic *V/S* will be more sensitive to "shifts in variance" than *KPSS* and will have more power than both the *R/S* and *KPSS* statistics against long memory. This is confirmed by simulation experiments.

The above asymptotic distributions and mutual relationships between the *R/S*, *KPSS*, and *V/S* statistics are basically the same as for the corresponding statistics based on the empirical distribution function $\hat{F}_n(u)$, used by Kuiper (1960) and Watson (1961) in the context of goodness-of-fit testing on the circle. The classical Kolmogorov and von Mises statistics are well suited for real-valued observations. To adjust them to test for the goodness-of-fit of observations on a circle, Kuiper (1960) used the modified classical Kolmogorov statistic

$$K_n = \sup_{0 \leq u \leq 1} (\hat{F}_n(u) - F(u)) - \inf_{0 \leq u \leq 1} (\hat{F}_n(u) - F(u)),$$

where $\hat{F}_n(u)$ is the empirical d.f. of the variables $X_1, \cdots, X_n$ taking values on a circle of unit length, and $F(u)$, $0 \leq u \leq 1$, is the d.f. The Watson statistic W_n is analogous to the von Mises statistic and has the form

$$W_n = n \left[\int_0^1 (\hat{F}_n(u) - F(u))^2 dF(u) - \left[\int_0^1 (\hat{F}_n(u) - F(u)) dF(u) \right]^2 \right].$$

The K_n- and W_n-tests have been shown to be powerful when we are more interested in the discrimination against shifts in variance than against shifts in mean.

Testing for long memory. Below we assume that $\{X_j\}$ is an I(d_0) process with $0 \leq d_0 < 1/2$. *R/S*, *KPSS*, and *V/S* statistics are used to differentiate between I(0) (short memory) and I(d_0) (long memory, $d_0 \in (0, 1/2)$) processes $\{X_j\}$. Under short memory we impose Assumption W, and derive the asymptotic distributions. Under long memory, we show the consistency of the tests. In the latter case Assumption W is not required.

Let $B^0(t) = B(t) - tB(1)$ denote a Brownian bridge, corresponding to Brownian motion $B(t) := B_{1/2}(t)$.

Proposition 9.5.2. *Under H_0, Assumption W, and (9.5.9),*

$$R/S(n) \to_D U_{R/S}, \quad U_{R/S} = \max_{0 \leq v \leq 1} B^0(v) - \min_{0 \leq v \leq 1} B^0(v). \quad (9.5.11)$$

Under H_1 (long memory), $R/S(n) \to_p \infty$.

Proof. Write

$$R_n = n^{1/2+d_0} \Big(\sup_{0 \leq v \leq 1} Y_n(v) - \inf_{0 \leq v \leq 1} Y_n(v) \Big).$$

Since Y_n satisfies Assumption W, by the continuous mapping theorem

$$n^{-1/2-d_0} R_n \to_D s_\varepsilon \big(\sup_{0 \leq v \leq 1} W_{d_0}(v) - \inf_{0 \leq v \leq 1} W_{d_0}(v) \big) \quad (9.5.12)$$

$$=: s_\varepsilon U_{R/S}(d_0).$$

Under H_0, $d_0 = 0$, (9.5.12) together with assumption (9.5.9) implies (9.5.11). Under H_1, $0 < d_0 < 1/2$, to prove $R/S(n) \to_p \infty$, use the fact that by auxiliary Lemma 9.5.1 below, for an $I(d_0)$ process $\{X_j\}$, both HAC and MAC estimators have the property $E\bar{s}_\varepsilon^2 = o(n^{2d_0})$. Thus, $\bar{s}_\varepsilon/n^{d_0} \to_p 0$ and by (9.5.12), for any $c > 0$,

$$P\big(R/S(n) \geq c\big) = P\Big(\frac{n^{-1/2}R_n}{\bar{s}_\varepsilon} \geq c\Big) = P(n^{-1/2-d_0}R_n \geq c\bar{s}_\varepsilon n^{-d_0})$$

$$\to P(U_{R/S} > 0) = 1.$$

Test. Let c_α denote the upper-α percentile of the limit distribution $U_{R/S}$, defined by $P(U_{R/S} > c_\alpha) = \alpha$. Proposition 9.5.2 implies that the test that rejects H_0 in favor of H_1, when $R/S(n) > c_\alpha$, is of asymptotic size α, and consistent against H_1, i.e.,

$$P\big(R/S(n) > c_\alpha\big) \to \alpha, \qquad d_0 = 0,$$

$$P\big(R/S(n) > c_\alpha\big) \to 1, \qquad 0 < d_0 < 1/2.$$

Remark 9.5.1. The distribution function of the limiting r.v. $U_{R/S}$ is the same as in Kuiper (1960):

$$F_{U_{R/S}}(x) = 1 + 2\sum_{k=1}^{\infty} (1 - 4k^2 x^2)e^{-2k^2 x^2}, \quad x \geq 0.$$

It can be shown that

$$EU_{R/S} = \sqrt{\pi/2}, \quad \text{Var}(U_{R/S}) = \pi(\pi - 3)/6.$$

For $\alpha = .025, .05, .1$, the upper-α R/S percentiles are $c_{.025} = 1.862$, $c_{.05} = 1.747$, and $c_{.1} = 1.620$.

Proposition 9.5.3. *Under H_0, Assumption W, and (9.5.9),*

$$KPSS(n) \to_D U_{KPSS}, \quad U_{KPSS} = \int_0^1 (B^0(u))^2 du. \quad (9.5.13)$$

Under H_1,

$$KPSS(n) \to_p \infty. \quad (9.5.14)$$

Remark 9.5.2. The distribution function of the random variable U_{KPSS} has a series expansion in terms of special functions, which converges very fast and can be used to tabulate the distribution of U_{KPSS} and can be found in Kiefer (1959). Rosenblatt (1952) showed that the random variable U_{KPSS} has a series representation

$$U_{KPSS} = \frac{1}{\pi^2} \sum_{j=1}^{\infty} Z_j^2/j^2,$$

where the Z_j's are independent standard normal variables. From this formula, we readily find

$$EU_{KPSS} = 1/6, \quad \mathrm{Var}(U_{KPSS}) = 1/45.$$

For $\alpha = .01, .05$, and $.10$, the upper-α percentiles of U_{KPSS} are 0.739, 0.463, and 0.347, respectively.

Next, let

$$U_{V/S} := \int_0^1 \left(B^0(u) - \int_0^1 B^0(v)dv\right)^2 du.$$

We have:

Proposition 9.5.4. *Under H_0, Assumption W, and (9.5.9),*

$$V/S(n) \to_D U_{V/S}. \quad (9.5.15)$$

Under H_1,

$$V/S(n) \to_D \infty. \quad (9.5.16)$$

Remark 9.5.3. The distribution function of the random variable $U_{V/S}$ is given by the formula

$$F_{U_{V/S}}(x) = 1 + 2\sum_{k=1}^{\infty}(-1)^k e^{-2k^2\pi^2 x}, \quad x \geq 0,$$

established by Watson (1961) in the context of goodness-of-fit tests on a circle. Note that $F_{U_{V/S}}(x) = F_K(\pi\sqrt{x})$, where F_K is the asymptotic distribution function of the standard Kolmogorov statistic $\sup_{0<v<1} \sqrt{n}(\hat{F}_n(v)-v)$. Hence, after a simple transformation the asymptotic distribution of V/S coincides with the limiting distribution of the standard Kolmogorov statistic. Moreover, $U_{V/S}$ has a representation

$$U_{V/S} = \frac{1}{4\pi^2} \sum_{j=1}^{\infty} \frac{(Z_{2j-1}^2 + Z_{2j}^2)}{j^2},$$

where the Z_j's are independent standard normal variables. Hence, $U_{V/S} = (U_1 + U_2)/4$, where U_1 and U_2 are two independent copies of the random variable U_{KPSS} of (9.5.13). It follows that

$$EU_{V/S} = 1/12, \quad \text{Var}(U_{V/S}) = 1/360.$$

Thus, the V/S statistic has a smaller variance than the *KPSS* statistic.

For $\alpha = .05$ and .1 the upper-α V/S percentiles are 0.190 and 0.153, respectively.

Proof of Propositions 9.5.3 and 9.5.4. Using the partial-sum process $Y_n(v)$, and noting that $S'_n = 0$, we rewrite,

$$n^{-2-2d_0} \sum_{k=1}^{n} S_k'^2 = \int_0^1 \left(Y_n(v) - \frac{[nv]}{n}Y_n(1)\right)^2 dv,$$

$$n^{-2-2d_0} \sum_{k=1}^{n} (S'_k - \bar{S}')^2 = \int_0^1 \left(Y_n(v) - \frac{[nv]}{n}Y_n(1)\right)^2 dv$$
$$- \left(\int_0^1 (Y_n(u) - \frac{[nu]}{n}Y_n(1))du\right)^2.$$

Then, by Assumption W and the continuous mapping theorem,

$$n^{-2-2d_0} \sum_{k=1}^{n} S_k'^2 \to_D s_\varepsilon^2 \int_0^1 \left(W_{d_0}(v) - vW_{d_0}(1)\right)^2 dv, \qquad (9.5.17)$$

$$n^{-2-2d_0} \sum_{k=1}^{n} (S'_k - \bar{S}')^2 \to_D s_\varepsilon^2 \left[\int_0^1 \left(W_{d_0}(v) - vW_{d_0}(1)\right)^2 dv\right.$$
$$\left. - \left(\int_0^1 (W_{d_0}(u) - uW_{d_0}(1))du\right)^2\right].$$

If H_0 holds, then $d_0 = 0$ and the convergence above is

$$n^{-2} \sum_{k=1}^{n} S_k'^2 \to_D s_\varepsilon^2 U_{KPSS}, \quad n^{-2} \sum_{k=1}^{n} (S'_k - \bar{S}')^2 \to_D s_\varepsilon^2 U_{V/S},$$

which together with (9.5.9) proves (9.5.13) and (9.5.15).

Consistency results (9.5.14) and (9.5.16) follow using Lemma 9.5.1 and (9.5.17) by the same argument as in the proof of Proposition 9.5.2.

Lemma 9.5.1. *Let $\{X_j\}$ be as in (9.3.1). Assume that $\{\varepsilon_j\}$ follows an I(d_0) model with $d_0 \in [0, 1/2)$, and $\bar{s}_\varepsilon^2$ is either a HAC or MAC estimator as in (9.5.7) and (9.5.8). Then*

$$E\bar{s}_\varepsilon^2 = o(n^{2d_0}), \quad d_0 \in (0, 1/2), \tag{9.5.18}$$

$$E\bar{s}_\varepsilon^2 = O(1), \quad d_0 = 0. \tag{9.5.19}$$

Proof. *HAC case.* Let $\bar{s}_\varepsilon^2$ be the HAC estimator (9.5.7). By (9.4.7), $\bar{s}_\varepsilon^2 \sim s_\varepsilon^2 q^{2d_0}$, for $q \leq Cn^\gamma$, $0 < \gamma < 1$, which implies (9.5.18) and (9.5.19).

MAC case. Let $\bar{s}_\varepsilon^2$ be the MAC estimator (9.5.8). By assumption (9.2.2), $f(u) \sim c|u|^{-2d_0}$, as $u \to 0$. Then, by Theorem 5.2.1(v),

$$EI_n(u_j)/f(u_j) \leq C, \quad 0 < u_j \leq u_m, \; u_m \to 0,$$

$$E[\bar{s}_\varepsilon^2] = 2\pi m^{-1} \sum_{j=1}^{m} EI_n(u_j) \leq Cm^{-1} \sum_{j=1}^{m} f(u_j)$$

$$\leq Cm^{-1} \sum_{j=1}^{m} (j/n)^{-2d_0} \leq C(n/m)^{2d_0},$$

which proves (9.5.18) and (9.5.19) because $m \to \infty$.

Testing for breaks. In various applications, e.g., analyzing financial and economic data, it is useful to perform the test for I(0) covariance stationarity as a *null hypothesis* against instabilities or breaks in the mean.

We shall assume that under H_0, $X_1, \cdots, X_n$ is a sample from a short memory I(0) (covariance) stationary process, whereas the alternatives H_1', $X_1, \cdots, X_n$ are generated by the non-stationary trend model

$$X_j = \mu + g_n(j) + \varepsilon_j, \tag{9.5.20}$$

where $g_n(j)$ is a deterministic trend and $\{\varepsilon_j\}$ is a second-order stationary I(0) process. The scheme covers a wide class of deterministic trends, such as multiple change points, non-parametric regression, and polynomial functions.

It is well known that in practice it is difficult to distinguish small trends graphically, and that tests designed to detect long memory often mistake long memory for a "spurious trend" and vice versa. We shall see below that the V/S test can be used not only to detect long memory, but also "small changes", e.g., multiple change points and deterministic trends. For short memory I(0) errors $\{\varepsilon_j\}$, the V/S test has exceptionally good power properties.

The following assumption for the trend functions $g_n(j)$ describes the class of deterministic trends that can be detected by the V/S statistic.

Assumption G. There exist constants $\gamma > -1/2$ and $0 < C < \infty$ and a function $g_\infty \in L_2[0,1]$ such that

$$n^{-\gamma}|g_n([nu])| < C, \qquad n \geq 1, \quad 0 \leq u \leq 1,$$

$$n^{-\gamma}g_n([nu]) \to g_\infty(u), \quad \text{a.e.}, \quad u \in [0,1], \quad n \to \infty.$$

We assume that $g_\infty(t)$ is not constant in $L_2[0,1]$.

Examples that satisfy this assumption include polynomial trends and non-parametric regression models. A polynomial trend with

$$g_n(j) = cj^\beta + o(j^\beta), \quad c \neq 0, \ \beta \geq 0$$

satisfies Assumption G with $\gamma = \beta$.

A non-parametric regression model where

$$g_n(k) = \mu(k/n), \quad k = 1, \cdots, n,$$

with $\mu(u)$, $u \in [0, 1]$, being a continuous function, also satisfies Assumption G with $\gamma = 0$. This class of trends also covers structural breaks in the mean (single and multiple change points) of the type

$$g_n(k) = \mu_1, \quad \text{for } 0 \leq k \leq [n\tau],$$

$$= \mu_2, \quad \text{for } [n\tau] < k \leq n, \text{ and some } 0 < \tau < 1.$$

Set $G(v) = \int_0^v g_\infty(u)du$, $0 \leq v \leq 1$. Then

$$B = \int_0^1 \left(G(v) - vG(1) - \int_0^1 (G(u) - uG(1))du \right)^2 dv > 0.$$

Proposition 9.5.5. *Let $\{X_j\}$ be as in (9.5.20). Assume that $\{\varepsilon_j\}$ follows the short memory model I(0) and satisfies Assumption W with $d_0 = 0$.*
Under H_0 and (9.5.9),

$$V/S(n) \to_D U_{V/S}. \tag{9.5.21}$$

Under H_1' and $B > 0$, $V/S(n) \to_p \infty$.

Let c_α denote the upper-α percentile of the limit distribution $U_{V/S}$. A consequence of the above proposition is that the test that rejects H_0, $d_0 = 0$, in favor of H_1' (a deterministic trend), whenever $V/S(n) > c_\alpha$, is of asymptotic size α and consistent. Here, $c_{.05} = 0.190$ and $c_{.1} = 0.153$.

The V/S test for I(0) covariance stationarity is a pretest in the sense that the rejection of the null hypothesis does not provide information about the alternative. Consequently, differentiation between a deterministic trend and long memory requires further investigation. The assumptions are weak and therefore the test might be applicable for linear and non-linear models such as conditionally heteroscedastic time series.

Proof. Under H_0, (9.5.21) follows from (9.5.15).

Assume that H_1' holds. Write $V/S(n) = V_n/\bar{s}_\varepsilon^2$, where

$$V_n = n^{-2}\sum_{k=1}^{n}(S_k^* - \bar{S}^*)^2, \quad S_k^* = \sum_{j=1}^{k}(X_j - \mu) = \sum_{j=1}^{k}(\varepsilon_j + g_n(j)).$$

Under H_1', $\{\varepsilon_j\}$ satisfies Assumption W with $d_0 = 0$. Hence, for $\gamma > -1/2$, $n^{-1-\gamma}\sum_{j=1}^{[n\cdot]}\varepsilon_j \Rightarrow 0$, in $\mathcal{D}[0,1]$. Since the $g_n(j)$'s satisfy Assumption G, by the DCT, $\forall v \in (0,1)$,

$$n^{-1-\gamma}\sum_{j=1}^{[nv]}g_n(j) \sim \int_0^v n^{-\gamma}g_n([nu])du \to \int_0^v g_\infty(u)du = G(v).$$

Thus, $Y_n^*(\cdot) \equiv n^{-1-\gamma}\sum_{j=1}^{[n\cdot]}(\varepsilon_j + g_n(j)) \Rightarrow G(\cdot)$, which, as in (9.5.17), implies

$$n^{-1-2\gamma}V_n \sim \int_0^1 \left(Y_n^*(v) - vY_n^*(1) - \int_0^1(Y_n^*(u) - uY_n^*(1))du\right)^2 dv$$
$$\to_p B > 0.$$

We shall show below that

$$n^{-1-2\gamma}E\bar{s}_\varepsilon^2 \to_p 0, \tag{9.5.22}$$

which yields $n^{-1-2\gamma}\bar{s}_\varepsilon^2 = o_p(1)$. Then, for any $0 < C < \infty$,

$$P(V/S(n) \geq C) = P\left(n^{-1-2\gamma}V_n \geq Cn^{-1-2\gamma}\bar{s}_\varepsilon^2\right) \to P(B > 0) = 1,$$

which proves the consistency $V/S(n) \to_p \infty$.

Proof of (9.5.22). Assume first that $\bar{s}_\varepsilon^2$ is the HAC estimator $\bar{s}_q^2(0)$ of (9.5.7). Let $\bar{s}_{X,q}^2 = q^{-1}\sum_{j,k=1}^{q}\hat{\gamma}_{|j-k|}$ stand for $\bar{s}_q^2(0)$, computed using

observations $X_1, \cdots, X_n$. Use $\hat{\gamma}_j = \int_\Pi e^{iju} I_X^*(u) du$, $|j| \le n$, $I_X^*(u) :=$
$(2\pi n)^{-1} |\sum_{k=1}^n e^{iku}(X_k - \bar{X})|^2$, to write

$$\bar{s}_{X,q}^2 = \int_\Pi b_n(u) I_X^*(u) du, \quad b_n(u) := q^{-1} |D_q(u)|^2.$$

Since $X_j - \mu = \varepsilon_j + g_n(j)$,

$$I_X^*(u) \le 2(I_\varepsilon^*(u) + I_g^*(u)), \quad s_{X,q}^2 \le 2(s_{\varepsilon,q}^2 + s_{g,q}^2). \tag{9.5.23}$$

By (9.5.19), $E s_{\varepsilon,q}^2 = O(1)$ and $n^{-1-2\gamma} E s_{\varepsilon,q}^2 \to 0$. Thus, to prove (9.5.22),
it remains to show that $n^{-1-2\gamma} s_{g,q}^2 \to 0$.

To bound $s_{g,q}^2$, observe that by Assumption G and the DCT,

$$n^{-1-2\gamma} \sum_{j=1}^n (g_n(j) - \bar{g}_n)^2 \le n^{-1-2\gamma} \sum_{j=1}^n g_n^2(j) \to \int_0^1 g_\infty^2(u) du. \tag{9.5.24}$$

Therefore, $|D_q(u)| \le q$, for all $u \in \Pi$, implies

$$s_{g,q}^2 \le q \int_\Pi I_g^*(u) du = q (2\pi n)^{-1} \sum_{j=1}^n g_n^2(j) \le C(\frac{q}{n}) n^{1+2\gamma} = o(n^{1+2\gamma}),$$

since $q = o(n)$, which implies (9.5.22) for the HAC estimator.

Let $\bar{s}_\varepsilon^2$ be the MAC estimator (9.5.8). By (9.5.23),

$$\bar{s}_\varepsilon^2 \le \frac{4\pi}{m} \left(\sum_{j=1}^m I_\varepsilon(u_j) + \sum_{j=1}^m I_g(u_j) \right) =: 2(\bar{s}_{\varepsilon,1}^2 + \bar{s}_{\varepsilon,2}^2).$$

By (9.5.19), $E\bar{s}_{\varepsilon,1}^2 = O(1) = o(n^{1+2\gamma})$. Since $\sum_{j=1}^n e^{iu_j k} = \sum_{j=1}^n e^{iju_k} = 0$
for $1 \le k \le n-1$, then

$$\bar{s}_{\varepsilon,2}^2 \le 4\pi m^{-1} \sum_{j=1}^n I_g(u_j) \le 2(mn)^{-1} \sum_{t,s=1}^n \sum_{j=1}^n e^{iu_j(t-s)} g_n(t) g_n(s)$$

$$= 2m^{-1} \sum_{t=1}^n g_n^2(t) \le Cm^{-1} n^{1+2\gamma} = o(n^{1+2\gamma}),$$

because of (9.5.24) and $m \to \infty$, which implies (9.5.22) for the MAC estimator.

Note: The above discussion benefited from Giraitis *et al.* (2003) and Giraitis, Leipus, and Philippe (2006).

Empirical size and power. (1) The simulation study examines the finite sample performance of the test procedures developed in the previous sections. The goal is to compare the $t(m)$, Lobato–Robinson t_{LR}, R/S, $KPSS$, and V/S tests. Sample sizes $n = 500$ and $n = 1000$ are used, repeated 10,000 times. The tables show the rate of rejection of a short memory null hypothesis H_0. In the R/S, $KPSS$, and V/S tests the long-run variance is estimated using the HAC estimator $\bar{s}_\varepsilon$, (9.5.7). The tests $t(m)$ and $t_{LR}(m)$ are computed with $m = \hat{m}^* = \min(\tilde{m}_{opt}, 0.8n^{4/5})$, where $\tilde{m}_{opt}$ is as in (8.5.49). The use of the automatic bandwidth selection rule is critical here, since fixed bandwidth choices may lead to significant size distortions.

Table 9.5.1 shows the empirical sizes of the tests for the AR(1) model $X_j = \rho X_{j-1} + \zeta_j$, where the ζ_j's are i.i.d. $\mathcal{N}(0,1)$ r.v.'s and ρ takes values $0, 0.5, 0.8$. It shows that when parameter ρ increases, the increase in short range dependence starts affecting the size. For short memory $d = 0$, the results for the HAC estimator derived in Section 9.4 suggest using the rate optimal bandwidth $q_{opt} = n^{1/3}$. For $n = 500, 1000$, this gives $q_{opt} = 8, 10$. The simulation uses values $q = 5, 10, 20$.

Table 9.5.2 compares the power of the tests under three long memory alternatives. The I(d) process $\{\varepsilon_j\}$ is modeled by the fractional ARFIMA$(0, d, 0)$ model with $d = 0.2$, $d = 0.3$, and $d = 0.4$. The simulations show that the V/S test has slightly higher power than the $KPSS$ test. The power of the R/S, $KPSS$, and V/S tests decreases when q increases. With an increase of q, the power of the R/S test deteriorates and it becomes less powerful than the $KPSS$ test. Table 9.5.2 shows that q should not be chosen too large, since under the long memory alternative, as q increases, these statistics have a strong bias towards accepting the null hypothesis, i.e., they show "spurious" short memory.

The simulations indicate that the V/S test achieves a somewhat better balance of size and power than the R/S and $KPSS$ tests, the R/S test suffers from size distortion and small power for large q, and the $KPSS$ test has almost uniformly smaller power. These tests can be used also for non-linear processes. For the linear models studied here, the Lobato–Robinson test distinguishes between short and long memory hypotheses with very high precision and should clearly be used in conjunction with the R/S-type tests. Besides long memory testing, the *KPSS* and *V/S* tests might be applicable to unit root testing; see Kwiatkowski *et al.* (1992), and Giraitis *et al.* (2006).

(2) It is natural to expect that the *V/S* test may exhibit better power properties detecting deterministic trends and breakpoints than detecting long memory. This is confirmed by the simulations. Consider the stationary short memory AR(1) model $X_j = \rho X_{j-1} + \varepsilon_j$, where $|\rho| < 1$ and ε_j are standard normal i.i.d. variables. The empirical size of the *V/S* test for different values of ρ is shown in Figure 9.5.2. The long-run variance is estimated by the HAC estimator with the bandwidth $q = n^{1/2}$. We observe the distortion of the size when the parameter $\rho > 0.7$ is close to the non-stationarity region $\rho = 1$. The test is moderately conservative (undersized) for $300 \le n \le 500$ and has good size for large $n > 500$.

Figure 9.5.1 illustrates the power of the *V/S* test under three types of alternatives:

(a) the unit root model $X_j = X_{j-1} + \xi_j$ with MA(1) errors $\xi_j = a\varepsilon_{j-1} + \varepsilon_j$, where ε_j are standard normal i.i.d. variables, and the parameter a varies from -1 to 0 where $a = -1$ with initialization $X_0 = 0$ corresponds to a short memory I(0) process,

(b) the linear trend model $X_j = cj + \varepsilon_j$,

(c) the change point model $X_j = \Delta I_{\{j > [n\tau]\}} + \varepsilon_j$.

The alternatives have parameters a, c, and Δ, respectively. In all cases the *V/S* test has good power properties. The top graph shows that the unit root test is powerful even for moderate sample sizes $n \ge 300$. The middle graph shows that the linear trend cj is effectively detected even for very small $c \ge 0.005$ with sample size $n \ge 300$ and the power tends to 1 when c increases. The bottom graph shows that small changes in the mean can be detected with probability 0.75 when the magnitude of the jump $\Delta \ge 0.35$ and $n \ge 500$.

n	q	$\rho = 0$		$\rho = 0.5$		$\rho = 0.8$	
		10%	5%	10%	5%	10%	5%
		MODIFIED R/S TEST					
500	5	6.88	2.94	14.06	7.46	43.41	30.87
	10	6.24	2.24	8.25	3.40	18.32	10.07
	20	5.12	1.38	4.83	1.39	6.32	2.01
1000	5	7.61	3.74	16.64	9.31	50.28	37.55
	10	7.19	3.50	10.62	5.03	24.54	15.19
	20	6.88	2.93	7.42	3.36	10.98	5.22
		$KPSS$ TEST					
500	5	9.71	5.00	15.99	8.90	35.46	24.11
	10	9.82	4.90	12.82	6.78	22.72	13.35
	20	9.71	4.56	11.27	5.38	15.04	8.12
1000	5	10.16	4.99	16.55	9.29	36.84	25.08
	10	10.31	4.85	13.13	6.73	23.42	14.20
	20	9.97	4.70	11.60	5.74	15.86	8.61
		V/S TEST					
500	5	9.64	4.55	18.58	10.47	48.51	35.27
	10	9.29	4.24	13.59	6.64	28.51	17.56
	20	8.35	3.44	10.43	4.44	16.89	7.92
1000	5	9.84	5.05	18.85	11.06	49.22	36.01
	10	9.94	4.87	14.19	7.46	28.94	18.56
	20	9.64	4.49	11.83	5.73	17.73	10.06
		$t(\widehat{m}^*)$ TEST					
500	$\widehat{m}^*$	10.93	5.45	24.73	16.08	28.68	21.17
1000	$\widehat{m}^*$	10.14	4.79	23.42	14.27	26.44	17.58
		$t_{LR}(\widehat{m}^*)$ TEST					
500	$\widehat{m}^*$	9.21	4.58	14.78	8.03	10.67	5.09
1000	$\widehat{m}^*$	9.35	4.53	16.67	9.43	13.2	7.09

Table 9.5.1: Empirical sizes (in %) of the modified R/S, $KPSS$, V/S, $t(\widehat{m}^*)$, and $t_{LR}(\widehat{m}^*)$ tests under the null hypothesis for the AR(1) model, $X_j = \rho X_{j-1} + \zeta_j$, with $\zeta_j \sim \mathcal{N}(0,1)$.

n	q	$d = 0.2$		$d = 0.3$		$d = 0.4$	
		10%	5%	10%	5%	10%	5%
		MODIFIED R/S TEST					
500	5	53.13	41.53	75.66	65.33	88.31	82.28
	10	37.90	26.80	56.47	44.01	71.62	61.03
	20	21.92	10.66	32.38	19.40	43.93	29.05
1000	5	70.29	59.51	89.63	83.97	96.98	94.48
	10	55.77	43.97	77.41	67.94	89.81	84.12
	20	39.21	27.28	57.67	45.82	72.73	62.50
		$KPSS$ TEST					
500	5	47.95	35.57	66.78	55.09	81.01	72.04
	10	39.28	27.67	55.35	43.27	69.41	57.78
	20	31.01	20.38	43.44	31.46	55.04	42.58
1000	5	57.54	45.34	79.01	68.91	90.55	84.52
	10	48.59	36.60	68.49	56.56	82.52	73.81
	20	39.51	27.98	55.80	43.91	70.06	59.04
		V/S TEST					
500	5	58.04	45.68	78.07	68.27	90.01	84.04
	10	46.82	34.86	65.24	53.74	78.74	69.40
	20	35.71	23.08	50.06	36.91	62.64	50.10
1000	5	69.44	58.40	89.08	82.11	96.67	93.73
	10	58.84	46.98	79.03	69.80	90.67	85.17
	20	46.96	34.48	65.32	54.12	79.46	70.10
		$t(\widehat{m}^*)$ TEST					
500	$\widehat{m}^*$	95.08	91.16	99.07	98.08	99.62	99.2
1000	$\widehat{m}^*$	99.61	99.06	99.98	99.91	100	99.98
		$t_{LR}(\widehat{m}^*)$ TEST					
500	$\widehat{m}^*$	92.94	87.55	97.97	95.82	98.84	97.43
1000	$\widehat{m}^*$	99.34	98.67	99.92	99.86	99.97	99.9

Table 9.5.2: Empirical power of the tests (in %) based on the modified R/S, $KPSS$, V/S, $t(\widehat{m}^*)$, and $t_{LR}(\widehat{m}^*)$ tests. The alternatives considered are ARFIMA(0,d,0) with $d = 0.2$, $d = 0.3$, and $d = 0.4$. The innovations ζ_j are standard normal. Each row is based on 10,000 replications.

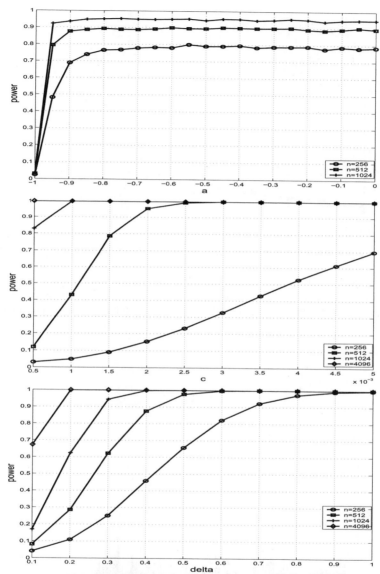

Fig. 9.5.1: Empirical power of the V/S test under (top) the unit root model $X_j = X_{j-1} + \xi_j$, where $\xi_j = a\zeta_{j-1} + \zeta_j$ with a varying from -1 to 0; (middle) the linear-trend model $X_j = cj + \zeta_j$ with c varying from 0.0005 to 0.005; (bottom) the change point model $X_j = \Delta I_{\{j>[n/2]\}} + \zeta_j$ with Δ varying from 0.1 to 1. The innovations ζ_j are i.i.d. $\mathcal{N}(0,1)$. The significance level $\alpha = 5\%$, the sample size $n = 256$, 512, 1024 in the top graph and $n = 256$, 512, 1024, 4096 in the two bottom graphs, all repeated 5,000 times.

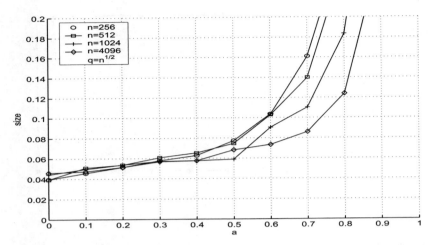

Fig. 9.5.2: Empirical size of the V/S test under the AR(1) model $X_j = \rho X_{j-1} + \zeta_j$, where the ζ_j are i.i.d. $\mathcal{N}(0,1)$. The significance level $\alpha = 5\%$ and the sample size $n = 256$, 512, 1024, 4096. Estimations are based on 5,000 independent replications.

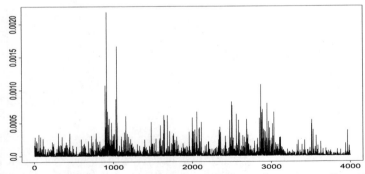

Fig. 9.5.3: Squared returns for the daily pound/dollar exchange rate. Four thousand observations ending January 21, 1997.

Example 9.5.1. Application to exchange rate data. We illustrate the theory and simulation by applying the tests to the squared returns r_j^2 of exchange rate data. To make the theoretical results comparable with the simulation, we divided a series of 4,000 daily returns for the pound/dollar exchange rate, shown in Figure 9.5.3, into four blocks of equal length. The corresponding p-values are displayed in Table 9.5.3. The evidence against the short memory null hypothesis in favor of the long memory alternative is strong for all blocks except perhaps the second one. The p-values, say, for the first block, are so small that one is inclined to believe that the data exhibit some other forms of departure from the null than the alternative discussed here. This may be due to some form of non-stationarity, e.g., a change in the parameters of the conditional heteroscedasticity process, or due to heavy tails of marginal distributions of the r_j^2. The considered series of r_j^2 exhibits the main second-order feature of long memory processes: a hyperbolic rate of decay of the autocorrelation function (and a pole in the periodogram near the zero frequency). The LW estimator of the long memory parameter yields $\hat{d} = 0.3627$. However, Figure 9.5.3 shows that such series do not display local spurious trends, which are common in linear ARFIMA-type models with a degree of persistence.

	Modified R/S	KPSS	V/S
$q = 5$	0.001	0.000	0.000
	0.074	0.024	0.067
	0.015	0.014	0.008
	0.001	0.000	0.001
$q = 11$	0.017	0.002	0.001
	0.148	0.042	0.117
	0.055	0.032	0.024
	0.015	0.000	0.009

Table 9.5.3: p-values of the tests based on the modified R/S, KPSS, and V/S statistics. For each value of q the four entries are the p-values corresponding to the four blocks shown in Figure 9.5.3.

Note: A number of tools for detecting the presence of long memory in a time series are used in statistics and econometrics. The *KPSS* test was suggested by Kwiatkowski *et al.* (1992). Lobato and Robinson (1998) considered a non-parametric test for I(0) model against long memory alternatives. The rescaled variance test was introduced by Giraitis *et al.* (2003). A useful guide for the practical use of *R/S*-type, Lobato–Robinson, *KPSS*, and *V/S* tests is provided by the empirical studies of Kirman and Teyssière (2002). They show that the Lobato–Robinson test can outperform *R/S*-type tests for the class of linear models. A theoretical extension of the Lagrange multiplier-type test of Lobato and Robinson (1998) to non-linear models remains an open problem. For a study on this topic see Robinson and Henry (1999). Kwiatkowski *et al.* (1992) and Shin and Schmidt (1992) obtained asymptotic distribution of the *KPSS* statistic under the null hypothesis of trend stationarity and a weakly dependent error process. Lee and Amsler (1997) considered non-stationary long memory alternatives. Tsay (1998) discussed testing the difference stationarity against a stationary and non-stationary I(d) using a modified Durbin–Watson statistic.

The modified *R/S* statistic has been extensively used to detect long memory in financial time series. Following the work of Hurst (1951), Mandelbrot and Taqqu (1979), and Lo (1991) developed a non-parametric *R/S*-type test. Goetzmann (1993), Crato and de Lima (1994), and Teverovsky, Taqqu and Willinger (1998) investigated long memory in stock returns and Cheung (1993) in foreign-exchange rates. Bhattacharya, Gupta and Waymire (1983) applied the *R/S* test for detecting the presence of a deterministic trend in the data.

Teverovsky and Taqqu (1997) discussed testing for long memory in the presence of shifting mean. Tests for stationarity, stability, and structural changes were considered by Ploberger, Krämer and Kontrus (1989), Andrews (1993), Kwiatkowski *et al.* (1992), and Giraitis *et al.* (2006). Lavancier, Philippe and Surgailis (2010) developed a two-sample test for comparison of the memory parameters based on the ratio of two *V/S* statistics. Surgailis, Teyssière and Vaičiulis (2008) suggested a test of long memory based on the increment ratio (IR) statistic. They showed that the IR statistic is robust against trends and structural changes and is less likely to confuse trends with long memory compared to the *V/S* and other *R/S*-type statistics. Bardet and Surgailis (2011) extended the IR statistic to continuous time processes. They also discuss estimation of the Hurst parameter for such processes.

Chapter 10

Empirical Processes

10.1 Introduction

Let $\{\varepsilon_j, j \in \mathbb{Z}\}$, be a stationary process and, let

$$\hat{F}_n(x) := n^{-1} \sum_{j=1}^{n} I(\varepsilon_j \leq x), \quad F(x) := P(\varepsilon_0 \leq x), \quad x \in \mathbb{R}.$$

The empirical d.f. $\hat{F}_n$ plays a central role in statistical inference in the one-sample setting. For example, under the classical i.i.d. set-up, $\hat{F}_n(x)$ is a uniformly minimum variance unbiased estimator of the population d.f. $F(x)$, for a large class of d.f.'s $\{F\}$; see Lehmann (1986). An analysis of some of the asymptotic statistical properties of numerous inference procedures based on $\hat{F}_n$ can be carried out in a unified fashion by first analyzing the large sample properties of the empirical process $\{\hat{F}_n(x), x \in \mathbb{R}\}$.

Analogously, in the regression context an equally important role is played by the *weighted residual empirical processes*

$$U_n(x) := \sum_{j=1}^{n} \gamma_{nj} I(\varepsilon_j \leq x + \xi_{nj}), \qquad x \in \mathbb{R}, \qquad (10.1.1)$$

where $\{\gamma_{nj}, \xi_{nj}, 1 \leq j \leq n\}$ are arrays of known real numbers. For example, in a simple linear regression model one observes a data set $\{(x_j, Y_j), 1 \leq j \leq n\}$ such that for some $\beta \in \mathbb{R}$, $\varepsilon_j = Y_j - \beta x_j$ is a stationary process. Let ψ be a real-valued function with $E\psi(\varepsilon_j) = 0$, for all $1 \leq j \leq n$. The corresponding M estimator $\hat{\beta}$ of β is defined as a solution t of the equation $\sum_{j=1}^{n} x_j \psi(Y_j - tx_j) = 0$. Note that $\psi(y) \equiv y$ and $\psi(y) \equiv \text{sgn}(y)$ give, respectively, the least-squares and the least-absolute deviation estimators of β. Some of the asymptotic distributional properties of a large class of

M estimators under long memory of errors are studied in detail in Chapter 11 below. Here, we shall relate them to the weighted residual empirical processes in (10.1.1). To this end, let

$$V_n(t, y) := \sum_{j=1}^{n} x_j \, I(Y_j \le y + tx_j), \qquad y \in \mathbb{R}.$$

For each t, $V_n(t, y)$ has a bounded variation as a function of $y \in \mathbb{R}$, being a weighted sum of indicator functions, and the Lebesgue–Stieltjes integral $\int \psi(y) V_n(t, dy) = \sum_{j=1}^{n} x_j \int \psi(y) dI(Y_j - tx_j \le y) = \sum_{j=1}^{n} x_j \psi(Y_j - tx_j)$. Therefore, $\hat{\beta}$ can be defined as a solution t of the equation $\int \psi(y) V_n(dy, t) = 0$. It is convenient to change to the variable $s = t - \beta$ in the last equation and rewrite it as $\int \psi(y) V_n(s + \beta, dy) = 0$ having the solution $(\hat{\beta} - \beta)$, since this change of variables reduces the problem to a study of the functional $\int \psi(y) V_n(s + \beta, dy)$ in the vicinity of $s = 0$. Clearly, $V_n(s + \beta, y)$ is an example of $U_n(y)$ with $\gamma_{nj} \equiv x_j$, $\xi_{nj} \equiv sx_j$, and $\varepsilon_j \equiv Y_j - \beta x_j$.

Asymptotic analysis of the processes V_n, and more generally of U_n, provides a unified way of analyzing the asymptotic distributions of these estimators and numerous other inference procedures in general regression and autoregressive models. A detailed analysis of this approach when errors are independent appears in the monographs of Koul (1992a, 2002).

The classical Glivenko–Cantelli lemma and weak convergence results assert that under the i.i.d. set-up,

$$\sup_x |\hat{F}_n(x) - F(x)| \to 0, \quad \text{a.s.,} \tag{10.1.2}$$

$$n^{1/2}(\hat{F}_n - F) \Rightarrow B^\circ(F), \quad \text{in } \mathcal{D}(\bar{\mathbb{R}}), \ \forall \, F \text{ continuous,} \tag{10.1.3}$$

where the convergence in (10.1.3) is w.r.t. the uniform metric, and B° is the Brownian bridge on $[0, 1]$. Because of the ergodic Theorem 2.5.2(i), (10.1.2) still holds when $\{\varepsilon_j, \, j \in \mathbb{Z}\}$ is strictly stationary and ergodic, regardless of having short or long memory. The analog of the weak convergence result (10.1.3), where $B^\circ(F)$ is replaced by some continuous Gaussian process, also still holds under weak dependence assumptions of the underlying processes; see, e.g., Theorem 22.1 in Billingsley (1968).

But under long memory with memory parameter $0 < d < 1/2$, the analogous asymptotics is completely different. The fluctuations of $\hat{F}_n(x) - F(x)$ are usually of order $n^{d-1/2} L(n)$, and the weak limit of

$Z_n(x) := n^{1/2-d}L^{-1}(n)(\hat{F}_n(x) - F(x))$, if it exists, is a degenerate process in x. This is the so-called *uniform reduction principle* of Dehling and Taqqu (1989). In particular, for Gaussian $\{\varepsilon_j\}$, this principle says that uniformly in $x \in \mathbb{R}$, $Z_n(x)$ can be approximated by $-f(x)n^{1/2-d}L^{-1}(n)\sum_{j=1}^{n}\varepsilon_j$, in probability. More generally, the degeneracy of the weak limit $Z(x)$ of $Z_n(x)$ means that $Z(x) = g(x)Y$, where $g(x)$ is a non-random function and Y a r.v. not depending on x. Among other things, such degeneracy implies that the increments of the process Z_n over disjoint intervals, or over disjoint observation sets, are asymptotically completely correlated. This fact is in complete contrast to the fact that these increments are asymptotically independent and non-degenerate when $\{\varepsilon_j\}$ are independent or weakly dependent. Because of this, the sample median and the sample mean are asymptotically equivalent, up to the first order. More generally, so-called M, R, and L estimators of the regression parameters in regression models with long memory errors are asymptotically equivalent, up to the first order, to the least-squares estimator.

This chapter analyzes the weak convergence behavior of the empirical process $\hat{F}_n$, and more generally of U_n, in the presence of long memory. Section 10.2 below establishes the uniform reduction principle for the Gaussian case and the moving average case separately.

Because of these first-order degeneracy results, it is of interest to carry out a higher-order asymptotic analysis for these empirical processes. An asymptotic expansion of $\hat{F}_n$ that is similar in spirit to a Taylor expansion is derived in Section 10.3. This expansion is also important for obtaining higher-order approximations of certain statistical functionals of these empirical processes such as M estimators, as explored in the next chapter.

10.2 Uniform reduction principle

This section contains proofs of the uniform reduction principle (URP) for the weighted residual empirical processes U_n when the underlying process $\{\varepsilon_j, j \in \mathbb{Z}\}$ is either a long memory Gaussian or a long memory moving-average process. A reason for discussing the Gaussian case separately is to illustrate the relative transparency of the proof. The case when $\{\varepsilon_j, j \in \mathbb{Z}\}$

is subordinated to a square integrable function of a long memory Gaussian process is also discussed briefly at the end of the first subsection.

10.2.1 URP: Gaussian case

Let $\{\varepsilon_j, \, j \in \mathbb{Z}\}$ be a mean-zero unit-variance stationary Gaussian process with the autocorrelation function

$$\gamma(k) = k^{-1+2d}L(k), \quad \text{for some } 0 < d < 1/2, \, k \geq 1, \quad (10.2.1)$$

where $L \in SV$. With d as in (10.2.1), throughout this subsection we let

$$\tau_n := n^d |L(n)|^{1/2}, \quad A_n := n^{1/2}\tau_n, \quad 0 < d < 1/2. \quad (10.2.2)$$

With ϕ denoting the standard Gaussian density, define, for $x \in \mathbb{R}$,

$$S_n(x) := \tau_n^{-1}\Big\{ U_n(x) - EU_n(x) + \sum_{j=1}^{n} \gamma_{nj}\phi(x + \xi_{nj})\varepsilon_j \Big\} \quad (10.2.3)$$

$$= \tau_n^{-1} \sum_{j=1}^{n} \gamma_{nj}\big[I(\varepsilon_j \leq x + \xi_{nj}) - \Phi(x + \xi_{nj}) + \phi(x + \xi_{nj})\varepsilon_j \big].$$

The following theorem gives the URP for the process $U_n(x)$.

Theorem 10.2.1. *Let $\{\varepsilon_j, \, j \in \mathbb{Z}\}$ be a mean-zero unit-variance stationary Gaussian process with the autocorrelation function given by (10.2.1), and assume*

$$(a) \quad n^{1/2} \max_{1 \leq j \leq n} |\gamma_{nj}| = O(1), \qquad (b) \quad \max_{1 \leq j \leq n} |\xi_{nj}| = O(1). \quad (10.2.4)$$

Then there exists an $\eta > 0$ depending on d and a constant C such that

$$P\Big(\sup_{x \in \mathbb{R}} |S_n(x)| > \delta \Big) \leq C n^{-\eta}\delta^{-2}, \quad \forall 0 < \delta \leq 1. \quad (10.2.5)$$

Moreover, for any $\epsilon > 0$,

$$\sup_{x \in \mathbb{R}} |S_n(x)| = O_p(n^{-d \wedge (\frac{1}{2} - d) + \epsilon}). \quad (10.2.6)$$

The constants $\gamma_{nj} \equiv n^{-1/2}$ a priori satisfy (10.2.4)(a). Moreover, the corresponding process U_n with $\xi_{nj} \equiv 0$ reduces to $n^{1/2}\hat{F}_n$. The above theorem is thus a generalization of Theorem 3.1 of Dehling and Taqqu (1989) to weighted empirical processes of the Gaussian r.v.'s. In application to regression models the variables γ_{nj}, ξ_{nj} are related to design variables.

The proof of this theorem is facilitated by the following lemma that provides a bound on the variance of the increments of the $S_n(x)$ process. Unless mentioned otherwise, in the proofs below C stands for a universal constant, which may be different in different contexts, but never depends on n. For any function g on $\bar{\mathbb{R}}$, let $g(x,y) := g(y) - g(x)$, $x, y \in \bar{\mathbb{R}}$. Let

$$\mu(x,y) := \int_x^y (1 + |z|)\phi(z/\sqrt{2})dz, \quad x \leq y. \tag{10.2.7}$$

Note that the measure μ is finite and continuous on $\mathbb{R}$.

We also need to define the Hermite rank of a family of functions. Let h_y, $y \in \mathbb{R}$, be a family of functions in $L_2(\mathbb{R}, \Phi)$. Let $J_k(y) := Eh_y(Z)H_k(Z)$. The H-rank of this family is defined to be

$$\text{H-rank}(h_y; y \in \mathbb{R}) := \inf\{k \geq 1 : J_k(y) \neq 0, \text{ for some } y \in \mathbb{R}\}. \tag{10.2.8}$$

Note that the H-rank of the indicator family of functions $\{h_y(x) := I(x \leq y), x \in \mathbb{R}\}$, $y \in \mathbb{R}$, is 1.

Lemma 10.2.1. *For any* $-\infty < x < y < \infty$ *and any* $\xi \in \mathbb{R}$,

$$\Phi(x+\xi, y+\xi) + |\phi(x+\xi, y+\xi)| \leq e^{\xi^2/2}(|\xi| + 2)\mu(x,y). \tag{10.2.9}$$

Moreover, under the conditions of Theorem 10.2.1, $\forall 0 < a < 2d \wedge (1-2d)$, *there exists a constant C such that* $\forall -\infty \leq x \leq y \leq \infty$, $n \geq 1$,

$$E(S_n(x,y))^2 \leq Cn^{-a}\mu(x,y). \tag{10.2.10}$$

Proof. Let $\xi \in \mathbb{R}$. Since $z^2 + \xi^2 - 2\xi z = (1/2)z^2 + (1/2)(z-2\xi)^2 - \xi^2 \geq (1/2)z^2 - \xi^2$, we have $\phi(z-\xi) \leq (1/\sqrt{2\pi})e^{-(1/4)z^2 + \xi^2/2}$ for all $z \in \mathbb{R}$, and

$$\Phi(x+\xi, y+\xi) = \int_x^y \phi(z-\xi)dz \leq \frac{e^{\xi^2/2}}{\sqrt{2\pi}} \int_x^y e^{-(1/4)z^2} dz$$

$$= e^{\xi^2/2} \int_x^y \phi(z/\sqrt{2})dz \leq e^{\xi^2/2}\mu(x,y).$$

Next, because $|\dot{\phi}(z-\xi)| \leq (|z| + |\xi|)\phi(z-\xi)$,

$$|\phi(x+\xi, y+\xi)| \leq \int_x^y |\dot{\phi}(z-\xi)|dz \leq \int_x^y (|z| + |\xi|)\phi(z-\xi)dz$$

$$\leq \frac{e^{\xi^2/2}}{\sqrt{2\pi}} \int_x^y (|z| + |\xi|)e^{-z^2/4}dz$$

$$\leq e^{\xi^2/2}(|\xi| + 1)\mu(x,y).$$

This proves the claim (10.2.9).

It remains to prove (10.2.10). In the proof below we shall omit the subscript n in the triangular arrays γ_{nj}, ξ_{nj}. Also let $\xi_{(n)} := \max_{1 \leq j \leq n} |\xi_j|$. Because of (10.2.4),

$$C_0 := \max_{n \geq 1} \left[4e^{\xi_{(n)}^2/2}(|\xi_{(n)}| \vee 1)n^{1/2} \max_{1 \leq j \leq n} |\gamma_j| \right] < \infty. \qquad (10.2.11)$$

Let $h_y(z) := I(z \leq y) - \Phi(y) + \phi(y)z$ and $h_{x,y}(z) := h_y(z) - h_x(z)$, $x, y, z \in \mathbb{R}$. Then, $\phi^2 \leq \phi$ and (10.2.9) imply that for all $x < y, \xi \in \mathbb{R}$,

$$\mathrm{Var}(h_{x+\xi,y+\xi}(Z)) \leq 2[\Phi(x+\xi, y+\xi) + |\phi(x+\xi, y+\xi)|] \qquad (10.2.12)$$
$$\leq 4e^{\xi^2/2}(|\xi| \vee 1)\mu(x,y).$$

From (10.2.8) it follows that the H-rank of the family of functions $\{h_y(x), y \in \mathbb{R}\}$ is 2. Inequalities (2.4.18) and (10.2.12) imply

$$\tau_n^2 E(S_n(x,y))^2$$
$$= \sum_{j,k=1}^{n} \gamma_{nj}\gamma_{nk}\mathrm{Cov}(h_{x+\xi_j,y+\xi_j}(\varepsilon_j), h_{x+\xi_k,y+\xi_k}(\varepsilon_k))$$
$$\leq \sum_{j,k=1}^{n} |\gamma_{nj}\gamma_{nk}|\gamma^2(j-k)\left\{ \mathrm{Var}(h_{x+\xi_j,y+\xi_j}(Z))\mathrm{Var}(h_{x+\xi_k,y+\xi_k}(Z)) \right\}^{1/2}$$
$$\leq C_0 n^{-1} \sum_{j,k=1}^{n} \gamma^2(j-k)\,\mu(x,y).$$

Applying (3.3.8) with $p = 2(1 - 2d)$ shows that

$$\sum_{j,k=1}^{n} \gamma^2(j-k) = O(n^{4d}L^2(n)), \quad 1/4 < d < 1/2,$$
$$= O(nL_0(n)), \qquad d = 1/4,$$
$$= O(n), \qquad\qquad 0 < d < 1/4,$$

where $L_0 \in SV$. Hence, using the fact that for any $\lambda > 0$ and any $L_i \in SV$, $i = 1, 2$, $L_1(n)/L_2(n) \leq Cn^\lambda$, see (2.3.3), we obtain

$$E(S_n(x,y))^2 \leq C\,n^{-\{[2d \wedge (1-2d)] - \lambda\}}\mu(x,y),$$

thereby proving (10.2.10) since $\lambda > 0$ can be chosen arbitrarily small.

Proof of Theorem 10.2.1. Because of (10.2.4), without loss of generality assume that $n^{1/2}|\gamma_{nj}| \leq 1$ and $|\xi_{nj}| \leq 1$, for all j. With $A_n = n^{1/2}\tau_n$, let

$$K = K_{a,n} := 1 + \text{integer part of } \log_2(n^{1+a}/A_n),$$

where $a > 0$ is the same as in (10.2.10). Define the refining partition of $\bar{\mathbb{R}}$ as follows. Let μ be as in (10.2.7), $\nu(x) := \mu(-\infty, x)$, and let ν^{-1} denote the inverse of the monotone increasing function $\nu(x)$. Define

$$\pi_{j,k} := \nu^{-1}(j2^{-k}), \quad j = 0, 1, \cdots, 2^k, \ k = 0, 1, \cdots, K. \quad (10.2.13)$$

Clearly, by (10.2.13), for each $k = 0, 1, \cdots, K$,

$$-\infty = \pi_{0,k} < \pi_{1,k} < \cdots < \pi_{2^k-1,k} < \pi_{2^k,k} = \infty, \quad (10.2.14)$$

$$\mu(\pi_{j,k}, \pi_{j+1,k}) = 2^{-k}, \quad j = 0, 1, \cdots, 2^k - 1.$$

Now, for any $x \in \bar{\mathbb{R}}$ and $k \in \{0, 1, \cdots, K\}$ define j_k^x by

$$\pi_{j_k^x,k} \leq x < \pi_{j_k^x+1,k}.$$

Define a chain linking $-\infty$ to each point $x \in \bar{\mathbb{R}}$ by

$$-\infty = \pi_{j_0^x,0} \leq \pi_{j_1^x,1} \leq \cdots \leq \pi_{j_K^x,K} \leq x < \pi_{j_K^x+1,K}.$$

Then we have the decomposition

$$S_n(x) = S_n(\pi_{j_0^x,0}, \pi_{j_1^x,1}) + S_n(\pi_{j_1^x,1}, \pi_{j_2^x,2}) \quad (10.2.15)$$

$$+ \cdots + S_n(\pi_{j_{K-1}^x,K-1}, \pi_{j_K^x,K}) + S_n(\pi_{j_K^x,K}, x).$$

First, consider the last term in the above decomposition. Let

$$B_n(x) := A_n^{-1} \sum_{j=1}^{n} \Big[\Phi(\pi_{j_K^x,K} + \xi_{nj}, \pi_{j_K^x+1,K} + \xi_{nj})$$

$$+ |\varepsilon_j| \, |\phi(\pi_{j_K^x,K} + \xi_{nj}, \pi_{j_K^x+1,K} + \xi_{nj})| \Big].$$

Because $e^{\xi^2/2}(|\xi| + 2) \leq C$, for all $|\xi| \leq 1$, by (10.2.9),

$$B_n(x) \leq C\mu(\pi_{j_K^x,K}, \pi_{j_K^x+1,K})A_n^{-1} \sum_{j=1}^{n}(1 + |\varepsilon_j|) \quad (10.2.16)$$

$$= C2^{-K}A_n^{-1} \sum_{j=1}^{n}(1 + |\varepsilon_j|) =: B_n.$$

Let $S_n^*(x)$ be the empirical process in (10.2.3) corresponding to $\gamma_j \equiv n^{-1/2}$. Obviously, Lemma 10.2.1 and (10.2.10) apply to S_n^*. We have

$$|S_n(\pi_{j_K^x,K}, x)| \quad (10.2.17)$$

$$\leq C_0 A_n^{-1} \sum_{j=1}^{n} I(\pi_{j_K^x,K} + \xi_{nj}, \pi_{j_K^x+1,K} + \xi_{nj}) + C_0 B_n(x)$$

$$\leq |S_n^*(\pi_{j_K^x,K}, \pi_{j_K^x+1,K})| + 2B_n.$$

Now assume without loss of generality that $C_0 \leq 1$. By (10.2.15), (10.2.17), and because $\delta/2 \geq \sum_{k=0}^{\infty} \delta/(k+3)^2 \geq \sum_{k=0}^{K} \delta/(k+3)^2$,

$$P(\sup_x |S_n(x)| > \delta)$$

$$\leq \sum_{k=0}^{K-1} P\left(\sup_x |S_n(\pi_{j_k^x,k}, \pi_{j_{k+1}^x,k+1})| > \frac{\delta}{(k+3)^2}\right) + P(B_n > \frac{\delta}{4})$$

$$+ P\left(\sup_x |S_n^*(\pi_{j_K^x,K}, \pi_{j_K^x+1,K})| > \frac{\delta}{(K+3)^2}\right).$$

By definition and the monotonicity property of the partitions in (10.2.13), any interval $[\pi_{j_k^x,k}, \pi_{j_{k+1}^x,k+1})$ is either empty, or else $[\pi_{j_k^x,k}, \pi_{j_{k+1}^x,k+1})$ equals $[\pi_{j-1,k+1}, \pi_{j,k+1})$, for some $j = 0, \cdots, 2^{k+1} - 1$. Hence, for any fixed $k \in \{0, 1, \cdots, K-1\}$,

$$\sup_{x \in \mathbb{R}} |S_n(\pi_{j_k^x,k}, \pi_{j_{k+1}^x,k+1})| = \max_{0 \leq j \leq 2^{k+1}-1} |S_n(\pi_{j-1,k+1}, \pi_{j,k+1})|.$$

Hence, by Chebyshev's inequality, (10.2.14), and by (10.2.10),

$$P\left(\sup_x |S_n(\pi_{j_k^x,k}, \pi_{j_{k+1}^x,k+1})| > \frac{\delta}{(k+3)^2}\right)$$

$$\leq \sum_{j=0}^{2^{k+1}-1} P\left(|S_n(\pi_{j,k+1}, \pi_{j+1,k+1})| > \frac{\delta}{(k+3)^2}\right)$$

$$\leq \frac{(k+3)^4}{n^a \delta^2} \sum_{j=0}^{2^{k+1}-1} \mu(\pi_{j,k+1}, \pi_{j+1,k+1}) = \frac{(k+3)^4}{n^a \delta^2}.$$

Similarly,

$$P\left(\sup_x |S_n^*(\pi_{j_{K-1}^x,K-1}, \pi_{j_K^x,K})| > \frac{\delta}{(K+3)^2}\right) \leq \frac{(K+3)^4}{n^a \delta^2}$$

and, by (10.2.16) and Chebyshev's inequality,

$$P(B_n > \delta/4) \leq (4C/\delta)2^{-K} A_n^{-1} n(1 + E|\varepsilon_0|) = C\delta^{-1} n^{-a}.$$

Therefore,

$$P(\sup_x |S_n(x)| > \delta) \leq Cn^{-a}\delta^{-2} \sum_{k=0}^{K}(k+3)^4 + Cn^{-a}\delta^{-1} \qquad (10.2.18)$$

$$\leq Cn^{-a}\left(\delta^{-2}(K+3)^5 + \delta^{-1}\right).$$

Since $(K+3)^5 = O(\log(n))$, this proves statement (10.2.5) of Theorem 10.2.1 with arbitrary $\eta < a$.

Relation (10.2.6) is immediate from (10.2.18) and the fact that $a < 2d \wedge (1 - 2d)$ is arbitrary (see Lemma 10.2.1), by taking $\delta = \delta' n^{(\epsilon - a)/2}$ and $a = 2d \wedge (1 - 2d) - \epsilon$ in (10.2.18), for $\delta' > 0$ fixed and ϵ small enough. Theorem 10.2.1 is proved.

Note that the constant C in (10.2.5) and (10.2.10) depends only on the C_0 of (10.2.11) and the slowly varying function L appearing in the autocovariance function $\gamma(k)$.

Remark 10.2.1. (*URP for a Gaussian subordinate process*). Suppose $\varepsilon_j = G(\eta_j)$, where η_j, $j \in \mathbb{Z}$, is a stationary standardized Gaussian process and G is some function in $L_2(\mathbb{R}, \Phi)$. Then, using the above method, we can prove the following theorem. Let F denote the d.f. of ε_0, and let, for $k \geq 1$,

$$\tau_{n,k} := n^{k\,d + \frac{1-k}{2}} L^{\frac{k}{2}}(n), \quad J_k(x) := E\varepsilon_0 H_k(\eta_0) I(\varepsilon_0 \leq x),$$

$$S_{n,k}(x) := \tau_{n,k}^{-1} \sum_{j=1}^{n} \gamma_{nj} \Big[I(\varepsilon_j \leq x + \xi_{nj}) - F(x + \xi_{nj})$$

$$- \frac{J_k(x + \xi_{nj})}{k!} H_k(\eta_j) \Big].$$

Note that $S_{n,k}$ is an analog of S_n of (10.2.3). Let $\mathcal{I} := \{x \in \mathbb{R}; 0 < F(x) < 1\} := (a_0, a_1)$, ν denote the H-rank of the family of functions $I(G(\eta) \leq x) - F(x)$, $x \in \mathbb{R}$, and let $\mu_\nu(x) := E\varepsilon_0 | H_\nu(\eta_0)| I(\varepsilon_0 \leq x)$, $x \in \mathbb{R}$.

Theorem 10.2.2. *Let $\{\varepsilon_j = G(\eta_j), \ j \in \mathbb{Z}\}$ be the above subordinate Gaussian process. Suppose (10.2.4) and the following conditions hold:*

(i) $\nu < 1/(1 - 2d)$,

(ii)   *the d.f. F has uniformly continuous density f on $\mathcal{I}$ and $f > 0$ a.e.,*

(iii)   *the functions J_ν and μ_ν are continuously differentiable and*

$$\dot{J}_\nu(x) \to 0, \ \dot{\mu}_\nu(x) \to 0 \text{ as } x \to a_0 \text{ and } x \to a_1.$$

Then $\sup_{x \in \bar{\mathbb{R}}} |S_{n,\nu}(x)| = o_p(1)$.

Theorem 10.2.1 corresponds to the identity function $G(y) \equiv y$ and Hermite rank $\nu = 1$. Theorem 10.2.2 is thus an extension of this theorem and that of Theorem 3.1 of Dehling and Taqqu (1989) to the subordinate Gaussian process and to the weighted residual empirical processes. The proof of this theorem can be found in Koul and Mukherjee (1993).

10.2.2   URP: moving-average case

Now assume the process $\{\varepsilon_j\}$ is the moving-average:

$$\varepsilon_j = \sum_{k=0}^{\infty} a_k \zeta_{j-k}, \quad j \in \mathbb{Z}, \tag{10.2.19}$$

where $\{\zeta_s, s \in \mathbb{Z}\}$ are i.i.d., with zero mean and finite variance σ^2. The coefficients a_j will be assumed to satisfy

$$a_j = j^{-1+d} L(j), \quad 0 < d < 1/2, \ j \geq 1, \tag{10.2.20}$$

for some function $L \in SV$. By definition, L takes positive values for sufficiently large j. However, if the coefficients a_j in (10.2.20) are such that $L(j) < 0$ and the function $L_1 := -L \in SV$, we can rewrite ε_j in (10.2.19) as $\varepsilon_j = \sum_{k=0}^{\infty} \tilde{a}_k \tilde{\zeta}_{j-k}$ where $\tilde{a}_j = j^{-1+d} L_1(j)$, $L_1 \in SV$, and $\tilde{\zeta}_s := -\zeta_s, s \in \mathbb{Z}$, are i.i.d. r.v.'s with zero mean and variance σ^2. Since the conditions on the innovations imposed below do not depend on their sign, the assumption that $L \in SV$ for (10.2.20) does not restrict the generality of the subsequent discussion.

By (2.3.3) and (2.3.4),

$$\sum_{j=0}^{\infty} |a_j| = \infty, \qquad \sum_{j=0}^{\infty} a_j^2 < \infty,$$

$$\mathrm{Cov}(\varepsilon_j, \varepsilon_{j+k}) = k^{-1+2d} \tilde{L}(k), \quad \forall k \geq 1,$$

where $\tilde{L} \in SV$ and such that

$$\lim_{k \to \infty} \frac{\tilde{L}(k)}{L^2(k)} = \sigma^2 B(d, 1-2d).$$

The rate of decay of the $a_j, j \geq 1$, determines the degree of dependence between the past and the future of the time series (10.2.19).

Our aim in this section is to establish the analog of Theorem 10.2.1. For this we need the following additional conditions:

$$|Ee^{iu\zeta_0}| \leq C(1 + |u|)^{-\delta}, \qquad \exists\, C, \delta > 0, \tag{10.2.21}$$

$$E|\zeta_0|^{2r} < \infty, \qquad \exists\, r > 1. \tag{10.2.22}$$

We will show later that (10.2.21) implies that F is infinitely differentiable. Let f denote the marginal probability density of the ε_j's and define

$$R_j(x) := I(\varepsilon_j \leq x) - F(x) + f(x)\,\varepsilon_j, \quad j \in \mathbb{Z}, \tag{10.2.23}$$

$$S_n(x) := \tau_n^{-1} \sum_{j=1}^{n} \gamma_{nj} R_j(x + \xi_{nj}).$$

This $\mathcal{S}_n$ is an analog of the S_n of (10.2.3). We are now ready to state

Theorem 10.2.3. *Suppose $\{\varepsilon_j\}$ is a linear process as in (10.2.19) with the a_j's and innovations $\{\zeta_j\}$ satisfying (10.2.20), (10.2.21), and (10.2.22). In addition, suppose γ_{nj} and ξ_{nj} satisfy (10.2.4).*

Then $\mathcal{S}_n(x)$ of (10.2.23) satisfies (10.2.5). Moreover, if (10.2.22) holds with $r = 3/2$, i.e., if $E|\zeta_0|^3 < \infty$, then $\mathcal{S}_n(x)$ satisfies the bound (10.2.6).

Theorem 10.2.3 implies the URP for the process $\mathcal{S}_n$ of (10.2.23) and its analog $\mathcal{S}_n^*$ where $\xi_{nj} \equiv 0$. Its statement is relatively simpler when $r = 3/2$ in (10.2.22) and when the slowly varying factor in (10.2.20) is asymptotically a constant, i.e., there exists a $c_0 > 0$ such that

$$a_j \sim c_0 j^{-1+d}, \quad j \to \infty. \tag{10.2.24}$$

For easy reference this is stated as a corollary below. Note that in this case $\operatorname{Var}(\mathcal{S}_n(x)) = O\big(n^{-(2d)\wedge(1-2d)}\big)$, for any $x \in \mathbb{R}$.

Corollary 10.2.1. *Suppose the conditions of Theorem 10.2.3 hold with $r = 3/2$ in (10.2.22) and a_j satisfying (10.2.24). Let*

$$\mathcal{S}_n(x) := n^{-d} \sum_{j=1}^n \gamma_{nj} R_j(x + \xi_{nj}), \quad \mathcal{S}_n^*(x) := n^{-d} \sum_{j=1}^n \gamma_{nj} R_j(x),$$

with $R_j(x)$ defined by (10.2.23). Then, for all $0 < d < 1/2$,

$$\sup_{x \in \bar{\mathbb{R}}} |\mathcal{S}_n(x)| = o_p(1) = \sup_{x \in \bar{\mathbb{R}}} |\mathcal{S}_n^*(x)|. \tag{10.2.25}$$

Remark 10.2.2. The rate of convergence in (10.2.6) in Theorems 10.2.1 and 10.2.3 is sharp. Consider, for example the ordinary empirical process obtained from U_n of (10.1.1) when $\gamma_{nj} \equiv n^{-1/2}$ and $\xi_{nj} \equiv 0$. Note that $\frac{1}{2} - d + d \wedge (\frac{1}{2} - d) = (1 - 2d) \wedge \frac{1}{2}$ for $0 < d < 1/2$. Then, from (10.2.6), we obtain

$$\sup_{x \in \mathbb{R}} |\hat{F}_n(x) - F(x) + f(x)\bar{\varepsilon}_n| = o_p(n^{-(1-2d)\wedge(1/2)+\epsilon}), \tag{10.2.26}$$

where $\epsilon > 0$ is arbitrarily small. However, from the asymptotic expansion of the empirical process given in Section 10.3, it follows that (10.2.26) is not true if ϵ is negative (and $|\epsilon|$ is arbitrarily small). The same rate of convergence in (10.2.26) applies both in the Gaussian as well as the moving-average case, under the third moment condition $E|\zeta_0|^3 < \infty$ and the regularity condition (10.2.21) on innovations.

Equation (10.2.26) and Corollary 4.4.1 imply the functional central limit theorem for the ordinary empirical process $\hat{F}_n$:

$$L(n)^{-1}n^{\frac{1}{2}-d}(\hat{F}_n(x) - F(x)) \Rightarrow s_{d,\gamma}f(x)\,Z,$$

where $Z \sim \mathcal{N}(0,1)$ and $s_{d,\gamma} = \left(E(\zeta_0^2)B(d, 1 - 2d)/d(1 + 2d)\right)^{1/2}$.

Remark 10.2.3. For $r > 1$ and $0 < d < 1/2$, let

$$\tau_1 := \min\left(1 - 2d,\ 2(1 - d)(r - 1)\right) > 0, \qquad (10.2.27)$$
$$t_1 := 1 - 2d + \tau_1.$$

Note that $r \geq 3/2$ and $0 < d < 1/2$ imply that $2(1 - d)(r - 1) \geq 1 - 2d$, and hence $\tau_1 = 1 - 2d$. This in turn implies $t_1 = 2(1 - 2d)$. This definition of t_1 is used repeatedly below.

Proof of Theorem 10.2.3. This follows from the chaining argument of the proof of Theorem 10.2.1 and Lemma 10.2.2 below, which is an analog of Lemma 10.2.1.

Lemma 10.2.2. *Under the conditions of Theorem 10.2.3, there exists a finite continuous measure μ on the real line, such that for any $-\infty < x < y < \infty$ and any $|\xi| \leq 1$,*

$$F(x + \xi, y + \xi) + |f(x + \xi, y + \xi)| \leq \mu(x, y). \qquad (10.2.28)$$

Moreover, for any $0 < a < \min\left(2d, t_1 - (1 - 2d)\right)$, with t_1 as in (10.2.27), there exists a constant C such that

$$E(S_n(x, y))^2 \leq Cn^{-a}\mu(x, y), \quad \forall\ -\infty \leq x \leq y \leq \infty. \qquad (10.2.29)$$

The proof of Lemma 10.2.2 is more complicated than the proof of Lemma 10.2.1 because the method of Hermite expansions used in the latter proof is not applicable. It is facilitated by Lemmas 10.2.3, 10.2.4, and 10.2.5. Lemmas 10.2.3 and 10.2.4 are auxiliary. Lemma 10.2.5 is the most important of these and gives a crucial bound on $\mathrm{Cov}(R_j(x, y), R_0(x, y))$, $x < y$, from which Lemma 10.2.2 easily follows. Lemma 10.2.5 uses a powerful method of martingale decomposition (originally introduced in Ho and Hsing (1996)), which, for linear processes, replaces the method of Hermite expansions. Lemmas 10.2.3 and 10.2.4 are also used in subsequent sections.

For $\gamma > 1$, set

$$h_\gamma(x) := (1 + |x|)^{-\gamma}, \quad \mu_\gamma(x, y) := \int_x^y h_\gamma(z)dz, \ x < y. \qquad (10.2.30)$$

Note that for each $\gamma > 1$, μ_γ is a continuous and finite measure on $\mathbb{R}$.

Lemma 10.2.3. *Let $h(x)$, $x \in \mathbb{R}$, be a real-valued function such that*

$$|h(x)| \leq Ch_\gamma(x), \quad |h(x) - h(y)| \leq C|x - y|h_\gamma(x), \qquad (10.2.31)$$

for any $x, y \in \mathbb{R}, |x - y| \leq 1$, and some $C < \infty$ and $1 < \gamma \leq 2$. Then there exists a constant C_γ depending only on γ and C of (10.2.31), such that for any $x, y, v, z \in \mathbb{R}$,

$$|h(x + y)| \leq C_\gamma h_\gamma(x)(1 \vee |y|^\gamma), \qquad (10.2.32)$$

$$\left| \int_0^y h(x + w)dw \right| \leq C_\gamma h_\gamma(x)(|y| \vee |y|^\gamma), \qquad (10.2.33)$$

$$\left| \int_x^y (h(\xi + v) - h(\xi))d\xi \right| \leq C_\gamma \mu_\gamma(x, y)(|v| \vee |v|^\gamma), \qquad (10.2.34)$$

$$\left| \int_0^v dw \int_x^y (h(\xi + w - z) - h(\xi - z))d\xi \right| \qquad (10.2.35)$$

$$\leq C_\gamma \mu_\gamma(x, y)|v|^\gamma (1 \vee |z|^\gamma).$$

Proof. *Proof of* (10.2.32). Note the elementary inequality

$$(1 + |z|) \leq (1 + |z + a|)(1 + |a|), \quad \forall z, a \in \mathbb{R}.$$

This inequality and (10.2.31) imply that

$$|h(x + y)| \leq Ch_\gamma(x + y) \leq Ch_\gamma(x)(1 \vee |y|^\gamma), \quad x, y \in \mathbb{R}.$$

Proof of (10.2.33). For $|y| \leq 1$, (10.2.33) is immediate by (10.2.32). Also, for $|y| \geq 1$ and $|x| \leq 1$, (10.2.33) is immediate by the integrability of h_γ. Consider the remaining case where $|y| \geq 1$ and $|x| \geq 1$. Assume $y \geq 1$; the case $y \leq -1$ follows similarly. By (10.2.31),

$$\left| \int_0^y h(x + w)dw \right| \leq C\left(\int_0^1 h_\gamma(x + w)dw + \int_1^y h_\gamma(x + w)dw \right).$$

By (10.2.32), the first integral in this bound is bounded above by

$$Ch_\gamma(x) \leq Ch_\gamma(x)|y|^\gamma.$$

To bound the second integral, consider the case $y \geq |x|/2$. Then, clearly,

$$\int_1^y h_\gamma(x + w)dw \leq \int_\mathbb{R} h_\gamma(x + w)dw =: C_1$$

$$\leq C_1(2y/|x|)^\gamma \leq C_\gamma h_\gamma(x)|y|^\gamma,$$

proving (10.2.33) in this case. Now, consider the case $1 \leq y \leq |x|/2$. Then, for $1 \leq w \leq y$, $|x + w| \geq |x| - |w| \geq |x|/2$, which in turn implies $h_\gamma(x + w) \leq C_\gamma h_\gamma(x)$. Hence,

$$\int_1^y h_\gamma(x + w)dw \leq C_\gamma h_\gamma(x)y^\gamma.$$

This completes the proof of (10.2.33).

Proof of (10.2.34). For $|v| \leq 1$, (10.2.34) is immediate by (10.2.31) and the definition (10.2.30) of μ_γ. Let $|v| \geq 1$. Then, by (10.2.31) and (10.2.32), the l.h.s. of (10.2.34) is bounded above by

$$\int_x^y (|h(z + v)| + |h(z)|)dz \leq C_\gamma(1 + |v|^\gamma) \int_x^y h_\gamma(z)dz$$
$$\leq C_\gamma |v|^\gamma \mu_\gamma(x, y).$$

Proof of (10.2.35). Consider first $|v| \leq 1$. Then, by (10.2.31) and (10.2.32), the l.h.s. of (10.2.35) does not exceed:

$$C \int_{-|v|}^{|v|} |w|dw \int_x^y h_\gamma(\xi - z)d\xi \; \leq C\mu_\gamma(x, y)v^2(1 \vee |z|^\gamma)$$
$$\leq C\mu_\gamma(x, y)|v|^\gamma(1 \vee |z|^\gamma).$$

For the case $|v| \geq 1$, the l.h.s. of (10.2.35) is bounded above by

$$C \int_{-|v|}^{|v|} dw \int_x^y (h_\gamma(\xi + w - z) + h_\gamma(\xi - z)) \, d\xi$$
$$\leq C(1 \vee |z|^\gamma) \int_x^y d\xi \int_{-|v|}^{|v|} (h_\gamma(\xi + w) + h_\gamma(\xi))dw$$
$$\leq C(1 \vee |z|^\gamma)\mu_\gamma(x, y)(|v|^\gamma + |v|) \leq \; C\mu_\gamma(x, y)|v|^\gamma(1 \vee |z|^\gamma),$$

where the middle inequality used (10.2.33) with $h = h_\gamma$. This completes the proof of Lemma 10.2.3.

Next, we decompose $\varepsilon_j = \varepsilon_{j,m} + \tilde{\varepsilon}_{j,m}$, $j \in \mathbb{Z}$, where

$$\varepsilon_{j,m} := \sum_{i=0}^m a_i \zeta_{j-i}, \qquad \tilde{\varepsilon}_{j,m} := \sum_{i=m+1}^\infty a_i \zeta_{j-i}, \qquad (10.2.36)$$

$$F_m(x) := P(\varepsilon_{j,m} \leq x), \qquad m \geq 0,$$

$$\tilde{\varepsilon}_{j,-1} := \varepsilon_j, \qquad F_{-1}(x) := I(x \geq 0).$$

The following lemma proves the smoothness properties of F and F_m. It also establishes a bound (10.2.39) on the bivariate probability density of

$(\varepsilon_0, \varepsilon_t)$, which is used in Section 13.2. These results do not require the regular asymptotics of the coefficients a_j in (10.2.20) but hold under the weak assumptions that $\sum_{j=0}^{\infty} a_j^2 < \infty$ and that $a_j \neq 0$ for infinitely many j's (these assumptions are implicit in the formulation of Lemma 10.2.4). For a smooth function g from $\mathbb{R}$ to $\mathbb{R}$, let $g^{(k)}$ denote its kth derivative, $k \geq 1$.

Lemma 10.2.4. *Assume (10.2.21) and* $E\zeta_0^2 < \infty$. *Then the following hold:*

(i) *For any* $p = 0, 1, \cdots$, *there exist* $m_0 \geq 1$ *and a constant* C *such that for any* $m \geq m_0$ *the distribution functions* $F(x)$ *and* $F_m(x)$ *are* $(p+1)$-*times continuously differentiable, and for any* $m \geq m_0, x \in \mathbb{R}$,

$$|F^{(p+1)}(x)| + |F_m^{(p+1)}(x)| \leq C(1 + |x|)^{-2}, \qquad (10.2.37)$$

$$|F_m^{(p+1)}(x) - F_{m+1}^{(p+1)}(x)| \leq Ca_m^2(1 + |x|)^{-2}. \qquad (10.2.38)$$

(ii) *If, in addition,* $E\zeta_0^4 < \infty$, *then there exists* $t_0 \geq 1$ *such that the bivariate probability density* f_t *of* $(\varepsilon_0, \varepsilon_t)$ *exists for all* $t \geq t_0$. *Moreover, for some* $C > 0$ *and all* $x, y \in \mathbb{R}$, *and all* $t \geq t_0$,

$$f_t(x, y) \leq C(1 + x^2)^{-1}(1 + y^2)^{-1}. \qquad (10.2.39)$$

Proof. (i) The densities $f(x) = dF(x)/dx$ and $f_m(x) = dF_m(x)/dx$ exist if the characteristic functions $\hat{f}(u) = Ee^{iu\varepsilon_0}$ and $\hat{f}_m(u) = Ee^{iu\varepsilon_{0,m}}$ are absolutely integrable on the real line. Then (10.2.37) and (10.2.38) follow by integration by parts if we show that for any $p \geq 0$ there exist m_0 and a constant C such that for any $m \geq m_0, u \in \mathbb{R}$,

$$|(u^p \hat{f}(u))^{(k)}| + |(u^p \hat{f}_m(u))^{(k)}| \leq C(1 + u^2)^{-1}, \qquad (10.2.40)$$

$$|(u^p(\hat{f}_m(u) - \hat{f}_{m-1}(u)))^{(k)}| \leq Ca_m^2(1 + u^2)^{-1}, \qquad (10.2.41)$$

for $k = 0$ and $k = 2$. Indeed, for some $C \in \mathbb{R}$,

$$\int e^{-iux} u^p \hat{f}(u)du = C\left(\int e^{-iux} \hat{f}(u)du\right)_x^{(p)} = C f^{(p)}(x),$$

and, similarly,

$$\int e^{-iux}(u^p \hat{f}(u))^{(2)} du = (ix)^2 \int e^{-iux} u^p \hat{f}(u)du = C x^2 f^{(p)}(x).$$

Hence, by (10.2.40),

$$|f^{(p)}(x)| \leq \int |u^p \hat{f}(u)|du \leq \int (1 + u^2)^{-1}du \leq C,$$

$$|x^2 f^{(p)}(x)| \leq \int |(u^p \hat{f}(u))^{(2)}|du \leq \int (1 + u^2)^{-1}du \leq C,$$

which clearly yield the bound (10.2.37) for $F^{(p+1)}(x) = f^{(p)}(x)$. The remaining bounds in (10.2.37) and (10.2.38) follow from (10.2.40) and (10.2.41) in a similar way.

We shall prove (10.2.40) for $\hat{f}_m$ and $k = 2$. We have

$$(u^p \hat{f}_m(u))^{(2)} = p(p-1)u^{p-2}\hat{f}_m(u) + 2pu^{p-1}\hat{f}_m^{(1)}(u) + u^p \hat{f}_m^{(2)}(u). \quad (10.2.42)$$

Put $\phi(u) := Ee^{iu\zeta_0}$, and for a subset $I \subset \{0, 1, 2, \cdots\}$, let

$$\Phi_I(u) := E \exp\left\{iu \sum_{r \in I} a_r \zeta_r\right\} = \prod_{r \in I} \phi(ua_r).$$

Then $\hat{f}_m(u) = \Phi_J(u)$ $(J := \{0, 1, \cdots, m\})$ and

$$\hat{f}_m^{(1)}(u) = \sum_{r \in J} \phi^{(1)}(ua_r)a_r\Phi_{J\setminus\{r\}}(u), \quad (10.2.43)$$

$$\hat{f}_m^{(2)}(u) = \sum_{r \in J} \phi^{(2)}(ua_r)a_r^2\Phi_{J\setminus\{r\}}(u) \quad (10.2.44)$$

$$+ \sum_{r_1, r_2 \in J, r_1 \neq r_2} \phi^{(1)}(ua_{r_1})\phi^{(1)}(ua_{r_2})a_{r_1}a_{r_2}\Phi_{J\setminus\{r_1,r_2\}}(u).$$

Because $|\phi^{(1)}(u)| \leq |u|$ and $|\phi^{(2)}(u)| \leq 1$ and the a_j^2's are summable,

$$\sum_{r \in J}|\phi^{(1)}(ua_r)a_r| \leq C|u|, \quad \sum_{r \in J}|\phi^{(2)}(ua_r)|a_r^2 \leq C, \quad (10.2.45)$$

where the constant C does not depend on u or m. Moreover, by (10.2.21), the characteristic functions $\Phi_J(u)$, $\Phi_{J\setminus\{r\}}(u)$, and $\Phi_{J\setminus\{r_1,r_2\}}(u)$ decay polynomially fast as $|u| \to \infty$, uniformly in r, r_1, r_2, provided m (the cardinality of J) is large enough. More precisely, for any given $p \geq 0$ there exist m_0 and a constant $C < \infty$, such that for all $m \geq m_0, u \in \mathbb{R}$, and any $r, r_1, r_2 \in J$,

$$|\Phi_J(u)| + |\Phi_{J\setminus\{r\}}(u)| + |\Phi_{J\setminus\{r_1,r_2\}}(u)| \leq C(1 + u^2)^{-p-4}. \quad (10.2.46)$$

The relations (10.2.42) to (10.2.46) easily imply (10.2.40) for $\hat{f}_m$ and $k = 2$. The remaining bounds in (10.2.40) and (10.2.41) follow analogously, thereby completing the proof of part (i).

(ii) Let $\hat{f}_t(u, v) = Ee^{i(u\varepsilon_0 + v\varepsilon_t)}$ denote the characteristic function of the random vector $(\varepsilon_0, \varepsilon_t)$. Similarly as in part (i), it suffices to show the inequality

$$\left|\frac{\partial^{k+l}\hat{f}_t(u, v)}{\partial u^k \partial v^l}\right| \leq \frac{C}{(1 + u^2)(1 + v^2)}, \quad (10.2.47)$$

for $t \geq t_0$ and for pairs of indices $(k, \ell) = (0,0), (2,0), (0,2)$, and $(2,2)$, with C independent of t. Indeed, using integration by parts,

$$
\int \int e^{-i(ux+vy)} \left[\hat{f}_t(u,v) - \frac{\partial^2 \hat{f}_t(u,v)}{\partial u^2} - \frac{\partial^2 \hat{f}_t(u,v)}{\partial v^2} + \frac{\partial^4 \hat{f}_t(u,v)}{\partial u^2 \partial v^2} \right] du\, dv
$$
$$
= C f_t(x,y)[1 - (ix)^2 - (iy)^2 + (ix)^2 (iy)^2]
$$
$$
= C f_t(x,y)(1 + x^2)(1 + y^2),
$$

as in part (i). Since the double integral on the l.h.s. is bounded by $C \int \int (1+u^2)^{-1}(1+v^2)^{-1} du\, dv < \infty$, (10.2.47) implies (10.2.39).

It remains to show (10.2.47). We have

$$
\hat{f}_t(u,v) = \prod_{s \in \mathbb{Z}} \phi(u a_s + v a_{t+s}),
$$

where $a_s := 0$, $s < 0$. For a subset $A \subset J := \{-t, -t+1, \cdots\}$, let

$$
\Phi_A(u,v;t) := E \exp \left\{ iu \sum_{s \in A} (u a_s + v a_{t+s}) \zeta_s \right\} = \prod_{s \in A} \phi(u a_s + v a_{t+s}).
$$

Note $\hat{f}_t(u,v) = \Phi_J(u,v;t)$. We first derive a bound similar to (10.2.46). That is, there exist $t_0 \geq 1$ and a constant $C < \infty$ such that for all $t \geq t_0$, $(u,v) \in \mathbb{R}^2$, and any subset $A \subset J$ with cardinality $|A| \leq 4$,

$$
|\Phi_{J \setminus A}(u,v;t)| \leq C(1+u^2)^{-3}(1+v^2)^{-3}. \tag{10.2.48}
$$

So, let $J_0 := \{-t, -t+1, \cdots, -1\}$ and $J_1 := \{0, 1, \cdots\}$ where $J_0 \cup J_1 = J$ and write $|\Phi_{J \setminus A}(u,v;t)| = \Phi_{J_0 \setminus A}(u,v;t) \Phi_{J_1 \setminus A}(u,v;t)$. Note $\Phi_{J_0 \setminus A}(u,v;t) = \prod_{s \in J_0 \setminus A} \phi(v a_{t+s})$ does not depend on u. As with (10.2.46), we find that there exist $t_0 \geq 1$ and a constant $C < \infty$ such that for all $t \geq t_0$, $(u,v) \in \mathbb{R}^2$, and $A \subset J, |A| \leq 4$,

$$
|\Phi_{J_0 \setminus A}(u,v;t)| \leq C(1+v^2)^{-6}. \tag{10.2.49}
$$

Next, consider

$$
|\Phi_{J_1 \setminus A}(u,v;t)| \leq \prod_{s=0, s \notin A}^{K} |\phi(u a_s + v a_{t+s})|,
$$

where $0 \leq K < \infty$ is fixed and specified below. Without loss of generality, assume $a_s \neq 0$, for every $s \geq 0$, and let $\bar{a}_K := \inf_{0 \leq s \leq K} |a_s| > 0$. Since $a_s \to 0$ as $s \to \infty$, there exists $t_1 \geq 1$ such that $|a_{t+s}| < \bar{a}_K/2$, $\forall t \geq t_1$, $s = 0, 1, \cdots, K$. Hence,

$$
|u + v \frac{a_{t+s}}{a_s}| \geq |u| - |v| \frac{|a_{t+s}|}{|a_s|} \geq \frac{|u|}{2}, \qquad \text{on } \{|v| \leq |u|\}.
$$

By (10.2.21), for any $A \subset J$, $|A| \leq 4$,

$$|\Phi_{J_1 \setminus A}(u, v; t)| I(|v| \leq |u|) \leq C(K) \prod_{s=0, s \notin A}^{K} (1 + \frac{1}{2}\bar{a}_K|u|)^{-\delta} \quad (10.2.50)$$

$$\leq C(1 + u^2)^{-3},$$

provided K is chosen so that $C = C(K)(\frac{\bar{a}_K}{2}) \wedge 1)^{-\delta(K+1)} < \infty$ and $K > 3 + (6/\delta)$. Combining (10.2.49) and (10.2.50) results in

$$|\Phi_{J \setminus A}(u, v; t)| \leq C(1 + v^2)^{-6}(1 + u^2)^{-3}I(|v| \leq |u|)$$
$$+ C(1 + v^2)^{-6}I(|v| > |u|)$$
$$\leq C(1 + u^2)^{-3}(1 + v^2)^{-3},$$

proving (10.2.48).

Let us return to the proof of (10.2.47). For $(k, \ell) = (0, 0)$, it is immediate from (10.2.48). Next, we prove (10.2.47) for $(k, \ell) = (2, 2)$. The proof of the remaining cases $(k, \ell) = (2, 0)$ and $(k, \ell) = (0, 2)$ follows similarly and will be omitted. As for (10.2.44), we have

$$\frac{\partial^4 \hat{f}_t(u, v)}{\partial u^2 \partial v^2} = \sum_s a_s^2 a_{t+s}^2 \phi^{(4)}(ua_s + va_{t+s})\Phi_{J \setminus \{s\}}(u, v; t)$$

$$+ \sum_{s_1, s_2}' a_{s_1}^2 a_{t+s_1} a_{t+s_2} \phi^{(3)}(ua_{s_1} + va_{t+s_1})$$

$$\times \phi^{(1)}(ua_{s_2} + va_{t+s_2})\Phi_{J \setminus \{s_1, s_2\}}(u, v; t)$$

$$+ \cdots$$

$$+ \sum_{s_1, \cdots, s_4}' a_{s_1} a_{s_2} a_{t+s_3} a_{t+s_4} \prod_{i=1}^{4} \phi^{(1)}(ua_{s_i} + va_{t+s_i})$$

$$\times \Phi_{J \setminus \{s_1, \cdots, s_4\}}(u, v; t),$$

where the sums $\sum_s, \sum_{s_1, s_2}', \cdots, \sum_{s_1, \cdots, s_4}'$ are taken over the corresponding disjoint indices from the set J. Using the bound in (10.2.48), the convergence $\sum_{s=0}^{\infty} a_s^2 < \infty$ and because $|\phi^{(1)}(u)| \leq |u|$ and $|\phi^{(i)}(u)| \leq E|\zeta_0^i| < C$, $i = 2, 3, 4$, as in part (i), we obtain that for all $t \geq t_0$,

$$\left|\frac{\partial^4 \hat{f}_t(u, v)}{\partial u^2 \partial v^2}\right| \leq C \frac{u^4 + v^4}{(1 + u^2)^3(1 + v^2)^3} \leq \frac{C}{(1 + u^2)(1 + v^2)}.$$

This completes the proof of Lemma 10.2.4.

Proof of Lemma 10.2.2. The proof uses Lemmas 10.2.3 and 10.2.4. Let $\bar{\mu}(x, y) := \mu_{r \wedge 2}(x, y)$. We shall first derive the bound (10.2.28) with $\mu := C\bar{\mu}$. Apply (10.2.37) with $p = 0, 1$, to obtain, for any $x < y$, $|\xi| \leq 1$,

$$F(x + \xi, y + \xi) + \int_{x+\xi}^{y+\xi} |f^{(1)}(z)| dz \leq C\mu_2(x + \xi, y + \xi)$$

$$\leq C\mu_2(x, y) \leq C\bar{\mu}(x, y).$$

The second inequality follows from (10.2.32) with $\gamma = 2$ and $h = h_2$.

It remains to show (10.2.29). Recall (10.2.27). Let $t_1 < 1$ and $t_1 - \epsilon < 1$. Use (10.2.51), as in the proof of Lemma 10.2.1, to obtain

$$E(S_n(x, y))^2 \leq C\bar{\mu}(x, y)\frac{1}{n^{1+2d}L^2(n)} \sum_{j,k=1}^{n} |j - k|_+^{-t_1 + \epsilon} \leq C\bar{\mu}(x, y)n^{-a},$$

with $a := t_1 - (1 - 2d) - 2\epsilon$ and $\epsilon > 0$ arbitrarily small. When $t_1 - \epsilon > 1$, the statement of Lemma 10.2.2 follows similarly.

The next lemma gives bounds on the covariances of the increments of $R_j(x, y) := R_j(y) - R_j(x)$, $x < y, j \in \mathbb{Z}$. These bounds are essential for bounding the second moment of the increments of the process S_n in (10.2.23). Let

$$k_+^{-b} := k^{-b}I(k > 0) + I(k = 0), \quad b > 0, k \geq 0.$$

Lemma 10.2.5. *Under the conditions of Theorem 10.2.3, the following statements hold. For any $\epsilon > 0$, there exists a constant $C < \infty$ such that for all $j \in \mathbb{Z}$, $x < y$, $x' < y'$,*

$$\left|\text{Cov}(R_0(x, y), R_j(x', y'))\right| \tag{10.2.51}$$

$$\leq C\big(\bar{\mu}(x, y)\bar{\mu}(x', y')\big)^{1/2} \begin{cases} |j|_+^{-t_1 + \epsilon}, & t_1 < 1, \\ |j|_+^{-(1+t_1)/2 + \epsilon}, & t_1 > 1, \end{cases}$$

$$\left|\text{Cov}(I(x < \varepsilon_0 \leq y), R_j(x', y'))\right|$$

$$\leq C\big(\bar{\mu}(x, y)\bar{\mu}(x', y')\big)^{1/2} \begin{cases} |j|_+^{-(t_1+1-2d)/2 + \epsilon}, & t_1 < 1, \\ |j|_+^{d-1}, & t_1 > 1, \end{cases}$$

$$\left|\text{Cov}(I(x < \varepsilon_0 \leq y), I(x' < \varepsilon_j \leq y'))\right|$$

$$\leq C\big(\bar{\mu}(x, y)\bar{\mu}(x', y')\big)^{1/2}|j|_+^{2d-1},$$

where $\bar{\mu}(x, y) := \mu_{r \wedge 2}(x, y)$ is a finite continuous measure on $\mathbb{R}$ and $t_1 > 1 - 2d$ is defined in (10.2.27).

Remark 10.2.4. Besides providing an indispensable tool for proving the uniform reduction principle, Lemma 10.2.5 also shows that for an ordinary empirical process $n^{1/2}\hat{F}_n$, i.e., when $\gamma_{nj} \equiv n^{-1/2}$ and $\xi_{nj} \equiv 0$, in the long memory case, the covariances of $\{R_j(x)\}$ decay faster than the covariances of $\{\varepsilon_j\}$, and therefore the limit laws of $\hat{F}_n(x) - F(x)$ and $f(x)\bar{\varepsilon}_n$ are the same. In contrast, in the i.i.d. case, $\{R_j(x)\}$ is again i.i.d. and hence uncorrelated and its covariance decays at the same rate as the covariance of $\{I(\varepsilon_j \leq x)\}$ and $\{\varepsilon_j\}$, and therefore $\hat{F}_n(x) - F(x)$ is not approximated by $f(x)\bar{\varepsilon}_n$.

Remark 10.2.5. Let ψ be a real-valued function of bounded variation on $\mathbb{R}$, $E\psi(\varepsilon_j) = 0$, $\lambda_1 := -\int f(y)d\psi(y)$, and $R_j(\psi) := \psi(\varepsilon_j) + \lambda_1\varepsilon_j$. Integration by parts gives $\psi(\varepsilon_j) = \psi(\infty) - \int I(\varepsilon_j \leq x)d\psi(x)$ and $E\psi(\varepsilon_j) = \psi(\infty) - \int F(x)d\psi(y)$, and hence

$$R_j(\psi) = -\int [I(\varepsilon_j \leq x) - F(x) + f(x)\varepsilon_j]d\psi(x) = -\int R_j(x)d\psi(x),$$

where $R_j(x)$ is as in (10.2.23). Then

$$\mathrm{Cov}(R_0(\psi), R_j(\psi)) = \int\int \mathrm{Cov}(R_0(x), R_j(y))d\psi(x)d\psi(y).$$

From (10.2.51) we find that, under the assumptions of Theorem 10.2.3, there exists $\kappa = \kappa(d, r) > 0$ such that

$$\mathrm{Cov}(R_0(\psi), R_j(\psi)) = O(j^{-(1-2d)-\kappa}), \tag{10.2.52}$$

$$\mathrm{Cov}(\psi(\varepsilon_0), \psi(\varepsilon_j)) = \lambda_1^2 \mathrm{Cov}(\varepsilon_0, \varepsilon_j)(1 + o(1)).$$

Indeed, the first relation in (10.2.52) follows from (10.2.51) and the bound $|\mathrm{Cov}(R_0(\psi), R_j(\psi))| \leq C \sup_{y,y'\in\mathbb{R}} |\mathrm{Cov}(R_0(y), R_j(y'))|$, with

$$\kappa = \begin{cases} t_1 - (1 - 2d) - \epsilon & = \tau_1 - \epsilon, & t_1 < 1, \\ (1 + t_1)/2 - (1 - 2d) - \epsilon = d + \tau_1/2 - \epsilon, & t_1 > 1, \end{cases}$$

where $\tau_1 > 0$ is defined by (10.2.27) and $\epsilon > 0$ is arbitrarily small.

Before proceeding with the proof of this lemma, we state an elementary inequality: for any $\alpha > 0$, $\beta > 0$, and $\alpha + \beta > 1$,

$$\sum_{i=1}^{\infty} i^{-\alpha}(j + i)^{-\beta} \leq C \begin{cases} j_+^{-\alpha-\beta+1}, & 0 < \alpha < 1, \\ j_+^{-\beta}, & \alpha > 1, \\ j_+^{-\beta}\log(1 + j), & \alpha = 1. \end{cases} \tag{10.2.53}$$

Proof of Lemma 10.2.5. Because of Remark 10.2.3, t_1 and the upper bounds in (10.2.51) do not depend on r for $r \geq 3/2$. Therefore, we shall assume $1 < r \leq 3/2$ in the rest of the proof, without loss of generality.

Recall the decomposition (10.2.36). Let $\mathcal{F}_k = \sigma\text{-field}\{\zeta_i : i \leq k\}$ be the past σ-field. Note that ε_j is $\mathcal{F}_j$ measurable and $\mathcal{F}_{-\infty}$ is the trivial σ-field. We have the telescoping identity

$$I(\varepsilon_j \leq x) - F(x) = \sum_{i=0}^{\infty} \left\{ P(\varepsilon_j \leq x | \mathcal{F}_{j-i}) - P(\varepsilon_j \leq x | \mathcal{F}_{j-i-1}) \right\}. \quad (10.2.54)$$

Because $\varepsilon_{j,i}$ and $\tilde{\varepsilon}_{j,i}$ are independent for each i, j,

$$P(\varepsilon_j \leq x | \mathcal{F}_{j-i}) = F_{i-1}(x - \tilde{\varepsilon}_{j,i-1}), \quad \forall i, j. \quad (10.2.55)$$

Let, for $j \in \mathbb{Z}$, $i = 0, 1, \cdots$,

$$U_{j,i}(x) := F_{i-1}(x - \tilde{\varepsilon}_{j,i-1}) - F_i(x - \tilde{\varepsilon}_{j,i}) + f(x)a_j\zeta_{j-i}. \quad (10.2.56)$$

By the well-known property of nested conditional expectations,

$$EU_{j,i}(x)U_{j',i'}(x') = 0, \quad \forall j - i \neq j' - i', x, x'. \quad (10.2.57)$$

Now, rewrite

$$R_j(x) = \sum_{i=0}^{\infty} U_{j,i}(x).$$

By (10.2.57), this series converges in mean square, and

$$|ER_j(x,y)R_0(x',y')| = \left| \sum_{i=0}^{\infty} EU_{j,j+i}(x,y)U_{0,i}(x',y') \right| \quad (10.2.58)$$

$$\leq \sum_{i=0}^{\infty} E^{1/2}(U_{j,j+i}(x,y))^2 E^{1/2}(U_{0,i}(x',y'))^2.$$

We will show that for all $x < y$, $j \in \mathbb{Z}$, $i \geq 0$,

$$E(U_{j,i}(x,y))^2 \leq C \mu_r(x,y) i_+^{-1-t_1+\epsilon}, \quad (10.2.59)$$

where μ_r is defined by (10.2.30). Clearly, (10.2.58) and (10.2.59) imply Lemma 10.2.5, because for any $j \geq 0$,

$$|\text{Cov}(R_j(x,y), R_0(x',y'))|$$

$$\leq C\{\mu_r(x,y)\mu_r(x',y')\}^{1/2} \sum_{i=0}^{\infty} (i(j+i))_+^{-(1+t_1)/2+(\epsilon/2)}$$

$$\leq C\{\mu_r(x,y)\mu_r(x',y')\}^{1/2} \begin{cases} j_+^{-t_1+\epsilon}, & t_1 < 1, \\ j_+^{-(1+t_1)/2+\epsilon}, & t_1 > 1. \end{cases}$$

The last inequality above follows from (10.2.53), provided $\epsilon > 0$ is chosen small enough.

To prove (10.2.59), take $i \geq m_0$ sufficiently large so that F_i is differentiable and the conclusions of Lemma 10.2.4 hold. With $f_i = F_i^{(1)}$, let

$$U_{j,i}^{(1)}(x) := F_i(x - \tilde{\varepsilon}_{j,i-1}) - F_i(x - \tilde{\varepsilon}_{j,i}) + a_i\zeta_{j-i}f_i(x - \tilde{\varepsilon}_{j,i}), \quad (10.2.60)$$

$$U_{j,i}^{(2)}(x) := F_{i-1}(x - \tilde{\varepsilon}_{j,i-1}) - F_i(x - \tilde{\varepsilon}_{j,i-1}),$$

$$U_{j,i}^{(3)}(x) := a_i\zeta_{j-i}\Big(f(x) - f(x - \tilde{\varepsilon}_{j,i})\Big),$$

$$U_{j,i}^{(4)}(x) := a_i\zeta_{j-i}\Big(f(x - \tilde{\varepsilon}_{j,i}) - f_j(x - \tilde{\varepsilon}_{j,i})\Big).$$

Then we can rewrite

$$U_{j,i}(x) = \sum_{\tau=1}^{4} U_{j,i}^{(\tau)}(x).$$

Clearly, (10.2.59) follows from

$$E\big(U_{j,i}(x,y)\big)^2 \leq C\mu_r(x,y), \qquad\qquad 1 \leq i \leq m_0, \quad (10.2.61)$$

$$E\big(U_{j,i}^{(\tau)}(x,y)\big)^2 \leq C\mu_r(x,y)i^{-1-t_1+\epsilon}, \qquad i \geq m_0, \quad (10.2.62)$$

for $\tau = 1, \cdots, 4$.

We shall first prove (10.2.62). Rewrite

$$U_{j,i}^{(1)}(x,y) = \int_0^{-a_j\zeta_{j-i}} du \int_x^y [f_i^{(1)}(w + u - \tilde{\varepsilon}_{j,i}) - f_i^{(1)}(w - \tilde{\varepsilon}_{j,i})]dw.$$

Apply (10.2.37) and (10.2.35), with $v = -a_i\zeta_{j-i}$ and $z = \tilde{\varepsilon}_{j,i}$, to obtain, for any $i \geq m_0$, $x < y$, and $\gamma \in (1, 2]$, with probability 1,

$$\big|U_{j,i}^{(1)}(x,y)\big| \leq C\mu_\gamma(x,y)|a_i|^\gamma|\zeta_{j-i}|^\gamma(1 \vee |\tilde{\varepsilon}_{j,i}|^\gamma). \quad (10.2.63)$$

By Rosenthal's inequality, see (2.5.5), for any $1 \leq \gamma \leq r$,

$$E\big|\tilde{\varepsilon}_{j,i}\big|^{2\gamma} \leq C\sum_{k=i+1}^{\infty}|a_k|^{2\gamma} + C\Big(\sum_{k=i+1}^{\infty}a_k^2\Big)^\gamma \leq Ci^{-\gamma(1-2d)+\epsilon}, \quad (10.2.64)$$

where the constant C depends only on $\epsilon > 0$ and where we used (10.2.20) and the property (2.3.3) of slowly varying functions. In particular,

$$E\big(1 \vee |\tilde{\varepsilon}_{j,i}|^{2r}\big) \leq C, \quad (10.2.65)$$

where C does not depend on j and i. Upon applying this inequality and using (10.2.20), (10.2.63), and the independence of ζ_{j-i} and $\tilde{\varepsilon}_{j,i}$, we find

$$E\big(U_{j,i}^{(1)}(x,y)\big)^2 \le C\mu_r(x,y)|a_i|^{2r} \tag{10.2.66}$$
$$\le C\mu_r(x,y)\big|i^{-(1-d)}L(i)\big|^{2r} \le C\mu_r(x,y)i^{-1-t_1+\epsilon},$$

as $2r(1-d) \ge 2(1-d) + \tau_1 = 1 + t_1$ by the definitions of τ_1 and t_1 in (10.2.27).

Consider $U_{j,i}^{(2)}$. By (10.2.32) and (10.2.38),

$$\big|U_{j,i}^{(2)}(x,y)\big| = \left| \int_x^y [f_{i-1}(u - \tilde{\varepsilon}_{j,i-1}) - f_i(u - \tilde{\varepsilon}_{j,i-1})]du \right|$$
$$\le Ca_i^2 \int_x^y (1 + |u - \tilde{\varepsilon}_{j,i-1}|)^{-2} du$$
$$\le C\mu_r(x,y)a_i^2(1 \vee |\tilde{\varepsilon}_{j,i}|)^r, \qquad \forall i \ge m_0.$$

Together with (10.2.65), this yields

$$E\big(U_{j,i}^{(2)}(x,y)\big)^2 \le C\mu_r(x,y)a_i^4 \le C\mu_r(x,y)i^{-1-t_1+\epsilon}, \tag{10.2.67}$$

since $4(1-d) \ge 1 + t_1$ by the definitions in (10.2.27).

Next, consider $U_{j,i}^{(3)}(x,y) = a_i\zeta_{j-i}\int_x^y(f(u) - f(u - \tilde{\varepsilon}_{j,i})du$. From (10.2.37) and (10.2.34), it follows that for any $x < y$ and $\gamma \in (1,2]$, with probability 1,

$$\big|U_{j,i}^{(3)}(x,y)\big| \le C\mu_\gamma(x,y)|a_i||\zeta_{j-i}|\big(|\tilde{\varepsilon}_{j,i}| + |\tilde{\varepsilon}_{j,i}|^\gamma\big). \tag{10.2.68}$$

By (10.2.64) and (10.2.68) with $\gamma = r$, we obtain

$$E\big(U_{j,i}^{(3)}(x,y)\big)^2 \le C\mu_r(x,y)a_i^2 i^{-(1-2d)+\epsilon/2} \tag{10.2.69}$$
$$\le C\mu_r(x,y)i^{-1-t_1+\epsilon},$$

since $2(1-d) + 1 - 2d \ge 1 + t_1$.

Next, consider $U_{j,i}^{(4)}(x,y)$. Observe by (10.2.38), for any $x \in \mathbb{R}$ and any $i \ge m_0$ sufficiently large,

$$|f(x) - f_i(x)| = \left| \sum_{k=i}^\infty (f_{k+1}(x) - f_k(x)) \right|$$
$$\le \frac{C}{(1 + |x|)^2} \sum_{k=i}^\infty a_k^2 \le \frac{C}{(1 + |x|)^2} i^{-(1-2d)+\epsilon}.$$

Then by (10.2.32), with probability 1, for any $1 < \gamma \le 2$,

$$\left| \int_x^y (f(u - \tilde{\varepsilon}_{j,i}) - f_i(u - \tilde{\varepsilon}_{j,i})du \right| \le C\mu_\gamma(x,y)i^{-(1-2d)+\epsilon}(1 \vee |\tilde{\varepsilon}_{j,i}|^\gamma).$$

This bound with $\gamma = r$ and (10.2.64) gives

$$E\big(U_{j,i}^{(4)}(x,y)\big)^2 \le C\mu_r(x,y)i^{-2(1-d)-2(1-2d)+4\epsilon} \le C\mu_r(x,y)i^{-1-t_1},$$

provided $\epsilon > 0$ was chosen small enough. The claim (10.2.62) now follows from this bound, (10.2.66), (10.2.67), and (10.2.69).

It remains to prove (10.2.61). But, because of (10.2.56), it follows from

$$E(F_j(x - \tilde{\varepsilon}_{j,i}, y - \tilde{\varepsilon}_{j,i}))^2 \le C\mu_r(x,y), \quad 0 \le i \le m_0,$$
$$(f(x,y))^2 \le C\mu_r(x,y).$$

The last claim follows from (10.2.37), while the first follows from the inequality

$$E(F_j(x - \tilde{\varepsilon}_{j,i}, y - \tilde{\varepsilon}_{j,i}))^2 \le EF_j(x - \tilde{\varepsilon}_{j,i}, y - \tilde{\varepsilon}_{j,i})$$
$$= F(x,y) = \int_x^y f(z)dz \le C\mu_2(x,y),$$

and (10.2.37). This proves (10.2.61) and also Lemma 10.2.5.

Note: This section has its roots in Dehling and Taqqu (1989), Koul (1992b), Koul and Mukherjee (1993), Ho and Hsing (1996), Giraitis, Koul and Surgailis (1996), and Giraitis and Surgailis (1999).

10.3   Expansion of the empiricals

In light of the URP of the previous section it is natural to explore the higher-order asymptotics of the empirical processes. This section describes a higher-order asymptotic expansion of the empirical process when $\{\varepsilon_j\}$ form a discrete-time Gaussian or moving-average process.

The next subsection discusses the case when $\{\varepsilon_j\}$ is a Gaussian process. In this case, a complete asymptotic expansion of the empirical process is derived via the standard techniques of Hermite expansions, as in the previous section. In Subsection 10.3.2 we formulate and discuss an asymptotic expansion of empirical processes of long memory moving averages (10.2.19). The proof of this expansion is preceded by a heuristic explanation. Lemmas 10.2.3 to 10.2.5 of the previous section are found useful in the proofs here also. For notational simplicity, in this section we assume that the slowly varying factors are asymptotically constant.

10.3.1 *Expansion of empiricals: Gaussian case*

Let $\{\varepsilon_j, j \in \mathbb{Z}\}$ be a stationary Gaussian process with zero mean, unit variance, and covariance function $\gamma(j) = E\varepsilon_0\varepsilon_j$. Recall the definition of the Hermite polynomials from (2.4.1) and the Hermite expansion of the centered indicator function:

$$I(\varepsilon_j \le x) - \Phi(x) = \sum_{k=1}^{\infty} \frac{(-1)^k}{k!}\Phi^{(k)}(x)H_k(\varepsilon_j), \qquad (10.3.1)$$

where $\Phi^{(k)}$ is the kth derivative of the standard normal d.f. Φ.

Here we shall assume that the covariance function $\gamma(j)$ satisfies

$$\gamma(j) \sim c_\gamma j^{-1+2d}, \qquad c_\gamma > 0, \; 0 < d < 1/2, \; j \to \infty. \qquad (10.3.2)$$

By the property (2.4.15) of Hermite polynomials,

$$\sum_{j=1}^{\infty} |\mathrm{Cov}(H_k(\varepsilon_0), H_k(\varepsilon_j))| = k! \sum_{j=1}^{\infty} |\gamma(j)|^k \begin{cases} = \infty, & k(1-2d) < 1, \\ < \infty, & k(1-2d) > 1. \end{cases}$$

Let $[x]$ denote the integer part of $x \in \mathbb{R}$ and let

$$k^* := \left[\frac{1}{1-2d}\right], \qquad 0 < d < 1/2. \qquad (10.3.3)$$

Accordingly, the r.h.s. of (10.3.1) can be split into a long memory part

$$\sum_{k=1}^{k^*} \frac{(-1)^k}{k!}\Phi^{(k)}(x)H_k(\varepsilon_j),$$

and a short memory part

$$R_j(x) := I(\varepsilon_j \le x) - \Phi(x) - \sum_{k=1}^{k^*} \frac{(-1)^k}{k!}\Phi^{(k)}(x)H_k(\varepsilon_j). \qquad (10.3.4)$$

The asymptotic distributions of the sums of the long memory terms in (10.3.1) were described as *non-central limit theorems* in Section 4.7. Let

$$Z_n^{(k)} := n^{k(1/2-d)-1} \sum_{j=1}^{n} \frac{H_k(\varepsilon_j)}{k!}, \qquad k = 1, 2, \cdots. \qquad (10.3.5)$$

Then

$$(Z_n^{(k)}, 1 \le k \le k^*) \to_D (Z^{(k)}, 1 \le k \le k^*), \qquad (10.3.6)$$

where the r.v.'s $Z^{(k)}$ can be written using the multiple Wiener–Itô integral

$$Z^{(k)} := \frac{c_\gamma^{k/2}}{k! B^{k/2}(d, 1 - 2d)} \tag{10.3.7}$$

$$\times \int_{\mathbb{R}^k} \left\{ \int_0^1 \prod_{i=1}^k (s - x_i)_+^{d-1} ds \right\} W(dx_1) \cdots W(dx_k)$$

w.r.t. Gaussian white-noise $W(ds)$, $E(W(ds))^2 = ds$. $s_+^{-\lambda} := s^{-\lambda}$ if $s > 0$, and is 0 otherwise. This r.v. coincides up to a multiplicative constant with the Hermite process $\mathcal{H}_k(t)$ at time $t = 1$ of (14.3.37), for which $E\mathcal{H}_k^2(1) = 1$. Accordingly, from (14.3.39), for $0 < d < 1/2$, $\mathrm{Var}(Z^{(k)}) = c_k^2(d)$, where

$$c_k^2(d) := \frac{2c_\gamma^k}{k! \big(1 - k(1 - 2d)\big)\big(2 - k(1 - 2d)\big)}.$$

A general definition of these integrals and some additional properties are given in Section 14.3 below.

Under the additional assumption that $\{\varepsilon_j\}$ is a Gaussian long memory moving average as in (10.2.19) and (10.2.24), the convergence in (10.3.6) follows as in the proof of Theorem 4.6.1. In the general case, the proof of (10.3.6) requires a multiple Wiener–Itô integral with respect to a spectral (complex-valued) Gaussian random measure, which we do not discuss in this book to avoid too many mathematical technicalities. An interested reader may consult Dobrushin and Major (1979) for details.

Next, let

$$\Psi_1 := \{\psi : \psi \text{ a real-valued non-decreasing, right continuous}$$
$$\text{and bounded function on } \mathbb{R}\}.$$

For $\psi \in \Psi_1$ and with $R_j(x)$ as in (10.3.4), let

$$S_{n,\psi} := n^{-1} \sum_{j=1}^n \psi(\varepsilon_j), \qquad R_j(\psi) := -\int R_j(x) d\psi(x), \tag{10.3.8}$$

$$Q_{n,\psi} := -\int Q_n(x) d\psi(x), \qquad Q_n(x) := n^{-1/2} \sum_{j=1}^n R_j(x),$$

$$\lambda(x) := E\psi(\varepsilon_0 - x) = \int \psi(y - x) d\Phi(y), \qquad x \in \mathbb{R},$$

$$\lambda_k := \lambda^{(k)}(0) = -\int \Phi^{(k)}(y) d\psi(y), \qquad k \geq 0.$$

We are now ready to state:

Theorem 10.3.1. *Let $\{\varepsilon_j\}$ be a stationary Gaussian process satisfying the above conditions including (10.3.2) for some $0 < d < 1/2$. Assume $1/(1 - 2d)$ is not an integer. Then the following asymptotic expansion of the empirical d.f. holds:*

$$\hat{F}_n(x) - \Phi(x) = \sum_{1 \leq k \leq k^*} (-1)^k n^{-k(1/2-d)} \Phi^{(k)}(x) Z_n^{(k)} \qquad (10.3.9)$$

$$+ n^{-1/2} Q_n(x),$$

where $Z_n^{(k)}$, $1 \leq k \leq k^*$, satisfy (10.3.6), and $\{Q_n(x), x \in \mathbb{R}\}$ converges weakly, in the Skorokhod space $\mathcal{D}(\bar{\mathbb{R}})$ with the sup-topology, to a Gaussian process $\{Q(x), x \in \mathbb{R}\}$ with zero mean and the covariance

$$\mathrm{Cov}(Q(x), Q(y)) = \sum_{j \in \mathbb{Z}} \mathrm{Cov}(R_0(x), R_j(y)).$$

Moreover, for every $\psi \in \Psi_1$ the following expansion holds for $S_{n,\psi}$:

$$S_{n,\psi} - ES_{n,\psi} \qquad (10.3.10)$$

$$= \sum_{1 \leq k \leq k^*} (-1)^k n^{-k(1/2-d)} \lambda_k Z_n^{(k)} + n^{-1/2} Q_{n,\psi},$$

where $Q_{n,\psi} \to_D \mathcal{N}(0, \sigma_\psi^2)$ with $\sigma_\psi^2 := \sum_{j \in \mathbb{Z}} \mathrm{Cov}(R_0(\psi), R_j(\psi))$.

Remark 10.3.1. Note that $\psi(x) = \sum_{k=0}^{\infty} (-1)^k (\lambda_k/k!) H_k(x)$ is the Hermite expansion of ψ. The Hermite rank of ψ is defined as H-rank$(\psi) = \inf\{k \geq 1; \lambda_k \neq 0\}$. Expansion (10.3.10) implies that the limit distribution of $S_{n,\psi}$ is determined by the H-rank of ψ: for $2 \leq$ H-rank$(\psi) \leq k^*$, this distribution is non-Gaussian and given by that of a corresponding multiple integral $Z^{(k)}$, while for H-rank$(\psi) = 1$ or H-rank$(\psi) > k^*$, the limit distribution of $S_{n,\psi}$ is Gaussian. These conclusions clearly agree with Theorems 4.6.1 and 4.7.1 of this text, which apply to arbitrary square integrable functions of a long memory Gaussian process.

Proof of Theorem 10.3.1. Note that the set of the discontinuity points of $\psi \in \Psi_1$ has the Lebesgue measure zero. Since the distribution of ε_j is absolutely continuous, so $\int \psi(x) dI(\varepsilon_j \leq x) = \psi(\varepsilon_j)$ a.s., for any $j \in \mathbb{Z}$, implying $S_{n,\psi} = \int \psi d\hat{F}_n$, a.s.

Claim (10.3.10) follows from (10.3.9), integration by parts, and

$$S_{n,\psi} - \int \psi d\Phi = \int \psi d(\hat{F}_n - \Phi) = - \int (\hat{F}_n - \Phi) d\psi, \quad \text{a.s.}$$

We proceed to prove (10.3.9). Let

$$h_x(z) := I(z \le x) - \sum_{0 \le k \le k^*} \frac{(-1)^k}{k!} \Phi^{(k)}(x) H_k(z), \quad z \in \mathbb{R}.$$

For each $x \in \mathbb{R}$, $h_x \in L_2(\Phi)$ has Hermite rank greater than k^* and $\sum_{j=0}^{\infty} |\gamma(j)|^{k^*+1} < \infty$, by (10.3.2) and the definition of k^*. Now, rewrite

$$Q_n(x) = n^{-1/2} \sum_{j=1}^{n} R_j(x) = n^{-1/2} \sum_{j=1}^{n} h_x(\varepsilon_j).$$

By Theorem 4.6.1, these facts imply that for every $x \in \mathbb{R}$, $Q_n(x) \to_D Q(x)$, where $Q(x)$ is a $\mathcal{N}(0, \sigma_x^2)$ r.v., with

$$\sigma_x^2 := \sum_{j \in \mathbb{Z}} \text{Cov}(h_x(\varepsilon_0), h_x(\varepsilon_j)) = \sum_{j \in \mathbb{Z}} \text{Cov}(R_0(x), R_j(x)).$$

A similar conclusion applies to any linear combination $\sum_{k=1}^{m} a_k Q_n(x_k)$, $a_k, x_k \in \mathbb{R}$, implying the finite-dimensional convergence $Q_n \to_{fdd} Q$, by the Cramér–Wold device.

It remains to show the tightness. The proof below uses the ideas of Csörgő and Mielniczuk (1996). Its main ingredient is the following inequality. Let $\langle x \rangle = x - [x] \in [0, 1)$ be the fractional part of $x \in \mathbb{R}$, and $\varepsilon = \varepsilon_0 \sim \mathcal{N}(0, 1)$ be the marginal distribution of $\{\varepsilon_j\}$.

Lemma 10.3.1. (Taqqu, 1977). *Let $n, m, p \ge 2, 1 \le q \le p$, be integers, and suppose $\bar{\gamma}_n := \max\{|\gamma(i)| : 1 \le i \le n\} < 1/(p-1)$. If $G_1, \cdots, G_p$ are functions in $L_2(\Phi)$ such that at least q of them have Hermite rank greater than or equal to m, then*

$$\sideset{}{'}\sum |E(G_1(\varepsilon_{t_1}) \cdots G_p(\varepsilon_{t_p}))| \tag{10.3.11}$$

$$\le K_p n^{p-q/2} \left(\sum_{j=1}^{n} |\gamma(j)|^m \right)^{q/2} \left(n^{-1} \sum_{j=1}^{n} |\gamma(j)|^m \right)^{\langle mq/2 \rangle / m},$$

where the summation $\sum'$ is over all different indices $1 \le t_1, \cdots, t_p \le n$ ($t_i \ne t_j$, $i \ne j$), and

$$K_p \le C_p \left\{ E(G_1^2(\varepsilon)) \cdots E(G_p^2(\varepsilon)) \right\}^{1/2}, \tag{10.3.12}$$

with $C_p := 2^p / (1 - \{(p-1)\bar{\gamma}_n\}^{1/2})^p$.

We also need the following elementary fact.

Lemma 10.3.2. Let $\{\varepsilon_j, j \in \mathbb{Z}\}$ be a stationary Gaussian process with zero mean and covariance $\gamma(j)$, with $\bar{\gamma} := \sup_{j \geq 1} |\gamma(j)| < \gamma(0) = 1$. Then for any $x_1 < x_2$, $x_3 < x_4$ and $j \geq 1$,

$$P(x_1 < \varepsilon_0 \leq x_2, x_3 < \varepsilon_j \leq x_4) \leq \bar{\mu}(x_1, x_2)\bar{\mu}(x_3, x_4), \quad (10.3.13)$$

where

$$\bar{\mu}(x, y) := \left(\sqrt{2\pi}(1 - \bar{\gamma})^{1/4}\right)^{-1} \int_x^y e^{-\frac{u^2}{2(1+\bar{\gamma})}} du \quad (10.3.14)$$

is a finite and continuous measure on $\mathbb{R}$.

Proof. We have

$$P(x_1 < \varepsilon_0 \leq x_2, x_3 < \varepsilon_k \leq x_4)$$
$$= \frac{1}{2\pi(1 - \gamma^2(k))^{1/2}} \int_{x_1}^{x_2} \int_{x_3}^{x_4} e^{-\frac{u_1^2 + u_2^2 - 2u_1 u_2 \gamma(k)}{2(1 - \gamma^2(k))}} du_1 du_2.$$

The lemma now follows from the following inequality:

$$\frac{u_1^2 + u_2^2 - 2u_1 u_2 \gamma}{2(1 - \gamma^2)} \geq \frac{u_1^2 + u_2^2}{2(1 + \bar{\gamma})}, \quad \forall |\gamma| \leq \bar{\gamma} < 1, u_1, u_2 \in \mathbb{R}.$$

In order to apply the above lemmas, assume first that

$$\bar{\gamma} := \sup_{i \geq 1} |\gamma(i)| < 1/3. \quad (10.3.15)$$

We shall verify the following inequality for Q_n. For any $x_1 < x_2 < x_3$,

$$E\big(Q_n(x_2) - Q_n(x_1)\big)^2 \big(Q_n(x_3) - Q_n(x_2)\big)^2 \leq C(\nu(x_1, x_3))^{3/2}, \quad (10.3.16)$$
$$E\big(Q_n(x_2) - Q_n(x_1)\big)^4 \leq C\big((\nu(x_1, x_2))^{3/2} + n^{-1}\nu(x_1, x_2)\big), \quad (10.3.17)$$

where

$$\nu(x, y) := \mu(x, y) + \bar{\mu}(x, y), \quad \mu(x, y) := \sum_{1 \leq k \leq k^* + 1} \int_x^y |\Phi^{(k)}(z)| dz$$

is a finite continuous measure on $\mathbb{R}$, with $\bar{\mu}$ defined by (10.3.14). Then the tightness of Q_n will follow from Lemma 4.4.1 applied to $\tilde{Q}_n(u) := Q_n(F^{-1}(u))$, where $F(x) := \nu(-\infty, x)/\nu(\mathbb{R})$.

Write $G_1(z) := h_{x_2}(z) - h_{x_1}(z)$ and $G_2(z) := h_{x_3}(z) - h_{x_2}(z)$. Then

$$n^2 \big(Q_n(x_2) - Q_n(x_1)\big)^2 \big(Q_n(x_3) - Q_n(x_2)\big)^2 \quad (10.3.18)$$
$$= \Sigma_1 + \Sigma_{21} + 2\Sigma_{22} + 2\Sigma_{23} + 2\Sigma_{24} + \Sigma_{31} + \Sigma_{32} + 2\Sigma_{33} + \Sigma_4,$$

where

$$\Sigma_1 := {\sum}' G_1^2(\varepsilon_t)G_2^2(\varepsilon_t), \quad \Sigma_{21} := {\sum}' G_1^2(\varepsilon_{t_1})G_2^2(\varepsilon_{t_2}),$$

$$\Sigma_{22} := {\sum}' G_1(\varepsilon_{t_1})G_2(\varepsilon_{t_1})G_1(\varepsilon_{t_2})G_2(\varepsilon_{t_2}),$$

$$\Sigma_{23} := {\sum}' G_1^2(\varepsilon_{t_1})G_2(\varepsilon_{t_1})G_2(\varepsilon_{t_2}), \quad \Sigma_{24} := {\sum}' G_1(\varepsilon_{t_1})G_1(\varepsilon_{t_2})G_2^2(\varepsilon_{t_2}),$$

$$\Sigma_{31} := {\sum}' G_1(\varepsilon_{t_1})G_1(\varepsilon_{t_2})G_2^2(\varepsilon_{t_3}), \quad \Sigma_{32} := {\sum}' G_2(\varepsilon_{t_1})G_2(\varepsilon_{t_2})G_1^2(\varepsilon_{t_3}),$$

$$\Sigma_{33} := {\sum}' G_1(\varepsilon_{t_1})G_2(\varepsilon_{t_2})G_1(\varepsilon_{t_3})G_2(\varepsilon_{t_3}),$$

$$\Sigma_4 := {\sum}' G_1(\varepsilon_{t_1})G_1(\varepsilon_{t_2})G_2(\varepsilon_{t_3})G_2(\varepsilon_{t_4}),$$

and where the sums $\sum'$ are taken over different summation indices as in (10.3.11). From the definition of h_x, it follows that for any $v \geq 1$, there is a constant $C < \infty$ depending on v alone such that, $\forall z \in \mathbb{R}, i = 1, 2,$

$$|G_i(z)|^v \leq C\Big[I(x_i < z \leq x_{i+1}) + \sum_{1 \leq k \leq k^*+1} \Big(\int_{x_i}^{x_{i+1}} |\Phi^{(k)}(u)|du\,|H_k(z)|\Big)^v\Big]$$

$$\leq C\Big[I(x_i < z \leq x_{i+1}) + \mu(x_i, x_{i+1}) \sum_{1 \leq k \leq k^*+1} |H_k(z)|^v\Big].$$

Hence, use the fact that $EH_k^{2v}(\varepsilon) < \infty$, $v \geq 0$, to conclude that for any even integers $v, w \geq 0, v + w \geq 2$, there exists a constant $C < \infty$ such that

$$E\{|G_1(\varepsilon)|^v|G_2(\varepsilon)|^w\} \tag{10.3.19}$$

$$\leq CE\Big\{\Big[I(x_1 < \varepsilon \leq x_2) + \mu(x_1, x_2)\sum_{k=1}^{k^*+1}|H_k(\varepsilon)|^v\Big]$$

$$\times \Big[I(x_2 < \varepsilon \leq x_3) + \mu(x_2, x_3)\sum_{k=1}^{k^*+1}|H_k(\varepsilon)|^w\Big]\Big\} \leq C\mu(x_1, x_3).$$

We shall now prove (10.3.16). We shall bound the expectation of the sums on the r.h.s. of (10.3.18), using Lemma 10.3.1 above.

To bound $E|\Sigma_4|$, use Lemma 10.3.1 with $p = q = 4$ and $m = k^* + 1$. As $\sum_{j \in \mathbb{Z}} |\gamma(j)|^{k_*+1} < \infty$, by (10.3.11), (10.3.12), and (10.3.19),

$$E|\Sigma_4| \leq Cn^2 EG_1^2(\varepsilon)EG_2^2(\varepsilon) \tag{10.3.20}$$

$$\leq Cn^2\mu(x_1, x_2)\mu(x_2, x_3) \leq Cn^2(\mu(x_1, x_3))^2.$$

Next, consider $E|\Sigma_{31}|$. In this case, use Lemma 10.3.1 with $p = 3, q = 2,$ and $m = k^* + 1$, to obtain

$$E|\Sigma_{31}| \leq Cn^2 EG_1^2(\varepsilon)(EG_2^4(\varepsilon))^{1/2}$$

$$\leq Cn^2\mu(x_1, x_2)(\mu(x_2, x_3))^{1/2} \leq Cn^2(\mu(x_1, x_3))^{3/2}.$$

A similar argument yields similar bounds for $E|\Sigma_{32}|$ and $E|\Sigma_{33}|$. Hence,

$$E|\Sigma_{3j}| \leq Cn^2(\mu(x_1, x_3))^{3/2}, \qquad j = 1, 2, 3. \qquad (10.3.21)$$

Consider the terms $E|\Sigma_{2j}|$, $j = 1, \cdots, 4$. Here, Lemma 10.3.1 cannot be applied since the Hermite rank of the product functions $G_i G_j$, $i, j = 1, 2$, could be zero. Instead, we use Lemma 10.3.2. Accordingly, for any $t_1 \neq t_2$,

$$EG_1^2(\varepsilon_{t_1})G_2^2(\varepsilon_{t_2}) \leq \prod_{i=1}^2 \int G_i^2(u)\bar{\mu}(du) \leq C(\bar{\mu}(x_1, x_3))^2,$$

as for (10.3.19), with $C < \infty$ independent of $t_1 \neq t_2$. In a similar way, for any $t_1 \neq t_2$,

$$E|G_1(\varepsilon_{t_1})G_2(\varepsilon_{t_1})G_1(\varepsilon_{t_2})G_2(\varepsilon_{t_2})| \leq \left(\int |G_1(u)G_2(u)|\bar{\mu}(du)\right)^2$$
$$\leq C(\bar{\mu}(x_1, x_3))^2,$$
$$E|G_1^2(\varepsilon_{t_1})G_2(\varepsilon_{t_1})G_2(\varepsilon_{t_2})| \leq \int |G_1^2(u)G_2(u)|\bar{\mu}(du) \int |G_2(u)|\bar{\mu}(du)$$
$$\leq C(\bar{\mu}(x_1, x_3))^2,$$
$$E|G_1(\varepsilon_{t_1})G_1(\varepsilon_{t_2})G_2^2(\varepsilon_{t_2})| \leq C(\bar{\mu}(x_1, x_3))^2.$$

This leads to the bounds

$$E|\Sigma_{2j}| \leq Cn^2(\bar{\mu}(x_1, x_3))^2, \qquad j = 1, 2, 3, 4. \qquad (10.3.22)$$

Finally, we obtain an upper bound for $E|\Sigma_1|$. Akin to (10.3.19), we have

$$E\{G_1^2(\varepsilon)G_2^2(\varepsilon)\} \qquad (10.3.23)$$
$$\leq C(\mu^{1/2}(x_1, x_2)\mu(x_2, x_3) + \mu(x_1, x_2)\mu^{1/2}(x_2, x_3)) \leq C(\mu(x_1, x_3))^{3/2},$$

using the definition of G_i and because $EI(x_1 < \varepsilon \leq x_2)I(x_2 < \varepsilon \leq x_3) = 0$ and $EI(x_i < \varepsilon \leq x_{i+1})H_k^2(\varepsilon) \leq E^{1/2}I(x_i < \varepsilon \leq x_{i+1}) E^{1/2}H_k^4(\varepsilon) \leq C(\mu(x_i, x_{i+1}))^{1/2}$, $i = 1, 2$. This yields

$$E|\Sigma_1| \leq Cn(\mu(x_1, x_3))^{3/2}. \qquad (10.3.24)$$

The claim (10.3.16) now follows from (10.3.18) and the bounds in (10.3.20) to (10.3.24).

The proof of (10.3.17) is very similar. As in (10.3.18), decompose

$$n^2\big(Q_n(x_2) - Q_n(x_1)\big)^4$$
$$= \tilde{\Sigma}_1 + \tilde{\Sigma}_{21} + 2\tilde{\Sigma}_{22} + 2\tilde{\Sigma}_{23} + 2\tilde{\Sigma}_{24} + \tilde{\Sigma}_{31} + \tilde{\Sigma}_{32} + 2\tilde{\Sigma}_{33} + \tilde{\Sigma}_4,$$

where $\tilde{\Sigma}_1, \cdots, \tilde{\Sigma}_4$ are defined as (10.3.18) with the only difference that the function G_2 is replaced by G_1. Then the bounds (10.3.20) to (10.3.22) are valid since they do not use the fact that the intervals $(x_1, x_2]$ and $(x_2, x_3]$ are disjoint; however, (10.3.23) and (10.3.24) must be changed to $EG_1^4(\varepsilon) \leq C\mu(x_1, x_3)$ and $E|\tilde{\Sigma}_1| \leq Cn\mu(x_1, x_3)$, respectively. This change contributes the extra term $n^{-1}\nu(x_1, x_2)$ in (10.3.17) and concludes the proof of (10.3.16) and (10.3.17).

Consider now the tightness of Q_n in the general case. Since $\lim_{j \to \infty} \gamma(j) = 0$, there exists a $k_0 \geq 1$ such that $\bar{\gamma}_{k_0} := \sup_{i \geq k_0} |\gamma(i)| < 1/3$. Assume for simplicity that $(n-1)/k_0 > 1$ is an integer, then $Q_n(x) = \sum_{j=1}^{k_0} Q_{nj}(x)$, where

$$Q_{nj}(x) := n^{-1/2} \sum_{i=0}^{[(n-j)/k_0]} h_x(\varepsilon_{j+ik_0}).$$

For any j, the "decimated" process $\{\varepsilon_{j+ik_0}, i \in \mathbb{Z}\}$ satisfies (10.3.15), in the sense that $\sup_{i \geq 1} |E(\varepsilon_j \varepsilon_{j+ik_0})| \leq \bar{\gamma}_{k_0} < 1/3$, and therefore Q_{nj} satisfies (4.4.1), namely

$$\lim_{\delta \to 0} \limsup_n P\left(\sup_{|x-y|<\delta} |Q_{nj}(x) - Q_{nj}(y)| \geq \epsilon \right) = 0, \qquad (10.3.25)$$

according to Lemma 4.4.1, as proved above. Obviously,

$$\left\{ \sup_{|x-y|<\delta} |Q_n(x) - Q_n(y)| > \epsilon \right\} \subset \bigcup_{j=1}^{k_0} \left\{ \sup_{|x-y|<\delta} |Q_{nj}(x) - Q_{nj}(y)| > \epsilon/k_0 \right\},$$

thus proving (10.3.25) with Q_{nj} replaced by Q_n. Therefore, the sequence $Q_n, n \geq 1$, is tight in $\mathcal{D}(\bar{\mathbb{R}})$ with uniform metric, see Theorem 4.4.1. This completes the proof of Theorem 10.3.1.

10.3.2   *Expansion of empiricals: moving-average case*

Let $\{\varepsilon_j, j \in \mathbb{Z}\}$ be the moving-average process of (10.2.19) with the coefficients a_j as in (10.2.24) and $\{\zeta_j\} \sim \text{IID}(0,1)$. Then, by Proposition 3.2.1, the autocovariances $\gamma(j)$ satisfy (10.3.2) with $c_\gamma = c_0^2 B(d, 1-2d)$.

Recall from (10.3.3) that $k^* = [1/(1-2d)]$ is an integer. The expansion of the empirical process of non-Gaussian long memory moving-average observations (10.2.19) is very similar to the expansion (10.3.9) in the Gaussian case except for the definition of $Z_n^{(k)}, 1 \leq k \leq k^*$. There are two ways to define the $Z_n^{(k)}$'s. The first way is to define $Z_n^{(k)}$ by replacing the Hermite polynomials $H_k(x)$ in (10.3.5) by $A_k(x)$ defined by (2.4.19), and use the finite *Appell expansion* of the indicator function (2.4.26), whose form is completely similar to that of the Hermite expansion in (10.3.1). As shown in Surgailis (1982) and Avram and Taqqu (1987) and discussed in Chapter 14, under the conditions (10.2.20) and $E\zeta_0^{2k} < \infty$, the limit distribution of these $Z_n^{(k)}$'s is again given by the same multiple Wiener–Itô integrals (10.3.7) as in the Gaussian case. The corresponding expansion of the empirical process appears in Koul and Surgailis (1997).

However, as mentioned in Section 2.4, although Appell polynomials coincide with Hermite polynomials when $\{\varepsilon_j\}$ is a Gaussian process, they lack the orthogonality property of the latter and therefore $Z_n^{(k)}$ and $Z_n^{(\ell)}$ for $k \neq \ell$ are not orthogonal. Moreover, the use of such $Z_n^{(k)}$'s in the asymptotic expansion of the empirical processes requires the existence of the $2k^*$th moments of ζ_0; see Koul and Surgailis (1997). For these reasons, it is more convenient to use, instead of $Z_n^{(k)}$, the sums

$$\tilde{Z}_n^{(k)} := n^{-1+k(\frac{1}{2}-d)} \sum_{j=1}^{n} \varepsilon_j^{(k)}, \tag{10.3.26}$$

where $\varepsilon_j^{(0)} := 1$, and

$$\varepsilon_j^{(k)} := \sum_{s_k < \cdots < s_1 \leq j} a_{j-s_1} \cdots a_{j-s_k} \zeta_{s_1} \cdots \zeta_{s_k}. \tag{10.3.27}$$

The series (10.3.27) converges in mean square for each $k \geq 1$, under the conditions $\sum_{k=0}^{\infty} a_k^2 < \infty$ and $E\zeta_0^2 < \infty$ alone, and (10.3.27) defines a strictly stationary process in $j \in \mathbb{Z}$ with zero mean and covariance

$$E\varepsilon_j^{(k)}\varepsilon_0^{(k)} = \sum_{0 \leq i_1 < \cdots < i_k} a_{j+i_1} a_{i_1} \cdots a_{j+i_k} a_{i_k}. \tag{10.3.28}$$

From (4.8.2) and (4.8.3) we conclude that for $j \neq i$,

$$E\varepsilon_j^{(k)}\varepsilon_i^{(\ell)} = 0, \qquad k, \ell = 0, 1, \cdots, \quad k \neq \ell. \tag{10.3.29}$$

It follows from (10.3.2) and (10.3.28) that for each $k \geq 1$,

$$|E\varepsilon_j^{(k)}\varepsilon_0^{(k)}| \leq \frac{1}{k!} \left(\sum_{i=0}^{\infty} |a_{j+i}a_i|\right)^k = O(j^{-k(1-2d)}), \quad j \to \infty, \tag{10.3.30}$$

$$E\left(\sum_{j=1}^{n} \varepsilon_j^{(k)}\right)^2 = O(n^{2-k(1-2d)}), \qquad k(1-2d) < 1, \tag{10.3.31}$$

$$= O(n), \qquad k(1-2d) > 1.$$

The properties (10.3.28) to (10.3.31) are very similar to those of the normalized Hermite polynomials $(k!)^{-1}H_k(\varepsilon_j)$ when $\{\varepsilon_j\}$ is Gaussian. In fact, the polynomial form (10.3.27) is the "off-diagonal" part, up to the factor $(k!)^{-1}$, of the Appell polynomial $A_k(\varepsilon_j)$. For example, when $k = 2$,

$$A_2(\varepsilon_j) = \varepsilon_j^2 - E\varepsilon_j^2 = 2\varepsilon_j^{(2)} + \sum_{i=0}^{\infty} a_i^2(\zeta_{j-i}^2 - E\zeta_{j-i}^2),$$

where the difference $\{A_2(\varepsilon_j) - 2\varepsilon_j^{(2)}\}$ is the moving average of the i.i.d. r.v.'s $\{A_2(\zeta_s) = \zeta_s^2 - E\zeta_s^2\}$ with summable weights $b_i^2, i \geq 0$, and therefore this difference is a short memory process. For general k, the relation between $A_k(\varepsilon_j)$ and $\varepsilon_j^{(k)}$ follows from the multinomial formula for Appell polynomials of Avram and Taqqu (1987); see (14.2.2) below. The limit distribution of $\tilde{Z}_n^{(k)}$ of (10.3.26) is the same as that of $Z_n^{(k)}$, but its existence does not require additional moments of ζ_0.

From Theorem 4.8.2, we readily obtain, under (10.2.24),

$$(\tilde{Z}_n^{(k)}, 0 \leq k \leq k^*) \to_D (Z^{(k)}, 0 \leq k \leq k^*), \qquad (10.3.32)$$

where

$$Z^{(k)} := \frac{c_0^k}{k!} \int_{\mathbb{R}^k} \left\{ \int_0^1 \prod_{i=1}^{k} (s - x_i)_+^{d-1} \, ds \right\} W(dx_1) \cdots W(dx_k). \quad (10.3.33)$$

Note the above definition agrees with (10.3.7) since $c_\gamma = c_0^2 B(d, 1 - 2d)$. From (14.3.39) we obtain

$$c_k^2(d) := E(Z^{(k)})^2 = \frac{2c_0^{2k} B^k(d, 1 - 2d)}{k!(1 - k(1 - 2d))(2 - k(1 - 2d))}.$$

Moreover, for all $1 \leq k < 1/(1 - 2d)$,

$$E(\tilde{Z}_n^{(j)} \tilde{Z}_n^{(k)}) \to c_k^2(d), \quad j = k,$$
$$= 0, \qquad j \neq k.$$

To describe the asymptotic expansions for $\hat{F}_n(x)$ and $S_{n,\psi} = n^{-1}\sum_{j=1}^{n} \psi(\varepsilon_j)$, we need some more notation. Akin to (10.3.8), define

$$\lambda(x) := E\psi(\varepsilon_0 - x) = \int_{\mathbb{R}} \psi(y - x)dF(y), \quad x \in \mathbb{R}, \qquad (10.3.34)$$

$$\lambda_k := \lambda^{(k)}(0) = -\int F^{(k)}(y)d\psi(y), \qquad \psi \in \Psi_1.$$

We are now ready to formulate one of the main results of this section.

Theorem 10.3.2. *Let $\{\varepsilon_j\}$ be a long memory moving-average (10.2.19) satisfying (10.2.24). Suppose that the d.f. of ζ_0 satisfies (10.2.21) and*

$$E\zeta_0^4 < \infty. \tag{10.3.35}$$

Let $1/(1-2d)$ be not an integer and $k^* := [1/(1-2d)] \geq 1$. Then

$$\hat{F}_n(x) - F(x) \tag{10.3.36}$$

$$= \sum_{1 \leq k \leq k^*} (-1)^k n^{-k(1/2-d)} F^{(k)}(x)\tilde{Z}_n^{(k)} + n^{-1/2}\tilde{Q}_n(x),$$

$$\sup_x |\tilde{Q}_n(x)| = O_p(n^\delta), \quad \forall \delta > 0. \tag{10.3.37}$$

Moreover, for every $\psi \in \Psi_1$, we have

$$S_{n,\psi} - ES_{n,\psi} = \sum_{k=1}^{k^*}(-1)^k \lambda_k n^{-k(1/2-d)}\tilde{Z}_n^{(k)} + n^{-1/2}\tilde{Q}_{n,\psi}, \tag{10.3.38}$$

where $\tilde{Q}_{n,\psi} = -\int \tilde{Q}_n(x)d\psi(x)$ is asymptotically normal:

$$\tilde{Q}_{n,\psi} \to_D \mathcal{N}(0, \tilde{\sigma}_\psi^2), \quad \tilde{\sigma}_\psi^2 := \sum_{j \in \mathbb{Z}}\text{Cov}(\tilde{R}_0(\psi), \tilde{R}_j(\psi)), \tag{10.3.39}$$

$$\tilde{R}_j(\psi) := \psi(\varepsilon_j) - \sum_{k=0}^{k^*}(-1)^k \lambda_k \varepsilon_j^{(k)}.$$

Remark 10.3.2. The fourth moment assumption (10.3.35) for the noise distribution is crucial for the expansions (10.3.37) and (10.3.38). Surgailis (2004) showed that if the fourth moment assumption is violated but instead ζ_0 is assumed to have regularly varying tails with index $2 < \alpha < 4$, the limit distribution of $S_{n,\psi}$ may be $\alpha(1-d)$-stable for (bounded) ψ with A-rank 2 and $\alpha(1-d) < 2$; in other words, sums of bounded functions of long memory moving averages with finite variance may tend to a limit law with infinite variance. Obtaining a complete asymptotic expansion of the empirical functionals $\hat{F}_n$ and $S_{n,\psi}$ in the case of an infinite fourth moment is an open question at the present time.

Remark 10.3.3. Because, by Lemma 10.2.4, F is infinitely differentiable and its derivatives are integrable, λ_k in (10.3.34) are well defined. Then the Appell rank of ψ can be defined as A-rank$(\psi) = \inf\{k \geq 1; \lambda_k \neq 0\}$. Similarly as in the Gaussian case (see Remark 10.3.1), from (10.3.38) we find that the limit distribution of $S_{n,\psi}$ is determined by the A-rank of ψ.

We can conclude from the above discussion that the asymptotic behavior of sums $S_{n,\psi}$ for long memory moving averages $\{\varepsilon_j\}$ with finite fourth moment is very similar to the Gaussian case.

Before proceeding with a detailed proof, we will give a heuristic argument of (10.3.36). Recall the decomposition (10.2.36) with its definition of F_m. For fixed m and j, the summands $\varepsilon_{j,m}$ and $\tilde{\varepsilon}_{j,m}$ are independent; moreover, given m, then $\{\varepsilon_{j,m}\}$ and $\{\tilde{\varepsilon}_{j,m}\}$ are strictly stationary moving averages in j, although they are not independent as processes. When m is large, the second expression in (10.2.36) is small, because $E\tilde{\varepsilon}_{j,m}^2 = \sum_{k=m+1}^{\infty} a_k^2 \to 0$ as $m \to \infty$. Under (10.2.21), Lemma 10.2.4 implies that

$$\sup_{x\in\mathbb{R}} |F_m^{(p)}(x) - F^{(p)}(x)| \to 0, \quad p = 0, 1, \cdots, \quad \text{as } m \to \infty. \qquad (10.3.40)$$

This means that for large m, $\varepsilon_{j,m}$ is the dominating term of the decomposition (10.2.36) in a distributional sense. On the other hand, $\{\tilde{\varepsilon}_{j,m}\}$ is the dominating term in the sense of "memory", which follows from the fact that, for fixed m, the process $\{\varepsilon_{j,m} = \varepsilon_j - \tilde{\varepsilon}_{j,m}\}$ is m-dependent, while $\{\tilde{\varepsilon}_{j,m}\}$ has long memory. As $j \to \infty$,

$$\text{Cov}(\tilde{\varepsilon}_{0,m}, \tilde{\varepsilon}_{j,m}) \sim \text{Cov}(\varepsilon_0, \varepsilon_j) \sim c_\gamma j^{-(1-2d)},$$

where $c_\gamma = c_0^2 B(d, 1 - 2d)$. This claim follows from (10.2.19) and (10.2.24).

Recall the identity (10.2.54) and equation (10.2.55). Also recall from (10.2.36) that $F_{-1}(x - \tilde{\varepsilon}_{j,-1}) \equiv I(\varepsilon_j \le x)$. We thus obtain

$$I(\varepsilon_j \le x) - F(x) = \sum_{i=0}^{\infty} [F_{i-1}(x - \tilde{\varepsilon}_{j,i-1}) - F_i(x - \tilde{\varepsilon}_{j,i})] \qquad (10.3.41)$$

$$\approx \sum_{i=0}^{\infty} [F_i(x - \tilde{\varepsilon}_{j,i-1}) - F_i(x - \tilde{\varepsilon}_{j,i})] \qquad \text{(by (10.3.40))}$$

$$= \sum_{i=0}^{\infty} [F_i(x - a_i\zeta_{j-i} - \tilde{\varepsilon}_{j,i}) - F_i(x - \tilde{\varepsilon}_{j,i})]$$

$$\approx -\sum_{i=0}^{\infty} a_i\zeta_{j-i} f_i(x - \tilde{\varepsilon}_{j,i}) \qquad \text{(by a Taylor expansion)}$$

$$= -F^{(1)}(x)\varepsilon_j + \sum_{i=0}^{\infty} a_i\zeta_{j-i}[F^{(1)}(x) - F_i^{(1)}(x - \tilde{\varepsilon}_{j,i})].$$

Because of (10.3.40), the second term in the r.h.s. of the last equation is negligible compared to the first term $-f(x)\varepsilon_j$. Therefore, $I(\varepsilon_j \le x) -$

$F(x) \approx -f(x)\varepsilon_j$, in the "first approximation", yielding

$$\hat{F}_n(x) \approx F(x) - f(x)\bar{\varepsilon}_n,$$

or the first-order expansion of $\hat{F}_n(x)$. In a similar way, one can formally obtain the second term of the expansion (10.3.36). With this goal in mind, first note that because $EF_i(x - \tilde{\varepsilon}_{j,i}) = F(x)$,

$$EF_i^{(1)}(x - \tilde{\varepsilon}_{j,i}) = \frac{d(EF_i(x - \tilde{\varepsilon}_{j,i}))}{dx} = F^{(1)}(x).$$

Now, expand $F_i^{(1)}(x - \tilde{\varepsilon}_{j,i})$ in (10.3.41) as above, to obtain

$$F_i^{(1)}(x - \tilde{\varepsilon}_{j,i}) - F^{(1)}(x)$$

$$= \sum_{k=1}^{\infty}[F_{i+k-1}^{(1)}(x - \tilde{\varepsilon}_{j,i+k-1}) - F_{i+k}^{(1)}(x - \tilde{\varepsilon}_{j,i+k})]$$

$$\approx \sum_{k=1}^{\infty}[F_{i+k}^{(1)}(x - \tilde{\varepsilon}_{j,i+k-1}) - F_{i+k}^{(1)}(x - \tilde{\varepsilon}_{j,i+k})]$$

$$= \sum_{k=1}^{\infty}[F_{i+k}^{(1)}(x - a_{i+k}\zeta_{j-i-k} - \tilde{\varepsilon}_{j,i+k}) - F_{i+k}^{(1)}(x - \tilde{\varepsilon}_{j,i+k})]$$

$$\approx -\sum_{k=1}^{\infty} a_{i+k}\zeta_{j-i-k}F_{i+k}^{(2)}(x - \tilde{\varepsilon}_{j,i+k})$$

$$= -F^{(2)}(x)\sum_{k=1}^{\infty} a_{i+k}\zeta_{j-i-k} + \sum_{k=1}^{\infty} a_{k+i}\zeta_{j-i-k}[F^{(2)}(x) - F_{i+k}^{(2)}(x - \tilde{\varepsilon}_{j,i+k})].$$

Dropping the last sum in this approximation, and together with (10.3.41), gives the second-order expansion of the indicator function:

$$I(\varepsilon_j \leq x) - F(x)$$

$$\approx -F^{(1)}(x)\sum_{i=0}^{\infty} a_i\zeta_{j-i} + F^{(2)}(x)\sum_{i=0}^{\infty}\sum_{k=1}^{\infty} a_i a_{i+k}\zeta_{j-i}\zeta_{j-i-k} + \cdots,$$

and, consequently, the two first terms of the expansion of $\hat{F}_n$:

$$n(\hat{F}_n(x) - F(x)) = -F^{(1)}(x)\sum_{j=1}^{n} \varepsilon_j^{(1)} + F^{(2)}(x)\sum_{j=1}^{n} \varepsilon_j^{(2)} + \cdots.$$

Proceeding in a similar way, we can derive further terms of the expansion (10.3.36), until the last expansion term is "short memory", or $(k+1)(1 - 2d) > 1$.

The above heuristic argument is based on the approximation of F_{i-1} by F_i given in Lemma 10.2.4, the Taylor expansion, and the iteration step $k - 1 \to k$, i.e., finding the next expansion term by induction on k. The following details provide a rigorous proof for this heuristic argument.

Proof of Theorem 10.3.2. Essentially, expansion (10.3.36) reduces to the statement (10.3.37) for $\tilde{Q}_n(x) := \tilde{Q}_{n,k^*}(x)$ where k^* is defined in (10.3.3), and where

$$\tilde{Q}_{n,k}(x) := n^{-1/2} \sum_{t=1}^{n} \tilde{R}_{j,k}(x), \tag{10.3.42}$$

$$\tilde{R}_{j,k}(x) := I(\varepsilon_j \le x) - \sum_{p=0}^{k} (-1)^p F^{(p)}(x)\varepsilon_j^{(p)}.$$

As in the proof of the URP, the control of $\sup_{x \in \mathbb{R}} |\tilde{Q}_{n,k}(x)|, 1 \le k \le k^*$, is largely facilitated by the fact that the indicator $I(\varepsilon_j \le x)$ is a monotone function of x, and reduces by the chaining argument, to obtaining an estimate for the variance of the increments $\tilde{Q}_{n,k}(x, y)$, $x < y$. To that end, we need to estimate the decay rate of the covariance $\text{Cov}(\tilde{R}_{0,k}(x, y), \tilde{R}_{j,k}(x, y))$ as $j \to \infty$. The relevant bound is provided by Lemma 10.3.3, which is the most difficult part of the proof of Theorem 10.3.2. This lemma is an extension of Lemma 10.2.5 and is suitable for obtaining a higher-order expansion. It is also of independent interest and is stated and proved here in more generality than is necessary for the proof of Theorem 10.3.2, as it refers to any $1 \le k \le k^*$ and not only to $k = k^*$. Moreover, this lemma is proved under weaker conditions than those of Theorem 10.3.2, in particular, with the aim of weakening the fourth moment condition (10.3.35), by assuming only (10.2.22). If $r < 2$, the number of terms in the expansion of $\hat{F}_n(x)$ may be less than k^*, depending on both r and d (also further terms of a different nature may appear; see Remark 10.3.2); in particular, as for the URP, for the first-order asymptotic expansion $\hat{F}_n(x) = F(x) - f(x)\bar{\varepsilon}_n + o_p(n^{d-1/2})$, it suffices to have condition (10.2.22) with $r > 1$, but arbitrarily close to 1.

Let $\varepsilon_{j,i}$, $\tilde{\varepsilon}_{j,i}$, and $F_i(x)$ be defined as in (10.2.36), and let $\tilde{\varepsilon}_{j,i}^{(0)} := 1$, and

$$\tilde{\varepsilon}_{j,i}^{(k)} := \sum_{i_k > \cdots > i_1 > i} a_{i_1} \cdots a_{i_k} \zeta_{j-i_1} \cdots \zeta_{j-i_k}, \quad k \ge 1. \tag{10.3.43}$$

For each $k = 1, 2, \cdots$, the r.v.'s $\varepsilon_j^{(k)}$ of (10.3.27) satisfy the relation

$$\varepsilon_j^{(k)} = \sum_{i=0}^{\infty} a_i \zeta_{j-i} \tilde{\varepsilon}_{j,i}^{(k-1)}. \tag{10.3.44}$$

Also, for $x \in \mathbb{R}$, let

$$\tilde{R}_{j,i,k}(x) := F_i(x - \tilde{\varepsilon}_{j,i}) - \sum_{q=0}^{k} (-1)^q F^{(q)}(x)\tilde{\varepsilon}_{j,i}^{(q)}. \qquad (10.3.45)$$

Lemma 10.3.3 includes the bound (10.3.48) for the variance of an increment of the sth derivative of (10.3.45), which is used in the proof of the main bound (10.3.47) by induction on k. As before, we shall assume (10.3.3). With r as in (10.2.22), let $r_* = r \wedge 2$, $\mu_r(x, y) = \int_x^y (1 + |z|)^{-r} dz$, and

$$\tau_k := (1 - 2d) \wedge \big(2(1 - d)r - 1 - k(1 - 2d)\big) > 0,$$
$$t_k := k(1 - 2d) + \tau_k, \qquad k \geq 0.$$

Lemma 10.3.3. *Let $\{\varepsilon_j\}$ be as in (10.2.19) with the a_j's as in (10.2.24). Let $0 \leq k \leq k^*$ be a given integer. Suppose the distribution of ζ_0 satisfies conditions (10.2.21) and (10.2.22) with r satisfying*

$$r > \max\left(1, \frac{1 + k(1 - 2d)}{1 + (1 - 2d)}\right). \qquad (10.3.46)$$

Then there exists a constant $C < \infty$ such that for any $j \geq 0$, $x' < y'$ and $x < y$,

$$\left|\mathrm{Cov}(\tilde{R}_{0,k}(x', y'), \tilde{R}_{j,k}(x, y))\right| \qquad (10.3.47)$$

$$\leq C\Big(\mu_{r_*}(x', y')\mu_{r_*}(x, y)\Big)^{1/2} j_+^{-t_k}, \qquad t_k < 1,$$

$$\leq C\Big(\mu_{r_*}(x, y)\mu_{r_*}(x', y')\Big)^{1/2} j_+^{-(1+t_k)/2}, \qquad t_k > 1.$$

Moreover, for any integer $p \geq 0$, there exist m_0 and a constant $C < \infty$ such that for any $i \geq m_0$ and $x < y$,

$$\mathrm{Var}(\tilde{R}_{0,i,k}^{(p)}(x, y)) \leq C\mu_{r_*}(x, y)i^{-t_k}, \qquad t_k < 1, \qquad (10.3.48)$$

$$\leq C\mu_{r_*}(x, y)i^{-(t_k+1)/2}, \qquad t_k > 1.$$

Remark 10.3.4. Let $r = 2$ in (10.2.22), i.e., the innovations have a finite fourth moment. Then (10.3.46) is satisfied for any $1 \leq k \leq k^*$, by the definition of k^*. Note that, in this case, $r_* = 2, \tau_k = 1 - 2d$, and $t_k = (1 + k)(1 - 2d)$, $1 \leq k \leq k^*$, in particular, $t_{k^*} > 1$. For the covariances of the increments of the short memory term $\tilde{R}_{j,k^*}(x)$ from (10.3.47) we obtain the summable decay rate:

$$\left|\mathrm{Cov}(\tilde{R}_{0,k^*}(x, y), \tilde{R}_{j,k^*}(x, y))\right| \leq C\mu_2(x, y)j_+^{-(1+(1+k^*)(1-2d))/2}.$$

Next, consider the case $r = 3/2$ and $k = 1$. Then (10.3.46) is satisfied with $r_* = 3/2, \tau_1 = 1 - 2d$, and $t_1 = 2(1 - 2d)$. From (10.3.47) we obtain

$$\left| \text{Cov}(\tilde{R}_{0,1}(x', y'), \tilde{R}_{j,1}(x, y)) \right| \leq C \Big(\mu_{3/2}(x', y') \mu_{3/2}(x, y) \Big)^{1/2} j_+^{-\rho},$$

with $\rho = (2 - 4d) \wedge (3/2 - 2d)$, which is similar to Lemma 10.2.5.

Proof of Lemma 10.3.3. We can see that the r.h.s. of (10.3.46) is less than 2, and $\tau_k = 1 - 2d$ for $r \geq 2$. (Indeed, $2(1 - d)r - 1 - k(1 - 2d) \geq 2(1-d)r - 2 \geq 4(1-d) - 2 = 2(1-2d) \geq 1 - 2d$ for $r \geq 2$ and $(1-2d)k \leq 1$.) Therefore, in the following we may assume $r \leq 2$ and $r_* = r$, without loss of any generality.

The proof below uses induction on $k \geq 0$. Accordingly, we need to verify the lemma for $k = 0$ and then to prove the induction step $k - 1$ to k. However, the two arguments are very similar and we shall start by discussing the general case $k \geq 0$ and later we will distinguish between the cases $k = 0$ and $k \geq 1$.

By analogy with (10.2.56), let

$$U_{j,i}(x) := F_{i-1}(x - \tilde{\varepsilon}_{j,i-1}) - F_i(x - \tilde{\varepsilon}_{j,i}) \tag{10.3.49}$$

$$- a_i \zeta_{j-i} \sum_{q=1}^{k} (-1)^q F^{(q)}(x) \tilde{\varepsilon}_{j,i}^{(q-1)},$$

with $\tilde{\varepsilon}_{j,i}^{(q-1)}$ defined by (10.3.43). For $k = 0$, the last sum is replaced by zero. By the telescoping identity (10.2.54) we can rewrite $\tilde{R}_{j,k}(x)$ as

$$\tilde{R}_{j,k}(x) = \sum_{i=0}^{\infty} U_{j,i}(x). \tag{10.3.50}$$

We shall show that for r, k, as in the statement of the lemma, there exists a constant $C < \infty$ such that for all $i \geq 0$, $x < y$,

$$E(U_{j,i}(x,y))^2 \leq C\mu_r(x,y) i_+^{-2(1-d)\nu}, \tag{10.3.51}$$

where

$$\nu := \min\left(r, \frac{1 + (k+1)(1 - 2d)}{1 + (1 - 2d)} \right) = \frac{1 + k(1 - 2d) + \tau_k}{2(1 - d)} \tag{10.3.52}$$

$$= \frac{1 + t_k}{2(1 - d)}.$$

The bound (10.3.51) and the orthogonality property (10.2.57) of the summands in the expansion (10.3.50) imply (10.3.47):

$$\left| \text{Cov}(\tilde{R}_{0,k}(x',y'), \tilde{R}_{j,k}(x,y)) \right|$$

$$= \left| \sum_{i=0}^{\infty} E U_{0,i}(x',y') U_{j,j+i}(x,y) \right|$$

$$\leq \sum_{i=0}^{\infty} \left(E(U_{0,i}(x',y'))^2 \right)^{1/2} \left(E(U_{j,j+i}(x,y))^2 \right)^{1/2}$$

$$\leq C \left(\mu_r(x',y') \mu_r(x,y) \right)^{1/2} \sum_{i=0}^{\infty} i_+^{-(1-d)\nu} (j+i)_+^{-(1-d)\nu}$$

$$\leq C \left(\mu_r(x,y) \mu_r(x',y') \right)^{1/2} \begin{cases} j_+^{-t_k}, & t_k < 1, \\ j_+^{-(1+t_k)/2}, & t_k > 1. \end{cases}$$

The last line above follows from the inequality (10.2.53) and the fact that $(1-d)\nu = (1+t_k)/2 < 1$ is equivalent to $t_k < 1$.

To prove the claim (10.3.51), assume first $k \geq 1$ and $i \geq m_0$ sufficiently large so that F_i is differentiable and the conclusions of Lemma 10.2.4 hold. Let $U_{j,i}^{(1)}(x)$ and $U_{j,i}^{(2)}(x)$ be as in (10.2.60), and define

$$U_{j,i}^{(3)}(x) := -a_i \zeta_{j-i} \left(F_i^{(1)}(x - \tilde{\varepsilon}_{j,i}) - \sum_{q=1}^{k} (-1)^{q-1} F^{(q)}(x) \tilde{\varepsilon}_{j,i}^{(q-1)} \right).$$

Then we can rewrite

$$U_{j,i}(x) = \sum_{\tau=1}^{3} U_{j,i}^{(\tau)}(x).$$

Thus, (10.3.51) will follow from the following inequalities.

$$E(U_{j,i}(x,y))^2 \leq C \mu_r(x,y), \qquad 1 \leq i \leq m_0, \qquad (10.3.53)$$

$$E(U_{j,i}^{(\tau)}(x,y))^2 \leq C \mu_r(x,y) i^{-2(1-d)\nu}, \quad i \geq m_0, \ \tau = 1,2,3. \quad (10.3.54)$$

For $\tau = 1$ and $\tau = 2$, (10.3.54) follows from (10.2.66) and (10.2.67), respectively, and the fact that $\nu \leq r \leq 2$ (see the definition of ν in (10.3.52)).

Consider $U_{j,i}^{(3)}(x)$. Rewrite it as

$$U_{j,i}^{(3)}(x) = -a_i \zeta_{j-i} \tilde{R}_{j,i,k-1}^{(1)}(x).$$

Therefore, by the inductive assumption (10.3.48), with k replaced by $k-1$,

$$E(U_{j,i}^{(3)}(x,y))^2 = a_i^2 E(\tilde{R}_{0,i,k-1}^{(1)}(x,y))^2 \qquad (10.3.55)$$

$$\leq C \mu_r(x,y) a_i^2 i^{-(k-1)(1-2d)-\tau_{k-1}}$$

$$\leq C \mu_r(x,y) i^{-2(1-d)\nu}.$$

(Indeed, note that $t_{k-1} = (k-1)(1-2d) + \tau_{k-1} \leq k(1-2d) < 1$ and $\tau_{k-1} \geq \tau_k$, implying $2(1-d) + (k-1)(1-2d) + \tau_{k-1} \geq 1 + t_k = 2(1-d)\nu$.) This completes the proof of the claim (10.3.54).

Next, we prove the claim (10.3.53). Note that by (10.2.37) and the fact that $\mu_2(x,y) \leq \mu_r(x,y)$ for $r \leq 2$, we readily obtain

$$\left(F^{(q)}(x,y)\right)^2 E\left(\tilde{\varepsilon}_{j,i}^{(q-1)}\right)^2 \leq C\mu_r(x,y), \quad 1 \leq q \leq k.$$

Again, for the same reasons we also have

$$E(F_i(x - \tilde{\varepsilon}_{j,i}, y - \tilde{\varepsilon}_{j,i}))^2 \leq EF_i(x - \tilde{\varepsilon}_{j,i}, y - \tilde{\varepsilon}_{j,i})$$
$$= F(x,y) = \int_x^y f(z)dz \leq C\mu_r(x,y).$$

These bounds complete the proof of (10.3.53), and also of (10.3.51) and (10.3.47) for $k \geq 1$.

The above argument applies also to the case $k = 0$, with some changes. Namely, the definition of $U_{j,i}^{(\tau)}(x)$ has to be modified to

$$U_{j,i}^{(1)}(x) := F_i(x - \tilde{\varepsilon}_{j,i-1}) - F_i(x - \tilde{\varepsilon}_{j,i}),$$
$$U_{j,i}^{(2)}(x) := F_{i-1}(x - \tilde{\varepsilon}_{j,i-1}) - F_i(x - \tilde{\varepsilon}_{j,i}), \quad U_{j,i}^{(3)}(x) := 0.$$

Write $U_{j,i}^{(1)}(x,y) = \int_x^y (F_i^{(1)}(u - a_i\zeta_{j-i} - \tilde{\varepsilon}_{j,i}) - F_i^{(1)}(u - \tilde{\varepsilon}_{j,i}))du$ and use (10.2.32), (10.2.34), and (10.2.37) to obtain the bound

$$|U_{j,i}^{(1)}(x,y)| \leq C\mu_\gamma(x,y)(|a_i\zeta_{j-i}| + |a_i\zeta_{j-i}|^\gamma)(1 + |\tilde{\varepsilon}_{j,i}|^\gamma), \quad \gamma \in (1,2].$$

Taking $\gamma = r$ and using (10.2.65) yields $E(U_{j,i}^{(1)}(x,y))^2 \leq C\mu_\gamma(x,y)|a_i|^2$. This proves (10.3.54) for $k = 0$ and $\tau = 1$, since $\tau_0 = 1 - 2d$ and $\nu = 1$; see the definition (10.3.52). The proof of (10.3.54) for $k = 0$ and $\tau = 2$ is analogous to the case $k \geq 1$ discussed above.

Next, we shall prove (10.3.48). The proof mimics that of (10.3.47) given above. Fix s and $m_0 \geq 1$ such that the inequalities (10.2.37) and (10.2.38) hold for all $p \leq s + k^*$ and all $m \geq m_0$, with the constant C independent of m. Note that for any $j \geq m_0$ and $i \geq 1$,

$$EF_j^{(s)}(x - \tilde{\varepsilon}_{0,j}) = F^{(s)}(x),$$
$$E[F_j^{(s)}(x - \tilde{\varepsilon}_{0,j})|\mathcal{F}_{-i-j}] = F_{i+j-1}^{(s)}(x - \tilde{\varepsilon}_{0,i+j-1}).$$

Write

$$F_j(x - \tilde\varepsilon_{0,j})$$

$$= \sum_{i=1}^{\infty} \Big(E[F_j(x - \tilde\varepsilon_{0,j})|\mathcal{F}_{-i-j}] - E[F_j(x - \tilde\varepsilon_{0,j})|\mathcal{F}_{-i-j-1}] \Big)$$

$$= \sum_{i=1}^{\infty} (F_{i+j-1}(x - \tilde\varepsilon_{0,i+j-1}) - F_{i+j}(x - \tilde\varepsilon_{0,i+j})).$$

Hence, the following expansion for $\tilde R_{0,j,k}^{(s)}(x - \tilde\varepsilon_{0,j})$ akin to (10.3.50) holds:

$$\tilde R_{0,j,k}^{(s)}(x - \tilde\varepsilon_{0,j}) = \sum_{i=1}^{\infty} \tilde U_{0,i}(x),$$

where

$$\tilde U_{0,i}(x) := F_{i+j-1}^{(s)}(x - \tilde\varepsilon_{0,i+j-1}) - F_{i+j}^{(s)}(x - \tilde\varepsilon_{0,i+j})$$

$$- a_{i+j}\zeta_{-i-j} \sum_{q=1}^{k} (-1)^q F^{(q+s)}(x)\tilde\varepsilon_{0,i+j}^{(q-1)}.$$

Then (10.3.48) follows from

$$E(\tilde U_{0,i}^{(\tau)}(x,y))^2 \le C\mu_r(x,y)(i+j)^{-2(1-d)\nu}, \quad i \ge 1, \tau = 1,2,3, \quad (10.3.56)$$

where ν is defined by (10.3.52) and where $\tilde U_{0,i}(x) = \sum_{\tau=1}^{3} \tilde U_{0,i}^{(\tau)}(x)$ is the decomposition analogous to that of $U_{j,i}(x)$ given in (10.3.49):

$$\tilde U_{0,i}^{(1)}(x) := F_{i+j}^{(s)}(x - \tilde\varepsilon_{0,i+j-1}) - F_{i+j}^{(s)}(x - \tilde\varepsilon_{0,i+j})$$

$$+ a_{i+j}\zeta_{-i-j} F_{i+j}^{(s+1)}(x - \tilde\varepsilon_{0,i+j}),$$

$$\tilde U_{0,i}^{(2)}(x) := F_{i+j-1}^{(s)}(x - \tilde\varepsilon_{0,i+j-1}) - F_{i+j}^{(s)}(x - \tilde\varepsilon_{0,i+j-1}),$$

$$\tilde U_{0,i}^{(3)}(x) := -a_{i+j}\zeta_{-i-j} \Big\{ F_{i+j}^{(s+1)}(x - \tilde\varepsilon_{0,i+j})$$

$$- \sum_{q=1}^{k} (-1)^{q-1} F^{(s+q)}(x)\tilde\varepsilon_{0,i+j}^{(q-1)} \Big\}$$

$$= -a_{i+j}\zeta_{-i-j} \tilde R_{0,i+j,k-1}^{(s+1)}(x).$$

The proof of (10.3.56) is completely analogous to that of (10.3.51). It uses the inductive assumption of (10.3.48) with k and s replaced by $k - 1$ and $s + 1$, respectively. It is important to note, however, that (10.3.56) holds for all $i \ge 1$, and that, contrary to (10.3.51), the bounds of Lemma 10.2.4

needed to prove (10.3.56) do not depend on $i \geq 1$, as they are guaranteed alone by choosing $j \geq m_0$ sufficiently large. Lemma 10.3.3 is proved.

Proof of Theorem 10.3.2 (continued). As noted at the beginning of the proof, the expansion (10.3.36) reduces to relation (10.3.37) for the remainder term $\tilde{Q}_n(x) \equiv \tilde{Q}_{n,k^*}(x)$. Clearly, the conditions of Lemma 10.3.3 hold with $r = 2$. Let us first check the bound

$$E(\tilde{Q}_n(x,y))^2 \leq \mu(x,y), \qquad x < y, \tag{10.3.57}$$

where $\mu(x,y)$ is a finite continuous measure independent of n. Indeed, by the definition of $\tilde{Q}_n$,

$$E(\tilde{Q}_n(x,y))^2 = n^{-1} \sum_{j_1,j_2=1}^{n} \mathrm{Cov}(\tilde{R}_{j_1,k^*}(x,y), \tilde{R}_{j_2,k^*}(x,y)).$$

According to (10.3.47) of Lemma 10.3.3, see Remark 10.3.4,

$$|\mathrm{Cov}(R_{j_1,k^*}(x,y), R_{j_2,k^*}(x,y))| \leq C\mu_2(x,y)|j_1 - j_2|_+^{-\alpha}, \tag{10.3.58}$$

with $\alpha = (1/2) + (1 - 2d)(k^* + 1)/2 > 1$, from the definition of k^*. This proves (10.3.57). The rest of the proof follows by the chaining argument similar to that in the proof of Theorem 10.2.1.

Consider the expansion (10.3.10) of $S_{n,\psi} - ES_{n,\psi} = -\int(\hat{F}_n - F)d\psi$. It reduces to the central limit theorem in (10.3.39) for the "remainder term" $\tilde{Q}_{n,\psi} = -\int \tilde{Q}_n(x)d\psi(x)$. The proof below uses an approximation of $\tilde{Q}_{n,\psi}$ by a stationary martingale-difference sequence and the martingale central limit theorem given in Lemma 2.5.1. The possibility of such an approximation is based on the covariance estimates in Lemma 10.3.3.

Recall (10.3.42) and the decomposition in (10.3.50). We have $\tilde{Q}_{n,\psi} = n^{-1/2}\sum_{j=1}^{n} \tilde{R}_{j,k^*}(\psi)$, where $\tilde{R}_{j,k^*}(\psi) = -\int R_{j,k^*}(x)d\psi(x)$ and hence the decomposition

$$\tilde{R}_{j,k^*}(\psi) = \sum_{i=0}^{\infty} U_{j,i}(\psi), \qquad U_{j,i}(\psi) := -\int U_{j,i}(x)d\psi(x).$$

By (10.3.51), (10.3.52), and the orthogonality property (10.2.57),

$$E(U_{j,i}(\psi))^2 = \int\int EU_{j,i}(x)U_{j,i}(y)d\psi(x)d\psi(y) \leq Ci_+^{-(1+t_{k^*})}, \tag{10.3.59}$$

$$EU_{j_1,i_1}(\psi)U_{j_2,i_2}(\psi) = 0, \qquad j_1 - i_1 \neq j_2 - i_2.$$

For a large K, let

$$\tilde{R}_{j,k^*}(\psi; K) := \sum_{i=0}^{K} U_{j,i}(\psi), \qquad T_{n1} := n^{-1/2}\sum_{j=1}^{n} \tilde{R}_{j,k^*}(\psi; K). \tag{10.3.60}$$

Then $\tilde{Q}_{n,\psi} = T_{n1} + T_{n2}$, where

$$
\begin{aligned}
ET_{n2}^2 &= \sum_{i_1,i_2>K} n^{-1} \sum_{1\leq j_1,j_2\leq n:j_1-i_1=j_2-i_2} EU_{j_1,i_1}(\psi)U_{j_2,i_2}(\psi) \\
&\leq C \sum_{i_1,i_2>K} i_1^{-(1+t_{k*})/2} i_2^{-(1+t_{k*})/2} \\
&\leq CK^{-(t_{k*}-1)} \to 0, \qquad K \to \infty, \tag{10.3.61}
\end{aligned}
$$

uniformly for $n \geq 1$ because $a_{k^*} > 1$; see Remark 10.3.4.

Next, decompose T_{n1} as

$$
T_{n1} = n^{-1/2} \sum_{j=1}^{n} \sum_{i=0}^{K} U_{j+i,i}(\psi) + N_n,
$$

$$
N_n := n^{-1/2} \sum_{i=0}^{K} \left(\sum_{j=1}^{i} U_{j,i}(\psi) - \sum_{j=n+1}^{n+i} U_{j,i}(\psi) \right).
$$

Observe that, for every fixed K, $n^{1/2}N_n$ is a sum of a finite number of stationary r.v.'s whose second moment is finite by (10.3.59). Thus, for every fixed $K < \infty$,

$$
EN_n^2 = O(n^{-1}). \tag{10.3.62}
$$

Furthermore, for any $0 \leq i \leq K$

$$
U_{j+i,i}(\psi) = E[\psi(\varepsilon_{j+i})|\mathcal{F}_j] - E[\psi(\varepsilon_{j+i})|\mathcal{F}_{j-1}] - a_i\zeta_j \sum_{q=1}^{k^*}(-1)^{q-1}\lambda_q \tilde{\varepsilon}_{j+i,i}^{(q-1)},
$$

see (10.3.42) and (10.3.44), where $\tilde{\varepsilon}_{j+i,i}^{(q-1)}$ is $\mathcal{F}_{j-1}$-measurable. Therefore, for each i and K fixed, $\{U_{j+i,j}(\psi), \; j \in \mathbb{Z}\}$ and

$$
\left\{ M_j(K) := \sum_{i=0}^{K} U_{j+i,i}(\psi), \; j \in \mathbb{Z} \right\}
$$

are square integrable martingale-difference sequences, satisfying

$$
E[U_{j+i,i}(\psi)|\mathcal{F}_{j-1}] = E[M_j(K)|\mathcal{F}_{j-1}] = 0.
$$

Moreover, these martingale-difference sequences are stationary and ergodic, being subordinated to the i.i.d. sequence $\{\zeta_j\}$. Therefore, by the martingale central limit theorem, Lemma 2.5.1,

$$
\tilde{T}_{n1} := T_{n1} - N_n = n^{-1/2} \sum_{j=1}^{n} M_j(K) \to_D \mathcal{N}(0, \tilde{\sigma}_{\psi,K}^2), \tag{10.3.63}
$$

where, by stationarity,

$$\tilde{\sigma}_{\psi,K}^2 := \mathrm{Var}(M_0(K)) = \sum_{i_1,i_2=0}^{K} EU_{0,i_1}(\psi)U_{i_2-i_1,i_2}(\psi)$$

$$= \sum_{j\in\mathbb{Z}} \sum_{i_1,i_2=0}^{K} EU_{0,i_1}(\psi)U_{j,i_2}(\psi), \qquad \text{(by orthogonality (10.3.59))}$$

$$= \sum_{j\in\mathbb{Z}} \mathrm{Cov}(\tilde{R}_{0,k^*}(\psi;K), \tilde{R}_{j,k^*}(\psi;K)),$$

see (10.3.60). It is clear from (10.3.59) that for each $j \in \mathbb{Z}$,

$$\mathrm{Cov}(\tilde{R}_{0,k^*}(\psi;K), \tilde{R}_{j,k^*}(\psi;K)) \rightarrow \mathrm{Cov}(\tilde{R}_{0,k^*}(\psi), \tilde{R}_{j,k^*}(\psi)),$$

as $K \to \infty$, and that these covariances are absolutely summable in $j \in \mathbb{Z}$ and satisfy a similar bound to (10.3.58), uniformly in $K \geq 1$. Therefore,

$$\tilde{\sigma}_{\psi,K}^2 \rightarrow \tilde{\sigma}_\psi^2, \quad K \to \infty,$$

where $\tilde{\sigma}_\psi^2$ is given by the convergent series in (10.3.39). Relations (10.3.61) and (10.3.62) imply that $\mathrm{Var}(\tilde{Q}_{n,\psi} - \tilde{T}_{n1}) \leq 2(\mathrm{Var}(T_{n2}) + \mathrm{Var}(N_n))$ can be made arbitrarily small by first, making K, and then n, large enough. These facts together with the central limit theorem in (10.3.63) easily yield the central limit theorem (10.3.39) of Theorem 10.3.2; see the proof of Proposition 4.6.2. This completes the proof of Theorem 10.3.2.

Note: Theorem 10.3.2 is based on Ho and Hsing (1996) and Koul and Surgailis (1997, 2002). Wu (2003) discusses the expansion of $\hat{F}_n$ under somewhat different conditions. In Koul and Surgailis (2000a) we also find analogous expansions for the weighted residual processes

$$\mathcal{U}_n(x) := n^{-1}\sum_{j=1}^{n} h_{nj} I(\varepsilon_j \leq x + \xi_{nj}), \quad x \in \mathbb{R},$$

where $h_{nj}, \xi_{nj}, 1 \leq j \leq n$, are arrays of real numbers, $\{\varepsilon_j\}$ is the moving-average process (10.2.19) with the a_j's as in (10.2.24). This expansion is useful in obtaining a higher-order expansion of a class of M estimators in regression models with a class of non-random designs as illustrated in Koul and Surgailis (2000a). We do not include these expansions in this text partly because of the complicated nature of their description and partly because to date they are proved only for simple linear regression models with special designs such as polynomials. A more useful result from the applications point of view would be to obtain similar expansions for regression models with random designs having short or long memory. Koul and Surgailis (2001) and Wu (2007) analyzed the large-sample behavior of empirical functionals and M estimators of linear processes with infinite variance and long memory.

Chapter 11

Regression Models

11.1 Introduction

A classical problem in statistics is to use a vector X of p variables to explain the one-dimensional response Y, $p \geq 1$. This is often done in terms of the conditional mean function $\mu(x) := E(Y|X = x)$, $x \in \mathbb{R}^p$, assuming it exists. A primary reason for this is that whenever Y is square integrable, $\mu(X)$ is the best mean-squared error predictor, i.e., among all square integrable functions $g(X)$ of X, the conditional mean function $\mu(X)$ minimizes the mean-squared error $E(Y - g(X))^2$.

The function $\mu(x)$ is also called the regression function in statistics. A model that formulates inference problems in terms of the regression function is called a regression model. Regression models are known to be important and useful in statistical analysis. From the 1980s examples from numerous social and physical sciences indicate a need for understanding the behavior of various inference procedures in regression models when design or error variables have long memory. This chapter will discuss the asymptotic distribution theory of several inference procedures in linear and non-linear regression models in the presence of long memory in errors, and when X is either a vector of known constants or r.v.'s that possibly have long memory.

A regression model is called linear if $\mu(x)$ is linear in x, non-linear parametric if $\mu(x)$ is non-linear in x and depends on some Euclidean parameter, and non-parametric if μ is completely unspecified but known to belong to an infinite dimensional class of functions.

In parametric regression models with random designs we are given an array of $p \times 1$ random vectors X_{ni}, $1 \leq i \leq n$, a family of functions

352

$m(x, \beta)$, $x \in \mathbb{R}^p$, $\beta \in \Theta \subset \mathbb{R}^q$, and we observe an array of random variables $\{Y_{ni}, 1 \leq i \leq n\}$ such that for some $\beta \in \mathbb{R}^q$,

$$Y_{ni} = m(X_{ni}, \beta) + \varepsilon_i, \quad 1 \leq i \leq n, \tag{11.1.1}$$

where the ε_i are the error r.v.'s with $E(\varepsilon_i | X_{ni}) = 0$, for all $1 \leq i \leq n$, and where p and q are known positive integers. Thus, we stipulate that for the design vectors X_{ni}, the regression function satisfies $\mu(X_{ni}) = m(X_{ni}, \beta)$, for some $\beta \in \mathbb{R}^q$, and for all $1 \leq i \leq n$. The model (11.1.1) is called a linear regression model with random design if $q = p$ and

$$m(x, \beta) = x'\beta, \quad x \in \mathbb{R}^p, \beta \in \mathbb{R}^p. \tag{11.1.2}$$

The model (11.1.1) is called a non-random design parametric regression model if $X_{ni} \equiv x_{ni}$ where the x_{ni}, $1 \leq i \leq n$, are arrays of $p \times 1$ vectors of known real numbers. The model (11.1.1) together with (11.1.2) and a non-random design is called a linear regression model with non-random design. In both cases, $E(\varepsilon_i) = 0$ for all i.

In this chapter we shall first focus on the large sample analysis of some inference procedures in non-random linear regression models with long memory errors. The long memory will be modeled by either a long memory Gaussian process or by a long memory moving average with an unknown innovation distribution. Among the inference procedures, we will discuss the asymptotic distributions of the least-squares, and the M and R estimators of the underlying regression parameters. Before proceeding further we need the following definition.

Definition 11.1.1. Let $\{S_n\}$ and $\{T_n\}$ be sequences of r.v.'s such that $ET_n^2 < \infty$ for all $n \geq 1$, and such that $T_n / \sqrt{\text{Var}(T_n)} = O_p(1)$. Then S_n is said to be asymptotically first-order equivalent to the sequence T_n, if

$$\frac{|S_n - T_n|}{\sqrt{\text{Var}(T_n)}} = o_p(1). \tag{11.1.3}$$

A surprising result from the 1990s is that in parametric regression models with stationary long memory moving-average errors a large class of the so-called robust estimators of the regression parameters are asymptotically first-order equivalent to the least-squares (LS) estimator in the sense of (11.1.3). This class of estimators includes the so-called M, R, and minimum-distance estimators. Thus, for example, asymptotically there is no difference up to the first order between the inference based on the least absolute deviation, or the Huber M, or the Hodges–Lehmann estimators and the LS estimator. This is in complete contrast to the i.i.d. or weakly dependent settings, where it is known that the first-order difference between these estimators and the LS estimator converges weakly to a normal distribution.

Section 11.2 below discusses these results at some length and indicates primary reasons for this phenomenon. Section 11.3 discusses the second-order asymptotics of a class of M estimators in some location models. To implement inference procedures based on M, R, or LS estimators we need to have $\log(n)$-consistent estimators of the long memory parameter d and consistent estimators of the other parameters appearing in the covariance structure of the error process. Section 11.4 discusses the asymptotic distribution of the Whittle estimators of these parameters when the design process is either non-random or random.

The asymptotic distributions of LS and M estimators in random design non-linear parametric regression models are discussed in Section 11.5. An important related problem of obtaining the asymptotic distributions of the sums $\sum_{j=1}^{n} h(X_j)\psi(\varepsilon_j)$ is also discussed in this section, where h and ψ are known square integrable functions, $\{X_j\}$ has short or long memory, and $\{\varepsilon_j\}$ has long memory.

Because the above class of M estimators are asymptotically first-order equivalent to each other, it is of interest to see if there is any finite sample difference between their standard error and how the values of the long memory parameters of the design and error processes affect these entities. In Section 11.6 we give the results of a finite sample simulation comparing the least-absolute deviation (LAD) and LS estimators in a simple linear regression model when both design and error variables have long memory.

11.2   First-order asymptotics of M and R estimators

This section discusses the first-order asymptotic behavior of the so-called M and R estimators of the regression parameters in non-random design linear regression models with either long memory Gaussian or long memory moving-average errors. There are three subsections. The first subsection discusses the asymptotic normality of the LS estimator under long memory errors while the second and the third subsections study the first-order asymptotics of M and R estimators, respectively.

First, we need to define the estimators. In the linear regression model (11.1.1) and (11.1.2) with non-random design, we are given an array x_{ni}, $1 \leq i \leq n$, of $p \times 1$ vectors of known real numbers, and we observe an array of r.v.'s $\{Y_{ni}, 1 \leq i \leq n\}$ satisfying

$$Y_{ni} = x'_{ni}\beta + \varepsilon_i, \quad E\varepsilon_i = 0, \quad 1 \leq i \leq n, \tag{11.2.1}$$

for some $\beta \in \mathbb{R}^p$. The precise additional assumptions about the probability structure of the errors ε_i will be given in the following subsections.

Let $\mathbf{X}_n$ denote the $n \times p$ matrix whose ith row is x'_{ni}. We shall assume

$$\mathbf{X}'_n\mathbf{X}_n = \sum_{i=1}^{n} x_{ni}x'_{ni} \text{ is non-singular.} \tag{11.2.2}$$

The LS estimator in the model (11.2.1) is defined as

$$\hat{\beta}_{LS} = \text{argmin}\Big\{ \sum_{i=1}^{n}(Y_{ni} - x'_{ni}t)^2 : t \in \mathbb{R}^p \Big\}.$$

The corresponding minimizer solves for t the system of linear equations

$$\sum_{i=1}^{n} x_{ni} (Y_{ni} - x'_{ni}t) = 0,$$

giving the solution

$$\hat{\beta}_{LS} = (\mathbf{X}'_n\mathbf{X}_n)^{-1} \sum_{i=1}^{n} x_{ni}Y_{ni}.$$

Observe that alternatively, this estimator may also be defined by the relation

$$\hat{\beta}_{LS} := \text{argmin}\Big\{ \Big\| D_n^{-1} \sum_{i=1}^{n} x_{ni} (Y_{ni} - x'_{ni} t) \Big\|^2 : t \in \mathbb{R}^p \Big\}, \tag{11.2.3}$$

$$D_n := (\mathbf{X}'_n\mathbf{X}_n)^{1/2},$$

where $\| \cdot \|$ is the Euclidean norm.

M estimators. Because the least-squares estimator $\hat{\beta}_{LS}$ depends linearly on observations, it is sensitive to the "outliers" in the errors. The idea behind the method of M estimation is to discount such "untypical" observations by replacing the residuals $Y_{ni} - x'_{ni}t$ in (11.2.3) by $\psi(Y_{ni} - x'_{ni}t)$, where

$\psi(y)$, $y \in \mathbb{R}$, is a *score function* typically bounded or with the property that $\psi(y) = o(|y|)$ as $|y| \to \infty$. Accordingly, let

$$\mathcal{M}(t; \psi) := D_n^{-1} \sum_{i=1}^{n} x_{ni} \psi(Y_{ni} - x'_{ni} t).$$

Then a class of M estimators, one for each ψ, of β is defined to be

$$\hat{\beta}_M := \operatorname{argmin}\{\|\mathcal{M}(t; \psi)\|^2 : t \in \mathbb{R}^p\}.$$

Important examples of these estimators correspond to the Huber score $\psi(x) \equiv x\, I(\|x\| \le c) + c\,(x/\|x\|) I(\|x\| > c)$, where c is a known positive constant, and the least absolute deviation score $\psi(x) = \operatorname{sgn}(x)$. The latter estimator is referred to as the LAD estimator, an analog of the *median* in the location case. It is also equivalent to

$$\hat{\beta}_{LAD} := \operatorname{argmin}\left\{ \sum_{i=1}^{n} |Y_{ni} - x'_{ni} t| : t \in \mathbb{R}^p\right\}.$$

R estimators. Another way to obtain estimators that are not sensitive to the extremes is to use ranks. Let R_{it} denote the rank of $Y_{ni} - x'_{ni} t$ among $Y_{nj} - x'_{nj} t$, $1 \le j \le n$, i.e.,

$$R_{it} := \sum_{j=1}^{n} I(Y_{nj} - x'_{nj} t \le Y_{ni} - x'_{ni} t), \quad 1 \le i \le n,\, t \in \mathbb{R}^p.$$

Let $\mathbf{X}_{cn}$ denote the $n \times p$ matrix whose ith row is $x'_{ni} - \overline{x}'_n$, $1 \le i \le n$, where $\overline{x}'_n = n^{-1} \sum_{i=1}^{n} x'_{ni}$. Here we shall assume

$$\mathbf{X}'_{cn}\mathbf{X}_{cn} = \sum_{i=1}^{n} (x_{ni} - \overline{x}_n)(x_{ni} - \overline{x}_n)' \text{ is non-singular.} \quad (11.2.4)$$

Let φ be a real-valued non-decreasing right continuous function on $[0, 1]$ and define

$$\mathcal{J}(t; \varphi) := \sum_{i=1}^{n} \varphi\left(\frac{R_{it}}{n+1}\right)(Y_{ni} - x'_{ni} t),$$

$$\mathcal{S}(t; \varphi) := D_{cn}^{-1} \sum_{i=1}^{n} (x_{ni} - \overline{x}_n)\varphi\left(\frac{R_{it}}{n+1}\right), \quad D_{cn} := (\mathbf{X}'_{cn}\mathbf{X}_{cn})^{1/2},$$

$$\hat{\beta}_R := \operatorname{argmin}\{\|\mathcal{S}(t; \varphi)\|^2 : t \in \mathbb{R}^p\}.$$

The estimator $\hat{\beta}_R$ is called an R (rank) estimator. The estimators $\hat{\beta}_R$ corresponding to $\varphi(u) \equiv u$ and $\varphi(u) = \Phi^{-1}(u)$, the inverse of the $\mathcal{N}(0, 1)$

d.f., are respectively the well-known Wilcoxon and van der Waard or normal scores estimators.

For i.i.d. errors with continuous distribution, Jaeckel (1972) showed that if (11.2.4) holds and if

$$\sum_{i=1}^{n} \varphi(\frac{i}{n+1}) = 0,$$

then the Jaeckel's dispersion $\mathcal{J}(t; \varphi)$ is convex in t with its derivative equal to $-D_{cn}\mathcal{S}(t; \varphi)$, a.e. This is purely an analytical property of $\mathcal{J}(t; \varphi)$ and holds under any kind of probability structure of the regression model that guarantees no ties among the ranks. In particular, this holds under any kind of long memory dependence that guarantees the continuity of the joint distributions of the observations $\{Y_{ni}, 1 \leq i \leq n\}$. Using the asymptotic uniform linearity of the scores $\mathcal{S}(\beta + D_{cn}^{-1}s; \varphi)$ in s, Jaeckel (1972) also showed that $D_{cn}(\hat{\beta}_R - \beta)$ is asymptotically equivalent, in the first order, to the minimizer of $\mathcal{J}(\beta + D_{cn}^{-1}s; \varphi)$ with respect to s. As we shall see later, the asymptotic uniform linearity result for a suitably standardized $\mathcal{S}$, and hence this first-order equivalence, holds even for long memory errors as long as the estimators are standardized to account for the assumed dependence.

In the classical case of i.i.d. errors, it is well known that several of these estimators are robust against outliers in the errors and have high asymptotic relative efficiencies (AREs) at various error distributions. For example, at the standard Gaussian error distribution, the ARE of the Wilcoxon estimator, relative to the LS estimator, is 0.955. In other words, the Wilcoxon estimator needs only about five percent of observations more than the LS estimator for estimating β with the same accuracy, for large samples. But at the same time it is robust against outliers in the errors for a large class of error distributions including Gaussian, while the LS estimator is nowhere robust. The ARE of the normal scores estimator, relative to the LS estimator, is at least 1 at all error d.f.'s with finite variance. For more on this subject see, for example, Huber (1981), Lehmann (1983), and Koul (2002). For comparable results in the case of weakly dependent errors, see Gastwirth and Rubin (1975) and Koul (1977).

But under long memory, the situation is completely different. If the error process is either long memory Gaussian or moving average, then all of these estimators are asymptotically first-order equivalent to the least-squares estimator and the above mentioned efficiency properties do not hold. Moreover, the asymptotic distributions of suitably standardized robust or LS estimators may not be normal when errors are subordinated to a long memory Gaussian process.

To make all these statements precise, we need to derive the asymptotic distributions of these estimators. There are two basic steps. In the first step, we establish the asymptotic uniform quadraticity of a suitably standardized defining dispersion over compact sets. In the second step, we prove that suitably standardized corresponding estimators are bounded in probability. These two steps together imply that the standardized minimizers are asymptotically like those of the corresponding asymptotic quadratic forms. Since the dispersions corresponding to M and R estimators involve squares of some kind of scores, the asymptotic uniform quadraticity of these dispersions is implied by the asymptotic uniform linearity of the underlying scores.

Before discussing any particular estimator, we will give a general lemma from Koul (1992a) that gives sufficient conditions for a suitably standardized estimator obtained by minimizing a dispersion to be bounded in probability. Accordingly, let $\xi_n = (\zeta_{n1}, \cdots, \zeta_{nn})'$ be a vector of n observations, $T_n(\theta) \equiv T_n(\xi_n, \theta)$, $\theta \in \Theta \subset \mathbb{R}^q$ be a real-valued objective function of the data where the parameter of interest is θ, herein called the *dispersion*. Let θ_0 denote the true parameter value. Define the minimum dispersion estimator of θ_0 to be

$$\hat{\theta}_n = \operatorname{argmin}\{T_n(t), \ t \in \Theta\}. \tag{11.2.5}$$

Our aim here is to provide sufficient conditions under which

$$\|\delta_n(\theta_0)(\hat{\theta}_n - \theta_0)\| = O_p(1), \tag{11.2.6}$$

for a sequence of positive definite $q \times q$ matrices $\delta_n(\theta_0)$, with $\|\delta_n(\theta_0)\| \to \infty$.

Consider the following two conditions. For every $\epsilon > 0$, there exists $0 < z_\epsilon < \infty$ and $N_{1\epsilon}$, such that

$$P(|T_n(\theta_0)| \leq z_\epsilon) \geq 1 - \epsilon, \quad \forall \ n \geq N_{1\epsilon}. \tag{11.2.7}$$

There exists a sequence of positive definite $q \times q$ real matrices $\delta_n = \delta_n(\theta_0)$ such that $\|\delta_n\| \to \infty$ and such that for every $\epsilon > 0$ and every $0 < \alpha < \infty$, there exists $N_{2\epsilon}$ and b (depending on ϵ and α) such that

$$P\left(\inf_{\|\delta_n(\theta - \theta_0)\| > b} T_n(\theta) \geq \alpha\right) \geq 1 - \epsilon, \quad \forall \ n \geq N_{2\epsilon}. \tag{11.2.8}$$

Here, P denotes the joint distribution of ξ_n under θ_0. The following lemma shows the sufficiency of these two conditions for (11.2.6). In other words, these two conditions imply that $\hat{\theta}_n$ is consistent for θ_0 at the rate δ_n^{-1}.

Lemma 11.2.1. *Under (11.2.7) and (11.2.8), (11.2.6) holds, for every minimizer $\hat{\theta}_n$ of (11.2.5).*

Proof. Let z_ϵ be as in (11.2.7). Choose $\alpha > z_\epsilon$ in (11.2.8). Then

$$\left\{ |T_n(\theta_0)| \leq z_\epsilon, \ \inf_{\|h\|>b} T_n(\theta_0 + \delta_n^{-1}h) \geq \alpha \right\}$$
$$\subset \left\{ \inf_{\|h\|\leq b} T_n(\theta_0 + \delta_n^{-1}h) \leq z_\epsilon, \ \inf_{\|h\|>b} T_n(\theta_0 + \delta_n^{-1}h) \geq \alpha \right\}$$
$$\subset \left\{ \inf_{\|h\|>b} T_n(\theta_0 + \delta_n^{-1}h) > \inf_{\|h\|\leq b} T_n(\theta_0 + \delta_n^{-1}h) \right\}.$$

Therefore, by (11.2.7) and (11.2.8), for any $\epsilon > 0$ there exists $0 \leq b < \infty$ (now depending only on ϵ) such that $\forall \, n \geq N_{1\epsilon} \vee N_{2\epsilon}$,

$$P\left(\inf_{\|h\|>b} T_n(\theta_0 + \delta_n^{-1}h) > \inf_{\|h\|\leq b} T_n(\theta_0 + \delta_n^{-1}h) \right) \geq 1 - \epsilon.$$

This in turn ensures the validity of (11.2.6), thereby completing the proof.

Remark 11.2.1. Clearly, (11.2.7) is equivalent to saying that $T_n(\theta_0)$ is bounded in probability. It is usually verified by an application of Markov's inequality in the case when $E|T_n(\theta_0)| = O(1)$. In some applications $T_n(\theta_0)$ converges weakly to a r.v., which also implies (11.2.7).

Roughly speaking, assumption (11.2.8) ensures that the smallest value of $T_n(\theta)$ for θ outside of a $\|\delta_n^{-1}\|$-neighborhood of θ_0 can be made asymptotically arbitrarily large with arbitrarily large probability. Often the verification of (11.2.8) is rendered easy by an application of a variant of the C–S inequality as will be seen below. See Koul (2002) for several other examples of applications of this methodology.

11.2.1   *CLT for the LS estimator*

Before proceeding to prove the asymptotic first-order equivalence of the M, R, and LS estimators we shall first establish the asymptotic normality of $\hat{\beta}_{LS}$ under long memory moving-average errors $\{\varepsilon_j\}$:

$$\varepsilon_j = \sum_{k=0}^{\infty} a_k \zeta_{j-k}, \quad j \in \mathbb{Z}, \tag{11.2.9}$$

where $\{\zeta_s, s \in \mathbb{Z}\}$ are i.i.d., with zero mean and finite variance σ^2, and

$$a_k = k^{-1+d}L(k), \quad 0 < d < 1/2, \quad L \in SV, \tag{11.2.10}$$

are regularly decaying coefficients. As in Chapter 10, (10.2.19) and (10.2.20), the condition $L \in SV$ does not restrict the subsequent discussion, and all results of this chapter apply to moving-average errors (11.2.9) and (11.2.10) with $-L \in SV$, or $L(j)$ taking negative values for sufficiently large j.

With $\tau_n := n^d|L(n)|$, let

$$\Sigma_n := \text{Cov}\big(\tau_n^{-1}D_n(\hat{\beta}_{LS} - \beta)\big) := \tau_n^{-2}E\big((\hat{\beta}_{LS} - \beta)D_n D_n'(\hat{\beta}_{LS} - \beta)\big)'.$$

The following theorem establishes the asymptotic normality of $\hat{\beta}_{LS}$ for long memory errors.

Theorem 11.2.1. *Suppose $\{\varepsilon_j\}$ in the linear regression model (11.2.1) satisfies (11.2.9) and (11.2.10). Assume, moreover, that (11.2.2) holds and*

$$n^{1/2} \max_{1 \le i \le n} \big\|D_n^{-1}x_{ni}\big\| = O(1). \tag{11.2.11}$$

Then

$$\tau_n^{-2}E\|D_n(\hat{\beta}_{LS} - \beta)\|^2 = O(1). \tag{11.2.12}$$

Moreover, if $\lim_{n\to\infty} \Sigma_n = \Sigma$ exists, then

$$\tau_n^{-1}D_n(\hat{\beta}_{LS} - \beta) \to_D \mathcal{N}_p(0, \Sigma). \tag{11.2.13}$$

Proof. Let

$$c_{ni} := D_n^{-1}x_{ni}. \tag{11.2.14}$$

Rewrite

$$\tau_n^{-1}D_n(\hat{\beta}_{LS} - \beta) = \tau_n^{-1}\sum_{i=1}^n c_{ni}\varepsilon_i.$$

Since, by (11.2.11), $n^{1/2} \max_{1 \le i \le n} \|c_{ni}\| \le C$, so

$$\tau_n^{-2}E\|D_n(\hat{\beta}_{LS} - \beta)\|^2 \le n^{-2d}L(n)^{-2}\sum_{i,j=1}^n \|c_{ni}\|\|c_{nj}\| \, |E\varepsilon_i\varepsilon_j|$$

$$\le Cn^{-2d-1}L(n)^{-2}\sum_{i,j=1}^n |i-j|_+^{-1+2d}L^2(|i-j|)$$

$$\le C,$$

by (3.3.8). This proves (11.2.12). The statement (11.2.13) is immediate from Theorem 4.3.2.

11.2.2 First-order asymptotics of M estimators

We shall now turn to the first-order asymptotic behavior of the above M and R estimators in linear parametric regression models with non-random design and long memory errors. We will establish their asymptotic first-order equivalence to $\hat{\beta}_{LS}$ and their asymptotic normality. Many technical details below are similar to those used in the i.i.d. errors case appearing in Koul (1992a, 2002).

First we shall focus on M estimators. From the above discussion for the LS estimator, a suitable standardization for an M estimator is $\tau_n^{-1}D_n$. Let

$$\tilde{\mathcal{M}}(s;\psi) := \mathcal{M}(\beta + D_n^{-1}\tau_n s;\psi), \quad s := \tau_n^{-1}D_n(t - \beta) \in \mathbb{R}^p. \quad (11.2.15)$$

Thus, $\tau_n^{-1}D_n(\hat{\beta}_M - \beta)$ is a minimizer of $\tilde{\mathcal{M}}(s;\psi)$ with respect to $s \in \mathbb{R}^p$.

A crucial result needed for the first-order analysis of these estimators is the so-called asymptotic uniform linearity of $\tau_n^{-1}\tilde{\mathcal{M}}(s;\psi)$ in s belonging to a compact set. To establish this result, assume that $\psi \in \Psi$, where

$$\Psi := \{\psi; \psi \text{ is a non-decreasing right continuous bounded}$$
$$\text{function from } \mathbb{R} \text{ to } \mathbb{R}, \text{ and } \int \psi dF = 0\}.$$

Introduce the weighted empirical process

$$V_n(s;y) := \frac{1}{\tau_n}\sum_{i=1}^{n}d_{ni}I(\varepsilon_i \leq y + \tau_n c_{ni}'s), \quad y \in \mathbb{R}, \ s \in \mathbb{R}^p, \quad (11.2.16)$$

where $\{d_{ni}, 1 \leq i \leq n\}$ and $\{c_{ni}, 1 \leq i \leq n\}$ are arrays of non-random vectors $d_{ni} \in \mathbb{R}^q, c_{ni} \in \mathbb{R}^p$. Note that, with $d_{ni} = c_{ni}$ of (11.2.14) and using the notation in (11.2.15) and (11.2.16), $\tau_n^{-1}\tilde{\mathcal{M}}(s;\psi)$ can be rewritten as

$$\tau_n^{-1}\tilde{\mathcal{M}}(s;\psi) = \tau_n^{-1}\sum_{i=1}^{n}c_{ni}\psi(Y_{ni} - x_{ni}'(\beta + D_n^{-1}\tau_n s)) \quad (11.2.17)$$

$$= \tau_n^{-1}\sum_{i=1}^{n}c_{ni}\psi(\varepsilon_i - \tau_n c_{ni}'s) = \int \psi(y)V_n(s;dy)$$

$$= \psi(\infty)\tau_n^{-1}\sum_{i=1}^{n}c_{ni} - \int V_n(s;y)\psi(dy).$$

Since the first term here does not depend on s, the asymptotic uniform linearity of $\tau_n^{-1}\tilde{\mathcal{M}}(s;\psi)$ will follow from a similar property of the empirical process $V_n(s;y)$. Assume

$$n^{1/2}\max_{1 \leq i \leq n}\|d_{ni}\| = O(1). \quad (11.2.18)$$

Then applying Theorem 10.2.3 with $\gamma_{ni} = a' d_{ni}$, $a \in \mathbb{R}^q$, and $\xi_{ni} = 0$, gives uniformly in $y \in \mathbb{R}$,

$$V_n(\mathbf{0}; y) = \frac{1}{\tau_n} \sum_{i=1}^{n} d_{ni} I(\varepsilon_i \leq y) \tag{11.2.19}$$

$$= \frac{F(y)}{\tau_n} \sum_{i=1}^{n} d_{ni} - \frac{f(y)}{\tau_n} \sum_{i=1}^{n} d_{ni}\varepsilon_i + o_p(1).$$

A similar expansion follows from Theorem 10.2.3 for the empirical process $V_n(s; y)$ for any $s \in \mathbb{R}^p$ *fixed*. However, to show a *uniform* asymptotic linearity, we need the following additional lemma.

Lemma 11.2.2. *Suppose $\{\varepsilon_j\}$ is a long memory moving average (11.2.9) and (11.2.10) with innovations $\{\zeta_j\}$ satisfying the conditions of Theorem 10.2.3. Assume also that (11.2.18) holds and*

$$n^{1/2} \max_{1 \leq i \leq n} \|c_{ni}\| = O(1). \tag{11.2.20}$$

Then, for every $0 \leq b < \infty$,

$$\sup_{\|s\| \leq b, \, y \in \mathbb{R}} \left\| V_n(s; y) - V_n(\mathbf{0}; y) - f(y) \sum_{i=1}^{n} d_{ni} c'_{ni} s \right\| = o_p(1). \tag{11.2.21}$$

In the particular case when $p = q$ and $d_{ni} \equiv c_{ni}$ of (11.2.14),

$$\sup_{\|s\| \leq b, \, y \in \mathbb{R}} \left\| V_n(s; y) - V_n(\mathbf{0}; y) - s f(y) \right\| = o_p(1), \tag{11.2.22}$$

provided (11.2.11) holds.

Proof. Note that (11.2.18) and (11.2.20) imply $\sum_{i=1}^{n} \|d_{ni} c_{ni}\| = O(1)$. Clearly, (11.2.22) follows from (11.2.21) since, from (11.2.2) and the definition of c_{ni} in (11.2.14), it follows that

$$\sum_{i=1}^{n} c_{ni} c'_{ni} = I_{p \times p}.$$

We now prove (11.2.21). Recall the definition of the weighted empirical process $S_n(x) = S_n^{(\xi)}(x)$ from (10.2.3).

$$S_n^{(\xi)}(x) = \frac{1}{\tau_n} \sum_{j=1}^{n} \gamma_{nj} [I(\varepsilon_j \leq x + \xi_{nj}) - F(x + \xi_{nj}) + f(x + \xi_{nj})\varepsilon_j],$$

where $\{\gamma_{nj}, 1 \leq j \leq n\}$ and $\{\xi_{nj}, 1 \leq j \leq n\}$ are given deterministic arrays of real numbers. Write $S_n^{(0)}(x)$ for the corresponding process with $\xi_{nj} \equiv 0$. Assume γ_{nj} satisfies (10.2.4)(a) and

$$\max_{1 \leq j \leq n} |\xi_{nj}| = O(n^{-1/2}\tau_n) = o(1). \qquad (11.2.23)$$

Also define

$$\mathcal{D}_n^{(\xi)}(x) := \frac{1}{\tau_n} \sum_{j=1}^{n} \gamma_{nj}(I(\varepsilon_j \leq x + \xi_{nj}) - I(\varepsilon_j \leq x) - f(x)\xi_{nj}).$$

We will show that Theorem 10.2.3 and (11.2.23) imply

$$\sup_{x \in \mathbb{R}} |\mathcal{D}_n^{(\xi)}(x)| = o_p(1). \qquad (11.2.24)$$

Indeed, $\mathcal{D}_n^{(\xi)}$ can be rewritten as

$$\mathcal{D}_n^{(\xi)} = S_n^{(\xi)}(x) - S_n^{(0)}(x) + \Gamma_{n1}(x) + \Gamma_{n2}(x),$$

$$\Gamma_{n1}(x) := \frac{1}{\tau_n} \sum_{j=1}^{n} \gamma_{nj}(F(x + \xi_{nj}) - F(x) - f(x)\xi_{nj}),$$

$$\Gamma_{n2}(x) := \frac{1}{\tau_n} \sum_{j=1}^{n} \gamma_{nj}(f(x) - f(x + \xi_{nj}))\varepsilon_j.$$

According to Theorem 10.2.3,

$$\sup_{x \in \mathbb{R}} |S_n^{(\xi)}(x)| = o_p(1) = \sup_{x \in \mathbb{R}} |S_n^{(0)}(x)|,$$

so that (11.2.24) follows from

$$\sup_{x \in \mathbb{R}} |\Gamma_{ni}^{(\xi)}(x)| = o_p(1), \qquad i = 1, 2. \qquad (11.2.25)$$

Use (10.2.4)(a), (11.2.23), the mean-value theorem, and the boundedness of $\dot{f}(x)$, guaranteed by Lemma 10.2.4, to obtain

$$\sup_{x \in \mathbb{R}} |\Gamma_{n1}^{(\xi)}(x)| \leq Cn^{1/2}\tau_n^{-1} \max_{1 \leq j \leq n} |\xi_{nj}|^2 \leq C\tau_n n^{-1/2} = o(1),$$

proving (11.2.25) for $i = 1$. Next, since $\tau_n^{-2}n^{-1}\sum_{i,j=1}^{n}|E\varepsilon_i\varepsilon_j| = O(1)$,

$$E(\Gamma_{n2}^{(\xi)}(x))^2 \leq C \max_{1 \leq i \leq n} (f(x) - f(x + \xi_{ni}))^2 = O(\max_{1 \leq i \leq n} \xi_{ni}^2) = o(1),$$

uniformly in $x \in \mathbb{R}$. Similarly, using the fact that $\max_{1 \leq j \leq n} |\xi_{nj}| \leq 1$ for all n large enough, we obtain

$$E(\Gamma_{n2}^{(\xi)}(x) - \Gamma_{n2}^{(\xi)}(y))^2 \leq C[\sup_{|\xi| \leq 1} |f(x + \xi) - f(y + \xi)|]^2$$

$$\leq C[\mu(x, y)]^2,$$

where μ is the finite continuous measure on $\mathbb{R}$ of Lemma 10.2.2, which can be used to totally bound $\mathbb{R}$. Thus, (11.2.25) for $i = 2$ follows from these facts and a routine argument. This proves the claim in (11.2.24).

Next, let $a \in \mathbb{R}^q$ and let $\gamma_{ni} := a'd_{ni}$. Decompose $\gamma_{ni} = \gamma_{ni}^+ - \gamma_{ni}^-$, where $\gamma_{ni}^+ := \max(\gamma_{ni}, 0) \geq 0$ and $\gamma_{ni}^- := \gamma_{ni}^+ - \gamma_{ni} \geq 0$, $1 \leq i \leq n$. Let

$$U_n^\pm(s; y) := \frac{1}{\tau_n} \sum_{i=1}^n \gamma_{ni}^\pm I(\varepsilon_i \leq y + \tau_n c_{ni}'s).$$

Note that

$$a'V_n(s; y) = U_n^+(s; y) - U_n^-(s; y), \qquad \forall\, y \in \mathbb{R},\ s \in \mathbb{R}^p.$$

We shall prove that for every $0 \leq b < \infty$,

$$\sup_{\|s\| \leq b;\, y \in \mathbb{R}} \left| U_n^\pm(s; y) - U_n^\pm(\mathbf{0}; y) - \sum_{i=1}^n \gamma_{ni}^\pm c_{ni}'sf(y) \right| = o_p(1). \qquad (11.2.26)$$

This result implies (11.2.21).

The result (11.2.26) is obvious for $b = 0$ and $a = \mathbf{0}$. Let $0 < b < \infty$ and $a \in \mathbb{R}^q$ with $\|a\| > 0$. Write $\mathcal{W}^\pm(s; y)$ for the expression inside the absolute values of the l.h.s. of (11.2.26). Below, we restrict the proof of (11.2.26) to $\mathcal{W}^+$ since the proof for $\mathcal{W}^-$ is analogous.

For $s \in \mathbb{R}^p$, $z, y \in \mathbb{R}$, let

$$W^+(s, z; y) := \frac{1}{\tau_n} \sum_{i=1}^n \gamma_{ni}^+ \Big[I\Big(\varepsilon_i \leq y + \tau_n(c_{ni}'s + \|c_{ni}\|z) \Big)$$

$$- I(\varepsilon_i \leq y) - f(y)\tau_n(c_{ni}'s + \|c_{ni}\|z) \Big].$$

Note that $W^+(s, 0; y) = \mathcal{W}^+(s; y)$. Apply (11.2.24) with

$$\xi_{ni} = \tau_n(c_{ni}'s + \|c_{ni}\|z)$$

and use (11.2.20) to show that for every $s \in \mathbb{R}^p$ and $z \in \mathbb{R}$ fixed,

$$\sup_{y \in \mathbb{R}} \left| W^+(s, z; y) \right| = o_p(1). \qquad (11.2.27)$$

Because the ball $\mathcal{K}_b := \{t \in \mathbb{R}^p; \|t\| \leq b\}$ is compact, for any $\delta > 0$ there exist $k = k(\delta, b)$ and points $s_1, \cdots, s_k \in \mathcal{K}_b$ such that for any $t \in \mathcal{K}_b$, $\|t - s_j\| < \delta$, for some $j = 1, \cdots, k$. Hence, by the triangle inequality,

$$\sup_{\|t\| \leq b,\, y \in \mathbb{R}} |\mathcal{W}^+(t; y)| \qquad (11.2.28)$$

$$\leq \max_{1 \leq j \leq k} \Big[\sup_{y \in \mathbb{R}} |\mathcal{W}^+(s_j; y)| + \sup_{\|t - s_j\| < \delta,\, y \in \mathbb{R}} \big| \mathcal{W}^+(t; y) - \mathcal{W}^+(s_j; y) \big| \Big].$$

Now fix an $s \in \mathcal{K}_b$. Note that for any $t \in \mathbb{R}^p$ such that $\|t - s\| < \delta$,

$$c'_{ni}s - \|c_{ni}\|\delta \leq c'_{ni}t \leq c'_{ni}s + \|c_{ni}\|\delta, \quad \forall 1 \leq i \leq n.$$

Hence, by monotonicity of the indicator function, for any s, t as above and for all $1 \leq i \leq n$,

$$\begin{aligned}
&|I(\varepsilon_i \leq y + \tau_n c'_{ni}t) - I(\varepsilon_i \leq y + \tau_n c'_{ni}s)| \\
&\leq |I(\varepsilon_i \leq y + \tau_n(c'_{ni}s + \|c_{ni}\|\delta)) - I(\varepsilon_i \leq y)| \\
&\quad + |I(\varepsilon_i \leq y + \tau_n(c'_{ni}s - \|c_{ni}\|\delta)) - I(\varepsilon_i \leq y)|,
\end{aligned}$$

and

$$|\mathcal{W}^+(t; y) - \mathcal{W}^+(s; y)| \tag{11.2.29}$$

$$\leq |W^+(s, \delta; y)| + |W^+(s, -\delta; y)| + 2\sum_{i=1}^n |\gamma'_{ni}c_{ni}|f(y)\,\delta$$

$$\leq |W^+(s, \delta; y)| + |W^+(s, -\delta; y)| + C\,\delta,$$

for some C independent of n, y, s, t, and δ. In obtaining this bound we also used the boundedness of f, (10.2.4), (11.2.20), and the fact that the γ_{ni}^+'s are positive. Equations (11.2.28) and (11.2.29) imply

$$\sup_{\|t\| \leq b, y \in \mathbb{R}} |\mathcal{W}^+(t; y)| \leq C\delta + \max_{1 \leq j \leq k} \Big[\sup_{y \in \mathbb{R}} |W^+(s_j, 0; y)|$$

$$+ \sup_{y \in \mathbb{R}} |W^+(s_j, \delta; y)| + \sup_{y \in \mathbb{R}} |W^+(s_j; -\delta; y)| \Big].$$

Together with (11.2.27) and the arbitrariness of $\delta > 0$, this entails (11.2.26). The proof of Lemma 11.2.2 can now be completed by applying (11.2.26) p-times, the ith time with γ_{nj} equal to the ith coordinate of d_{nj}, $i = 1, \cdots, p$.

We are now ready to prove the first-order asymptotic equivalence of an M estimator to the LS estimator. Recall that $\lambda_1 = -\int f(y)d\psi(y)$. By Remark 10.3.3, $\lambda_1 \neq 0$ means that the Appell rank of ψ is 1. Let

$$\Psi_2 := \Big\{ \psi \in \Psi; \lambda_1 \neq 0 \Big\}.$$

Theorem 11.2.2. *Suppose the non-random design linear regression model (11.2.1) holds with $\{\varepsilon_j\}$ as in (11.2.9) and (11.2.10). In addition, suppose the innovations $\{\zeta_i\}$ satisfy (10.2.21) and (10.2.22), and $\{x_{ni}\}$ satisfy (11.2.2) and (11.2.11). Then every M estimator corresponding to $\psi \in \Psi_2$ is asymptotically first-order equivalent to the LS estimator:*

$$\tau_n^{-1} D_n(\hat{\beta}_M - \beta) = \tau_n^{-1} D_n(\hat{\beta}_{LS} - \beta) + o_p(1). \tag{11.2.30}$$

Proof. Recall that $\tau_n^{-1}D_n(\hat\beta_M - \beta)$ is a minimizer of $\tilde{\mathcal{M}}(s; \psi)$ of (11.2.15). From (11.2.22) and (11.2.17) we readily obtain that for every $0 \le b < \infty$,

$$\sup_{\|s\|\le b}\left\|\tau_n^{-1}[\tilde{\mathcal{M}}(s; \psi) - \tilde{\mathcal{M}}(\mathbf{0}; \psi)] - \lambda_1 s\right\| = o_p(1). \qquad (11.2.31)$$

To establish (11.2.30), we shall first prove

$$\tau_n^{-1}\|D_n(\hat\beta_M - \beta)\| = O_p(1). \qquad (11.2.32)$$

We use Lemma 11.2.1 with $T_n(t) = \|\tau_n^{-1}\mathcal{M}(t; \psi)\|^2$, $\delta_n(\theta_0) = \tau_n^{-1}D_n$, and $\theta_0 = \beta$. Clearly, condition (11.2.7) follows from

$$\tau_n^{-1}\|\tilde{\mathcal{M}}(\mathbf{0}; \psi)\| = O_p(1). \qquad (11.2.33)$$

From (11.2.17) and (11.2.19) we obtain

$$\tau_n^{-1}\tilde{\mathcal{M}}(\mathbf{0}; \psi) = \tau_n^{-1}\sum_{i=1}^{n}c_{ni}\left(\psi(\infty) - \int F(y)d\psi(y)\right)$$
$$+ \tau_n^{-1}\sum_{i=1}^{n}c_{ni}\varepsilon_i\int f(y)d\psi(y) + o_p(1).$$

Here, $\psi(\infty) - \int F(y)d\psi(y) = 0$, because $E\psi(\varepsilon_0) = \int \psi dF = 0$. Therefore,

$$\tau_n^{-1}\tilde{\mathcal{M}}(\mathbf{0}; \psi) = -\lambda_1\tau_n^{-1}D_n(\hat\beta_{LS} - \beta) + o_p(1). \qquad (11.2.34)$$

Hence, (11.2.33) and (11.2.7) follow from the consistency rate of the LS estimator in (11.2.12).

Next, we verify condition (11.2.8) of Lemma 11.2.1. Note that $\tau_n^{-1}\|D_n(t - \beta)\| > b$ is equivalent to $\|s\| > b$, for t, s related as in (11.2.15). With this notation, condition (11.2.8) can be rewritten as follows: $\forall \epsilon > 0$ and $0 < \alpha < \infty$, $\exists N_\epsilon$ and b (depending on ϵ and α) such that

$$P\left(\inf_{\|s\|>b}\left\|\tau_n^{-1}\tilde{\mathcal{M}}(s; \psi)\right\|^2 \ge \alpha\right) \ge 1 - \epsilon, \quad \forall n \ge N_\epsilon. \qquad (11.2.35)$$

To prove (11.2.35), let $\epsilon > 0$ and $0 < \alpha < \infty$. By (11.2.33), there exists $n_{\epsilon 1}$ and $k_\epsilon > \sqrt{\alpha/4}$ such that

$$P\left(\tau_n^{-1}\|\tilde{\mathcal{M}}(\mathbf{0}; \psi)\| \le k_\epsilon\right) > 1 - \epsilon/2, \quad \forall n > n_{\epsilon 1}. \qquad (11.2.36)$$

Next, write $s \in \mathbb{R}^p$ as $s = re$, $r \ge 0$, $e \in \mathbb{R}^p$, and $\|e\| = 1$. Consider the function $g_n(\cdot; e) : [0, \infty) \to \mathbb{R}$ defined by

$$g_n(r; e) := \tau_n^{-1}e'\tilde{\mathcal{M}}(re; \psi) = \tau_n^{-1}\sum_{i=1}^{n}e'c_{ni}\psi(\varepsilon_i - \tau_n re'c_{ni}).$$

By the monotonicity of ψ, $g_n(r; e)$ is non-increasing in r, for every e. Thus,

$$|g_n(r; e)|^2 I(g_n(b; e) \leq 0) \geq |g_n(b; e)|^2 I(g_n(b; e) \leq 0), \quad \forall r \geq b.$$

Moreover,

$$\left\| \tau_n^{-1} \tilde{\mathcal{M}}(re; \psi) \right\|^2 \geq |\tau_n^{-1} e' \tilde{\mathcal{M}}(re; \psi)|^2 = |g_n(r; e)|^2.$$

Hence,

$$\inf_{\|s\|>b} \left\| \tau_n^{-1} \tilde{\mathcal{M}}(s; \psi) \right\|^2 \geq \inf_{r>b, \|e\|=1} |g_n(r; e)|^2 I(g_n(b; e) \leq 0)$$

$$\geq \inf_{\|e\|=1} |g_n(b; e)|^2 I(g_n(b; e) \leq 0).$$

Thus, it suffices to prove that, for any given $\epsilon > 0$ and $\alpha > 0$, there exist N_ϵ and b (depending on ϵ and α) such that for all $n \geq N_\epsilon$,

$$P\Big(\inf_{\|e\|=1} |g_n(b; e)|^2 I(g_n(b; e) \leq 0) \geq \alpha \Big) \geq 1 - \epsilon. \quad (11.2.37)$$

To show this, decompose

$$g_n(b; e) = [g_n(b; e) - g_n(0; e) - \lambda_1 b] + g_n(0; e) + \lambda_1 b.$$

Note that $\lambda_1 < 0$ implies

$$|\lambda_1| b = 4k_\epsilon, \quad \sup_{\|e\|=1} |g_n(0; e)| \leq k_\epsilon, \quad \sup_{\|e\|=1} |g_n(b; e) - g_n(0; e) - \lambda_1 b| \leq k_\epsilon,$$

which in turn implies

$$|g_n(b; e)| \geq 2k_\epsilon \quad \text{and} \quad g_n(b; e) \leq -2k_\epsilon < 0, \quad \forall \|e\| = 1.$$

Here, $P(\sup_{\|e\|=1} |g_n(0; e)| \leq k_\epsilon) \geq P(\tau_n^{-1} \|\tilde{\mathcal{M}}(0; \psi)\| \leq k_\epsilon)$ is bounded from below as in (11.2.36), while

$$P\Big(\sup_{\|e\|=1} |g_n(b; e) - g_n(0; e) - \lambda_1 b| \leq k_\epsilon \Big)$$

$$= P\Big(\sup_{\|e\|=1} \left\| e'\big(\tau_n^{-1}[\tilde{\mathcal{M}}(be; \psi) - \tilde{\mathcal{M}}(0; \psi)] - \lambda_1 be \big) \right\| \leq k_\epsilon \Big)$$

$$\geq P\Big(\sup_{\|s\|\leq b} \left\| \tau_n^{-1}[\tilde{\mathcal{M}}(s; \psi) - \tilde{\mathcal{M}}(0; \psi)] - \lambda_1 s \right\| \leq k_\epsilon \Big) \geq 1 - \epsilon/2,$$

for all $n > n_{\epsilon 2}$ according to (11.2.31). Therefore, for $b = 4k_\epsilon/|\lambda_1|$, $k_\epsilon > \sqrt{\alpha}/4$, and any $n > n_{\epsilon 1} \vee n_{\epsilon 2} =: N_\epsilon$, relation (11.2.37) holds. This proves (11.2.35) and (11.2.32) by Lemma 11.2.1.

We are now ready to complete the proof of (11.2.30). Let

$$\hat{s}_M := \tau_n^{-1} D_n(\hat{\beta}_M - \beta), \quad \hat{s}_{LS} := \tau_n^{-1} D_n(\hat{\beta}_{LS} - \beta),$$

$$\hat{s}_{M0} := -\lambda_1^{-1} \tau_n^{-1} \tilde{\mathcal{M}}(0), \quad \mathcal{N}(s) := \tau_n^{-1} \tilde{\mathcal{M}}(0) + s\lambda_1.$$

$\tilde{\mathcal{M}}(s) \equiv \tilde{\mathcal{M}}(s; \psi)$. Then $\hat{s}_M$ is a minimizer of $\|\tilde{\mathcal{M}}(s)\|^2$, $\hat{s}_{M0}$ is the minimizer of $\|\mathcal{N}(s)\|^2$, $\mathcal{N}(\hat{s}_{M0}) = 0$, and $\hat{s}_{LS} = \hat{s}_{M0} + o_p(1)$; see (11.2.34). Therefore, (11.2.30) is equivalent to

$$\hat{s}_M = \hat{s}_{M0} + o_p(1). \tag{11.2.38}$$

From these definitions, we have the following implications:

$$\left\{ \|\hat{s}_M - \hat{s}_{M0}\| > \epsilon, \|\hat{s}_M\| \le b, \|\hat{s}_{M0}\| \le b \right\}$$

$$\subset \left\{ \inf_{\|s - \hat{s}_{M0}\| \le \epsilon, \|s\| \le b} \|\tilde{\mathcal{M}}(s)\| \ge \inf_{\|s - \hat{s}_{M0}\| > \epsilon, \|s\| \le b} \|\tilde{\mathcal{M}}(s)\|, \right.$$

$$\left. \|\hat{s}_M\| \le b, \|\hat{s}_{M0}\| \le b \right\}$$

$$\subset \left\{ \sup_{\|s - \hat{s}_{M0}\| \le \epsilon, \|s\| \le b} \|\tau_n^{-1}[\tilde{\mathcal{M}}(s) - \tilde{\mathcal{M}}(0)] - \lambda_1 s\| \right.$$

$$\ge \inf_{\|s - \hat{s}_{M0}\| > \epsilon, \|s\| \le b} \|\mathcal{N}(s)\|$$

$$- \sup_{\|s - \hat{s}_{M0}\| > \epsilon, \|s\| \le b} \|\tau_n^{-1}[\tilde{\mathcal{M}}(s) - \tilde{\mathcal{M}}(0)] - \lambda_1 s\|,$$

$$\left. \|\hat{s}_M\| \le b, \|\hat{s}_{M0}\| \le b \right\}$$

$$\subset \left\{ 2 \sup_{\|s\| \le b} \|\tau_n^{-1}[\tilde{\mathcal{M}}(s) - \tilde{\mathcal{M}}(0)] - \lambda_1 s\| \ge \inf_{\|s - \hat{s}_{M0}\| > \epsilon, \|s\| \le b} \|\mathcal{N}(s)\| \right\}.$$

By (11.2.31), the l.h.s. in the last inequality is $o_p(1)$ and the r.h.s. is $|\lambda_1|\epsilon$ for $0 < \epsilon < b/2$ and $\|\hat{s}_{M0}\| \le b/2$, according to the definitions of $\mathcal{N}(s)$ and $\hat{s}_{M0}$. Since $\|\hat{s}_{M0}\| = O_p(1)$, we find that for any $b > 0$ and $\epsilon > 0$, the probability $P\left(\|\hat{s}_M - \hat{s}_{M0}\| > \epsilon, \|\hat{s}_M\| \le b, \|\hat{s}_{M0}\| \le b\right) \to 0$ $(n \to \infty)$. Together with the fact that $\|\hat{s}_M\|$ and $\|\hat{s}_{M0}\|$ are bounded in probability, this proves (11.2.38) and Theorem 11.2.2 as well.

Remark 11.2.2. Rewrite (11.2.38) as

$$\tau_n^{-1} D_n(\hat{\beta}_M - \beta) = -\lambda_1^{-1} \tau_n^{-1} D_n^{-1} \sum_{i=1}^{n} x_{ni} \psi(\varepsilon_i) + o_p(1). \tag{11.2.39}$$

Huber (1981) showed that for classical i.i.d. errors we also have an expansion analogous to (11.2.39), without τ_n^{-1}. Then an application of the Lindeberg–Feller CLT to the sum on the r.h.s. gives the asymptotic normality of the sequence $D_n(\hat{\beta}_M - \beta)$. But, under long memory there is no "ready-made" theorem one can apply to obtain an asymptotic distribution of the leading term on the r.h.s. of (11.2.39). However, Theorems 11.2.1 and 11.2.2 immediately give the following:

Corollary 11.2.1. *In addition to the conditions of Theorem 11.2.2, assume* $\Sigma = \lim_{n\to\infty} \mathrm{Cov}(\tau_n^{-1} D_n(\hat{\beta}_{LS} - \beta))$ *exists. Then* $\tau_n^{-1} D_n(\hat{\beta}_M - \beta) \to_D$ $\mathcal{N}_p(0, \Sigma)$. *Moreover,*

$$\tau_n^{-1} D_n^{-1} \sum_{i=1}^{n} x_{ni}\psi(\varepsilon_i) = -\lambda_1^{-1}\tau_n^{-1} D_n^{-1} \sum_{i=1}^{n} x_{ni}\varepsilon_i + o_p(1)$$

$$\to_D \mathcal{N}(0, \lambda_1^{-2}\Sigma).$$

Note that $\lambda_1 \neq 0$ means that the Appell rank of ψ equals 1.

11.2.3 First-order asymptotics of R estimators

We shall now establish the first-order asymptotic equivalence of R estimators to the LS estimator for the linear regression model (11.2.1) with non-random design and moving-average errors $\{\varepsilon_j\}$. Let f denote the density of ε_0 and let

$$\Phi := \{\varphi; \ \varphi \ \text{a right continuous d.f. on } [0,1]\}.$$

Theorem 11.2.3. *Assume the conditions of Theorem 11.2.2 hold and that* $(\varepsilon_1, \cdots, \varepsilon_n)$ *have a joint probability density for all* $n \geq 1$, *with* $f > 0$, *a.e. In addition, suppose the design variables satisfy (11.2.4) and*

$$n^{1/2} \max_{1 \leq i \leq n} \|D_{cn}^{-1}(x_{ni} - \bar{x}_n)\| = O(1). \tag{11.2.40}$$

Then every R estimator corresponding to $\varphi \in \Phi$ *is asymptotically first-order equivalent to the LS estimator:*

$$\tau_n^{-1} D_{cn}(\hat{\beta}_R - \beta) = \tau_n^{-1} D_{cn}(\hat{\beta}_{LS} - \beta) + o_p(1). \tag{11.2.41}$$

Proof. Again, the method of proof is the same as used for independent errors as in Koul (1992a, 2002). The basic tool for proving the theorem is the asymptotic uniform linearity of the score $\mathcal{S}(\beta + \tau_n D_{cn}^{-1} s; \varphi)$ in s in a compact set; see (11.2.44). Combined with a monotonicity property of the score in Lemma 11.2.3, below, and the general Lemma 11.2.1, it will allow us to establish that the consistency rate of the estimator $\hat{\beta}_R$ is $O_p(\|\tau_n^{-1} D_{cn}\|)$, and then the asymptotic equivalence in (11.2.41), as for M estimators in Theorem 11.2.2.

Define

$$\mathcal{Z}(t; u) := D_{cn}^{-1} \sum_{i=1}^{n} (x_{ni} - \bar{x}_n) I(R_{it} \leq nu), \qquad 0 \leq u \leq 1.$$

Because the weights in $\mathcal{S}$ are centered at their means,

$$\mathcal{S}(t; \varphi) = \int_0^1 \varphi\left(\frac{nu}{n+1}\right) \mathcal{Z}(t; du) = -\int_0^1 \mathcal{Z}(t; u) d\varphi\left(\frac{nu}{n+1}\right),$$

for all $\varphi \in \Phi$. For simplicity, we shall assume below that the regressors x_{ni} are centered, i.e., that

$$\bar{x}_n := n^{-1} \sum_{i=1}^{n} x_{ni} = 0. \tag{11.2.42}$$

Then $D_{cn} = D_n$ and let $c_{ni} := D_n^{-1} x_{ni}$ as in (11.2.14).

We shall prove below that for any $b < \infty$,

$$\sup_{\|s\| \leq b, 0 \leq u \leq 1} \left\| \tau_n^{-1} \mathcal{Z}(\beta + \tau_n D_{cn}^{-1} s; u) - T_n(u) - sq(u) \right\| = o_p(1), \tag{11.2.43}$$

$$\sup_{\|s\| \leq b} \left\| \tau_n^{-1} \mathcal{S}(\beta + \tau_n D_{cn}^{-1} s; \varphi) + T_n(\varphi) + sq(\varphi) \right\| = o_p(1), \tag{11.2.44}$$

where $q(u) := f(F^{-1}(u))$, $F^{-1}(u) := \inf\{y : F(y) \geq u\}$, and

$$T_n(u) := \tau_n^{-1} \sum_{i=1}^{n} c_{ni} [I(\varepsilon_i \leq F^{-1}(u)) - u],$$

$$T_n(\varphi) := \int_0^1 T_n(u) d\varphi(u), \qquad q(\varphi) := \int_0^1 q(u) d\varphi(u).$$

Note that (11.2.44) follows from (11.2.43), $\delta_n(T) := \int_0^1 T_n(u) d\left(\varphi\left(\frac{nu}{n+1}\right) - \varphi(u)\right) = o_p(1)$, and $\delta_n(q) := \int_0^1 q(u) d\left(\varphi\left(\frac{nu}{n+1}\right) - \varphi(u)\right) = o(1)$. The latter follows from the uniform continuity of q on the interval $[0, 1]$ and the fact that $q(1) = \lim_{u \to 1} q(u) = 0$. Relation $\delta_n(T) = o_p(1)$ follows similarly, using Theorem 10.2.3.

The method of the proof of (11.2.43) is the same as for i.i.d. errors. It relies on the asymptotic uniform linearity result in Lemma 11.2.2.

For $y \in \mathbb{R}$, $s \in \mathbb{R}^p$, and $0 \leq u \leq 1$, define

$$\tilde{\mathcal{Z}}(s; u) := \mathcal{Z}(\beta + \tau_n D_n^{-1} s; u), \quad F_n(s; y) := n^{-1} \sum_{i=1}^{n} I(\varepsilon_i \leq y + \tau_n c'_{ni} s),$$

$$F_{ns}^{-1}(u) := \inf\{y : F_n(s; y) \geq u\}.$$

Apply (11.2.21) with $q = 1$ and $d_{ni} \equiv n^{-1/2}$ and use the centering condition (11.2.42) to conclude that for every $0 \le b < \infty$,

$$\sup_{\|s\| \le b, \, y \in \mathbb{R}} n^{1/2} \tau_n^{-1} |F_n(s; y) - F_n(0; y)| = o_p(1). \qquad (11.2.45)$$

Now, let $\tilde{R}_{is} \in \{1, \cdots, n\}$ denote the rank of $Y_{ni} - x'_{ni}(\beta + \tau_n D_n^{-1} s)$. Note that $\tilde{R}_{is} = n F_n(s; \varepsilon_i - \tau_n c'_{ni} s)$, for all i. Thus,

$$\tilde{Z}(s; u) = \sum_{i=1}^{n} c_{ni} I\big(F_n(s; \varepsilon_i - \tau_n c'_{ni} s) \le u\big).$$

Recall the definition of V_n in (11.2.16), where now $d_{ni} \equiv c_{ni}$. We have

$$V_n(s; F_{ns}^{-1}(u)) = \tau_n^{-1} \sum_{i=1}^{n} c_{ni} I\big(\varepsilon_i \le F_{ns}^{-1}(u) + \tau_n c'_{ni} s\big).$$

Then, uniformly in $s \in \mathbb{R}^p$ and $0 \le u \le 1$,

$$\|\tau_n^{-1} \tilde{Z}(s; u) - V_n(s; F_{ns}^{-1}(u))\| \le \frac{2 \max_{1 \le i \le n} \|c_{ni}\|}{\tau_n} = o(1), \qquad (11.2.46)$$

by (11.2.40) and because $0 < d < 1/2$. Write

$$V_n(s; F_{ns}^{-1}(u)) - T_n(u) - sq(u) = \sum_{k=1}^{4} L_{nk},$$

where

$$L_{n1} := \frac{1}{\tau_n} \sum_{i=1}^{n} c_{ni} \big[I(\varepsilon_i \le F_{ns}^{-1}(u) + \tau_n c'_{ni} s) - I(\varepsilon_i \le F_{ns}^{-1}(u))$$
$$- \tau_n c'_{ni} s f(F_{ns}^{-1}(u)) \big],$$

$$L_{n2} := \frac{1}{\tau_n} \sum_{i=1}^{n} c_{ni} \big[I(\varepsilon_i \le F_{ns}^{-1}(u)) - F(F_{ns}^{-1}(u)) + f(F_{ns}^{-1}(u)) \varepsilon_i \big],$$

$$L_{n3} := \frac{(-1)}{\tau_n} \sum_{i=1}^{n} c_{ni} \big[I(\varepsilon_i \le F^{-1}(u)) - F(F^{-1}(u)) + f(F^{-1}(u)) \varepsilon_i \big],$$

$$L_{n4} := \frac{1}{\tau_n} \big[f(F_{ns}^{-1}(u)) - f(F^{-1}(u)) \big] \sum_{i=1}^{n} c_{ni} \varepsilon_i,$$

and where we used the centering condition (11.2.42) again. Hence from (11.2.46), we see that (11.2.43) is implied by

$$\sup_{\|s\| \le b, \, 0 \le u \le 1} \|L_{ni}\| = o_p(1) \quad \text{for any } 0 \le b < \infty, \; i = 1, \cdots, 4. \qquad (11.2.47)$$

For $i = 1$, (11.2.47) follows from (11.2.22) and for $i = 2, 3$ it follows from the URP Theorem 10.2.3. Finally, relation (11.2.47) for $i = 4$ follows from $\sum_{i=1}^{n} c_{ni}\varepsilon_i = O_p(\tau_n)$, and

$$\sup_{\|s\| \leq b} |\theta_n(s)| = o_p(1), \quad \text{for any } 0 \leq b < \infty, \tag{11.2.48}$$

where $\theta_n(s) := \sup_u |f(F_{ns}^{-1}(u)) - f(F^{-1}(u))|$. Since f is continuous on the extended line $\bar{\mathbb{R}} = [-\infty, +\infty]$ and F is strictly increasing, this implies that $q(u) = f(F^{-1}(u))$ is uniformly continuous on $[0, 1]$. Rewrite

$$\begin{aligned}\theta_n(s) &= \sup_{u \in [0,1]} |f(F^{-1}FF_{ns}^{-1}(u)) - f(F^{-1}(u))| \\ &= \sup_{u \in [0,1]} |q(F(F_{ns}^{-1}(u))) - q(u)|.\end{aligned}$$

Next, by $F(F_{ns}^{-1}(u)) - u = F(F_{ns}^{-1}(u)) - F_{ns}(F_{ns}^{-1}(u)) + O(n^{-1})$, a.s., and (11.2.45),

$$\begin{aligned}\sup_{0 \leq u \leq 1, \|s\| \leq b} &|F(F_{ns}^{-1}(u)) - u| \\ &= \sup_{y \in \mathbb{R}, \|s\| \leq b} |F_{ns}(y) - F(y)| + O(n^{-1}) \\ &\leq \sup_{y \in \mathbb{R}, \|s\| \leq b} |F_{ns}(y) - F_{no}(y)| + \sup_{y \in \mathbb{R}} |F_{no}(y) - F(y)| + O(n^{-1}) \\ &= O_p(\tau_n n^{-1/2}) = o_p(1).\end{aligned}$$

This together with the uniform continuity of the function $q(u)$ proves (11.2.48) and concludes the proof of the asymptotic uniform linearity results in (11.2.43) and (11.2.44).

Next, we prove the $\tau_n^{-1} D_{cn}$-consistency of $\hat{\beta}_R$, namely

$$\tau_n^{-1} \|D_{cn}(\hat{\beta}_R - \beta)\| = O_p(1), \tag{11.2.49}$$

see (11.2.32). So as in the proof of (11.2.32), we use Lemma 11.2.1 with $\theta_0 = \beta$, $\delta_n(\theta_0) = \tau_n^{-1} D_{cn}$, and $T_n(t) = \|\tau_n^{-1} S(t; \varphi)\|^2$. Therefore, condition (11.2.7) of this lemma follows from

$$\tau_n^{-1} S(\beta; \varphi) = -q(\varphi)\tau_n^{-1} D_{cn}(\hat{\beta}_{LS} - \beta) + o_p(1) \tag{11.2.50}$$

and Theorem 11.2.1. But, according to (11.2.44), $\tau_n^{-1} S(\beta; \varphi) = T_n(\varphi) + o_p(1)$ and hence (11.2.50) follows from Theorem 10.2.3.

Let $\tilde{S}(s; \varphi) := S(\beta + \tau_n D_{cn}^{-1} s; \varphi)$. Then condition (11.2.8) is in this case: $\forall \epsilon > 0$ and $0 < \alpha < \infty$, $\exists N_\epsilon$ and b (depending on ϵ and α) such that

$$P\left(\inf_{\|s\| > b} \|\tau_n^{-1}\tilde{S}(s; \varphi)\|^2 \geq \alpha\right) \geq 1 - \epsilon, \quad \forall n \geq N_\epsilon. \tag{11.2.51}$$

The verification of (11.2.51) is as for the M estimators; see (11.2.35), and uses (11.2.50) and (11.2.44), together with the fact that $q(\varphi) > 0$. As in the proof of (11.2.35), for $s \in \mathbb{R}^p$ let $s = re$, $r \geq 0$, $e \in \mathbb{R}^p$, $\|e\| = 1$, and consider the function $g_n(\cdot; e) : [0, \infty) \to \mathbb{R}$ defined by

$$g_n(r; e) := \tau_n^{-1} e' \tilde{S}(re; \varphi) = \tau_n^{-1} \sum_{i=1}^n e' c_{ni} \varphi \left(\frac{\tilde{R}_{ie}r}{n+1} \right),$$

where $\tilde{R}_{is} \in \{1, \cdots, n\}$ denotes the rank of $Y_{ni} - x'_{ni}(\beta + \tau_n D_{cn}^{-1}s) = \varepsilon_i - \tau_n c'_{ni}s$, as above. To proceed, we need the following result due to Hájek (1969, Theorem II.7E, p. 35). See also Lehmann (1966).

Lemma 11.2.3. *Let* $c_1, c_2, \cdots, c_n$; $d_1, d_2, \cdots, d_n$; *and* $v_1, v_2, \cdots, v_n$ *be real numbers such that not all* c_i *and* d_i *are the same and no two* v_i *are the same. Let* r_{iu} *denote the rank of* $v_i - ud_i$ *among* $\{v_j - ud_j; 1 \leq j \leq n\}$, $u \in \mathbb{R}$. *Let* $\{b(i); 1 \leq i \leq n\}$ *be a set of non-decreasing real numbers and set*

$$J(u) := \sum_{i=1}^n c_i b(r_{iu}), \quad u \in \mathbb{R}.$$

Assume

$$(c_i - c_j)(d_i - d_j) \geq 0, \quad \forall 1 \leq i \leq j \leq n. \tag{11.2.52}$$

Then the step function $J(u)$ *is non-increasing for all* $u \in \mathbb{R}$ *for which there are no ties among* $\{v_j - ud_j; 1 \leq j \leq n\}$, *i.e., for which no two differences among* $\{v_j - ud_j; 1 \leq j \leq n\}$ *are the same.*

Apply this lemma with $u = r, v_i = \varepsilon_i$, and $c_i = d_i = \tau_n^{-1} e'_i c_{ni}$, to see that $e'\tilde{S}(re; \varphi)$ is a.s. non-increasing in r, for every $e \in \mathbb{R}^p, \|e\| = 1$. The remaining details of the proof of (11.2.51) are as for (11.2.35) and we omit them. This proves the desired consistency rate of $\hat{\beta}_R$ in (11.2.49).

The asymptotic equivalence of $\hat{\beta}_R$ and $\hat{\beta}_{LS}$ as in (11.2.41) now follows from (11.2.44), (11.2.49), and (11.2.50), analogously as for the M estimators; see the end of the proof of Theorem 11.2.2. This completes the proof of Theorem 11.2.3.

Proof of Lemma 11.2.3. The following proof is essentially as in Hájek (1969). We reproduce it for the sake of convenience. Let $u_1 < u_2$ be such that there are no ties among $\{v_j - u_1 d_j; 1 \leq j \leq n\}$ or among $\{v_j - u_2 d_j; 1 \leq j \leq n\}$. We wish to show that $J(u_1) \geq J(u_2)$. The proof is by induction.

Let $R_1 = (r_{1u_1}, \cdots, r_{nu_1})$ and $R_2 = (r_{1u_2}, \cdots, r_{nu_2})$. Suppose there is only one change between the rank sets R_1 and R_2, i.e., suppose for some $i, k = 1, \cdots, n$, that $r_{ku_1} = r_{iu_2}$, $r_{iu_1} = r_{ku_2}$, and $r_{ju_1} = r_{ju_2}$, $j \neq i, k$. Then

$$
\begin{aligned}
J(u_1) - J(u_2) &= c_i b(r_{iu_1}) + c_k\, b(r_{ku_1}) - c_i b(r_{iu_2}) - c_k b(r_{ku_2}) \\
&= c_i [b(r_{iu_1}) - b(r_{iu_2})] + c_k\, [b(r_{ku_1}) - b(r_{ku_2})] \\
&= (c_i - c_k)[b(r_{ku_2}) - b(r_{ku_1})].
\end{aligned}
$$

Now suppose without loss of generality $r_{ku_2} > r_{iu_2}$. Then $r_{iu_1} > r_{ku_1}$ and

$$
\begin{aligned}
& v_k - u_2 d_k > v_i - u_2 d_i, \quad v_i - u_1 d_i > v_k - u_1 d_k \\
\Rightarrow\ & v_k - v_i > u_2(d_k - d_i), \quad v_i - v_k > u_1(d_i - d_k) \\
\Rightarrow\ & (u_2 - u_1)(d_k - d_i) < 0, \ \Rightarrow\ d_i > d_k, \text{ since } u_2 > u_1, \\
\Rightarrow\ & c_i > c_k, \quad \text{by (11.2.52).}
\end{aligned}
$$

Hence, $J(u_1) \geq J(u_2)$, because $b(i)$ is non-decreasing in i.

Now suppose there are m changes in the rank sets R_1 and R_2. Then write $u_2 - u_1 = \sum_{k=1}^m \theta_k$, such that for each $\theta_k > 0$, each pair of sets of ranks of the following variables

$$
\begin{aligned}
& \{v_j - (u_1 + \theta_1)\, d_j\}_{j=1}^n && \&\ \ \{v_j - u_1\, d_j\}_{j=1}^n, \\
& \{v_j - (u_1 + \theta_1 + \theta_2)\, d_j\}_{j=1}^n && \&\ \ \{v_j - (u_1 + \theta_1) d_j\}_{j=1}^n,
\end{aligned}
$$

$$\cdots$$

$$
\{v_j - u_2\, d_j\}_{j=1}^n \qquad \&\ \ \{v_j - (u_1 + \delta_{m-1})\, d_j\}_{j=1}^n
$$

will have exactly one change between them, where $\delta_{m-1} := \sum_{k=1}^{m-1} \theta_k$. Then

$$
\begin{aligned}
J(u_2) - J(u_1) &= [J(u_1 + \theta_1) - J(u_1)] + [J(u_1 + \theta_1 + \theta_2) - J(u_1 + \theta_1)] \\
&\quad + \cdots + [J(u_2) - J(u_1 + \theta_1 + \cdots + \theta_{m-1})].
\end{aligned}
$$

By the first step of the above proof, each of these summands is non-positive. This completes the proof of the lemma.

Remark 11.2.3. (*Errors subordinated to a Gaussian process*). Analogs of the above results when errors are subordinated to a Gaussian long memory process were obtained by Koul and Mukherjee (1993). They considered linear regression models with $\varepsilon_j = G(\eta_j)$, where $\{\eta_j\}$ is a stationary long memory Gaussian process with zero mean, unit variance, and covariance

$E\eta_0\eta_j = L^2(j)j^{1-2d}$, $0 < d < 1/2$, $L \in SV$, and where $G \in L^2(\mathbb{R}, \Phi)$ is a square integrable function having H-rank $\kappa \geq 1$. In this case, the limit distribution of the sum on the r.h.s. of (11.2.39) depends on the Hermite rank ν of the composite function $\psi \circ G$ and $\tau_n = L(n)n^d$ in (11.2.39) must be replaced by $\tau_{n,\nu} := L(n)n^{d\nu + \frac{1-\nu}{2}}$, for $1 \leq \nu < 1/(1 - 2d)$. For M estimators, Koul and Mukherjee (1993) showed that if the design variables satisfy (11.2.2) and (11.2.11), then, with $c_{ni} = D_n^{-1}x_{ni}$,

$$\tau_{n,\nu}^{-1}D_n(\hat{\beta}_M - \beta) = -\frac{J_\nu(\psi(G))}{\nu!\lambda_1}\tau_{n,\nu}^{-1}\sum_{i=1}^{n} c_{ni}H_\nu(\eta_i) + o_p(1), \quad (11.2.53)$$

where $J_\nu(\psi(G)) := E\psi(G(\eta))H_\nu(\eta)$ is the corresponding Hermite coefficient. Similar results are obtained for R estimators also.

A particular case $\psi(x) = x$ of (11.2.53) corresponds to the first-order approximation of the LS estimator under subordinate Gaussian errors. Note that apart from the difference in the H-ranks ($\nu \neq \kappa$ in general), the coefficients $J_\nu(\psi(G))$ and $J_\kappa(G)$ are also different. However, suppose $G(x) = x$, i.e., the errors are Gaussian, then $\kappa = \nu = 1$ and $J_1(\psi) = \int \psi(x)x\phi(x)dx = \int \phi(x)d\psi(x) = -\lambda_1$ so that (11.2.53) coincides with (11.2.30).

An example where $\kappa = \nu = 1$ and the errors are non-Gaussian is obtained when G is strictly increasing. Indeed, $J_1(G) = \int \phi(x)dG(x) \neq 0$ and, similarly, $J_1(\psi(G)) = \int \phi(x)d\psi(G(x)) \neq 0$. In this case it makes sense to compare the asymptotic variances of the LS estimator and an M estimator, which are proportional to the squares of the coefficients on the r.h.s. of (11.2.53). Thus, the ratio

$$\rho := \left(\frac{J_1(\psi(G))}{\lambda_1 J_1(G)}\right)^2 = \left(\frac{\int \phi(x)d\psi(G(x))}{\int f(x)d\psi(x) \int \phi(x)dG(x)}\right)^2$$

may be used to assess the ARE of the M estimator corresponding to the given ψ relative to the LS estimator. The values of $\rho < 1$ would imply that the standardized M estimator is asymptotically more efficient than the LS estimator.

Note: Much of Section 11.2 is based on Giraitis *et al.* (1996), Koul (1992b), and Koul and Mukherjee (1993). Beran (1992b) and Koul (1992b) were the first to establish the first-order equivalence of M estimators in the one sample location and linear regression models, respectively, when errors are long memory Gaussian.

Mukherjee (1994, 1999) studied the asymptotic distribution of certain classes of L and minimum-distance estimators of regression parameters in linear regression models when errors are subordinated to a long memory Gaussian process, while Li (2003) contains similar results for a class of minimum-distance estimators when errors form a long memory moving average.

11.3   Higher-order asymptotics of M estimators

Because of the first-order asymptotic equivalence of robust estimators to the LS estimator discussed in the previous section it is desirable to investigate their higher-order asymptotic behavior. This section discusses the higher-order asymptotic expansion of a class of M estimators of the location parameter in the one-sample location model without any proofs.

As a consequence of these higher-order expansions, we conclude that a suitably standardized second-order difference between the sample median and the sample mean has a non-degenerate limiting distribution. The nature of the limiting distribution depends on the range of the dependence parameter d. If $0 < d < 1/3$, then it has a normal distribution provided the common error distribution is symmetric. For very long memory, where $1/3 < d < 1/2$, the corresponding limiting distribution is non-normal.

Consider the one-sample location model where we observe $\{Y_j\}$, which obeys the relation

$$Y_j = \mu + \varepsilon_j, \quad 1 \le j \le n, \tag{11.3.1}$$

for some $\mu \in \mathbb{R}$. The errors $\varepsilon_j, j \in \mathbb{Z}$, are assumed to form the moving average (10.2.19) with the weights a_j satisfying (10.2.24). Obviously, (11.3.1) is a particular case of the general linear regression model in (11.2.1) with $p = 1$, $x_{ni} \equiv 1$, and $\beta = \mu \in \mathbb{R}^1$. Note that $\mathbf{X}'_n \mathbf{X}_n = n$ and $D_n = n^{1/2}$.

Given function $\psi \in \Psi$, let

$$S_{n,\psi}(x) := \frac{1}{n} \sum_{j=1}^{n} \psi(Y_j - x), \quad \lambda(x) := E\psi(Y_0 - x), \ x \in \mathbb{R}.$$

The corresponding M estimator of μ is defined to be

$$T_n = \operatorname{argmin}\{|S_{n,\psi}(x)| : x \in \mathbb{R}\}.$$

Recall that for $\psi \in \Psi_2$, $\lambda(\mu) = \int \psi dF = 0$ and $\lambda_1 \ne 0$ where F is the d.f. of ε_0 and $\lambda_k := \lambda^{(k)}(0) = -\int F^{(k)} d\psi$.

In order to describe the higher-order expansion of these estimators we need to introduce the following polynomials. Let $z_0 := 1$, $Q_0 = 0$, and $Q_1(z_1) := z_1$. For $k \geq 2$, define Q_k by the following iterative relation, using

$$\mathbf{P}_k^{(p)}(z_1, \cdots, z_{k-p+1}) = \sum \binom{k}{j_1, \cdots, j_p} \prod_{s=1}^{p} Q_{j_s}(z_1, \cdots, z_{j_s}),$$

where the sum is taken over all integers $j_1, \cdots, j_p \geq 1$ such that $j_1 + \cdots + j_p = k$; $k \geq p \geq 1$. Then, in terms of the $\mathbf{P}_k$'s,

$$Q_k(z_1, \cdots, z_k) = \frac{1}{\lambda_1} \sum_{j=2}^{k} \frac{\lambda_j}{j!} \sum_{r=0}^{j-1} (-1)^{r-1} \binom{j}{r} \frac{k!}{(k-r)!} z_r$$

$$\times \mathbf{P}_{k-r}^{(j-r)}(z_1, \cdots, z_{k-j+1}) + (-1)^{k-1} \frac{\lambda_k}{\lambda_1} z_k.$$

Note that $Q_k(z_1, \cdots, z_k)$ is a polynomial in the real variables $z_1, \cdots, z_k$ with coefficients given by $\lambda_i, 1 \leq i \leq k$. All polynomials $\mathbf{P}_{k-r}^{(j-r)}$ on the r. h. s. of the above equation are expressed in terms of Q_j, $1 \leq j \leq k-1$. Also, note that $\mathbf{P}_k^{(1)} = Q_k$ and $\mathbf{P}_k^{(k)} = k! Q_1^k$, $\forall k \geq 1$. For convenience, we describe the first few product polynomials $\mathbf{P}_k^{(p)}$:

$$\mathbf{P}_2^{(2)} = 2Q_1^2, \quad \mathbf{P}_3^{(2)} = 6Q_1 Q_2, \quad \mathbf{P}_3^{(3)} = 6Q_1^3,$$

$$\mathbf{P}_4^{(2)} = 4!\left(\frac{2}{1!3!} Q_1 Q_3 + \frac{1}{4} Q_2^2\right), \quad \mathbf{P}_4^{(3)} = 4! \frac{3}{1!1!2!} Q_1^2 Q_2, \quad \mathbf{P}_4^{(4)} = 4! Q_1^4.$$

Using these, we have

$$Q_2(z_1, z_2) = \frac{\lambda_2}{\lambda_1}(z_1^2 - z_2), \quad Q_3(z_1, z_2, z_3) = \frac{\lambda_3}{\lambda_1}(2z_1^3 - 3z_1 z_2 + z_3),$$

$$Q_4(z_1, z_2, z_3, z_4) = \frac{3\lambda_2(2\lambda_1\lambda_3 - \lambda_2^2)}{\lambda_1^3}(z_1^2 - z_2)^2$$

$$+ \frac{\lambda_4}{\lambda_1}\{3z_1^4 - 6z_1^2 z_2 + 4z_1 z_3 - z_4\}.$$

It is clear from the previous section that the asymptotic expansion of M estimators should rely on similar expansions for the empirical function $S_{n,\psi}(T_n)$ and the Taylor expansion of $\lambda(T_n)$ around μ. In fact, we can establish the following asymptotic expansion. Recall the definitions of k^*, $\tilde{Z}_n^{(k)}$, and $Z^{(k)}$ from (10.3.3), (10.3.26), and (10.3.33). Let

$$\Lambda_j(x) := \psi(\varepsilon_j - x) - \sum_{k=0}^{k^*} (-1)^k \lambda^{(k)}(x)\varepsilon_j^{(k)}, \quad x \in \mathbb{R}.$$

We are now ready to state the higher-order expansion result.

Theorem 11.3.1. *Suppose $\{Y_j\}$ follows the location model (11.3.1) with long memory errors $\{\varepsilon_j\}$ as in (10.2.19) and (10.2.24) and innovations $\{\zeta_j\}$ satisfying (10.2.21). In addition, suppose $1/(1-2d)$ is not an integer and $E\zeta_0^4 < \infty$.*

Then, for every $\psi \in \Psi_2$ for which $\lambda^{(k)}(x) = \int \psi(y-x) f^{(k)}(y) dy$, $k = 0, 1, \cdots$, $x \in \mathbb{R}$, the following holds:

$$T_n - \mu = \bar{Y}_n - \mu + \sum_{2 \leq k \leq k^*} \frac{n^{-k(1/2-d)}}{k!} Q_k(1! \tilde{Z}_{n,1}, \cdots, k! \tilde{Z}_{n,k}) \quad (11.3.2)$$

$$+ n^{-1/2} \lambda_1^{-1} W_n(\psi).$$

Furthermore, $\forall \, 1 \leq k \leq k^*$,

$$Q_k(1! \tilde{Z}_n^{(1)}, \cdots, k! \tilde{Z}_n^{(k)}) \to_D Q_k(1! Z^{(1)}, \cdots, k! Z^{(k)}), \quad (11.3.3)$$

while $W_n(\psi) \to_D \mathcal{N}(0, \sigma_\psi^2)$ where $\sigma_\psi^2 = \sum_{j \in \mathbb{Z}} \mathrm{Cov}(\Lambda_0(0), \Lambda_j(0))$.

A proof of this theorem appears in Koul and Surgailis (1997). Because of its technical nature we do not present it here. Instead, we shall discuss a statistical application of the above theorem, considering second-order asymptotic variance comparisons of these estimators.

To that effect, suppose, additionally, that f is symmetric around 0, ψ is skew-symmetric: $\psi(-x) \equiv -\psi(x)$, and $\mu = 0$. Then $f^{(k)}(y) \equiv (-1)^k f^{(k)}(-y)$ and integration by parts shows that $\lambda_k = \int_0^\infty \psi(y) [1 - (-1)^k] f^{(k)}(y) \, dy$ so that

$$\lambda_k = 0, \qquad \qquad \text{if } k = 2, 4, \cdots, \quad (11.3.4)$$

$$= 2 \int_0^\infty \psi(y) \, f^{(k)}(y) dy, \qquad \text{if } k = 1, 3, \cdots.$$

Thus $\lambda_2 = 0$ and $\lambda_1 \neq 0$. Now, $k^* = 1, 2$ or ≥ 3 depending on whether $1/2 < (1-2d) < 1$, $1/3 < (1-2d) < 1/2$, or $0 < (1-2d) < 1/3$. In particular, if $\lambda_3 \neq 0$, then (11.3.2) and (11.3.3) give, for $d \neq 1/4$,

$$n^{3(1-2d)/2}(T_n - \bar{Y}_n) \qquad \qquad (11.3.5)$$

$$\to_D \frac{\lambda_3}{3! \lambda_1} \big(2(Z^{(1)})^3 - 3Z^{(1)} Z^{(2)} + Z^{(3)} \big), \quad 1/3 < d < 1/2,$$

$$n^{1/2}(T_n - \bar{Y}_n) \to_D \mathcal{N}_1(0, \sigma_\psi^2/\lambda_1^2), \qquad \qquad 0 < d < 1/3.$$

We will now write $T_{n,\psi}$, $\lambda_{k,\psi}$, etc. to emphasize the dependence on ψ.

We will consider two cases. *Case 1:* $1/3 < d < 1/2$. This case may be thought of as *very long memory*. Note that here the the asymptotic distribution of $T_{n,\psi}$ depends on ψ only through $C_\psi := \lambda_{3,\psi}/\lambda_{1,\psi}$. It is thus natural to define, for any two score functions ψ_1, ψ_2 of the above type, the *second-order asymptotic relative efficiency* (ARE) of T_{n,ψ_1}, relative to T_{n,ψ_2}, as the ratio

$$e_{\psi_1,\psi_2} := \{C_{\psi_1}/C_{\psi_2}\}^2.$$

The three interesting ψ functions are

$$\psi_1(x) := (1/2)\mathrm{sgn}(x), \qquad \psi_2(x) := 2F(x) - 1,$$
$$\psi_h(x) := h\,\mathrm{sgn}(x)I(|x| > h) + xI(|x| \le h), \ h > 0.$$

The estimator T_{n,ψ_1} is the sample median. When errors are i.i.d., it is well known that this estimator is asymptotically first-order optimal for the double exponential errors while T_{n,ψ_2} has the same property for the logistic errors. All three scores yield robust estimators against heavy tail errors.

Now consider the case of Gaussian errors $\{\varepsilon_j\}$ with zero mean and covariance $\gamma(k) = E\varepsilon_0\varepsilon_k$, $\gamma(0) = \tau^2$, for some $\tau > 0$. Then $F(x) \equiv \Phi(x/\tau)$. With ϕ denoting the density of the standard normal distribution d.f. Φ, let

$$C_{\psi_1} = -1/\tau^2, \quad C_{\psi_2} = -1/2\tau^2, \quad C_{\psi_h} = \frac{-(h/\tau^3)\phi(h/\tau)}{\big(\Phi(h/\tau) - \Phi(0)\big)}.$$

Then

$$e_{\psi_1,\psi_2} = 4, \qquad e_{\psi_h,\psi_1} = \left(\frac{(h/\tau)\phi(h/\tau)}{\Phi(h/\tau) - \Phi(0)}\right)^2,$$
$$e_{\psi_h,\psi_2} = 4\left(\frac{(h/\tau)\phi(h/\tau)}{\Phi(h/\tau) - \Phi(0)}\right)^2.$$

Thus in terms of the second-order ARE, T_{n,ψ_2} is four times as efficient as the sample median. Also note that $e_{\psi_h,\psi_2} \to 4$ as $\tau \to \infty$ while $e_{\psi_h,\psi_2} \to 0$ as $\tau \to 0$. Thus for an arbitrarily large error variance τ^2, T_{n,ψ_2} is preferable to T_{n,ψ_h} for all $h > 0$. Moreover, using the fact that the function $\Phi(x) - 1/2 - 2x\phi(x)$, $x \in \mathbb{R}$, is positive for $x \ge 1.4$, we can say that the estimator T_{n,ψ_h} is preferable to T_{n,ψ_2} for all those h for which $h \ge 1.4\tau$ while the opposite is true otherwise.

Observe that in all of the above three situations $\lambda_3 \ne 0$. Note that $\lambda_k = \tau^{-1-k}\int \psi(x)\phi^{(k)}(x/\tau)dx$, for all $k \ge 1$ and any ψ with $\int \psi^2(x)\phi(x/\tau)dx <$

∞. The Hermite expansion of $\psi(\varepsilon)$ is thus

$$\psi(\varepsilon) = \sum_{k=0}^{\infty} \frac{(-1)^k \lambda_k \tau^k}{k!} H_k(\varepsilon/\tau).$$

Notice that $\lambda_3 = 0$ if and only if

$$\int_{|x|<\tau} (\tau^2 - x^2)\phi(x/\tau)d\psi(x) = \int_{|x|>\tau} (x^2 - \tau^2)\phi(x/\tau)d\psi(x).$$

In particular, $\lambda_3 = 0$ if the measure $d\psi$ is purely atomic with atoms only at $\pm\tau$.

Next, consider *Case 2:* $0 < d < 1/3$. In this case the asymptotic distribution of $T_{n,\psi}$ in (11.3.5) depends on ψ only through $K_\psi := (\sigma_\psi/\lambda_{1,\psi})^2$, where now

$$\sigma_\psi^2 = \sum_{j\in\mathbb{Z}} E(\psi(\varepsilon_0) + \lambda_{1,\psi}\varepsilon_0)(\psi(\varepsilon_j) + \lambda_{1,\psi}\varepsilon_j).$$

In general, it is hard to obtain a closed expression for this. But some simplification is possible for the above type of Gaussian errors, where we shall now let $\tau = 1$, for convenience. Let $\rho_k := \sum_{j\in\mathbb{Z}} \gamma^k(j)$. Because $\gamma(j) \sim c_\gamma j^{-1+2d}$, $j \to \infty$, ρ_k converges for $k \geq 3$ in this case. Now, using (11.3.4) and the orthogonality of Hermite polynomials (2.4.15),

$$\sigma_\psi^2 = \sum_{k=1}^{\infty} \frac{\lambda_{2k+1,\psi}^2 \, \rho_{2k+1}}{(2k+1)!}, \qquad K_\psi = \sum_{k=1}^{\infty} \left(\frac{\lambda_{2k+1,\psi}}{\lambda_{1,\psi}}\right)^2 \times \frac{\rho_{2k+1}}{(2k+1)!}.$$

This expression for K_ψ can be used to minimize it over a given class of functions ψ. For example, consider the class of M estimators corresponding to the class of functions $\{\psi_h, \ h > 0\}$. This class includes the median and the sample mean as the limiting cases $h \to 0$ and $h \to \infty$, respectively. In this case, the Hermite coefficients $\lambda_k \equiv \lambda_{k,h}$ can be explicitly found: $\lambda_{k,h} = -2H_{k-2}(h)\phi(h)$, $k = 3, 5, \cdots$, and $\lambda_{1,h} = 1 - 2\Phi(h)$, resulting in

$$K_{\psi_h} = \sum_{k=1}^{\infty} \left(\frac{H_{2k-1}(h)\phi(h)}{\Phi(h) - \Phi(0)}\right)^2 \times \frac{\rho_{2k+1}}{(2k+1)!}.$$

The function $h \mapsto K_{\psi_h}$ is continuous and strictly positive on $(0,\infty)$, $K_{\psi_h} \to 0$, $h \to \infty$, and $\lim_{h\to 0} K_{\psi_h} = K_{\psi_1}$ is the corresponding constant for the median. It follows that K_{ψ_h} has a well-defined minimum on each compact interval, which depends on the covariance function $\{\gamma(j), j \in \mathbb{Z}\}$. By assuming some specific form of the latter, e.g., taking it to be the covariance of fractional Gaussian noise or fractional ARFIMA errors, the corresponding minimization problem for K_{ψ_n} can be dealt with numerically.

Suppose that the errors are i.i.d. with variance τ^2. Then, under minimal conditions on ψ and F, we know, say from Huber (1981),

$$n^{1/2}(T_n - \mu) = -n^{-1/2}\lambda_1^{-1}\sum_{i=1}^n \psi(\varepsilon_i) + o_p(1),$$

$$n^{1/2}(T_n - \bar{Y}_n) = -n^{-1/2}\lambda_1^{-1}\sum_{i=1}^n \{\psi(\varepsilon_i) + \lambda_1\varepsilon_i\} + o_p(1)$$

$$\to_D \mathcal{N}(0, \varphi^2), \qquad \varphi^2 := \lambda_1^{-2}\mathrm{Var}\Big(\psi(\varepsilon_0) + \lambda_1\varepsilon_0\Big).$$

Observe that φ^2 is precisely the first term in the asymptotic variance σ_ψ^2 of Theorem 11.3.1. It thus follows that the asymptotic variance of $n^{1/2}(T_n - \bar{Y}_n)$ under the above type of long memory dependent errors is larger than its counterpart for independent errors.

Finally, we state the following conjecture. Suppose that the convergence in (11.3.2) extends to the variances on both sides. Then we conjecture that the following holds even without the assumption of Gaussianity: in the very long memory region $1/3 < d < 1/2$,

$$E(T_n - \mu)^2 = E(\bar{Y}_n - \mu)^2 + n^{-2(1-2d)}\frac{\lambda_3}{\lambda_1}c_{13}(d)(1 + o(1)),$$

where

$$c_{13}(d) := EZ^{(1)}[2(Z^{(1)})^3 - 6Z^{(1)}Z^{(2)} + 6Z^{(3)}]/3$$
$$= 2(E(Z^{(1)})^2)^2 - 2E\{(Z^{(1)})^2Z^{(2)}\}.$$

The last equality is obtained by using the Gaussianity of $Z^{(1)}$ and that $EZ^{(1)}Z^{(3)} = 0$.

Assuming the above approximation to be true, it can be used to compare the mean-squared errors of T_n and $\bar{Y}_n$ at the second-order level as follows. First, using the definition (10.3.7) of the $Z^{(k)}$'s verify that

$$c_{13}(d) = \Big(\frac{B(d, 1-2d)}{d}\Big)^2 c_0^4\ell(d),$$

$$\ell(d) := \frac{2}{(1+2d)^2} - \frac{1}{(1+4d)} - B(1+2d, 1+2d).$$

The function ℓ is negative on $(0, 1/2)$ with values ranging from 0 to -0.004994776 so that $c_{13}(d) < 0$, $0 < d < 1/2$.

Thus if $\lambda_{3,\psi}/\lambda_{1,\psi} > 0\,(<0)$, then the mean-squared error of T_n is smaller (larger) than that of $\bar{Y}_n$, in the second-order sense, i.e., in the sense that

$$\lim_n n^{2(1-2d)}\{E(T_n - \mu)^2 - E(\bar{Y}_n - \mu)^2\} < 0\ (>0).$$

For example, for the Gaussian $Y_j \sim \mathcal{N}(0,1)$,

$$\lambda_{3,\psi}/\lambda_{1,\psi} = \int (x^2 - 1)\phi(x)d\psi(x) \Big/ \int \phi(x)d\psi(x) \ > 0,$$

if the support of $d\psi \subset (-\infty, -1) \cup (1, +\infty)$ while $\lambda_{3,\psi}/\lambda_{1,\psi} < 0$ if the support of $d\psi \subset (-1, +1)$. For example, the estimator corresponding to $\psi(x) := -I(x \leq -2) + (x + 1)I(-2 \leq x \leq -1) + (x - 1)I(1 \leq x \leq 2) + I(x \geq 2)$ would have a smaller mean-squared error in the second-order sense than the sample mean for long memory Gaussian errors!

Note: Theorem 11.3.1 is proved in Koul and Surgailis (1997). An analogous expansion in a simple linear regression model with non-random design is established in Koul and Surgailis (2000a). The problem of establishing similar expansions for these estimators in multiple linear regression models with random designs is open at the time of writing.

11.4 Whittle estimator in linear regression

As seen in the previous sections the asymptotic distributions of a large class of estimators of the regression parameter vector depends on the covariance structure of the errors. For parametric long memory moving-average errors as in (11.4.2) below, this covariance depends on the parameters in the moving-average coefficients, which also includes the long memory parameter d. Because the consistency rate of these estimators is typically $n^{d-1/2}$, to implement any inference procedure based on these estimators we must first have a $\log(n)$-consistent estimator of d and consistent estimators of any other parameters appearing in this covariance.

This section establishes the consistency and asymptotic normality of the Whittle estimators of the parameters σ^2 and θ appearing in the errors (11.4.2) based on the residuals for the two types of linear regression models. For the first type, Case 1, the design variables are non-random and for the second type, Case 2, they are random, which allows for long memory.

Booth, Kaen and Koveos (1982), Lo (1991), and Cheung (1993) provide examples from finance where both error and design variables have long memory.

In Case 1, we assume the linear regression model (11.2.1) holds while in Case 2, we see that (X_j, Y_j), $j = 1, \cdots, n$, satisfies

$$Y_j = X'_j \beta + \varepsilon_j, \quad 1 \le j \le n, \tag{11.4.1}$$

for some $\beta \in \mathbb{R}^p$. Here, X'_j is a p-dimensional row vector of a known $n \times p$ matrix $\mathbf{X}_n$ of design random variables. In both cases, the errors $\{\varepsilon_j\}$ are assumed to be the parametric moving-average process

$$\varepsilon_j = \sum_{k=0}^{\infty} a_k(\theta)\zeta_{j-k}, \quad a_0(\theta) = 1, \quad j \in \mathbb{Z}, \quad \theta \in \Theta, \tag{11.4.2}$$

with the innovations $\{\zeta_i\} \sim \text{IID}(0, \sigma^2)$ and finite fourth moment; see (8.2.1). Recall the definitions of $Q_Y(\theta)$ and the $b_j(\theta)$'s from (8.2.3) to (8.2.5). Let $\tilde{\beta}_n$ be a consistent estimator of β and define the residuals

$$\hat{\varepsilon}_j := Y_j - x'_{nj}\tilde{\beta}_n, \qquad \text{in Case 1,}$$
$$:= Y_j - X'_j\tilde{\beta}_n, \qquad \text{in Case 2,} \quad 1 \le j \le n,$$

and let $Q_{\hat{\varepsilon}}(\theta) := n^{-1} \sum_{j,k=1}^{n} b_{k-j}(\theta)\hat{\varepsilon}_j\hat{\varepsilon}_k$, accordingly. Throughout, θ_0 will denote the true parameter value and it is assumed to be in the interior of Θ. Whittle's estimator $(\hat{\theta}_n, \hat{\sigma}_n^2)$ of (θ_0, σ^2) is defined to be the minimizer of the quadratic form $\Lambda_n(\hat{\varepsilon}; \sigma, \theta) = (2\sigma^2)^{-1}Q_{\hat{\varepsilon}}(\theta) + \log \sigma$ with respect to $(\theta, \sigma) \in \Theta \times (0, \infty)$ as in Definition 8.2.1, giving

$$\hat{\theta}_n := \text{argmin}_{\theta \in \Theta} Q_{\hat{\varepsilon}}(\theta), \qquad \hat{\sigma}_n^2 := Q_{\hat{\varepsilon}}(\hat{\theta}_n). \tag{11.4.3}$$

For any function $g(\theta)$ from Θ to $\mathbb{R}$, let

$$\nabla g(\theta) = \left(\frac{\partial g(\theta)}{\partial \theta_j}\right)_{j=1,\cdots,q}, \quad \nabla^2 g(\theta) = \left(\frac{\partial^2 g(\theta)}{\partial \theta_i \partial \theta_j}\right)_{i,j=1,\cdots,q}.$$

Recall from (8.2.2) that $f_\theta(u) = \sigma^2 s_\theta(u)/2\pi$ is the spectral density of the error process. To prove the asymptotic normality of $\hat{\theta}_n$ in (11.4.3), we need to impose stronger conditions on the parametric form of $s_\theta(u)$ than the conditions (a0) to (a4) of Sections 8.2 and 8.3.1 dealing with the asymptotic properties of the parametric Whittle estimator in the fully observable case.

(a0′) The parameter space Θ is compact, $s_\theta(u)$, $u \in \Pi$, determines $\theta \in \Theta$ uniquely and $s_\theta^{-1}(u)$ is continuous in $(u, \theta) \in \Pi \times \Theta$. Moreover, the true value θ_0 lies in the interior of Θ and $\int_\Pi \log s_\theta(u)du = 0$, $\forall \theta \in \Theta$.

Furthermore, there exists a small compact ball $\Theta_0 \subset \Theta$ centered at θ_0 such that

(a1′) $\int_\Pi \log s_\theta(u)du$ is twice differentiable in $\theta \in \Theta_0$ under the integral sign.

(a2′) $s_\theta(u)$ is continuous at all (u, θ), $u \neq 0$, $\theta \in \Theta_0$, and $\nabla s_\theta^{-1}(u)$ and $\nabla^2 s_\theta^{-1}(u)$ are continuous on $\Pi \times \Theta_0$.

(a3′) There exist $0 < \alpha < 1$ and $0 \leq \eta < \alpha$, such that $\alpha - \eta < 1/2$ and

$$s_\theta(u) \leq C|u|^{-\alpha}, \quad \|\partial s_\theta(u)/\partial u\| \leq C|u|^{-\alpha-1},$$
$$\|\nabla s_\theta^{-1}(u)\| \leq C|u|^\eta, \quad \forall (u, \theta) \in \Pi \times \Theta_0.$$

(a4′) $s_\theta^{-1}(u) \leq C|u|^{\alpha/2}, \quad \forall (u, \theta) \in \Pi \times \Theta_0.$

(a5′) $\|\nabla^2 s_\theta^{-1}(u)\| \leq C|u|^{\alpha/2}, \quad \forall (u, \theta) \in \Pi \times \Theta_0.$

Assumptions (a1′) and (a2′) basically agree with (a1) and (a2) of Section 8.3.1. Also note that (a0′) implies that $s_\theta(u)$ is separated from zero on $\Pi \times \Theta$ and hence implies (a0). Assumption (a3′) is stronger than (a3), in the sense that it allows for positive values of α only and therefore excludes the negative dependence of errors. (a4′) and (a5′) (where α is the same as in (a3′)) are extra assumptions needed to deal with the quadratic form $Q_{\hat\varepsilon}(\theta)$ based on the residuals.

Consider Case 1, where the design is non-random. Here, we shall use the standardization

$$D_n := (\mathbf{X}_n'\mathbf{X}_n)^{1/2}, \quad V_j = V_{nj} = n^{1/2}D_n^{-1}x_{nj}, \quad 1 \leq t \leq n.$$

We make the following additional assumption.

There exist $\mathbb{R}^p$-valued continuously differentiable functions $\quad$ (11.4.4)
$v_n(t)$, $t \in (0, 1]$, $n \geq 1$, having uniformly bounded derivatives
on $(0, 1]$ such that $V_{nj} \equiv v_n(j/n)$.

In the following theorem all probability statements are to be understood to hold under the model (11.4.2) and when the true parameter in the spectral density is θ_0.

Theorem 11.4.1. *Suppose $\{Y_j\}$ follows the regression model (11.4.1) with parametric errors $\{\varepsilon_j\}$ as in (11.4.2) having the spectral density f_{θ_0} and that assumptions (a0′) to (a5′) hold. In addition, suppose the non-random design $\{x_{nj}\}$ satisfies (11.4.4) and an estimator $\tilde\beta_n$ of β satisfies*

$$n^{-1/2}D_n(\tilde\beta_n - \beta) = o_p(n^{\eta/2-1/4}). \quad (11.4.5)$$

Then $\hat{\sigma}_n^2 \to_p \sigma^2$ and

$$n^{1/2}(\hat{\theta}_n - \theta_0) \to_D \mathcal{N}_q(0, 4\pi W^{-1}(\theta_0)), \tag{11.4.6}$$

where $W(\theta_0)$ is the same as in Theorem 8.3.1.

Remark 11.4.1. A large class of deterministic designs, x_{nj}, satisfying condition (11.4.4) can be constructed as follows. Let $u_1(z), \cdots, u_p(z)$, $z \in [0, 1]$, be continuously differentiable real-valued functions on $[0, 1]$ and linearly independent in $L_2[0, 1]$. Let $u(z)' = (u_1(z), \cdots, u_p(z))$, and set $x_{nj} = u(j/n)$, $1 \leq j \leq n$. Note that the matrices $\mathbf{U}_n' \mathbf{U}_n := n^{-1} \mathbf{X}_n' \mathbf{X}_n$, with entries $n^{-1} \sum_{j=1}^n u_\ell(j/n) u_k(j/n)$, $1 \leq \ell, k \leq p$, converge as $n \to \infty$ to the non-degenerate matrix

$$\mathbf{U}'\mathbf{U} = \left(\int_0^1 u_\ell(z) u_k(z) dz \right)_{\ell, k = 1, \cdots, p}$$

and condition (11.4.4) holds with $v_n(z) = (\mathbf{U}_n' \mathbf{U}_n)^{-1/2} u(z)$, approaching uniformly in $z \in [0, 1]$, together with their derivatives, to $v(z) = (\mathbf{U}'\mathbf{U})^{-1/2} u(z)$. For example, we can take the u-functions $u_k(z) = \cos(2\pi(k-1)z)$, $1 \leq k \leq p$, or $u_k(z) = z^{k-1}$, $1 \leq k \leq p$.

Remark 11.4.2. Assumption (11.4.5) is rather a weak condition on the rate of consistency of $\tilde{\beta}_n$. Since condition (11.4.4) implies both (11.2.2) and (11.2.11), from Theorem 11.2.1 and Corollary 11.2.1 we see that for the above types of non-random designs and long memory moving-average errors, a large class of estimators $\tilde{\beta}_n$ satisfy

$$n^{-1/2} D_n(\tilde{\beta}_n - \beta) = O_p(n^{(\alpha-1)/2}).$$

This rate implies (11.4.5) since $1/2 > \alpha - \eta$ in (a3'). Thus, condition (11.4.5) is satisfied by this class of estimators, which includes the LS and LAD estimators.

Proof of Theorem 11.4.1. According to the definition,

$$Q_{\tilde{\varepsilon}}(\theta) = \frac{1}{n} \sum_{t,s=1}^n b_{t-s}(\theta)(Y_t - x_{nt}' \tilde{\beta}_n)(Y_s - x_{ns}' \tilde{\beta}_n).$$

Arguing as in the proof of Theorem 8.3.1, it suffices to show that

$$n^{1/2} \nabla Q_{\tilde{\varepsilon}}(\theta_0) \to_D \mathcal{N}_q(0, \sigma^4 \pi^{-1} W(\theta_0)), \tag{11.4.7}$$

and that, uniformly in θ in a neighborhood Θ_0 of θ_0,

$$\nabla^2 Q_{\hat{\varepsilon}}(\theta) = \sigma^2 (2\pi)^{-1} W(\theta_0) + o_p(1). \qquad (11.4.8)$$

Consider (11.4.7). Set $T_n := n^{-1/2} D_n(\tilde{\beta}_n - \beta)$. Then

$$Q_{\hat{\varepsilon}}(\theta) = Q_1(\theta) + 2Q_2(\theta) + Q_3(\theta), \quad \forall \theta \in \Theta, \qquad (11.4.9)$$

where

$$Q_1(\theta) := \frac{1}{n} \sum_{t,s=1}^{n} b_{t-s}(\theta)\varepsilon_t\varepsilon_s, \quad Q_2(\theta) := \frac{1}{n} \sum_{t,s=1}^{n} b_{t-s}(\theta)\varepsilon_t V_s' T_n,$$

$$Q_3(\theta) := \frac{1}{n} \sum_{t,s=1}^{n} b_{t-s}(\theta)V_t' T_n V_s' T_n.$$

Equations (11.4.7) and (11.4.8) for the quadratic form Q_1 instead of $Q_{\hat{\varepsilon}}$ follow from Lemmas 8.3.1 and 8.2.2; see (8.3.6). Accordingly, (11.4.7) and (11.4.9) will follow from

$$\nabla Q_i(\theta_0) = o_p(n^{-1/2}), \quad i = 2, 3, \qquad (11.4.10)$$

$$\nabla^2 Q_i(\theta) \to 0, \quad i = 2, 3, \qquad (11.4.11)$$

uniformly in θ in a neighborhood of θ_0, in probability.

Let $\nabla_j = \partial/\partial\theta_j$ and $\nabla_{ij}^2 = \partial^2/\partial\theta_i\partial\theta_j$, $i, j = 1, \cdots, q$, denote partial derivatives. From the definitions of $Q_i, i = 2, 3$, for $j = 1, \cdots, q$,

$$|\nabla_j Q_2(\theta)| \leq n^{-1} \left\| \sum_{t,s=1}^{n} \nabla_j b_{t-s}(\theta)\varepsilon_t V_s' \right\| \|T_n\|, \qquad (11.4.12)$$

$$|\nabla_j Q_3(\theta)| \leq n^{-1} \left\| \sum_{t,s=1}^{n} \nabla_j b_{t-s}(\theta)V_t V_s' \right\| \|T_n\|^2.$$

Clearly, (11.4.10) for $i = 3$ follows from (11.4.12), (11.4.5), and

$$\sum_{t,s=1}^{n} \nabla_j b_{t-s}(\theta)V_t V_s' = O(n^{1-\eta}),$$

with η as in (a3'), where the last equation follows from Lemma 11.4.1, below, with $\kappa = \eta$, and assumption (a3') about the derivative ∇s_θ^{-1}. Relation (11.4.10) for $i = 2$ follows similarly from Lemma 11.4.1 and the fact that the sum of the resulting exponents of n equals $-1 + (\frac{1+\alpha}{2} - \eta) \vee 0 + \frac{\eta}{2} - \frac{1}{4}$, which is strictly less than $-\frac{1}{2}$ because $\eta > \alpha - \frac{1}{2}$ in (a3').

Consider (11.4.11). Here, we apply Lemma 11.4.1 with $\kappa = \alpha/2$; see (a5'). Estimating $|\nabla_{jk}^2 Q_i(\theta)|$, $i = 2, 3$, as in (11.4.12) and using

(11.4.5), we obtain $\sup_{\theta \in \Theta_0} |\nabla^2_{jk} Q_2(\theta)| = O_p(n^{\frac{\eta}{2}-\frac{3}{4}} \log n) = o_p(1)$ and $\sup_{\theta \in \Theta_0} |\nabla^2_{jk} Q_3(\theta)| = O_p(n^{\eta-\frac{\alpha}{2}-\frac{1}{2}}) = o_p(1)$ since $\eta < \alpha < 1$. This completes the proof of (11.4.11) and also of Theorem 11.4.1.

Now consider Case 2, where the design is random. To allow for a wide range of dependence in design variables $X'_j = (X_{j,1}, \cdots, X_{j,p})$, where $X_{j,1} \equiv 1$, the remaining components $\mathcal{Y}_j := (X_{j,2}, \cdots, X_{j,p})'$ are assumed to have zero mean $E\mathcal{Y}_j = 0$ and a $(p-1) \times (p-1)$-matrix valued spectral density $g(u), u \in \Pi$, such that for some $C > 0$, $0 \leq \gamma < 1$,

$$\|g(u)\| \leq C|u|^{-\gamma} \qquad u \in \Pi. \tag{11.4.13}$$

We further assume that $\{\mathcal{Y}_j\}$ is independent of $\{\varepsilon_j\}$, implying the independence of the designs and the errors in (11.4.1). Condition (11.4.13) is very general and includes both short memory (in particular, i.i.d.) and long memory random designs.

Because of the difference between the consistency rates of the location parameter estimators and the slope parameter estimators, it is convenient to write $\beta' = (\beta_1, \beta'_2)$, where β_1 is a scalar and β_2 a $(p-1) \times 1$ vector. Similarly, write the estimator $\tilde{\beta}'_n = (\tilde{\beta}_{n1}, \tilde{\beta}'_{n2})$.

Theorem 11.4.2. *Suppose $\{Y_j\}$ follows the regression model (11.4.1) with errors $\{\varepsilon_j\}$ as in (11.4.2) having the spectral density f_{θ_0} and that assumptions (a0') to (a4') hold. Assume that the design process $X_j = (1, \mathcal{Y}'_j)', j \in \mathbb{Z}$, with $E\mathcal{Y}_j \equiv 0$, is independent of errors and (11.4.13) holds. In addition, suppose the regression parameter estimators $\tilde{\beta}_{n1}$ and $\tilde{\beta}_{n2}$ satisfy*

$$\tilde{\beta}_{n1} - \beta_1 = O_p(n^{-\gamma_1}), \quad \text{for some } \gamma_1 > (0 \vee \frac{1-2\eta}{4}), \tag{11.4.14}$$

$$\tilde{\beta}_{n2} - \beta_2 = o_p(n^{-1/4}). \tag{11.4.15}$$

Then the conclusion (11.4.6) holds.

Proof. As in the proof of the previous theorem, we need to verify (11.4.7) and (11.4.8) only. Let

$$\Delta_1 := \tilde{\beta}_{n1} - \beta_1, \quad \Delta_2 := \tilde{\beta}_{n2} - \beta_2, \quad \Delta := \tilde{\beta}_n - \beta = (\Delta_1, \Delta'_2)'.$$

The decomposition (11.4.9) is still valid with obvious modifications. In particular, we can rewrite

$$Q_2(\theta) = n^{-1} \sum_{t,s=1}^{n} b_{t-s}(\theta)\varepsilon_t X_s'(\tilde{\beta}_n - \beta)$$

$$= n^{-1} \sum_{t,s=1}^{n} b_{t-s}(\theta)\varepsilon_t \, \Delta_1 \; + \; n^{-1} \sum_{t,s=1}^{n} b_{t-s}(\theta)\varepsilon_t \mathcal{Y}_s' \Delta_2,$$

$$Q_3(\theta) = n^{-1}(\tilde{\beta}_n - \beta)' \sum_{t,s=1}^{n} b_{t-s}(\theta) X_t X_s' \; (\tilde{\beta}_n - \beta)$$

$$= n^{-1} \sum_{t,s=1}^{n} b_{t-s}(\theta) \, \Delta_1^2 \; + \; 2n^{-1}\Delta_1 \sum_{t,s=1}^{n} b_{t-s}(\theta)\mathcal{Y}_t' \Delta_2$$

$$+ n^{-1}\Delta_2' \sum_{t,s=1}^{n} b_{t-s}(\theta)\mathcal{Y}_t \mathcal{Y}_s' \, \Delta_2.$$

To show (11.4.7), as in the proof of Theorem 11.4.1, we need to prove (11.4.10) and (11.4.11). This in turn is facilitated by Lemmas 11.4.1 and 11.4.2, below, applied for a fixed $j = 1, \cdots, q$, with $q_t := \nabla_j b_t(\theta_0)$. With this notation and using a similar argument as in (11.4.12) together with (11.4.14) and (11.4.15) in mind, (11.4.10) will follow from the following five statements where γ_1 is as in (11.4.14).

$$\sum_{t,s=1}^{n} q_{t-s} = o(n^{(1/2)+2\gamma_1}), \tag{11.4.16}$$

$$\sum_{t,s=1}^{n} q_{t-s} X_{t,k} X_{s,m} = O_p(n), \quad k, m = 2, \cdots, p, \tag{11.4.17}$$

$$\sum_{t,s=1}^{n} q_{t-s} X_{t,k} = O_p(n^{(3/4)+\gamma_1}), \quad k = 2, \cdots, p, \tag{11.4.18}$$

$$\sum_{t,s=1}^{n} q_{t-s}\varepsilon_t X_{s,k} = O_p(n^{3/4}), \quad k = 2, \cdots, p, \tag{11.4.19}$$

$$\sum_{t,s=1}^{n} q_{t-s}\varepsilon_t = O_p(n^{(1/2)+\gamma_1}). \tag{11.4.20}$$

We shall apply Lemmas 11.4.1 and 11.4.2 with $\kappa = \eta$ of (a3').

Consider (11.4.16). By (11.4.24), $\sum_{t,s=1}^{n} q_{t-s} = O(n^{1-\eta})$ where $1 - \eta < (1/2) + 2\gamma_1$ because of the lower bound on γ_1 in (11.4.14).

Claim (11.4.17) is immediate from (11.4.28). Next, consider (11.4.18). Use (11.4.22) with $V_t \equiv 1, \varepsilon_t = X_{t,k}$, and $\alpha = \gamma$ to obtain

$\sum_{t,s=1}^{n} q_{t-s} X_{t,k} = O_p(n^{(\frac{1+\gamma}{2} - \eta) \vee 0}) = O_p(n^{1-\eta})$ since $\gamma < 1$. But $1 - \eta \leq (3/4) + \gamma_1$ by (11.4.14). This proves (11.4.18).

To prove (11.4.19), use (11.4.30) to obtain

$$\sum_{t,s=1}^{n} q_{t-s} \varepsilon_t X_{s,k} = O_p(n^{\frac{\alpha+1}{2} - \eta}) = O_p(n^{\frac{3}{4}}), \quad \eta < \alpha/2,$$

$$= O_p(n^{\frac{1}{2}}), \qquad \text{otherwise,}$$

since $\alpha - \eta < 1/2$, by (a3′).

Finally, (11.4.22) gives $\sum_{t,s=1}^{n} q_{t-s} \varepsilon_t = O_p(n^{(\frac{1+\alpha}{2} - \eta) \vee 0} \log n)$, which proves (11.4.20). Indeed, let $\eta < 1/2$. Then $(1/2) + \gamma_1 > (3/4) - \eta/2 > (1 + \alpha)/2 - \eta$, according to $\alpha - \eta < 1/2$; see (a3′). Moreover, if $\eta < 1/2$ then $(1/2) + \gamma_1 > 1/2 > \max((1 + \alpha)/2 - \eta, 0)$ since $2\eta \geq 1 > \alpha$. This proves (11.4.16) to (11.4.20) and hence (11.4.10).

It remains to show (11.4.11). Since $\Delta = o_p(1)$, (11.4.11) follows from the fact that the supremum over $\theta \in \Theta_0$ of the quadratic forms on the left hand sides of (11.4.16) to (11.4.20) grow at the rate $O_p(n)$. The last fact is immediate from (11.4.28) applied with $\kappa = 0$; see assumption (a2′). (Recall that Theorem 11.4.2 does not assume condition (a5′).) This proves (11.4.11) and also completes the proof of Theorem 11.4.2.

Remark 11.4.3. Theorem 11.4.2 does not assume any specific dependence conditions on the design except (11.4.13). Clearly, this condition allows a broad range of dependence structures, from i.i.d. to long memory.

The dependence structure of the design and the errors have a bearing on the rate of consistency of $\tilde{\beta}_{n2}$ stipulated in the assumption (11.4.15). Clearly, this condition is much weaker than the usual $n^{1/2}$-consistency rate. For example, as pointed out in Remark 11.5.2 below, for i.i.d. designs with zero mean, a class of M-estimators of β_2 is $n^{1/2}$-consistent, which is much faster than the rate $o_p(n^{-1/4})$ of (11.4.15). Robinson and Hidalgo (1997) give several estimators of β_2 that also obtain $n^{1/2}$-consistency rates under some conditions on the designs and long memory errors. For the sake of completeness we describe their estimators briefly. They are like weighted LS estimators: $\hat{\beta}_\varphi := \hat{A}_\varphi^{-1} \hat{b}_\varphi$, where, with $\varphi_r := (2\pi)^{-2} \int_\Pi \varphi(u) \cos(ru) \, du$, $r \in \mathbb{Z}$,

$$\hat{A}_\varphi := n^{-1} \sum_{j=1}^{n} \sum_{k=1}^{n} (X_j - \bar{X})(X_k - \bar{X})' \varphi_{j-k},$$

$$\hat{b}_\varphi := n^{-1} \sum_{j=1}^n \sum_{k=1}^n (X_j - \bar{X})(Y_k - \bar{Y})\varphi_{j-k}.$$

For example, $\varphi \equiv 1$ gives the LS estimator of β and $\varphi(\cdot) \equiv 1/f_{\hat{\theta}}(\cdot)$, where $\hat{\theta}$ is a consistent estimator of θ_0, gives a generalized LS estimator that is asymptotically efficient for Gaussian errors. Any of these estimators could provide a candidate for $\tilde{\beta}_{n2}$, under appropriate assumptions. Nielsen (2005) gives some further results in this direction.

In Section 11.2, we showed that condition (11.4.14) is satisfied by a large class of estimators of the location parameter; see (11.2.12) and (11.2.32).

We now state and prove the two lemmas of somewhat general interest. They are used in the proof of Theorems 11.4.1 and 11.4.2. To state these lemmas we need some more notation. Let Ω be a compact subset of a Euclidean space, $\hat{q}(x, \omega)$, $\omega \in \Omega$, be a family of functions in $L_2(\Pi)$, $q_j(\omega) := \int_\Pi e^{ijx} \hat{q}(x, \omega) dx$, $j \in \mathbb{Z}$, $V_j \equiv V_{nj}$, $j = 1, \cdots, n$ be as in (11.4.4) and let

$$R_1(\omega) := \sum_{t,s=1}^n q_{t-s}(\omega) V_t \varepsilon_s, \quad R_2(\omega) := \sum_{t,s=1}^n q_{t-s}(\omega) V_t V_s'.$$

Lemma 11.4.1. *Let $\{\varepsilon_j, j \in \mathbb{Z}\}$ be a covariance stationary process with zero mean and spectral density f. Suppose $q_j(\omega)$, $j \in \mathbb{Z}$, $\omega \in \Omega$, are real coefficients. Assume that there exist $0 \le \alpha < 1$, $0 \le \kappa < 1$ and constant $C > 0$, all independent of ω, such that*

$$|f(x)| \le C|x|^{-\alpha}, \quad \sup_{\omega \in \Omega} |\hat{q}(x, \omega)| \le C|x|^\kappa, \quad \forall x \in \Pi. \tag{11.4.21}$$

Then, for any $\omega \in \Omega$,

$$\|R_1(\omega)\| = \begin{cases} O_p(n^{\frac{1+\alpha}{2}-\kappa}), & 1 + \alpha - 2\kappa > 0, \\ O_p(\log n), & 1 + \alpha - 2\kappa \le 0. \end{cases} \tag{11.4.22}$$

Moreover,

$$\sup_{\omega \in \Omega} \|R_1(\omega)\| = \begin{cases} O_p(n^{\frac{1}{2}+(\frac{\alpha}{2}-\kappa)\vee 0}), & \frac{\alpha}{2} - \kappa \ne 0, \\ O_p(n^{\frac{1}{2}} \log n), & \frac{\alpha}{2} - \kappa = 0, \end{cases} \tag{11.4.23}$$

$$\sup_{\omega \in \Omega} \|R_2(\omega)\| = O(n^{1-\kappa}). \tag{11.4.24}$$

Proof. It suffices to prove the lemma for a scalar-valued $V_j \in \mathbb{R}$. Let

$$\hat{v}_n(x) := n^{-1} \sum_{t=1}^{n} e^{\mathrm{i}(t/n)x} v_n(t/n), \quad x \in \mathbb{R},$$

where $V_j = v_n(j/n)$ and $v_n(z), z \in [0,1]$, satisfy the conditions in (11.4.4). Clearly, the functions $\hat{v}_n$ are uniformly bounded on $\mathbb{R}$: $|\hat{v}_n(x)| \leq C$. Summation by parts with $\Delta v_n(t/n) := v_n(t/n) - v_n((t-1)/n)$ and $v_n((n+1)/n) = v_n(-1/n) := 0$ and the inequality $|1 - e^{\mathrm{i}x}| \geq C|x|, \forall x \in \Pi$, imply

$$|\hat{v}_n(x)| = \left| \frac{1}{n(1 - e^{\mathrm{i}x/n})} \sum_{t=1}^{n} \Delta v_n(t/n) e^{\mathrm{i}(t/n)x} \right| \leq C/|x|, \quad x \in n\Pi,$$

since

$$|\Delta v_n(t/n)| \leq C \begin{cases} 1, & t = 1 \text{ or } t = n+1, \\ n^{-1}, & t = 2, \cdots, n, \end{cases}$$

by assumption (11.4.4) on the design. Therefore, uniformly in x,

$$|\hat{v}_n(x)| \leq C/(1 + |x|), \quad x \in n\Pi. \tag{11.4.25}$$

Consider the claim (11.4.24). We have

$$R_2(\omega) = \int_{\Pi} \sum_{t,s=1}^{n} e^{\mathrm{i}(t-s)x} v_n(t/n) v_n(s/n) \hat{q}(x,\omega) dx$$

$$= n \int_{n\Pi} |\hat{v}_n(x)|^2 \hat{q}(x/n,\omega) dx.$$

Hence, by the bounds on $\hat{v}_n$ in (11.4.25) and $\hat{q}$ in (11.4.21),

$$n^{\kappa-1} \sup_{\omega \in \Omega} |R_2(\omega)| \leq C \int_{\mathbb{R}} \frac{|x|^{\kappa} dx}{(1 + |x|)^2} < \infty.$$

Next, consider (11.4.22). Rewrite

$$R_1(\omega) = \int_{n\Pi} \left(\sum_{t=1}^{n} e^{\mathrm{i}(t/n)y} \varepsilon_t \right) \hat{v}_n(y) \hat{q}(y/n) dy. \tag{11.4.26}$$

Therefore, as above, and using (2.1.8) for $|D_n(x)|$,

$$ER_1^2(\omega) = n^{-1} \int_{n\Pi} f(\frac{x}{n}) dx \left| \int_{n\Pi} \hat{q}(\frac{y}{n},\omega) \hat{v}_n(y) D_n(\frac{x+y}{n}) e^{\mathrm{i}y(n+1)/2n} dy \right|^2$$

$$\leq C n^{\alpha-2\kappa-1} \int_{n\Pi} |x|^{-\alpha} dx \left(\int_{n\Pi} \frac{|y|^{\kappa}}{1+|y|} |D_n(\frac{x+y}{n})| dy \right)^2$$

$$\leq C n^{1+\alpha-2\kappa} \int_{\mathbb{R}} |x|^{-\alpha} dx \left(\int_{\mathbb{R}} \frac{|y|^{\kappa}}{(1+|y|)(1+|x+y|)} dy \right)^2$$

$$=: C n^{1+\alpha-2\kappa} J.$$

Recall the elementary inequality: For all $\kappa \in [0, 1)$,

$$\int_{\mathbb{R}} |y|^\kappa (1 + |y|)^{-1} (1 + |x + y|)^{-1} dy \leq C(1 + |x|)^{\kappa - 1} (1 + \log_+ |x|).$$

This can be derived from Lemma 6.3.2. Hence, $J \leq C \int_{\mathbb{R}} |x|^{-\alpha}(1 + |x|)^{2\kappa - 2}(1 + \log_+ |x|)^2 dx < \infty$, if $1 + \alpha > 2\kappa$, which proves (11.4.22) for $1 + \alpha - 2\kappa > 0$.

Next, consider the case $1 + \alpha - 2\kappa \leq 0$. Without loss of generality, assume $\kappa = (1 + \alpha)/2$. Then the above calculations yield $ER_1^2(\omega) \leq C \int_{n\Pi} |x|^{-\alpha}(1 + |x|)^{\alpha - 1}(1 + \log_+ |x|)^2 dx \leq C \log^2 n$, thereby proving (11.4.22).

Finally, consider (11.4.23). Assumption (11.4.21) and equations (11.4.26) and (11.4.25) imply

$$\sup_{\omega \in \Omega} |R_1(\omega)| \leq C \int_{n\Pi} \Big| \sum_{t=1}^n e^{\mathrm{i}(t/n)y} \varepsilon_t \Big| \frac{|y/n|^\kappa dy}{1 + |y|}$$

$$= Cn \int_{\Pi} \Big| \sum_{t=1}^n e^{\mathrm{i}ty} \varepsilon_t \Big| \frac{|y|^\kappa dy}{1 + n|y|}.$$

Therefore,

$$E \sup_{\omega \in \Omega} |R_1(\omega)| \leq Cn \int_{\Pi} E^{1/2} \Big| \sum_{t=1}^n e^{\mathrm{i}ty} \varepsilon_t \Big|^2 \frac{|y|^\kappa dy}{1 + n|y|} \tag{11.4.27}$$

$$= Cn \int_{\Pi} \Big(\int_{\Pi} D_n^2(x + y) f(x) dx \Big)^{1/2} \frac{|y|^\kappa dy}{1 + n|y|}$$

$$\leq Cn^{\frac{1}{2} + \frac{\alpha}{2} - \kappa} J_n,$$

where

$$J_n := \int_{n\Pi} \Big(\int_{n\Pi} \frac{|x|^{-\alpha} dx}{1 + |x + y|^2} \Big)^{1/2} \frac{|y|^\kappa dy}{1 + |y|}$$

$$\leq C \int_{n\Pi} \frac{dy}{(1 + |y|)|y|^{\alpha/2 - \kappa}} =: C \tilde{J}_n.$$

Clearly,

$$\tilde{J}_n \leq C \begin{cases} 1, & \alpha/2 > \kappa, \\ \log n, & \alpha/2 = \kappa, \\ n^{\kappa - \alpha/2}, & \alpha/2 < \kappa. \end{cases}$$

Together with (11.4.27), this proves (11.4.23) and Lemma 11.4.1 too.

Lemma 11.4.2. *Let $\{\varepsilon_j\}$, f, and $\{q_j(\omega)\}$ be the same as in Lemma 11.4.1, satisfying (11.4.21). Let $\{X_{i,t}, t \in \mathbb{Z}\}, i = 1, 2$, be two covariance stationary processes. Consider the quadratic forms*

$$R_3(\omega) := \sum_{t,s=1}^{n} q_{t-s}(\omega) X_{1,t} \varepsilon_s, \quad R_4(\omega) := \sum_{t,s=1}^{n} q_{t-s}(\omega) X_{1,t} X_{2,s}.$$

Then

$$\sup_{\omega \in \Omega} |R_k(\omega)| = O_p(n), \quad k = 3, 4. \tag{11.4.28}$$

Suppose, in addition, $\{X_{1,t}\}$ is independent of $\{\varepsilon_s\}$ and has the spectral density g such that for some $C > 0$, $0 \le \gamma < 1$,

$$g(u) \le C|u|^{-\gamma}, \quad u \in \Pi.$$

Then, for any $\omega \in \Omega$,

$$R_3(\omega) = \begin{cases} O_p(n^{A/2+\epsilon}), & A := \alpha + \gamma - 2\kappa \ge 1, \\ O_p(n^{1/2}), & A < 1, \end{cases} \tag{11.4.29}$$

where $\epsilon > 0$ is arbitrarily small. In particular,

$$R_3(\omega) = \begin{cases} O_p(n^{\frac{\alpha+1}{2}-\kappa}), & \kappa < \alpha/2, \\ O_p(n^{\frac{1}{2}}), & \kappa \ge \alpha/2. \end{cases} \tag{11.4.30}$$

Proof. It suffices to prove (11.4.28) for $k = 4$. We have

$$\sup_{\omega \in \Omega} |R_4(\omega)| = \sup_{\omega \in \Omega} \left| \int_\Pi \left(\sum_{t,s=1}^{n} e^{\mathbf{i}(t-s)u} X_{1,t} X_{2,s} \right) \hat{q}(u, \omega) du \right|$$

$$\le \sup_{\omega \in \Omega} \int_\Pi \left| \sum_{t=1}^{n} e^{\mathbf{i}tu} X_{1,t} \right| \left| \sum_{s=1}^{n} X_{2,s} e^{\mathbf{i}su} X_{2,s} \right| |\hat{q}(u, \omega)| du$$

$$\le C \sum_{i=1}^{2} \int_\Pi \left| \sum_{t=1}^{n} e^{\mathbf{i}tu} X_{i,t} \right|^2 du,$$

since $\kappa \ge 0$ in (11.4.21).

Recall the spectral representation $\gamma_i(t - s) := \mathrm{Cov}(X_{i,s}, X_{i,t}) = \int_\Pi e^{\mathbf{i}(t-s)u} \varphi_i(du)$, where φ_i is a finite measure on Π, $i = 1, 2$. Thus, with $\mu_i := EX_{i,0}, i = 1, 2$,

$$E \int_\Pi \left| \sum_{t=1}^{n} e^{\mathbf{i}tu} X_{i,t} \right|^2 du$$

$$\le 2 \int_\Pi \left(|\mu_i|^2 \left| \sum_{t=1}^{n} e^{\mathbf{i}tu} \right|^2 + \sum_{t,s=1}^{n} e^{\mathbf{i}(t-s)u} \gamma_i(t - s) \right) du$$

$$= 2 \left(|\mu_i|^2 \int_\Pi D_n^2(u) du + \int_{\Pi^2} D_n^2(u + v) \varphi_i(dv) du \right).$$

Hence, (11.4.28) follows from $\sup_{v \in \Pi} \int_{\Pi} D_n^2(u+v)du \leq Cn$ and the finiteness of φ_i.

To show (11.4.29), note $ER_3(\omega) = 0$ and write the variance of $R_3(\omega)$ as the trace: $\mathrm{Var}(R_3(\omega)) = \mathrm{tr}(Q(\omega)R_{\varepsilon\varepsilon}Q(\omega)R_{X_1X_1})$, where

$$Q(\omega) = \big(q_{t-s}(\omega)\big)_{t,s=1,\cdots,n}, \quad R_{\varepsilon\varepsilon} := \big(E\varepsilon_t\varepsilon_s\big)_{t,s=1,\cdots,n},$$
$$R_{X_1X_1} := \big(\mathrm{Cov}(X_{1,t}, X_{1,s})\big)_{t,s=1,\cdots,n}$$

are Toeplitz matrices. Then (11.4.29) is a direct consequence of Lemma 14.4.1 and Corollary 14.4.1. Claim (11.4.30) follows from (11.4.29) since $(\alpha + \gamma - 2\kappa)/2 + \epsilon < (\alpha + 1)/2 - \kappa$, for $\gamma < 1$ and $\epsilon > 0$ small enough. Lemma 11.4.2 is proved.

Note: More details of the proofs of Theorems 11.4.1 and 11.4.2 can be found in Koul and Surgailis (2000b). Giraitis and Koul (1997) prove the asymptotic normality of maximum likelihood estimators of the underlying parameters in linear regression models with long memory Gaussian errors.

11.5 Non-linear regression with random design

The importance of non-linear regression models is well documented in the literature, see, e.g., Jennrich (1969), Hannan (1973), Bates and Watts (1988), and Seber and Wild (1989). However, in most of the literature the design variables are taken to be either non-random or i.i.d. and the errors are assumed to be either i.i.d. or weakly dependent.

In finance there are examples where both design and error variables have long memory. For example, in the forward premium anomaly we often regress the spot returns on forward premiums. Baillie and Bollerslev (1994) and Maynard and Phillips (2001) have noted the widespread property of forward premium having long memory. There also are instances where the errors could either have long memory or form an uncorrelated sequence.

We are thus motivated to investigate asymptotic distributions of a class of M estimators in non-linear regression models with various dependent probability structures on the explanatory variables and the errors. After proving asymptotic uniform linearity of M scores under fairly general conditions, we then discuss cases where the design has long memory and errors form either a long memory or a martingale-difference sequence.

Let p, q be fixed positive integers, Ω be an open subset of $\mathbb{R}^q$, and let h be a known function on $\mathbb{R}^p \times \Omega$ to $\mathbb{R}$, measurable in the first p coordinates and such that if $h(x, \beta_1) = h(x, \beta_2)$, for all $x \in \mathbb{R}^p$, then $\beta_1 = \beta_2$. In the non-linear regression (NLR) model considered here we observe (X_i, Y_i), $1 \leq i \leq n$, where $X_i := (X_{i1}, X_{i2}, \cdots, X_{ip})'$, $i \in \mathbb{Z}$, are p-dimensional stationary random vectors, and where for some $\beta \in \Omega$,

$$Y_i = h(X_i, \beta) + \varepsilon_i, \qquad i \in \mathbb{Z}, \tag{11.5.1}$$

with $\{\varepsilon_i\}$ being a stationary mean-zero process.

To proceed further consider the following assumption: There exists a q-vector $\dot{h}$ of functions on $\mathbb{R}^p \times \Omega$ and a sequence a_n of positive numbers tending to infinity, such that for every $b_1 \in \Omega$, $0 < k < \infty$,

$$\sup_{1 \leq i \leq n, \|a_n(b-b_1)\| \leq k} \left\{ \frac{|h(X_i, b) - h(X_i, b_1) - (b - b_1)'\dot{h}(X_i, b_1)|}{\|b - b_1\|} \right\} \tag{11.5.2}$$
$$= o_p(1).$$

This condition plays a similar role as the classical Cramér's conditions on the likelihood score function when discussing the asymptotics of maximum likelihood estimators; see Rao (1973). This condition is *a priori* satisfied by any function $h(x, b)$ that is linear in b, which includes polynomial regression models. In general, $\dot{h}$ need not be the derivative of h with respect to b nor does the differentiability of h alone imply this condition as is illustrated by taking $q = 1 = p$, $\Omega = (0, \infty)$, and $h(x, \beta) = \beta^2 x$. In this example, (11.5.2) is satisfied provided $\max_{1 \leq i \leq n} |X_i|/a_n = o_p(1)$, since the l.h.s. of (11.5.2) is bounded above by $k \max_{1 \leq i \leq n} |X_i|/a_n$.

Next, let ψ be a real-valued non-decreasing function on $\mathbb{R}$, $\dot{h}$ be as in (11.5.2), and define the corresponding M score and M estimator:

$$M(b) := \sum_{i=1}^{n} \dot{h}(X_i, b)\psi(Y_i - h(X_i, b)), \quad b \in \Omega, \tag{11.5.3}$$
$$\widehat{\beta} := \operatorname{argmin}_{b \in \Omega} \|M(b)\|.$$

Computational aspects of M estimators for various h were addressed in Seber and Wild (1989). As mentioned before, the choice of $\psi(x) = x$, $\psi(x) = xI(|x| \leq c) + c\operatorname{sgn}(x))I(|x| > c)$, and $\psi(x) = \operatorname{sgn}(x)$, respectively, give the LS, the Huber(c), and the LAD estimators. Note that the first two scores are continuous, in contrast to the last score. Theorem 11.5.1 given below is applicable when ψ is *smooth* in the sense of $(\psi 1)$ given below. The LAD estimator is discussed later.

Condition $(\psi 1)$. The score function ψ is absolutely continuous with $E\psi(\varepsilon_0) = 0$ and $E\psi^2(\varepsilon_0) < \infty$. Its a.e. derivative $\dot{\psi}$ is such that $E|\dot{\psi}(\varepsilon_0)| < \infty$, and $\forall \eta > 0$,

$$\lim_{z \to 0} E\left\{ \|\dot{h}(X_0, \beta)\| \left[\|\dot{h}(X_0, \beta)\| + \eta \right] \left| \dot{\psi}(\varepsilon_0 - z(\|\dot{h}(X_0, \beta)\| + \eta)) - \dot{\psi}(\varepsilon_0) \right| \right\} = 0.$$

We shall also assume the following where $\dot{h}_{ni}(u) := \dot{h}(X_i, \beta + a_n^{-1}u)$.

$$a_n \to \infty, \quad a_n = o(n), \quad b_n := n/a_n. \tag{11.5.4}$$

$$b_n^{-1}\|M(\beta)\| = O_p(1). \tag{11.5.5}$$

$$a_n^{-1} \max_{1 \le i \le n} \|\dot{h}(X_i, \beta)\| = o_p(1). \tag{11.5.6}$$

$$E\|\dot{h}(X_0, \beta)\|^2 < \infty. \tag{11.5.7}$$

$$\max_{1 \le i \le n, \|u\| \le k} \|\dot{h}_{ni}(u) - \dot{h}_{ni}(0)\| = o_p(1), \quad \forall 0 < k < \infty. \tag{11.5.8}$$

$$b_n^{-1}\left\| \sum_{i=1}^{n} \left[\dot{h}_{ni}(u) - \dot{h}_{ni}(0) \right] \psi(\varepsilon_i) \right\| = o_p(1), \quad \forall u \in \mathbb{R}^q. \tag{11.5.9}$$

For every $\epsilon > 0$, there exists $\delta > 0$ and $N < \infty$ such that $\forall k < \infty$, and for every $v \in N_k := \{z \in \mathbb{R}^q; \|z\| \le k\}$, $n > N$,

$$P\left(\sup_{\|u-v\| \le \delta} b_n^{-1}\left\| \sum_{i=1}^{n} \{\dot{h}_{ni}(u) - \dot{h}_{ni}(v)\}\psi(\varepsilon_i) \right\| > \epsilon \right) \le \epsilon. \tag{11.5.10}$$

Examples where the above assumptions (11.5.5) to (11.5.10) are satisfied are discussed in Remarks 11.5.1 to 11.5.3 below.

We are now ready to state a general result as far as the structure of the design and error variables is concerned. It reduces the question about the limit distribution of the M estimator $\hat{\beta}$ in the NLR model in (11.5.1) to a much simpler one about the limit distribution of the additive functional $M(\beta) = \sum_{i=1}^{n} \dot{h}(X_i, \beta)\psi(\varepsilon_i)$. The latter question for (independent) long memory errors and designs is discussed in Proposition 11.5.1 below.

Let $\lambda_1 := -E\dot{\psi}(\varepsilon_0)$. Consider the $q \times q$-matrix:

$$H := E\dot{h}(X_0, \beta)\dot{h}'(X_0, \beta). \tag{11.5.11}$$

Theorem 11.5.1. *Assume that $\{X_i, Y_i; i \in \mathbb{Z}\}$ is as in (11.5.1), X_i is independent of ε_i for each $i \in \mathbb{Z}$, and that $\{(X_i, \varepsilon_i); i \in \mathbb{Z}\}$ is strictly* *stationary and ergodic. In addition, suppose (11.5.2), (ψ1), and (11.5.4) to (11.5.10) hold. Then, with $\dot{h}$ as in (11.5.2),*

$$n^{-1} \sum_{i=1}^{n} \dot{h}(X_i, \beta) \dot{h}'(X_i, \beta) = H + o_p(1), \qquad (11.5.12)$$

and, for every $0 < k < \infty$,

$$\sup_{\|u\| \leq k} \left\| b_n^{-1} [M(\beta + a_n^{-1} u) - M(\beta)] + H u \lambda_1 \right\| = o_p(1). \qquad (11.5.13)$$

If, in addition, the matrix H is non-singular, $\lambda_1 > 0$, $E\psi(\varepsilon_0) = 0$, and $\widehat{\beta}$ of (11.5.3) satisfies

$$\|a_n(\widehat{\beta} - \beta)\| = O_p(1), \qquad (11.5.14)$$

then

$$a_n(\widehat{\beta} - \beta) = -\lambda_1^{-1} b_n^{-1} H^{-1} M(\beta) + o_p(1). \qquad (11.5.15)$$

Proof. Clearly, (11.5.12) follows from (11.5.7) and the ergodic Theorem 2.5.2(i). To prove (11.5.13), let

$$D_{ni}(u) := h(X_i, \beta + a_n^{-1} u) - h(X_i, \beta), \qquad (11.5.16)$$
$$\dot{D}_{ni}(u) := \dot{h}_{ni}(u) - \dot{h}_{ni}(0), \qquad u \in \mathbb{R}^q,$$
$$g(x) := k\left(\|\dot{h}(x, \beta)\| + \epsilon\right), \qquad x \in \mathbb{R}^p, \quad g_{ni} := a_n^{-1} g(X_i), \quad 1 \leq i \leq n,$$
$$A_n := \{\sup_{i,u} \|D_{ni}(u) - a_n^{-1} u' \dot{h}_i\| \leq k\epsilon a_n^{-1}\}, \quad \epsilon > 0, 0 < k < \infty.$$

Here, and in the proof below, the indices i and u in $\sup_{i,u}$ vary over $1 \leq i \leq n$, $u \in N_k := \{u \in \mathbb{R}^q; \|u\| \leq k\}$. Also write h_i, $\dot{h}_i$ for $h(X_i, \beta)$, $\dot{h}(X_i, \beta)$, respectively. By (11.5.2), there exists an n_1 such for all $n > n_1$,

$$P(A_n) > 1 - \epsilon.$$

On A_n,

$$\|D_{ni}(u)\| \leq g_{ni}, \quad \text{for all } 1 \leq i \leq n, \ u \in N_k. \qquad (11.5.17)$$

But, by (11.5.6), $\max_i g_{ni} = o_p(1)$, and hence

$$\sup_{i,u} \|D_{ni}(u)\| = o_p(1).$$

Now, let $Z_n(u) := b_n^{-1}[M(\beta + a_n^{-1}u) - M(\beta)]$. Rewrite

$$Z_n(u) = b_n^{-1} \sum_{i=1}^{n} \dot{D}_{ni}(u)\,\psi(\varepsilon_i - D_{ni}(u)) \qquad (11.5.18)$$

$$+ b_n^{-1} \sum_{i=1}^{n} \dot{h}_i\,\{\psi(\varepsilon_i - D_{ni}(u)) - \psi(\varepsilon_i)\},$$

$$= Z_{n1}(u) + Z_{n2}(u), \quad \text{say.}$$

Consider the second term first. Rewrite

$$Z_{n2}(u) = b_n^{-1} \sum_{i=1}^{n} \dot{h}_i\,\{\psi(\varepsilon_i - D_{ni}(u)) - \psi(\varepsilon_i) - D_{ni}(u)\dot{\psi}(\varepsilon_i)\}$$

$$+ b_n^{-1} \sum_{i=1}^{n} \dot{h}_i\{D_{ni}(u) - \dot{h}_i'a_n^{-1}\,u\}\dot{\psi}(\varepsilon_i) + b_n^{-1} \sum_{i=1}^{n} \dot{h}_i\dot{h}_i'a_n^{-1}\dot{\psi}(\varepsilon_i)u$$

$$= Z_{n21}(u) + Z_{n22}(u) + Z_{n23}\,u, \quad \text{say.}$$

The independence of X_i and ε_i, (11.5.4), (11.5.12), and the ET imply

$$Z_{n23} = n^{-1} \sum_{i=1}^{n} \dot{h}_i\dot{h}_i'\dot{\psi}(\varepsilon_i) + o_p(1) = H\,\lambda_1 + o_p(1).$$

Next, on A_n, for every $n > n_1$,

$$\sup_u \|Z_{n22}(u)\| \le k\epsilon n^{-1} \sum_{i=1}^{n} \|\dot{h}_i\|\,|\dot{\psi}(\varepsilon_i)|,$$

so that, by the ET, (11.5.7), and $(\psi 1)$,

$$\sup_u \|Z_{n22}(u)\| = \epsilon\,O_p(1).$$

To deal with $Z_{n21}(u)$, use the absolute continuity of ψ and a standard argument to obtain, in view of (11.5.17), that on A_n,

$$\sup_u \|Z_{n21}(u)\| \le b_n^{-1} \int_{-a_n^{-1}}^{a_n^{-1}} \sum_{i=1}^{n} \|\dot{h}(X_i, \beta)\|\,g(X_i)\,|\dot{\psi}(\varepsilon_i - zg(X_i)) - \dot{\psi}(\varepsilon_i)|dz,$$

for all $n > n_1$. But, by condition $(\psi 1)$ and (11.5.4), the expected value of this upper bound equals

$$b_n^{-1}n \int_{-a_n^{-1}}^{a_n^{-1}} E\Big\{\|\dot{h}(X_0, \beta)\|\,g(X_0)\,|\dot{\psi}(\varepsilon_0 - zg(X_0)) - \dot{\psi}(\varepsilon_0)|\Big\}dz = o(1).$$

This fact, the above derivations, and the arbitrariness of ϵ give

$$\sup_u \|Z_{n2}(u) - Hu\,\lambda_1\| = o_p(1).$$

It remains to show that $Z_{n1}(u)$ of (11.5.18) is negligible. Rewrite

$$Z_{n1}(u) = b_n^{-1} \sum_{i=1}^{n} \dot{D}_{ni}(u) \left\{ \psi(\varepsilon_i - D_{ni}(u)) - \psi(\varepsilon_i) \right\} + b_n^{-1} \sum_{i=1}^{n} \dot{D}_{ni}(u)\psi(\varepsilon_i)$$

$$= Z_{n11}(u) + Z_{n12}(u), \qquad \text{say.}$$

By (11.5.9), $Z_{n12}(u) = o_p(1)$, for every $u \in \mathbb{R}^q$. Also, $\forall\, u,\, v \in N_k$,

$$Z_{n12}(u) - Z_{n12}(v) = b_n^{-1} \sum_{i=1}^{n} \left\{ \dot{h}_{ni}(u) - \dot{h}_{ni}(v) \right\} \psi(\varepsilon_i).$$

Hence, (11.5.10), the compactness of N_k, and a routine argument imply

$$\sup_u \left\| Z_{n12}(u) \right\| = o_p(1). \tag{11.5.19}$$

Now, consider $Z_{n11}(u)$. Again, by the absolute continuity of ψ, on A_n,

$$\sup_u \left\| Z_{n11}(u) \right\| \le b_n^{-1} \sum_{i=1}^{n} \left\| \dot{D}_{ni}(u) \right\| \int_{-a_n^{-1}}^{a_n^{-1}} g(X_i)\, |\dot{\psi}(\varepsilon_i - sg(X_i))|\, ds,$$

$$\le \sup_{i,u} \left\| \dot{D}_{ni}(u) \right\| \cdot \frac{a_n}{n} \sum_{i=1}^{n} \int_{-a_n^{-1}}^{a_n^{-1}} g(X_i)|\dot{\psi}(\varepsilon_i - zg(X_i))|\, dz.$$

But, by condition $(\psi 1)$, the expected value of the second factor in this upper bound is $O(1)$. Hence, by (11.5.8),

$$\sup_u \left\| Z_{n11}(u) \right\| \le \sup_{i,u} \left\| \dot{D}_{ni}(u) \right\| O_p(1) = o_p(1).$$

This, together with the above derivations completes the proof of (11.5.13), and that of Theorem 11.5.1.

We emphasize that in this theorem $\{(X_i, \varepsilon_i),\ i \in \mathbb{Z}\}$ is a general stationary ergodic process and X_i is independent of ε_i for each $i \in \mathbb{Z}$. Neither of the two processes need to have either long or short memory.

For the LS score $\psi(x) \equiv x$, the assumptions $E\varepsilon_0 = 0$ and $E\varepsilon_0^2 < \infty$ imply condition $(\psi 1)$ trivially. In the case of the Huber(c) score, $(\psi 1)$ is satisfied as long as the d.f. F of ε_0 is continuous at c.

Remark 11.5.1. (*On condition* (11.5.14)). A sufficient condition for (11.5.14) to hold is as follows. For every $\delta > 0$ and $0 < \epsilon < \infty$, there exists an N_δ and a $K \equiv K_{\delta,\epsilon}$ such that

$$P(\inf_{\|u\|>K} \|b_n^{-1} M(\beta + a_n^{-1}u)\| \ge \epsilon) \ge 1 - \delta, \qquad n > N_\delta.$$

This condition is known to hold for a non-decreasing ψ in the models where β appears linearly in the regression function h; see, e.g., Subsection 11.2.2.

Remark 11.5.2. (*On conditions* (11.5.4), (11.5.8), (11.5.9), *and* (11.5.10)). The latter three conditions are vacuously satisfied by a linear regression model, i.e., when $p = q$ and $h(x, \beta) = x'\beta$, regardless of any probabilistic structure on $\{X_i, \varepsilon_i\}$. In general, the satisfaction of these conditions requires more information either on the underlying probability structure or on the regression function.

Consider, for example, the following assumptions where $\dot{D}_{ni}$ is as in (11.5.16): Suppose there exists a $q \times q$ matrix of real-valued functions $\ddot{h}$ defined on $\mathbb{R}^p \times \Omega$ such that $E\|\ddot{h}(X, \beta)\|^2 < \infty$, and for every $k < \infty$,

$$\sup_{1 \le i \le n, \|u\| \le k} \left\| \frac{\dot{D}_{ni}(u)}{a_n^{-1}} - u'\ddot{h}(X_i, \beta) \right\| = o_p(1). \tag{11.5.20}$$

Then, under (11.5.4) and the assumption $E\psi(\varepsilon_0) = 0$, both (11.5.9) and (11.5.10) hold. This follows from the ET, (11.5.4), and the decomposition

$$b_n^{-1} \sum_{i=1}^n \dot{D}_{ni}(u)\psi(\varepsilon_i) = b_n^{-1} \sum_{i=1}^n \left\{ \dot{D}_{ni}(u) - a_n^{-1}u'\ddot{h}(X_i, \beta) \right\} \psi(\varepsilon_i)$$

$$+ u'n^{-1} \sum_{i=1}^n \ddot{h}(X_i, \beta)\psi(\varepsilon_i).$$

If, in addition, $a_n^{-1} \max_{1 \le i \le n} \|\ddot{h}(X_i, \beta)\| = o_p(1)$, then (11.5.8) also holds.

Alternatively, suppose the errors $\{\varepsilon_i\}$ and ψ are such that $E\psi(\varepsilon_0) = 0$, $E\psi^2(\varepsilon_0) < \infty$, and $E\{\psi(\varepsilon_j)|(\varepsilon_{i-1}, X_i) : i \le j\} = 0$, for all $j \in \mathbb{Z}$. In addition, suppose $\{X_i\}$ obey the following long memory model: For some $0 < d_1 < 1/2$,

$$\Gamma_k := \text{Cov}(X_i, X'_{i+k}) = k^{-1+2d_1}\mathcal{L}(k), \quad i \in \mathbb{Z}, \, k \ge 1, \tag{11.5.21}$$

where $\mathcal{L}$ is a $p \times p$ matrix of slowly varying functions at infinity and $\mathcal{L}(k)$ is positive definite for all large k. Then the choice of $a_n = n^{1/2} = b_n$, and the assumptions (11.5.7), (11.5.8), and

$$E\|\dot{h}(X, \beta + a_n^{-1}u) - \dot{h}(X, \beta)\|^2 = o(1), \quad \forall u \in \mathbb{R}^q, \tag{11.5.22}$$

imply (11.5.5), (11.5.4), (11.5.6), and (11.5.9). Moreover, $n^{-1/2}M(\beta)$ is asymptotically normal with mean zero and the covariance matrix $H E\psi^2(\varepsilon_0)$. If, in addition (11.5.10) holds, then we obtain the classical result, $n^{1/2}(\hat{\beta} - \beta) \to_D \mathcal{N}_q(0, H^{-1} E\psi^2(\varepsilon_0)/(\int f d\psi)^2)$. In particular, these results hold when $\{X_i\}$ obey the long memory model (11.5.21) and $\{\varepsilon_i\}$

are i.i.d. Thus, the uncorrelatedness of the process $\{\psi(\varepsilon_i)\}$ overwhelms the long memory of the covariate process $\{X_i\}$. The model given by (11.5.1) and (11.5.21) includes NLR models with fractional Gaussian and fractional ARFIMA designs.

Remark 11.5.3. (*On conditions* (11.5.5), (11.5.4), *and* (11.5.9) *under long memory*). When $\{X_i, \varepsilon_i\}$ follow the long memory conditions (11.5.21), (10.2.19), and (10.2.20), the satisfaction of these conditions is not straightforward, unless h is assumed to be additionally smooth, e.g., as in (11.5.20) above. In general, the relevant problem from the current perspective is to assess the magnitude of sums of the type

$$S := \sum_{i=1}^{n} \varphi_n(X_i)\psi(\varepsilon_i), \tag{11.5.23}$$

where φ_n, $n \geq 1$, is a sequence of functions from $\mathbb{R}^p$ to $\mathbb{R}^q$ such that $E\|\varphi_n(X_0)\|^2 < \infty$, for all $n \geq 1$. Note that on taking $\varphi_n(X_i) \equiv \dot{D}_{ni}(u)$, S equals the sum inside the norm on the l.h.s. of (11.5.9), while the choice $\varphi_n(X_0) \equiv \dot{h}(X_0, \beta)$ gives S equal to $M(\beta)$. Assessing the magnitude of S ensures the satisfaction of (11.5.9) in the former case and that of (11.5.5) in the latter case.

We thus need to assess the magnitude of S under the following two conditions:

$$\text{(a)} \qquad E\|\varphi_n(X_0)\|^2 = o(1); \quad \text{(b)} \qquad \varphi_n(X_0) \equiv \varphi(X_0), \tag{11.5.24}$$

where φ is an $\mathbb{R}^q$-valued function with $E\|\varphi(X_0)\|^2 < \infty$.

To illustrate the essence of the argument, suppose $p = 1$ and that $\{X_j\}$ and $\{\varepsilon_j\}$ are standardized Gaussian moving averages

$$X_j = \sum_{i=0}^{\infty} g_i \xi_{j-i}, \quad \varepsilon_j = \sum_{i=0}^{\infty} w_i \zeta_{j-i}, \tag{11.5.25}$$

$$g_j = L_1(j) j^{d_1 - 1}, \qquad w_j = L(j) j^{d-1},$$

where $0 < d_1, d < 1/2$, $L_1, L \in SV$, and where $\{\xi_j\}$ and $\{\zeta_j\}$ are mutually independent i.i.d. $\mathcal{N}(0,1)$ r.v.'s so that the processes $\{X_j, j \in \mathbb{Z}\}$ and $\{\varepsilon_j, j \in \mathbb{Z}\}$ are independent. By (11.5.25), as $k \to \infty$,

$$\gamma_X(k) = O(k^{-1+2d_1} L_1^2(k)), \qquad \gamma_\varepsilon(k) = O(k^{-1+2d} L^2(k)). \tag{11.5.26}$$

Decompose $S = S_1 + S_2$, where $S_j := \sum_{i=1}^{n} U_{ij}$, $j = 1, 2$,

$$U_{i1} := (\varphi_n(X_i) - E\varphi_n(X_0))\psi(\varepsilon_i), \quad U_{i2} := \psi(\varepsilon_i)E\varphi_n(X_0).$$

Then

$$EU_{01}U_{k1} = \text{Cov}(\varphi_n(X_0), \varphi_n(X_k))\text{Cov}(\psi(\varepsilon_0), \psi(\varepsilon_k)) \qquad (11.5.27)$$
$$= O(\gamma_X(k)\gamma_\varepsilon(k)),$$
$$EU_{02}U_{k2} = (E\varphi_n(X_0))^2\text{Cov}(\psi(\varepsilon_0), \psi(\varepsilon_k)) = (E\varphi_n(X_0))^2 O(\gamma_\varepsilon(k)).$$

Note that $\{U_{i2}\}$ is a long memory process with memory parameter d, modulus a constant $E\varphi_n(X_0) \in \mathbb{R}^q$, which may vanish with n, while the process $\{U_{i1}\}$ has either long memory or absolutely summable correlations, depending on whether $d + d_1 > 1/2$ or $d + d_1 < 1/2$. Accordingly, let

$$b_n := n^{1/2}, \qquad\qquad\quad \text{if } E\varphi_n(X_0) \equiv 0, \ d + d_1 < 1/2, \quad (11.5.28)$$
$$:= n^{d+d_1}L(n)L_1(n), \quad \text{if } E\varphi_n(X_0) \equiv 0, \ d + d_1 > 1/2,$$
$$:= n^{1/2+d}L(n), \qquad\quad \text{if } E\varphi_n(X_0) \not\equiv 0.$$

The above derivation and $\|E\varphi_n(X_0)\|^2 \le E\|\varphi_n(X_0)\|^2$, imply

$$b_n^{-1}\|S\| = O_p(E^{1/2}\|\varphi_n(X_0)\|^2) \qquad\qquad\qquad (11.5.29)$$
$$= o_p(1), \quad \text{under } (11.5.24)(a),$$
$$= O_p(1), \quad \text{under } (11.5.24)(b).$$

Now, we apply this result to $\varphi_n(X_i) = \dot{D}_{ni}(u)$. Note $E\varphi_n(X_0) \equiv 0$ if $E(\dot{h}(X_0, \beta + v) - \dot{h}(X_0, \beta)) = 0$, for all $v \in \mathbb{R}^p$ in a sufficiently small neighborhood of $v = 0$, and $E\varphi_n(X_0) \not\equiv 0$ otherwise. Let

$$a_n := nb_n^{-1} \qquad\qquad\qquad\qquad\qquad\qquad\qquad\qquad (11.5.30)$$
$$= n^{1/2}, \qquad\qquad\qquad E\varphi_n(X_0) \equiv 0, \ d + d_1 < 1/2, \quad (11.5.31)$$
$$= n^{1-d_1-d}L_1^{-1}(n)L^{-1}(n), \quad E\varphi_n(X_0) \equiv 0, \ d + d_1 > 1/2,$$
$$= n^{(1/2)-d}L^{-1}(n), \qquad\quad E\varphi_n(X_0) \not\equiv 0.$$

Then, from (11.5.29), the assumption (11.5.22) readily implies (11.5.9).

Next, we apply (11.5.29) and the above argument to $\varphi(x) \equiv \dot{h}(x, \beta)$ independent of n. It is easy to see that in this case the magnitude of $S = M(\beta)$ and its asymptotic distribution depends on whether $E\dot{h}(X_0, \beta) \neq 0$ or $E\dot{h}(X_0, \beta) = 0$ and, in the latter case, whether $d+d_1 < 1/2$ or $d+d_1 > 1/2$. We thus have the following

Proposition 11.5.1. *Suppose $\{X_i, i \in \mathbb{Z}\}$ and $\{\varepsilon_i, i \in \mathbb{Z}\}$ are two independent univariate standardized long memory Gaussian processes as in*

(11.5.25). In addition, suppose $E\psi(\varepsilon_0) = 0$, $E\psi^2(\varepsilon_0) < \infty$, and (11.5.2) and (11.5.7) hold. Then the following conclusions hold:
(i) *If $E\dot{h}(X_0, \beta) = 0$ and $d + d_1 < 1/2$ then*

$$n^{-1/2}M(\beta) = O_p(1). \qquad (11.5.32)$$

(ii) *If $E\dot{h}(X_0, \beta) = 0$ and $d + d_1 > 1/2$ then*

$$n^{-(d+d_1)}(L(n)L_1(n))^{-1}M(\beta) \qquad (11.5.33)$$

$$= -J_1 c_1 n^{-(d+d_1)}(L(n)L_1(n))^{-1} \sum_{i=1}^{n} X_i \varepsilon_i + o_p(1),$$

where $J_1 := -E(\psi(\varepsilon_0)\varepsilon_0)$, $c_1 := E(\dot{h}(X_0, \beta)X_0)$.
(iii) *If $E\dot{h}(X_0, \beta) \neq 0$ then*

$$n^{-d-1/2}L^{-1}(n)M(\beta) = -J_1 c_0\, n^{-d-1/2}L^{-1}(n) \sum_{i=1}^{n} \varepsilon_i + o_p(1), \quad (11.5.34)$$

where J_1 is the same as in (ii) and $c_0 := E\dot{h}(X_0, \beta)$.

Proof. Part (i) follows immediately from (11.5.27). Parts (ii) and (iii) follow by noting that $c_0, c_1, J_0 = E\psi(\varepsilon_0) = 0$, and $-J_1$ are nothing other than the coefficients up to order ≤ 2 of the Hermite expansions of $\dot{h}(X_i, \beta)$ and $\psi(\varepsilon_i)$, respectively; see Remark 10.3.1. This implies that the covariances of the "remainders" $\{\dot{h}(X_i, \beta) - c_0 - c_1 X_i\}$ and $\{\psi(\varepsilon_i) + J_1 \varepsilon_i\}$ decay faster than the covariances of $\{X_i\}$ and $\{\varepsilon_i\}$, see Proposition 4.6.1, leading to the approximations in (11.5.33) and (11.5.34).

Notice that Proposition 11.5.1 extends to vector-valued regressors $X_j \in \mathbb{R}^p$, under similar assumptions for the covariance matrix $\gamma_X(k)$, and the limit distribution of $M(\beta)$ can be identified in each of the cases (i) to (iii). We do not include this generalization since its proof uses multivariate Hermite expansions and the discussion in Chapter 4 is limited to the univariate case and scalar-valued Gaussian processes $\{X_j\}$. Moreover, Proposition 11.5.1 can be extended to non-Gaussian linear processes $\{X_i\}$ and $\{\varepsilon_i\}$ with respective memory parameters $d, d_1 \in (0, 1/2)$ under regularity assumptions on $\dot{h}(\cdot, \beta)$ and ψ, and moment assumptions on the innovations of these linear processes. In the latter case, c_0, c_1, and λ_1 should be replaced by Appell coefficients $\lambda_1 = -\int f d\psi$, $c_0 = E\dot{h}(X_0, \beta)$, and $c_1 = \int g(x)\dot{h}(dx, \beta)$, where f, g are the probability densities of ε_0, X_0, respectively, assuming that these integrals exist. Again, such an extension to multivariate linear processes would be based on a generalization of the asymptotic results of Chapter 10, e.g., to vector-valued linear processes and unbounded functions.

However, for the *linear* regression model with long memory moving-average designs and errors, proofs of the analogs of the above results when $p > 1$ are relatively simple. Suppose now (11.5.1) and (11.5.21) hold with $p = q \geq 1$ and $h(x, \beta) = x'\beta$. In this case $M(\beta) = \sum_{i=1}^{n} X_i \psi(\varepsilon_i)$. Then (11.5.2) and (11.5.7) to (11.5.10) are all trivially satisfied as long as $E\|X_0\|^2 < \infty$. Moreover, for linear long memory regressors and errors, the analogs of (11.5.32) to (11.5.34) also hold.

More precisely, suppose the errors form a long memory linear process

$$\varepsilon_j = \sum_{k=0}^{\infty} w_k \zeta_{j-k}, \quad w_k = k^{-1+d} L(k), \ L \in SV, \ 0 < d < 1/2, \quad (11.5.35)$$

where $\{\zeta_j\} \sim \mathrm{IID}(0, 1)$ and satisfies the same conditions as needed for the URP in Subsection 10.2.2, namely,

$$|Ee^{iu\zeta_0}| \leq C(1 + |u|)^{-\delta}, \quad E|\zeta_0|^{2+\delta} < \infty, \quad \exists \, C, \delta > 0. \quad (11.5.36)$$

Assume that the regressors form a $\mathbb{R}^p$-valued long memory linear process

$$X_j = \mu + \sum_{k=0}^{\infty} g_k \xi_{j-k}, \quad j \in \mathbb{Z}, \quad (11.5.37)$$

where $\mu \in \mathbb{R}^p$, and $\{\xi_j = (\xi_{j,1}, \cdots, \xi_{j,r})'\}$ is an i.i.d. sequence of $(p-1)$-dimensional random vectors with values in $\mathbb{R}^{p-1}$, with zero mean $E\xi_{j,i} = 0$, and unit variance $E\xi_{j,i}^2 = 1$, $i = 1, \cdots, p-1$, and g_k are $p \times (p-1)$ matrices having entries of the form

$$g_{k,ij} = k^{-1+d_1} L_{ij}(k), \quad L_{ij} \in SV, \ 0 < d_1 < 1/2; \quad (11.5.38)$$

moreover, there exists $L_1 \in SV$ such that

$$\lim_{k \to \infty} L_{ij}(k)/L_1(k) < \infty, \quad i = 1, \cdots, p, \ j = 1, \cdots, p-1. \quad (11.5.39)$$

We also assume that the sequences $\{\zeta_j\}$ and $\{\xi_j\}$ are independent. Then for $M(\beta) = \sum_{i=1}^{n} X_i \psi(\varepsilon_i)$ we have the following statement.

Proposition 11.5.2. *In addition to (11.5.35) to (11.5.39), assume that ψ has bounded variation.*
(i) *If $\mu = EX_0 = 0$ and $d + d_1 < 1/2$ then $M(\beta) = O_p(n^{1/2})$.*

(ii) *If $\mu = EX_0 = 0$ and $d + d_1 > 1/2$ then*

$$\frac{1}{n^{d+d_1} L(n) L_1(n)} M(\beta) = -\frac{\lambda_1}{n^{d+d_1} L(n) L_1(n)} \sum_{i=1}^{n} X_i \varepsilon_i + o_p(1).$$

(iii) *If $\mu = EX_0 \neq 0$ then*

$$\frac{1}{n^{d+1/2} L(n)} M(\beta) = -\frac{\lambda_1 \mu}{n^{d+1/2} L(n)} \sum_{i=1}^{n} \varepsilon_i + o_p(1).$$

Proof. It suffices to discuss the case $p = 2$ only. Then the statements in (i) to (iii) follow by writing $\mathrm{Cov}(X_0 \psi(\varepsilon_0), X_j \psi(\varepsilon_j)) = \mu \mathrm{Cov}(\psi(\varepsilon_0), \psi(\varepsilon_j)) + \mathrm{Cov}(X_0, X_j) \mathrm{Cov}(\psi(\varepsilon_0), \psi(\varepsilon_j))$ and using the properties of the covariance of $\{\psi(\varepsilon_j)\}$ in Remark 10.2.5, as well as the form of the covariances of $\{X_j\}$ and $\{\varepsilon_j\}$; see (11.5.26). $\blacksquare$

Asymptotic distribution of $\widehat{\beta}$. From the discussion in Remark 11.5.3, we can now deduce the asymptotic distribution of $\widehat{\beta}$ for various situations as follows. Recall that $c_0 := E\dot{h}(X_0, \beta) \in \mathbb{R}^q$, $c_1 := E\dot{h}(X_0, \beta) X_0 \in \mathbb{R}^q$, $\lambda_1 = -E\dot{\psi}(\varepsilon_0) = -\int f d\psi \neq 0$, H is the $q \times q$ matrix of (11.5.11), and let $\tilde{\Sigma}_0 := \sum_{j \in \mathbb{Z}} E\psi(\varepsilon_0)\psi(\varepsilon_j) E\dot{h}(X_0, \beta)\dot{h}(X_j, \beta)'$.

Proposition 11.5.3. *Suppose (11.5.1) holds with $p = 1$, where $\{X_j\}$ and $\{\varepsilon_j\}$ are independent standardized univariate Gaussian long memory processes. In addition, suppose $E\psi(\varepsilon_0) = 0$, $E\psi^2(\varepsilon_0) < \infty$, (11.5.2), (11.5.7), and the conclusion (11.5.15) of Theorem 11.5.1 hold with a_n and b_n as in (11.5.30) and (11.5.28), respectively. Then the following conclusions hold:*
(i) *If $c_0 = 0$ and $d + d_1 < 1/2$ then*

$$n^{1/2}(\widehat{\beta} - \beta) \to_D \mathcal{N}_q(0, \Sigma_0),$$

where $\Sigma_0 := \lambda_1^{-2} H^{-1} \tilde{\Sigma}_0 H^{-1}$.
(ii) *If $c_0 = 0$ and $d + d_1 > 1/2$ then*

$$n^{1-d-d_1} (L(n) L_1(n))^{-1} (\widehat{\beta} - \beta) \to_D H^{-1} c_1 \mathcal{W},$$

where $\mathcal{W}$ is a double Wiener–Itô integral as defined in (11.5.48) below.
(iii) *If $c_0 \neq 0$ then*

$$n^{-d+1/2} L^{-1}(n)(\widehat{\beta} - \beta) \to_D \mathcal{N}_q(0, \Sigma_1),$$

where $\Sigma_1 := 2B(d, 1 - 2d) H^{-1} c_0 c_0' H^{-1}$.

Proof. Part (iii) is immediate from Proposition 11.5.1(iii) and Theorem 4.3.1. Part (ii) follows from Propositions 11.5.1(ii) and 11.5.6 below.

By Proposition 11.5.1(i), the proof of part (i) reduces to proving

$$n^{-1/2}M(\beta) = n^{-1/2}\sum_{j=1}^{n} \dot{h}(X_j, \beta)\psi(\varepsilon_j) \to_D \mathcal{N}(0, \tilde{\Sigma}_0). \qquad (11.5.40)$$

By Proposition 4.2.4, it suffices to prove (11.5.40) for $q = 1$. In the rest of the proof, we use the notation $\varphi(x) = \dot{h}(x, \beta)$ and $S_n := M(\beta) = \sum_{j=1}^{n}\varphi(X_j)\psi(\varepsilon_j)$ as in (11.5.23). Since $\{X_j\}$ and $\{\varepsilon_j\}$ are mutually independent standardized Gaussian r.v.'s and $E\varphi(X_0) = 0 = E\psi(\varepsilon_0)$, we can expand $\varphi(X_j)$ and $\psi(\varepsilon_j)$ using Hermite polynomials. Using the properties of Hermite polynomials, as in the proof of Theorem 4.6.1, it suffices to prove (11.5.40) for a finite sum of the Hermite polynomials, namely, for

$$\varphi(X_j)\psi(\varepsilon_j) = \sum_{k,i=1}^{K} \frac{(-1)^i c_k \lambda_i}{k!i!} H_k(X_j)H_i(\varepsilon_j), \quad \text{for an arbitrary } K < \infty,$$

$$c_k := E\varphi(X_0)H_k(X_0), \quad (-1)^j \lambda_j := E\psi(\varepsilon_0)H_j(\varepsilon_0), \quad k, j \geq 1.$$

To prove this CLT, we approximate S_n by a sum of m-dependent r.v.'s, as in the proof of Proposition 4.6.2. So, for a fixed $m \geq 1$, consider m-dependent approximations of the moving averages in (11.5.25):

$$X_{j,m} := \sum_{k=0}^{\infty} g_{k,m}\xi_{j-k}, \quad g_{k,m} := \kappa_{1,m}^{-1}g_k I(0 \leq k \leq m),$$

$$\varepsilon_{j,m} := \sum_{k=0}^{\infty} w_{k,m}\zeta_{j-k}, \quad w_{k,m} := \kappa_{2,m}^{-1}w_k I(0 \leq k \leq m),$$

where $\kappa_{1,m}^2 := \sum_{k=0}^{m} g_k^2$ and $\kappa_{2,m}^2 := \sum_{k=0}^{w} w_k^2$. Then $\{X_{j,m}\}$ and $\{\varepsilon_{j,m}\}$ are standardized stationary Gaussian processes and $\{\varphi(X_{j,m})\psi(\varepsilon_{j,m})\}$ is a stationary m-dependent process. Let $S_{n,m} := \sum_{j=1}^{n}\varphi(X_{j,m})\psi(\varepsilon_{j,m})$. By Proposition 4.2.3, for any $m < \infty$ fixed,

$$n^{-1/2}S_{n,m} \to_D \mathcal{N}(0, \tilde{\Sigma}_m), \qquad (11.5.41)$$

$$\tilde{\Sigma}_m := \sum_{j\in\mathbb{Z}} \text{Cov}\big(\varphi(X_{0,m})\psi(\varepsilon_{0,m}), \varphi(X_{j,m})\psi(\varepsilon_{j,m})\big).$$

Then (11.5.40) follows from (11.5.41) and

$$\limsup_{n\to\infty} n^{-1}\text{Var}(S_n - S_{n,m}) \to 0, \quad m \to \infty, \qquad (11.5.42)$$

$$\tilde{\Sigma}_{n,m} \to \tilde{\Sigma}, \quad m \to \infty.$$

The proof of (11.5.42) is similar to that of the analogous equation (4.6.31) in Proposition 4.6.2, we only sketch an outline of the proof here. By the orthogonality property (2.4.15) of Hermite polynomials and the independence of $\{X_j\}$ and $\{\varepsilon_j\}$, it suffices to show (11.5.42) for $\varphi = H_k$ and $\psi = H_j$, $k, j \geq 1$. Then $\mathrm{Var}(S_n - S_{n,m}) = \sum_{j,k=1}^{n} \rho_m(j-k)$, where

$$\rho_m(t) := \mathrm{Cov}\{H_k(X_0)H_j(\varepsilon_0) - H_k(X_{0,m})H_j(\varepsilon_{0,m}),$$
$$H_k(X_t)H_j(\varepsilon_t) - H_k(X_{t,m})H_j(\varepsilon_{t,m})\}$$

is dominated in the absolute value by a summable sequence

$$C\Big(\sum_{u=0}^{\infty} |g_u g_{u+t}|\Big)^k \Big(\sum_{v=0}^{\infty} |w_v w_{t+v}|\Big)^j$$
$$\leq C\sum_{u=0}^{\infty} |g_u g_{u+t}| \sum_{v=0}^{\infty} |w_v w_{t+v}| \leq CL_1^2(t)L^2(t)t^{2d+2d_1-2},$$

with C independent of m and t, and $\rho_m(t) \to 0$, $m \to \infty$, $\forall t \geq 1$ fixed. From these facts and the DCT, (11.5.42) follows as in the proof of Proposition 4.6.2. This completes the proof of Proposition 11.5.3.

From Propositions 11.5.1 and 11.5.3 we can conclude that, under a long memory Gaussian set-up as in (11.5.25), all M estimators of β in (11.5.1) are asymptotically first-order equivalent to each other, provided either $E\dot{h}(X_0, \beta) \neq 0$ or $E\dot{h}(X_0, \beta) = 0$ and $d + d_1 > 1/2$. The effect of the assumed model h appears in the above asymptotic distributions through the vectors c_0, c_1, and the matrix H. As mentioned earlier, analogous results hold for $p > 1$, but the proofs are technically more involved.

However, in the linear regression model, the situation is relatively easy, and the above result can be extended to $p > 1$ and to non-Gaussian regressors and errors as in (11.5.35) to (11.5.37), as follows. The restriction to $p = 1$ in Proposition 11.5.4 (ii) below is made purely for simplicity, since in the case $p > 1$ the description of the limiting r.v. $\mathcal{W}$ is more involved.

Proposition 11.5.4. *Suppose the regression model (11.5.1) holds with $p = q \geq 1$, $h(x, \beta) = x'\beta$, and errors as in (11.5.35). In addition, assume that ψ has bounded variation, $(\psi 1)$, (11.5.36) to (11.5.39) hold with a_n and b_n given by (11.5.30) and (11.5.28), the matrix $H = EXX'$ is positive definite, and that the innovations $\{\xi_j\}$ in (11.5.37) satisfy*

$$E|\xi_0|^r < \infty, \quad \exists\, r \geq 2 \text{ such that } n = O(a_n^r). \tag{11.5.43}$$

Then the following conclusions hold:

(i) *If $EX_0 = 0$ and $d + d_1 < 1/2$ then*

$$n^{1/2}(\widehat{\beta} - \beta) \to_D \mathcal{N}_q(0, \Sigma_0),$$

where $\Sigma_0 := \lambda_1^{-2} H^{-1} \sum_{i \in \mathbb{Z}} E\psi(\varepsilon_0)\psi(\varepsilon_i) EX_0 X_i' H^{-1}$.

(ii) *If $p = 1$, $EX_0 = 0$, and $d + d_1 > 1/2$ then*

$$n^{1-d-d_1}(L(n)L_1(n))^{-1}(\widehat{\beta} - \beta) \to_D H^{-1}\mathcal{W}.$$

(iii) *If $EX_0 \neq 0$ then*

$$n^{-d+1/2}L^{-1}(n)(\widehat{\beta} - \beta) \to_D \mathcal{N}_q(0, \Sigma_1).$$

In (ii) *and* (iii), *$\mathcal{W}$ and Σ_1 are the same as in Proposition 11.5.3, with $c_0 = EX_0$ and $L_1 = L_{11}$.*

Proof.   First, note that (11.5.43) implies (11.5.6) of Theorem 11.5.1, namely,

$$\max_{1 \leq i \leq n} |X_i| = o_p(a_n). \tag{11.5.44}$$

Indeed, by (2.5.5), $E|X_0|^r \leq C\left(1 + \sum_{j=0}^{\infty} |g_j|^r E|\xi_j|^r\right) < \infty$. Hence, by the stationarity of $\{X_i\}$ and Chebyshev's inequality, for any $\epsilon > 0$,

$$P(\max_{1 \leq i \leq n} |X_i| > \epsilon a_n) \leq nP(|X_0| > \epsilon a_n)$$

$$\leq na_n^{-r}\epsilon^{-r}E|X_0|^2 I(|X_0| > \epsilon a_n) = o(1),$$

proving (11.5.44). Since the other conditions of Theorem 11.5.1 are satisfied by the linearity of h and Proposition 11.5.2, we can apply the conclusion (11.5.15) of Theorem 11.5.1 and Proposition 11.5.2 to derive the convergences in (i) to (iii), as follows. In particular, (ii) and (iii) are immediate from (11.5.15) and Proposition 11.5.6, below, and Theorem 4.3.1, respectively.

It remains to prove (i), or

$$n^{-1/2} \sum_{j=1}^{n} X_j \psi(\varepsilon_j) \to_D \mathcal{N}_q(0, \tilde{\Sigma}_0), \tag{11.5.45}$$

where now $\tilde{\Sigma}_0 := \lambda^2 H\Sigma_0 H = \sum_{i \in \mathbb{Z}} E\psi(\varepsilon_0)\psi(\varepsilon_i) EX_0 X_i'$. The proof below uses an approximation of the moving average X_j in (11.5.37) by the m-dependent (truncated) moving average $X_{j,m}$. It does not depend on $p =$

$q \geq 1$. Thus, we shall assume $p = q = 1$ in the rest of the proof for notational simplicity.

So, let $X_{j,m} := \sum_{k=0}^{m} g_k \xi_{j-k}$ and $S_n := \sum_{j=1}^{n} X_j \psi(\varepsilon_j)$, $S_{n,m} := \sum_{j=1}^{n} X_{j,m} \psi(\varepsilon_j)$. Then (11.5.45) follows from

$$n^{-1/2} S_{n,m} \to_D \mathcal{N}(0, \tilde{\Sigma}_{0,m}), \quad n \to \infty, \ \forall m \geq 1, \tag{11.5.46}$$

$$\lim_{m \to \infty} \limsup_{n \to \infty} n^{-1} E(S_{n,m} - S_n)^2 = 0, \quad \tilde{\Sigma}_{0,m} \to \tilde{\Sigma}_0, \quad m \to \infty.$$

Let $\mathcal{F}^\varepsilon$ be the σ-field generated by $\{\varepsilon_j, j \in \mathbb{Z}\}$. Then, conditionally on $\mathcal{F}^\varepsilon$, $\{X_{j,m} \psi(\varepsilon_j)\}$ is an m-dependent process, with zero mean and bounded variance. Applying Proposition 4.2.3 for m-dependent sequences, we see that relation (11.5.46) is implied by

$$n^{-1} E[S_{n,m}^2 | \mathcal{F}^\varepsilon] \to_p \tilde{\Sigma}_{0,m} := \sum_{i \in \mathbb{Z}} E\psi(\varepsilon_0) \psi(\varepsilon_i) \, EX_{0,m} X_{i,m}. \tag{11.5.47}$$

The independence of $\{\varepsilon_j\}$ and $\{X_j\}$, the definition of $\{X_{j,m}\}$, and the ET, give

$$n^{-1} E[S_{n,m}^2 | \mathcal{F}^\varepsilon] = n^{-1} \sum_{1 \leq j,k \leq n: |j-k| \leq m} \psi(\varepsilon_j) \psi(\varepsilon_k) EX_{j,m} X_{k,m}$$

$$= \sum_{i=-m}^{m} EX_{0,m} X_{i,m} \, n^{-1} \sum_{j=1}^{n} \psi(\varepsilon_j) \psi(\varepsilon_{j+i}) + o_p(1)$$

$$= \sum_{i=-m}^{m} EX_{0,m} X_{i,m} E\psi(\varepsilon_0) \psi(\varepsilon_i) + o_p(1) = \tilde{\Sigma}_{0,m} + o_p(1),$$

proving (11.5.47) and the first relation in (11.5.46), too.

Consider

$$n^{-1} E(S_{n,m} - S_n)^2$$

$$= n^{-1} \sum_{j,k=1}^{n} E\psi(\varepsilon_j) \psi(\varepsilon_k) E(X_j - X_{j,m})(X_k - X_{k,m})$$

$$= \sum_{i=-n}^{n} \left(1 - \frac{|i|}{n}\right) E\psi(\varepsilon_0) \psi(\varepsilon_i) E(X_0 - X_{0,m})(X_i - X_{i,m}).$$

Here, according to Theorem 10.2.3 and Remark 10.2.5,

$$E\psi(\varepsilon_0) \psi(\varepsilon_i) = O(E\varepsilon_0 \varepsilon_i) = O(L^2(i) |i|^{-(1-2d)}),$$

while $|E(X_0 - X_{0,m})(X_i - X_{i,m})| \leq C L_1^2(i) |i|^{-1+2d_1}$, uniformly in $m \geq 1$ and $E(X_0 - X_{0,m})(X_i - X_{i,m}) \to 0$, $m \to \infty$, for any $i \in \mathbb{Z}$, fixed. Therefore, since $d + d_1 < 1/2$, the second claim in (11.5.46) follows by the DCT. The third claim of (11.5.46) follows similarly. Proposition 11.5.4 is proved.

To complete the discussion of the asymptotic distribution of $\widehat{\beta}$ in the random design case, we need to prove statements (ii) of Propositions 11.5.3 and 11.5.4. As mentioned above, these statements reduce to the limit distribution of sums $\sum_{j=1}^{n} X_j \varepsilon_j$ of products of long memory moving averages $\{X_j\}$ and $\{\varepsilon_j\}$ in the situation when $d + d_1 > 1/2$, that is, when the product process $\{X_j \varepsilon_j\}$ has long memory.

Hence, we need a generalization of the double Wiener–Itô integral, introduced in Section 4.7, for *independent* Brownian motions. Let $\{W_i(u), u \in \mathbb{R}\}$, $i = 1, 2$, be two independent Brownian motions on the real line with zero means and covariances $EW_i(u)W_i(v) = EW_i(-u)W_i(-v) = \min(u, v)$, $u, v \geq 0$. A double Wiener–Itô integral

$$\int_{\mathbb{R}^2} f(u_1, u_k) W_1(du_1) W_2(du_2) \equiv \int f\, dW_1 dW_2 \qquad (11.5.48)$$

is defined for any square integrable function $f \in L_2(\mathbb{R}^2)$, as a limit in mean square of the weighted double sums of the increments of $\{W_i(u)\}$, $i = 1, 2$, as in the case of single Brownian motion. Namely, if f is a "simple" function taking a finite number of non-zero constant values $f(j_1, j_2)$ on rectangles $(j_1/n, (j_1+1)/n] \times (j_2/n, (j_2+1)/n] \subset \mathbb{R}^2$, $(j_1, j_2) \in \mathbb{Z}^2$, the corresponding double integral coincides with the sum

$$\int f\, dW_1 dW_2 = \sum_{j_1, j_2 \in \mathbb{Z}} f(j_1, j_2)\, \Delta_n W_1(j_1) \Delta_n W_2(j_2),$$

where $\Delta_n W_i(j) := W_i((j+1)/n) - W_i(j/n)$, $j \in \mathbb{Z}$. Note that because of the independence of W_1 and W_2, the integrand f need not vanish on the diagonal $j_1 = j_2$, in contrast to the definition in Section 4.7. The integral in (11.5.48) has similar properties as the Wiener–Itô integral discussed in Chapter 14; namely, $E \int f\, dW_1 dW_2 = 0$ and

$$E\left(\int f\, dW_1 dW_2 \right)^2 = \int_{\mathbb{R}^2} f^2(u_1, u_2)\, du_1 du_2 = \|f\|^2.$$

We shall also need a generalization of the criterion for the convergence in the distribution of polynomial forms in i.i.d. r.v.'s given in Chapter 6. Consider a double sum

$$Q_2(g) := \sum g(j_1, j_2) \xi_{j_1} \zeta_{j_2}, \qquad (11.5.49)$$

where $\{\xi_j\} \sim \text{IID}(0,1)$ and $\{\zeta_j\} \sim \text{IID}(0,1)$ are independent sequences of standardized i.i.d. r.v.'s, with real square-summable coefficients $g(j_1, j_2)$. Obviously, $EQ_2^2(g) = \sum g^2(j_1, j_2)$. With any $g \in L_2(\mathbb{Z}^2)$ and any $n \geq 1$, we can associate its extension to $\mathbb{R}^2$, namely, $\tilde{g}_n(u_1, u_2) := ng([u_1 n], [u_2 n])$, $(u_1, u_2) \in \mathbb{R}^2$; see also (4.7.10). The proof of the next proposition is similar to that of Proposition 14.3.2 in Chapter 14, and hence omitted.

Proposition 11.5.5. *Let $Q_2(g_n)$, $n = 1, 2, \cdots$, be a sequence of double sums as in (11.5.49), with coefficients $g_n \in L_2(\mathbb{Z}^2)$. Suppose for some $f \in L_2(\mathbb{R}^2)$,*

$$\|\tilde{g}_n - f\| \to 0. \tag{11.5.50}$$

Then $Q_2(g_n) \to_D \int f dW_1 dW_2$.

Proposition 11.5.6. *Let $\varepsilon_j = \sum_{k=0}^{\infty} w_k \zeta_{j-k}$ and $X_j = \sum_{k=0}^{\infty} g_k \xi_{j-k}$ be moving averages with regularly decaying coefficients $g_k = k^{-1+d_1} L_1(k)$ and $w_k = k^{-1+d} L(k)$, $k \geq 1$, $L, L_1 \in SV$, $0 < d, d_1 < 1/2$, and innovations $\{\xi_j\} \sim \text{IID}(0,1)$ and $\{\zeta_j\} \sim \text{IID}(0,1)$. Assume $\{\xi_j\}$ and $\{\zeta_j\}$ are independent and $d + d_1 > 1/2$. Then*

$$n^{-d_1-d_2}(L_1(n)L_2(n))^{-1} \sum_{j=1}^{n} X_j \varepsilon_j \to_D \mathcal{W}, \tag{11.5.51}$$

where $\mathcal{W} = \int f dW_1 dW_2$ is a double Wiener–Itô integral as defined in (11.5.48) with integrand

$$f(u_1, u_2) := \int_0^1 (\tau - u_1)_+^{d_1-1}(\tau - u_2)_+^{d-1} d\tau. \tag{11.5.52}$$

Proof. Rewrite the l.h.s. of (11.5.51) as $Q_2(g_n)$, where $g_t = w_t := 0$, $t < 0$,

$$g_n(j_1, j_2) := n^{-d_1-d_2}(L_1(n)L_2(n))^{-1} \sum_{t=1}^{n} g_{t-j_1} w_{t-j_2}.$$

Then (11.5.51) follows from Proposition 11.5.5, provided (11.5.50) holds, with f given by (11.5.52). The verification of (11.5.50) uses the DCT and the standard properties of slowly varying functions and is completely analogous to the verification of the corresponding equation in the proof of Theorem 4.8.2. Proposition 11.5.6 is proved.

Remark 11.5.4. (*LAD estimator*). This estimator corresponds to the M estimator with $\psi(x) \equiv \text{sgn}(x)$. It is not covered by the above discussion, but a similar method of proof gives the following

Corollary 11.5.1. *Assume (11.5.1), (11.5.21), and (11.5.2) to (11.5.10) hold with a_n, b_n as in (11.5.30), (11.5.28), and with $\psi(x) \equiv \text{sgn}(x)$. In addition, suppose (11.5.36) is satisfied and (11.5.14) holds for $\psi(x) \equiv \text{sgn}(x)$. Then the conclusions of Propositions 11.5.3 and 11.5.4 also hold for the LAD estimator.*

Note: This section has its roots in the works of Koul (1996), Koul, Baillie and Surgailis (2004), and Surgailis (2003).

11.6   A simulation study

Because the above class of M estimators are asymptotically equivalent to each other, it is of interest to see if there is any finite sample difference between their standard errors and how the values of the long memory parameters d_1 and d affect these entities. The following simulation study, first given in Koul and Baillie (2003), sheds some light on these issues in terms of the LAD and LS estimators applied to the model (11.5.1) with $p = q = 1$, $h(x, \beta) \equiv x\beta$, and $\beta = 1$, where the X_j's and the ε_j's were chosen to be the two independent moving-average processes with long memory parameters d_1, d, respectively, and $\mathcal{N}(0, 1)$ innovations. The first $2,500$ observations were neglected to avoid any effects of initialization. The values of d and d_1 were chosen to be $0.1, 0.2, 0.3$ and 0.4. Five thousand replications were computed for each combination. Tables 11.6.1 through 11.6.4 give the results of the simulation for sample sizes $n = 50, 100, 200$, and 500.

In terms of the standard errors, the ratio $r := SE_{LAD}/SE_{LS}$ varies from 1.050 to 1.362, when considered as a function of the pair (d_1, d). The smallest value of r occurs for the pair $d_1 = 0.4$, $d = 0.3$, for all sample sizes considered. The largest value of r is for the pair $d_1 = 0.1$, $d = 0.4$ for sample sizes $50, 100, 200$, while for $n = 500$, the largest values is for the pair $d_1 = 0.1$, $d = 0.1$. For the sample sizes of 500, the largest value of r is 1.23. In terms of bias there is generally little to choose between the estimators for all chosen sample sizes.

d_1	d	$\hat{\beta}_{LS}$	$\hat{\beta}_{LAD}$	SE$_{LS}$	SE$_{LAD}$
0.1	0.1	1.0016	1.0017	0.1486	0.1840
0.1	0.2	1.0017	1.0017	0.1595	0.1979
0.1	0.3	1.0015	1.0013	0.1823	0.2288
0.1	0.4	1.0010	0.9992	0.2303	0.3137
0.2	0.1	1.0024	1.0028	0.1474	0.1814
0.2	0.2	1.0024	1.0037	0.1650	0.2006
0.2	0.3	1.0022	1.0025	0.1999	0.2433
0.2	0.4	1.0013	1.0003	0.2698	0.3482
0.3	0.1	1.0031	1.0049	0.1424	0.1742
0.3	0.2	1.0032	1.0050	0.1677	0.1997
0.3	0.3	1.0029	1.0044	0.2156	0.2536
0.3	0.4	1.0016	1.0029	0.3077	0.3724
0.4	0.1	1.0032	1.0043	0.1318	0.1585
0.4	0.2	1.0034	1.0050	0.1631	0.1882
0.4	0.3	1.0030	1.0059	0.2203	0.2480
0.4	0.4	1.0016	1.0023	0.3274	0.3749

Table 11.6.1: LS, LAD estimators, and SEs, $n = 50$.

d_1	d	$\hat{\beta}_{LS}$	$\hat{\beta}_{LAD}$	SE$_{LS}$	SE$_{LAD}$
0.1	0.1	1.0008	1.0016	0.1040	0.1274
0.1	0.2	1.0008	1.0004	0.1126	0.1362
0.1	0.3	1.0010	0.9999	0.1304	0.1612
0.1	0.4	1.0014	1.0000	0.1674	0.2220
0.2	0.1	1.0006	1.0019	0.1040	0.1253
0.2	0.2	1.0007	1.0002	0.1186	0.1409
0.2	0.3	1.0009	0.9993	0.1479	0.1757
0.2	0.4	1.0014	1.0002	0.2061	0.2608
0.3	0.1	1.0004	1.0013	0.1014	0.1208
0.3	0.2	1.0004	1.0001	0.1233	0.1416
0.3	0.3	1.0005	0.9988	0.1656	0.1882
0.3	0.4	1.0007	0.9978	0.2472	0.2950
0.4	0.1	1.0002	1.0008	0.0943	0.1104
0.4	0.2	1.0001	0.9997	0.1221	0.1373
0.4	0.3	0.9998	0.9987	0.1741	0.1917
0.4	0.4	0.9995	0.9977	0.2718	0.3061

Table 11.6.2: LS, LAD estimators, and SEs, $n = 100$.

d_1	d	$\hat{\beta}_{LS}$	$\hat{\beta}_{LAD}$	SE_{LS}	SE_{LAD}
0.1	0.1	1.0012	1.0014	0.0719	0.0888
0.1	0.2	1.0015	1.0026	0.0779	0.0947
0.1	0.3	1.0020	1.0032	0.0909	0.1099
0.1	0.4	1.0025	1.0035	0.1189	0.1530
0.2	0.1	1.0013	1.0021	0.0721	0.0875
0.2	0.2	1.0017	1.0033	0.0830	0.0982
0.2	0.3	1.0022	1.0036	0.1057	0.1222
0.2	0.4	1.0028	1.0039	0.1523	0.1859
0.3	0.1	1.0013	1.0022	0.0706	0.0844
0.3	0.2	1.0016	1.0032	0.0878	0.1005
0.3	0.3	1.0021	1.0035	0.1223	0.1360
0.3	0.4	1.0026	1.0038	0.1904	0.2199
0.4	0.1	1.0011	1.0019	0.0658	0.0773
0.4	0.2	1.0013	1.0024	0.0884	0.0985
0.4	0.3	1.0016	1.0022	0.1319	0.1424
0.4	0.4	1.0019	1.0022	0.2155	0.2371

Table 11.6.3: LS, LAD estimators, and SEs, $n = 200$.

d_1	d	$\hat{\beta}_{LS}$	$\hat{\beta}_{LAD}$	SE_{LS}	SE_{LAD}
0.1	0.1	1.0001	0.9999	0.0457	0.0572
0.1	0.2	1.0001	0.9996	0.0495	0.0608
0.1	0.3	1.0001	1.0002	0.0578	0.0703
0.1	0.4	1.0001	0.9995	0.0759	0.0946
0.2	0.1	0.9999	0.9998	0.0458	0.0562
0.2	0.2	0.9998	0.9996	0.0531	0.0630
0.2	0.3	0.9997	0.9996	0.0688	0.0800
0.2	0.4	0.9994	0.9990	0.1021	0.1208
0.3	0.1	0.9997	0.9995	0.0451	0.0540
0.3	0.2	0.9995	0.9993	0.0575	0.0654
0.3	0.3	0.9992	0.9992	0.0833	0.0919
0.3	0.4	0.9987	0.9983	0.1362	0.1535
0.4	0.1	0.9996	0.9996	0.0424	0.0493
0.4	0.2	0.9994	0.9990	0.0596	0.0651
0.4	0.3	0.9991	0.9993	0.0941	0.0997
0.4	0.4	0.9986	0.9980	0.1623	0.1755

Table 11.6.4: LS, LAD estimators, and SEs, $n = 500$.

Chapter 12

Non-parametric Regression

12.1 Introduction

The previous chapter dealt with parametric homoscedastic regression models. This chapter looks at non- and semi-parametric heteroscedastic regression models. We discuss large sample properties of kernel-type estimators of the regression and heteroscedasticity functions in non-parametric heteroscedastic regression models with uniform non-random design on the unit interval and moving-average errors that may have long memory. The consistency and asymptotic distributions of these estimators is established. The results thus obtained are useful when constructing large sample confidence bands for these functions.

12.2 Uniform design

Consider the non-parametric regression model

$$Y_j = \mu(\frac{j}{n}) + \sigma(\frac{j}{n})\varepsilon_j, \quad j = 1, 2, \cdots, n, \tag{12.2.1}$$

where μ is a real-valued function and σ is a positive function, both defined on $[0, 1]$. The errors $\{\varepsilon_j\}$ are assumed to be a linear mean-zero process with i.i.d. innovations; see Assumption 12.2.1 below, having either short memory:

$$\sum_{k \in \mathbb{Z}} |\gamma_\varepsilon(k)| < \infty, \quad \sum_{k \in \mathbb{Z}} \gamma_\varepsilon(k) > 0, \tag{12.2.2}$$

or long memory, satisfying:

$$\gamma_\varepsilon(k) \sim c_\gamma |k|^{-1+2d}, \quad k \to \infty, \ c_\gamma > 0, \ d \in (0, 1/2). \tag{12.2.3}$$

415

For short memory (12.2.2) we set $d = 0$. Clearly, $EY_j \equiv \mu(j/n)$ and $\mathrm{Var}(Y_j)$ $\equiv \sigma^2(j/n)\gamma_\varepsilon(0)$. We shall absorb the unknown $\gamma_\varepsilon(0)$ into σ^2, and in the rest of this chapter assume without loss of generality that $\gamma_\varepsilon(0) = 1$.

Now, let K and W be non-negative kernel functions on $\mathbb{R}$, and $b \equiv b_n$ and $\omega \equiv \omega_n$ be the two deterministic bandwidth sequences. The kernel estimators of μ and σ^2 to be investigated here are Nadaraya–Watson kernel-type estimators

$$\hat{\mu}(x) := \frac{\sum_{j=1}^n K((nx - j)/nb)Y_j}{\sum_{j=1}^n K((nx - j)/nb)}, \tag{12.2.4}$$

$$\hat{\sigma}^2(x) := \frac{\sum_{j=1}^n W((nx - j)/n\omega)[Y_j - \hat{\mu}(\frac{j}{n})]^2}{\sum_{j=1}^n W((nx - j)/n\omega)}, \qquad x \in [0, 1].$$

We shall establish the consistency and asymptotic distributions of these estimators. The range of values of d affects the asymptotic distribution of $\hat{\sigma}^2$ but not of $\hat{\mu}$. Theorem 12.2.2 below shows that finite-dimensional distributions of $(nb)^{1/2-d}(\hat{\mu}(x) - \mu(x))$ converge weakly to multivariate Gaussian distributions for all values of d. In comparison, Theorem 12.3.1 shows that a similar CLT holds for $\hat{\sigma}^2$ only for $0 \le d < 1/4$.

A similar phenomenon is seen to hold for consistency rates of these estimators. The consistency rate for $\hat{\sigma}^2(x)$ is affected by the range of values of d. As seen in Theorem 12.3.1(i) below, for a large class of innovations, the mean-squared error of $\hat{\sigma}^2(x)$ is of the order $(n\omega)^{-1} + \omega^4$, for $0 \le d < 1/4$, and $(n\omega)^{4d-2} + \omega^4$, for $1/4 < d < 1/2$.

Proposition 12.2.1 shows that the uniform consistency rate of $\hat{\mu}$ over any closed subinterval of $(0, 1)$ is $(nb)^{1/2-d-\delta}$ for all $0 \le d < 1/2$ and any $\delta > 0$, and for a class of innovations ζ_j that have all moments finite.

Because the standardizing sequence is unknown, we need consistent estimators of $\mathrm{Var}(\hat{\mu}(x))$ and $\mathrm{Var}(\hat{\sigma}^2(x))$ in order for these results to be applicable in practice. An estimator of $\mathrm{Var}(\hat{\mu}(x))$ that avoids the need to have a $\log(n)$-consistent estimator of d is given in Theorem 12.4.1 below.

For the sake of easy reference we shall now state several assumptions.

Assumption 12.2.1. The error process is

$$\varepsilon_j = \sum_{k=0}^\infty a_k \zeta_{j-k}, \quad j \in \mathbb{Z}, \quad \{\zeta_j\} \sim \mathrm{IID}(0, \sigma_\zeta^2),$$

with γ_ε satisfying either (12.2.2) or (12.2.3).

Assumption 12.2.2. The bandwidth sequences b_n and ω_n satisfy

$$b_n \to 0, \quad \omega_n \to 0, \quad nb_n \to \infty, \quad n\omega_n \to \infty. \qquad (12.2.5)$$

Definition 12.2.1. We say that $K \in \mathcal{K}$, if $K(u), u \in \mathbb{R}$, is an even differentiable function such that $K(0) > 0$,

$$|K(u)| \le C(1+|u|)^{-4}, \quad |\dot{K}(u)| \le C(1+|u|)^{-3}, \quad u \in \mathbb{R}. \quad (12.2.6)$$

Definition 12.2.2. We say that $\mu \in \mathcal{C}_2$ if $\mu(v)$, $v \in (0,1)$, has a bounded continuous second derivative.

Establishing the consistency and asymptotic distribution of $\hat{\mu}(x)$ is facilitated by the content of the next subsection.

12.2.1 *Some preliminaries for $\hat{\mu}$*

This section establishes asymptotic properties of the following kernel-type weighted sums

$$S_n(x) := \sum_{j=1}^{n} K\left(\frac{nx-j}{nb}\right)\sigma\left(\frac{j}{n}\right)\varepsilon_j, \qquad (12.2.7)$$

$$\bar{\mu}(x) := \frac{\sum_{j=1}^{n} K\left(\frac{nx-j}{nb}\right)\mu\left(\frac{j}{n}\right)}{K_{nx}}, \quad K_{nx} := \sum_{j=1}^{n} K\left(\frac{nx-j}{nb}\right), \quad x \in [0,1].$$

To proceed further, let $\|K\|_2^2 := \int_{\mathbb{R}} K^2(u)du$, and

$$v_{0,K}^2 := \|K\|_2^2 \sum_{k \in \mathbb{Z}} \gamma_\varepsilon(k),$$

$$v_{d,K}^2 := c_\gamma \int_{\mathbb{R}^2} K(u)K(v)|u-v|^{-1+2d}dudv, \quad d \in (0,1/2).$$

Let $\mathcal{I}$ be the $k \times k$ identity matrix. We are now ready to state and prove:

Theorem 12.2.1. *Assume that $\{\varepsilon_j\}$ satisfies Assumption 12.2.1. In addition, suppose $K(u)$, $u \in \mathbb{R}$, is continuous and*

$$|K(u)| \le C(1+|u|)^{-4}, \quad \forall u \in \mathbb{R}. \qquad (12.2.8)$$

Furthermore, suppose $\sigma(u), u \in [0,1]$, is positive and has a bounded derivative, and the bandwidth b satisfies (12.2.5).

Then, for any $k \ge 1$ and for any distinct $0 < x_1, \cdots, x_k < 1$,

$$\left(\frac{S_n(x_1)}{\sqrt{\mathrm{Var}(S_n(x_1))}}, \cdots, \frac{S_n(x_k)}{\sqrt{\mathrm{Var}(S_n(x_k))}}\right) \to_D Z \sim \mathcal{N}_k(0, \mathcal{I}), \quad (12.2.9)$$

$$\mathrm{Var}(S_n(x_i)) \sim (nb)^{1+2d}v_{d,K}^2\sigma^2(x_i), \quad i = 1, \cdots, k.$$

Proof. The proof uses Proposition 4.3.1 and Theorem 4.3.2. So, let, for any integer $1 \le j \le n$,

$$z_{nx,j} := K\left(\frac{nx - j}{nb}\right)\sigma\left(\frac{j}{n}\right), \quad 0 \le x \le 1, \tag{12.2.10}$$

$$S_n(x) = \sum_{j=1}^{n} z_{nx,j}\varepsilon_j, \quad 0 \le x \le 1.$$

Let $\|z_{nx}\|^2 := \sum_{j=1}^{n} z_{nx,j}^2$. By Lemmas 12.2.1 and 12.2.2 below, for any $x \in (0,1)$, $\|z_{nx}\|^2 \sim nb\,\sigma^2(x)\|K\|_2^2$, $\mathrm{Var}(S_n(x)) \sim (nb)^{1+2d}\sigma^2(x)v_{d,K}^2$, and for any $x, y \in (0,1)$, $\mathrm{Corr}(S_n(x), S_n(y)) \to I(x = y)$. Since K and σ are bounded functions, and $(nb)^{1+2d} \to \infty$, we readily find $\max_{1 \le j \le n} |z_{nx,j}| = O(1)$, and $\max_{1 \le j \le n} |z_{nx,j}| = o\left(\sqrt{\mathrm{Var}(S_n(x))}\right)$. Thus, for $0 \le d < 1/2$, the sums $S_n(x_i)$, $i = 1, \cdots, k$, satisfy condition (i) of Proposition 4.3.1. Therefore, (12.2.9) holds by Theorem 4.3.2.

In the following two lemmas, C is a universal constant not depending on $0 \le x, y \le 1$ and $n \ge 1$, and $I_K = \int_{\mathbb{R}} K(v)dv$.

Lemma 12.2.1. *Under the assumptions of Theorem 12.2.1,* $\forall x \in (0,1)$,

$$\|z_{nx}\|^2 \sim nb\,\sigma^2(x)\|K\|_2^2, \quad (nb)^{-1}\sum_{j=1}^{n} z_{nx,j} \to \sigma(x)I_K, \tag{12.2.11}$$

$$|z_{nx,1}| + |z_{nx,n}| + \sum_{j=2}^{n} |z_{nx,j} - z_{nx,j-1}| = o(\|z_{nx}\|).$$

Moreover, for any $p \ge 1/2$,

$$\sup_{0 \le x \le 1} \sum_{j=1}^{n} z_{nx,j}^p \le Cnb, \quad \sup_{0 \le x \le 1} \sum_{j=1}^{n} z_{nx,j}^2 \left(\frac{nx - j}{nb}\right)^2 \le Cnb. \tag{12.2.12}$$

Proof. Write

$$\sum_{j=1}^{n} z_{nx,j}^2 = \sum_{j=1-[nx]}^{n-[nx]} z_{nx,j+[nx]}^2 = \int_{1-[nx]}^{n+1-[nx]} g_{nx,\lfloor u \rfloor}^2 \, du,$$

where

$$g_{nx,u} := z_{nx,u+[nx]} = K\left(\frac{nx - [nx] - u}{nb}\right)\sigma\left(\frac{[nx] + u}{n}\right).$$

By (12.2.8), for every $u \in \mathbb{R}$,

$$|g_{nx,\lfloor unb \rfloor}| \le C(1 + |u|)^{-4}, \quad g_{nx,\lfloor unb \rfloor} \to K(u)\sigma(x). \tag{12.2.13}$$

Thus by the DCT,

$$(nb)^{-1}\sum_{j=1}^{n} z_{nx,j}^2 = \int_{(1-[nx])/nb}^{(n-[nx]+1)/nb} g_{nx,[unb]}^2 du \to \sigma^2(x)\int_{\mathbb{R}} K^2(u)du,$$

which proves the first claim of (12.2.11).

To prove the second claim of (12.2.11), note that for $K \in \mathcal{K}$, by the mean-value theorem,

$$|z_{nx,j} - z_{nx,j-1}|$$
$$\leq C\{|K(\frac{nx-j}{nb}) - K(\frac{nx-j+1}{nb})| + |\sigma(\frac{j}{n}) - \sigma(\frac{j-1}{n})|\}$$
$$\leq C(nb)^{-1}\big(1 + (\frac{nx-j}{nb})\big)^{-3} + Cn^{-1}.$$

Thus as above,

$$\sum_{j=1}^{n}|z_{nx,j} - z_{nx,j-1}| \leq C(nb)^{-1}\sum_{j=1}^{n}\big(1 + (\frac{nx-j}{nb})\big)^{-3} + O(1) = O(1),$$

which implies the second bound of (12.2.11), because $\|z_{nx}\| \sim C(nb)^{1/2} \to \infty$ and $\max_{j=1,\cdots,n} z_{nx,j} \leq \sup_{u\in\mathbb{R}} K(u) + \sup_{u\in[0,1]} \sigma(u) < \infty$.

The bounds in (12.2.12) can be shown using the same argument as used in proving (12.2.11). In particular, by (12.2.13), uniformly in $0 \leq x \leq 1$,

$$(nb)^{-1}\sum_{j=1}^{n} z_{nx,j}^p \leq C\int_{\mathbb{R}}(1+|u|)^{-4p}du < \infty.$$

This completes the proof of the lemma.

The next lemma analyzes the asymptotic properties of the covariances of $S_n(x)$ and $S_n(y)$. Let

$$v_n(x,y) := \mathrm{Cov}(S_n(x), S_n(y)), \quad 0 \leq x,y \leq 1.$$

Lemma 12.2.2. *Under the assumptions of Theorem 12.2.1,*

$$(bn)^{-1-2d}v_n(x,y) \to \sigma^2(x)v_{d,K}^2 I(x=y), \quad 0 < x \leq y < 1. \quad (12.2.14)$$

Moreover,

$$\sup_{0\leq x\leq 1} v_n(x,x) \leq C(bn)^{1+2d}, \quad (12.2.15)$$

$$\sum_{j,k=1}^{n} z_{nx,j}^p z_{nx,k}^p |\gamma_\varepsilon(j-k)| \leq C(bn)^{1+2d}, \quad p \geq 1/2,$$

$$|v_n(x,y)| \leq C\big(1 + \frac{|x-y|}{b}\big)^{-2} n^{2d} nb, \quad n \geq 1. \quad (12.2.16)$$

Proof. We shall first prove (12.2.16). Note that

$$\frac{1}{1+|u|}\frac{1}{1+|v|} \leq \frac{1}{1+|u \pm v|}, \quad u,v \in \mathbb{R}. \tag{12.2.17}$$

This together with (12.2.8) gives

$$K(u)K(v) \leq C\big((1+|u|)(1+|v|)\big)^{-4} \leq C(1+|u-v|)^{-4}, \quad u,v \in \mathbb{R}.$$

Hence,

$$(z_{nx,j}z_{ny,j})^{1/2} \leq C\big(1+\frac{|x-y|}{b}\big)^{-2}, \qquad (z_{nx,j}z_{ny,j})^{1/2} \leq z_{nx,j} + z_{ny,j},$$

$$z_{nx,j}z_{ny,j} \leq C\big(1+\frac{|x-y|}{b}\big)^{-2}(z_{nx,j}+z_{ny,j}).$$

Thus, for any $0 \leq x,y \leq 1$,

$$|v_n(x,y)| \leq \sum_{j,k=1}^{n} z_{nx,j}z_{ny,k}|\gamma_\varepsilon(j-k)|$$

$$\leq C\big(1+\frac{|x-y|}{b}\big)^{-2} \sum_{j,k=1}^{n} (z_{nx,j}+z_{ny,k})|\gamma_\varepsilon(j-k)|$$

$$\leq C\big(1+\frac{|x-y|}{b}\big)^{-2} \sum_{j=1}^{n}(z_{nx,j}+z_{ny,j}) \sum_{s=-n}^{n} |\gamma_\varepsilon(s)|.$$

By (12.2.2) and (12.2.3), $\sum_{s=-n}^{n} |\gamma_\varepsilon(s)| = O(n^{2d})$, which together with (12.2.12), gives

$$|v_n(x,y)| \leq C\big(1+\frac{|x-y|}{b}\big)^{-2}(bn)n^{2d},$$

and hence (12.2.16).

Proof of (12.2.14) and (12.2.15). If $x \neq y$ are fixed, then by (12.2.16),

$$|v_n(x,y)| \leq Cb(bn)n^{2d} = o((bn)^{1+2d}),$$

since $b \to 0$ and $0 \leq 2d < 1$, which proves (12.2.14) when $x \neq y$.

Now, let $x = y$. Consider first the case $d = 0$. Then $\{\varepsilon_j\}$ has a continuous spectral density f_ε, and $f_\varepsilon(0) > 0$, and by (12.2.11),

$$|z_{nx,1}| + |z_{nx,n}| + \sum_{j=2}^{n}|z_{nx,j} - z_{nx,j-1}| = o((\sum_{j=1}^{n} z_{nx,j}^2)^{1/2}).$$

Therefore, by (4.3.15) and (12.2.11),

$$v_n(x,x) \sim 2\pi f_\varepsilon(0) \sum_{j=1}^{n} z_{nx,j}^2 \sim \sigma^2(x)\|K\|^2 \sum_{k \in \mathbb{Z}} \gamma_\varepsilon(k),$$

because of the equality $\sum_{k\in\mathbb{Z}}\gamma_\varepsilon(k) = 2\pi f(0)$, which yields (12.2.14) for $d = 0$. To obtain the first bound of (12.2.15), by (12.2.12),

$$\sup_{0\le x\le 1}|v_n(x,x)| = \sup_{0\le x\le 1}\sum_{j,k=1}^{n} z_{nx,j}z_{nx,k}\gamma_\varepsilon(j-k)$$

$$\le \sup_{0\le x\le 1}\sum_{j=1}^{n} z_{nx,j}^2 \sum_{s=-n}^{n}|\gamma_\varepsilon(s)| \le C\,bn.$$

Consider the case $0 < d < 1/2$. Then

$$v_n(x,x) = \sum_{j,k=1}^{n} z_{nx,j}z_{nx,k}\gamma_\varepsilon(j-k)$$

$$= \sum_{j,k=1-[nx]}^{n-[nx]} z_{nx,j+[nx]}z_{nx,k+[nx]}\gamma_\varepsilon(j-k)$$

$$= \int\int_{1-[nx]}^{n-[nx]+1} g_{nx,\lfloor u\rfloor}g_{nx,\lfloor v\rfloor}\gamma_\varepsilon(\lfloor u\rfloor - \lfloor v\rfloor)dudv$$

$$= (nb)^{1+2d}\int\int_{(1-[nx])/nb}^{(n-[nx]+1)/nb} g_{nx,\lfloor unb\rfloor}g_{nx,\lfloor vnb\rfloor}$$

$$\times |u-v|^{-1+2d}r_n(u,v)dudv,$$

$$r_n(u,v) := (nb|u-v|)^{1-2d}\gamma_\varepsilon(\lfloor unb\rfloor - \lfloor vnb\rfloor).$$

By (12.2.3), for every $u,v \in \mathbb{R}$, $|r_n(u,v)| \le C$ and $|r_n(u,v)| \to c_\gamma$. These results together with (12.2.13), by the DCT, imply

$$(nb)^{-1-2d}v_n(x,x) \to \sigma^2(x)c_\gamma\int_{\mathbb{R}^2} K(u)K(v)|u-v|^{-1+2d}dudv = \sigma^2(x)v_{d,K}^2,$$

for every $x \in (0,1)$. Moreover,

$$(nb)^{-1-2d}\sup_{0\le x\le 1}v_n(x,x) \le C\int_{\mathbb{R}^2}(1+|u|)^{-4}(1+|v|)^{-4}|u-v|^{-1+2d}dudv < \infty.$$

This completes the proof of (12.2.14) and the first bound of (12.2.15) for $0 < d < 1/2$. The second bound of (12.2.15) for $0 \le d < 1/2$ is established similarly. This completes the proof of the lemma.

Deriving the asymptotic properties of the sum $\bar{\mu}_n(x)$ requires stronger assumptions on K and μ. Let $a_K := \int_{\mathbb{R}} v^2 K(v)dv/2$.

Lemma 12.2.3. *Suppose that $K \in \mathcal{K}$ and $\mu \in C_2$. Then, for any $x \in (0,1)$,*

$$\bar{\mu}(x) = \mu(x) + \frac{\ddot{\mu}(x)a_K}{I_K}b^2 + o(b^2) + O(n^{-1}). \qquad (12.2.18)$$

Moreover, for all $0 \leq x \leq 1$,

$$|\bar{\mu}(x) - \mu(x)| \leq C\left(b^2 + n^{-1} + b\left(\frac{b}{(1-x) \wedge x + b}\right)^2\right). \qquad (12.2.19)$$

Proof. Let

$$Q_{nx,1} := \dot{\mu}(x) \sum_{j=1}^{n} K\left(\frac{nx-j}{nb}\right)\left(\frac{j}{n} - x\right),$$

$$Q_{nx,2} := \frac{1}{2}\ddot{\mu}(x) \sum_{j=1}^{n} K\left(\frac{nx-j}{nb}\right)\left(\frac{j}{n} - x\right)^2,$$

$$Q_{nx,3} := \sum_{j=1}^{n} K\left(\frac{nx-j}{nb}\right)\left(\mu\left(\frac{j}{n}\right) - \mu(x)\right) - Q_{nx,1} - Q_{nx,2}.$$

Write

$$\bar{\mu}(x) - \mu(x) = K_{nx}^{-1} \sum_{j=1}^{n} K\left(\frac{nx-j}{nb}\right)\left(\mu\left(\frac{j}{n}\right) - \mu(x)\right)$$

$$= K_{nx}^{-1}(Q_{nx,1} + Q_{nx,2} + Q_{nx,3}).$$

By (12.2.11),

$$(bn)^{-1}K_{nx} \to I_K, \quad \forall x \in (0,1). \qquad (12.2.20)$$

We shall show that for some $c > 0$,

$$\inf_{0 \leq x \leq 1} (nb)^{-1}K_{nx} \geq c, \qquad \forall n \geq 1, \qquad (12.2.21)$$

$$(nb)^{-1}|Q_{nx,1}| \leq C\left\{b\left(\frac{b}{(1-x) \wedge x + b}\right)^2 + n^{-1}\right\}, \quad \forall n \geq 1, \qquad (12.2.22)$$

$$(b^3 n)^{-1}Q_{nx,2} \to \ddot{\mu}(x)a_K, \quad (b^3 n)^{-1}Q_{nx,3} \to 0, \quad \forall x \in (0,1), \qquad (12.2.23)$$

$$(b^3 n)^{-1} \sup_{0 \leq x \leq 1} (|Q_{nx,2}| + |Q_{nx,3}|) \leq C, \quad n \geq 1. \qquad (12.2.24)$$

These bounds yield (12.2.18) and (12.2.19) in a routine fashion.

Proof of (12.2.21). The continuity of K and $K(0) > 0$ imply there exists $c > 0$ and $\epsilon > 0$ such that $K(u) \geq c$ for $|u| \leq \epsilon$. To prove (12.2.21), it suffices to show that for any $0 \leq x \leq 1$, there are at least $[\epsilon b n]$ of j's in $\{1, \cdots, n\}$ such that

$$|j - xn|/bn \leq \epsilon. \qquad (12.2.25)$$

Since $b \to 0$, assume $b\epsilon \le 1/3$. Let $0 \le x \le 1/2$. Then $[nx] + [\epsilon b\, n] \le n$, and (12.2.25) holds for $j = [nx] + 1, \cdots, [nx] + \epsilon b\, n$. If $1/2 < x \le 1$, then $[nx] - [\epsilon bn] \ge 1$, and (12.2.25) holds for $j = [nx] - 1, \cdots, [nx] - [\epsilon b\, n]$.

Proof of (12.2.22). Let $0 < x < 1$. Write

$$\sum_{j=1}^{n} K\Big(\frac{nx-j}{nb}\Big)\Big(\frac{j}{n} - x\Big) = -b\sum_{j=1}^{n} K\Big(\frac{nx-j}{nb}\Big)\Big(\frac{nx-j}{nb}\Big) =: -bs_{nx}.$$

Since K is an even function,

$$I_{nx} := \int_{\mathbb{R}} K\Big(\frac{nx-u}{nb}\Big)\Big(\frac{nx-u}{nb}\Big)du = nb\int_{\mathbb{R}} K(v)v\,dv = 0.$$

Split I_{nx} into two:

$$I_{nx} = \int_{[0,\,n+1]}[\cdots]du + \int_{\mathbb{R}\backslash[0,\,n+1]}[\cdots]du := I_{nx,1} + I_{nx,2},$$

$$|s_{nx}| = |s_{nx} - I_{nx}| \le |s_{nx} - I_{nx,1}| + |I_{nx,2}|.$$

Let $G(u) := K(u)u$. By (12.2.6), $|G(u)| \le C(1 + |u|)^{-3}$ and $|\dot{G}(u)| \le C(1 + |u|)^{-2}$, $u \in \mathbb{R}$. Therefore,

$$|s_{n,1} - I_{nx,1}| \le \int_{0}^{n+1} \Big| K\Big(\frac{nx-[u]}{nb}\Big)\Big(\frac{nx-[u]}{nb}\Big)$$
$$- K\Big(\frac{nx-u}{nb}\Big)\Big(\frac{nx-u}{nb}\Big)\Big| du$$
$$\le C(bn)^{-1}\int_{\mathbb{R}}\Big(1 + \Big|\frac{nx-u}{nb}\Big|\Big)^{-2}du \le C,$$

where C does not depend on x and $n \ge 1$. In addition,

$$|I_{nx,2}| \le C\int_{n}^{\infty}\Big(1 + \frac{u-xn}{nb}\Big)^{-3}du + \int_{0}^{\infty}\Big(1 + \frac{u+xn}{nb}\Big)^{-3}du$$
$$\le Cbn\big((b/(1-x+b))^2 + (b/(x+b))^2\big).$$

The above bounds clearly yield (12.2.22).

Proof of (12.2.23). By the same argument used to prove (12.2.11),

$$(b^3 n)^{-1}Q_{nx,2} = \frac{1}{2}\ddot\mu(x)(bn)^{-1}\sum_{j=1}^{n}K\Big(\frac{nx-j}{nb}\Big)\Big(\frac{nx-j}{nb}\Big)^2 \qquad (12.2.26)$$

$$\to \frac{1}{2}\ddot\mu(x)\int_{\mathbb{R}}K(u)u^2du.$$

To bound $Q_{nx,3}$, let

$$h(v) := \sup_{\xi:|\xi-x|\le|v-x|}|\ddot\mu(\xi) - \ddot\mu(x)|, \quad 0 \le v \le 1.$$

Note that h is a continuous bounded function and $h(x) = 0$. Then, by a Taylor expansion,

$$\left| \mu(v) - \mu(x) - (v-x)\dot{\mu}(x) - \frac{1}{2}(v-x)^2 \ddot{\mu}(x) \right| \leq \frac{1}{2}(v-x)^2 h(v),$$

$$(b^3 n)^{-1}|Q_{nx,3}| \leq \frac{1}{2}(bn)^{-1} \sum_{j=1}^{n} K(\frac{nx-j}{nb})(\frac{nx-j}{nb})^2 h(\frac{j}{n}) \quad (12.2.27)$$

$$\to \frac{h(x)}{2} \int_{\mathbb{R}} K(u)u^2 du = 0,$$

by the same argument as used in proving (12.2.11).

Proof of (12.2.24). Since $\ddot{\mu}$ and h are bounded on $[0,1]$, (12.2.26) and (12.2.27), together with (12.2.12), give

$$(b^3 n)^{-1} \sup_{0 \leq x \leq 1} (|Q_{nx,2}| + |Q_{nx,3}|)$$

$$\leq C \sup_{0 \leq x \leq 1} (nb)^{-1} \sum_{j=1}^{n} K(\frac{nx-j}{nb})(\frac{nx-j}{nb})^2 \leq C.$$

This completes the proof of the lemma.

12.2.2 Asymptotic properties of $\hat{\mu}$

This section establishes the consistency, asymptotic normality, and uniform convergence rate of the kernel-type estimator $\hat{\mu}(x)$ of $\mu(x)$.

Using the notation of (12.2.7), rewrite

$$\hat{\mu}(x) = \bar{\mu}(x) + K_{nx}^{-1} S_n(x) = \mu(x) + [\bar{\mu}(x) - \mu(x)] + K_{nx}^{-1} S_n(x). \quad (12.2.28)$$

Let I_K and a_K be as in (12.2.11) and (12.2.18), respectively. Notice that $E\hat{\mu}(x) = \bar{\mu}(x)$ and $\hat{\mu}(x) - E\hat{\mu}(x) = K_{nx}^{-1} S_n(x)$. The decomposition (12.2.28), Lemma 12.2.3, and Theorem 12.2.1 are used in Theorem 12.2.2 below to describe the asymptotic behavior of the bias term $\bar{\mu}(x) - \mu(x)$ and the stochastic term $K_{nx}^{-1} S_n(x)$ in $\hat{\mu}(x) - \mu(x)$. Let $Z \sim \mathcal{N}_k(0, \mathcal{I})$ random vector.

Theorem 12.2.2. *Let $\{Y_j\}$ be as in (12.2.1), where $\mu \in C_2$, σ is a continuous and positive function on $[0,1]$, and $\{\varepsilon_i\}$ satisfies Assumption 12.2.1. Then, for every $K \in \mathcal{K}$ and for each $x \in (0,1)$,*

$$E(\hat{\mu}(x) - \mu(x))^2 = O(b^4 + (nb)^{-1+2d}), \quad (12.2.29)$$

$$E\hat{\mu}(x) = \mu(x) + \ddot{\mu}(x)(a_K/I_K)\, b^2 + o(b^2) + O(n^{-1}).$$

Moreover, for every $k \geq 1$ and for any distinct $x_1, \cdots, x_k \in (0,1)$,

$$\left(\frac{\hat{\mu}(x_1) - E\hat{\mu}(x_1)}{\sqrt{\operatorname{Var}(\hat{\mu}(x_1))}}, \cdots, \frac{\hat{\mu}(x_k) - E\hat{\mu}(x_k)}{\sqrt{\operatorname{Var}(\hat{\mu}(x_k))}} \right) \to_D Z, \qquad (12.2.30)$$

$$\operatorname{Var}(\hat{\mu}(x_i)) \sim (nb)^{-1+2d} (v_{d,K}^2 / I_K^2) \sigma^2(x_i), \quad i = 1, \cdots, k.$$

From the above theorem, the following corollary is immediate. Let $v_n^2(x) = \operatorname{Var}(S_n(x))$, $0 < x < 1$.

Corollary 12.2.1. *Suppose the assumptions of Theorem 12.2.2 hold, $0 \leq d < 1/2$ and*

$$(nb)^{1/2-d} b^2 \to 0. \qquad (12.2.31)$$

Then, for any $k \geq 1$ and distinct x_i, $i = 1, \cdots, k$, in $(0,1)$,

$$\left(\frac{K_{nx_i}}{v_n(x_i)} (\hat{\mu}(x_i) - \mu(x_i)), \quad i = 1, \cdots, k \right) \to_D Z, \qquad (12.2.32)$$

$$\frac{K_{nx_i}}{v_n(x_i)} \sim (nb)^{1/2-d} \frac{I_K}{v_{d,K} \sigma(x_i)}.$$

For $b = o(n^{-1/5})$, $(nb)^{1/2-d} b^2 = o(n^{-4d/5}) \to 0$, $\forall d \geq 0$.

Moreover, if for $0 < x < 1$ an estimate $\hat{v}_n^2(x)$ has the property

$$\hat{v}_n^2(x) = \operatorname{Var}(S_n(x))(1 + o_p(1)), \qquad (12.2.33)$$

then

$$\frac{K_{nx}}{\hat{v}_n(x)} (\hat{\mu}(x) - \mu(x)) \to_D Z. \qquad (12.2.34)$$

In Section 12.4 below, we describe a sequence of estimators $\hat{v}_n^2(x)$ of the variance $\operatorname{Var}(S_n(x))$ that satisfies (12.2.33) and allows the studentization (12.2.34). These estimators are easy to compute and do not require the estimation of the parameters d and $v_{d,K}$ appearing in the limit (12.2.32).

12.2.3 Uniform rate of convergence

In this section we shall give the uniform consistency rate of $\hat{\mu}$ for μ under somewhat stronger assumptions than needed for finite-dimensional weak convergence. We shall now assume that $K(u)$, $u \in \mathbb{R}$, has a second bounded derivative, such that for some $C < \infty$,

$$|\dot{K}(u)| \leq C(1+|u|)^{-4}, \quad |\ddot{K}(u)| \leq C(1+|u|)^{-4}, \quad u \in \mathbb{R}. \qquad (12.2.35)$$

We also exclude the small bandwidth case $b = o(n^{-1/2})$ by assuming that there exists $c > 0$ such that

$$b^2 n \geq c > 0, \qquad n \geq 1. \tag{12.2.36}$$

Proposition 12.2.1. *Suppose the conditions of Theorem 12.2.2, (12.2.35), and (12.2.36) hold and all moments of the innovations $\{\zeta_j\}$ are finite. Then*

$$\sup_{a \leq x \leq 1-a} |\hat{\mu}(x) - \mu(x)| = O_p\Big(b^2 + (nb)^{-1/2+d+\delta}\Big), \qquad \forall\, a, \delta > 0,$$

$$\sup_{0 \leq x \leq 1} |\hat{\mu}(x) - \mu(x)| = O_p\Big(b + (nb)^{-1/2+d+\delta}\Big).$$

The proof of this proposition is facilitated by the following lemma.

Lemma 12.2.4. *Suppose $\{\varepsilon_j\}$ is as in Assumption 12.2.1 with the innovations $\{\zeta_j\}$ having finite moments. In addition, assume that $K \in \mathcal{K}$ satisfies (12.2.35), $\sigma(u), u \in [0,1]$, is continuous and positive, and b satisfies (12.2.36). Then, with $S_n(x)$ as in (12.2.7),*

$$\sup_{0 \leq x \leq 1} |S_n(x)| = o_p((bn)^{1/2+d+\delta}), \qquad \forall\, \delta > 0. \tag{12.2.37}$$

Proof. Denote by m_n the l.h.s. of (12.2.37). Let $N = [bn]$, $h = 1/N$, and define $x_i = ih$, $i = 0, \cdots, N$. Then

$$m_n \leq \max_{j=1,\cdots,N} \Big(|S_n(x_i)| + \sup_{x_{i-1} < x \leq x_i} |S_n(x) - S_n(x_i)|\Big).$$

To prove (12.2.37), it suffices to show that with $k_n := N^{1/2+d+\delta}$,

$$\limsup_n P(m_n \geq a k_n) \to 0, \qquad \forall\, a > 0. \tag{12.2.38}$$

Set, for $i = 1, \cdots, N$,

$$p_{ni} := P(|S_n(x_i)| \geq a k_n), \quad p'_{ni} := P\Big(\sup_{x_{i-1} < x \leq x_i} |S_n(x) - S_n(x_i)| \geq a k_n\Big).$$

Then $P(m_n \geq a k_n) \leq \sum_{i=1}^{N} (p_{ni} + p'_{ni})$. We shall show that there exists an $\epsilon > 0$, such that

$$p_{ni} \leq C_a N^{-1-\epsilon}, \quad p'_{ni} \leq C_a N^{-1-\epsilon}, \quad i = 1, \cdots, N, \tag{12.2.39}$$

where C_a depends on a but not n. This clearly implies

$$P(m_n \geq a k_n) \leq 2 C_a N^{-\epsilon}, \qquad n \geq 1,$$

thereby proving (12.2.38) because $N = [bn] \to \infty$.

Observe that by (12.2.15),

$$\max_{i=1,\cdots,N} ES_n^2(x_i) \le C(bn)^{1+2d}, \quad n \ge 1,$$

where, here and in the rest of the proof, C does not depend on n or i. The weighted sum $S_n(x_i)$ can be rewritten as a moving-average process with innovations $\{\zeta_j\}$; see (4.3.13). By (2.5.4) and because all moments of ζ_0 are finite, for any integer $p \ge 1$, for all $i = 1, \cdots, N$,

$$E|S_n(x_i)|^{2p} \le C(E(S_n^2(x_i)))^p \le CN^{(1+2d)p}. \tag{12.2.40}$$

Hence,

$$p_{ni} \equiv P(|S_n(x_i)| \ge ak_n) \le \frac{E|S_n(x_i)|^{2p}}{(ak_n)^{2p}} \tag{12.2.41}$$

$$\le C\frac{N^{(1+2d)p}}{(aN^{1/2+d+\delta})^{2p}} = CN^{-2p\delta}.$$

Choosing the integer p such that $2p\delta > 1$, gives the first claim in (12.2.39). To bound p'_{ni}, let

$$s_{ni} := \sup_{x_{i-1} < x \le x_i} |S_n(x) - S_n(x_i)|, \qquad Q(u) := (1+u^2)^{-1}, \ u \in \mathbb{R},$$

$$k_{x,j} := K(\frac{xn-j}{nb}), \qquad q_{x,j} := Q(\frac{xn-j}{nb}), \qquad \sigma_j = \sigma(\frac{j}{n}),$$

$$\widetilde{S}_j(x) = \sum_{l=1}^{j} q_{x,l} \, \sigma_l \varepsilon_l, \qquad j = 1, \cdots, n, \qquad 0 < x < 1.$$

Then summation by parts (2.5.8) gives

$$S_n(x) - S_n(x_i)$$

$$= \sum_{j=1}^{n} (k_{x,j} - k_{x_i,j}) q_{x_i,j}^{-1} \{q_{x_i,j} \sigma_j \varepsilon_j\}$$

$$= \sum_{j=1}^{n-1} \{(k_{x,j} - k_{x_i,j}) q_{x_i,j}^{-1} - (k_{x,j+1} - k_{x_i,j+1}) q_{x_i,j+1}^{-1}\} \widetilde{S}_j(x_i)$$

$$+ (k_{x,n} - k_{x_i,n}) q_{x_i,n}^{-1} \widetilde{S}_n(x_i).$$

By assumption (12.2.36), $h/b \le A < \infty$ for some $A < \infty$. Notice that for all $|y| \le A$ and $u \in \mathbb{R}$,

$$(1 + |u+y|)^{-1} \le C(1 + |u|)^{-1} \le C(1 + u^2)^{-1/2}, \tag{12.2.42}$$

where C does not depend on u but depends on A. Therefore, by (12.2.35), the mean-value theorem, (12.2.42), and with $\dot{k}_{x,j} := \dot{K}((xn - j)/nb)$ and $\ddot{k}_{x,j} := \ddot{K}((xn - j)/nb)$, we have

$$\left|(k_{x,j} - k_{x_i,j})q_{x_i,j}^{-1} - (k_{x,j+1} - k_{x_i,j+1})q_{x_i,j+1}^{-1}\right|$$

$$\leq |k_{x,j} - k_{x_i,j}||q_{x_i,j}^{-1} - q_{x_i,j+1}^{-1}|$$

$$+ |k_{x,j} - k_{x,j+1} - (k_{x_i,j} - k_{x_i,j+1})|q_{x_i,j+1}^{-1}$$

$$\leq C(h/b)|\dot{k}_{\tilde{x},j}|(nb)^{-1}q_{x_i,j}^{-1/2} + C(h/b)|\dot{k}_{x^*,j} - \dot{k}_{x^*,j+1}|q_{x_i,j+1}^{-1}$$

$$\leq C(nb)^{-1}q_{x_i,j}^{-1} + C(nb)^{-1}|\ddot{k}_{x^*,j^*}|q_{x_i,j+1}^{-1}$$

$$\leq C(nb)^{-1}q_{x_i,j}^{-1}, \qquad \tilde{x}, x^* \in [x, x_j], \ j^* \in [j, j+1].$$

Similarly, $|k_{x,n} - k_{x_i,n}|q_{x_i,n}^{-1} \leq C(h/n)|\dot{k}_{x^*,n}|q_{x_i,n}^{-1} \leq C$, where C does not depend on x, x_i, j, or n. Thus

$$s_{ni} \leq C(nb)^{-1}\sum_{j=1}^{n-1} q_{x_i,j}|\widetilde{S}_j(x_i)| + |\widetilde{S}_n(x_i)|.$$

We shall show that for any integer $p \geq 1$,

$$Es_{ni}^{2p} \leq CN^{(1+2d)p}, \tag{12.2.43}$$

which by the same argument as in (12.2.41) proves (12.2.39) for p'_{ni}.

Note that by (12.2.12) and (12.2.15),

$$(nb)^{-1}\sum_{k=1}^{n} q_{x_i,k} \leq C,$$

$$E\left|\widetilde{S}_j(x_i)\right|^2 \leq \sum_{l,k=1}^{n} q_{x_i,l}q_{x_i,k}|\gamma_\varepsilon(l-k)| \leq C(nb)^{1+2d}, \quad 1 \leq j \leq n.$$

Thus, by the same argument as in the proof of (12.2.40),

$$\max_{1 \leq j \leq n} E\left|\widetilde{S}_j(x_i)\right|^{2p} \leq CN^{(1+2d)p}, \quad \forall i = 1, \cdots, N.$$

Therefore,

$$Es_{ni}^{2p} \leq C(nb)^{-2p}E\left(\sum_{j=1}^{n-1} q_{x_i,j}|\widetilde{S}_j(x_i)|\right)^{2p} + CE|\widetilde{S}_n(x_i)|^{2p}$$

$$\leq C(nb)^{-2p}\sum_{j_1,\cdots,j_{2p}=1}^{n-1} q_{x_i,j_1}\cdots q_{x_i,j_{2p}}E|\widetilde{S}_{j_1}(x_i)\cdots\widetilde{S}_{j_{2p}}(x_i)|$$

$$+ CN^{(1+2d)p}$$

$$\leq C\left((nb)^{-1}\sum_{k=1}^{n-1} q_{x_i,k}\right)^{2p}\max_{1 \leq j \leq n-1} E\widetilde{S}_j(x_i)^{2p} + CN^{(1+2d)p}$$

$$\leq CN^{(1+2d)p}, \qquad \forall i = 1, \cdots, N.$$

This completes the proof of (12.2.43) and of the lemma.

Remark 12.2.1. By Theorem 12.2.2, the asymptotic MSE, and the point-wise consistency rate of $\hat{\mu}(x)$, respectively, are

$$c_1(b^4 + c_2(nb)^{-1+2d}), \quad c_1 > 0, \ c_2 > 0, \tag{12.2.44}$$
$$\hat{\mu}(x) - \mu(x) = O_p(b^2 + (nb)^{-1/2+d}), \quad 0 < x < 1.$$

Here the dominating term in the squared bias is of the order b^4, while the variance is of the order $(nb)^{-1+2d}$. The optimal b that minimizes this MSE with respect to n is

$$b_{opt} = c_3 n^{-(1-2d)/(5-2d)}, \quad c_3 := \left(\frac{c_2(1-2d)}{4}\right)^{1/(5-2d)}. \tag{12.2.45}$$

This in turn yields the point-wise convergence rate, for each fixed $x \in (0,1)$,

$$\hat{\mu}(x) - \mu(x) = O_p(b_{opt}^2 + (nb_{opt})^{-(1/2-d)}) = O_p(n^{-(1-2d)/(5/2-d)}).$$

Hall and Hart (1990) showed that in non-parametric regression models (12.2.1) with $\sigma(x) \equiv \sigma$, a positive constant, and long memory moving-average Gaussian errors with autocovariances as in (12.2.3), the uniform convergence rate of a non-parametric estimator of the regression function μ on intervals $[a, 1-a]$, $0 < a < 1$, is the same as above, $n^{-(1/2-d)/(5/2-d)}$, and it is obtained with the bandwidth b as in (12.2.45).

In comparison, the uniform consistency rate given in Proposition 12.2.1 in the current set-up does not require the errors $\{\varepsilon_j\}$ to be Gaussian. Moreover, this rate is almost optimal in the following sense. In (12.2.44), $(nb)^{-1/2+d}$ can be replaced by $(nb)^{-1/2+d+\delta}$ for any $\delta > 0$. Furthermore, Guo and Koul (2007) showed that if the moving-average coefficients a_j of the error process $\{\varepsilon_j\}$ are non-increasing and its innovations $\{\zeta_j\}$ satisfy the Cramér condition, then the uniform consistency rate is $(nb)^{-1/2+d} \log n$.

12.3 Asymptotics of $\hat{\sigma}^2$

Let $\hat{\mu}$, $\hat{\sigma}^2$, b, W, and ω be as in (12.2.4). This section discusses the asymptotic behavior of the bias and variance of the estimator $\hat{\sigma}^2$, followed by its asymptotic distribution. Set $\mu_j := \mu(j/n)$, $\hat{\mu}_j := \hat{\mu}(j/n)$, $\sigma_j := \sigma(j/n)$, $e_j := \sigma_j \varepsilon_j$, $\hat{e}_j := Y_j - \hat{\mu}_j = \mu_j - \hat{\mu}_j + e_j$, and

$$w_{xj} := W(\frac{xn - j}{n\omega}), \quad 1 \le j \le n, \quad W_{nx} := \sum_{j=1}^{n} w_{xj}, \quad 0 \le x \le 1.$$

Recall that $Ee_j^2 = \sigma_j^2 E\varepsilon_j^2 = \sigma_j^2$. The following decomposition is useful in analyzing the MSE of $\hat{\sigma}^2(x)$ and deriving its asymptotic distribution.

$$\hat{\sigma}^2(x) = W_{nx}^{-1} \sum_{j=1}^{n} w_{xj}\hat{e}_j^2 = \bar{\sigma}^2(x) + W_{nx}^{-1}S_{n\varepsilon^2}(x) + W_{nx}^{-1}R_n(x), \quad (12.3.1)$$

$$\bar{\sigma}^2(x) := W_{nx}^{-1} \sum_{j=1}^{n} w_{xj}\sigma_j^2, \quad S_{n\varepsilon^2}(x) := \sum_{j=1}^{n} w_{xj}\sigma_j^2(\varepsilon_j^2 - 1),$$

$$R_n(x) := \sum_{j=1}^{n} w_{xj}(\hat{e}_j^2 - e_j^2).$$

The asymptotics of the bias term $\bar{\sigma}^2(x)$ follows from Lemma 12.2.3. In addition, we need to establish the asymptotic distribution of the stochastic term $S_{n\varepsilon^2}(x)$ and the upper bound for the remainder $R_n(x)$. All these results are facilitated by the content of the following subsection.

12.3.1 *Some preliminaries for $\hat{\sigma}^2$*

The following two lemmas are useful in the large sample analysis of $S_{n\varepsilon^2}(x)$ and $R_n(x)$. Below, we assume that $\{\varepsilon_j\}$ satisfies (12.2.2) and (12.2.3) with parameter $0 \le d < 1/2$. Let $\gamma_{\varepsilon^2}(k) := \text{Cov}(\varepsilon_k^2, \varepsilon_0^2)$, $k \ge 0$, and define

$$\tilde{d} := 0, \qquad\qquad 0 \le d < 1/4, \qquad\qquad\qquad (12.3.2)$$
$$:= 2d - 1/2, \qquad 1/4 < d < 1/2.$$

By Theorem 4.5.2 and Lemma 4.5.3,

$$\sum_{k\in\mathbb{Z}} |\gamma_{\varepsilon^2}(k)| < \infty, \qquad\qquad 0 \le d < 1/4, \qquad (12.3.3)$$

$$\gamma_{\varepsilon^2}(k) \sim \gamma_\varepsilon^2(k) \sim c_\gamma^2 |k|^{-1+2\tilde{d}}, \qquad 1/4 < d < 1/2.$$

Therefore, $\tilde{d}$ is the memory parameter of the process $\{\varepsilon_j^2\}$ with short memory for $0 \le d < 1/4$ and long memory for $1/4 < d < 1/2$. The corresponding asymptotic variances are

$$\tilde{v}_{0,W}^2 := \|W\|_2^2 \sum_{k\in\mathbb{Z}} \gamma_{\varepsilon^2}(k), \quad 0 \le d < 1/4.$$

$$\tilde{v}_{\tilde{d},W}^2 := c_\gamma^2 \int_{\mathbb{R}^2} W(u)W(v)|u - v|^{-1+2\tilde{d}}dudv, \quad 1/4 < d < 1/2.$$

When $1/4 < d < 1/2$, we shall need an additional assumption on the coefficients a_j appearing in Assumption 12.2.1:

$$a_j \sim c_a j^{-1+d}, \quad j \to 0, \qquad c_a > 0. \tag{12.3.4}$$

By Proposition 3.2.1, this assumption implies (12.2.3) with $c_\gamma = c_a^2 \sigma_\zeta^2$ $\times B(d, 1-2d)$. To describe the limiting distribution of $\hat{\sigma}^2(x)$ in this case we need to introduce the 2-tuple Wiener–Itô integral

$$\mathcal{R}_2 := \int_{\mathbb{R}^2} f(u_1, u_2) B(du_1) B(du_2), \tag{12.3.5}$$

$$f(u_1, u_2) := c_a^2 \int_{\mathbb{R}} W(v)(v - u_1)_+^{-1+d}(v - u_2)_+^{-1+d} dv, \quad u_1, u_2 \in \mathbb{R},$$

where $B(u)$ is the standard Brownian motion, as in (4.7.1), where it was denoted by $W(u)$. Note that $f \in L_2(\mathbb{R}^2)$.

Lemma 12.3.1. *Assume that* $\{\varepsilon_j\}$ *satisfies Assumption 12.2.1,* $W \in \mathcal{K}$, $\sigma \in \mathcal{C}_2$, $E\zeta_0^4 < \infty$, *and the bandwidth* ω *satisfies (12.2.5). In addition, if* $1/4 < d < 1/2$, *assume (12.3.4) holds.*

Then, for any $0 < x < 1$,

$$\frac{S_{n\varepsilon^2}(x)}{\sqrt{\text{Var}(S_{n\varepsilon^2}(x))}} \to_D Z \sim \mathcal{N}(0,1), \qquad 0 \le d < 1/4, \tag{12.3.6}$$

$$\to_D (\text{Var}(\mathcal{R}_2))^{-1/2} \mathcal{R}_2, \quad 1/4 < d < 1/2,$$

$$\text{Var}(S_{n\varepsilon^2}(x)) \sim (n\omega)^{1+2d} \tilde{v}_{d,W}^2.$$

Proof. Let $w_x := (w_{x1}, \cdots, w_{xn})'$. Write

$$S_{n\varepsilon^2}(x) = \sum_{j=1}^n w_{xj}(\varepsilon_j^2 - E\varepsilon_j^2), \quad 0 \le x \le 1.$$

By (12.3.3), for $0 \le d < 1/2$, $d \ne 1/4$, the second relation of (12.3.6) follows from (12.2.14), whereas (12.2.11) implies

$$(n\omega)^{-1}\|w_x\|^2 \to \|W\|_2^2, \tag{12.3.7}$$

$$|w_{x1}| + |w_{xn}| + \sum_{j=2}^n |w_{xj} - w_{xj-1}| = o(\|w_x\|). \tag{12.3.8}$$

Let $0 \le d < 1/4$. Then, by (12.3.3), the process $\{\varepsilon_j^2\}$ has short memory. Since the weights $\{w_{xj}\}$ of $S_{n\varepsilon^2}(x)$ satisfy (12.3.7) and $\max_{1 \le j \le n} w_{xj} = O(1) = o(\|w_x\|)$, the claim (12.3.6) regarding the asymptotic normality follows from Proposition 4.5.2.

Next, consider the case $1/4 < d < 1/2$. Setting $a_k = 0$, $k \leq 0$,

$$g_n(s,t) := \sum_{j=1}^{n} w_{xj}\, a_{j-s} a_{j-t}, \quad s,t \in \mathbb{Z},$$

we rewrite

$$\varepsilon_j^2 - E\varepsilon_j^2 = \sum_{s,t \in \mathbb{Z}} a_{j-s} a_{j-t}(\zeta_s \zeta_t - E\zeta_s \zeta_t),$$

$$S_{n\varepsilon^2}(x) = \sum_{s,t \in \mathbb{Z}} g_n(s,t)(\zeta_s \zeta_t - E\zeta_s \zeta_t)$$

$$= \sum_{s=t}[\cdots] + \sum_{s\neq t}[\cdots] = s_{n,1} + s_{n,2}.$$

We shall show that

$$(n\omega)^{-1/2-\tilde{d}} s_{n,1} = o_p(1), \quad (n\omega)^{-1/2-\tilde{d}} s_{n,2} \to_D \mathcal{R}_2, \tag{12.3.9}$$

which will complete the proof of (12.3.6) when $1/4 < d < 1/2$.

Since the ζ_s's are i.i.d. r.v.'s,

$$E s_{n,1}^2 = \mathrm{Var}(\zeta_0^2) \sum_{s \in \mathbb{Z}} g_n^2(s,s) \leq C \sum_{s \in \mathbb{Z}} \Big(\sum_{j=1}^{n} w_{xj}\, a_{j-s} a_{j-s}\Big)^2$$

$$= C \sum_{s \in \mathbb{Z}} \sum_{j,k=1}^{n} w_{xj} w_{xk}\, a_{j-s} a_{j-s} a_{k-s} a_{k-s}$$

$$\leq C \sum_{j=1}^{n} w_{xj}^2 \Big(\sum_{l \in \mathbb{Z}} a_l^2\Big)^2 \leq C\, n\omega,$$

by (12.2.11) because $\sum_{l \in \mathbb{Z}} a_l^2 < \infty$. This yields $s_{n,1} = O_p((n\omega)^{1/2})$, which implies (12.3.9) because $\tilde{d} > 0$.

Next, since $\{\zeta_j\} =_D \{\zeta_{j-[nx]}\}$,

$$s_{n,2} = \sum_{s\neq t} g_n(s,t)(\zeta_s \zeta_t - E\zeta_s \zeta_t)$$

$$=_D \sum_{s\neq t} g_n(s,t)(\zeta_{s-[nx]}\zeta_{t-[nx]} - E\zeta_{s-[nx]}\zeta_{t-[nx]})$$

$$= \sum_{s\neq t} g_n(s+[nx], t+[nx])(\zeta_s \zeta_t - E\zeta_s \zeta_t).$$

The sum on the r.h.s. is an off-diagonal polynomial form in the i.i.d. r.v.'s ζ_j; see (4.7.8). Hence, to show its convergence to the multiple Wiener–Itô integral $\mathcal{R}_2$ of (12.3.5), we can apply the criterion of convergence of discrete multiple integrals of Section 4.7. Let $N = [n\omega]$. Note that by (12.3.3), $1/2 + \tilde{d} = 2d$. The corresponding extension of g_n to $\mathbb{R}^2$, given by (4.7.10), for $N^{-1/2-\tilde{d}} s_{n,2}$ is

$$\tilde{g}_N(u_1, u_2) := N\{N^{-1/2-\tilde{d}} g_n(\lfloor Nu_1 \rfloor + [nx], \lfloor Nu_2 \rfloor + [nx])\},$$

$$= N^{1-2d} \sum_{j=1}^{n} w_{x,j}\, a_{j-\lfloor Nu_1 \rfloor - [nx]} a_{j-\lfloor Nu_2 \rfloor - [nx]}, \quad u_1, u_2 \in \mathbb{R},$$

where $n = n(N, \omega)$ is a function of N and ω. Note that the above extension $\tilde{g}_N$ is an analog of $\tilde{g}_n$ of (4.7.10) with n replaced by N.

By Corollary 4.7.1, to prove the second claim in (12.3.9), it suffices to prove

$$\|\tilde{g}_N - f\|_2 \to 0. \tag{12.3.10}$$

Write

$$\tilde{g}_N(u_1, u_2) = N^{2-2d} \int_{(1-[nx])/N}^{(n-[nx]+1)/N} w_{x,\lfloor Nv \rfloor + [nx]}$$
$$\times a(\lfloor Nv \rfloor - \lfloor Nu_1 \rfloor) a(\lfloor Nv \rfloor - \lfloor Nu_2 \rfloor) dv.$$

By assumption (12.3.4), for every $v, u \in \mathbb{R}$,

$$(N\,|v - u|)^{1-d} |a(\lfloor Nv \rfloor - \lfloor Nu \rfloor)| \le C, \quad \forall n \ge 1,$$
$$(N\,|v - u|)^{1-d} a(\lfloor Nv \rfloor - \lfloor Nu \rfloor) \to c_a,$$

whereas, because $W \in \mathcal{K}$,

$$w_{x,\lfloor Nv \rfloor + [nx]} = W\Big(\frac{nx - [nx] - \lfloor Nv \rfloor}{N}\Big) \to W(v),$$
$$|w_{x,\lfloor Nv \rfloor + [nx]}| \le C, \quad \forall n \ge 1.$$

Thus by the DCT, for every $u_1, u_2 \in \mathbb{R}$,

$$|\tilde{g}_N(u_1, u_2)| \le C f(u_1, u_2), \quad \tilde{g}_N(u_1, u_2) \to f(u_1, u_2),$$

with f as in (12.3.5), which, in turn, implies (12.3.10). This completes the proof of (12.3.9) and of the lemma.

The next lemma provides a bound for the remainder $W_{nx}^{-1} R_n(x)$ term in (12.3.1) where C does not depend on n or x.

Lemma 12.3.2. *Under the assumptions of Lemma 12.3.1, $\forall x \in (0, 1)$,*

$$E\big(R_n(x)/W_{nx}\big)^2 \le C\,\{b^8 + b^4\omega^4 + (nb)^{-2+4d} \tag{12.3.11}$$
$$+ (b^4 + b^2\omega^2)(n\omega)^{-1+2d}$$
$$+ (nb)^{-1+2d}(n\omega)^{-1+2d}\}.$$

In particular, if $d = 0$, then

$$E\big(R_n(x)/W_{nx}\big)^2 \le C\left(b^8 + b^4\omega^4 + (nb)^{-2} + (n\omega)^{-2}\right).$$

Proof. By (12.2.21),

$$W_{nx}^{-1} \le C(n\omega)^{-1}, \quad 0 \le x \le 1, \; n \ge 1, \qquad (12.3.12)$$

$$(n\omega)^{-1}W_{nx} \to I_W := \int_{\mathbb{R}} W(u)du, \quad 0 < x < 1.$$

Let, for $j, s = 1, \cdots, n$,

$$k_{sj} := K((s-j)/nb), \quad K_{n,j} = \sum_{i=1}^{n} k_{ji}, \quad S_j := S_n(j/n), \quad \bar{\mu}_j := \bar{\mu}(j/n).$$

By (12.2.28),

$$\hat{\mu}_j - \mu_j = (\bar{\mu}_j - \mu_j) + K_{n,j}^{-1}S_j =: m_j + z_j. \qquad (12.3.13)$$

Therefore, $\hat{e}_j = Y_j - \hat{\mu}_j = e_j - m_j - z_j$, and

$$\hat{e}_j^2 - e_j^2 = (m_j + z_j - e_j)^2 - e_j^2 = (m_j + z_j)^2 - 2m_j e_j - 2z_j e_j.$$

Hence,

$$R_n(x) = \sum_{j=1}^{n} w_{xj}(\hat{e}_j^2 - e_j^2)$$

$$= \sum_{j=1}^{n} w_{xj}(m_j + z_j)^2 - 2\sum_{j=1}^{n} w_{xj}m_j\varepsilon_j - 2\sum_{j=1}^{n} w_{xj}z_j\varepsilon_j$$

$$=: q_{n,1} - 2q_{n,2} - 2q_{n,3}.$$

We shall show that for a fixed $0 < x < 1$, and all $n \ge 1$,

$$|q_{n,1}| \le C\, n\omega\{b^4 + b^2\omega^2 + (nb)^{-1+2d}\}, \qquad (12.3.14)$$

$$Eq_{n,2}^2 \le C(n\omega)^{1+2d}\{b^4 + b^2\omega^2 + n^{-2}\}, \qquad (12.3.15)$$

$$Eq_{n,3}^2 \le C(n\omega)^2\{(nb)^{-2+4d} + (nb)^{-1+2d}(n\omega)^{-1+2d}\}, \qquad (12.3.16)$$

which together with (12.3.12) will imply (12.3.11).

Proof of (12.3.14). Use $(m_j + z_j)^2 \le 2m_j^2 + 2z_j^2$, to obtain

$$E|q_{n,1}| \le 2\sum_{j=1}^{n} w_{xj}m_j^2 + 2\sum_{j=1}^{n} w_{xj}Ez_j^2 =: q_{n,11} + q_{n,12}.$$

To bound $q_{n,11}$, first we show that, for any fixed $x \in (0, 1)$,

$$w_{xj}^{1/4}|m_j| \le C_x(b^2 + b\omega + n^{-1}), \quad j = 1, \cdots, n, \qquad (12.3.17)$$

where $C_x := (x \wedge (1-x))^{-1}C$. By (12.2.19),

$$|\bar{\mu}_j - \mu_j| \leq C(b^2 + n^{-1} + b(\frac{(n-j) \wedge j}{nb} + 1)^{-1}),$$

for all $j = 1, \cdots, n$, where $0 < C < \infty$ does not depend on j, and $n \geq 1$.
The assumption that $W \in \mathcal{K}$ yields $(W(u))^{1/4} \leq C(1 + |u|)^{-1}$, $u \in \mathbb{R}$. Use
these two bounds together with (12.2.17), to obtain

$$w_{xj}^{1/4}(1 + \frac{(n-j) \wedge j}{nb})^{-1} \leq (1 + \frac{|xn-j|}{n(b \vee \omega)})^{-1}(1 + \frac{(n-j) \wedge j}{n(b \vee \omega)})^{-1}$$

$$\leq C(\frac{(1-x) \wedge x}{b \vee \omega} + 1)^{-1} \leq C_x(b \vee \omega),$$

$$w_{xj}^{1/4}|m_j| = w_{xj}^{1/4}|\bar{\mu}_j - \mu_j| \leq C_x(b^2 + b\omega + n^{-1}), \quad j = 1, \cdots, n.$$

Thus, by (12.2.12),

$$q_{n,11} = \sum_{j=1}^{n} \sqrt{w_{xj}}|w_{xj}^{1/4}m_j|^2 \leq C(b^4 + b^2\omega^2 + n^{-2}) \sum_{j=1}^{n} \sqrt{w_{xj}}$$

$$\leq C(b^4 + b^2\omega^2 + n^{-2}) n\omega.$$

To bound $q_{n,12}$, by (12.2.21) and (12.2.15),

$$K_{n,j}^{-1} \leq C(nb)^{-1}, \quad ES_j^2 \leq C(nb)^{1+2d}, \quad 1 \leq j, l \leq n, \quad (12.3.18)$$

$$Ez_j^2 = E(K_{n,j}^{-1}S_j)^2 \leq C(nb)^{-1+2d}, \quad 1 \leq j, l \leq n,$$

where C does not depend on $1 \leq j \leq n$. By the last bound and (12.2.11),

$$q_{n,12} \leq C(nb)^{-1+2d} \sum_{j=1}^{n} w_{xj} \leq C n\omega(nb)^{-1+2d},$$

which completes the proof of (12.3.14) since $n^{-2} = o((nb)^{-1+2d})$.

Proof of (12.3.15). Observe that $Eq_{n,2} = 0$ and

$$Eq_{n,2}^2 = \sum_{j,l=1}^{n} w_{x\,j}\, m_j\, w_{x\,l}\, m_l\, \gamma_\varepsilon(j - l).$$

By (12.3.17),

$$|w_{xj}^{1/4}m_j w_{xl}^{1/4}m_l| \leq C(b^4 + b^2\omega^2 + n^{-2}), \quad 1 \leq j, l \leq n,$$

$$|Eq_{n,2}^2| \leq C(b^4 + b^2\omega^2 + n^{-2}) \sum_{j,l=1}^{n} w_{xj}^{3/4}w_{xl}^{3/4}|\gamma_\varepsilon(j - l)|$$

$$\leq C(b^4 + b^2\omega^2 + n^{-2})(n\omega)^{1+2d},$$

by (12.2.15), which completes the proof of (12.3.15).

Proof of (12.3.16). By the elementary equation $Eq_{n,3}^2 = E(q_{n,3} - Eq_{n,3})^2 + (Eq_{n,3})^2$, it suffices to show

$$|Eq_{n,3}| \leq C\, n\omega (nb)^{-1+2d}, \tag{12.3.19}$$

$$E(q_{n,3} - Eq_{n,3})^2 \leq C(n\omega)^2 (n\omega)^{-1+2d}(nb)^{-1+2d}.$$

But

$$Eq_{n,3} = \sum_{j=1}^n w_{xj} E[z_j \varepsilon_j] = \sum_{j,\,l=1}^n w_{xj} K_{n,j}^{-1} k_{j,\,l} \gamma_\varepsilon (j - l).$$

By (12.2.21), $K_{n,j}^{-1} \leq C(nb)^{-1}$, for all $1 \leq j \leq n$, and by (12.2.6),

$$|k_{j,l}| \leq C(1 + \frac{|j - l|}{nb})^{-4}, \quad 1 \leq j,\ l \leq n.$$

Therefore

$$|Eq_{n,3}| \leq C(nb)^{-1} \sum_{j,\,l=1}^n |w_{xj}|(1 + \frac{|j - l|}{nb})^{-4} |\gamma_\varepsilon (j - l)|$$

$$\leq C(nb)^{-1} \sum_{j=1}^n w_{xj}\, i_n, \quad i_n := \sum_{s=-n}^n (1 + \frac{|s|}{nb})^{-4} |\gamma_\varepsilon (s)|.$$

By (12.2.11), $\sum_{j=1}^n w_{xj} \leq Cn\omega$. For $d = 0$, by (12.2.2),

$$i_n \leq C \sum_{s \in \mathbb{Z}} |\gamma_\varepsilon (s)| < \infty,$$

whereas, for $0 < d < 1/2$, by (12.2.3),

$$i_n \leq C \sum_{s=-n}^n (1 + \frac{|s|}{nb})^{-4}(1 + |s|)^{-1+2d}$$

$$\leq C(nb)^{2d} \int_{\mathbb{R}} (1 + |u|)^{-4} |u|^{-1+2d} du \leq C(nb)^{2d}.$$

Thus $|Eq_{n,3}| \leq C\, n\omega (nb)^{-1+2d}$, which proves the first bound of (12.3.19).

Next, by definition,

$$z_j = K_{n,j}^{-1} \sum_{s=1}^n k_{j,s} e_s, \quad e_j = \sigma_j \varepsilon_j, \quad j = 1, \cdots, n,$$

are weighted sums of the linear process $\{\varepsilon_j\}$ with innovations $\{\zeta_j\}$. Therefore, the r.v.'s z_j and e_j, $j = 1, 2, \cdots, n$, can be written as moving averages of $\{\zeta_j\}$, as in (4.5.5). Thus, by Lemma 4.5.2,

$$E(q_{n,3} - Eq_{n,3})^2 \leq C \sum_{j,\,l=1}^n w_{xj} w_{xl} \text{Cov}(z_j, z_l) \text{Cov}(e_j, e_l).$$

Now

$$|\text{Cov}(z_j, z_l)| \leq \left(\text{Var}(z_j)\text{Var}(z_l)\right)^{1/2}$$

$$\leq (K_{n,j}K_{n,l})^{-1}\left(\text{Var}(S_j)\text{Var}(S_l)\right)^{1/2} \leq C(bn)^{-1+2d},$$

by (12.3.18), where C does not depend on j or n. Thus, using

$$|\text{Cov}(e_j, e_l)| = \sigma_j\sigma_l|\text{Cov}(\varepsilon_j, \varepsilon_l)| \leq C|\gamma_\varepsilon(j-l)|,$$

$$E(q_{n,3} - Eq_{n,3})^2 \leq C(nb)^{-1+2d}\sum_{j,l=1}^{n} w_{xj}w_{xl}|\gamma_\varepsilon(j-l)|$$

$$\leq C(nb)^{-1+2d}(\omega n)^{1+2d},$$

by (12.2.15), which proves the second bound of (12.3.19). This completes the proof of (12.3.16) and the lemma.

12.3.2   *MSE and asymptotic distribution of* $\hat{\sigma}^2$

We shall now discuss the large sample behavior of the MSE and the asymptotic distribution of $\hat{\sigma}^2$. Because of Lemmas 12.3.1 and 12.3.2, we need to consider two cases: the not so strong long memory case $0 \leq d < 1/4$, and the very strong long memory case $1/4 < d < 1/2$. Recall the definition of $\tilde{d}$ from (12.3.2).

Theorem 12.3.1. *Let* $0 \leq d < 1/2$*. Assume that* $\{\varepsilon_j\}$ *satisfies Assumption 12.2.1,* $W \in \mathcal{K}$*,* $\sigma \in \mathcal{C}_2$*, and* $E\zeta_0^4 < \infty$*. Let the bandwidths* b *and* ω *satisfy (12.2.5) and*

$$b = O(n^{-1/8}), \qquad \omega = O(b). \tag{12.3.20}$$

In addition, for $1/4 < d < 1/2$*, assume (12.3.4). Then the following results hold for every* $x \in (0,1)$*.*
(i) *The MSE satisfies*

$$E\left(\hat{\sigma}^2(x) - \sigma^2(x)\right)^2 = O\left(\omega^4 + (n\omega)^{-1+2\tilde{d}}\right).$$

(ii) *If* $\omega = o(n^{-1/5})$*, and if, for* $d \in (1/4, 1/2)$*,* $\omega = o(b)$*, then*

$$\frac{\hat{\sigma}^2(x) - \sigma^2(x)}{\sqrt{\text{Var}(\hat{\sigma}^2(x))}} \to_D Z \sim \mathcal{N}(0,1), \qquad 0 \leq d < 1/4, \tag{12.3.21}$$

$$\to_D (\text{Var}(\mathcal{R}_2))^{-1/2}\mathcal{R}_2, \quad 1/4 < d < 1/2,$$

$$\text{Var}(\hat{\sigma}^2(x)) = \frac{\text{Var}(S_{n\varepsilon^2}(x))}{W_{nx}^2} \sim (n\omega)^{-1+2\tilde{d}}\tilde{v}_{\tilde{d},W}^2/I_W^2.$$

Proof. Fix $0 < x < 1$. By (12.3.1),

$$\hat{\sigma}^2(x) - \sigma^2(x) = \bar{\sigma}^2(x) - \sigma^2(x) + S_{n\varepsilon^2}(x)/W_{nx} + R_n(x)/W_{nx}.$$

(i) We shall now bound the terms on the r.h.s. of the above equality. By Lemma 12.2.3, $\bar{\sigma}^2(x) - \sigma^2(x) = O(\omega^2 + n^{-1})$. By Lemma 12.3.1, $S_{n\varepsilon^2}(x)$ satisfies (12.3.6). In particular, $S_{n\varepsilon^2}(x) = O_p((n\omega)^{1/2+\tilde{d}})$, whereas by (12.3.12), $W_{nx} \sim n\omega I_W$, and thus

$$S_{n\varepsilon^2}(x)/W_{nx} = O_p((n\omega)^{-1/2+\tilde{d}}).$$

Finally, (12.3.20) implies $b^4 = O(n^{-1/2}) = o((n\omega)^{-1/2})$, and hence, by Lemma 12.3.2,

$$E(R_n(x)/W_{nx})^2 = O\big(b^8 + \omega^4 + (n\omega)^{-2+4d}\big), \qquad \text{if } \omega = O(b),$$
$$= O(b^8 + \omega^4) + o((n\omega)^{-2+4d}), \qquad \text{if } \omega = o(b).$$

These bounds prove (i) because by the definition of $\tilde{d}$, $(n\omega)^{-1+2\tilde{d}} \equiv (n\omega)^{-1}$ for $0 \le d < 1/4$, and $(n\omega)^{-1+2\tilde{d}} \equiv (n\omega)^{-2+4d}$ for $1/4 < d < 1/2$.

(ii) Claim (12.3.21) follows from the above bounds and Lemma 12.3.1, because $\omega = o(n^{-1/5})$ implies $\omega^4 = o((n\omega)^{-1}) = o((n\omega)^{-1+2\tilde{d}})$. This completes the proof of the theorem.

Remark 12.3.1. In particular, if $\mu(x)$ and $\sigma^2(x)$ are estimated using the same bandwidth $b = \omega$, then the MSE of $\hat{\sigma}^2(x)$ is as in Theorem 12.3.1(i). In addition, for $0 \le d < 1/4$, we can use $b = \omega = o_p(n^{-1/5})$ to obtain the normal approximation (12.3.21).

For short memory errors $\{\varepsilon_j\}$, $d = 0$, assumption $\omega = O(b)$ of (12.3.20) can be relaxed. Let $b = O(n^{-1/8})$. Then by Lemmas 12.3.1, 12.3.2, and 12.2.3,

$$\text{Var}(\hat{\sigma}^2(x)) = \frac{\tilde{v}_{0,W}^2}{n\omega\, I_W^2} + o\big(\frac{1}{n\omega}\big),$$
$$\hat{\sigma}^2(x) - \sigma^2(x) = O_p\big(\omega^2 + (nb)^{-1} + (n\omega)^{-1/2}\big).$$

If, in addition, $\omega = o(n^{-1/5})$ and $\omega = o(nb^2)$, then

$$\frac{\hat{\sigma}^2(x) - \sigma^2(x)}{\sqrt{\text{Var}(\hat{\sigma}^2(x))}} = Z + O_p\big(\frac{\sqrt{n\omega}}{nb}\big) + o_p(1)$$
$$\to_D Z.$$

In particular, the above CLT holds if we choose $\omega = n^{-\gamma_\omega}$ and $b = n^{-\gamma_b}$ with $1/5 < \gamma_\omega < 1$, and $1/8 \le \gamma_b < 1/2 + \gamma_\omega/2$.

12.4 Estimation of Var$(\hat{\mu}(x))$

Deriving the confidence intervals for $\mu(x)$, $0 \le x \le 1$, based on the asymptotic normality result (12.2.32), requires estimation of the unknown variance Var$(\hat{\mu}(x))$. In this section we shall provide estimators of this variance that do not require the estimation of the unknown parameters d and $v_{d,K}$ appearing in the limit variance (12.2.32).

Since Var$(\hat{\mu}(x)) = $ Var$(S_n(x))/K_{nx}^2$, it suffices to construct an estimator $\hat{v}_n^2(x)$ of $v_n^2(x) = $ Var$(S_n(x))$ satisfying (12.2.33). Let $\hat{\mu}(x)$ be an estimator of $\mu(x)$ of the model (12.2.1) computed with kernel K and bandwidth b. Recall the notation

$$k_{nx,j} = K(\frac{xn - j}{bn}), \quad \sigma_j = \sigma(j/n), \quad e_j = \sigma_j \varepsilon_j, \quad j = 1, \cdots, n.$$

Recall also the convention $\gamma_\varepsilon(0) = 1$, so that the autocorrelation $r_\varepsilon(j) \equiv \gamma_\varepsilon(j)/\gamma_\varepsilon(0) = \gamma_\varepsilon(j)$, $j \ge 0$. Note that

$$v_n^2(x) = \sum_{j,l=1}^{n} k_{nx,j} k_{nx,l} \mathrm{Cov}(e_j, e_l).$$

Let $\hat{\mu}_h$ denote the estimator $\hat{\mu}$ of (12.2.4) with $b = h$, a possibly different bandwidth sequence such that $h \to 0$, $hn \to \infty$, and set

$$\hat{e}_j := Y_j - \hat{\mu}_h(j/n), \quad j = 1, \cdots, n,$$

$$\hat{\gamma}_{\hat{e}}(k) := n^{-1} \sum_{j=1}^{n-|k|} \hat{e}_j \hat{e}_{j+|k|}, \quad \hat{r}_{\hat{e}}(k) := \frac{\hat{\gamma}_{\hat{e}}(k)}{\hat{\gamma}_{\hat{e}}(0)}, \quad |k| \le n - 1.$$

Our estimator of $v_n^2(x)$ is

$$\hat{v}_n^2(x) := \hat{\sigma}^2(x) \sum_{j,l=1}^{n} k_{nx,j} k_{nx,l} \hat{r}_{\hat{e}}(j - l), \quad 0 < x < 1. \tag{12.4.1}$$

In (12.4.1), the estimate $\hat{\sigma}^2(x)$ denotes any consistent estimate of $\sigma^2(x)$.

If $\sigma^2(x) \equiv \sigma^2$, then (12.4.1) needs to be replaced by

$$\hat{v}_n^2(x) := \sum_{j,l=1}^{n} k_{nx,j} k_{nx,l} \hat{\gamma}_{\hat{e}}(j - l). \tag{12.4.2}$$

Theorem 12.4.1. *Suppose the assumptions of Theorem 12.2.2 hold and $K \in \mathcal{K}$. Let $\hat{v}_n^2(x)$ be as in (12.4.1), with the $\hat{e}_j$'s obtained using kernel K and the bandwidth h, and $\hat{\sigma}^2(x)$ be an estimator of $\sigma^2(x)$ satisfying*

$$\hat{\sigma}^2(x) \to_p \sigma^2(x), \quad \forall x \in (0, 1). \tag{12.4.3}$$

Assume also that

$$nbh^3 \to 0, \qquad b = o(h). \tag{12.4.4}$$

Then, for every $0 < x < 1$,

$$\hat{v}_n^2(x) = v_n^2(x)(1 + o_p(1)). \tag{12.4.5}$$

Moreover, if $\sigma(x) \equiv \sigma$, $0 \le x \le 1$, is constant, then $\hat{v}_n^2(x)$ of (12.4.2) also satisfies (12.4.5).

Proof. Fix $x \in (0, 1)$. Let $c_\sigma := \int_0^1 \sigma^2(u)du$,

$$Q_{n,\hat{e}} := \sum_{j,l=1}^n k_{nx,j}k_{nx,l}\hat{\gamma}_{\hat{e}}(j-l), \quad \tilde{v}_n^2(x) := \sum_{j,l=1}^n k_{nx,j}k_{nx,l}\gamma_\varepsilon(j-l).$$

Write

$$\hat{v}_n^2(x) = \frac{\hat{\sigma}^2(x)Q_{n,\hat{e}}}{\widehat{\gamma}_{\hat{e}}(0)}.$$

We shall prove below that

$$\hat{\gamma}_{\hat{e}}(0) \to_p c_\sigma, \tag{12.4.6}$$

$$Q_{n,\hat{e}} = c_\sigma \tilde{v}_n^2(x) + o_p((nb)^{1+2d}). \tag{12.4.7}$$

This together with (12.4.3) proves (12.4.5):

$$\hat{v}_n^2(x) = \sigma^2(x)\tilde{v}_n^2(x)(1 + o_p(1)) = v_n^2(x)(1 + o_p(1)),$$

where the last equality holds because of (12.2.14),

$$\sigma^2(x)\tilde{v}_n^2(x) = v_n^2(x) + o((nb)^{1+2d}) = v_n^2(x)(1 + o(1)) \tag{12.4.8}$$
$$\sim \sigma^2(x)v_{d,K}^2(nb)^{1+2d}.$$

Proof of (12.4.6). Write

$$\hat{\gamma}_{\hat{e}}(0) = \hat{\gamma}_e(0) + (\hat{\gamma}_{\hat{e}}(0) - \hat{\gamma}_e(0)). \tag{12.4.9}$$

By $\hat{e}_j^2 - e_j^2 = 2(\hat{e}_j - e_j)e_j + (\hat{e}_j - e_j)^2$, and the C–S inequality,

$$|\hat{\gamma}_{\hat{e}}(0) - \hat{\gamma}_e(0)| \le n^{-1}\sum_{j=1}^n |\hat{e}_j^2 - e_j^2| \le 2(\gamma_{\hat{e}-e}(0)\gamma_e(0))^{1/2} + \gamma_{\hat{e}-e}(0).$$

We shall show that

$$\hat{\gamma}_e(0) \to_p c_\sigma, \quad \hat{\gamma}_{\hat{e}-e}(0) \to_p 0, \tag{12.4.10}$$

which, together with (12.4.9), will yield (12.4.6).

Recall that $E\varepsilon_j^2 = \gamma_\varepsilon(0) = 1$. Hence, (12.2.2) and (12.2.3) imply,

$$E\hat{\gamma}_e(0) = n^{-1} \sum_{j=1}^{n} \sigma_j^2 \to \int_0^1 \sigma^2(u)du = c_\sigma,$$

$$E\big(\hat{\gamma}_e(0) - E\hat{\gamma}_e(0)\big)^2 = n^{-2} \sum_{j,k=1}^{n} \sigma_j\sigma_k\gamma_\varepsilon(j-k)$$

$$\leq Cn^{-2} \sum_{j,k=1}^{n} |\gamma_\varepsilon(j-k)| \to 0,$$

which yields the first convergence of (12.4.10).

Next, consider $\hat{\gamma}_{\hat{e}-e}(0)$. Write $\hat{\mu}_j = \bar{\mu}_j + K_{n,j}^{-1}S_j$, where $\bar{\mu}_j := \bar{\mu}(j/n)$, and $K_{n,j}$ and S_j, are as in (12.2.28), but computed with the bandwidth h instead of b. Then

$$(\hat{e}_j - e_j)^2 = (\hat{\mu}_j - \mu_j)^2 \leq 2(\bar{\mu}_j - \mu_j)^2 + 2K_{n,j}^{-2}S_j^2.$$

By Lemma 12.2.3,

$$|\bar{\mu}_j - \mu_j| \leq C\{h^2 + n^{-1} + h\,(1 + \frac{(n-j)\wedge j}{nh})^{-2}\},$$

$$\sum_{j=1}^{n} |\bar{\mu}_j - \mu_j|^2 \leq Cn(h^4 + n^{-2}) + Ch^2 \sum_{j=1}^{n} (1 + \frac{j}{nh})^{-2}$$

$$\leq C\Big(nh^4 + n^{-1} + nh^3 \int_0^\infty (1+u)^{-2}du\Big)$$

$$\leq C(nh^3 + n^{-1}).$$

By (12.2.21) and (12.2.15),

$$\sum_{j=1}^{n} K_{n,j}^{-2}ES_j^2 \leq C \sum_{j=1}^{n} (nh)^{-1+2d} = Cn(nh)^{-1+2d}.$$

Thus

$$E\hat{\gamma}_{\hat{e}-e}(0) \equiv n^{-1} \sum_{j=1}^{n} E(\hat{e}_j - e_j)^2 \leq C(h^3 + n^{-2} + (nh)^{-1+2d}). \quad (12.4.11)$$

Since the r.h.s. of (12.4.11) tends to zero, this implies the second convergence of (12.4.10) and completes the proof of (12.4.6).

Proof of (12.4.7). Decompose

$$Q_{n,\hat{e}} = c_\sigma \tilde{v}_n^2(x) + (Q_{n,e} - EQ_{n,e}) + (EQ_{n,e} - c_\sigma\tilde{v}_n^2(x)) + (Q_{n,\hat{e}} - Q_{n,e}).$$

We shall show that

$$Q_{n,e} - EQ_{n,e} = o_p((nb)^{1+2d}), \qquad (12.4.12)$$

$$EQ_{n,e} - c_\sigma \tilde{v}_n^2(x) = o((nb)^{1+2d}), \qquad (12.4.13)$$

$$Q_{n,\hat{e}} - Q_{n,e} = o_p((nb)^{1+2d}), \qquad (12.4.14)$$

which yield (12.4.7). Moreover, by (12.4.8), (12.4.12), and (12.4.13),

$$Q_{n,e} = O_p((nb)^{1+2d}), \qquad (12.4.15)$$

which, in turn, will be used to prove (12.4.14). Let, for $v \in \Pi$,

$$\Psi(v) := \sum_{j=1}^{n} e^{ijv} k_{nx,j}, \quad I_{\hat{e}}(v) := (2\pi n)^{-1} \left| \sum_{j=1}^{n} e^{ivj} \hat{e}_j \right|^2.$$

Note that $\hat{\gamma}_{\hat{e}}(k) = \hat{\gamma}_{\hat{e}}(-k)$, for $|k| \leq n-1$, and

$$\hat{\gamma}_{\hat{e}}(k) = \int_\Pi e^{ikv} I_{\hat{e}}(v) dv, \quad |k| \leq n-1,$$

$$Q_{n,\hat{e}} := \sum_{j,l=1}^{n} k_{nx,j} k_{nx,l} \int_\Pi e^{i(j-l)v} I_{\hat{e}}(v) dv = \int_\Pi |\Psi(v)|^2 I_{\hat{e}}(v) dv,$$

$$Q_{n,e} = \int_\Pi |\Psi(v)|^2 I_e(v) dv.$$

Proof of (12.4.12). Let

$$\psi(s) := (2\pi)^{-1} \int_\Pi e^{isu} |\Psi(u)|^2 du \qquad (12.4.16)$$

$$= \sum_{j=1}^{n-|s|} k_{nx,j} k_{nx,j+|s|}, \quad |s| \leq n-1.$$

Then

$$Q_{n,e} = \int_\Pi |\Psi(u)|^2 I_e(u) du$$

$$= \frac{1}{2\pi n} \int_\Pi |\Psi(u)|^2 \sum_{t,s=1}^{n} e^{i(t-s)u} e_t e_s du = \frac{1}{n} \sum_{t,s=1}^{n} \psi(t-s) e_t e_s.$$

By Lemma 4.5.2,

$$\operatorname{Var}(Q_{n,e}) \leq C n^{-2} \sum_{j_1,k_1,j_2,k_2=1}^{n} |\psi(j_1-k_1)\psi(j_2-k_2)| \qquad (12.4.17)$$

$$\times |\gamma_\varepsilon(j_1-j_2)\gamma_\varepsilon(k_1-k_2)|.$$

First, consider the case when $0 \le d < 1/4$. Clearly,

$$|\psi(j_1 - k_1)\psi(j_2 - k_2)| \le \psi^2(j_1 - k_1) + \psi^2(j_2 - k_2),$$

$$\operatorname{Var}(Q_{n,e}) \le Cn^{-1} \sum_{s_1=-n}^{n} \psi^2(s_1) \sum_{s_2=-n}^{n} |\gamma_\varepsilon(s_2)| \sum_{s_3=-n}^{n} |\gamma_\varepsilon(s_3)|.$$

By (12.4.16) and (12.2.12), and by assumptions (12.2.2) and (12.2.3),

$$\sum_{s=-n}^{n} \psi^2(s) \le C(\sum_{j=1}^{n} k_{nx,j})^3 \le C(bn)^3, \qquad \sum_{s=-n}^{n} |\gamma_\varepsilon(s)| \le Cn^{2d}.$$

Thus

$$\operatorname{Var}(Q_{n,e}) \le Cn^{-1}(nb)^3 n^{4d} = (nb)^{2+4d} b^{1-4d} = o((nb)^{2+4d}),$$

which proves (12.4.12) for $0 \le d < 1/4$.

Consider the case $1/4 \le d < 1/2$. The inequality (12.4.17) together with the bound $|\gamma_\varepsilon(j_1 - j_2)\gamma_\varepsilon(k_1 - k_2)| \le \gamma_\varepsilon^2(j_1 - j_2) + \gamma_\varepsilon^2(k_1 - k_2)$ implies

$$\operatorname{Var}(Q_{n,e}) \le Cn^{-1} \left(\sum_{s=-n}^{n} \psi(s) \right)^2 \sum_{j=-n}^{n} \gamma_\varepsilon^2(j).$$

By (12.4.16) and (12.2.12),

$$\sum_{s=-n}^{n} \psi(s) \le C(\sum_{j=1}^{n} k_{nx,j})^2 \le C(bn)^2.$$

By (12.2.3), $\gamma_\varepsilon^2(j) \le C(1 + |j|)^{-2(1-2d)}$. Because $0 < 2(1 - 2d) \le 1$,

$$\sum_{j=-n}^{n} \gamma_\varepsilon^2(j) \le C \log n, \qquad d = 1/4,$$

$$\le Cn^{-1+4d}, \qquad 1/4 < d < 1/2.$$

Therefore, for $d = 1/4$,

$$\operatorname{Var}(Q_{n,e}) \le Cn^{-1}(nb)^4 \log n = (nb)^{2+4d} b \log n = o((nb)^{2(1+2d)}),$$

since by (12.4.4), $b \log n \to 0$, which implies (12.4.12).

Further, for $1/4 < d < 1/2$,

$$\operatorname{Var}(Q_{n,e}) \le C(nb)^4 n^{-2+4d} = (nb)^{2+4d} b^{2-4d} = o((nb)^{2(1+2d)}),$$

since $b^{2-4d} = o(1)$, which completes the proof of (12.4.12).

Proof of (12.4.13). By definition, $Q_{n,e} = \sum_{j,l=1}^{n} k_{nx,j} k_{nx,l} \hat{\gamma}_e(j-l)$. Use the equalities

$$\mathrm{Cov}(e_j, e_{j+k}) = \sigma_j \sigma_{j+k} \gamma_\varepsilon(k),$$

$$E\hat{\gamma}_e(k) = n^{-1} \sum_{j=1}^{n-|k|} \mathrm{Cov}(e_j, e_{j+|k|}) = \gamma_\varepsilon(k) n^{-1} \sum_{j=1}^{n-|k|} \sigma_j \sigma_{j+|k|}$$

$$=: \gamma_\varepsilon(k) \hat{\gamma}_\sigma(k), \qquad k = 0, \pm 1, \cdots, \pm(n-1),$$

to obtain

$$|EQ_{n,e} - c_\sigma \tilde{v}_n^2(x)| \tag{12.4.18}$$

$$\leq C \sum_{j,l=1}^{n} k_{nx,j} k_{nx,l} |\gamma_\varepsilon(j-l)| \left| \hat{\gamma}_\sigma(j-l) - c_\sigma \right|.$$

Since σ^2 is continuous and bounded, the DCT yields

$$|\hat{\gamma}_\sigma(k) - c_\sigma| \leq \sup_{0 \leq u \leq 1} \sigma^2(u) < \infty,$$

$$|\hat{\gamma}_\sigma(k) - c_\sigma| \to 0, \qquad \forall \text{ fixed } k = 0, \pm 1, \cdots,$$

$$|\hat{\gamma}_\sigma([unb]) - c_\sigma| \to 0, \qquad \forall \text{ fixed } u \geq 0.$$

Now, if $\{\varepsilon_j\}$ has long memory with parameter $0 < d < 1/2$, then (12.4.13) follows by bounding the r.h.s. of (12.4.18) as in the proof of (12.2.14) for $x = y$.

If $\{\varepsilon_j\}$ has short memory, bound the r.h.s. of (12.4.18) by

$$C \sum_{j=1}^{n} k_{nx,j}^2 \sum_{k=-n}^{n} |\gamma_\varepsilon(k)| |\hat{\gamma}_\sigma(k) - c_\sigma|.$$

By (12.2.11), the first sum in this product is of order $O(nb)$ while, by the DCT, the second sum tends to zero because $|\gamma_\varepsilon(k)|$ is summable. This proves (12.4.13) for the short memory case.

Proof of (12.4.14). Use

$$|I_{\hat{e}}(v) - I_e(v)|$$

$$\leq (2\pi n)^{-1} \{ 2 | \sum_{j=1}^{n} e^{ivj} (\hat{e}_j - e_j) | | \sum_{j=1}^{n} e^{ivj} e_j | + | \sum_{j=1}^{n} e^{ivj} (\hat{e}_j - e_j) |^2 \}$$

$$\leq 2 I_{\hat{e}-e}^{1/2}(v) I_e^{1/2}(v) + I_{\hat{e}-e}(v),$$

and the C–S inequality, to bound

$$|Q_{n,\hat{e}} - Q_{n,e}| \leq \int_{\Pi} |\Psi(v)|^2 |I_{\hat{e}}(v) - I_e(v)| dv$$

$$\leq 2\left(\int_{\Pi} |\Psi(v)|^2 I_{\hat{e}-e}(v) dv\right)^{1/2} \left(\int_{\Pi} |\Psi(v)|^2 I_e(v) dv\right)^{1/2}$$

$$+ \int_{\Pi} |\Psi(v)|^2 I_{\hat{e}-e}(v) dv = 2(Q_{n,\hat{e}-e})^{1/2}(Q_{n,e})^{1/2} + Q_{n,\hat{e}-e}.$$

By (12.4.15), $Q_{n,e} = O_p((nb)^{1+2d})$. To complete the proof of (12.4.14), it suffices to show that

$$Q_{n,\hat{e}-e} = o_p((nb)^{1+2d}). \tag{12.4.19}$$

By (12.2.12), $|\Psi(v)| \leq \sum_{j=1}^{n} k_{nx,j} \leq Cnb$. Hence,

$$Q_{n,\hat{e}-e} \leq C(nb)^2 \int_{\Pi} I_{\hat{e}-e}(v) dv \leq C(nb)^2 n^{-1} \sum_{j=1}^{n} (\hat{e}_j - e_j)^2.$$

Thus by (12.4.11) and assumption (12.4.4),

$$EQ_{n,\hat{e}-e} \leq C(nb)^2 \{h^3 + n^{-2} + (nh)^{-1+2d}\}$$

$$\leq C(nb)^{1+2d}\{nbh^3 + n^{-1} + (b/h)^{1-2d}\} = o((nb)^{1+2d}),$$

which yields (12.4.19). This completes the proof of the theorem for the estimate (12.4.1).

When $\sigma(x) \equiv \sigma$, a constant, rewrite estimate (12.4.2) as $\hat{v}_n^2(x) = Q_{n,\hat{e}}$. Therefore, (12.4.5) follows from (12.4.7) and (12.4.8), which completes the proof of the theorem.

Note: Robinson (1997) established the asymptotic distribution of $\hat{\mu}$ in the homoscedastic non-parametric regression model (12.2.1), where $\sigma^2(x) \equiv \sigma^2$, a constant. Some of the proofs above are influenced by Guo and Koul (2007), who obtained similar results to those discussed in this chapter under somewhat different conditions and using the CLT given in Robinson (1997). Csörgő and Mielniczuk (1995a, 1995b) discussed kernel-type estimators of the regression function in homoscedastic models where the errors are subordinated to a long memory Gaussian process. Fan and Yao (1998) proved the finite-dimensional asymptotic normality of the estimators $\hat{\sigma}^2$ for stationary and absolutely regular errors.

Chapter 13

Model Diagnostics

13.1 Introduction

In the previous two chapters we assumed the existence of a regression model and then analyzed the large sample behavior of various inference procedures pertaining to the model. In this chapter we discuss lack-of-fit tests of a given parametric regression model and the tests of a subhypothesis for linear regression models in the presence of long memory in errors or design variables. The asymptotic null distributions of some tests of the goodness-of-fit of a marginal error distribution are also discussed.

There are examples in finance where even the design variables are believed to have long memory. In international finance we often regress spot returns on the forward premium. Baillie and Bollerslev (1994, 2000) and Maynard and Phillips (2001) documented the long memory characteristics of the forward premium while Cheung (1993) found empirical evidence for long memory in some spot return series. The occurrence of long memory in both spot return and in the forward premium then gives rise to the possibility of a balanced regression. However, some authors promote the case of an unbalanced regression when spot returns are close to being uncorrelated, while the forward premium is a long memory process. This would then lead to fitting a regression model with long memory design but martingale-difference errors. This is the motivation for analyzing the statistical behavior of testing procedures for the lack-of-fit hypothesis of a regression model in the presence of long memory in design and when errors either form a martingale-difference sequence or a stationary long memory process.

446

The problem of fitting an error distribution is also important for many applications. Often statisticians use inference procedures that are valid when the underlying process is Gaussian. But, if in a given situation the hypothesis of Gaussianity of the marginal error distribution is rejected, then the validity of such an inference would be in doubt.

In Section 13.2 we study the asymptotic behavior of a test based on a particular marked empirical process for the lack-of-fit hypothesis when errors are homoscedastic while a similar analysis is carried out in Section 13.3 for non-parametric heteroscedastic situations. Sections 13.4 and 13.5 discuss the problem of testing a subhypothesis in linear regression models and goodness-of-fit testing, respectively.

13.2 Lack-of-fit tests

A classical problem in statistics is to assess the effect of an input variable X on the response variable Y. Assuming Y has finite expectation, this is often formulated in terms of the regression function $\mu(x) := E(Y|X = x)$. In practice we often have a parametric model $\mathcal{M} := \{m_\beta(x); \beta \in \Omega \subset \mathbb{R}^q, x \in \mathbb{R}\}$ available and we are interested in testing $\mathcal{H}_0 : \mu \in \mathcal{M}$, against the alternative that $\mathcal{H}_0$ is not true based on n observations (X_i, Y_i), $1 \le i \le n$, from the distribution of (X, Y).

Processes useful for this problem are

$$V(x) := \sum_{i=1}^{n}(Y_i - \mu(X_i))I(X_i \le x), \qquad (13.2.1)$$

$$V_\beta(x) := \sum_{i=1}^{n}(Y_i - m_\beta(X_i))I(X_i \le x), \quad \beta \in \Omega, \, x \in \mathbb{R}.$$

These are known as *marked residual empirical processes* with the marks $Y_i - \mu(X_i)$ and $Y_i - m_\beta(X_i)$, respectively. Tests of the simple hypothesis $\mu = \mu_0$, where μ_0 is a known regression function, are based on V^0, the V process with $\mu = \mu_0$, while tests of $\mathcal{H}_0$ are based on $V_{\beta_n}(x)$, where β_n is an estimator of β under $\mathcal{H}_0$.

Several well-known lack-of-fit tests are based on the analogs of the V_{β_n} process; see Hart (1997). Von Neumann (1941) considered the problem of testing for no design effect when $q = 1$ and $X_i = i/n$. In other words his testing problem was to test for $\mu(x) \equiv \mu$, a constant in x. Under this assumption, $\beta = \mu$, $m_\beta(x) \equiv \mu$ and a natural estimator of β is $\beta_n = \bar{Y} = \sum_{i=1}^n Y_i/n$. If we let $e_i := Y_i - \bar{Y}$, then we see that $V_{\beta_n}(x) = \sum_{i \leq [nx]} e_i$, $0 \leq x \leq 1$, is the partial-sum process of the residuals e_i, $1 \leq i \leq n$. Let $T_j := \sum_{i \leq j} e_i$, $s_n := \left(\sum_{i=1}^n (Y_i - Y_{i-1})^2/(n-1) \right)^{1/2}$, and $B_n := n^{-2} \sum_{j=1}^n T_j^2 = n^{-1} \int_0^1 V_{\beta_n}^2(x) dx$. The von Neumann test uses $\max_{1 \leq j \leq n} |T_j|/s_n \sqrt{n}$.

For the same problem, Buckley (1991) showed that the test based on $2B_n/s_n^2$ is locally the most powerful for a particular Bayesian model.

For the sake of clarity we shall focus on the special case where $m_\beta(x) = \beta' \ell(x)$, $\ell := (\ell_1, \cdots, \ell_q)'$, and where $\ell := (\ell_1, \cdots, \ell_q)'$ is a vector of measurable functions satisfying

$$E\|\ell(X_0)\|^2 < \infty, \quad \Lambda := E\ell(X_0)\ell(X_0)' \text{ positive definite.} \qquad (13.2.2)$$

The null hypothesis now becomes

$$H_0 : \mu(x) = \beta_0' \ell(x), \ \ \forall\, x \in \mathbb{R}, \text{ for some } \beta_0 \in \mathbb{R}^q, \qquad (13.2.3)$$

against the alternative H_0 is not true. This hypothesis, for example, covers the case of fitting a polynomial of degree q where $\ell_j(x) = x^{j-1}$, $j = 1, \cdots, q$. Under H_0, the Y_i's follow the regression model

$$Y_i = \beta_0' \ell(X_i) + \varepsilon_i, \quad i = 1, \cdots, n, \qquad (13.2.4)$$

where $\{\varepsilon_i\}$ is a mean-zero stationary process, independent of $\{X_j\}$.

The corresponding marked empirical process of interest here is

$$\mathcal{V}_\beta(x) := \sum_{i=1}^n (Y_i - \beta' \ell(X_i)) I(X_i \leq x), \ \ \beta \in \Omega, \ x \in \mathbb{R}.$$

With $\varepsilon_i := Y_i - \beta_0' \ell(X_i)$,

$$\mathcal{V}_{\beta_0}(x) := \sum_{i=1}^n \varepsilon_i I(X_i \leq x), \quad x \in \mathbb{R}.$$

In this section the design variables X_i, $i \in \mathbb{Z}$, are assumed to be stationary long memory moving averages of the form

$$X_j = m + \sum_{i=0}^\infty b_i \xi_{j-i}, \quad b_i \sim c_1 i^{-1+d_1}, \ i \to \infty, \qquad (13.2.5)$$

for some $0 < d_1 < 1/2$, $c_1 > 0$, $\{\xi_j\} \sim \text{IID}(0,1)$, and $m = EX_0$. We shall additionally assume that for some $C < \infty$, $\delta > 0$, and $p > 2$,

$$|Ee^{iu\xi_0}| \leq C(1 + |u|)^{-\delta}, \quad \forall\, u \in \mathbb{R}, \qquad (13.2.6)$$

$$E|\xi_0|^p < \infty. \qquad (13.2.7)$$

These conditions are analogous to (10.2.21) and (10.2.22). Let G denote the d.f. of X_0. By Lemma 10.2.4, (13.2.6) implies that G is infinitely differentiable.

In the next two subsections we shall discuss the weak convergence of the above processes when errors form martingale differences and when they have long memory, respectively.

13.2.1 *Testing for H_0: martingale-difference errors*

Let β_n be the LS estimator of β_0 under H_0. Write $\mathcal{V}_n$ for $\mathcal{V}_{\beta_n}$ and $\mathcal{V}$ for $\mathcal{V}_{\beta_0}$. Our proposed test of H_0 is based on the process $\mathcal{V}_n$. Its implementation for a large sample is facilitated by obtaining the weak convergence results for $\mathcal{V}$ and $\mathcal{V}_n$ under H_0.

This subsection discusses the asymptotic distributions of $\mathcal{V}$ and $\mathcal{V}_n$ under H_0 when the X_i's have long memory structure (13.2.5), and the errors ε_i of (13.2.4) are independent of the design process and satisfy

$$\{\varepsilon_i\} \text{ are mean-zero finite-variance martingale differences with} \quad (13.2.8)$$
$$E\left(\varepsilon_i^2|\mathcal{F}_{i-1}\right) \equiv \sigma_\varepsilon^2,$$

where $\mathcal{F}_i := \sigma\text{-field}\{\varepsilon_j, j \leq i\}$ and σ_ε^2 is a positive constant.

In order to proceed further we need some more notation. Let

$$\mathcal{Z}_n := \sum_{i=1}^n \ell(X_i)\varepsilon_i, \quad \Lambda_n := \sum_{i=1}^n \ell(X_i)\ell(X_i)', \quad \bar{\Lambda}_n := n^{-1}\Lambda_n, \quad (13.2.9)$$

$$\nu_n(x) := \sum_{i=1}^n \ell(X_i)I(X_i \leq x), \qquad \nu(x) := E\ell(X_0)I(X_0 \leq x),$$

$$\bar{\nu}_n(x) := n^{-1}\nu_n(x), \qquad G_n(x) := n^{-1}\sum_{i=1}^n I(X_i \leq x), \quad x \in \mathbb{R}.$$

We find the following preliminary result useful in the proofs below. For a measurable real-valued function g, where $E|g(X_0)| < \infty$,

$$\sup_{x \in \mathbb{R}} \left|n^{-1}\sum_{i=1}^n g(X_i)I(X_i \leq x) - Eg(X_0)I(X_0 \leq x)\right| \to_{a.s.} 0. \quad (13.2.10)$$

For a non-negative g this follows by the ergodic Theorem 2.5.2(i) and the classical Glivenko–Cantelli argument where $\bar{\mathbb{R}}$ is partitioned such that the oscillation of the measure $Eg(X)I(X_0 \leq x)$ is small. This result applied with $g^{\pm} := (\pm)g \vee 0$, $g = g^{+} - g^{-}$, and the triangle inequality proves (13.2.10) for a general g.

Assume ℓ satisfies (13.2.2) and

$$\Lambda_n \text{ is almost surely positive definite for all } n \geq q. \qquad (13.2.11)$$

Then, by the ET and (13.2.10), $\bar{\Lambda}_n \to \Lambda$, a.s., and

$$\bar{\Lambda}_n^{-1} \to \Lambda^{-1}, \qquad \sup_{x \in \bar{\mathbb{R}}} \|\bar{\nu}_n(x) - \nu(x)\| \to 0, \quad \text{a.s.} \qquad (13.2.12)$$

Under (13.2.5), (13.2.8), and because $\{X_i\}$ is independent of $\{\varepsilon_i\}$,

$$\text{Cov}(\mathcal{V}(x), \mathcal{V}(y)) = n\,\sigma_\varepsilon^2\, G(x \wedge y), \qquad E\mathcal{Z}_n \mathcal{Z}_n' = n\,\sigma_\varepsilon^2\, \Lambda.$$

Note that $\mathcal{V}(-\infty) = 0 = \mathcal{V}_n(-\infty)$ and

$$\mathcal{V}(\infty) := \sum_{i=1}^{n} \varepsilon_i, \qquad \mathcal{V}_n(\infty) := \sum_{i=1}^{n}(Y_i - \beta_n' \ell(X_i)).$$

The next lemma is useful in proving the tightness of the process $\mathcal{V}$. It follows from Lemma 10.2.5 applied to the X_i process with $r = 2$. Let

$$\mu(x, y] := \int_x^y \frac{1}{(1 + u^2)}\, du, \quad -\infty \leq x \leq y \leq \infty. \qquad (13.2.13)$$

Lemma 13.2.1. *Under the assumptions (13.2.5) to (13.2.7) with $p = 4$, there exists a constant C such that for all $-\infty \leq x_1 \leq x_2 \leq x_3 \leq \infty$,*

$$\left| \text{Cov}\big(I(x_1 < X_0 \leq x_2), I(x_2 < X_j \leq x_3)\big) \right|$$
$$\leq C j^{2d_1 - 1}\, \mu^{1/2}(x_1, x_2]\, \mu^{1/2}(x_2, x_3], \qquad \forall\, j \geq 1.$$

We are now ready to prove the weak convergence of $\mathcal{V}$ and $\mathcal{V}_n$ in the Skorokhod space $\mathcal{D}(\bar{\mathbb{R}})$, w.r.t. the uniform metric.

Lemma 13.2.2. *Assume that $X_i, i \in \mathbb{Z}$, satisfies (13.2.5) to (13.2.7) with $p = 4$ and the errors satisfy (13.2.8) with $\max_i E\varepsilon_i^4 < \infty$. Then*

$$n^{-1/2}\mathcal{V} \Rightarrow \sigma_\varepsilon\, B \circ G, \quad \text{in } \mathcal{D}(\bar{\mathbb{R}}), \qquad (13.2.14)$$

$$n^{-1/2}\mathcal{V}_n \Rightarrow W_G, \quad \text{in } \mathcal{D}(\bar{\mathbb{R}}), \qquad (13.2.15)$$

where $B \circ G(x)$ is a continuous Brownian motion on $\mathbb{R}$ w.r.t. time $G(x)$, and W_G is a continuous mean-zero Gaussian process on $\mathbb{R}$ with $W_G(-\infty) = 0$, and the covariance function

$$K(x, y) := \sigma_\varepsilon^2 [G(x \wedge y) - \nu(x)' \Lambda^{-1} \nu(y)], \qquad x, y \in \mathbb{R}.$$

Proof. Let $V_n(x) := n^{-1/2}\mathcal{V}(x) = n^{-1/2}\sum_{i=1}^{n}\varepsilon_i I(X_i \le x)$, $x \in \bar{\mathbb{R}} :=$ $[-\infty, \infty]$. Apply Lemma 2.5.1 to show that the finite-dimensional distributions of V_n converge weakly to those of $\sigma_\varepsilon B \circ G$, under the assumed conditions.

To prove the tightness of V_n in the uniform metric, we shall first prove that for any $-\infty \le x_1 < x_2 < x_3 \le \infty$, with $\eta = 1 - 2d_1 > 0$,

$$E[V_n(x_1) - V_n(x_2)]^2[V_n(x_2) - V_n(x_3)]^2 \le C\big(\mu(x_1, x_3]\big)^2, \qquad (13.2.16)$$

$$E[V_n(x_1) - V_n(x_2)]^4 \le C\{(\mu(x_1, x_2])^2 + n^{-\eta}\mu(x_1, x_2]\}. \qquad (13.2.17)$$

Let $\alpha_i := I(x_1 < X_i \le x_2)$ and $\beta_i := I(x_2 < X_i \le x_3)$, $i \in \mathbb{Z}$. By Lemma 10.2.4(ii), there exist $k_0 \ge 1$ and $C > 0$ such that for all $k \ge k_0$ and $-\infty \le x_1 < x_2 < x_3 \le \infty$,

$$E(\alpha_0\beta_k) = P(x_1 < X_0 \le x_2, x_2 < X_k \le x_3) \qquad (13.2.18)$$
$$\le C\mu(x_1, x_2]\mu(x_2, x_3],$$

where μ is defined by (13.2.13). Assume first that $k_0 = 1$; the general case $k_0 \ge 1$, requires minor changes and is briefly discussed at the end of the proof. By the independence of $\{\varepsilon_i\}$ and $\{X_i\}$, the l.h.s. of (13.2.16) equals

$$n^{-2}\sum_{i,j,k,l}E(\varepsilon_i\varepsilon_j\varepsilon_k\varepsilon_l)E(\alpha_i\alpha_j\beta_k\beta_l).$$

Now, if the largest index among i, j, k, l is not matched by any other, then $E(\varepsilon_i\varepsilon_j\varepsilon_k\varepsilon_l) = 0$. Hence, this in turn equals

$$\frac{1}{n^2}\sum_{i,j<k}E(\varepsilon_i\varepsilon_j\varepsilon_k^2)\big[E(\alpha_i\alpha_j\beta_k) + E(\beta_i\beta_j\alpha_k)\big],$$

where we also used $\alpha_i\beta_i = 0$, $\alpha_i^2 = \alpha_i$, and $\beta_i^2 = \beta_i$. By (13.2.8), $E(\varepsilon_i\varepsilon_j\varepsilon_k^2) = E\{\varepsilon_i\varepsilon_j E(\varepsilon_k^2|\mathcal{F}_{k-1})\} = \sigma_\varepsilon^2 E\{\varepsilon_i\varepsilon_j\} = 0$ and $E(\varepsilon_j^2\varepsilon_k^2) = \sigma_\varepsilon^4$ for all $i \ne j < k$. These equations together with (13.2.18) give

$$E[V_n(x_1) - V_n(x_2)]^2[V_n(x_2) - V_n(x_3)]^2$$
$$= \frac{\sigma_\varepsilon^4}{n^2}\sum_{i<k}\big(E(\alpha_i\beta_k) + E(\beta_i\alpha_k)\big) \le C\mu(x_1, x_2]\mu(x_2, x_3] \le C\big(\mu(x_1, x_3]\big)^2,$$

thereby proving (13.2.16).

The proof of (13.2.17) is similar. We have

$$E[V_n(x_1) - V_n(x_2)]^4 = \frac{3}{n^2}\sum_{i,j<k}E(\varepsilon_i\varepsilon_j\varepsilon_k^2)E(\alpha_i\alpha_j\alpha_k^2)$$

$$+ \frac{3}{n^2}\sum_{i<k}E(\varepsilon_i\varepsilon_k^3)E(\alpha_i\alpha_k) + \frac{1}{n^2}\sum_{k}E(\varepsilon_k^4)E(\alpha_k).$$

Here,

$$\sum_{i,j<k} E(\varepsilon_i\varepsilon_j\varepsilon_k^2)E(\alpha_i\alpha_j\alpha_k) = \sigma_\varepsilon^4 \sum_{i<k} E(\alpha_i\alpha_k)$$

$$= \sigma_\varepsilon^4 \sum_{i<k} \big(\mathrm{Cov}(\alpha_i,\alpha_k) + (E\alpha_i)^2\big),$$

where $|E\alpha_i| = |G(x_1) - G(x_2)| \le C\mu(x_1,x_2]$. Therefore, by Lemma 13.2.1,

$$n^{-2}\Big| \sum_{i,j<k} E(\varepsilon_i\varepsilon_j\varepsilon_k^2)E(\alpha_i\alpha_j\alpha_k)\Big|$$

$$\le Cn^{-2} \sum_{1\le i<k\le n} \Big\{(k-i)^{-\eta}\mu(x_1,x_2] + \mu(x_1,x_2]^2\Big\}$$

$$\le C\big(n^{-\eta}\mu(x_1,x_2] + \mu(x_1,x_2]^2\big).$$

In a similar way, using $|E\varepsilon_i\varepsilon_k^3| \le E^{1/4}\varepsilon_i^4 E^{3/4}\varepsilon_k^4 = E\varepsilon_0^4$,

$$n^{-2}\sum_{i<k}|E(\varepsilon_i\varepsilon_k^3)E(\alpha_i\alpha_k)| \le Cn^{-2} \sum_{i<k} \big(\mathrm{Cov}(\alpha_i,\alpha_k) + (E\alpha_i)^2\big)$$

$$\le C\big(n^{-\eta}\mu(x_1,x_2] + \mu(x_1,x_2]^2\big).$$

Finally, $n^{-2}\sum_k E(\varepsilon_k^4)E(\alpha_k) \le Cn^{-1}\mu(x_1,x_2]$. Because $n^{-1} \le n^{-\eta}$, this proves (13.2.17).

Now let $J(x) := \mu(-\infty,x]/\mu(\bar{\mathbb{R}})$, where μ is as in (13.2.13). Because J is a uniformly continuous strictly increasing d.f. on $\bar{\mathbb{R}}$, proving the tightness of $V_n(x), x \in \bar{\mathbb{R}}$, is equivalent to proving that of $U_n(t) := V_n(J^{-1}(t)), t \in [0,1]$. The bounds (13.2.16) and (13.2.17) verify the tightness criterion (4.4.3) and (4.4.4) for the U_n process. Hence, by Lemma 4.4.1, U_n satisfies (4.4.1) when (13.2.18) holds with $k_0 = 1$.

Consider the general case when (13.2.18) holds with some $k_0 \ge 1$. Assume that $(n-1)/k_0 > 1$ is an integer. Then $V_n(x) = \sum_{j=1}^{k_0} V_{nj}(x)$, where

$$V_{nj}(x) := n^{-1/2} \sum_{i=0}^{[(n-j)/k_0]} \varepsilon_{j+ik_0}I(X_{j+ik_0} \le x).$$

For any j, the "decimated" process $\{X_{j+ik_0}, i \in \mathbb{Z}\}$ satisfies (13.2.18) for any $k \ge 1$ and therefore V_{nj} satisfies (4.4.1), namely,

$$\lim_{\delta\to 0} \limsup_n P\Big(\sup_{|x-y|<\delta} |V_{nj}(x) - V_{nj}(y)| \ge \epsilon\Big) = 0, \qquad (13.2.19)$$

as proved above. Obviously,

$$\Big\{ \sup_{|x-y|<\delta} |V_n(x) - V_n(y)| > \epsilon\Big\} \subset \bigcup_{j=1}^{k_0} \Big\{ \sup_{|x-y|<\delta} |V_{nj}(x) - V_{nj}(y)| > \epsilon/k_0\Big\},$$

proving (13.2.19) with V_{nj} replaced by V_n. This proves (13.2.14).

To prove (13.2.15), because β_n satisfies $\beta_n - \beta_0 = \Lambda_n^{-1} Z_n$,

$$\mathcal{V}_n(x) = \sum_{i=1}^{n} (Y_i - \beta_n' \ell(X_i)) \, I(X_i \leq x) = \mathcal{V}(x) - Z_n' \, \bar{\Lambda}_n^{-1} \, \bar{\nu}_n(x),$$

where $\mathcal{V}(x) \equiv \mathcal{V}_{\beta_0}(x) \equiv \sum_{i=1}^{n} \varepsilon_i I(X_i \leq x)$. Hence, by (13.2.12),

$$n^{-1/2} \mathcal{V}_n(x) = n^{-1/2} [\mathcal{V}(x) - Z_n' \Lambda^{-1} \nu(x)]$$

$$- n^{-1/2} Z_n' \, [\bar{\Lambda}_n^{-1} \, \bar{\nu}_n(x) - \Lambda^{-1} \, \nu(x)]$$

$$= n^{-1/2} \mathcal{V}(x) - n^{-1/2} Z_n' \Lambda^{-1} \nu(x) + u_p(1).$$

The tightness in (13.2.15) now follows from the tightness of $n^{-1/2}\mathcal{V}$ and that of the degenerate processes $n^{-1/2} Z_n' \Lambda^{-1} \nu(x)$, because $\|n^{-1/2} Z_n\| = O_p(1)$ and $\nu(x)$ is uniformly continuous on $\mathbb{R}$. This completes the proof of (13.2.15) and hence that of Lemma 13.2.2.

Testing for H_0. From (13.2.15) it follows that the asymptotic null distribution of $\mathcal{V}_n$ is the same as in the case of i.i.d. design and errors. It depends on the null model and the estimator β_n of β_0 in a complicated fashion and is generally unknown. We shall now describe a transformation $\mathcal{T}_n$ of $\mathcal{V}_n$ that converges weakly to $\sigma_\varepsilon B(G)$, where B is the standard Brownian motion on $[0, \infty)$. Consequently, tests based on a suitably normalized process $\mathcal{T}_n$ will be asymptotically distribution free. We shall also give a computation formula for the test based on the supremum norm of $\mathcal{T}_n$. Let $\Lambda(x) := E\ell(X_0)\ell(X_0)'I(X_0 \geq x)$. Suppose $\Lambda(x)$ is positive definite for every $-\infty \leq x < \infty$. For a real-valued function ψ, define for $x < \infty$,

$$\mathcal{T}\psi(x) := \psi(x) - \int_{y \leq x} \ell(y)' \Lambda^{-1}(y) \Big\{ \int \ell(z) \, I(z \geq y) \psi(dz) \Big\} G(dy),$$

$$\mathcal{T}_n(x) := \mathcal{V}_n(x) - \int_{y \leq x} \ell(y)' \bar{\Lambda}_n^{-1}(y) \Big\{ \int \ell(z) \, I(z \geq y) \, \mathcal{V}_n(dz) \Big\} G_n(dy),$$

where G_n is the empirical d.f. of X_i, $1 \leq i \leq n$, see (13.2.9). Note that $\mathcal{T}\mathcal{V}_n$ with G replaced by G_n is equal to $\mathcal{T}_n$. Since $\mathcal{T}$ is a linear functional, $\mathcal{T}B(G)$ is a centered Gaussian process. Using Lemma 13.2.2 and the method used in the proof in Stute, Thies and Zhu (1998) and Koul and Stute (1999), we can prove that for every $y < \infty$, $n^{-1/2}\mathcal{T}_n$ converges weakly to $\sigma_\varepsilon B(G)$ on $\mathcal{D}[-\infty, y]$ with uniform metric. This implies that for any $y < \infty$,

$$\sup_{x \leq y} n^{-1/2} |\mathcal{T}_n(x)| \to_D \sigma_\varepsilon \sup_{x \leq y} |B(G(x))|$$

$$= \sigma_\varepsilon \sup_{0 \leq u \leq G(y)} |B(u)| =_D \sigma_\varepsilon G(y)^{1/2} \sup_{0 \leq u \leq 1} |B(u)|,$$

where the last equality follows by self-similarity of Brownian motion. The above form of the limit distribution suggests dividing the statistic $\mathcal{T}_n$ by estimates $\hat{s}_n := (n^{-1}\sum_{i=1}^n \hat{\varepsilon}_i^2)^{1/2}$, $\hat{\varepsilon}_i := Y_i - \beta'_n \ell(X_i)$, and $G_n^{1/2}(y)$ of σ_ε and $G^{1/2}(y)$, respectively, in order to have the limiting distribution free of any parameters. Thus

$$\sup_{x\leq y} |\mathcal{T}_n(x)|/[n\hat{s}_n^2 G_n(y)]^{1/2} \to_D \sup_{0\leq u\leq 1} |B(u)|, \quad \forall\, y < \infty.$$

In applications we use this result with $y = \max_{1\leq j\leq n} X_j$, where $G_n \equiv 1$.

We can derive a computational formula for $\mathcal{T}_n$ as follows. Let $X_{(j)}$, $1 \leq j \leq n$, denote the ordered X_i's in ascending order and $\hat{\eta}_i$'s denote the corresponding $\hat{\varepsilon}_i$'s. Also, let $\Lambda_{in} := \sum_{j=i}^n \ell(X_{(j)})\ell(X_{(j)})'$, $1 \leq i, j \leq n$. Then

$$\mathcal{T}_n(x) = \sum_{i=1}^n \left[I(X_{(i)} \leq x) - \sum_{j=1}^n \ell(X_{(j)})'\Lambda_{jn}^{-1}\ell(X_{(i)})\, I(X_{(j)} \leq X_{(i)} \wedge x) \right] \hat{\eta}_i.$$

Hence, with $X_{(n+1)} = \infty$, if $X_{(j)} \leq x < X_{(j+1)}$ for some $1 \leq j \leq n$, then $\mathcal{T}_n(x) \equiv \mathcal{S}_j$, where

$$\mathcal{S}_j = \sum_{i=1}^j \left[1 - \sum_{s=1}^i \ell(X_{(s)})'\Lambda_{sn}^{-1}\ell(X_{(i)}) \right]\hat{\eta}_i - \sum_{i=j+1}^n \sum_{s=1}^j \ell(X_{(s)})'\Lambda_{sn}^{-1}\ell(X_{(i)})\hat{\eta}_i.$$

Let

$$\mathcal{D}_n := \frac{\max_{1\leq j\leq n} |\mathcal{S}_j|}{n^{1/2}\hat{s}_n}. \tag{13.2.20}$$

Then the test that rejects H_0 whenever $\mathcal{D}_n > q_\alpha$ is of the asymptotic size $0 < \alpha < 1$, where q_α is such that $P(\sup_{0\leq u\leq 1} |B(u)| > q_\alpha) = \alpha$.

For ease of reference in Table 13.2.1 we give a few values of q_α reported in Khmaladze and Koul (2004). For more detail see the web site *http://www.mcs.vuw.ac.nz/~ray/Brownian/#one-dim-lim.*

α	0.1	0.05	0.025	0.01
q_α	1.9599	2.2414	2.4977	2.8070

Table 13.2.1: Selected values of q_α.

On page 553 of Resnick (1992) we find that

$$P(\sup_{0\leq t\leq 1} |B(t)| < q) \tag{13.2.21}$$

$$= P(B(1) < q) + 2\sum_{i=1}^\infty (-1)^i P\big((2i-1)q < B(1) < (2i+1)q\big).$$

Formula (13.2.21) may be used to obtain any other value of q_α. It can be used to check the accuracy of the Table 13.2.1. For example, $|P(\sup_{0 \leq t \leq 1} |B(t)| < 2.2414) - 0.95| \leq (0.1)^6$.

Extension: Testing for $\mathcal{H}_0$. Let $\{m_\beta(x); \beta \in \Omega \subset \mathbb{R}^q, x \in \mathbb{R}\}$ be a given family of parametric regression models. We shall now describe an extension of the above test for testing $\mathcal{H}_0$: $\mu(x) = m_{\beta_0}(x)$, $x \in \mathbb{R}$, and for some $\beta_0 \in \Omega$. Suppose $m_\beta(x)$ is twice continuously differentiable in β with $\dot{m}_\beta := \partial m_\beta / \partial \beta$ satisfying $E\|\dot{m}_\beta(X)\|^2 < \infty$. Let $\mathcal{B}_{\beta_0}(x) := E\dot{m}_{\beta_0}(X_0)\dot{m}_{\beta_0}(X_0)'I(X_0 \geq x)$, with $\mathcal{B}_{\beta_0} := \mathcal{B}(-\infty) = E\dot{m}_{\beta_0}(X_0)\dot{m}_{\beta_0}(X_0)'$. Unlike when testing for H_0, these entities depend on the null parameter β_0.

Now, let $\tilde{\beta}_n$ be the LS estimator of β_0 under $\mathcal{H}_0$, $\tilde{\mathcal{V}}_n$ denote the $V_{\tilde{\beta}_n}$ of (13.2.1), $\tilde{m}_{ni} := \dot{m}_{\tilde{\beta}_n}(X_{(i)})$, $\hat{s}_n := \{\sum_{i=1}^n (Y_i - \tilde{m}_n(X_i))^2/n\}^{1/2}$, and let

$$\tilde{\nu}_n(x) := n^{-1}\sum_{i=1}^n \tilde{m}_{ni}I(X_i \leq x), \quad \tilde{\mathcal{B}}_n(x) := n^{-1}\sum_{i=1}^n \tilde{m}_{ni}\tilde{m}_{ni}'I(X_i \geq x),$$

$$\tilde{\mathcal{B}}_{ni} := \tilde{\mathcal{B}}_n(X_{(i)}), \quad 1 \leq i \leq n, \, x \in \mathbb{R}.$$

Let $\tilde{\eta}_i$ denote the residual corresponding to $X_{(i)}$ among the residuals $\{Y_j - m_{\tilde{\beta}_n}(X_j), 1 \leq j \leq n\}$, $1 \leq i \leq n$. Then an analog of the statistic $\mathcal{D}_n$ of (13.2.20) is $\tilde{\mathcal{D}}_n := (n^{1/2}\hat{s}_n)^{-1}\max_{1 \leq j \leq n} |\tilde{\mathcal{T}}_j|$, where

$$\tilde{\mathcal{T}}_j := \sum_{i=1}^j \Big[1 - \sum_{k=1}^i \tilde{m}_{nk}'\tilde{\mathcal{B}}_{nk}^{-1}\tilde{m}_{ni}\Big]\tilde{\eta}_i - \sum_{i=j+1}^n \sum_{k=1}^j \tilde{m}_{nk}'\tilde{\mathcal{B}}_{nk}^{-1}\tilde{m}_{ni}\tilde{\eta}_i.$$

We can verify that under (13.2.5), (13.2.8), and smoothness conditions on $\mathcal{M}$, which would also guarantee that $\tilde{\beta}_n$ is $n^{1/2}$-consistent for β_0, under $\mathcal{H}_0$, $\tilde{\mathcal{D}}_n \to_D \sup_{0 \leq u \leq 1} |B(u)|$, so that the test based on this statistic is asymptotically distribution free.

Note: The contents of Subsection 13.2.1 have roots in Stute *et al.* (1998), Koul and Stute (1999), and Khmaladze and Koul (2004) where we can find more detail about the proofs pertaining to $\mathcal{D}_n$ and $\tilde{\mathcal{D}}_n$ tests and their consistency.

13.2.2 *Lack-of-fit tests: LM moving-average errors*

Now we will discuss the asymptotic null distribution and power properties of the test of H_0 based on $\sup_x \|\mathcal{V}_n(x)\|$ when errors form a long memory moving-average process:

$$\varepsilon_i = \sum_{k=0}^{\infty} a_k \zeta_{i-k}, \quad i \in \mathbb{Z}, \quad a_k \sim c_0 k^{-1+d}, \quad k \to \infty, \quad (13.2.22)$$

for some $c_0 > 0$, $0 < d < 1/2$, with $\{\zeta_k\} \sim \text{IID}(0,1)$, and the design process is a long memory moving average as in (13.2.5). By Proposition 3.2.1, assumptions (13.2.22) and (13.2.5) imply that the respective auto-covariances decay with $k \to \infty$ as

$$\gamma_\varepsilon(k) \sim c_{\varepsilon,\gamma} k^{-1+2d}, \qquad \gamma_X(k) \sim c_{X,\gamma} k^{-1+2d_1}, \qquad (13.2.23)$$
$$c_{\varepsilon,\gamma} := c_0^2 B(d, 1-2d), \qquad c_{X,\gamma} := c_1^2 B(d_1, 1-2d_1).$$

In addition, we shall assume (13.2.6) and (13.2.7) and the independence of $\{X_i\}$ and $\{\varepsilon_i\}$.

To proceed further, we first need to determine the magnitude of $\mathcal{V}_n$. This in turn depends on the nature of the $\ell(X_0)$ and on whether $d + d_1 > 1/2$ or $d + d_1 < 1/2$. We shall first state some general rate results, which are needed for determining the magnitude of $\mathcal{V}_n$. Let

$$\mathcal{W}_n := \sum_{i=1}^{n} [\ell(X_i) - E\ell(X_i)] \, \varepsilon_i, \quad \mu := E\ell(X_0), \qquad (13.2.24)$$

$$\mathcal{U}_n := \sum_{i=1}^{n} (X_i - m)\varepsilon_i, \quad S_n := \sum_{i=1}^{n} \varepsilon_i.$$

Let g denote the Lebesgue density of G, and define

$$\mu(u) := E\ell(X_0 + u) = \int_{\mathbb{R}} \ell(x) g(x - u) dx, \quad \mu := \mu(0),$$

$$\nu(u; x) := E\ell(X_0 + u) I(X_0 + u \le x) = \int_{-\infty}^{x} \ell(y) g(y - u) dy,$$

$$\nu(x) := \nu(0, x), \quad x, u \in \mathbb{R}.$$

By Lemma 10.2.4, under (13.2.6), G has an infinitely differentiable density g. This together with (13.2.25) below implies the infinite differentiability of the functions $\mu(\cdot)$ and $\nu(\cdot\,; x)$ with

$$\dot{\nu}(x) := \frac{\partial \nu(u; x)}{\partial u}\bigg|_{u=0} = -\int_{-\infty}^{x} \ell(y) \dot{g}(y) dy,$$

$$\dot{\mu} := \frac{\partial \mu(u)}{\partial u}\bigg|_{u=0} = -\int_{\mathbb{R}} \ell(x) \dot{g}(x) dx.$$

We need the following lemma, which can be proved by using the method of proof of (10.3.38) in Theorem 10.3.2 above. In this lemma, $\delta := d + d_1$. Moreover, for any sequence $\delta_n > 0$, $n = 1, 2, \cdots$, $U_p(\delta_n)$ and $u_p(\delta_n)$ will stand for a sequence of stochastic processes whose absolute value is uniformly $O_p(\delta_n)$ and $o_p(\delta_n)$, respectively.

Lemma 13.2.3. *Suppose (10.2.19), (13.2.5), (13.2.6), and (13.2.7) with $p > 2$ hold. Let $\ell(x)$, $x \in \mathbb{R}$, be any function with values in $\mathbb{R}^q$ such that*

$$\|\ell(x)\| \leq C(1 + |x|)^\lambda, \text{ for some } 0 \leq \lambda < (p-2)/2. \qquad (13.2.25)$$

Then there exists $\kappa > 0$ such that

$$\mathcal{W}_n = \mu \mathcal{U}_n + O_p(n^{\delta - \kappa}), \qquad \delta > 1/2, \qquad (13.2.26)$$
$$= O_p(n^{1/2}), \qquad \delta < 1/2.$$

$$\bar{\nu}_n(x) = \nu(x) + \dot{\nu}(x)(\bar{X}_n - m) + U_p(n^{d_1 - 1/2 - \kappa}). \qquad (13.2.27)$$

$$\mathcal{V}(x) = G(x)S_n - g(x)\mathcal{U}_n + U_p(n^{\delta - \kappa}), \qquad \delta > 1/2, \qquad (13.2.28)$$
$$= G(x)S_n + U_p(n^{1/2}), \qquad \delta < 1/2.$$

Moreover,

$$S_n = O_p(n^{d+1/2}), \qquad \bar{X}_n - m = O_p(n^{d_1 - 1/2}). \qquad (13.2.29)$$

$$\mathcal{U}_n = O_p(n^\delta), \qquad \delta > 1/2, \qquad (13.2.30)$$
$$= O_p(n^{1/2}), \qquad \delta < 1/2.$$

By Corollary 4.4.1, under conditions (10.2.19), (10.2.20), and (13.2.5),

$$n^{-d-1/2}S_n \to_D Y, \qquad n^{-d_1+1/2}(\bar{X}_n - m) \to_D Y_1, \qquad (13.2.31)$$

where Y and Y_1 are two independent normal r.v.'s with zero means and respective long-run variances $s^2_{\varepsilon,d}$ and s^2_{X,d_1}, where

$$s^2_{\varepsilon,d} = c_{\varepsilon,\gamma} d/(1 + 2d), \qquad s^2_{X,d_1} = c_{X,\gamma} d_1/(1 + 2d_1). \qquad (13.2.32)$$

To state the next result, let

$$J(x) := G(x) - \mu' \Lambda^{-1} \nu(x), \qquad x \in \mathbb{R}.$$

Lemma 13.2.4. *Under the conditions of Lemma 13.2.3,*

$$\sup_{x \in \mathbb{R}} \left| n^{-d-1/2}\{\mathcal{V}_n(x) - J(x) S_n\} \right| = o_p(1).$$

Consequently,

$$n^{-d-1/2}\mathcal{V}_n(x) \Rightarrow J(x)Y, \qquad (13.2.33)$$

in $\mathcal{D}(\bar{\mathbb{R}})$ with uniform metric.

Proof. With the notation of (13.2.24), rewrite

$$\mathcal{V}_n(x) = \mathcal{V}(x) - (\mu S_n + \mathcal{W}_n)' \bar{\Lambda}_n^{-1} \bar{\nu}_n(x). \qquad (13.2.34)$$

First, suppose $d + d_1 > 1/2$. Use (13.2.34) and Lemma 13.2.3, (13.2.26) to (13.2.28), (13.2.30), to obtain

$$\mathcal{V}_n(x) = S_n \big(G(x) - \mu' \bar{\Lambda}_n^{-1} \nu(x) \big) + U_p(n^{d+d_1}). \qquad (13.2.35)$$

Now (13.2.33) clearly follows from (13.2.12), (13.2.31), and (13.2.35) as $d + 1/2 > d + d_1$. A similar argument with small changes applies also in the case $d + d_1 = 1/2$.

Next, consider the case $d + d_1 < 1/2$. In this case we will use the following identity:

$$\mathcal{V}_n(x) = S_n(G(x) - \mu'\Lambda^{-1}\nu(x)) + S_n\Big[\mu'\Lambda^{-1}\nu(x) - \mu'\bar{\Lambda}_n^{-1}\bar{\nu}_n(x)\Big] \quad (13.2.36)$$

$$+ \sum_{i=1}^{n} \varepsilon_i \Big[I(X_i \le x) - G(x) - (\ell(X_i) - \mu)'\Lambda^{-1}\nu(x) \Big]$$

$$- \Big(\sum_{i=1}^{n} \varepsilon_i \big(\ell(X_i) - \mu\big) \Big)' \Big[\bar{\Lambda}_n^{-1}\bar{\nu}_n(x) - \Lambda^{-1}\nu(x) \Big].$$

By (13.2.12), $\mu'\Lambda^{-1}\nu(x) - \mu'\bar{\Lambda}_n^{-1}\bar{\nu}_n(x) = u_p(1)$. Hence, the second term on the r.h.s. of (13.2.36) is $u_p(|S_n|)$, and the fourth term is $U_p(|\mathcal{W}_n|) = U_p(n^{1/2}) = u_p(|S_n|)$; see (13.2.27) and (13.2.29). Finally, for the third term, by (13.2.26) and (13.2.28),

$$\sum_{i=1}^{n} \varepsilon_i(I(X_i \le x) - G(x) - (\ell(X_i) - \mu)'\Lambda^{-1}\nu(x))$$

$$= \mathcal{V}(x) - G(x)S_n - \mathcal{W}_n'\Lambda^{-1}\nu(x)$$

$$= U_p(|\mathcal{U}_n|) + U_p(n^{1/2}) = U_p(n^{1/2}) = u_p(|S_n|),$$

thereby completing the proof of Lemma 13.2.4.

Application of Lemma 13.2.4 to testing of H_0. We now discuss some important consequences of the above lemma to the testing problem. First, note that unlike the situation with i.i.d. or martingale-difference errors, the weak limit of $\mathcal{V}_n$ under H_0 given by (13.2.33) is the degenerate process $J(x)Y$. This result is analogous to the URP discussed in Section 10.2. It implies that if $\sup_x |J(x)| \ne 0$, and if we can estimate $J(x)$ uniformly consistently, the long-run variance $s_{\varepsilon,\gamma}^2$ consistently, and the long memory parameter $d \log(n)$-consistently, then we can easily implement tests based on continuous functions of $\mathcal{V}_n$, using standard normal distribution percentiles.

A uniformly consistent estimator of $J(x)$ is given by

$$\hat{J}_n(x) := \frac{1}{n}\sum_{i=1}^n \{I(X_i \le x) - \ell(X_i)'\bar{\Lambda}_n^{-1}\bar{\nu}_n(x)\}, \quad x \in \bar{\mathbb{R}}.$$

Let $\hat{d}$ denote a local Whittle estimator of d of Section 8.5 and $\hat{s}_{\varepsilon,d}^2$ a HAC estimator of $s_{\varepsilon,d}^2$ based on $Y_i - \beta_n'\ell(X_i)$, $1 \le i \le n$. Their required consistency is proved using the method of Sections 8.6 and 9.4. Consequently, the test that rejects H_0 whenever

$$\widehat{\mathcal{D}}_n := \frac{\sup_x |\mathcal{V}_n(x)|}{n^{\hat{d}+1/2}\hat{s}_{\varepsilon,d}\sup_x |\hat{J}_n(x)|} \ge z_{\alpha/2}, \tag{13.2.37}$$

will be of the asymptotic size α.

An example where $J(x) \ne 0$ for some x, is given by $\ell(x) = (x, x^2, \cdots, x^q)$, $q \ge 2$, i.e., when fitting a polynomial regression of order two or higher through the origin. Because of assumption (13.2.25), the degree of the polynomial will depend on the existence of the moments of ξ_0. If (13.2.7) holds with $p = q + 1$, then the above test would be good for fitting a polynomial through the origin of degree at most q.

Also, if $E\ell(X_0) = 0$ then $J(x) \equiv G(x)$, and the test that rejects H_0 whenever $(n^{\hat{d}+1/2}\hat{s}_{\varepsilon,d})^{-1}\sup_x |\mathcal{V}_n(x)| \ge z_{\alpha/2}$, would also be of the asymptotic size α.

Asymptotic power. We now briefly address the question of consistency and asymptotic power against local alternatives of the above test. Let h be a real-valued function on $\mathbb{R}$ satisfying

$$0 < Eh^2(X_0) < \infty, \tag{13.2.38}$$

$$\Delta(x) := Eh(X_0)I(X_0 \le x) - \nu'(x)\Lambda^{-1}E\ell(X_0)h(X_0) \ne 0, \tag{13.2.39}$$

for some $x \in \mathbb{R}$. Then the above test (13.2.37) is consistent against any fixed alternative $\mu(x) = \beta_0'\ell(x) + h(x)$, assuming that the errors $\varepsilon_i \equiv Y_i - \beta_0'\ell(x) - h(x)$ continue to be a long memory moving average as in (13.2.22).

The test (13.2.37), valid when $\sup_x |J(x)| \ne 0$ and under (13.2.5) and (13.2.22), has a non-trivial asymptotic power against the sequence of alternatives $\mu(x) = \beta_0'\ell(x) + n^{d-1/2}h(x)$. In fact, the asymptotic power of this test against these alternatives is

$$P\big(J(x)Z > z_{\alpha/2} \sup_y |J(y)| + \Delta(x), \text{ for some } x \in \mathbb{R}\big)$$

$$+ P\big(J(x)Z < -z_{\alpha/2} \sup_y |J(y)| + \Delta(x), \text{ for some } x \in \mathbb{R}\big).$$

An example where the conditions (13.2.38) and (13.2.39) hold is when X_0 has six finite moments, $\ell(x)' = (x, x^2)$, $h(x) = \alpha x^3$, $\alpha \in \mathbb{R}$. Another class of examples is given by those functions h that satisfy (13.2.38), have $E\ell(X_0)h(X_0) = 0$, and $\sup_x |Eh(X)I(X \le x)| \ne 0$. An example where (13.2.39) does not hold is when $h(x) = a'\ell(x)$, $a \in \mathbb{R}^q$.

Now consider the case $J(x) \equiv 0$. Rewrite

$$J(x) = \int_{-\infty}^{x} (1 - \mu'\Lambda^{-1}\ell(y))dG(y).$$

Clearly, $J(x) \equiv 0$ implies $\mu'\Lambda^{-1}\ell(y) = 1$, a.e. w.r.t. measure G (G-a.e.). Assume $\Lambda^{-1}\mu \ne 0$. Then the functions $1, \ell_1(x), \cdots, \ell_q(x)$, are *linearly dependent* G-a.e. In other words, there exist some constants $k_0, k_1, \cdots, k_q$ not all of them identically zero and such that

$$k_0 + \sum_{i=1}^{q} k_i\ell_i(x) = 0, \qquad G\text{-a.e.} \tag{13.2.40}$$

In particular, (13.2.40) holds if one of the functions $\ell_1, \cdots, \ell_q$ is a constant G-a.e., and $\ell_1, \cdots, \ell_q$ are linearly independent. Assume in the following that $\ell_1(x) \equiv 1$. Then by Lemma 13.2.3,

$$\mathcal{V}_n(x) = G(x)S_n - g(x)\mathcal{U}_n + U_p(n^{d+d_1-\kappa})$$
$$- S_n\mu'\Big[\Lambda^{-1} + (\bar{\Lambda}_n^{-1} - \Lambda^{-1})\Big]$$
$$\times \Big[\nu(x) + \dot\nu(x)(\bar{X}_n - m) + U_p(n^{d_1-1/2-\kappa})\Big]$$
$$- \Big[\dot\mu\mathcal{U}_n + O_p(n^{d+d_1-\kappa})\Big]'\Big[\Lambda^{-1} + o_p(1)\Big]\Big[\nu(x) + u_p(1)\Big].$$

After cancelation of the main term $(G(x) - \mu'\Lambda^{-1}\nu(x))S_n = J(x)S_n$, this gives

$$\mathcal{V}_n(x) \tag{13.2.41}$$
$$= -g(x)\mathcal{U}_n - S_n\mu'(\bar{\Lambda}_n^{-1} - \Lambda^{-1})\nu(x) - S_n(\bar{X}_n - \theta)\mu'\Lambda^{-1}\dot\nu(x)$$
$$\quad - \mathcal{U}_n\dot\mu'H^{-1}\nu(x) + O_p(n^{d+d_1-\kappa}) + o_p(|\mathcal{U}_n|)$$
$$= S_n\mu'\Big[\Lambda^{-1} - \bar{\Lambda}_n^{-1}\Big]\nu(x) - S_n(\bar{X}_n - \theta)\mu'\Lambda^{-1}\dot\nu(x)$$
$$\quad - \mathcal{U}_n\Big[g(x) + \dot\mu'\Lambda^{-1}\nu(x)\Big] + O_p(n^{d+d_1-\kappa}) + o_p(|\mathcal{U}_n|).$$

This fact is useful in studying the asymptotic null distribution of $\mathcal{V}_n$. We illustrate this by applying it to the case $q = 2$ in the next lemma. The results of this lemma are, in particular, useful in arriving at the asymptotic null distribution of tests based on $\mathcal{V}_n$ when fitting a simple linear regression model with a non-zero intercept, provided $\{X_i\}$ is a non-Gaussian process. Let $\varphi(X_0)$ be a r.v. with finite variance,

$$K(x) := E\varphi(X_0)I(X_0 \le x), \quad \mu_1 := E\varphi(X_0), \quad \sigma_\varphi^2 := \text{Var}(\varphi(X_0)),$$
$$\dot{\mu}_1 := \partial E\varphi(X_0 + u)/\partial u\big|_{u=0},$$
$$\mathcal{J}(x) := g(x) + \frac{\dot{\mu}_1}{\sigma_\varphi^2}\left(K(x) - \mu_1 G(x)\right), \quad x \in \mathbb{R}.$$

Lemma 13.2.5. *Assume (13.2.5), (13.2.6), (13.2.7) with $p > 2$, and (13.2.22) hold, and that $d + d_1 > 1/2$. Let φ be a measurable function with*

$$\sigma_\varphi^2 > 0, \quad |\varphi(x)| \le C(1 + |x|)^\lambda, \quad x \in \mathbb{R}, \text{ and for some } \lambda < (p-2)/4.$$

Assume $q = 2$, and let

$$\ell(x)' = (1, \varphi(x)), \quad x \in \mathbb{R}. \tag{13.2.42}$$

Then there exists $\kappa > 0$, such that

$$\mathcal{V}_n(x) = (S_n(\bar{X}_n - m) - \mathcal{U}_n)\mathcal{J}(x) + U_p(n^{d+d_1-\kappa}). \tag{13.2.43}$$

When $\varphi(x) = x$, (13.2.43) is valid provided the moment condition (13.2.7) holds for some $p > 4$.

Proof. Recall (13.2.41). We shall first identify the expressions $\mu'\Lambda^{-1}\dot{\nu}(x)$ and $\dot{\mu}'\Lambda^{-1}\nu(x)$ for the case (13.2.42). Here, $\mu' = (1, \mu_1)$, $\mu_2 := E\varphi^2(X)$, and

$$\Lambda = \begin{pmatrix} 1 & \mu_1 \\ \mu_1 & \mu_2 \end{pmatrix}, \quad \Lambda^{-1} = \sigma_\varphi^{-2}\begin{pmatrix} \mu_2 & -\mu_1 \\ -\mu_1 & 1 \end{pmatrix}.$$

Let $\dot{K}(x) \equiv -\int_{-\infty}^x \varphi(y)\dot{g}(y)dy$. Then $\dot{\nu}(x)' = (-g(x), \dot{K}(x))$, and

$$\mu'\Lambda^{-1}\dot{\nu}(x) \tag{13.2.44}$$
$$= -\frac{1}{\sigma_\varphi^2}\left(\mu_2 g(x) - \mu_1^2 g(x) + \mu_1\dot{K}(x) - \mu_1\dot{K}(x)\right) = -g(x).$$

Also,

$$\dot{\mu}' = \left(0, -\int_{\mathbb{R}}\varphi(x)\dot{g}(x)dx\right) = (0, \dot{\mu}_1), \quad \nu(x)' = (G(x), K(x)),$$

implying

$$\dot{\mu}'\Lambda^{-1}\nu(x) = \frac{\dot{\mu}_1}{\sigma_\varphi^2}\big[K(x) - \mu_1 G(x)\big]. \tag{13.2.45}$$

To complete the proof of the lemma, we need to examine the behavior of $\bar{\Lambda}_n^{-1} - \Lambda^{-1}$. Let

$$\overline{\varphi} := \frac{1}{n}\sum_{i=1}^n \varphi(X_i), \qquad \overline{\varphi^2} := \frac{1}{n}\sum_{i=1}^n \varphi^2(X_i).$$

Then

$$\bar{\Lambda}_n^{-1} = \frac{1}{\tau_n^2}\begin{pmatrix} \overline{\varphi^2} & -\overline{\varphi} \\ -\overline{\varphi} & 1 \end{pmatrix}, \qquad \tau_n^2 := \det(\bar{\Lambda}_n) = \overline{\varphi^2} - (\overline{\varphi})^2,$$

and $\bar{\Lambda}_n^{-1} - \Lambda^{-1} = Q_{1n} + Q_{2n}$, where

$$Q_{1n} := \frac{1}{\tau_n^2}\begin{pmatrix} \overline{\varphi^2} - \mu_2 & -\overline{\varphi} + \mu_1 \\ -\overline{\varphi} + \mu_1 & 0 \end{pmatrix}, \qquad Q_{2n} := \frac{\sigma_\varphi^2 - \tau_n^2}{\tau_n^2}\Lambda^{-1}.$$

The limiting behavior of $\overline{\varphi^2} - \mu_2$ will depend on the first Appell coefficient

$$\dot{\mu}_2 := \frac{\partial E\varphi^2(X_0 + u)}{\partial u}\bigg|_{u=0} = -\int_{\mathbb{R}}\varphi^2(x)\dot{g}(x)dx.$$

By Lemma 13.2.3, under condition (13.2.2), there exists a $\kappa > 0$ such that

$$\overline{\varphi^2} - \mu_2 = n^{-1}\sum_{i=1}^n(\varphi^2(X_i) - E\varphi^2(X_0)) = \dot{\mu}_2(\bar{X}_n - m) + O_p(n^{d_1 - 1/2 - \kappa}),$$

$$\overline{\varphi} - \mu_1 = n^{-1}\sum_{i=1}^n(\varphi(X_i) - E\varphi(X_0)) = \dot{\mu}_1(\bar{X}_n - m) + O_p(n^{d_1 - 1/2 - \kappa}).$$

Hence,

$$Q_{1n} = \frac{\bar{X}_n - m}{\tau_n^2}\begin{pmatrix} \dot{\mu}_2 & -\dot{\mu}_1 \\ -\dot{\mu}_1 & 0 \end{pmatrix} + O_p(n^{d_1 - 1/2 - \kappa}).$$

We also have

$$\tau_n^2 - \sigma_\varphi^2 = n^{-1}\sum_{i=1}^n \varphi^2(X_i) - \Big(n^{-1}\sum_{i=1}^n \varphi(X_i)\Big)^2 - \sigma_m^2$$

$$= n^{-1}\sum_{i=1}^n(\varphi^2(X_i) - E\varphi^2(X_i)) - (\mu_1 + (\overline{\varphi} - \mu_1))^2 + \mu_1^2$$

$$= \dot{\mu}_2(\bar{X}_n - m) - 2\mu_1\dot{\mu}_1(\bar{X}_n - m) + O_p(n^{d_1 - 1/2 - \kappa}),$$

implying $\tau_n^{-2} - \sigma_\varphi^{-2} = \sigma_\varphi^{-4}(2\mu_1\dot\mu_1 - \dot\mu_2)(\bar{X}_n - m) + O_p(n^{d_1 - 1/2 - \kappa})$. Hence,

$$\bar{\Lambda}_n^{-1} - \Lambda^{-1} = \frac{(\bar{X}_n - m)}{\sigma_\varphi^2}\left\{\begin{pmatrix} \dot\mu_2 & -\dot\mu_1 \\ -\dot\mu_1 & 0 \end{pmatrix}\right.$$

$$\left. + \frac{2\mu_1\dot\mu_1 - \dot\mu_2}{\sigma_\varphi^2}\begin{pmatrix} \mu_2 & -\mu_1 \\ -\mu_1 & 1 \end{pmatrix}\right\} + O_p(n^{d_1 - 1/2 - \kappa}).$$

After some algebra and numerous cancelations, this result gives that for some $\kappa > 0$,

$$S_n\mu'(\Lambda^{-1} - \bar{\Lambda}_n^{-1})\nu(x) = \frac{S_n(\bar{X}_n - m)}{\sigma_\varphi^2}\{\dot\mu_1[K(x) - \mu_1 G(x)]\} + U_p(n^{d + d_1 - \kappa}).$$

The lemma follows from this, (13.2.41), (13.2.44), and (13.2.45).

The following corollary describes the limiting distribution of $\mathcal{V}_n$.

Corollary 13.2.1. *Under the conditions of Lemma 13.2.5,*

$$n^{-d - d_1}\mathcal{V}_n(x) \Rightarrow \mathcal{J}(x)\,(YY_1 - U),$$

where Y and Y_1 are the same as in (13.2.31) and

$$U := \int_{-\infty}^1 \int_{-\infty}^1 \left\{\int_0^1 (s - x)_+^{-(1-d)}(s - x_1)_+^{-(1-d_1)}ds\right\}B(dx)B_1(dx_1),$$

with B and B_1 being mutually independent Brownian motions, defined on the same probability space as Y and Y_1.

The above result is useful for testing H_0 as long as $\mathcal{J}(x) \neq 0$, for some $x \in \mathbb{R}$. It is interesting to note that in the case $\ell(x)' = (1, x)$, the condition $\mathcal{J}(x) \equiv 0$ is equivalent to X_0 being a Gaussian r.v. For a further discussion on the limiting null distribution of $\mathcal{V}_n$ when $\{X_i\}$ is Gaussian, see Koul, Baillie and Surgailis (2004).

The next lemma describes the case $d + d_1 < 1/2$. Let

$$\rho_i(x, y) := \text{Cov}\Big(I(X_0 \leq x) - G(x) - (\ell(X_0) - \mu)'\Lambda^{-1}\nu(x),$$

$$I(X_i \leq y) - G(y) - (\ell(X_i) - \mu)'\Lambda^{-1}\nu(y)\Big).$$

Lemma 13.2.6. *Assume ℓ is of the form (13.2.42) and satisfies the conditions of Lemma 13.2.5, and $d + d_1 < 1/2$. Then $n^{-1/2}\mathcal{V}_n(x) \Rightarrow \mathcal{G}(x)$, where $\mathcal{G}(x), x \in \mathbb{R}$, is a Gaussian process with zero mean and covariance*

$$\text{Cov}(\mathcal{G}(x), \mathcal{G}(y)) = \sum_{i=-\infty}^{\infty} \text{Cov}(\varepsilon_0, \varepsilon_i)\rho_i(x, y).$$

Note: The contents of Subsection 13.2.2 have roots in Koul *et al.* (2004), where we can also find the proofs of Corollary 13.2.1 and Lemma 13.2.6.

13.3　Long memory design and heteroscedastic errors

In this section we will propose a test of H_0 of (13.2.3) when the errors form a heteroscedastic long memory process, the heteroscedasticity function $\sigma^2(x)$ is non-parametric, and the design forms a long memory Gaussian process. So, we will first establish the consistency of the kernel and cross validation-type estimators of $\sigma^2(x)$ under the assumed conditions. These results are then applied subsequently in Subsections 13.3.1 to 13.3.4 below for implementing lack-of-fit tests based on the V_n-process.

Consider the regression model

$$Y_i = \beta_0' \ell(X_i) + \sigma(X_i)\varepsilon_i, \quad i \in \mathbb{Z}, \tag{13.3.1}$$

where $\sigma(x)$ is a positive function defined on $\mathbb{R}$, and $\ell := (\ell_1, \cdots, \ell_q)'$ is a vector of measurable functions satisfying (13.2.2). The process $\{\varepsilon_i\}$ is a long memory moving average as in (13.2.22) with memory parameter $0 < d < 1/2$ and the design process $\{X_i\}$ is a long memory moving-average Gaussian process with mean m, memory parameter $0 < d_1 < 1/2$, and covariance γ_X decaying as in (13.2.23), independent of the ε_i's. In addition, we shall suppose that the innovations ζ_j's of the error process in (13.2.22) have finite fourth moment.

Estimation of $\sigma^2(x)$. To define the kernel-type estimator of $\sigma^2(x)$, let W be a density function on $[-1, 1]$, $\omega = \omega_n$ be a sequence of positive numbers, ϕ denote the standard normal density, $\tau := \gamma_X^{1/2}(0)$, and

$$\bar{X} := \frac{1}{n} \sum_{i=1}^n X_i, \quad s^2 := \frac{1}{n} \sum_{i=1}^n (X_i - \bar{X})^2, \tag{13.3.2}$$

$$\tilde{\phi}(x) := \tau^{-1}\phi((x-m)/\tau), \quad \tilde{\phi}_n(x) := s^{-1}\phi((x-\bar{X})/s).$$

Let $W_\omega(x) \equiv W(x/\omega)/\omega$, $W_{\omega i}(x) := W_\omega(x - X_i)$, and define the kernel-type estimator of $\sigma^2(x)$ to be

$$\hat{\sigma}^2(x) := \frac{1}{n\tilde{\phi}_n(x)} \sum_{i=1}^n W_{\omega i}(x)\, \hat{e}_i^2, \qquad \hat{e}_i := Y_i - \beta_n' \ell(X_i),$$

where β_n is the LS estimator of β in (13.3.1).

The following assumption about the window width $\omega = \omega_n$ is standard:

$$\omega \to 0, \quad n\omega \to \infty. \tag{13.3.3}$$

We shall make the following assumption regarding W:

    Kernel W is an even positive probability density on $(-1, 1)$,   (13.3.4)
    vanishing off $(-1, 1)$.

The final set of assumptions refer to functions σ and ℓ in the regression model (13.3.1).

    Function σ is strictly positive and twice differentiable on $\mathbb{R}$.   (13.3.5)
    Moreover, $E\sigma^4(X_0) < \infty, E\|\ell(X_0)\|^4 < \infty$, and the functions
    $\sigma^4(x)\tilde{\phi}(x), \|\ell(x)\|^4\tilde{\phi}(x)$, and $\partial^2(\sigma^2(x)\tilde{\phi}(x))/\partial^2 x$ are bounded on $\mathbb{R}$.

The following theorem gives the rates of consistency of $\hat{\sigma}^2(x)$ for $\sigma^2(x)$, where $\tau(n, d) := 1$ if $d \neq 1/4$, and $\tau(n, d) := \log^{1/2}(n)$ if $d = 1/4$.

Theorem 13.3.1. *Suppose (13.2.22) and (13.2.23) hold with $E\zeta_0^4 < \infty$, and that $\{X_i\}$ is Gaussian and independent of $\{\varepsilon_i\}$. In addition, assume (13.2.2), (13.3.1), (13.3.3), (13.3.4), and (13.3.5) hold. Then, $\forall x \in \mathbb{R}$,*

$$\hat{\sigma}^2(x) - \sigma^2(x) = O(\omega^2) + O_p\left(\frac{1}{\sqrt{\omega n}}\right) + O_p\left(n^{d_1 - 1/2}\right) \qquad (13.3.6)$$

$$+ O_p\left(\frac{\tau(n, d)}{n^{(1-2d)\wedge(1/2)}}\right).$$

Remark 13.3.1. Consider $\omega = O(n^{-\delta})$ for some $\delta > 0$. From (13.3.6) we easily obtain the convergence rate

$$\hat{\sigma}^2(x) - \sigma^2(x) = O_p\left(n^{d_1 - 1/2}\right), \qquad 2d < 1/2 + d_1, \qquad (13.3.7)$$

$$\delta > (1 - 2d_1)/4,$$

$$\hat{\sigma}^2(x) - \sigma^2(x) = O_p(n^{2d - 1}), \qquad 2d \geq 1/2 + d_1, \qquad (13.3.8)$$

$$\delta > 1/2 - d.$$

A closer look at the remainder terms in the proof of Theorem 13.3.1 below reveals that the limit distribution of the estimator $\hat{\sigma}^2(x)$ can be non-Gaussian, depending on the long memory parameters d and d_1. We do not pursue this here further but relevant results can be found in Guo and Koul (2008). The problem of obtaining the limiting distribution of these estimators for general non-Gaussian X_j's, analogous to (12.3.21), is open at the time of writing. Even for the Gaussian X_j's this problem is not solved completely. Another open problem is obtaining uniform consistency rates of $\hat{\sigma}^2(x)$ for random or non-random designs and long memory errors.

Proof of Theorem 13.3.1. For convenience, assume in this proof that $m = 0$ and $\tau = 1$ so that $X_0 \sim \mathcal{N}(0,1)$. Consider the r.v. $W_\omega(x - X_0)\sigma^2(X_0) = \omega^{-1}W((x - X_0)/\omega)\sigma^2(X_0)$ indexed by $x \in \mathbb{R}$. According to (13.3.4) and (13.3.5),

$$\lambda_\omega(x) := E\Big(W_\omega(x - X_0)\sigma^2(X_0)\Big)^2 \tag{13.3.9}$$

$$= \omega^{-1}\int_{-1}^{1} W^2(y)\sigma^4(x + \omega y)\phi(x + \omega y)dy$$

$$\leq C\omega^{-1}\int_{-1}^{1} W^2(y)dy \leq C\omega^{-1},$$

uniformly in $x \in \mathbb{R}$. Let

$$J_{k\omega}(x) := EH_k(X)W_\omega(x - X)\sigma^2(X), \quad k = 0, 1, \cdots$$

be the coefficients of the Hermite expansion of $W_\omega(x - X_0)\sigma^2(X_0)$, and let $k_* \geq 1$ be the smallest integer such that $(k_* + 1)(1 - 2d_1) > 1$. Then, as above,

$$|J_{k\omega}(x)| = \Big|\int_{-1}^{1} W(y)H_k(x + \omega y)\sigma^2(x + \omega y)\phi(x + \omega y)dy\Big| \tag{13.3.10}$$

$$\leq C\int_{-1}^{1} W(y)|H_k(x + \omega y)|\phi^{1/2}(x + \omega y)dy$$

$$\leq C\int_{-1}^{1} W(y)dy \leq C, \quad k = 1, \cdots, k_*,$$

uniformly in $x \in \mathbb{R}$. Let

$$\widetilde{\sigma}^2(x) := \frac{1}{n\phi(x)}\sum_{i=1}^{n} W_{\omega i}(x)\,\sigma^2(X_i)\varepsilon_i^2.$$

Decompose

$$\widehat{\sigma}^2(x) - \sigma^2(x) = A + B + C,$$

where

$$A := \widetilde{\sigma}^2(x) - \sigma^2(x), \quad B := \widehat{\sigma}^2(x) - \frac{\phi(x)}{\phi_n(x)}\widetilde{\sigma}^2(x), \quad C := \Big[\frac{\phi(x)}{\phi_n(x)} - 1\Big]\widetilde{\sigma}^2(x).$$

Clearly, (13.3.6) follows from

$$A = O_p\Big(\frac{\tau(n, d)}{n^{(1-2d)\wedge(1/2)}}\Big) + O_p(n^{d_1-1/2}) \tag{13.3.11}$$

$$+ O_p\Big(\frac{1}{\sqrt{\omega n}}\Big) + O(\omega^2),$$

$$B = O_p(n^{2d-1}) + O_p\Big(\frac{n^{d-1/2}}{\sqrt{\omega n}}\Big), \tag{13.3.12}$$

$$C = O_p(n^{d_1-1/2}). \tag{13.3.13}$$

Proof of (13.3.11). Split $A = \sum_{j=1}^{4} A_j$, where

$$A_1 := \frac{J_{0\omega}(x)}{n\phi(x)} \sum_{i=1}^{n} (\varepsilon_i^2 - E\varepsilon_0^2), \quad A_2 := \frac{1}{n\phi(x)} \sum_{i=1}^{n} \varepsilon_i^2 \sum_{k=1}^{k_*} H_k(X_i) \frac{J_{k\omega}(x)}{k!},$$

$$A_3 := \frac{1}{n\phi(x)} \sum_{i=1}^{n} \varepsilon_i^2 \Big[W_{\omega i}(x)\sigma^2(X_i) - \sum_{k=0}^{k_*} H_k(X_i) \frac{J_{k\omega}(x)}{k!} \Big],$$

$$A_4 := \frac{J_{0\omega}(x)}{\phi(x)} - \sigma^2(x),$$

and where we used $E\varepsilon_0^2 = 1$. By Lemma 4.5.3,

$$\text{Cov}(\varepsilon_0^2, \varepsilon_i^2) = 2 \Big(\sum_{k=0}^{\infty} a_k a_{i+k} \Big)^2 + [E\zeta_0^4 - 3] \sum_{k=0}^{\infty} a_k^2 a_{k+i}^2 \qquad (13.3.14)$$

$$= O(i^{2(2d-1)}) + O(i^{2(d-1)}) = O(i^{2(2d-1)}), \quad i \to \infty.$$

Therefore, using (13.3.10),

$$EA_1^2 = \frac{J_{0\omega}^2(x)}{n^2\phi^2(x)} \sum_{i,j=1}^{n} \text{Cov}(\varepsilon_i^2, \varepsilon_j^2) \leq Cn^{-(2-4d)\wedge 1} \tau^2(n, d),$$

implying

$$A_1 = O_p\big(n^{-(1-2d)\wedge(1/2)} \tau(n, d)\big). \qquad (13.3.15)$$

Use the independence of the $\{X_i\}$'s and $\{\varepsilon_j\}$'s and (2.4.15) to obtain

$$EA_2^2 \leq \frac{E\varepsilon_0^4}{n^2\phi(x)^2} \sum_{k=1}^{k_*} \sum_{i,j=1}^{n} |\text{Cov}(X_i, X_j)|^k \frac{J_{k\omega}^2(x)}{k!}$$

$$\leq Cn^{-2} \sum_{i,j=1}^{n} |\text{Cov}(X_i, X_j)| \leq Cn^{2d_1-1},$$

implying

$$A_2 = O_p\big(n^{d_1-1/2}\big). \qquad (13.3.16)$$

Next, using (13.3.9), the independence of the design and errors and (2.4.18), with $\mathcal{Y}_i(x) := W_{\omega i}(x)\sigma^2(X_i) - \sum_{k=0}^{k_*} H_k(X_i) \frac{J_{k\omega}(x)}{k!}$,

$$EA_3^2 = \frac{1}{n^2\phi^2(x)} \sum_{i,j=1}^{n} \text{Cov}\big(\mathcal{Y}_i(x), \mathcal{Y}_j(x)\big) E[\varepsilon_i^2 \varepsilon_j^2]$$

$$\leq \frac{C(E\varepsilon_0^4)\lambda_\omega(x)}{n^2\phi^2(x)} \sum_{i,j=1}^{n} |\text{Cov}(X_i, X_j)|^{k_*+1} \leq C\omega^{-1} n^{-1},$$

implying

$$A_3 = O_p\big((\omega n)^{-1/2}\big). \tag{13.3.17}$$

Finally, rewrite

$$A_4 = \phi(x)^{-1} \int_{-1}^{1} W(y)\big[\sigma^2(x+\omega y)\phi(x+\omega y) - \sigma^2(x)\phi(x)\big]dy.$$

Use $\int_{-1}^{1} yW(y)dy = 0$, assumption (13.3.5), and a Taylor expansion of $\sigma^2(x)\phi(x)$ up to two terms, to verify

$$|A_4| \leq C\omega^2. \tag{13.3.18}$$

Clearly, (13.3.15) to (13.3.18) prove (13.3.11).

Proof of (13.3.12). Let $\widehat{\Delta} := \beta_n - \beta_0$ and

$$\Sigma_n := \frac{1}{n\phi(x)} \sum_{i=1}^{n} W_{\omega i}(x)\ell(X_i)\ell(X_i)',$$

$$S_n := \frac{1}{n\phi(x)} \sum_{i=1}^{n} W_{\omega i}(x)\sigma(X_i)\ell(X_i)\varepsilon_i.$$

Then

$$B = \frac{1}{n\phi_n(x)} \sum_{i=1}^{n} W_{\omega i}(x) \left(\big(Y_i - \beta_n'\ell(X_i)\big)^2 - \sigma^2(X_i)\varepsilon_i^2 \right)$$

$$= \frac{\phi(x)}{\phi_n(x)} \left[\widehat{\Delta}'\Sigma_n\widehat{\Delta} - 2\widehat{\Delta}'S_n \right].$$

The LS estimator β_n satisfies

$$\widehat{\Delta} = \Lambda_n^{-1} n^{-1} \sum_{i=1}^{n} \ell(X_i)\sigma(X_i)\varepsilon_i = O_p(n^{d-1/2}), \tag{13.3.19}$$

where we used (13.2.23) and the fact that $\Lambda_n := n^{-1}\sum_{i=1}^{n}\ell(X_i)\ell(X_i)'$ $\to_p \Lambda$, a non-singular matrix by assumption. In a similar way, because $E|W_{\omega i}(x)|\|\ell(X)\|^2 = \int_{-1}^{1}|W(y)|\|\ell(x+\omega y)\|^2\phi(x+\omega y)dy \leq C$,

$$\Sigma_n = O_p(1). \tag{13.3.20}$$

Assessing the order of S_n is similar to that of A in (13.3.11). Let $\widetilde{J}_{k\omega}(x) := EH_k(X_0)W_\omega(x - X_0)\sigma(X_0)\ell(X_0)$. Consider the expansion

$$W_{\omega i}(x)\sigma(X_i)\ell(X_i) = \sum_{k=0}^{k_*} H_k(X_i)\frac{\widetilde{J}_{k\omega}(x)}{k!} + \widetilde{\mathcal{Y}}_i(x).$$

The coefficients $\widetilde{J}_{k\omega}$ satisfy

$$|\widetilde{J}_{k\omega}(x)| \le C, \quad k = 1, \cdots, k_*. \tag{13.3.21}$$

Accordingly, decompose $S_n = S_{n1} + S_{n2} + S_{n3}$, where

$$S_{n1} := \frac{\widetilde{J}_{0\omega}(x)}{n\phi(x)} \sum_{i=1}^{n} \varepsilon_i, \quad S_{n2} := \frac{1}{n\phi(x)} \sum_{i=1}^{n} \varepsilon_i \sum_{k=1}^{k_*} H_k(X_i) \frac{\widetilde{J}_{k\omega}(x)}{k!},$$

$$S_{n3} := \frac{1}{n\phi(x)} \sum_{i=1}^{n} \varepsilon_i \widetilde{\mathcal{Y}}_i(x).$$

Using (13.3.21), as in the proof of (13.3.11), we obtain

$$S_{n1} = O_p(n^{d-1/2}), \quad S_{n2} = O_p\Big(\frac{\tau(n, (d+d_1)/2)}{n^{(1-d-d_1)\wedge(1/2)}}\Big) = o_p(n^{d-1/2}),$$

$$S_{n3} = O_p\Big(\frac{1}{\sqrt{\omega n}}\Big).$$

Upon combining (13.3.19) to (13.3.22) and using $\phi(x)/\phi_n(x) = O_p(1)$, see the proof of (13.3.13) below, claim (13.3.12) follows.

Proof of (13.3.13). Recall that $m = EX_0 = 0$, $\tau^2 = \mathrm{Var}(X) = 1$, and the definition of s^2 from (13.3.2). We shall first prove

$$s - 1 = O_p\Big(\frac{\tau(n, d_1)}{n^{(1-2d_1)\wedge(1/2)}}\Big). \tag{13.3.22}$$

Let $s_1^2 = n^{-1} \sum_{i=1}^{n} X_i^2$. By (13.2.23) and the Gaussianity of $\{X_i\}$,

$$E(s_1^2 - 1)^2 = n^{-2} E\Big(\sum_{i=1}^{n}(X_i^2 - 1)\Big)^2 = n^{-2} \sum_{i,j=1}^{n} \mathrm{Cov}(X_i^2, X_j^2)$$

$$= 2n^{-2} \sum_{i,j=1}^{n} \big(\mathrm{Cov}(X_i, X_j)\big)^2 = O\Big(\frac{\tau^2(n, d_1)}{n^{(2-4d_1)\wedge 1}}\Big).$$

Because, $s^2 = s_1^2 - \bar{X}^2$ and because $E\bar{X}^2 = O(n^{-1+2d_1})$, we have $(s^2 - 1) = O_p\big(\frac{\tau(n,d_1)}{n^{(1-2d_1)\wedge(1/2)}}\big)$. Hence, (13.3.22) follows by writing $s-1 = (s^2-1)/(s+1)$ and using the fact that $1/(s+1) \to_p 1/2$. Claim (13.3.13) follows by taking a Taylor expansion of $\phi_n(x) = s^{-1}\phi((x - \bar{X})/s)$ around $\bar{X} = 0, s = 1$ and using (13.3.22) and the fact that $\bar{X} = O_p(n^{d_1-1/2})$, which decays at a slower rate than $s-1$ in (13.3.22), so that $\phi(x)/\phi_n(x) - 1 = O_p(\bar{X}) = O_p(n^{d_1-1/2})$. This completes the proof of Theorem 13.3.1.

Next, we shall describe a cross validation-type estimator of $\sigma^2(x)$ and prove its consistency. This is needed later when carrying out a test of H_0.

For this purpose we assume the density of X_0 is known, i.e., m and τ are known, and take them to be 0 and 1, respectively, without loss of generality. Let

$$\hat{\Sigma}_{-i}(x) := \left(\frac{1}{n-1} \sum_{j=1, j\neq i}^n W_{\omega j}(x)\hat{e}_j^2\right)^{1/2}, \qquad (13.3.23)$$

$$V_i(x) := \hat{\Sigma}_{-i}(x)\phi^{-1/2}(x), \quad i = 1, \cdots, n,$$

$$\tilde{\Sigma}_{-i}(x) := \left(\frac{1}{n-1} \sum_{j=1, j\neq i}^n W_{\omega j}(x)\sigma^2(X_j)\varepsilon_j^2\right)^{1/2}.$$

The statistic $n^{-1} \sum_{i=1}^n V_i(x)$ is a cross validation-type estimator of $\sigma(x)$. The following theorem establishes a consistency type result for this estimator, which is used in Subsection 13.3.1 below.

Theorem 13.3.2. *Suppose the model (13.3.1) holds with $\{\varepsilon_i\}$ as in (13.2.22) and where $\{X_i\}$ is a stationary Gaussian process having zero mean and unit variance, independent of $\{\varepsilon_i\}$, and γ_X satisfying (13.2.23). In addition, assume $E\zeta_0^4 < \infty$, and (13.2.2), (13.3.1), (13.3.3), and (13.3.4) hold. Furthermore, assume that there exist $C > 0, p > 4$ such that*

$$\big(\|\ell(x)\|^p + \sigma^p(x)\big)\phi(x) \leq C, \quad \forall x \in \mathbb{R}. \qquad (13.3.24)$$

Then

$$n^{-1} \sum_{i=1}^n \big|V_i(X_i) - \sigma(X_i)\big| \|\ell(X_i)\| \to_p 0. \qquad (13.3.25)$$

Proof. It suffices to show that

$$n^{-1} \sum_{i=1}^n \big|\hat{\Sigma}_{-i}(X_i) - \tilde{\Sigma}_{-i}(X_i)\big| \|\ell(X_i)\|\phi^{-1/2}(X_i) = o_p(1), \qquad (13.3.26)$$

$$n^{-1} \sum_{i=1}^n \big|\tilde{\Sigma}_{-i}(X_i)\phi^{-1/2}(X_i) - \sigma(X_i)\big| \|\ell(X_i)\| = o_p(1). \qquad (13.3.27)$$

By Hölder's inequality, the expectation of the l.h.s. of (13.3.27) is bounded above by

$$n^{-1} \sum_{i=1}^n E^{1/4}\{\tilde{\Sigma}_{-i}(X_i) - \sigma(X_i)\phi^{1/2}(X_i)\}^4 E^{3/4}\{\|\ell(X_0)\|^{4/3}\phi^{-2/3}(X_0)\}.$$

Since, by (13.3.24),

$$E\|\ell(X_0)\|^{4/3}\phi^{-2/3}(X_0) = \int \|\ell(x)\|^{4/3}\phi^{1/3}(x)dx < \infty,$$

the claim (13.3.27) follows from Lemma 13.3.2 below. Similarly, by the C–S inequality, the l.h.s. of (13.3.26) is bounded above by

$$\frac{1}{n}\left\{\sum_{i=1}^{n}\left|\hat{\Sigma}_{-i}(X_i) - \tilde{\Sigma}_{-i}(X_i)\right|^2\phi^{-1/2}(X_i)\right\}^{1/2}\left\{\sum_{i=1}^{n}\|\ell(X_i)\|^2\phi^{-1/2}(X_i)\right\}^{1/2}.$$

The claim (13.3.26) follows from this bound, Lemma 13.3.1 below, and the fact that $n^{-1}\sum_{i=1}^{n}\|\ell(X_i)\|^2\phi^{-1/2}(X_i) \to_{a.s.} E\|\ell(X_0)\|^2\phi^{-1/2}(X_0) < \infty$. Theorem 13.3.2 is proved.

It remains to prove Lemmas 13.3.1 and 13.3.2.

Lemma 13.3.1. *Under the assumptions of Theorem 13.3.2,*

$$n^{-1}\sum_{i=1}^{n}\left|\hat{\Sigma}_{-i}(X_i) - \tilde{\Sigma}_{-i}(X_i)\right|^2\phi^{-1/2}(X_i) \to_p 0. \tag{13.3.28}$$

Proof. Let R_n denote the l.h.s. of (13.3.28) and $\hat{\Delta} := \beta_n - \beta_0$. Using the elementary inequality $|a^{1/2} - b^{1/2}|^2 \le |a - b|$, $a, b \ge 0$, we can see that $R_n \le R_{n1}\|\hat{\Delta}\|^2 + 2R_{n2}\|\hat{\Delta}\|$, where

$$R_{n1} := \frac{1}{n(n-1)}\sum_{i=1}^{n}\sum_{j\neq i}^{n}W_{\omega j}(X_i)\|\ell(X_j)\|^2\phi^{-1/2}(X_i),$$

$$R_{n2} := \frac{1}{n(n-1)}\sum_{i=1}^{n}\sum_{j\neq i}^{n}W_{\omega j}(X_i)\|\ell(X_j)\|\,\sigma(X_j)\phi^{-1/2}(X_i)\,|\varepsilon_j|.$$

By (13.3.29) below, $E[R_{n1} + R_{n2}] \le C$, and by (13.3.19), $\|\hat{\Delta}\| = o_p(1)$. Hence, $R_n = o_p(1)$.

It remains to prove that for all $1 \le i \neq j \le n$,

$$\sup_{j\neq i} E\left\{|W_{\omega j}(X_i)|[\|\ell(X_j)\|^2 + \sigma^2(X_j)]\phi^{-1/2}(X_i)\right\} \le C. \tag{13.3.29}$$

Let (X, Y) be bivariate Gaussian r.v.'s with zero means, unit variances, and correlation coefficient $\rho \in (-1, 1)$. Let $\phi(y|x) := (2\pi(1 - \rho^2))^{-1/2}\exp\{-(y - \rho x)^2/2(1 - \rho^2)\}$ denote the corresponding conditional density. Then $\phi(y|x) \le C(1 - \rho^2)^{-1/2} \le C$ and

$$\omega^{-1} E\Big\{ W\Big(\frac{X-Y}{\omega}\Big) \|\ell(Y)\|^2 \phi^{-1/2}(X) \Big\} \tag{13.3.30}$$

$$= \omega^{-1} \int_{\mathbb{R}} \phi^{1/2}(x) \int_{\mathbb{R}} \phi(y|x) \|\ell(y)\|^2 W\Big(\frac{x-y}{\omega}\Big) dx dy$$

$$\leq C\omega^{-1} \int_{\mathbb{R}} \phi^{1/2}(x) \int_{\mathbb{R}} \|\ell(y)\|^2 W\Big(\frac{x-y}{\omega}\Big) dx dy$$

$$\leq C \int_{-1}^{1} du \int_{\mathbb{R}} \frac{\phi^{1/2}(x)}{\phi^{2/p}(x+\omega u)} dx$$

$$= C \int_{-1}^{1} e^{(\omega u)^2/(p-4)} du \int_{\mathbb{R}} e^{-a(x-(\omega u/ap))^2} dx \leq C,$$

where $a = 1/4 - 1/p > 0$. The second inequality in the above bounds follows from (13.3.24) and because W is bounded. By a similar argument we can obtain a similar bound for the second summand, involving $\sigma^2(X_j)$, in the l.h.s. of (13.3.29), thereby completing the proof of (13.3.29), and hence of Lemma 13.3.1.

Lemma 13.3.2. *Under the assumptions of Theorem 13.3.2,*

$$\max_{1 \leq i \leq n} E\big| \tilde{\Sigma}_{-i}(X_i) - \sigma(X_i)\phi^{1/2}(X_i) \big|^4 \to 0. \tag{13.3.31}$$

Proof. Claim (13.3.31) follows from

$$\max_{1 \leq i \leq n} E\big\{ \tilde{\Sigma}^2_{-i}(X_i) - \sigma^2(X_i)\phi(X_i) \big\}^2 \to 0, \tag{13.3.32}$$

and the elementary inequality $|a^{1/2} - b^{1/2}|^2 \leq |a-b|$, $a, b \geq 0$. Because the argument below does not depend on i, we give details for $i = 1$ only. Write

$$\tilde{\Sigma}^2_{-1}(X_1) - \sigma^2(X_1)\phi(X_1) = \frac{1}{n-1} \sum_{j=2}^{n} \Gamma_j,$$

$$\Gamma_j := W_{\omega j}(X_1)\sigma^2(X_j)\varepsilon_j^2 - \sigma^2(X_1)\phi(X_1).$$

Arguing as for (13.3.30), verify that $\max_{2 \leq j \leq n} E\Gamma_j^2 \leq \omega^{-1} C$.

For $K \geq 1$, let

$$\mathcal{A}_K := \{ (j,k) \in \{2,n\}^2 : j \geq K, k \geq K, |j-k| \geq K \}, \quad \mathcal{A}_K^c := \{2,n\}^2 \backslash \mathcal{A}_K,$$

$$U_{1,K} := \frac{1}{(n-1)^2} \sum_{(j,k)\in\mathcal{A}_K} E\{\Gamma_j\Gamma_k\}, \quad U_{1,K^c} := \frac{1}{(n-1)^2} \sum_{(j,k)\in\mathcal{A}_K^c} E\{\Gamma_j\Gamma_k\}.$$

Then, for any $K \geq 1$,

$$(n-1)^{-2} E \Big(\sum_{j=2}^{n} \Gamma_j \Big)^2 = \frac{1}{(n-1)^2} \sum_{j,k=2}^{n} E\{\Gamma_j \Gamma_k\} = U_{1,K} + U_{1,K^c}.$$

Clearly, the set $\mathcal{A}_K^c$ has at most $3K(n-1)$ elements and therefore

$$|U_{1,K^c}| \leq \frac{3K}{n-1} \max_{2 \leq j \leq n} E\{\Gamma_j^2\} \leq \frac{CK}{n\omega} \to 0, \qquad (13.3.33)$$

for any fixed $K < \infty$. We will show that for any $\epsilon > 0$ we can find $n_0 < \infty$ and K such that

$$U_{1,K} \leq \epsilon, \qquad \forall n > n_0. \qquad (13.3.34)$$

Statement (13.3.32) follows from (13.3.33) and (13.3.34).

To prove (13.3.34), we need the following result from Soulier (2001). (Later, we shall apply this result with $\nu = 2$ and 3 only.) Let $\boldsymbol{\xi} = (\xi_1, \cdots, \xi_\nu)'$ be a ν-dimensional Gaussian vector with $E\xi_i = 0, E\xi_i^2 = 1, i = 1, \cdots, \nu$, and non-degenerate covariance matrix $B = (b_{ij})_{i,j=1,\cdots,\nu}$. Let $\varrho \geq 0$ denote the largest eigenvalue of the matrix

$$B - I = \begin{pmatrix} 0 & b_{12} & b_{13} & \cdots & b_{1\nu} \\ b_{21} & 0 & b_{23} & \cdots & b_{2\nu} \\ \cdots & & & & \\ b_{\nu 1} & b_{\nu 2} & b_{\nu 3} & \cdots & 0 \end{pmatrix}.$$

Then

$$\varrho \leq \max_{i=1,\cdots,\nu} \sum_{j=1, j \neq i}^{\nu} |b_{ij}|. \qquad (13.3.35)$$

Let E_ν^0 denote the expectation with respect to which $\boldsymbol{\xi}$ has a standard Gaussian distribution $\mathcal{N}_\nu(0, I)$. For a square integrable function $g = g(\boldsymbol{\xi})$, $\|g\|_0 := (E_\nu^0 g^2(\boldsymbol{\xi}))^{1/2} < \infty$, the Hermite rank $h(g)$ of $g(\boldsymbol{\xi}) - E_\nu^0 g(\boldsymbol{\xi})$ is defined by

$$h(g) := \min \{k \geq 1 : J_{i_1,\cdots,i_\nu}(g) \neq 0, i_1 + \cdots + i_\nu = k\}, \qquad (13.3.36)$$

where

$$J_{i_1,\cdots,i_\nu}(g) := E_\nu^0 \{g(\boldsymbol{\xi}) H_{i_1}(\xi_1) \cdots H_{i_\nu}(\xi_\nu)\}, \qquad (13.3.37)$$

$i_1, \cdots, i_\nu = 0, 1, \cdots$, and H_i, $i = 0, 1, \cdots$, are Hermite polynomials.

Theorem 2.1 of Soulier (2001) states that there exists a constant $C < \infty$ such that for all sufficiently small ϱ and any square integrable function g, the following inequality holds:

$$\left| Eg(\boldsymbol{\xi}) - E_\nu^0 g(\boldsymbol{\xi}) \right| \leq C\|g\|_0 \varrho^{h(g)/2}. \qquad (13.3.38)$$

To apply the above result, write

$$E\{\Gamma_k\Gamma_j\} = E\{W_{\omega j}(X_1)W_{\omega k}(X_1)\sigma^2(X_j)\sigma^2(X_k)\} E\{\varepsilon_k^2\varepsilon_j^2\} \qquad (13.3.39)$$
$$- E\{W_{\omega j}(X_1)\sigma^2(X_j)\sigma^2(X_1)\phi(X_1)\}$$
$$- E\{W_{\omega k}(X_1)\sigma^2(X_k)\sigma^2(X_1)\phi(X_1)\}$$
$$+ E\{\sigma^4(X_1)\phi^2(X_1)\}.$$

We will show that all four terms on the r.h.s. of (13.3.39) tend to the same limit (the last term) as $\omega \to 0$ and the indices $1, j, k$ tend to be far apart, and therefore the expectation on the l.h.s. of (13.3.39) tends to zero. In particular, assuming that X_1, X_j, and X_k are independent, we have

$$E_3^0\{W_{\omega j}(X_1)W_{\omega k}(X_1)\sigma^2(X_j)\sigma^2(X_k)\}$$
$$= \int_{[-1,1]^2} W(u_1)W(u_2)du_1 du_2$$
$$\times \int_{\mathbb{R}} \sigma^2(x + \omega u_1)\sigma^2(x + \omega u_2)\phi(x + \omega u_1)\phi(x + \omega u_2)\phi(x)dx$$
$$= \int_{\mathbb{R}} \sigma^4(x)\phi^3(x)dx = E\{\sigma^4(X_1)\phi^2(X_1)\} + o(1),$$

by the DCT and the bound (13.3.24). In a similar way,

$$E_2^0\{W_{\omega j}(X_1)\sigma^2(X_j)\sigma^2(X_1)\phi(X_1)\} = E\{\sigma^4(X_1)\phi^2(X_1)\} + o(1).$$

Moreover, by (13.3.14), $\mathrm{Cov}(\varepsilon_j^2, \varepsilon_k^2) = O(|j - k|^{2(2d-1)}) \to 0$, implying $E\{\varepsilon_k^2\varepsilon_j^2\} \to 1$ as $|k - j| \to \infty$. Hence, for any $\epsilon > 0$ there exist n_0 and K such that

$$\left| E_3^0\{\Gamma_k\Gamma_j\} \right| < \epsilon, \quad \forall n > n_0, \ \forall j, k \in \mathbb{Z}, \ |j - k| > K. \qquad (13.3.40)$$

Next, we evaluate the difference $E\{\Gamma_k\Gamma_j\} - E_3^0\{\Gamma_k\Gamma_j\}$. We have

$$\left| E\{\Gamma_k\Gamma_j\} - E_3^0\{\Gamma_k\Gamma_j\} \right| \leq T_{jk} + T_j + T_k,$$

where

$$T_{jk} := \left| E\{W_{\omega j}(X_1)W_{\omega k}(X_1)\sigma^2(X_j)\sigma^2(X_k)\} \right.$$
$$\left. - E_3^0\{W_{\omega j}(X_1)W_{\omega k}(X_1)\sigma^2(X_j)\sigma^2(X_k)\} \right|,$$
$$T_j := \left| E\{W_{\omega j}(X_1)\sigma^2(X_j)\sigma^2(X_1)\phi(X_1)\} \right.$$
$$\left. - E_2^0\{W_{\omega j}(X_1)\sigma^2(X_j)\sigma^2(X_1)\phi(X_1)\} \right|,$$

and T_k is defined similarly. Let $\xi_1 := X_1$, $\xi_2 := X_j$, $\xi_3 := X_k$, $\boldsymbol{\xi} = (\xi_1, \xi_2, \xi_3)'$, and let

$$g(\boldsymbol{\xi}) := W_{\omega j}(X_1) W_{\omega k}(X_1) \sigma^2(X_j) \sigma^2(X_k).$$

For any $q = 2, 3, \cdots$, rewrite $g(\boldsymbol{\xi}) = g_1(\boldsymbol{\xi}) + g_2(\boldsymbol{\xi})$, where

$$g_1(\boldsymbol{\xi}) := \sum_{1 \leq i_1 + i_2 + i_3 \leq q-1} \frac{J_{i_1, i_2, i_3}(g)}{i_1! i_2! i_3!} H_{i_1}(\xi_1) H_{i_2}(\xi_2) H_{i_3}(\xi_3),$$

$$g_2(\boldsymbol{\xi}) := g(\boldsymbol{\xi}) - g_1(\boldsymbol{\xi}).$$

According to the definition of the Hermite rank in (13.3.36),

$$h(g_1) \geq 1, \quad h(g_2) \geq q. \tag{13.3.41}$$

Clearly,

$$T_{jk} \leq T_{jk,1} + T_{jk,2}, \quad T_{jk,i} := \left| E\{g_i(\boldsymbol{\xi})\} - E_3^0\{g_i(\boldsymbol{\xi})\} \right|, \quad i = 1, 2.$$

Note $E_3^0\{g_1(\boldsymbol{\xi})\} = 0$, $\|g_1\|_0 \leq \|g\|_0 \leq C/\omega$, and

$$\|g_1\|_0^2 = \sum_{1 \leq i_1 + i_2 + i_3 \leq q-1} \frac{J_{i_1, i_2, i_3}^2(g)}{i_1! i_2! i_3!} \leq C,$$

by the definition of $J_{i_1, i_2, i_3}(g)$ in (13.3.37) and the orthogonality property (2.4.15) of Hermite polynomials. The above facts together with (13.3.38) give the bounds

$$T_{jk,1} \leq C\varrho(j, k)^{1/2}, \quad T_{jk,2} \leq C\omega^{-1}\varrho(j, k)^{q/2}, \tag{13.3.42}$$

where $0 \leq \varrho(j, k)$ equals the largest eigenvalue of the 3×3 matrix

$$\begin{pmatrix} 0 & \gamma_X(j-1) & \gamma_X(k-1) \\ \gamma_X(j-1) & 0 & \gamma_X(j-k) \\ \gamma_X(k-1) & \gamma_X(j-k) & 0 \end{pmatrix}.$$

By (13.3.35),

$$\varrho(j, k) \leq C(|\gamma_X(j-1)| + |\gamma_X(k-1)| + |\gamma_X(k-j)|).$$

Let $q \geq 2$ be the smallest integer such that $q(1 - 2d_1) > 2$. Then $\sum_{k \in \mathbb{Z}} |\gamma_X(k)|^{q/2} < \infty$ and from (13.3.42) and (13.3.35), we obtain

$$\sum_{(j,k) \in A_K} T_{jk,1} \leq Cn^{2 - (1 - 2d_1)/2}, \quad \sum_{(j,k) \in A_K} T_{jk,2} \leq C\omega^{-1}n.$$

Similarly, applying (13.3.38) with $\nu = 2$, $\boldsymbol{\xi} := (X_1, X_j)'$, $g(\boldsymbol{\xi}) := W_{\omega j}(X_1)$ $\sigma^2(X_j)\sigma^2(X_1)\phi(X_1)$, and using $\|g\|_0 \leq C\omega^{-1/2}$ we obtain

$$\sum_{(j,k)\in A_K} T_j \leq C\big(n^{2-(1-2d_1)/2} + \omega^{-1/2}n\big).$$

Because $d_1 < 1/2$, the above bounds give as $\omega n \to \infty$,

$$\frac{1}{(n-1)^2} \sum_{(j,k)\in A_K} \big|E\{\Gamma_k\Gamma_j\} - E_3^0\{\Gamma_k\Gamma_j\}\big| \leq C\big(\frac{1}{n^{1/2-d_1}} + \frac{1}{\omega n}\big) = o(1).$$

Together with (13.3.40), this proves (13.3.34) and hence (13.3.32), too, thereby completing the proof of Lemma 13.3.2.

13.3.1   *Lack-of-fit tests: heteroscedastic LMMA errors*

We are now ready to discuss the problem of testing H_0 when the errors are heteroscedastic and have long memory, i.e., under H_0, the model (13.3.1) holds for some positive function $\sigma(x)$, where the ε_i are as in (13.2.22) and are independent of the X_i's. We also assume that $\sigma^2(x) = \text{Var}[(Y - \beta_0'\ell(X_0))|X_0 = x]$, and that the process $\{X_i\}$ is a long memory Gaussian process as in (13.2.5). The function ℓ is assumed to satisfy (13.2.2) and (13.2.11). Let β_n denote the LS estimator of β under H_0.

Tests of H_0 are to be based on $\widetilde{\mathcal{V}}_n$, an analog of the $\mathcal{V}_n$ process:

$$\widetilde{\mathcal{V}}_n(x) = \sum_{i=1}^{n} \big(Y_i - \beta_n'\ell(X_i)\big)I(X_i \leq x), \quad x \in \mathbb{R}.$$

To describe its asymptotic distribution, recall the definitions of ν and $\bar{\nu}_n$ from (13.2.9) and let

$$J_\sigma(x) := E\Big([\sigma(X_0) - E\{\sigma(X_0)\ell(X_0)\}'\Lambda^{-1}\ell(X_0)]I(X_0 \leq x)\Big),$$

$$Z_n := \sum_{i=1}^{n} \ell(X_i)\sigma(X_i)\varepsilon_i, \quad \mathcal{L}(x) := E\sigma^2(X_0)I(X_0 \leq x),$$

$$F_\sigma(x) := E\sigma(X_0)I(X_0 \leq x), \quad \widetilde{\mathcal{V}}(x) := \sum_{i=1}^{n} \sigma(X_i)\varepsilon_i I(X_i \leq x),$$

$$\mathcal{U}_n(x) := \sum_{i=1}^{n} \varepsilon_i\{\sigma(X_i)I(X_i \leq x) - F_\sigma(x)\}, \quad x \in \mathbb{R}.$$

Also let $S_n := \sum_{i=1}^{n} \varepsilon_i$. Akin to Lemma 13.2.4, the following lemma gives the weak convergence of the $\widetilde{\mathcal{V}}_n$ process.

Lemma 13.3.3. *Suppose the model (13.3.1) holds with ℓ satisfying (13.2.2) and (13.2.11), $\{\varepsilon_i\}$ is as in (13.2.22), and $\{X_i\}$ is a long memory moving-average Gaussian process, independent of $\{\varepsilon_i\}$. Then*

$$\sup_{x \in \mathbb{R}} \left| n^{-d-1/2} \{ \widetilde{\mathcal{V}}_n(x) - J_\sigma(x) S_n \} \right| = o_p(1).$$

Consequently, under H_0, $n^{-d-1/2} \widetilde{\mathcal{V}}_n(x) \Rightarrow J_\sigma(x) \, s_{\varepsilon,d} \, Z$ in $\mathcal{D}(\bar{\mathbb{R}})$ with uniform metric, where $Z \sim \mathcal{N}(0,1)$ and $s_{\varepsilon,d}$ is as in (13.2.32).

Proof. Note $\mathcal{U}_n(x) = \widetilde{\mathcal{V}}(x) - F_\sigma(x) S_n$. Using the inequality (2.4.18), for any $i, j \in \mathbb{Z}$ and any $x < y$ we obtain

$$\left| \mathrm{Cov}\left(\sigma(X_i) I(x < X_i \leq y), \sigma(X_j) I(x < X_j \leq y) \right) \right|$$

$$\leq |\gamma_X(i-j)| \mathrm{Var}\left(\sigma(X_0) I(x < X_0 \leq y) \right) \leq |\gamma_X(i-j)| \, |\mathcal{L}(y) - \mathcal{L}(x)|.$$

Therefore, from (13.2.23) we obtain

$$E\left\{ n^{-d-1/2} [\mathcal{U}_n(x) - \mathcal{U}_n(y)] \right\}^2$$

$$\leq n^{-2d-1} |\mathcal{L}(y) - \mathcal{L}(x)| \sum_{i,j=1}^{n} |\gamma_X(i-j) \gamma_\varepsilon(i-j)|$$

$$\leq C n^{-\kappa} |\mathcal{L}(y) - \mathcal{L}(x)|, \quad \forall x, y \in \mathbb{R},$$

for any $0 < \kappa < (1-2d_1) \wedge (2d)$. This inequality and the chaining argument used in the proof of Theorem 10.2.1 give $\sup_{x \in \bar{\mathbb{R}}} |\mathcal{U}_n(x)| = o_p(1)$; see also the proof of Lemma 13.2.2 above. Hence,

$$n^{-d-1/2} \widetilde{\mathcal{V}}(x) = F_\sigma(x) S_n + u_p(1). \tag{13.3.43}$$

Using the representation of $\beta_n - \beta_0$ in (13.3.19), rewrite $\widetilde{\mathcal{V}}_n(x)$ as

$$\widetilde{\mathcal{V}}_n(x) = \widetilde{\mathcal{V}}(x) - Z_n' \bar{\Lambda}_n^{-1} \bar{\nu}_n(x).$$

Decomposing $Z_n = \mu_{\ell\sigma} S_n + \sum_{i=1}^{n} (\ell(X_i) \sigma(X_i) - E\ell(X_i) \sigma(X_i)) \varepsilon_i$ where $\mu_{\ell\sigma} := E\{\sigma(X)\ell(X)\}$, and using (2.4.18) and (13.2.23), as above, gives

$$n^{-d-1/2} Z_n = \mu_{\ell\sigma} n^{-d-1/2} S_n + o_p(1).$$

Notice that $J_\sigma(x) = F_\sigma(x) - \mu_{\ell\sigma}' \Lambda^{-1} \nu(x)$, $\bar{\Lambda}_n = \Lambda + o_p(1)$, and $\bar{\nu}_n(x) = \nu(x) + u_p(1)$; see (13.2.12). From these equations and (13.3.43) we readily obtain

$$n^{-d-1/2} \widetilde{\mathcal{V}}_n(x) = J_\sigma(x) n^{-d-1/2} S_n + u_p(1).$$

This, the uniform continuity of J_σ, and (13.2.31) complete the proof.

In order to implement the above result, we need a uniformly consistent estimator $\widetilde{J}_n(x)$ of $J_\sigma(x)$. With V_i as in (13.3.23), define

$$n\widetilde{J}_n(x) = \sum_{i=1}^{n} V_i(X_i)I(X_i \le x) - \sum_{i=1}^{n} \ell(X_i)V_i(X_i)\bar{\Lambda}_n^{-1}\bar{\nu}_n(x).$$

Its uniform consistency is given by the following lemma.

Lemma 13.3.4. *Under the assumptions of Theorem 13.3.2,*

$$\sup_{x\in\bar{\mathbb{R}}} |\widetilde{J}_n(x) - J_\sigma(x)| = o_p(1).$$

Proof. The proof is in two parts. First,

$$\sup_{x} \frac{1}{n}\left| \sum_{i=1}^{n} V_i(X_i)I(X_i \le x) - F_\sigma(x) \right|$$

$$\le \sup_{x} \frac{1}{n}\left| \sum_{i=1}^{n} V_i(X_i)I(X_i \le x) - \sigma(X_i)I(X_i \le x) \right|$$

$$+ \sup_{x} \frac{1}{n}\left| \sum_{i=1}^{n} \sigma(X_i)I(X_i \le x) - F_\sigma(x) \right|$$

$$\le \frac{1}{n}\sum_{i=1}^{n} \left| V_i(X_i) - \sigma(X_i) \right| + \sup_{x} \frac{1}{n}\left| \sum_{i=1}^{n} \sigma(X_i)I(X_i \le x) - F_\sigma(x) \right|.$$

The first term in this bound tends to zero in probability by (13.3.25) applied with $\ell(x) \equiv 1$. The second term tends to zero almost surely by (13.2.10). Second,

$$\sup_{x}\left| \frac{1}{n}\sum_{i=1}^{n} \left(V_i(X_i)\ell(X_i)'\bar{\Lambda}_n\bar{\nu}_n(x) - E\sigma(X_0)\ell(X_0)'\Lambda^{-1}\nu(x) \right) \right|$$

$$\le \left\| \frac{1}{n}\sum_{i=1}^{n} \left(V_i(X_i) - \sigma(X_i) \right)\ell(X_i) \right\| \|\bar{\Lambda}_n\| \sup_{x} \|\bar{\nu}_n(x)\|$$

$$+ \frac{1}{n}\sum_{i=1}^{n} \sigma(X_i)\|\ell(X_i)\| \|\bar{\Lambda}_n - \Lambda^{-1}\| \sup_{x} \|\bar{\nu}_n(x)\|$$

$$+ \frac{1}{n}\sum_{i=1}^{n} \sigma(X_i)\|\ell(X_i)\| \|\Lambda^{-1}\| \sup_{x} \|\bar{\nu}_n(x) - \nu(x)\|$$

$$+ \left\| \frac{1}{n}\sum_{i=1}^{n}(\sigma(X_i)\ell(X_i) - E(\sigma(X_0)\ell(X_0))) \right\| \|\Lambda^{-1}\| \sup_{x} \|\nu(x)\|.$$

This bound tends to zero in probability, by (13.3.25), the Theorem 2.5.2(i), and because $\sup_x |\bar{\nu}_n(x) - \nu(x)| \to_{a.s.} 0$ and $\sup_x \|\nu(x)\| \leq \sqrt{E\|\ell(X_0)\|^2} < \infty$.

Similarly as for the diagnostic test $\hat{D}_n$ of (13.2.37) used for homoscedastic errors, to implement the test based on $\widetilde{V}_n(x)$ in the heteroscedastic case we need consistent and $\log(n)$-consistent estimators $\widetilde{s}^2_{\varepsilon,\gamma}$ and $\widetilde{d}$ of $s^2_{\varepsilon,\gamma}$ and d, respectively, based on the residuals $Y_i - \beta'_n \ell(X_i), i = 1, \cdots, n$. These estimators can be constructed using the techniques of Sections 8.5 and 9.4. Guo and Koul (2008) provide one such construction.

A consequence of the above results is that whenever $\sup_x |J_\sigma(x)| \neq 0$, the test that rejects H_0, whenever

$$\mathcal{T}_n := \frac{1}{n^{\widetilde{d}+1/2} \widetilde{s}_{\varepsilon,d} \sup_x |\widetilde{J}_n(x)|} \sup_x |\widetilde{V}_n(x)| \geq z_{\alpha/2}, \qquad (13.3.44)$$

is of the asymptotic size α.

In a simple linear regression model with non-zero intercept, i.e., when $\ell(x) = (1, x)'$, $J_\sigma(x) \equiv 0$ if and only if $\sigma(x)$ is constant in x. For a polynomial regression through the origin, $\sup_x |J_\sigma(x)|$ never vanishes. In particular, the above test is applicable when fitting a heteroscedastic polynomial of a known degree.

13.3.2 *A Monte Carlo simulation*

In this subsection we report the findings of a finite-sample simulation study on the level and power of the test (13.3.44) based on $\mathcal{T}_n$ first reported in Guo and Koul (2008). In this simulation $\ell(x) = (1, x)'$, $\beta_0 = 0$, $\beta_1 = 2$, and $\sigma^2(x) = 1 + x^2$. The errors $\{\varepsilon_i\}$ are taken to be ARFIMA$(0, d, 0)$ with standardized Gaussian innovations and $\{X_i\}$ is taken to be fractional Gaussian noise with the LM parameter d_1. The values of d and d_1 range in the interval $[0.1, 0.45]$ with increments of 0.05. The processes were generated using the algorithm given in Beran (1994, Chapter 12).

We first concentrate on the properties of the LS estimator β_{1n} of β_1 and $\widetilde{d}$. Table 13.3.1 shows the root mean square errors (RMSE) of the LS estimator β_{1n} with sample size 500 and 2,000 replications. As can be seen from this table, when $d + d_1$ increases, so does the RMSE of β_{1n}. Typically, when $d + d_1 < 1/2$, the RMSE is small.

$d \setminus d_1$	0.10	0.15	0.20	0.25	0.30	0.35	0.40	0.45
0.10	.0087	.0086	.0088	.0098	.0105	.0115	.0135	.0192
0.15	.0084	.0095	.0107	.0117	.0123	.0135	.0176	.0247
0.20	.0104	.0101	.0114	.0135	.0146	.0176	.0215	.0341
0.25	.0108	.0121	.0135	.0155	.0194	.0227	.0304	.0465
0.30	.0123	.0139	.0177	.0192	.0244	.0333	.0479	.0735
0.35	.0141	.0178	.0218	.0283	.0362	.0488	.0704	.1254
0.30	.0186	.0237	.0310	.0398	.0540	.0834	.1201	.2087
0.45	.0257	.0341	.0519	.0647	.1137	.1762	.2738	.4962

Table 13.3.1:   RMSE of the LS estimator β_{1n}, $n = 500$.

$d \setminus d_1$	0.1	0.15	0.20	0.25	0.30	0.35	0.40	0.45
0.10	.0396	.0387	.0393	.0387	.0377	.0394	.0399	.0394
0.15	.0428	.0427	.0440	.0426	.0433	.0451	.0416	.0432
0.20	.0481	.0475	.0477	.0482	.0504	.0486	.0475	.0469
0.25	.0536	.0548	.0554	.0547	.0538	.0508	.0529	.0494
0.30	.0623	.0630	.0623	.0592	.0582	.0597	.0581	.0551
0.35	.0708	.0720	.0708	.0670	.0658	.0631	.0622	.0577
0.40	.0833	.0832	.0809	.0780	.0769	.0724	.0678	.0651
0.45	.11288	.1110	.1089	.1030	.0946	.0852	.0782	.0672

Table 13.3.2:   RMSE of $\widetilde{d}$ based on $Y_i - \beta_{1n}X_i$, $n = 500$.

$d \setminus d_1$	0.15	0.25	0.35	0.45
0.15	(a) (.175, .3)	(a) (.125, .5)	(a) (.075, .3)	(a) (.025, .1)
0.25	(a) (.175, .3)	(a) (.125, .5)	(a) (.075, .3)	(a) (.025, .1)
0.35	(b) (.150, .3)	(a) (.125, .5)	(a) (.075, .3)	(a) (.025, .1)
0.45	(b) (.050, .3)	(b) (.050, .5)	(b) (.050, .3)	(a) (.025, .1)

Table 13.3.3:   Ranges for δ of the bandwidths for estimating σ.

Table 13.3.2 shows the RMSEs of the local Whittle estimator $\widetilde{d}$ of d based on pseudo-residuals $\widetilde{\varepsilon}_i = Y_i - \beta_{1n}X_i$ without estimating the variance function $\sigma^2(x)$ and on the samples of size 500 with 1,000 replications. From this table, it is observed that for $d \leq 0.35$, the overall RMSE is less than 0.072 and stable regardless of the values of d_1.

Next, to assess the finite sample behavior of $\widehat{\sigma}^2$, the estimator $\widehat{\sigma}^2(x)$ was simulated for the values of x in the grid $x_1 = -1.50$, $x_2 = -1.49, \cdots,$

h \ Summary	Bandwidth	Q1	Median	Mean	Q3
0.15	$3.0n^{-0.200}$	.0261	.0369	.0424	.0512
0.25	$3.5n^{-0.200}$	.0256	.0383	.0432	.0557
0.35	$4.0n^{-0.200}$	.0273	.0417	.0595	.0617
0.45	$1.5n^{-0.099}$	.0366	.0663	.1138	.1058

Table 13.3.4: Summary of $\mathrm{ASE}(\hat{\sigma}^2)$ for $d = 0.15$.

d_1 \ Summary	Bandwidth	Q1	Median	Mean	Q3
0.15	$4n^{-0.200}$	.0442	.0711	.0887	.1127
0.25	$4n^{-0.200}$	.0465	.0652	.0888	.1076
0.35	$4n^{-0.200}$	.0467	.0774	.1043	.1252
0.45	$2n^{-0.099}$	.0627	.0995	.2190	.1902

Table 13.3.5: Summary of $\mathrm{ASE}(\hat{\sigma}^2)$ for $d = 0.25$.

d_1 \ Summary	Bandwidth	Q1	Median	Mean	Q3
0.15	$4.5n^{-0.200}$	.1562	.2724	.5402	.5584
0.25	$6.0n^{-0.200}$	.1594	.3113	.5449	.6330
0.35	$5.0n^{-0.200}$	.1625	.3252	.5475	.6103
0.45	$2.5n^{-0.099}$	.1704	.3155	.7092	.6235

Table 13.3.6: Summary of $\mathrm{ASE}(\hat{\sigma}^2)$ for $d = 0.35$.

d_1 \ Summary	Bandwidth	Q1	Median	Mean	Q3
0.15	$6.0n^{-0.200}$	1.153	3.214	16.24	11.83
0.25	$7.0n^{-0.200}$	1.137	3.078	14.75	11.25
0.35	$13.5n^{-0.200}$	1.018	2.611	12.77	11.59
0.45	$4.5n^{-0.099}$	1.136	3.374	12.57	11.85

Table 13.3.7: Summary of $\mathrm{ASE}(\hat{\sigma}^2)$ for $d = 0.45$.

$x_{301} = 1.50$, and for $0.15 \leq d, d_1 \leq 0.45$. The calculations used the built-in smoothing function of the R program with normal kernel and sample size 500, repeated 500 times. The ranges for δ in the bandwidths $c = Cn^{-\delta}$ are given in Table 13.3.3, using Remark 13.3.1. The symbols (a) and (b) indicate conditions (13.3.7) and (13.3.8), respectively. Based on Table 13.3.3, for convenience we used $\delta = 0.2$ and $c = Cn^{-0.2}$ in our simulations for all values of d and d_1 except when $d_1 = 0.45$, where $\delta = 0.099$ was used. The constant C was adjusted for different values of d and d_1, so as to minimize the average-squared errors: $\text{ASE} := \sum_{k=1}^{301}(\widehat{\sigma}^2(x_k)/\sigma^2(x_k) - 1)^2/301$. The summary statistics ASE are shown in Tables 13.3.4 to 13.3.7. It can be seen that the estimator $\widehat{\sigma}^2(x)$ is relatively stable for the values of d, $d_1 \leq 0.35$. Similar results are observed when we replace the normal kernel by the kernel function $W(x) = 0.5(1 + \cos(x\pi))I(|x| \leq 1)$ or the uniform kernel.

Although Guo and Koul (2008) used a different value of δ when $d_1 = 0.45$, because of (13.3.7) and (13.3.8), the choice $\omega = Cn^{-0.2}$ is suitable for all values of d and d_1.

13.3.3 *Application to the forward premium anomaly*

Here we give an application of the tests (13.2.20) and (13.2.37) to the well-known forward premium anomaly in financial economics and international finance, which was first reported in Koul *et al.* (2004). The anomaly refers to the widespread empirical finding that the returns on most freely floating nominal exchange rates up until the early 1990s appear to be negatively correlated with the lagged forward premium or forward discount. Engel (1996) provides a good survey of the literature on the forward premium anomaly.

Let s_i denote the log spot exchange rate at time i, and f_i the log of the forward exchange rate at time i, for delivery at time $i+1$. All the rates use the US dollar as the numeraire currency. The uncovered interest rate parity theory assumes rational expectations, risk neutrality, free capital mobility, and the absence of taxes on capital transfers. A common test of the theory is to estimate the regression

$$s_{i+1} - s_i = \beta_0 + \beta_1(f_i - s_i) + u_{i+1}, \tag{13.3.45}$$

and to test the hypothesis that $\beta_0 = 0$ and $\beta_1 = 1$, when the u_{i+1} are serially uncorrelated.

To illustrate the above testing procedures, we now consider instances where both the lagged forward premiums $f_i - s_i$ and spot returns $s_{i+1} - s_i$ appear to have long memory characteristics. Baillie and Bollerslev (2000) and Maynard and Phillips (2001) discuss the case of an unbalanced regression when spot returns are close to being uncorrelated, while the forward premium is a long memory process. However, a number of authors including

Booth *et al.* (1982), Cheung (1993), and Lo (1991) have argued that asset and exchange rate returns in general may exhibit long memory behavior, in which case the regression may not be unbalanced. In particular, Cheung (1993) argues that slow adjustments to purchasing power parity can cause the long memory property of spot returns, as was seen for some currencies in the initial part of the post Bretton Woods era of floating exchange rates.

The regression model (13.3.45) was fitted to a very similar sample period and currencies to that of Cheung (1993) where some spot returns appeared to exhibit long memory. The first example uses monthly observations of the Canadian dollar to US dollar spot and the one-month forward rate from January 1974 through December 1991, with a total of $n = 215$ observations. As expected, there is strong evidence for long memory in the forward premium: the Ljung–Box statistic for the first 20 lags is $Q(20) = 803.52$, and the first eleven autocorrelation coefficients of the forward premium series are $0.89, 0.74, 0.63, 0.55, 0.52, 0.50, 0.49, 0.47, 0.43, 0.36$, and 0.28, respectively. The local Whittle estimator for the long memory parameter in the forward premium is 0.25.

The monthly Canadian dollar versus US dollar spot returns over the same period also show evidence of long memory: the d estimate in an ARFIMA$(1, d, 0)$ model is 0.12 and the robust t statistic is significant at the .05 level. We test the hypothesis H_0 with $\ell(X_i) = (1, X_i)'$, with X_i representing the lagged forward premium $f_i - s_i$ and Y_i the variable $s_{i+1} - s_i$. The LS estimates of β were $(-2.606, -1.432)'$ with standard errors of $(0.811, 0.407)'$. Hence, the regression gives rise to the usual negative slope coefficient that is consistent with the anomaly. Ljung–Box tests on the autocorrelations of the standardized residuals and also their squares failed to reject the hypothesis that the residuals were serially uncorrelated and without ARCH effects. It thus seemed reasonable to assume that the errors form a martingale-difference sequence. This justifies the application of the test statistic $\mathcal{D}_n$ of (13.2.20) to fit the model (13.3.45) to this data, resulting in $\mathcal{D}_n = 5.82$, indicating a clear rejection of this model.

A further example of the methodology is provided by the monthly British pound to US dollar rate over the same time period. The Whittle estimator of the long memory parameter from an ARFIMA model for the spot returns gives a value of 0.293 with standard error of 0.080, which indicates long memory in spot returns. For the British forward premium, the Whittle estimator of the long memory parameter turns out to be 0.45 with a standard error of 0.124. Estimation of the traditional forward premium anomaly regression (13.3.45) confirmed the standard anomalous negative slope coefficient. Moreover, there is also evidence of substantial, persistent autocorrelation in the residuals, which can also be represented as a long memory process. Variables associated with the risk premium may give rise to the long memory process in the residuals.

This suggests fitting the second-degree polynomial to the British pound to US dollar data with long memory errors, i.e., testing for H_0 with $\ell(x) = (x, x^2)'$, where again x represents the lagged forward premium. The LS estimates of β were $(-2.414, -0.244)'$ with standard errors of $(0.835, 0.129)'$. To test for further misspecification of this regression, the statistic $\widehat{\mathcal{D}}_n$ in (13.2.37) was calculated to be 6.72, where, in the error process, the Whittle estimator of the long memory parameter d equals 0.262. Thus again the regression appears to be misspecified.

A possible implication of the above data analysis is that the usual linear regression model (13.3.45) to forecast future exchange rates may be misspecified, even if we allow for the long memory behavior of both spot returns and forward premiums. Such a misspecification might provide a partial explanation of the forward premium anomaly. The second data example suggests that fitting a second-order polynomial instead of a linear function in the exchange data does not improve the situation and that further economic variable terms associated with a risk premium should be included in the relationship (13.3.45).

13.3.4 *Application to a foreign exchange data set*

In this section we shall give the results of a real data example discussed in Guo and Koul (2008). In this example the lack-of-fit test $\mathcal{T}_n$ of (13.3.44) is applied to fit a simple linear regression model with heteroscedastic errors to some currency exchange rate data obtained from *www.federalreserve.gov/releases/H10/hist/*. The data are noon-buying rates in New York for cable transfers payable in foreign currencies.

In this example, currency exchange rates for British pound to US dollar and Swiss franc to US dollar from January 4, 1971 to December 2,

2005 are used. Missing values were deleted leaving about 437 monthly observations. The symbols $X = $ dlUK and $Y = $ dlSZ stand for the differenced log exchange rate of British pound to US dollar and Swiss franc to US dollar, respectively. In Figure 13.3.1, we see that these two sequences appear to be stationary. Also, $\bar{X} = -0.0001775461$, $s_X = 0.001701488$, $\bar{Y} = -0.00004525129$, and $s_Y = 0.001246904$.

Local Whittle estimated values of the long memory parameters d_X and d_Y are $\widehat{d}_X = 0.1610273$ and $\widehat{d}_Y = 0.2147475$. In computing these estimators the choice of the smoothing parameter m is crucial. Taqqu and Teverovsky (1997) recommend $m = n/4$ and $m = n/32$ for sample sizes $n = 100$ and $n = 10,000$, respectively. The sample size of 437 being in between these two, $m = n/8$ was chosen to obtain our estimates.

Comparing the X-process with a simulated fractional Gaussian noise with $d_1 = \widehat{d}_X = 0.1610273$ and $n = 437$, Figure 13.3.2 suggests that the marginal distribution of X is Gaussian.

Next, we regressed Y on X, using a non-parametric kernel regression estimator and parametric simple linear regression model of Y upon X. Both of these estimates are depicted in Figure 13.3.3, which also describes scatter plot of the data. They display a negative association between X and Y. The estimated linear equation is $\widetilde{Y} = -0.000118775 - 0.4141107\,X$, with a residual standard error of 0.00102992.

Figure 13.3.4 shows the non-parametric kernel estimator of $\sigma(x)$ when regressing Y on X with $W(x) = 0.5(1 + \cos(x\pi))I(|x| \leq 1)$.

The estimators of d based on the residuals $\widetilde{\varepsilon} = Y - \beta'_n X$ and $\widehat{\varepsilon} = (Y - \beta'_n X)/\widehat{\sigma}(X)$ are equal to 0.1046235 and 0.1246576, respectively. This again suggests the presence of long memory in the error process.

Finally, to check if the regression of Y on X is simple linear, we computed $\mathcal{T}_n = 0.4137897$, which corresponds to the asymptotic p-value 66%. As expected, this test fails to reject the null hypothesis that there exists a linear relationship between these two processes.

Note: Other works that propose and analyze lack-of-fit tests based on the process V_n under independence and in a time-series context include An and Cheng (1991), Stute (1997), Stute et al. (1998), Koul and Stute (1999), Stute and Zhu (2002), Khmaladze and Koul (2004), and the review article of Koul (2006). Some of the above arguments used for proving the tightness of $V_n := n^{-1/2}\mathcal{V}$ are similar to arguments in Koul and Stute (1999). Li (2004) contains several other lack-of-fit tests.

Csörgő and Mielniczuk (2000) discuss the asymptotic behavior of a class of kernel-type estimators of the regression function in random-design non-parametric regression models with errors given by an unknown function of a long memory moving-average process and the i.i.d. explanatory random vectors. Masry and Mielniczuk (1999) obtained the joint asymptotic distribution of the local linear estimators of a multivariate regression function and its derivatives in the same regression model as in the previous paper except the errors are now a function of a long memory stationary Gaussian process and independent explanatory random vectors.

Sections 13.2 and 13.3.3 are based on the work of Koul *et al.* (2004) while Sections 13.3, 13.3.2, and 13.3.4 are based on the work of Guo and Koul (2008).

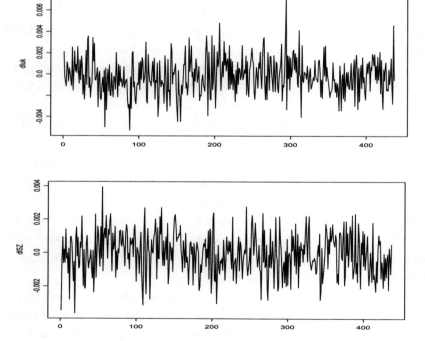

Fig. 13.3.1: Time series plots of dlUK and dlSZ.

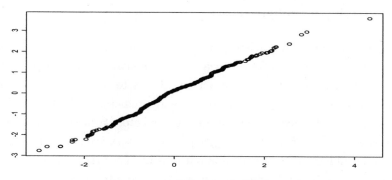

Fig. 13.3.2: QQ-plot of dlUK.

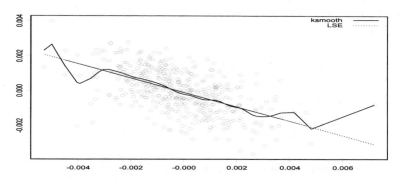

Fig. 13.3.3: Kernel estimation of $r(x)$.

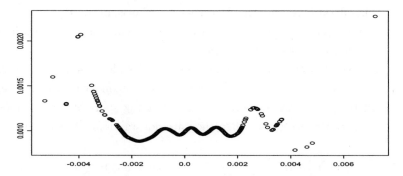

Fig. 13.3.4: Kernel estimation of $\sigma(x)$.

13.4   Testing a subhypothesis

A classical problem in statistics is to see whether, among a given set of predictor variables, a subset of variables is significant or not for predicting a response variable. One way to achieve this is to first stipulate a linear regression model between the response and predictor variables and then to test for the absence of a subset of the predictor variables in the model by testing that the corresponding slope parameters are zero. This is the so-called problem of testing a subhypothesis in linear regression.

More precisely, let k and p be known positive integers with $k \leq p$. Let $\boldsymbol{X}_{i1}, (\boldsymbol{X}_{i2})$ be $k \times 1, ((p-k) \times 1)$ random vectors and Y_i denote the response variable, $1 \leq i \leq n$. Consider the regression model where for some $\beta_1 \in \mathbb{R}^k$ and $\beta_2 \in \mathbb{R}^{p-k}$,

$$Y_i = \beta_1' \boldsymbol{X}_{i1} + \beta_2' \boldsymbol{X}_{i2} + \varepsilon_i, \quad i = 1, \cdots, n. \tag{13.4.1}$$

The problem of interest is to test

$$H_0 : \beta_2 = 0, \quad \text{vs.} \quad H_1 : \beta_2 \neq 0.$$

The problem with independent errors and when the design variables are either non-random or random and i.i.d., has been thoroughly studied; see Rao (1973) and references therein. A classical testing procedure is the likelihood ratio test when the errors are Gaussian, or the analysis of variance-type tests via the least-squares theory, which are asymptotically valid without the Gaussianity assumption.

Here we shall discuss an analog of the generalized least-squares-type test for H_0 when the errors $\{\varepsilon_i\}$ form a parametric long memory moving average, and when the covariates are either random having short or long memory, independent of the errors, or when they are non-random. Accordingly, let

$$\varepsilon_i = \sum_{j=0}^{\infty} a_j(\theta) \zeta_{i-j}, \quad i \in \mathbb{Z}, \quad \text{for some } \theta \in \Theta \subset \mathbb{R}^q, \tag{13.4.2}$$

with $\{\zeta_j\} \sim \text{IID}(0, \sigma^2)$ and having finite fourth moment $E\zeta_0^4 < \infty$. The spectral density f_{θ, σ^2} of $\{\varepsilon_i\}$ is unbounded at the origin and satisfies the regularity conditions of Sections 8.3.1 and 11.4. Let $s_\theta(u)$ and $b_j(\theta)$ be defined as in (8.2.2) and (8.2.4), namely,

$$f_{\theta,\sigma^2}(u) = \frac{\sigma^2}{2\pi} s_\theta(u), \quad s_\theta(u) := \left| \sum_{k=0}^{\infty} a_k(\theta) e^{iku} \right|^2, \quad u \in \Pi, \quad (13.4.3)$$

$$b_j(\theta) := \frac{1}{2\pi} \int_\Pi \frac{e^{iju}}{s_\theta(u)} du, \quad j \in \mathbb{Z}, \ \theta \in \Theta.$$

Let $\boldsymbol{X}_i' := (\boldsymbol{X}_{i1}', \boldsymbol{X}_{i2}')$ and $\beta' := (\beta_1', \beta_2')$ and define

$$\Lambda_n(\theta, \beta) := \sum_{i,j=1}^{n} b_{i-j}(\theta)(Y_i - \boldsymbol{X}_i'\beta)(Y_j - \boldsymbol{X}_j'\beta), \quad \beta \in \mathbb{R}^p,$$

$$(\theta_n, \beta_n) := \operatorname{argmin}_{(\theta,\beta) \in \Theta \times \mathbb{R}^p} \Lambda_n(\theta, \beta),$$

$$\Lambda_{n1}(\theta, \beta_1) := \sum_{i,j=1}^{n} b_{i-j}(\theta)(Y_i - \boldsymbol{X}_{i1}'\beta_1)(Y_j - \boldsymbol{X}_{j1}'\beta_1), \quad \beta_1 \in \mathbb{R}^k,$$

$$(\theta_{n1}, \beta_{n1}) := \operatorname{argmin}_{(\theta,\beta_1) \in \Theta \times \mathbb{R}^k} \Lambda_{n1}(\theta, \beta_1).$$

Note that $\Lambda_n(\theta, \beta)$ and $\Lambda_{n1}(\theta, \beta_1)$ are the Whittle quadratic forms of (8.2.4) (modulus factor n^{-1}) in which the unobserved error process $\varepsilon_i, i = 1, \cdots, n$, is replaced by observations $Y_i - \boldsymbol{X}_i'\beta$ and $Y_i - \boldsymbol{X}_{i1}'\beta_1$, $i = 1, \cdots, n$, respectively. Minimizing the above forms over (θ, β) and (θ, β_1) leads to Whittle estimates (θ_n, β_n) and $(\theta_{n1}, \beta_{n1})$ of these parameters under the respective hypotheses H_1 and H_0.

The analog of the likelihood ratio test for H_0 is based on

$$\mathcal{Q}_n := -2[\Lambda_n(\theta_n, \beta_n) - \Lambda_{n1}(\theta_{n1}, \beta_{n1})]. \quad (13.4.4)$$

Strictly speaking, the exact likelihood ratio test for Gaussian errors would have the elements of the inverse of the covariance matrix as weights in the quadratic forms Λ_n and Λ_{n1}, instead of the $b_{i-j}(\theta)$'s. The quadratic forms Λ_n and Λ_{n1} are their Whittle approximations.

For classical i.i.d. errors, the asymptotic null distribution of $\mathcal{Q}_n$ is known to be a chi-square distribution. The purpose of this section is to extend this result to a more general situation where the errors or design can exhibit long memory.

For this purpose, we need to impose conditions (a0′) to (a5′) of Section 11.4 on the parametric form $s_\theta(u)$. These conditions are similar but stronger than conditions (a0) to (a4) of Subsection 8.3.1, which discusses the asymptotic normality of the Whittle estimator for a completely observable process. For convenience we reproduce below the last three conditions.

In the following, β_0 will denote the true parameter value of β in the model (13.4.1), and $\Theta_0 \subset \Theta$ will be a small ball centered at the true value θ_0 of the parameter θ in (13.4.3).

(a3') There exist $0 < \alpha < 1$ and $0 \le \eta < \alpha$ such that $\alpha - \eta < 1/2$ and

$$s_\theta(u) \le C|u|^{-\alpha}, \quad \|\partial s_\theta(u)/\partial u\| \le C|u|^{-\alpha-1},$$

$$\|\nabla s_\theta^{-1}(u)\| \le C|u|^\eta, \quad \forall\, (u, \theta) \in \Pi \times \Theta_0.$$

(a4') $s_\theta^{-1}(u) \le C|u|^{\alpha/2}, \qquad \forall\, (u, \theta) \in \Pi \times \Theta_0.$

(a5') $\|\nabla^2 s_\theta^{-1}(u)\| \le C|u|^{\alpha/2}, \quad \forall\, (u, \theta) \in \Pi \times \Theta_0.$

We shall first consider random design.

13.4.1 *Random design*

By *random design* we mean a zero-mean vector-valued stationary process $\{\boldsymbol{X}_i, i \in \mathbb{Z}\}$, independent of the errors $\{\varepsilon_j\}$. More specifically, about the covariates $\boldsymbol{X}_i' = (X_{i,1}, \cdots, X_{i,p})$ we shall assume that *all* of the components form the moving-average processes

$$X_{i,k} = \sum_{j=1}^p \sum_{s=0}^\infty \psi_{kj}(s)\xi_{i-s,j}, \quad i \in \mathbb{Z}, \tag{13.4.5}$$

$$\sum_{s=0}^\infty \psi_{kj}^2(s) < \infty, \quad k, j = 1, \cdots, p,$$

where $\xi_s := (\xi_{s,1}, \cdots, \xi_{s,p})'$, $s \in \mathbb{Z}$, are i.i.d. vector-valued r.v.'s with zero mean and identity covariance matrix, and independent of the innovations $\zeta_s, s \in \mathbb{Z}$ of $\{\varepsilon_j\}$. The last assumption implies the independence of $\{X_{i,k}\}$ and $\{\varepsilon_i\}$ in (13.4.1). The above form for the $\{X_{i,k}\}$'s covers various dependence structures. It allows us to consider the case when the $X_{i,k}$, $k = 1, \cdots, p$, are mutually independent, which corresponds to independent innovation sequences $\{\xi_{s,k}\}, k = 1, \cdots, p$, and "diagonal" weights $\psi_{kj} \equiv 0\, (k \neq j)$, as well as the case when the r.v.'s are mutually dependent moving averages in a common innovation sequence, which happens when $\psi_{kj} \equiv 0, j = 2, \cdots, p$.

 We also impose a weak condition (g1), below, on the spectral density of $\{\boldsymbol{X}_i\}$, which is similar to condition (11.4.13). These designs include both the stationary short memory (in particular, i.i.d.) and the stationary long memory random designs. The above assumptions obviously imply

$E\boldsymbol{X}_0 = 0$ and therefore the model (13.4.1) does not allow for a non-zero intercept parameter.

The linear form in (13.4.5) imposes a structural restriction on the design, which is not present in Section 11.4. The reason for this restriction is that in the present section, the covariates play an important part in the limit distribution of Q_n while in Section 11.4 they play the role of a "nuisance parameter" whose effect is asymptotically eliminated from the Whittle likelihood using mild consistency assumptions on regression estimates.

Let $\mathbf{g}(u)$ be the $p \times p$ matrix-valued spectral density of $\{\boldsymbol{X}_i\}$ and let

$$V(\theta) := \int_{\Pi} \mathbf{g}(u)s_\theta^{-1}(u)du, \quad W(\theta) := \int_{\Pi} s_\theta(u)\nabla^2_{\theta\theta}s_\theta^{-1}(u)du. \quad (13.4.6)$$

Write

$$\mathbf{g}(u) = \begin{pmatrix} \mathbf{g}_{11}(u) & \mathbf{g}_{12}(u) \\ \mathbf{g}_{12}(u)' & \mathbf{g}_{22}(u) \end{pmatrix}, \quad V(\theta) = \begin{pmatrix} V_{11}(\theta) & V_{12}(\theta) \\ V_{12}(\theta)' & V_{22}(\theta) \end{pmatrix},$$

where $\mathbf{g}_{11}(u)$ and $\mathbf{g}_{22}(u)$ are matrix-valued (cross-)spectral densities of $\{\boldsymbol{X}_{i1}\}$ and $\{\boldsymbol{X}_{i2}\}$ having dimensions $k \times p$ and $(p-k) \times (p-k)$, respectively, and where $V_{ij}(\theta)$ is defined similarly to $V(\theta)$ above with $\mathbf{g}(u)$ replaced by $\mathbf{g}_{ij}(u)$, $i,j = 1,2$. We also need the following assumption for $\mathbf{g}$:

(g1) There exists $0 \le \gamma < 1$ such that $\|\mathbf{g}(u)\| \le C|u|^{-\gamma}, \forall u \in \Pi$.

We are now ready to state the following theorem:

Theorem 13.4.1. *Suppose the regression model (13.4.1) along with (a0')
to (a5'), (g1), and (13.4.5) hold and that $V(\theta_0)$ is positive definite. Then*

$$Q_n \to_D 2\sigma^2 \chi^2_{p-k}, \quad under\ H_0.$$

This theorem allows us to construct the following test of H_0. Let $\sigma_n^2 := n^{-1}\Lambda_{n1}(\theta_{n1}, \beta_{n1})$. Using arguments as in Subsection 8.3.1, we can verify that $\sigma_n^2 \to_p \sigma^2$ under H_0. Hence, for a given $0 < \nu < 1$, the test that rejects H_0 whenever $Q_n/(2\sigma_n^2) > \chi^2_{\nu,p-k}$ has asymptotic level ν. Here $\chi^2_{\nu,p-k}$ is the $(1-\nu)$th percentile of the χ^2_{p-k} distribution.

The proof of the above theorem is facilitated by the following lemma. Let Z_{θ_0} and Z_{β_0} denote two independent random vectors with $\mathcal{N}_q(0, \sigma^4\pi^{-1}W(\theta_0))$ and $\mathcal{N}_p(0, \sigma^2 V(\theta_0))$ distributions, respectively. Then

$(Z'_{\theta_0}, Z'_{\beta_0})'$ has the $\mathcal{N}_{q+p}(0, \Gamma)$ distribution, where

$$\Gamma := \begin{pmatrix} \sigma^4 \pi^{-1} W(\theta_0) & 0 \\ 0 & \sigma^2 V(\theta_0) \end{pmatrix}.$$

Let $Z_{\beta_0 1}$ denote the vector of the first k components of Z_{β_0}, and set

$$\mathcal{Q} := 2[Z'_{\beta_0} V(\theta_0)^{-1} Z_{\beta_0} - Z'_{\beta_0 1} V_{11}(\theta_0)^{-1} Z_{\beta_0 1}].$$

We also need to define

$$\Lambda_{n0}(\theta) := \Lambda_n(\theta, \beta_0) \equiv \sum_{t,s=1}^{n} b_{t-s}(\theta) \varepsilon_t \varepsilon_s, \quad Z_{n\theta_0} := n^{-1/2} \nabla_\theta \Lambda_{n0}(\theta_0),$$

$$T_n(\theta) := \sum_{t,s=1}^{n} b_{t-s}(\theta) \varepsilon_t \boldsymbol{X}_s, \quad Z_{n\beta_0} := n^{-1/2} T_n(\theta_0),$$

$$T_{n1}(\theta) := \sum_{t,s=1}^{n} b_{t-s}(\theta) \varepsilon_t \boldsymbol{X}_{s1}, \quad Z_{n\beta_0 1} := n^{-1/2} T_{n1}(\theta_0),$$

$$A_n(\theta) := \sum_{t,s=1}^{n} b_{t-s}(\theta) \boldsymbol{X}_t \boldsymbol{X}'_s, \quad \beta'_0 := (\beta'_{01}, \beta'_{02}).$$

The following lemma is needed in the proof of Theorem 13.4.1.

Lemma 13.4.1. *Under the conditions of Theorem 13.4.1,*

$$\Lambda_n(\theta_n, \beta_n) = \Lambda_n(\theta_0, \beta_0) - (\frac{\pi}{\sigma^2}) Z'_{n\theta_0} W(\theta_0)^{-1} Z_{n\theta_0} \tag{13.4.7}$$
$$- Z'_{n\beta_0} V(\theta_0)^{-1} Z_{n\beta_0} + o_p(1),$$

$$\Lambda_{n1}(\theta_{n1}, \beta_{n1}) = \Lambda_{n1}(\theta_0, \beta_{01}) - (\frac{\pi}{\sigma^2}) Z'_{n\theta_0} W(\theta_0)^{-1} Z_{n\theta_0} \tag{13.4.8}$$
$$- Z'_{n\beta 1} V_{11}(\theta_0)^{-1} Z_{n\beta 1} + o_p(1),$$

$$(Z'_{n\theta_0}, Z'_{n\beta_0}) \to_D (Z'_{\theta_0}, Z'_{\beta_0}). \tag{13.4.9}$$

The proof of this lemma is given later. It is used in the following proof.
Proof of Theorem 13.4.1. Since, under H_0, $\Lambda_n(\theta_0, \beta_0)$ is equivalent to $\Lambda_{n1}(\theta_0, \beta_{01})$, (13.4.7) and (13.4.8) give

$$\mathcal{Q}_n = 2[Z'_{n\beta_0} V(\theta_0)^{-1} Z_{n\beta_0} - Z'_{n\beta_0 1} V_{11}(\theta_0)^{-1} Z_{n\beta_0 1}] + o_p(1).$$

By (13.4.9), $\mathcal{Q}_n \to_D \mathcal{Q}$. Now the claim follows from the following.
Let $Z \sim \mathcal{N}_p(0, \Sigma)$ where Σ is a positive definite covariance matrix, and

let Z_1 denote its first k components with covariance matrix Σ_1. Then the distribution of $Z'\Sigma^{-1}Z - Z_1'\Sigma_1^{-1}Z_1$ is χ^2_{p-k}.

A way to prove this is to use the Cholesky decomposition: $Z = GU$, where $U \sim \mathcal{N}_p(0, I)$ is a standard p-dimensional normal and G is a lower triangular matrix. Then $\Sigma = EZZ' = EGUU'G' = GEUU'G' = GG'$ and $Z'\Sigma^{-1}Z = U'G'\Sigma^{-1}GU = U'U$. Similarly, write $Z_1 = G_1U_1$, where $U_1 \in \mathbb{R}^k$ is formed by the first k coordinates of U and the matrix G_1 by the first k upper-left entries of G, implying $Z_1'\Sigma_1^{-1}Z_1 = U_1'U_1$ and $Z'\Sigma^{-1}Z - Z_1'\Sigma_1^{-1}Z_1 = U'U - U_1'U_1$. Obviously, the last difference has a χ^2_{p-k}-distribution.

Proof of Lemma 13.4.1. *Proof of (13.4.7).* Assume that (13.4.9) holds, and without loss of generality assume $\sigma^2 = 1$. By arguing as in Section 8.2 above, we can verify that under the assumed set-up, (θ_n, β_n) is consistent for (θ_0, β_0). In other words, for all sufficiently large n, $\theta_n \in \Theta_0$, and $\beta_n \in B_0$ with a probability arbitrarily close to 1, where B_0 is an arbitrarily small neighborhood of β_0. For the sake of brevity, let $\delta_n := \theta_n - \theta_0$ and $\Delta_n := \beta_n - \beta_0$. A Taylor expansion around (θ_0, β_0) gives

$$\Lambda_n(\theta_n, \beta_n) = \Lambda_n(\theta_0, \beta_0) + \delta_n'\nabla_\theta\Lambda_n(\theta_0, \beta_0) + \Delta_n'\nabla_\beta\Lambda_n(\theta_0, \beta_0) \quad (13.4.10)$$
$$+ \frac{1}{2}\delta_n'\nabla^2_{\theta\theta}\Lambda_n(\theta_n^*, \beta_n^*)\delta_n + \Delta_n'\nabla^2_{\theta\beta}\Lambda_n(\theta_n^*, \beta_n^*)\delta_n$$
$$+ \frac{1}{2}\Delta_n'\nabla^2_{\beta\beta}\Lambda_n(\theta_n^*, \beta_n^*)\Delta_n,$$

where $(\theta_n^*, \beta_n^*) \in \Theta \times \mathbb{R}^p$ are some random vectors such that

$$\|\theta_n^* - \theta_0\| \le \|\Delta_n\| = o_p(1), \quad \|\beta_n^* - \beta_0\| \le \|\delta_n\| = o_p(1). \quad (13.4.11)$$

We shall establish below the following asymptotic relations:

$$n^{1/2}\delta_n = -(2\pi/\sigma^2)W(\theta_0)^{-1}Z_{n\theta_0} + o_p(1), \quad (13.4.12)$$
$$n^{1/2}\Delta_n = V(\theta_0)^{-1}Z_{n\beta_0} + o_p(1). \quad (13.4.13)$$

Now, a Taylor expansion of $\Lambda_n(\theta, \beta)$ in the β variable around β_0 gives

$$\Lambda_n(\theta, \beta) = \Lambda_{n0}(\theta) - 2(\beta - \beta_0)'T_n(\theta) + (\beta - \beta_0)'A_n(\theta)(\beta - \beta_0). \quad (13.4.14)$$

The rest of the proof is facilitated by the following lemma. Its proof is given later.

Lemma 13.4.2. *Under the assumptions of Theorem 13.4.1, the following conclusions hold:*

$$n^{-1}\nabla^2_{\theta\theta}\Lambda_{n0}(\theta) \to (\sigma^2/2\pi)W(\theta), \qquad (13.4.15)$$

$$n^{-1}\nabla_\theta T_n(\theta) \to 0, \qquad (13.4.16)$$

$$n^{-1}A_n(\theta) \to V(\theta), \qquad (13.4.17)$$

uniformly in $\theta \in \Theta_0$, a.s. Moreover,

$$\sup_{\theta\in\Theta_0}\left(\|\nabla_\theta A_n(\theta)\| + \|\nabla_{\theta\theta} A_n(\theta)\|\right) = O_p(n), \qquad (13.4.18)$$

$$\sup_{\theta\in\Theta_0}\left(\|\nabla_\theta T_n(\theta)\| + \|\nabla_{\theta\theta} T_n(\theta)\|\right) = O_p(n).$$

Note that (13.4.11) and (13.4.14) to (13.4.18) imply the convergence of the second derivatives in (13.4.10) (in probability):

$$n^{-1}\nabla^2_{\theta\theta}\Lambda_n(\theta^*_n, \beta^*_n) \to_p (\sigma^2/2\pi)W(\theta_0), \qquad (13.4.19)$$

$$n^{-1}\nabla^2_{\theta\beta}\Lambda_n(\theta^*_n, \beta^*_n) \to_p 0,$$

$$n^{-1}\nabla^2_{\beta\beta}\Lambda_n(\theta^*_n, \beta^*_n) \to_p 2V(\theta_0).$$

The expansion (13.4.7) in turn follows from these three results, (13.4.9), (13.4.10), (13.4.12), and (13.4.13). Claims (13.4.9), (13.4.12), and (13.4.13) are proved next.

Proof of the CLT in (13.4.9). Recall the definitions of ε_i and X_i from (13.4.2) and (13.4.5). In the following, we shall not exhibit θ_0 in terms such as $a_i(\theta_0)$ of (13.4.2) and elsewhere. Thus, for example, we will write b_j for $b_j(\theta_0)$ and $\nabla_\theta h(u)$ for $\nabla_\theta h_\theta(u)\big|_{\theta=\theta_0}$, for any function $h_\theta(u)$. Many details of the following proof of (13.4.9) are similar to those in Section 6.3.

For notational simplicity, we shall assume that $\sigma = q = p = 1$, in particular, that $X_i = \sum_{s=0}^{\infty}\psi(s)\xi_{i-s}$ is a scalar process, where $\xi_s \equiv \xi_{s,1}$ and $\psi(s) \equiv \psi_{11}(s)$ of (13.4.5). Similarly as for Theorem 6.3.1, we shall first approximate the quadratic forms $Z_{n\theta_0}, Z_{n\beta_0}$ by the corresponding "diagonal" forms $\tilde{Z}_{n\theta_0}, \tilde{Z}_{n\beta_0}$ defined as follows. Let

$$\hat{a}(u) := \frac{1}{2\pi}\sum_{j=0}^{\infty} a_j e^{-iju}, \qquad \hat{\psi}(u) := \frac{1}{2\pi}\sum_{j=0}^{\infty}\psi(j)e^{-iju}, \qquad (13.4.20)$$

$$\hat{b}(u) := (2\pi s(u))^{-1}, \qquad \hat{\nu}_1(u) := (2\pi)^{1/2}\hat{a}(u)|\nabla_\theta \hat{b}(u)|^{1/2},$$

$$\hat{\nu}_2(u) := (2\pi)^{1/2}\hat{a}(u)|\nabla_\theta \hat{b}(u)|^{1/2}\mathrm{sgn}(\nabla_\theta \hat{b}(u)),$$

$$\hat{\nu}_3(u) := (2\pi)^{1/2}\hat{a}(u)(\hat{b}(u))^{1/2}, \qquad \hat{\nu}_4(u) := (2\pi)^{1/2}\hat{\psi}(u)(\hat{b}(u))^{1/2},$$

$$\nu_j(t) := \int_\Pi e^{itu} \hat{\nu}_j(u) du, \quad t \in \mathbb{Z}, \; j = 1, 2, 3, 4.$$

(To avoid confusion, note that the above Fourier transforms $v_j(t)$ differ from those defined in Chapter 2 by a factor $1/2\pi$.) We also need

$$\tilde{\varepsilon}_{j,i} := \sum_{s=0}^\infty \nu_j(s)\zeta_{i-s}, \; j = 1,2,3; \quad \tilde{X}_i := \sum_{s=0}^\infty \nu_4(s)\xi_{i-s}. \quad (13.4.21)$$

Then the approximating quadratic forms are

$$\tilde{Z}_{n\theta_0} := n^{-1/2} \sum_{i=1}^n (\tilde{\varepsilon}_{1,i}\tilde{\varepsilon}_{2,i} - E\tilde{\varepsilon}_{1,i}\tilde{\varepsilon}_{2,i}), \quad \tilde{Z}_{n\beta_0} := n^{-1/2} \sum_{i=1}^n \tilde{\varepsilon}_{3,i}\tilde{X}_i.$$

The processes $\{\tilde{\varepsilon}_{j,i}\}$ and $\{\tilde{X}_i\}$, are analogs of $\{Y_{j,i}\}$ in (6.3.7) with weights $\nu_j(i), j = 1, 2, 3, 4$, defined through their Fourier transforms as in (6.3.8).

By assumption (a2'), $\hat{b} = s^{-1}/2\pi$ and $\nabla_\theta \hat{b}$ are bounded, which in turn implies $\sum_{i=0}^\infty \nu_j^2(i) < \infty$. Note also that $s \geq 0$ implies $\hat{b} \geq 0$. In particular, the process $\{\tilde{X}_i\}$ in (13.4.21) is a well-defined moving-average process independent of processes $\{\tilde{\varepsilon}_{j,i}\}, j = 1, 2, 3$. Clearly, for $q = p = 1$, (13.4.9) follows from the following two results.

$$(\tilde{Z}_{n\theta_0}, \tilde{Z}_{n\beta_0}) \to_D \mathcal{N}_2(0, \Gamma), \quad (13.4.22)$$

$$E\left(Z_{n\theta_0} - \tilde{Z}_{n\theta_0}\right)^2 = o(1), \quad E\left(Z_{n\beta_0} - \tilde{Z}_{n\beta_0}\right)^2 = o(1). \quad (13.4.23)$$

The first claim in (13.4.23) follows from Lemma 6.3.1 and because $EZ_{n,\theta_0} = o_p(1)$; see (8.3.8). To prove the second claim, write

$$E\left(Z_{n\beta_0} - \tilde{Z}_{n\beta_0}\right)^2 = \text{Var}(Z_{n\beta_0}) - 2\text{Cov}(Z_{n\beta_0}, \tilde{Z}_{n\beta_0}) + \text{Var}(\tilde{Z}_{n\beta_0}).$$

Analogous to (14.4.6) below, the above expectations can be written as traces of Toeplitz matrices:

$$\text{Var}(Z_{n\beta_0}) = n^{-1}\text{tr}(BR_{\varepsilon\varepsilon}BR_{XX}), \quad (13.4.24)$$

$$\text{Cov}(Z_{n\beta_0}, \tilde{Z}_{n\beta_0}) = n^{-1}\text{tr}(BR_{\varepsilon\tilde{\varepsilon}_3}\delta R_{X\tilde{X}}),$$

$$\text{Var}(\tilde{Z}_{n\beta_0}) = n^{-1}\text{tr}(\delta R_{\tilde{\varepsilon}_3\tilde{\varepsilon}_3}\delta R_{\tilde{X}\tilde{X}}),$$

where $\delta := (\delta_{t-s})_{t,s=1,\cdots,n}$ is the unit matrix, $B := (b_{t-s})_{t,s=1,\cdots,n}$ is the coefficient matrix of the quadratic form $Z_{n\beta_0}$, and $R_{\varepsilon\varepsilon}, \cdots, R_{\tilde{X}\tilde{X}}$ are the corresponding (cross-)covariance matrices:

$$R_{\varepsilon\varepsilon} := (E\varepsilon_t\varepsilon_s)_{t,s=1,\cdots,n}, \qquad R_{XX} := (EX_tX_s)_{t,s=1,\cdots,n},$$

$$R_{\varepsilon\tilde{\varepsilon}_3} := (E\varepsilon_t\tilde{\varepsilon}_{3,s})_{t,s=1,\cdots,n}, \qquad R_{X\tilde{X}} := (EX_t\tilde{X}_s)_{t,s=1,\cdots,n},$$

$$R_{\tilde{\varepsilon}_3\tilde{\varepsilon}_3} := (E\tilde{\varepsilon}_{3,t}\tilde{\varepsilon}_{3,s})_{t,s=1,\cdots,n}, \qquad R_{\tilde{X}\tilde{X}} := (E\tilde{X}_t\tilde{X}_s)_{t,s=1,\cdots,n}.$$

By the definitions in (13.4.20), the corresponding (cross-)spectral densities of these processes are

$$f_{\varepsilon\varepsilon} = f, \quad g_{XX} = g, \quad f_{\varepsilon\tilde{\varepsilon}_3} = 2\pi\,\hat{\nu}_3\overline{\hat{a}} = (2\pi)^{1/2}f\,\hat{b}^{1/2}, \qquad (13.4.25)$$

$$g_{X\tilde{X}} = 2\pi\,\hat{\nu}_4\overline{\hat{\psi}} = (2\pi)^{1/2}g\,\hat{b}^{1/2}, \quad f_{\tilde{\varepsilon}_3\tilde{\varepsilon}_3} = 2\pi|\hat{\nu}_3|^2 = 2\pi f\,\hat{b},$$

$$g_{\tilde{X}\tilde{X}} = 2\pi|\hat{\nu}_4|^2 = 2\pi g\,\hat{b}.$$

By Lemma 14.4.1 and using $\hat{\delta} = 1/2\pi$, we see that the limits of all three traces in (13.4.24) are the same:

$$n^{-1}\mathrm{tr}(BR_{\varepsilon\varepsilon}BR_{XX}) \to \int_\Pi (2\pi)^3 f_{\varepsilon\varepsilon}(u)g_{XX}(u)(\hat{b}(u))^2 du, \qquad (13.4.26)$$

$$n^{-1}\mathrm{tr}(BR_{\varepsilon\tilde{\varepsilon}_3}\delta R_{X\tilde{X}}) \to \int_\Pi (2\pi)^2 f_{\varepsilon\tilde{\varepsilon}_3}(u)g_{X\tilde{X}}(u)\hat{b}(u)du,$$

$$n^{-1}\mathrm{tr}(\delta R_{\tilde{\varepsilon}_3\tilde{\varepsilon}_3}\delta R_{\tilde{X}\tilde{X}}) \to \int_\Pi (2\pi)f_{\tilde{\varepsilon}_3\tilde{\varepsilon}_3}(u)g_{\tilde{X}\tilde{X}}(u)du,$$

since each of these integrals equals $\int_\Pi g(u)s^{-1}(u)du$. The crucial condition $A := \alpha_1 + \alpha_2 + \beta_1 + \beta_2 < 1$ of Lemma 14.4.1 is verified for all three traces in (13.4.26) by using the expressions in (13.4.25) and assumptions (a3′), (a4′), and (g1). In particular, since $|f_{\varepsilon\tilde{\varepsilon}_3}(u)| \le C|u|^{-\alpha+\alpha/4}$ and $|g_{X\tilde{X}}(u)| \le C|u|^{-\gamma+\alpha/4}$, for the second trace in (13.4.26), Lemma 14.4.1 applies with $\alpha_1 := 3\alpha/4$, $\alpha_2 := (\gamma - \alpha/4) \vee 0$, $\beta_1 := -\alpha/2$, $\beta_2 := 0$, and $A = \max(\gamma, \alpha/4) < 1$.

It remains to prove the CLT in (13.4.22). First we verify the convergence of the corresponding covariance matrices. Since $\mathrm{Cov}(\tilde{Z}_{n\theta_0}, \tilde{Z}_{n\beta_0}) = 0$, by the independence of errors and designs, this claim follows from

$$\mathrm{Var}(\tilde{Z}_{n\theta_0}) \to \pi^{-1}\int_\Pi s(u)\nabla_{\theta\theta}^2 s^{-1}(u)du, \qquad (13.4.27)$$

$$\mathrm{Var}(\tilde{Z}_{n\beta_0}) \to \int_\Pi g(u)s^{-1}(u)du;$$

see the definitions in (13.4.6). The first claim in (13.4.27) follows from Theorem 4.5.2 while the second from the final lines in (13.4.24) and (13.4.26).

The rest of the proof of (13.4.22) is similar to the proof of Theorem 4.5.2. Accordingly, we define below "2M-dependent" ($M < \infty$) approximations:

$$\tilde{\varepsilon}_{j,i}^M := \sum_{|s-i|\le M} \nu_j^M(i-s)\zeta_s, \quad j = 1, 2, 3, \qquad (13.4.28)$$

$$\tilde{X}_i^M := \sum_{|s-i|\le M} \nu_4^M(i-s)\xi_s, \quad i \in \mathbb{Z},$$

to (13.4.21), and the corresponding quadratic forms:

$$\tilde{Z}_{n\theta_0}^M := \frac{1}{\sqrt{n}} \sum_{i=1}^{n} (\tilde{\varepsilon}_{1,i}^M \tilde{\varepsilon}_{2,i}^M - E\tilde{\varepsilon}_{1,i}^M \tilde{\varepsilon}_{2,i}^M), \quad \tilde{Z}_{n\beta_0}^M := \frac{1}{\sqrt{n}} \sum_{i=1}^{n} \tilde{\varepsilon}_{3,i}^M \tilde{X}_i^M.$$

The approximating processes, i.e.,, the weight functions $\nu_j^M(t)$, are defined through the respective Fourier transforms $\hat{\nu}_j^M$. In particular, $\hat{\nu}_j^M$, $j = 1, 2$, is a trigonometric polynomial of degree M such that

$$\|\hat{\nu}_j - \hat{\nu}_j^M\|_{L_4(\Pi)} < \epsilon, \qquad j = 1, 2, \tag{13.4.29}$$

where $\epsilon = \epsilon(M) > 0$ tends to zero with $M \to \infty$; see also (4.5.12). Claim (13.4.22) follows from

$$(\tilde{Z}_{n\theta_0}^M, \tilde{Z}_{n\beta_0}^M) \to_D \mathcal{N}_2(0, \Gamma^M), \quad \forall M < \infty, \tag{13.4.30}$$

where Γ^M is a covariance matrix, and

$$\lim_{M \to \infty} \|\Gamma^M - \Gamma\| = 0, \tag{13.4.31}$$

$$\lim_{M \to \infty} \limsup_{n \to \infty} E(\tilde{Z}_{n\theta_0}^M - \tilde{Z}_{n\theta_0})^2 = 0, \tag{13.4.32}$$

$$\lim_{M \to \infty} \limsup_{n \to \infty} E(\tilde{Z}_{n\beta_0}^M - \tilde{Z}_{n\beta_0})^2 = 0; \tag{13.4.33}$$

see (4.5.13) to (4.5.15) in the proof of Theorem 4.5.2. Claim (13.4.32) follows from (13.4.29) as for (4.5.15). The CLT in (13.4.30) is a consequence of the fact that the approximating processes in (13.4.28) are all stationary and $2M$-dependent.

We shall now show that there exist $\nu_j^M(t), j = 3, 4$, which satisfy (13.4.33). Introduce the truncated versions

$$\nu_{j,K}(t) := \int_\Pi e^{itx} \hat{\nu}_j(x) I(|\hat{\nu}_j(x)| \le K) dx, \quad j = 3, 4, \tag{13.4.34}$$

$$\tilde{\varepsilon}_{3,t,K} := \sum_{s \in \mathbb{Z}} \nu_{3,K}(s) \zeta_{t-s}, \quad \tilde{X}_{t,K} := \sum_{s \in \mathbb{Z}} \nu_{4,K}(s) \xi_{t-s}.$$

Then, for every $K < \infty$, as in (13.4.26),

$$n^{-1} E \Big(\sum_{t=1}^{n} (\tilde{\varepsilon}_{3,t} \tilde{X}_t - \tilde{\varepsilon}_{3,t,K} \tilde{X}_{t,K}) \Big)^2 \tag{13.4.35}$$

$$\to (2\pi)^3 \int_\Pi |\hat{\nu}_3(z) \hat{\nu}_4(z)|^2 I(|\hat{\nu}_3(z)| > K \text{ or } |\hat{\nu}_4(z)| > K) dz.$$

Next, using the boundedness of the truncated functions $\hat{\nu}_j(x) I(|\hat{\nu}_j(x)| \le K)$ by K in (13.4.34), we can approximate them in $L_4(\Pi)$ by trigonometric polynomials of degree M, as in (13.4.29) and (4.5.12) in the proof of Theorem 4.5.2, and the corresponding moving averages by $2M$-dependent moving averages. More precisely, for any $K, \epsilon > 0$ there exist $M > 0$ and trigonometric polynomials $\hat{\nu}_j^M$, $j = 3, 4$, such that the Fourier coefficients $\nu_j^M(t) = \int_\Pi e^{itx} \hat{\nu}_j^M(x) dx$, $j = 3, 4$, vanish for $|t| > M$ and such that

$$\int_\Pi \left| \hat{\nu}_j(x) I(|\hat{\nu}_j(x)| \leq K) - \hat{\nu}_j^M(x) \right|^4 dx \ < \epsilon, \quad j = 3, 4. \qquad (13.4.36)$$

As for (13.4.35),

$$n^{-1} E \left(\sum_{t=1}^n (\tilde{\varepsilon}_{3,t}^M \tilde{X}_t^M - \tilde{\varepsilon}_{3,t,K} \tilde{X}_{t,K}) \right)^2 \qquad (13.4.37)$$

$$\to \ (2\pi)^3 \int_\Pi \left| \hat{\nu}_3(x) \hat{\nu}_4(x) I(|\hat{\nu}_3(x)| \leq K, |\hat{\nu}_4(x)| \leq K) - \hat{\nu}_3^M(x) \hat{\nu}_4^M(x) \right|^2 dx.$$

Clearly, by (13.4.36), the above limit in (13.4.37) does not exceed $C_K \epsilon^{1/2}$, i.e., it can be made arbitrarily small by an appropriate choice of M and the approximating polynomials. Together with (13.4.35) and the fact that the r.h.s. of (13.4.35) tends to zero as $K \to \infty$, this concludes the proof of (13.4.33). Claim (13.4.31) follows from (13.4.27), (13.4.35), and (13.4.37). The proof of the CLT in (13.4.9) is now complete.

Proof of (13.4.12) and (13.4.13). First, we prove the $n^{1/2}$-consistency of the estimator (θ_n, β_n). By definition, (θ_n, β_n) is a solution in $\Theta_0 \times B_0$ of the equations

$$\sum_{t,s=1}^n \nabla_\theta b_{t-s}(\theta_n)(Y_t - \boldsymbol{X}_t' \beta_n)(Y_s - \boldsymbol{X}_s' \beta_n) = 0, \qquad (13.4.38)$$

$$\sum_{t,s=1}^n b_{t-s}(\theta_n) \nabla_\beta (Y_t - \boldsymbol{X}_t' \beta_n)(Y_s - \boldsymbol{X}_s' \beta_n) = 0. \qquad (13.4.39)$$

Equation (13.4.39) gives

$$\beta_n \ = \ \beta_0 + \bar{A}_n^{-1} \hat{T}_n, \qquad (13.4.40)$$

where $\bar{A}_n := A_n(\theta_n)$ and $\hat{T}_n := T_n(\theta_n)$. Equation (13.4.38) defines the minimizer $\theta_n = \operatorname{argmin}_\theta \tilde{\Lambda}_n(\theta)$, where the quadratic form $\tilde{\Lambda}_n(\theta) := \Lambda_n(\theta, \beta_n)$ is the same as the quadratic form $nQ_{\tilde{\varepsilon}}(\theta)$ in the proof of Theorem 11.4.1 of Section 11.4, with x_{nt} and $\tilde{\beta}_n$ there replaced by $\boldsymbol{X}_t$ and β_n. Therefore we can use the results in that section to estimate $\|\delta_n\|$ given a rate of $\|\Delta_n\|$

and *vice versa*. Namely, expanding $\nabla_\theta \tilde{\Lambda}_n$ as in Section 11.4, the proof of Theorem 11.4.2 gives

$$0 = \nabla_\theta \tilde{\Lambda}_n(\theta_n) = \nabla_\theta \tilde{\Lambda}_n(\theta_0) + \nabla_{\theta\theta}^2 \tilde{\Lambda}_n(\theta_n^*)\delta_n. \qquad (13.4.41)$$

Using the consistency of (θ_n, β_n), as in (13.4.19), it follows that

$$n^{-1}\nabla_{\theta\theta}^2 \tilde{\Lambda}_n(\theta_n^*) = (\sigma^2/2\pi)W(\theta_0) + o_p(1).$$

Therefore, according to (13.4.41),

$$\|\delta_n\| = O_p(\|n^{-1}\nabla_\theta \tilde{\Lambda}_n(\theta_0)\|). \qquad (13.4.42)$$

But

$$\nabla_\theta \tilde{\Lambda}_n(\theta_0) = n^{1/2}Z_{n\theta_0} + \nabla_\theta \Lambda_{n2}(\theta_0)\Delta_n + \Delta_n' \nabla_\theta \Lambda_{n3}(\theta_0)\Delta_n, \qquad (13.4.43)$$

where $Z_{n\theta_0} = O_p(1)$ by (13.4.9) and, analogous to (11.4.17) and (11.4.19),

$$\nabla_\theta \Lambda_{n3}(\theta_0) := \sum_{t,s=1}^{n} \nabla_\theta b_{t-s}(\theta_0)\boldsymbol{X}_t\boldsymbol{X}_s' = O_p(n), \qquad (13.4.44)$$

$$\nabla_\theta \Lambda_{n2}(\theta_0) := \sum_{t,s=1}^{n} \nabla_\theta b_{t-s}(\theta_0)\varepsilon_t\boldsymbol{X}_s' = O_p(n^{3/4}).$$

From (13.4.42) to (13.4.44) we conclude

$$\|\delta_n\| = O_p(n^{-1/2}) + O_p(n^{-1/4}\|\Delta_n\|) + O_p(\|\Delta_n\|^2) \qquad (13.4.45)$$
$$= O_p\big(\max\big(\|\Delta_n\|^2, n^{-1/2}\big)\big).$$

Next, we shall obtain a rate for $\|\Delta_n\|$ given a rate for $\|\delta_n\|$. Equation (13.4.40) gives

$$\|\Delta_n\| \qquad\qquad\qquad\qquad\qquad\qquad\qquad\qquad\qquad (13.4.46)$$
$$= \left\| \bar{A}_n^{-1} \sum_{t,s=1}^{n} \big(b_{t-s}(\theta_n) - b_{t-s}(\theta_0)\big)\boldsymbol{X}_t\varepsilon_s + \bar{A}_n^{-1} \sum_{t,s=1}^{n} b_{t-s}(\theta_0)\boldsymbol{X}_t\varepsilon_s \right\|$$
$$\leq \|\delta_n\| \sup_{\theta'\in\Theta_0} \|nA_n^{-1}(\theta')\| \sup_{\theta''\in\Theta_0} \|n^{-1}\nabla_\theta T_n(\theta'')\|$$
$$+ n^{-1/2} \sup_{\theta\in\Theta_0} \|nA_n^{-1}(\theta)\| \, \|Z_{n\beta_0}\|.$$

By (13.4.9), (13.4.17), and (13.4.18),

$$Z_{n\beta_0} = O_p(1), \quad \sup_{\theta\in\Theta_0} \|nA_n^{-1}(\theta)\| = O_p(1), \quad \sup_{\theta''\in\Theta_0} \|n^{-1}\nabla_\theta T_n(\theta'')\| = O_p(1).$$

Hence, from (13.4.46),

$$\|\Delta_n\| = O_p\big(\max\big(\|\delta_n\|, n^{-1/2}\big)\big). \tag{13.4.47}$$

Substituting (13.4.45) into (13.4.47) and using $\|\delta_n\| = o_p(1)$ gives

$$\|\Delta_n\| = O_p\big(n^{-1/2}\big), \tag{13.4.48}$$

which in turn, together with (13.4.45), gives

$$\|\delta_n\| = O_p\big(n^{-1/2}\big). \tag{13.4.49}$$

Now we are ready to prove (13.4.12) and (13.4.13). The consistency of θ_n and (13.4.17) imply $n\bar{A}_n^{-1} = V(\theta_0)^{-1} + o_p(1)$. This and (13.4.40) imply $n^{1/2}\Delta_n = V(\theta_0)^{-1}Z_{n\beta_0} + O_p\big(n^{-1/2}\|T_n(\theta_n) - T_n(\theta_0)\|\big) + o_p(1)$, where

$$\begin{aligned}
\|T_n(\theta_n) - T_n(\theta_0)\| &= \|\delta'_n \nabla_\theta T_n(\theta_0) + (1/2)\delta'_n \nabla^2_{\theta\theta} T_n(\theta_n^*)\delta_n\| \\
&\leq n\|\delta_n\| \|n^{-1}\nabla_\theta T_n(\theta_0)\| \\
&\quad + n\|\delta_n\|^2 \sup_{\theta\in\Theta_0} \|n^{-1}\nabla^2_{\theta\theta} T_n(\theta)\| = o_p(n^{1/2}),
\end{aligned}$$

according to (13.4.49), (13.4.16), and (13.4.18). This proves (13.4.13). Claim (13.4.12) follows from Theorem 11.4.2 and (13.4.48).

Next, we prove (13.4.15) to (13.4.17). For any two processes $\eta_i = \{\eta_{it}, t \in Z\}, i = 1, 2$, define

$$\mathcal{S}_n(\eta_1, \eta_2, \theta) := n^{-1} \sum_{t,s=1}^{n} b_{t-s}(\theta)\eta_{1t}\eta_{2s}, \quad \mathcal{S}_n(\eta_i, \theta) \equiv \mathcal{S}_n(\eta_i, \eta_i, \theta), \quad i = 1, 2.$$

By symmetry $b_{t-s}(\theta) = b_{s-t}(\theta)$,

$$\mathcal{S}_n(\eta_1, \eta_2, \theta) = (1/2)(\mathcal{S}_n(\eta_1 + \eta_2, \theta) - \mathcal{S}_n(\eta_1 - \eta_2, \theta)). \tag{13.4.50}$$

A similar identity holds for $\nabla_\theta \mathcal{S}_n(\eta_1, \eta_2, \theta)$ and $\nabla^2_\theta \mathcal{S}_n(\eta_1, \eta_2, \theta)$. Write $\varepsilon = \{\varepsilon_j, j \in Z\}$ and $X_k = \{X_{j,k}, j \in Z\}, k = 1, \cdots, p$.

Recall from (a0′) and (a2′) that $s_\theta^{-1}(u)$, $\nabla_\theta s_\theta^{-1}(u)$, and $\nabla^2_\theta s_\theta^{-1}(u)$ are jointly continuous on $\Pi \times \Theta$. Then (13.4.15) or $\nabla^2_\theta \mathcal{S}_n(\varepsilon, \theta) \to (\sigma^2/2\pi)W(\theta)$, uniformly in $\theta \in \Theta$, follows immediately from Lemma 8.2.2.

Consider (13.4.16), or $\nabla_\theta \mathcal{S}_n(\varepsilon, X_k, \theta) \to 0$, uniformly in $\theta \in \Theta$, $k = 1, \cdots, p$. Because of (13.4.50), this equation follows from

$$\nabla_\theta \mathcal{S}_n(\varepsilon + X_k, \theta) \to \int_\Pi \nabla_\theta s_\theta^{-1}(u)[f_{\theta,\sigma^2}(u) + g_k(u)]du, \tag{13.4.51}$$

$$\nabla_\theta \mathcal{S}_n(\varepsilon - X_k, \theta) \to \int_\Pi \nabla_\theta s_\theta^{-1}(u)[f_{\theta,\sigma^2}(u) + g_k(u)]du,$$

uniformly in $\theta \in \Theta$, where $f_{\theta,\sigma^2}(u) + g_k(u)$ is the common spectral density of stationary processes $\varepsilon + X_k$ and $\varepsilon - X_k$. Since these processes are stationary and ergodic, both results in (13.4.51) follow from Lemma 8.2.2.

The proof of (13.4.17) is similar to that of (13.4.16), since (13.4.50) reduces the problem to the convergences of positive quadratic forms $\mathcal{S}_n(X_k \pm X_j, \theta)$, $k, j = 1, \cdots, p$, which can be shown with the help of Lemma 8.2.2 as above. This completes the proof of claims (13.4.15) to (13.4.17).

Finally, because of assumption (a2') and Lemma 11.4.2, (13.4.18) follows from $\sup_{(\theta,u) \in \Theta_0 \times \Pi} \left(\left\| \nabla_\theta s_\theta^{-1}(u) \right\| + \left\| \nabla_{\theta\theta} s_\theta^{-1}(u) \right\| \right)$ being bounded. This concludes the proof of Lemma 13.4.1.

13.4.2 *Non-random design*

Here we shall discuss the null behavior of the $\mathcal{Q}_n$ statistic in (13.4.4) when the design is non-random as described next. Let $V_1, \cdots, V_p$ be p continuously differentiable real-valued functions on $[0, 1]$ and linearly independent in $L_2[0, 1]$. Let $\mathbf{V}_1' = (V_1, \cdots, V_k)$, $\mathbf{V}_2' = (V_{k+1}, \cdots, V_p)$, and $\mathbf{V} = (\mathbf{V}_1', \mathbf{V}_2')'$. Assume $\mathbf{X}_t = \mathbf{V}(\frac{t}{n})$ so that $\mathbf{X}_{t1} = \mathbf{V}_1(\frac{t}{n})$ and $\mathbf{X}_{t2} = \mathbf{V}_2(\frac{t}{n})$, $t = 1, \cdots, n$. This form of non-random design is very similar to that discussed in Section 11.4; see Remark 11.4.1 and condition (11.4.4). As in the previous sections, let θ_0 and β_0 denote the true values of θ and β, and let $\alpha_0 := \alpha(\theta_0)$. We also need the following notation:

$$\hat{\mathbf{V}}(x) := \int_0^1 e^{\mathbf{i}xz}\mathbf{V}(z)dz, \quad x \in \mathbb{R}, \tag{13.4.52}$$

$$\Gamma_V(\theta) := \int_{\mathbb{R}^3} |x_1|^{\alpha(\theta)}|x_2|^{\alpha(\theta)}|y|^{-\alpha(\theta)}\hat{\mathbf{V}}(x_1)\overline{\hat{\mathbf{V}}(x_2)}'$$
$$\times \frac{(e^{\mathbf{i}(y-x_1)} - 1)(e^{-\mathbf{i}(y-x_2)} - 1)}{(y - x_1)(y - x_2)}dx_1 dx_2 dy,$$

$$R_V(\theta) := \int_{\mathbb{R}} |x|^{\alpha(\theta)}\hat{\mathbf{V}}(x)\overline{\hat{\mathbf{V}}(x)}' dx.$$

Also, let $R_{Vij}(\theta)$ and $\Gamma_{Vij}(\theta)$ be defined as above with $\mathbf{V}$ and $\hat{\mathbf{V}}$ replaced by $\mathbf{V}_i$ and $\hat{\mathbf{V}}_j$, respectively, $i, j = 1, 2$, where $\hat{\mathbf{V}}_j(x) := \int_0^1 e^{\mathbf{i}xz}\mathbf{V}_j(z)dz$, $j = 1, 2$. Note that the matrix $R_V(\theta)$ is positive definite for all θ because $y'R_V(\theta)y = \int |x|^{\alpha(\theta)}|y'\hat{\mathbf{V}}(x)|^2 dx = 0$, for some $y \in \mathbb{R}^p$, which implies $y'\mathbf{V}(z) \equiv 0$ for $z \in [0, 1]$, which contradicts the assumption of linear independence of the components of $\mathbf{V}$. Similarly, we conclude that $\Gamma_V(\theta)$, $R_{Vjj}(\theta)$, and $\Gamma_{Vjj}(\theta)$, $j = 1, 2$, are positive definite matrices.

For non-random design, the errors are assumed to follow a moving-average process as in (13.4.2) however, assumptions (a0′) to (a5′) need to be strengthened as follows:

(a6′) $\hat{a}(u, \theta)$ is continuous for all $(u, \theta), u \neq 0, \theta \in \Theta_0$; $\hat{a}^{-1}(u, \theta)$ is continuous on $\Pi \times \Theta_0$; and there exist strictly positive and continuous functions $\alpha(\theta)$ and $C_a(\theta)$, $\theta \in \Theta_0$, such that $\alpha/2 \leq \alpha(\theta) \leq \alpha$ for any $\theta \in \Theta_0$ and such that

$$\lim_{u \to 0} |u|^{\alpha(\theta)/2} \hat{a}(u, \theta) = C_a(\theta), \quad \text{uniformly in } \theta \in \Theta_0.$$

By the definitions in (13.4.3), assumption (a6′) implies

$$s_\theta(u) \sim (2\pi)^2 C_a^2(\theta)|u|^{-\alpha(\theta)}, \quad u \to 0. \tag{13.4.53}$$

Hence, it implies (a4′) and the first bound in (a3′). In particular, α_0 satisfies $\alpha/2 \leq \alpha_0 \leq \alpha$, where $\alpha \in (0, 1)$ is the same as in (a3′) to (a5′).

Let $Z_{\beta_0 v} = (Z'_{\beta_{01} v}, Z'_{\beta_{02} v})'$ be a random vector with an $\mathcal{N}_p(0, \Gamma_V(\theta_0))$ distribution, where $Z_{\beta_{01} v}$ and $Z_{\beta_{02} v}$ denote the vectors of the first k and the last $(p - k)$ components of $Z_{\beta_0 v}$, respectively. We are now ready to state the following theorem.

Theorem 13.4.2. *Assume the regression model (13.4.1) with non-random design* $\mathbf{X}$ *as above and errors* $\{\varepsilon_i\}$ *as in (13.4.2), and that (a0′) to (a6′) are satisfied. Then, under H_0,*

$$Q_n \to_D Q := \frac{\sigma^2}{2\pi^2} [Z'_{\beta_0 v} R_V^{-1}(\theta_0) Z_{\beta_0 v} - Z'_{\beta_{01} v} R_{V11}^{-1}(\theta_0) Z_{\beta_{01} v}].$$

Remark 13.4.1. Observe that the r.v. Q is a positive definite quadratic form of the standardized Gaussian r.v.'s. To see this, let

$$\mathcal{R} := R_{V22} - R'_{V12} R_{V11}^{-1} R_{V12}, \quad \mathcal{Z} := Z_{\beta_{02}} - R_{V21} R_{V11}^{-1} Z_{\beta_{01}},$$
$$\Sigma := \Gamma_{V22} - 2\Gamma_{V21} R_{V11}^{-1} R_{V12} + R_{V21} R_{V11}^{-1} \Gamma_{V11} R_{V11}^{-1} R_{V12},$$

where $R_{Vij} = R_{Vij}(\theta_0)$ and $\Gamma_{Vij} = \Gamma_{Vij}(\theta_0)$. Note that $\Sigma = E\mathcal{Z}\mathcal{Z}'$. The positive definiteness of $R_V = R_V(\theta_0)$ and $\Gamma_V = \Gamma_V(\theta_0)$ implies that of $\mathcal{R}$ and Σ. Now, write $R_V = (R_{Vij})_{i,j=1,2}$, and use the fact that

$$R_V^{-1} = \begin{pmatrix} R_{V11}^{-1} + R_{V11}^{-1} R_{V12} \mathcal{R}^{-1} R_{V21} R_{V11}^{-1} & -R_{V11}^{-1} R_{V12} \mathcal{R}^{-1} \\ -\mathcal{R}^{-1} R_{V21} R_{V11}^{-1} & \mathcal{R}^{-1} \end{pmatrix},$$

to obtain $\mathcal{Q} = (\sigma^2/2\pi^2)\mathcal{Z}'\mathcal{R}^{-1}\mathcal{Z}$. Note that $\mathcal{Z} \sim \mathcal{N}_{p-k}(0, \Sigma)$. Since Σ is positive definite, we can define $\mathcal{Y} := \Sigma^{-1/2}\mathcal{Z} \sim \mathcal{N}_{p-k}(0, I)$, and rewrite $\mathcal{Q} = (\sigma^2/2\pi^2)\mathcal{Y}'(\Sigma^{1/2}\mathcal{R}_V^{-1}\Sigma^{1/2})\mathcal{Y}$, which is an a.s. positive quadratic form of the standardized Gaussian r.v.'s and thus has a weighted chi-square distribution; see Rao (1973). This is unlike the situation in random design of the previous section where we had a chi-square distribution in the limit.

Remark 13.4.2. The limiting quadratic form $\mathcal{Q}$ in Theorem 13.4.2 depends only on α_0 and the design function $\mathbf{V}$ through the matrix integrals $R_V(\theta_0) = \tilde{R}_V(\alpha(\theta_0))$ and $\Gamma_V(\theta_0) = \tilde{\Gamma}_V(\alpha(\theta_0))$ where $\tilde{R}_V(\tilde{\alpha})$ and $\tilde{\Gamma}_V(\tilde{\alpha})$ denote the integrals in (13.4.52) with $\alpha(\theta)$ replaced by $\tilde{\alpha}$. Write $\mathcal{Q}(\theta_0)$ for $\mathcal{Q}$. Note that the maps $\tilde{\alpha} \mapsto \tilde{R}_V(\tilde{\alpha})$, $\tilde{\alpha} \mapsto \tilde{R}_V^{-1}(\tilde{\alpha})$, and $\tilde{\alpha} \mapsto \tilde{\Gamma}_V(\tilde{\alpha})$ are continuous functions of $\tilde{\alpha}$ in the interval $(0, 1)$. Since $\theta \mapsto \alpha(\theta)$ is also continuous, see (a6'), it thus follows that the distribution of $\mathcal{Q}$ is weakly continuous on Θ_0. Consequently, $\mathcal{Q}(\tilde{\theta}_n) \to_p \mathcal{Q}(\theta_0)$ for every consistent estimator $\tilde{\theta}_n$ of θ_0.

Remark 13.4.3. For a polynomial design where $\mathbf{V}(t) = (1, t, \cdots, t^{p-1})'$, the elements of $\tilde{R}_V = (\tilde{R}_V^{ij})_{i,j=1,\cdots,p}$ can be explicitly computed using elementary functions of the parameter $\tilde{\alpha} \in (0, 1)$:

$$\tilde{R}_V^{ij} = d_{\tilde{\alpha}} \frac{(i - \tilde{\alpha})B(i, 1 - \tilde{\alpha}) + (j - \tilde{\alpha})B(j, 1 - \tilde{\alpha})}{i + j - 1 - \tilde{\alpha}},$$

where $d_{\tilde{\alpha}} := \sin(\pi\tilde{\alpha}/2)\Gamma(\tilde{\alpha})$ and B is the beta function. The integrals $\tilde{\Gamma}_V^{ij}$ are more cumbersome but can also be explicitly found in terms of hypergeometric functions. In particular,

$$\tilde{\Gamma}_V^{11} = 2c_{\tilde{\alpha}}d_{\tilde{\alpha}}^2 \left\{ \frac{3B(\tilde{\alpha}, 1 - \tilde{\alpha})}{1 - \tilde{\alpha}} + \frac{{}_3F_2(\tilde{\alpha}, 1, 1; 1 + \tilde{\alpha}, 2; 1)}{\tilde{\alpha}} \right\},$$

where $c_{\tilde{\alpha}} := 2\sin(\pi\tilde{\alpha}/2)\Gamma(1 - \tilde{\alpha})$ and ${}_3F_2$ is a hypergeometric function; see, e.g., Abramowitz and Stegun (1972), p. 556.

The proof of Theorem 13.4.2 is essentially similar to that of Theorem 13.4.1 above. The starting point again is the Taylor expansion in (13.4.10). We need to establish the analogs of Lemmas 13.4.1 and 13.4.2 for the non-random design. The primary reason why the weak limit of $\mathcal{Q}_n$ in Theorem 13.4.2 is different from the one for the random design of the previous section is the limiting behavior of $Z_{n\beta_0 v}$. For random design the quadratic form

$$T_n(\theta_0) = n^{1/2}Z_{n\beta_0} = \sum_{t,s=1}^{n} b_{t-s}(\theta_0)\varepsilon_t \mathbf{X}_s$$

has a limiting Gaussian distribution Z_{β_0} under normalization $n^{1/2}$ and its variance grows as $O(n)$. However, for non-random design the corresponding quadratic form

$$T_{nv}(\theta_0) = n^{(1-\alpha_0)/2} Z_{n\beta_0 v} = \sum_{t,s=1}^{n} b_{t-s}(\theta_0) \varepsilon_t \mathbf{V}(s/n)$$

grows at a much slower rate $O_p(n^{(1-\alpha_0)/2}) = o_p(n^{1/2})$ and has a limit Gaussian distribution $Z_{\beta_0 v}$ under normalization $n^{(1-\alpha_0)/2} \ll n^{1/2}$.

The above behavior is a characteristic of negative memory and can be explained by the fact that the sum of the coefficients of $T_{nv}(\theta_0)$ is zero:

$$\sum_{t \in \mathbb{Z}} b_t(\theta_0) = 0, \tag{13.4.54}$$

and the design $\mathbf{V}(s/n)$ changes very slowly. Equation (13.4.54) is the consequence of $\hat{b}(0, \theta_0) = 0$; see (13.4.53). Moreover, under additional regularity conditions on s_θ, it follows that, with $\gamma_\varepsilon(j) = E\varepsilon_0\varepsilon_j$,

$$b_j(\theta_0) = O(|j|^{-1-\alpha_0}), \quad \gamma_\varepsilon(j) = O(|j|^{-1+\alpha_0}), \quad |j| \to \infty. \tag{13.4.55}$$

To see the quantitative effect of (13.4.54) and (13.4.55) on the rate of growth of $T_{nv}(\theta_0)$, consider the case $\mathbf{V} \equiv 1$. In this case we can rewrite $T_{nv}(\theta_0) = -\sum_{t=1}^{n}(c_{n1}(t) + c_{n2}(t))\varepsilon_t$, where, by (13.4.54) and (13.4.55),

$$c_{n1}(t) = \sum_{s=-\infty}^{0} b_{t-s}(\theta_0) = O(t^{-\alpha_0}),$$

$$c_{n2}(t) = \sum_{s=n+1}^{\infty} b_{t-s}(\theta_0) = O((n-t)^{-\alpha_0}), \quad 1 \leq t \leq n.$$

Therefore, by substituting $t_i = n\tau_i$, $i = 1, 2$, in the second double sum and using the fact that the dominating integral

$$\int_0^1 \int_0^1 (\tau_1^{-\alpha_0} + (1-\tau_1)^{-\alpha_0})(\tau_2^{-\alpha_0} + (1-\tau_2)^{-\alpha_0})|\tau_1 - \tau_2|^{-1+\alpha_0} d\tau_1 d\tau_2 < \infty,$$

$$\mathrm{Var}(T_{nv}(\theta_0))$$

$$= \sum_{t_1,t_2=1}^{n} (c_{n1}(t_1) + c_{n2}(t_1))(c_{n1}(t_2) + c_{n2}(t_2))\gamma_\varepsilon(t_1 - t_2)$$

$$\leq C \sum_{t_1,t_2=1}^{n} (t_1^{-\alpha_0} + (n-t_1)^{-\alpha_0})(t_2^{-\alpha_0} + (n-t_2)^{-\alpha_0})$$

$$\times (1 + |t_1 - t_2|)^{-1+\alpha_0} = O(n^{1-\alpha_0}).$$

Needless to say, the above argument is not applicable to $\mathrm{Var}(T_n(\theta_0))$ whose asymptotics are obtained from Lemma 14.4.1; see (13.4.24) and (13.4.26).

Note: Section 13.4 is based on Koul and Surgailis (2008, 2009). In the latter paper the proposed test is shown to be consistent against a class of fixed alternatives and has a non-trivial asymptotic power against local alternatives that converge to H_0 at the rate $n^{(-1+\alpha_0)/2}$. A thorough discussion about the consistency of the $\mathcal{Q}_n$-test for random and non-random designs other than those considered above is a desirable and an open problem at the time of writing this monograph.

13.5 Goodness-of-fit of a marginal distribution function

In this section we shall discuss the asymptotic behavior of some goodness-of-fit tests pertaining to the marginal d.f. of a strictly stationary long memory moving-average process in five different cases. In the first case the process is completely observable, while in the other four cases the process is observable up to an unknown location parameter, a scale parameter, location and scale parameters, or regression parameters, respectively.

Because of the URP, unlike in the i.i.d. set-up, classical tests based on empirical process are relatively easy to implement in the first case. For the same reason, in the cases where the process is observed up to an unknown location parameter or unknown location-scale parameters, the null weak limit of the first-order difference between the residual empirical process and the null model is degenerate at zero. Hence, it cannot be used to asymptotically distinguish between two marginal distributions that differ only in their means or in their means and variances. A similar phenomena holds when fitting an error d.f. in a regression model with a non-zero intercept. Another important implication of the URP is that the first-order residual empirical process tests are robust against not knowing the error variance when fitting an error distribution up to an unknown error variance. We will end this section by analyzing a test based on the second-order difference between the residual empirical and the null d.f.'s in the location model.

Goodness-of-fit tests when $\varepsilon_i, i \in \mathbb{Z}$, **is observable.** Suppose we observe ε_i, $i = 1, \cdots, n$, from a long memory moving-average process

$\{\varepsilon_i, \, i \in \mathbb{Z}\}$ of (13.2.22). Let F denote the d.f. of ε_0 and let F_0 be a known d.f., and consider the problem of testing

$$H_0 : F = F_0, \quad \text{vs.} \quad H_1 : F \neq F_0.$$

This problem is of interest in applications. For example, as discussed in Tsay (2002), for the value at risk analysis, various probability calculations are based on the assumption that the underlying process is a Gaussian process. If we reject the null hypothesis that a marginal distribution is Gaussian then such an analysis would be in doubt.

Let

$$\hat{F}_n(x) := n^{-1} \sum_{j=1}^{n} I(\varepsilon_j \leq x), \qquad K_n := \sup_{x \in \mathbb{R}} |\hat{F}_n(x) - F_0(x)|,$$

$$C_{np} := \int |\hat{F}_n(x) - F_0(x)|^p dx, \quad \mathcal{I}_{n0} := \int [\hat{F}_n(x) - F_0(x)]^2 dF_0(x).$$

We shall analyze the asymptotic behavior of omnibus tests based on these statistics. The classical Kolmogorov test is based on K_n while C_{np} and $\mathcal{I}_{n0}$ are Cramér–von Mises-type test statistics.

We shall now describe the asymptotic null distribution of these statistics. Assume $E|\zeta_0|^3 < \infty$ and that (10.2.21) holds. Then, under H_0, Lemma 10.2.4(i) implies that F_0 is infinitely differentiable with a smooth and bounded Lebesgue density f_0. On taking $\gamma_{nj} \equiv n^{-1/2}$ in Corollary 10.2.1, we obtain $\mathcal{S}_n^*(x) \equiv n^{1/2-d}\{\hat{F}_n(x) - F_0(x) + f_0(x)n^{-1} \sum_{i=1}^{n} \varepsilon_i\}$, and by (10.2.25),

$$\sup_{x \in \mathbb{R}} \left| n^{1/2-d}(\hat{F}_n(x) - F_0(x)) + f_0(x) \, n^{-1/2-d} \sum_{i=1}^{n} \varepsilon_i \right| = o_p(1). \quad (13.5.1)$$

But, because by Corollary 4.4.1,

$$s_{\varepsilon,d}^{-1} n^{-1/2-d} \sum_{i=1}^{n} \varepsilon_i \to_D Z, \quad (13.5.2)$$

we obtain

$$n^{\frac{1}{2}-d}(\hat{F}_n(x) - F_0(x)) \Rightarrow s_{\varepsilon,d} \, f_0(x) Z, \quad (13.5.3)$$

where $s_{\varepsilon,d}$ is as in (13.2.32). Consequently,

$$n^{\frac{1}{2}-d} K_n \to_D s_{\varepsilon,d} |Z| \|f_0\|_\infty, \quad \|f_0\|_\infty := \sup_{x \in \mathbb{R}} f_0(x),$$

$$n^{p(\frac{1}{2}-d)} C_{np} \to_D s_{\varepsilon,d}^p |Z|^p \int f_0^p(x) dx, \quad n^{1-2d} \mathcal{I}_{n0} \to_D s_{\varepsilon,d}^2 Z^2 \int f_0^3(x) dx.$$

From these results we readily obtain the following, where $\vartheta := (s_{\varepsilon,d}, d)$. Under H_0,

$$\mathcal{K}_n(\vartheta) := \frac{n^{1/2-d}K_n}{s_{\varepsilon,d}\|f_0\|_\infty} \to_D |Z|, \qquad T_{n1}(\vartheta) := \frac{n^{1/2-d}C_{n1}}{s_{\varepsilon,d}} \to_D |Z|,$$

$$T_{n2}(\vartheta) := \frac{n^{1-2d}C_{n2}}{s_{\varepsilon,d}^2 \int f_0^2(x)dx} \to_D Z^2, \qquad \mathcal{I}_{n0}(\vartheta) := \frac{n^{1-2d}\mathcal{I}_{n0}}{s_{\varepsilon,d}^2 \int f_0^3(x)dx} \to_D Z^2.$$

where $\vartheta := (s_{\varepsilon,d}, d)$. Implementation of any one of these tests requires consistent and $\log(n)$-consistent estimators of $s_{\varepsilon,d}$ and d. The semiparametric estimators $\hat{d}$ and $\hat{s}_{\varepsilon,d}^2$ of Sections 8.5 and 9.4 were shown to satisfy these conditions. Let $\hat{\vartheta} := (\hat{s}_{\varepsilon,d}, \hat{d})'$. Any of the above statistics with ϑ replaced by $\hat{\vartheta}$ may be used to test for H_0.

Clearly, the determination of the large sample critical values of these tests is trivial, relative to their analog in the i.i.d. situation. For example, the Kolmogorov test that rejects H_0 whenever $\mathcal{K}_n(\hat{\vartheta}) > z_{\alpha/2}$ would be of asymptotic size α, $0 < \alpha < 1$. Similarly, the Cramér–von Mises test that rejects H_0 whenever $\mathcal{I}_{n0}(\hat{\vartheta}) > \chi_{1-\alpha}^2$ would have the same property, where χ_α^2 is the αth percentile of the chi-square distribution with one degree of freedom.

Note that T_{n1} does not depend on F_0. It is thus an asymptotically distribution-free test, i.e., no matter which F_0 we are trying to fit within the class that satisfies the assumed conditions, this statistic has the same limiting distribution. To remove the dependence of the other statistics on F_0, we can replace the entities involving f_0 by their consistent estimators.

We shall now analyze the asymptotic power of these tests. Consider the Kolmogorov test. Let $F \neq F_0$ be a marginal d.f. of ε_0 of (13.2.22) such that (10.2.21) and (10.2.22) with $r = 3/2$ are satisfied and where a local Whittle estimator $\hat{d}$ is $\log(n)$-consistent for d, and where the long-run variance estimator $\hat{s}_{\varepsilon,d}^2$ of Section 9.4 is consistent for $s_{\varepsilon,d}$. Then (10.2.25) holds and, arguing as above, F has a smooth Lebesgue density f and

$$\mathcal{K}_n(\hat{\vartheta}) = \frac{\sup_{x\in\mathbb{R}} |s_{\varepsilon,d}f(x)Z + n^{1/2-d}(F(x) - F_0(x))|}{s_{\varepsilon,d}\|f_0\|_\infty} + o_p(1).$$

From this we readily see that the above Kolmogorov test is consistent for this F. Moreover, it is consistent for all those sequences $\{F\}$ satisfying the above conditions and for which $\sup_x n^{1/2-d}|F(x) - F_0(x)| \to \infty$. It has a trivial asymptotic power against sequences of alternatives for which

$\sup_x n^{1/2-d} |F(x) - F_0(x)| \to 0$. Thus, this test cannot distinguish the $n^{1/2}$-neighborhoods of F_0, i.e., this test has asymptotic power α against those $\{F\}$ in the class of d.f. satisfying the above assumed conditions and for which $n^{1/2} \sup_x |F(x) - F_0(x)| = O(1)$.

The asymptotic power of this test against a local alternative $F = F_0 + n^{-1/2+d} \Delta$, where Δ is an absolutely continuous function with the a.e. derivative $\dot{\Delta}$ bounded, equals

$$P\left(\sup_x | - f_0(x)vZ + \Delta(x)| > z_{\alpha/2} s_{\varepsilon,d} \|f_0\|_\infty \right).$$

A similar conclusion is available for the other tests proposed above.

Fitting an error d.f. in a location model. A more interesting and, at the same time, surprisingly challenging problem is to test

$$H_0^{loc} : F(x) = F_0(x - \mu), \quad \forall x \in \mathbb{R}, \text{ for some } \mu \in \mathbb{R}; \text{ vs.}$$
$$H_1^{loc} : H_0^{loc} \text{ is not true.}$$

This problem is equivalent to testing for H_0 based on the observations

$$Y_i = \mu + \varepsilon_i, \quad i = 1, \cdots, n, \quad \text{for some } \mu \in \mathbb{R}, \tag{13.5.4}$$

where ε_i, $i \in \mathbb{Z}$, are now unobservable moving-average errors of (13.2.22). Here the tests are based on

$$\bar{F}_n(x) := n^{-1} \sum_{i=1}^n I(Y_i - \bar{Y}_n \le x), \quad \bar{Y}_n := n^{-1} \sum_{i=1}^n Y_i, \quad x \in \mathbb{R}.$$

Let $\bar{\varepsilon}_n := \sum_{i=1}^n (Y_i - \mu)/n$. Note that $\bar{F}_n(x) = \hat{F}_n(x + \bar{\varepsilon}_n)$. The following proposition gives an approximation to this empirical process.

Proposition 13.5.1. (URP for $\bar{F}_n$). *Consider the location model (13.5.4) with errors as in (13.2.22) and such that (10.2.21) holds and $E|\zeta_0|^3 < \infty$. Then*

$$\sup_{x \in \mathbb{R}} n^{1/2-d} |\bar{F}_n(x) - F_0(x)| = o_p(1). \tag{13.5.5}$$

Proof. From (13.5.1) and by the mean-value theorem,

$$n^{1/2-d}(\bar{F}_n(x) - F_0(x))$$
$$= n^{1/2-d} \big[\hat{F}_n(x + \bar{\varepsilon}_n) - F_0(x + \bar{\varepsilon}_n) + \bar{\varepsilon}_n f_0(x + \bar{\varepsilon}_n) \big]$$
$$+ \int_x^{x+\bar{\varepsilon}_n} (f_0(u) - f_0(x + \bar{\varepsilon}_n)) du$$
$$= u_p(1) + O_p(\|\dot{f}_0\|_\infty n^{1/2-d} |\bar{\varepsilon}_n|^2) = u_p(1).$$

According to Proposition 13.5.1, the first-order difference

$$n^{1/2-d}\big(\bar{F}_n(x) - F_0(x)\big)$$

cannot distinguish between two marginal distributions of a long memory moving-average process that differ only in their means. This finding is in sharp contrast to the i.i.d. errors situation where the first-order difference $n^{1/2}[\bar{F}_n(x) - F_0(x)]$ converges weakly to a time-transformed Brownian bridge with a drift under this hypothesis; see Durbin (1973). Some tests based on the second-order difference for this problem are presented towards the end of this section. The study of the asymptotic behavior of these tests is facilitated by the higher-order approximations of the empirical processes discussed in Section 10.3.

Fitting an error d.f. in a scale model. Another problem of interest is to test the equivalence of F to F_0 up to a scale parameter, i.e., to test

$$H_0^{sc} : F(x) = F_0(x/\sigma), \quad \forall\, x \in \mathbb{R}, \text{ for some } \sigma > 0; \text{ vs.}$$
$$H_1^{sc} : H_0^{sc} \text{ is not true.}$$

This problem is equivalent to testing for H_0 based on the observations

$$Y_i = \sigma\varepsilon_i, \quad i = 1,\cdots,n, \quad \text{for some } \sigma > 0, \tag{13.5.6}$$

where again ε_i, $i \in \mathbb{Z}$, are the moving-average errors as in (13.2.22). Here tests will be based on

$$\tilde{F}_n(x) := \frac{1}{n}\sum_{i=1}^{n} I(\frac{Y_i}{\tilde{\sigma}_n} \le x) = \hat{F}_n(\frac{x\tilde{\sigma}_n}{\sigma}), \ x \in \mathbb{R}, \quad \tilde{\sigma}_n^2 := n^{-1}\sum_{i=1}^{n} Y_i^2.$$

The following proposition establishes a URP result for $\tilde{F}_n$.

Proposition 13.5.2. (URP for $\tilde{F}_n$). *Consider the scale model (13.5.6) with errors as in (13.2.22) and such that (10.2.21) holds and $E\zeta_0^4 < \infty$. Then*

$$\sup_{x\in\mathbb{R}} \big|n^{1/2-d}[\tilde{F}_n(x) - F_0(x)] + n^{1/2-d}\bar{\varepsilon}_n\, f_0(x)\big| = o_p(1).$$

Proof. Without loss of generality, assume $E\varepsilon_0^2 = \sum_{j=0}^{\infty} b_j^2 = 1$. We have

$$E(\tilde{\sigma}_n^2 - \sigma^2)^2 = \sigma^4 n^{-2} E\Big(\sum_{i=1}^{n}(\varepsilon_i^2 - 1)\Big)^2 = \sigma^4 n^{-2} \sum_{i,j=1}^{n} \text{Cov}(\varepsilon_i^2, \varepsilon_j^2).$$

By (13.3.14),

$$\text{Cov}(\varepsilon_i^2, \varepsilon_j^2) = O((\text{Cov}(\varepsilon_i, \varepsilon_j))^2) = O(|i-j|^{-2(1-2d)}) \ \text{ as } |i-j| \to \infty.$$

Hence,

$$\frac{\tilde{\sigma}_n}{\sigma} - 1 = O_p(n^{-1/2}), \qquad\qquad 0 < d < 1/4, \qquad (13.5.7)$$

$$= O_p((\log(n)/n)^{1/2}), \qquad d = 1/4,$$

$$= O_p(n^{-1+2d}), \qquad\qquad 1/4 < d < 1/2.$$

Now, let $\Delta_n := (\tilde{\sigma}_n/\sigma - 1)$. Recall from Lemma 10.2.4 that under (10.2.21), F_0 is infinitely smooth with density f_0 satisfying

$$\sup_{x \in \mathbb{R}} (1 + x^2)(f_0(x) + |\dot{f}_0(x)|) < \infty.$$

This bound and (10.2.33) applied with $\gamma = 2$ imply

$$|F_0(x + x\Delta_n) - F_0(x)| = \left| \int_0^{x\Delta_n} f_0(x+u)du \right| \qquad (13.5.8)$$

$$\leq C(1+x^2)^{-1}(|x\Delta_n| + x^2\Delta_n^2) \leq C(|\Delta_n| + \Delta_n^2),$$

$$|f_0(x + x\Delta_n) - f_0(x)| = \left| \int_0^{x\Delta_n} \dot{f}_0(x+u)du \right|$$

$$\leq C(1+x^2)^{-1}(|x\Delta_n| + x^2\Delta_n^2) \leq C(|\Delta_n| + \Delta_n^2).$$

Hence,

$$n^{1/2-d}\left| \tilde{F}_n(x) - F_0(x) + \bar{\varepsilon}_n f_0(x) \right|$$

$$\leq n^{1/2-d}\left| \hat{F}_n(x\tilde{\sigma}/\sigma) - F_0(x\tilde{\sigma}/\sigma) + \bar{\varepsilon}_n f_0(x\tilde{\sigma}/\sigma) \right|$$

$$+ n^{1/2-d}\bar{\varepsilon}_n\left| f_0(x) - f_0(x + x\Delta_n) \right| + n^{1/2-d}\left| F_0(x + x\Delta_n) - F_0(x) \right|$$

$$\leq u_p(1) + Cn^{1/2-d}(|\Delta_n| + \Delta_n^2).$$

The statement of the proposition now follows from this bound, (13.5.7), and the fact that $0 < d < 1/2$.

Note the URP for $\mathbb{F}_n$ is precisely the same as for $\hat{F}_n$ given in (13.5.1). In other words, the asymptotic null distribution of tests based on $n^{1/2-d}(\mathbb{F}_n - F_0)$ for testing H_0^{sc} is the same as those of tests based on $n^{1/2-d}(\hat{F}_n - F_0)$ for testing H_0. This is a kind of robustness property of these tests against the unknown error variance. It is also unlike the above situation in the location model, and unlike the i.i.d. situation, where $n^{1/2}(\mathbb{F}_n - F_0)$ weakly converges to a Brownian bridge with a drift; see, e.g., Durbin (1973).

Fitting an error d.f. up to unknown location+scale parameters. Now consider the problem of testing for the hypothesis

$$\mathcal{H}_0 : F(x) = F_0(\frac{x - \mu}{\sigma}), \quad \forall\, x \in \mathbb{R} \text{ and for some } \mu \in \mathbb{R}, \sigma > 0,$$

$$\mathcal{H}_1 : \mathcal{H}_0 \text{ is not true.}$$

This is equivalent to testing for $\mathcal{H}_0$ based on the observations

$$Y_i = \mu + \sigma\varepsilon_i, \quad i = 1, \cdots, n, \quad \text{for some } \mu \in \mathbb{R}, \sigma > 0,$$

where the errors ε_i, $i \in \mathbb{Z}$, are as in (13.2.22). Here tests will be based on

$$\mathcal{F}_n(x) := n^{-1} \sum_{i=1}^{n} I(Y_i - \bar{Y}_n \le x\hat{\sigma}_n), \quad x \in \mathbb{R}; \quad \hat{\sigma}_n^2 := n^{-1} \sum_{i=1}^{n}(Y_i - \bar{Y}_n)^2.$$

Assume, as in the scale problem, that $E\zeta_0^4 < \infty$. Let $\delta_n := (\hat{\sigma}_n - \sigma)/\sigma$, $\varepsilon_i = (Y_i - \mu)/\sigma$, and $\bar{\varepsilon}_n := (\bar{Y}_n - \mu)/\sigma$. Then, with $\bar{F}_n$ the same as in Proposition 13.5.1,

$$\mathcal{F}_n(x) = n^{-1} \sum_{i=1}^{n} I(Y_i - \bar{Y}_n \le x\hat{\sigma}_n) = n^{-1} \sum_{i=1}^{n} I(\varepsilon_i \le x + x\delta_n + \bar{\varepsilon}_n)$$

$$= \bar{F}_n(x + x\delta_n).$$

Moreover, the bound (13.5.7) holds with $\tilde{\sigma}_n$ replaced by $\hat{\sigma}_n$ under these conditions. Use these equations, (13.5.5), and an argument like the one used for deriving the bounds in (13.5.8), to obtain, under $\mathcal{H}_0$,

$$n^{1/2-d} \sup_{x} |\mathcal{F}_n(x) - F_0(x)|$$

$$\le n^{1/2-d} \Big\{ \sup_{x} |\bar{F}_n(x) - F_0(x)| + \sup_{x} |F_0(x + x\delta_n) - F_0(x)| \Big\}$$

$$= o_p(1) + C n^{1/2-d}(|\delta_n| + \delta_n^2) = o_p(1).$$

Thus, here the location parameter dominates in the sense that the first-order difference $n^{1/2-d}[\mathcal{F}_n(x) - F_0(x)]$, $x \in \mathbb{R}$, is not useful for fitting a marginal d.f. up to the unknown location and scale parameters.

Fitting an error d.f. in a regression model. Consider the linear regression model (11.2.1) with errors as in (13.2.22) and the design matrix $\boldsymbol{X}_n$ assumed to satisfy (11.2.2). Now F and f denote the marginal d.f. and density of the error process, respectively. Consider the problem of testing the above $\mathcal{H}_0$ vs. $\mathcal{H}_1$ based on $\{x_{ni}, Y_{ni}\}, 1 \le i \le n$, satisfying (11.2.1).

Here tests are to be based on the residual empirical process

$$\mathbb{F}_n(x) := n^{-1} \sum_{i=1}^{n} I(Y_{ni} - \tilde{\beta}'_n x_{ni} \leq x), \quad x \in \mathbb{R},$$

where $\tilde{\beta}_n$ is the LS estimator of β. The following proposition is useful in obtaining the weak null limit of this process. Let x'_{ni} denote the ith row of $\boldsymbol{X}_n$, $D_n := (\boldsymbol{X}'_n \boldsymbol{X}_n)^{1/2}$, and

$$\mathcal{Z}_n := n^{-1/2-d} \sum_{i=1}^{n} [x'_{ni}(\tilde{\beta}_n - \beta) - 1]\varepsilon_i.$$

Proposition 13.5.3. (URP for $\mathbb{F}_n$). *Assume for $\{(x_{ni}, Y_{ni}), 1 \leq i \leq n\}$ and the linear regression model (11.2.1) that $\{\varepsilon_i\}$ satisfies (13.2.22) and that (10.2.21) holds, and that $\{x_{ni}\}$ satisfy (11.2.2) and (11.2.11). Then, under H_0,*

$$\sup_{x \in \mathbb{R}} \left| n^{1/2-d}[\mathbb{F}_n(x) - F_0(x)] - f_0(x)\mathcal{Z}_n \right| = o_p(1).$$

Proof. With $c_{ni} := D_n^{-1} x_{ni}$ and $s = n^{-d} D_n(\tilde{\beta}_n - \beta)$ write

$$\mathbb{F}_n(x) := \frac{1}{n} \sum_{i=1}^{n} I(Y_{ni} \leq x + \tilde{\beta}'_n x_{ni}) = \frac{1}{n} \sum_{i=1}^{n} I(\varepsilon_i \leq x + n^d c'_{ni} s).$$

On taking $\gamma_{ni} \equiv n^{-1/2}$ and $\xi_{ni} \equiv n^d c'_{ni} s$ in Corollary 10.2.1 and arguing as in the proof of Lemma 11.2.2, we obtain,

$$\sup_{x \in \mathbb{R}} \left| n^{1/2-d}[\mathbb{F}_n(x) - \hat{F}_n(x)] - n^{-1/2-d} \sum_{i=1}^{n} x'_{ni}(\tilde{\beta}_n - \beta) f_0(x) \right| = o_p(1).$$

This together with (13.5.1) implies, under H_0,

$$n^{1/2-d}[\mathbb{F}_n(x) - F_0(x)]$$

$$= n^{1/2-d}[\hat{F}_n(x) - F_0(x)] + n^{-1/2-d} \sum_{i=1}^{n} x'_{ni}(\tilde{\beta}_n - \beta) f_0(x) + u_p(1)$$

$$= f_0(x)[-n^{-1/2-d} \sum_{i=1}^{n} \varepsilon_i + n^{-1/2-d} \sum_{i=1}^{n} x'_{ni}(\tilde{\beta}_n - \beta)\varepsilon_i] + u_p(1)$$

$$= \mathcal{Z}_n + u_p(1).$$

This concludes the proof of the proposition.

Let $v_{ni} := n^{1/2} D_n^{-1} x_{ni}$ and $\bar{v}_n := n^{-1} \sum_{i=1}^{n} v_{ni}$. Because $D_n(\tilde{\beta}_n - \beta) = D_n^{-1} \sum_{i=1}^{n} x_{ni} \varepsilon_i$, rewrite

$$\mathcal{Z}_n = n^{-1/2-d} \sum_{i=1}^{n} [\bar{v}_n' v_{ni} - 1] \varepsilon_i.$$

Note that if the regression model (11.2.1) is the one sample location model (13.5.4), i.e., if in (11.2.1) $p = 1$, $x_{ni} \equiv 1$, and $\beta = \mu$ then $v_{ni} \equiv 1$, the LS estimator is $\bar{Y}_n$, $\mathbb{F}_n(x) = \bar{F}_n(x)$, and $\mathcal{Z}_n = 0$, so that we again obtain the conclusion (13.5.5).

More generally, $\mathcal{Z}_n = 0$, for all $n \geq 1$, whenever there is a non-zero intercept parameter in the model (11.2.1). To see this and to keep the exposition transparent, consider the model (11.2.1) with $p = 2$ and $x_{ni}' = (1, a_i)$, where $a_1, \cdots, a_n$ are some known constants with $\sigma_a^2 := \sum_{i=1}^{n}(a_i - \bar{a})^2 > 0$, and $\bar{a} := n^{-1} \sum_{i=1}^{n} a_i$. Then

$$\bar{v}_n' v_{ni} = n(1, \bar{a})(\boldsymbol{X}_n' \boldsymbol{X}_n)^{-1} \begin{pmatrix} 1 \\ a_i \end{pmatrix}$$

$$= \frac{n}{\sigma_a^2}(1, \bar{a}) \begin{pmatrix} n^{-1} \sum_{j=1}^{n} a_j^2 & -\bar{a} \\ -\bar{a} & 1 \end{pmatrix} \begin{pmatrix} 1 \\ a_i \end{pmatrix} = 1, \quad \forall i = 1, \cdots, n.$$

Thus, here also the first-order difference $n^{1/2-d}(\mathbb{F}_n - F_0)$ cannot be used to fit F_0 to the error d.f. in the linear regression model (11.2.1) with a non-zero intercept parameter.

Next, if in (11.2.1) $\sum_{i=1}^{n} x_{ni} = 0$ then $\mathcal{Z}_n = -n^{-1/2-d} \sum_{i=1}^{n} \varepsilon_i$, and by (13.5.2), we again obtain the analog of (13.5.1) for $\mathbb{F}_n$. In other words, if the design vectors corresponding to the slope parameters are orthogonal to the p vector of 1's, then the asymptotic null distribution of $n^{1/2-d}(\mathbb{F}_n - F_0)$ is not affected by not knowing the slope parameters, and is the same as in (13.5.3). This conclusion is also applicable for the classical i.i.d. errors; see Koul (2002, Chapter 6).

In general, the asymptotic distribution of $\mathcal{Z}_n$ will depend on the design variables and ϑ through $s_n^2(\vartheta) := \text{Var}(\mathcal{Z}_n)$. Note that $\mathcal{Z}_n$ is a weighted sum of long memory moving-average errors with weights $c_{ni} \equiv \bar{v}_n' v_{ni} - 1$, which are, because of (11.2.11), uniformly bounded. Assuming $\lim_n s_n^2(\vartheta) = s^2(\vartheta)$ exists, by Proposition 4.3.1(i), $\mathcal{Z}_n \to_D \mathcal{N}(0, s^2(\vartheta))$.

As an example of a design where this limit exists consider the first-degree polynomial regression through the origin where $p = 1$ and $x_{ni} \equiv i/n$. Then

$v_{ni} := a_n\,(i/n)$ and $\bar{v}_n = a_n(n+1)/2n$ where $a_n := n^{3/2}6^{1/2}/(n(n+1)(2n+1))^{1/2} \sim 3^{1/2}$. Note that $\bar{v}_n v_{ni} = a_n^2[(n+1)/2n](i/n) \sim (3/2)(i/n)$. Hence, with $c_{\varepsilon,\gamma}$ as in (13.2.23),

$$s_n^2(\vartheta) \sim 2c_{\varepsilon,\gamma}n^{-1-2d} \sum_{1 \le i < j \le n} [\frac{3}{2}(\frac{i}{n}) - 1][\frac{3}{2}(\frac{j}{n}) - 1](j - i)^{-(1-2d)}$$

$$\sim 2c_{\varepsilon,\gamma} \int_0^1 \int_0^v (1.5u - 1)(1.5v - 1)(v - u)^{2d-1}\,du\,dv$$

$$= s_{\varepsilon,d}^2 \frac{3 - 4d^2 - d}{2(2d+2)(2d+3)} =: s^2(\vartheta).$$

Let $\mathcal{Z}_\vartheta$ be a $\mathcal{N}(0, s^2(\vartheta))$ r.v. Consequently, the analog of the Kolmogorov test is based on

$$\tilde{\mathcal{K}}_n := \frac{n^{1/2-\tilde{d}}\sup_x |\mathbb{F}_n(x) - F_0(x)|}{s(\tilde{\vartheta})\|f_0\|_\infty}.$$

Here $\tilde{\vartheta} = (\tilde{s}_{\varepsilon,d}^2, \tilde{d})$, where $\tilde{s}_{\varepsilon,d}^2$ is the long-run variance estimator of $s_{\varepsilon,d}^2$ and $\tilde{d}$ is the local Whittle estimator of d, based either on Y_{ni}'s or on the residuals $\tilde{\varepsilon}_i$'s. Other tests may be modified in a similar fashion.

Tests for H_0^{loc} based on the second-order difference. Because of (13.5.5), it is desirable to base the tests on a second-order difference of $\bar{F}_n$ and F_0. The nature of a limiting distribution of such a second-order difference depends on whether $0 < d < 1/4$ or $1/4 < d < 1/2$. The result in the former case when the underlying process is Gaussian is especially simple to describe. Recall the equations (10.3.3) to (10.3.7) and Theorem 10.3.1.

Suppose $\{\varepsilon_j,\ j \in \mathbb{Z}\}$ is a long memory Gaussian process and $k^* = 1$ or $0 < d < 1/4$. Let

$$R_j(x) := I(\varepsilon_j \le x) - F_0(x) + f_0(x)\varepsilon_j, \quad j \in \mathbb{Z}, \qquad (13.5.9)$$

$$Q_n(x) := n^{-1/2} \sum_{j=1}^n R_j(x)$$

$$= n^{1/2}\{\hat{F}_n(x) - F_0(x) + f_0(x)\bar{\varepsilon}_n\}, \quad x \in \mathbb{R}.$$

Then Theorem 10.3.1 gives the following:

Proposition 13.5.4. *Let $\{\varepsilon_j\}$ be a Gaussian long memory moving-average satisfying (13.2.22) with $0 < d < 1/4$. Then, under H_0^{loc}, $Q_n(x) \Rightarrow Q(x)$,* *where* $\{Q(x), x \in \mathbb{R}\}$ *is a continuous Gaussian process with zero mean and covariance function*

$$\text{Cov}(Q(x), Q(y)) = \sum_{j \in \mathbb{Z}} \text{Cov}(R_0(x), R_j(y)).$$

Moreover,

$$n^{1/2}\{\bar{F}_n(x) - F_0(x)\} \Rightarrow Q(x). \tag{13.5.10}$$

Proof. The first claim is exactly Theorem 10.3.1 when $0 < d < 1/4$. We need only to prove (13.5.10). Using the definition of $Q_n(x)$ in (13.5.9), rewrite $n^{1/2}(\bar{F}_n(x) - F_0(x)) = Q_n(x + \bar{\varepsilon}_n) + U_n(x)$, where $U_n(x) := n^{1/2}[F_0(x + \bar{\varepsilon}_n) - F_0(x) - f_0(x + \bar{\varepsilon}_n)\bar{\varepsilon}_n]$. By a Taylor expansion,

$$\sup_x |U_n(x)| \leq n^{1/2}(\bar{\varepsilon}_n)^2 \sup_x |\dot{f}_0(x)| = O_p(n^{2d-1/2}) = o_p(1).$$

Therefore, (13.5.10) follows from (13.5.9) and

$$\sup_x |Q_n(x + \bar{\varepsilon}_n) - Q_n(x)| = o_p(1). \tag{13.5.11}$$

For any $\delta_1, \delta_2 > 0$,

$$P(\sup_x |Q_n(x + \bar{\varepsilon}_n) - Q_n(x)| > \delta_1)$$
$$\leq P(|\bar{\varepsilon}_n| \geq \delta_2) + P(\sup_{|x-y|<\delta_2} |Q_n(x) - Q_n(y)| > \delta_1).$$

Since the sequence $Q_n, n \geq 1$, is tight in the sup-topology on $\bar{\mathbb{R}}$, the last probability in the above bound can be made arbitrarily small for all n large enough, by choosing δ_2 small enough. As $|\bar{\varepsilon}_n| = o_p(1)$, this proves (13.5.11) and completes the proof of Proposition 13.5.4.

Proposition 13.5.4 implies that the large sample critical values of the test that rejects H_0^{loc} whenever $\sup_{x \in \mathbb{R}} n^{1/2}|\bar{F}_n(x) - F_0(x)|$ is large can be determined from the distribution of $\|Q\|_\infty$. This is similar to the "usual" asymptotics of the Kolmogorov test for short memory errors based on $\hat{F}_n$, except that the percentiles of $\|Q\|_\infty$ are relatively harder to obtain. However, we can use a resampling method to estimate them. For this purpose it may be necessary to use an estimate of $\text{Cov}(Q(x), Q(y))$ to generate the sampling distribution of the process Q.

Now consider the general case of moving-average processes, not necessarily Gaussian. Recall (10.3.7), (10.3.26), (10.3.27), (10.3.32), and Theorem 10.3.2, which gives the higher-order expansion for $\hat{F}_n$. As discussed for Theorem 10.3.2, while we can show the weak convergence of all finite-dimensional distributions of the remainder process $\{Q_n(x),\ x \in \bar{\mathbb{R}}\}$ to those of a continuous Gaussian process, proving the tightness of this process remains an open technical problem. Because of this, when $0 < d < 1/4$ or $k^* = 1$, the expansion (10.3.37) yields only the URP of (13.5.5) and is not useful in deriving the limiting distribution of the second-order difference of $\hat{F}_n - F_0$. However, for $1/4 < d < 1/2$ the situation is different. Let $\mathcal{Y} := Z^{(2)} - 2^{-1}(Z^{(1)})^2$ and $\tilde{Z}_n^{(k)}$ be as in (10.3.26).

Proposition 13.5.5. *Assume the same conditions as in Theorem 10.3.2, and let $1/4 < d < 1/2$. Then, under H_0^{loc},*

$$\left|n^{1-2d}\{\bar{F}_n(x) - F_0(x)\} - \dot{f}_0(x)\big[\tilde{Z}_n^{(2)} - 2^{-1}(\tilde{Z}_n^{(1)})^2\big]\right| = u_p(1), \quad (13.5.12)$$

$$n^{1-2d}\{\bar{F}_n(x) - F_0(x)\} \Rightarrow \dot{f}_0(x)\mathcal{Y}. \quad (13.5.13)$$

Moreover, with $s_{\varepsilon,d}^2$ as in (13.2.32),

$$E\mathcal{Y} = -2^{-1}s_{\varepsilon,d}^2, \quad (13.5.14)$$

$$E\mathcal{Y}^2 = s_{\varepsilon,d}^4 \frac{(1+2d)^2}{2d^3}\left\{\frac{1}{2(4d-1)} + \frac{1}{2d(1+2d)^2}\right. \quad (13.5.15)$$
$$\left. - \frac{1}{d(4d+1)} - \frac{B(1+2d, 1+2d)}{d}\right\}.$$

Proof. Note that $d > 1/4$ implies $2d - 1/2 > 0$ and $k^* \geq 2$. Combine these with (10.3.36) and (10.3.37) and with $\delta < 2d - 1/2$ to obtain the second-order expansion under H_0:

$$\sup_{x\in\mathbb{R}}\left|n^{1-2d}\{\hat{F}_n(x) - F_0(x) + f_0(x)\bar{\varepsilon}_n\} - \dot{f}_0(x)\tilde{Z}_n^{(2)}\right| = o_p(1). \quad (13.5.16)$$

The decomposition

$$\bar{F}_n(x) - F_0(x) = \hat{F}_n(x + \bar{\varepsilon}_n) - F_0(x + \bar{\varepsilon}_n) + F_0(x + \bar{\varepsilon}_n) - F_0(x)$$

and (13.5.16) give

$$\sup_{x\in\mathbb{R}}\left|n^{1-2d}\{\bar{F}_n(x) - F_0(x) + \bar{\varepsilon}_n\, f_0(x + \bar{\varepsilon}_n)\} - \dot{f}_0(x + \bar{\varepsilon}_n)\tilde{Z}_n^{(2)}\right.$$
$$\left. - n^{1-2d}\{F_0(x + \bar{\varepsilon}_n) - F_0(x)\}\right| = o_p(1).$$

Use a Taylor expansion for f_0, $\dot{f}_0$, and the boundedness of $\ddot{f}_0$, to obtain

$$n^{1-2d}\{\bar{F}_n(x) - F_0(x)\}$$

$$= -n^{1-2d}\bar{\varepsilon}_n f_0(x + \bar{\varepsilon}_n) + \dot{f}_0(x + \bar{\varepsilon}_n)\tilde{Z}_n^{(2)}$$

$$- n^{1-2d}\{F_0(x + \bar{\varepsilon}_n) - F_0(x)\} + u_p(1)$$

$$= - n^{1-2d}\bar{\varepsilon}_n\left[f_0(x) + \bar{\varepsilon}_n \dot{f}_0(x) + \frac{(\bar{\varepsilon}_n)^2}{2}\ddot{f}_0(x + \xi_n)\right]$$

$$+ \left[\dot{f}_0(x) + \bar{\varepsilon}_n \ddot{f}_0(x + \xi_n)\right]\tilde{Z}_n^{(2)}$$

$$+ n^{1-2d}\left[\bar{\varepsilon}_n f_0(x) + \frac{(\bar{\varepsilon}_n)^2}{2}\dot{f}_0(x) + \frac{(\bar{\varepsilon}_n)^3}{6}\ddot{f}_0(x + \xi_n)\right] + u_p(1)$$

$$= \dot{f}_0(x)\left[\tilde{Z}_n^{(2)} - \frac{(\bar{\varepsilon}_n)^2}{2}n^{1-2d}\right] + O_p\left(n^{1-2d}(\bar{\varepsilon}_n)^3\right) + u_p(1),$$

where ξ_n is a sequence of r.v.'s with $|\xi_n| \le |\bar{\varepsilon}_n|$, $(\bar{\varepsilon}_n)^2 n^{1-2d} = (\tilde{Z}_n^{(1)})^2 \to_D (Z^{(1)})^2$, and $n^{1-2d}(\bar{\varepsilon}_n)^3 = o_p(n^{d-1/2}) = o_p(1)$. This proves (13.5.12). Claim (13.5.13) follows from (13.5.12) and (10.3.32).

Proof of (13.5.14) and (13.5.15). Let $C(d) := c_0^4 B^2(d, 1 - 2d) = s_{\varepsilon,d}^4 (1 + 2d)^2/4d^2$. From definition (10.3.7) and the diagram formula of Wiener–Itô integrals in Chapter 14 below, we find $EZ^{(2)} = 0$, and

$$E(Z^{(1)})^2 = c_0^2 \int_{\mathbb{R}} \left\{\int_0^1 (u - s)_+^{d-1}du\right\}^2 ds = \frac{c_0^2 B(d, 1 - 2d)}{d(1 + 2d)} = s_{\varepsilon,d}^2,$$

$$E(Z^{(1)})^4 = 3\big(E(Z^{(1)})^2\big)^2 = 3s_{\varepsilon,d}^4,$$

$$E(Z^{(2)})^2 = 2^{-1}c_0^4 \int_{\mathbb{R}} \left\{\int_0^1 (u - s_1)_+^{d-1}(u - s_2)_+^{d-1}du\right\}^2 ds$$

$$= \frac{C(d)}{4d(4d - 1)},$$

$$EZ^{(2)}(Z^{(1)})^2 = c_0^4 \int_{\mathbb{R}} \int_{\mathbb{R}} ds_1 ds_2 \int_0^1 (u - s_1)_+^{d-1}(u - s_2)_+^{d-1} du$$

$$\times \int_0^1 (v - s_1)_+^{d-1} dv \int_0^1 (w - s_2)_+^{d-1} dw$$

$$= C(d) \int_0^1 \int_0^1 \int_0^1 |u - v|^{2d-1}|u - w|^{2d-1} \, du \, dv \, dw$$

$$= \frac{C(d)}{2d^2}\left(\frac{1}{4d + 1} + B(2d + 1, 2d + 1)\right).$$

This completes the proof of Proposition 13.5.5.

Now, let $\hat{d}$ be a $\log(n)$-consistent estimator of d under H_0^{loc} and

$$\mathcal{K}_n := \sup_{x \in \mathbb{R}} n^{1-2\hat{d}} |\bar{F}_n(x) - F_0(x)|.$$

From Proposition 13.5.5, we readily find that the asymptotic null distribution of $\mathcal{K}_n$ is the same as that of $\|\dot{f}_0\|_\infty \mathcal{Y}$. Consequently, the critical values of the test that rejects H_0^{loc} whenever, $\mathcal{K}_n / \|\dot{f}_0\|_\infty$ is large may be determined from the distribution of $\mathcal{Y}$. Unfortunately this distribution depends on $s_{\varepsilon,d}$ and d in a complicated fashion, and is not easy to track. However, we can use a resampling method to simulate the distribution of this r.v. corresponding to $\hat{s}_{\varepsilon,d}$ and $\hat{d}$. No theoretical properties of any bootstrap methodology are known at the time of writing.

Higher-order moments of the limiting r.v. $\mathcal{Y}$ can be computed using the diagram formula for Wiener–Itô integrals given in Proposition 14.3.6 below, although the resulting formulas are rather cumbersome. Rosenblatt (1961) obtained the characteristic function of the r.v. $Z^{(2)}$ and its representation as an infinite weighted sum of chi-squares. See also Taqqu (1975, (6.2)).

Using the techniques of Chapter 10 above it should be possible to obtain the higher-order expansion for $\mathbb{F}_n$ in the regression situation. Such an expansion would then yield a test for fitting an error d.f. in linear regression models with non-zero intercepts based on second- or higher-order terms of this expansion.

Note: This chapter has roots in Koul *et al.* (2004), Guo and Koul (2008), and Koul and Surgailis (2008, 2009, 2010). At the time of writing the problems of proving the consistency of the tests discussed in Sections 13.3 and 13.4.1 are unsolved.

Chapter 14

Appendix

This chapter contains some results from probability theory that are of general interest and not necessarily directly relevant to long memory processes only. It provides a definition of *Wick products* and formulas for their moments and cumulants. Wick products are multivariate generalizations of Appell and Hermite polynomials. They are useful in deriving limit distributions of sums $S_n = \sum_{j=1}^{n} A_k(X_j)$ of Appell polynomials of a linear process $\{X_j\}$. This chapter also gives a definition and some basic properties of Wiener–Itô integrals. The chapter ends with a lemma describing the limit of the trace of a product of matrices.

14.1 Appell polynomials, Wick products, and diagram formulas

Direct computation of the moments and cumulants of a function $h(X_j)$ of a Gaussian or a linear process $\{X_j\}$ is a difficult task. As seen in the proof of Theorem 4.6.1, this task was facilitated by a Hermite expansion and the diagram formula of Proposition 14.3.6 below. Computation of the moments and cumulants of $h(X_j, X_{j-1}, \cdots)$ in terms of those of X_j's is enabled by expansions using multivariate Hermite and Appell polynomials, which are the so-called Wick products. These products are also used to describe the main terms in asymptotic expansions of sums of transformed linear processes; see Chapters 10 and 11.

Let X_i, $1 \leq i \leq m$, be a set of r.v.'s satisfying $E|X_i|^m < \infty, i = 1, \cdots, m$. The *Wick product* of the r.v.'s $X_1, \cdots, X_m$ is defined by

$$:X_1 \times \cdots \times X_m: = \frac{(-i)^m \partial^m}{\partial u_1 \cdots \partial u_m} \Big(\frac{\exp\{i\sum_{j=1}^m u_j X_j\}}{E \exp\{i\sum_{j=1}^m u_j X_j\}} \Big)\Big|_{u_1=\cdots=u_m=0}.$$

In the following, a Wick product is also called a *multivariate Appell polynomial*.

A Wick product $:X_1 \times \cdots \times X_m:$ is an mth-order polynomial in the variables $X_1, \cdots, X_m$ with coefficients defined by their joint cumulants; see Proposition 14.1.1 below. For Gaussian r.v.'s, Wick products are called *multivariate Hermite polynomials*.

Wick multiplication is different from the usual multiplication of r.v.'s, although it enjoys some of the properties of the latter (see below). In particular, Wick multiplication is commutative so exchanging the order of the r.v.'s does not affect the Wick product. Moreover,

$$:X^m: \equiv\ :X \times \cdots \times X: = A_m(X), \qquad (14.1.1)$$
$$= H_m(X), \text{ if } X \sim \mathcal{N}(0,1),$$

where $A_m(X)$ is the mth Appell polynomial of the r.v. X as defined in (2.4.19). By definition, it follows that if $X_1, \cdots, X_m$ are independent, then for any $k_1, \cdots, k_m \geq 1$,

$$:X_1^{k_1} \times \cdots \times X_m^{k_m}: = \prod_{j=1}^m :X_j^{k_j}: = \prod_{j=1}^m A_{k_j}(X_j), \qquad (14.1.2)$$

$$:X_1 \times \cdots \times X_m: = \prod_{j=1}^m X_j, \quad \text{if } EX_1 = \cdots = EX_m = 0.$$

The next proposition provides a rule for the computation of the coefficients of Wick products. Let $T = \{1, \cdots, m\}$. For any subset $U \subset T$, let $|U|$ denote the cardinality of U, $X^U = \prod_{j \in U} X_j$, and $\text{Cum}(X^U) = \text{Cum}(X_j, j \in U)$. In the following $:X^U:$ will denote the Wick product of the r.v.'s $X_j, j \in U$, in particular, $:X^T: = :X_1 \times \cdots \times X_m:$. Set $:X^\emptyset: = 1$.

Proposition 14.1.1.

$$X^T = \sum_{U \subset T} :X^U: EX^{T \setminus U}, \qquad (14.1.3)$$

$$:X^T: = \sum_{U \subset T} X^U \sum_{T \setminus U} (-1)^r \text{Cum}(X^{V_1}) \cdots \text{Cum}(X^{V_r}), \qquad (14.1.4)$$

$$E:X^T: = 0 \quad \text{if } T \neq \emptyset, \qquad (14.1.5)$$

where the sum $\sum_{U \subset T}$ is taken over all subsets $U \subset T$ including $U = \emptyset$, and the sum $\sum_{T \setminus U}$ is taken over all partitions $(V_1, \cdots, V_r)$ of $T \setminus U$ into non-empty disjoint subsets.

Proof. For $U \subset T$, let $\nabla^U := \prod_{j \in U} \partial/\partial u_j$. By the Leibnitz rule for differentiation, for any functions f and g,

$$\nabla^T(fg) = \sum_{U \subset T} (\nabla^U f)(\nabla^{T \setminus U} g).$$

Let $\phi_1 := \exp\{i \sum_{j=1}^m u_j X_j\}$ and $\phi_2 := E \exp(i \sum_{j=1}^m u_j X_j)$. Observe that for any $U \subset T$, and by the definition of a Wick product,

$$\nabla^U \phi_1|_{u=0} = i^{|U|} X^U, \quad \nabla^U \frac{\phi_1}{\phi_2}\Big|_{u=0} = i^{|U|} :X^U:, \tag{14.1.6}$$

$$\nabla^U \phi_2\Big|_{u=0} = i^{|U|} E X^U, \quad \nabla^U \log(\phi_2)\big|_{u=0} = i^{|U|} \mathrm{Cum}(X^U).$$

This and the following equation for the partial derivatives

$$\nabla^T \phi_1 \equiv \nabla^T[(\frac{\phi_1}{\phi_2})\phi_2] = \sum_{U \subset T} \nabla^U(\frac{\phi_1}{\phi_2})\nabla^{T \setminus U}\phi_2,$$

computed at $u_1 = \cdots = u_m = 0$, implies (14.1.3).

To prove (14.1.4), use the equality

$$\nabla^T(\frac{\phi_1}{\phi_2}) = \sum_{U \subset T} \nabla^U \phi_1 \nabla^{T \setminus U}(\frac{1}{\phi_2}).$$

Because of the above proof of (14.1.3), it suffices to show that for any $W \subset T$,

$$\nabla^W(\frac{1}{\phi_2})\Big|_{u=0} = i^{|W|} \sum_W (-1)^r \mathrm{Cum}(X^{V_1}) \cdots \mathrm{Cum}(X^{V_r}), \tag{14.1.7}$$

where the sum $\sum_W$ is defined in the same way as the sum $\sum_{T \setminus U}$ in (14.1.4). This can be shown by induction on $|W|$. Suppose $|W| = 1$. Write $W = \{i\}$. Then

$$\frac{\partial}{\partial u_i}(\frac{1}{\phi_2}) = -\frac{1}{\phi_2} \frac{\partial}{\partial u_i} \log(\phi_2),$$

$r = 1$, $\log(\phi_2)|_{\tilde{u}} = \sum_{k=0}^\infty \frac{(iu_i)^k}{k!} \mathrm{Cum}_k(X_i)$, $\tilde{u} = (0, \cdots, 0, u_i, 0, \cdots, 0)$, and

$$\nabla^W(\frac{1}{\phi_2})\big|_{u=0} = -i \, \mathrm{Cum}_1(X_i) = (-1)i \, \mathrm{Cum}(X^W).$$

This proves the equality (14.1.7).

Next we show that if (14.1.7) is valid for $|W| \leq k_0 - 1$, $k_0 \geq 2$, then it also holds for $|W| = k_0$. Write $W = (W\backslash\{i\}) \cup \{i\} =: W' \cup \{i\}$. Then $\nabla^W(\frac{1}{\phi_2}) = -\nabla^{W'}(\frac{1}{\phi_2}\frac{\partial \log(\phi_2)}{\partial u_i})$, and (14.1.7) follows applying the Leibnitz rule, the induction assumption for $\nabla^{W'}$, and (14.1.6).

To prove (14.1.5), notice that $E[: X^T :] = 0$ if $|T| = 1$. For $|T| \geq 2$, taking expectations on both sides of (14.1.3), by induction we find $EX^T = E[: X^T:] + EX^T$ which implies (14.1.5). This completes the proof of the proposition.

As noted above, some properties of the Wick products are similar to those of the usual products of r.v.'s.

(Symmetry): The polynomial $:X_1 \times \cdots \times X_m:$ is invariant under permutations of $X_1, \cdots, X_m$.

(The multilinearity): If $X_i = \sum_{j=1}^n a_{ij}\xi_j$, where $\xi_j, j = 1, \cdots, m$, are r.v.'s and $a_{ij} \in \mathbb{R}$, then

$$:X_1 \times \cdots \times X_m: = \sum_{j_1, \cdots, j_m=1}^n a_{1j_1} \cdots a_{mj_m} :\xi_{j_1} \times \cdots \times \xi_{j_m}: . \quad (14.1.8)$$

The multilinearity property follows from (14.1.4), because products and cumulants are multilinear functions. We also note that $:c := 0$ for all $c \in \mathbb{R}$.

Example 14.1.1. Proposition 14.1.1 gives practical rules for the computation of the coefficients of Wick products, and Appell and Hermite polynomials. Suppose that $X \sim \mathcal{N}(0, 1)$. Then, by (14.1.3) and (14.1.4),

$$X^3 = H_3(X) + 3H_1(X), \quad X^4 = H_4(X) + 6H_2(X) + 3,$$

$$H_3(X) = X^3 - 3X, \quad H_4(X) = X^4 - 6X^2 + 3.$$

Suppose $\{X_j\}$ is the linear process (4.3.1) with i.i.d. innovations $\{\zeta_k\}$ such that $E|\zeta_k|^m < \infty$. The multilinearity property (14.1.8) allows us to write the Appell polynomial $A_m(X_j)$, $m \geq 1$ in terms of Wick products of i.i.d. noise ζ_k:

$$A_m(X_j) = :\left(\sum_{k \leq j} a_{j-k}\zeta_k\right)^m: \quad (14.1.9)$$

$$= \sum_{k_1, \cdots, k_m \leq j} a_{j-k_1} \cdots a_{j-k_m} :\zeta_{k_1} \times \cdots \times \zeta_{k_m}:$$

where the Wick products $:\zeta_{k_1} \times \cdots \times \zeta_{k_m}:$ have a simple structure, and their moments and cumulants can be easily computed. For example,

$$:\zeta_{k_1}\zeta_{k_2}: = \zeta_{k_1}\zeta_{k_2}, \quad \text{if } k_1 \neq k_2,$$

$$= :\zeta_{k_1}^2 := A_2(\zeta_{k_1}) = \zeta_{k_1}^2 - E\zeta_{k_1}^2, \quad \text{if } k_1 = k_2.$$

Notice, that the Wick polynomials $:X_t X_s:$ and $:\zeta_t \zeta_s:$ are different, since they correspond to different random variables and therefore they have different coefficients.

Diagram formulas for moments and cumulants. The diagram method allows us to compute the moments and cumulants of Appell and Hermite polynomials and Wick products.

Consider a table $T = \{T_i\}_{i=1,\cdots,k}$ with rows

$$T_i = \{(i,1), \cdots, (i,k_i)\}, \quad k_i \geq 1, \quad i = 1, \cdots, k,$$

and a set of r.v.'s X_{ij}, $(i,j) \in T$, such that $E|X_{ij}|^m < \infty$, $m = \sum_{i=1}^{k} k_i$.

Definition 14.1.1. (i) A partition or diagram $\{V\} = (V_1, \cdots, V_r)$ of T divides T into $r \geq 1$ non-empty disjoint subsets (edges) V_j.

(ii) An edge V_j is called *flat* if $V_j \subset T_i$ for some $i = 1, \cdots, k$.

(iii) A partition $\{V\}$ is called *connected* if the rows $T_1, \cdots, T_k$ of the table T cannot be divided into two groups each of which is partitioned by $\{V\}$ separately.

(iv) A partition $\{V\} = (V_1, \cdots, V_r)$ is called *Gaussian* if $|V_1| = \cdots = |V_r| = 2$ and $m = 2r$ is even.

For example, the trivial partition (T) consisting of the single set T is connected, the partition $(T_1, \cdots, T_k)$ of T into rows $T_1, \cdots, T_k$ is not connected, whereas the partition $((1,1), T\backslash(1,1))$ by two subsets $(1,1)$ and $T\backslash(1,1)$ is connected if $k_1 \geq 2$ and is not connected if $k_1 = 1$.

We need the following classes of partitions $\{V\}$ of T. Let

P_T be all partitions, $\quad P_T^c$ be partitions without flat edges, (14.1.10)

Γ_T be all connected partitions,

Γ_T^c be all connected partitions without flat edges.

As before, for $V \subset T$,

$$EX^T = E\prod_{(i,j)\in V} X_{ij}, \quad \mathrm{Cum}(X^V) = \mathrm{Cum}(X_{ij}, (i,j) \in V).$$

For a given partition $\{V\} = (V_1, \cdots, V_r)$, define

$$I_V = \mathrm{Cum}(X^{V_1}) \times \cdots \times \mathrm{Cum}(X^{V_r}). \tag{14.1.11}$$

The next theorem provides diagram formulas for the moments and cumulants of the usual products and the Wick products of random variables.

Theorem 14.1.1. (Diagram formula). *Let T be a table with rows T_1, $\cdots$, T_k and let X_{ij}, $(i,j) \in T$, be r.v.'s with $E|X_{ij}|^{|T|} < \infty$. Then*

$$E[X^{T_1} \times \cdots \times X^{T_k}] = \sum_{\{V\} \in P_T} I_V, \qquad (14.1.12)$$

$$\text{Cum}(X^{T_1}, \cdots, X^{T_k}) = \sum_{\{V\} \in \Gamma_T} I_V, \qquad (14.1.13)$$

$$E[:X^{T_1}: \times \cdots \times :X^{T_k}:] = \sum_{\{V\} \in P_T^c} I_V, \qquad (14.1.14)$$

$$\text{Cum}(:X^{T_1}:, \cdots, :X^{T_k}:) = \sum_{\{V\} \in \Gamma_T^c} I_V. \qquad (14.1.15)$$

Proof. Recall the notation ∇^U and ϕ_2 from the proof of Proposition 14.1.1. To prove (14.1.12), it suffices to show that

$$\nabla^T \phi_2 = \phi_2 \sum_{\{V\} \in P_T} \nabla^{V_1}(\log \phi_2) \cdots \nabla^{V_r}(\log \phi_2), \qquad (14.1.16)$$

which, on setting $u_1 = \cdots = u_n = 0$ implies (14.1.12). For $|T| = 1$, (14.1.16) follows, since $(\partial/\partial u_1)\phi_2 = \phi_2(\partial/\partial u_1)\log(\phi_2)$. For $|T| \geq 2$, it follows by induction, applying the Leibnitz rule, as for (14.1.4).

The proof of (14.1.13) is based on induction in k. Let $Y_i := X^{T_i}$, $i = 1$, $\cdots$, k. Suppose $k = 2$. Then, using (14.1.12),

$$\text{Cum}(Y_1, Y_2) = EY_1Y_2 - EY_1EY_2$$

$$= \sum_{\{V\} \in P_T} I_V - \sum_{\{V\} \in P_{T_1}} I_V \sum_{\{V\} \in P_{T_2}} I_V = \sum_{\{V\} \in \Gamma_T} I_V.$$

We shall show that if (14.1.13) holds for $k \geq k_0 - 1$, $k_0 \geq 2$, then it holds for $k = k_0$. Setting $A = \{1, \cdots, k\}$, (14.1.12) gives

$$E[Y_1 \times \cdots \times Y_k] = \sum_{\{V\} \subset A} \text{Cum}(Y^{V_1}) \cdots \text{Cum}(Y^{V_r}) \qquad (14.1.17)$$

$$= \text{Cum}(Y^A) + \sum_{\{V\} \subset A: r \geq 2} \text{Cum}(Y^{V_1}) \cdots \text{Cum}(Y^{V_r}),$$

$$\sum_{\{V\} \in P_T} I_V = \text{Cum}(Y_1, \cdots, Y_k) + \sum_{\{V\} \in P_T/\Gamma_T} I_V,$$

where the second equality in (14.1.17) follows by using relation (14.1.12) for the l.h.s. and an induction argument on the r.h.s., which proves (14.1.13) for $k = k_0$.

To prove (14.1.14), using (14.1.4), write

$$:X^{T_i}: = \sum_{U \subset T_i} X^U Q_{T_i \setminus U}, \quad i = 1, \cdots, k,$$

$$Q_{T_i \setminus U} \equiv \sum_{T \setminus U} (-1)^r \text{Cum}(X^{V_1}) \cdots \text{Cum}(X^{V_r}).$$

Multiplying, we find

$$\prod_{i=1}^{k} :X^{T_i}: = \prod_{i=1}^{k} \sum_{U_i \subset T_i} X^{U_i} Q_{T_i \setminus U_i} = \sum_{U \subset T} X^U \prod_{i=1}^{k} Q_{T_i \setminus U}.$$

Apply the rule (14.1.12), to verify that $\forall U \subset T$,

$$EX^U = \sum_{\{V\} \in P_U} I_V = \sum_{W \subset U} \sum_{\{V\} \in P_W^c} I_V \prod_{i=1}^{k} EX^{T_i \cap (U \setminus W)}.$$

The second equality follows from $P_U = \cup_{W \subset U} P_W^c \cup \{\cup_{i=1}^{k} P_{T_i \cap (U \setminus W)}\}$, and because $\sum_{\{V\} \in P_{T_i \cap (U \setminus W)}} I_V = EX^{T_i \cap (U \setminus W)}$, which holds by (14.1.12). Therefore,

$$E \prod_{i=1}^{k} :X^{T_i}: = \sum_{W \subset T} \sum_{\{V\} \in P_W^c} I_V H_W,$$

$$H_W := \sum_{U : W \subset U \subset T} \prod_{i=1}^{k} EX^{T_i \cap (U \setminus W)} Q_{T_i \setminus U}.$$

Let $S := U \setminus W$. Then $S \subset T \setminus W$, $S \cap W = \emptyset$, $T_i \setminus U = (T_i \setminus W) \setminus S$, and, by (14.1.4),

$$H_W = \prod_{i=1}^{k} \left[\sum_{S \subset T_i \setminus W} EX^{(T_i \setminus W) \cap S} Q_{(T_i \setminus W) \setminus S} \right] = \prod_{i=1}^{k} E[: X^{T_i \setminus W} :].$$

By (14.1.5), $E[: X^{T_i \setminus W} :] = 0$ if $T_i \setminus W \neq \emptyset$. Thus, $E \prod_{i=1}^{k} : X^{T_i} : = \sum_{\{V\} \in P_T^c} I_V$, which completes the proof of (14.1.14).

The proof of (14.1.15) is obtained by induction on k. As for the proof of (14.1.13), set $Y_i =: X^{T_i}:$, $i = 1, \cdots, k$. Suppose $k = 2$. Then $EY_1 = EY_2 = 0$, and (14.1.15) follows from $\mathrm{Cum}(Y_1, Y_2) = EY_1 Y_2$, because of (14.1.14). Assuming that (14.1.15) is valid for $k \leq k_0 - 1$, $k_0 \geq 2$, we show that it is valid for $k = k_0$. Then (14.1.17) holds, and applying (14.1.14) to the l.h.s. and the inductive assumption to the sum on the r.h.s., we find

$$\sum_{\{V\} \in P_T^c} I_V = \mathrm{Cum}(Y_1, \cdots, Y_k) + \sum_{\{V\} \in P_T^c / \Gamma_T^c} I_V.$$

This in turn implies (14.1.15) for $k = k_0$ and completes the proof of the theorem.

Equalities (14.1.12) and (14.1.13) are known as the Leonov–Shiryaev formulas; see Leonov and Shiryaev (1959), and (14.1.13) is Theorem 2.3.2 in Brillinger (1981).

In the Gaussian case, the diagram formula simplifies. Observe that

$$\text{Cum}(X^V) = EX_{ij}, \qquad \text{if } V = (i,j),$$
$$= \text{Cov}(X_{ij}, X_{i'j'}), \qquad \text{if } V = \{(i,j),(i',j')\},$$
$$= 0, \qquad \text{if } |V| \geq 3 \text{ and } X_{ij}, (i,j) \in V, \text{ are Gaussian.}$$

Let G_T be the set of *Gaussian partitions* $\{V\} = (V_1, \cdots, V_{|T|/2})$ such that $|V_s| = 2$, for any $s = 1, \cdots, |T|/2$. For $V_s = \{(i_s, j_s), (i'_s, j'_s)\}$, $s = 1, \cdots, |T|/2$ and $|T|$ even,

$$I_V = \prod_{s=1}^{|T|/2} \text{Cov}(X_{i_s}, X_{i'_s}). \qquad (14.1.18)$$

If $|T|$ is odd, then G_T is empty and $I_V = 0$.

Corollary 14.1.1. *Suppose $X_{i,j}, (i,j) \in T$, are jointly Gaussian r.v.'s with $EX_{i,j} \equiv 0$. Then the following hold:*

(i) If $|T|$ is even, then summation in (14.1.12) to (14.1.15) is restricted, in addition, to partitions $\{V\} \in G_T$. If $|T|$ is odd, then the corresponding moments and cumulants are equal to zero.

(ii) Wick products satisfy the orthogonality property:

$$E[:X^{T_1}: \times :X^{T_2}:] = 0, \qquad |T_1| \neq |T_2|,$$
$$E[:X_1^m: \times :X_2^p:] = 0, \qquad m \neq p,$$
$$= m! \text{Cov}^m(X_1, X_2), \qquad m = p.$$

The simplest case of the diagram formula is a table having two rows: $T = T_1 \cup T_2$ where $|T_1| = m$ and $|T_2| = p$. Clearly, if $m \neq p$, the class of Gaussian diagrams over such a table is empty. However, if $m = p$, the number of Gaussian diagrams equals the number $m!$ of permutations of $\{1, \cdots, m\}$.

Theorem 14.1.1 is useful in providing computational formulas for moments, covariances, and cumulants of Appell and Hermite polynomials, which we summarize as follows.

Corollary 14.1.2. *For the Appell polynomials $A_{k_i}(X_i)$, $i = 1, \cdots, m$, and a given table $T = \{(i,1), \cdots, (i,k_i)\}$, $i = 1, \cdots, m$, the following holds:*

$$E \prod_{i=1}^m A_{k_i}(X_i) = \sum_{\{V\} \in P_T^c} I_V, \qquad (14.1.19)$$
$$\text{Cum}(A_{k_1}(X_1), \cdots, A_{k_m}(X_m)) = \sum_{\{V\} \in \Gamma_T^c} I_V.$$

If, in addition, $X_i \sim \mathcal{N}(0,1)$, $i = 1, \cdots, m$, are jointly Gaussian, then

$$E \prod_{i=1}^{m} H_{k_i}(X_i) = \sum_{\{V\} \in P_T^c \cap G_T} I_V, \qquad (14.1.20)$$

$$\text{Cum}(H_{k_1}(X_1), \cdots, H_{k_m}(X_m)) = \sum_{\{V\} \in \Gamma_T^c \cap G_T} I_V,$$

where summation is over Gaussian partitions with I_V as in (14.1.18).

Remark 14.1.1. The Wick product of $X_1, \cdots, X_m$, can be also understood as

$$:X_1 \times \cdots \times X_m: = P_{X_1, \cdots, X_m}(x_1, \cdots, x_m)\Big|_{x_1 = X_1, \cdots, x_n = X_m},$$

where $P_{X_1, \cdots, X_m}(x_1, \cdots, x_m)$ is a polynomial in the real variables $x_1, \cdots,$ x_m whose coefficients are defined by the joint cumulants of $X_1, \cdots, X_m$ by rule (14.1.4). Similarly, $A_n(X) = A_{n,X}(x)\Big|_{x=X}$, where $A_{n,X}(x)$ is an Appell polynomial and has coefficients defined by (2.4.19).

Polynomials $A_{n,X}(x)$ and $P_{X_1, \cdots, X_n}(x_1, \cdots, x_m)$ have a number of interesting analytical properties. For example, the multilinearity property (14.1.8) implies that $\forall\, a, b, x \in \mathbb{R}$,

$$A_{n,X+a}(x+a) = A_{n,X}(x), \quad A_{n,aX}(ax) = a^n A_{n,X}(x),$$

$$P_{aX_1, bX_2}(ax_1, bx_2) = ab P_{X_1, X_2}(x_1, x_2).$$

Example 14.1.2. Let $\{X_j\}$ be a zero-mean stationary process with finite fourth moment and $A_2(X_j)$ denote the second-order Appell polynomial. By (14.1.1), (14.1.14), and (14.1.15),

$$\begin{aligned}
\text{Cov}(A_2(X_j), A_2(X_s)) &= \text{Cum}(:X_jX_j:, :X_sX_s:) \\
&= 2\text{Cum}^2(X_j, X_s) + \text{Cum}(X_j, X_j, X_s, X_s), \\
\text{Cov}(X_j, A_2(X_s)) &= \text{Cum}(:X_j:, :X_sX_s:) \\
&= \text{Cum}(X_j, X_s, X_s), \\
\text{Cov}(X_j^2, X_jX_s) &= \text{Cum}(X_j^2, X_jX_s) \\
&= 2\text{Cov}(X_j, X_j)\text{Cov}(X_j, X_s) \\
&\quad + \text{Cum}(X_j, X_j, X_s, X_s).
\end{aligned}$$

Note: The above discussion is based on Giraitis and Surgailis (1986). For more on Wick products and multivariate Appell and Hermite polynomials see Feinsilver (1978).

14.2 Sums of Appell polynomials

Let $\{X_j\}$ be the linear process (4.3.1) and $A_{k,X}$, $k \geq 1$, denote an Appell polynomial corresponding to the r.v. X_0 as defined in (2.4.19). Suppose $E|\zeta_0|^{2k} < \infty$. Then, by Corollary 2.5.1, $E|X_0|^{2k} < \infty$, and

$$Y_j := A_{k,X}(X_j), \ j \in \mathbb{Z},$$

is a stationary process with zero mean and finite variance.

Below we will show that the sums

$$S_n := \sum_{j=1}^{n} A_{k,X}(X_j)$$

of the kth Appell polynomial $A_k(X_j)$ have the same asymptotic properties as the sums of the Hermite polynomial $H_k(X_j)$ of a Gaussian process $\{X_j\}$, discussed in Proposition 4.6.2 and Theorem 4.7.1. As in Proposition 4.6.2, the CLT will be derived under condition (4.6.28) for the a_j's.

Theorem 14.2.1. *Let $\{X_j\}$ be a stationary linear process (4.3.1) with $\{\zeta_k\} \sim \mathrm{IID}(0,1)$. Assume that the a_j's satisfy (4.6.28) and $E\zeta_0^{2k} < \infty$. Then $\sum_{j\in\mathbb{Z}} |\gamma_Y(j)| < \infty$ and*

$$n^{-1/2}S_n \to_D \mathcal{N}(0,\sigma^2), \quad \sigma^2 = \sum_{j\in\mathbb{Z}} \gamma_Y(j). \qquad (14.2.1)$$

Proof. First we decompose $A_{k,X}(X_j)$ into a linear combination of polynomial forms. By (14.1.9),

$$A_{k,X}(X_j) = \sum_{s_1,\cdots,s_k \leq j} a_{j-s_1} \cdots a_{j-s_k} : \zeta_{s_1} \times \cdots \times \zeta_{s_k} : . \qquad (14.2.2)$$

By (14.1.2), $: \zeta_{s_1}^{k_1} \times \cdots \times \zeta_{s_m}^{k_m} : = A_{k_1,\zeta}(\zeta_{s_1}) \cdots A_{k_m,\zeta}(\zeta_{s_m})$ for $s_m < \cdots < s_1$, where $A_{p,\zeta}$ is the pth-order Appell polynomial corresponding to the r.v. ζ_0.

In the following, $I(k)$, $k \geq 1$, denotes the set of all indices $(k)_m := (k_1, \cdots, k_m)$, of the positive integers k_j, such that $k_1 + \cdots + k_m = k$ for some $m = 1, \cdots, k$. Define a polynomial form:

$$Q_j^{(k)m} = \sum_{-\infty < s_m < \cdots < s_1 \leq j} a_{j-s_1}^{k_1} \cdots a_{j-s_m}^{k_m} A_{k_1,\zeta}(\zeta_{s_1}) \cdots A_{k_m,\zeta}(\zeta_{s_m}).$$

Then the r.h.s. of (14.2.2) can be written as

$$Y_j = \sum_{m=1}^{k} \sum_{(k)_m \in I(k)} C_{(k)_m} Q_j^{(k)m}, \qquad (14.2.3)$$

with some constants $C_{(k)_m}$ that do not depend on j, and, for $(k)_m = (k_1, \cdots, k_m) \in I(k)$, $(t)_p = (t_1, \cdots, t_p) \in I(k)$, and $1 \le m, p \le k$,

$$\text{Cov}(Q_j^{(k)_m}, Q_0^{(t)_p}) = 0, \quad \forall j, \quad \text{if } m \ne p,$$

$$|\text{Cov}(Q_j^{(k)_m}, Q_0^{(t)_m})| \le C \prod_{i=1}^{m} \left(\sum_{s_i=0}^{\infty} |a_{j+s_i}|^{k_i} |a_{s_i}|^{t_i} \right) \quad (14.2.4)$$

$$=: Cr_{(k)_m,(t)_m}(j), \quad \forall j \ge 0,$$

where C does not depend on j. Note that

$$\gamma_Y(j) = \sum_{m=1}^{k} \sum_{(k)_m,(t)_m \in I(k)} C_{(k)_m} C_{(t)_m} \text{Cov}(Q_0^{(k)_m}, Q_j^{(t)_m}).$$

The following lemma allows us to prove the first claim of the theorem.

Lemma 14.2.1. *Suppose $\sum_{k=0}^{\infty} a_k^2 < \infty$. Then*

$$\sum_{j=0}^{\infty} r_{(k)_m,(t)_m}(j) < \infty, \quad \forall (k)_m, (t)_m \in I(k), \quad 1 \le m < k. \quad (14.2.5)$$

If, in addition, (4.6.28) holds, then (14.2.5) is satisfied for $m = k$.

Proof. Let B denote the l.h.s. of (14.2.5). When $m = k$, $B < \infty$ follows from (14.2.4) and (4.6.28). If $m < k$, then either (a) $k_i \wedge t_i \ge 2$ for some $i = 1, \cdots, m$; or (b) $k_i \ge 2, t_i = 1$ and $k_j = 1, t_j \ge 2$ for some $i, j = 1, \cdots, m, i \ne j$.

In case (a), without loss of generality, assume that both $k_1, t_1 \ge 2$. Then $\alpha_p := \sum_{s=0}^{\infty} |a_s|^p < \infty$, for $p \ge 2$, and by the C-S inequality,

$$B \le \sum_{s_1=0}^{\infty} \sum_{j=0}^{\infty} |a_{j+s_1}|^{k_1} |a_{s_1}|^{t_1} \prod_{i=2}^{m} \alpha_{2k_i}^{1/2} \alpha_{2t_i}^{1/2} \le \alpha_{k_1} \alpha_{t_1} \prod_{i=2}^{m} \alpha_{2k_i}^{1/2} \alpha_{2t_i}^{1/2} < \infty.$$

In case (b) it suffices to consider $k_1 \ge 2, t_1 = 1$ and $k_2 = 1, t_2 \ge 2$. Then

$$B \le \sum_{j=0}^{\infty} \sum_{s_1,s_2=0}^{\infty} |a_{j+s_1}|^{k_1} |a_{s_1}| |a_{j+s_2}| |a_{s_2}|^{t_2} \prod_{i=3}^{m} \alpha_{2k_i}^{1/2} \alpha_{2t_i}^{1/2}$$

$$\le \alpha_{k_1} \alpha_2 \alpha_{t_2} \prod_{i=3}^{m} \alpha_{2k_i}^{1/2} \alpha_{2t_i}^{1/2} < \infty.$$

Clearly, the above lemma implies that, under the assumptions of Theorem 14.2.1, $\sum_{j \in \mathbb{Z}} |\gamma_Y(j)| < \infty$. Similarly as in the proof of Theorem 4.8.1, it allows us to prove (14.2.1) by approximating the polynomial forms $Q_j^{(k)_m}$

and the r.v.'s Y_j by m-dependent variables. The rest of the proof follows the proof of Theorem 4.8.1 and therefore is omitted.

The conditions of the following non-central limit theorem combine assumptions on the weights a_j of the linear process $\{X_j\}$ and the order $k \geq 1$ of an Appell polynomial:

$$a_j \sim c_a j^{-1+d_X}, \quad j \to \infty, \quad 0 < d_X < 1/2, \quad c_a > 0, \qquad (14.2.6)$$
$$k(1 - 2d_X) < 1.$$

Let $2d_Y = 1 - k(1 - 2d_X)$, $c_Y := c_a^{2k} B^k(d_X, 1 - 2d_X)$, and

$$S_n(\tau) = \sum_{j=1}^{[n\tau]} A_{k,X}(X_j), \qquad 0 \leq \tau \leq 1,$$

$$Z_j := k! \sum_{-\infty < s_k < \cdots < s_1 \leq j} a_{j-s_1} \cdots a_{j-s_k} \zeta_{s_1} \cdots \zeta_{s_k}, \quad j \in \mathbb{Z}.$$

Theorem 14.2.2. *Let $\{X_j\}$ be a stationary linear process (4.3.1) with $\{\zeta_j\} \sim \text{IID}(0, 1)$. Assume that the a_j's satisfy (14.2.6) and $E\zeta_0^{2k} < \infty$. Then $Y_j := A_{k,X}(X_j)$ can be written as $Y_j = Z_j + R_j$ such that*

$$\gamma_Z(j) \sim c_Y j^{-1+2d_Y}, \qquad \sum_{j \in \mathbb{Z}} |\gamma_R(j)| < \infty, \qquad (14.2.7)$$

$$n^{-1/2-d_Y} S_n(\tau) \Rightarrow c_a^k I_k(g_\tau), \text{ in } \mathcal{D}[0,1] \text{ with uniform metric}, \quad (14.2.8)$$

where g_τ and $I_k(g_\tau)$ are the same as in Theorem 4.8.2.

Proof. Equations (14.2.2) and (14.2.3) allow us to write Y_j as

$$Y_j = Z_j + \sum_{(k)_m \in I(k):\, m < k} C_{(k)_m} Q_j^{(k)_m} =: Z_j + R_j.$$

By (14.2.4) and Lemma 14.2.1, $\sum_{j \in \mathbb{Z}} |\gamma_R(j)| < \infty$. Since $d_Y > 0$, this in turn gives

$$n^{-1} \text{Var}\Big(\sum_{j=1}^{n} R_j\Big) \leq \sum_{j \in \mathbb{Z}} |\gamma_R(j)| < \infty, \quad \forall n \geq 1, \qquad (14.2.9)$$

$$n^{-1/2-d_Y} \sum_{j=1}^{[n\tau]} R_j \to_p 0.$$

The asymptotic properties of the partial-sum process $n^{-1/2-d_Y} \sum_{j=1}^{[n\tau]} Z_j$ of the polynomial form Z_j were discussed in Theorem 4.8.2, which implies (14.2.7) for $\gamma_Z(j)$ and, together with (14.2.9), gives (14.2.8).

14.3 Multiple Wiener–Itô integrals

Multiple Wiener–Itô integrals are r.v.'s that describe limit distributions in the non-CLT for the sums of functions of a dependent Gaussian process and appear as the limits of some polynomials of i.i.d. r.v.'s. This section will define these integrals and discuss some of their properties.

Let $\{W(dx), x \in \mathbb{R}\}$ be a real-valued Gaussian random signed measure on $\mathbb{R}$ with the properties

$$EW(dx) = 0, \quad E(W(dx))^2 = dx, \quad x \in \mathbb{R}, \qquad (14.3.1)$$

$$E\{W(dx)W(dy)\} = 0, \quad x \neq y.$$

It is convenient to view it as a stochastic process $W(\cdot)$ indexed by a set $A \subset \mathbb{R}$ such that on an interval $(a, b]$, $a < b$, $W(a, b] = W(b) - W(a)$, where $\{W(x), x \in \mathbb{R}\}$ is a Wiener process on $\mathbb{R}$. Such a measure is a.s. additive but not σ-additive, and $E\{W(A_1)W(A_2)\} = \int_{A_1 \cap A_2} dx$, for any intervals A_1, A_2. Conversely, this random measure can be constructed from a Wiener process $\{W(x), x \in \mathbb{R}\}$ by putting $W(A) := W(b) - W(a)$ on the interval $A = (a, b]$ and then extending it to an arbitrary bounded Borel set $A \subset \mathbb{R}$ in the usual way.

Let $L_2(\mathbb{R}^k)$ denote the space of all complex-valued functions $f : \mathbb{R}^k \to \mathbb{C}$ with a finite norm $\|f\| = \left(\int_{\mathbb{R}^k} |f(x)|^2 d^k x \right)^{1/2} < \infty$, $d^k x := dx_1 \cdots dx_k$. Consider the symmetrization operator:

$$(\operatorname{sym} f)(x_1, \cdots, x_k) := \frac{1}{k!} \sum_{(\pi)_k} f(x_{\pi_1}, \cdots, x_{\pi_k}),$$

where the sum is taken over all permutations $(\pi)_k = (\pi_1, \cdots, \pi_k)$ of the integers $1, \cdots, k$. If f is symmetric, i.e., invariant under all permutations of its arguments, then $f = (\operatorname{sym} f)$.

We shall give a constructive definition of the *Wiener–Itô integral*

$$I_k(f) \equiv \int_{\mathbb{R}^k} f(x_1, \cdots, x_k) W(dx_1) \cdots W(dx_k), \quad k \geq 0, \qquad (14.3.2)$$

for $f \in L_2(\mathbb{R}^k)$. This is done by first defining it for simple functions and then using an approximation of f using the simple functions.

Let $(\Delta) \equiv \Delta_1 \times \cdots \times \Delta_k \subset \mathbb{R}^k$ be a "cube" in $\mathbb{R}^k$. Write $(\Delta) \in \{\Delta_M\}$, to indicate that the edges $\Delta_1, \cdots, \Delta_k \in \mathcal{J}_M$ where

$$\mathcal{J}_M := \{(\tfrac{j}{M}, \tfrac{j+1}{M}], \quad j \in \mathbb{Z}\}, \quad M = 1, 2, \cdots.$$

The set $\{\Delta_M\}$ is a partition of $\mathbb{R}^k$ into cubes with edges of size $1/M$. Let $\{\Delta_M^{\text{diag}}\}$ be the set of cubes intersecting the diagonals $\{x = (x_1, \cdots, x_k) \in \mathbb{R}^k : x_i = x_j$ for some $i \neq j\}$: $(\Delta) \in \{\Delta_M^{\text{diag}}\}$, if $\Delta_i = \Delta_j$ for some $i \neq j$.

Let $S_M(\mathbb{R}^k)$ denote the class of *simple* functions $f \in L_2(\mathbb{R}^k)$, which have constant values on a finite number of cubes (Δ) and are equal to zero on the diagonals and elsewhere, i.e., $f \in S_M(\mathbb{R}^k)$ satisfies

$$f(x) = f^{\Delta_1, \cdots, \Delta_k}, \quad x \in (\Delta) = \Delta_1 \times \cdots \times \Delta_k, \ (\Delta) \in \{\Delta_M\};$$
$$= 0, \qquad\qquad x \in (\Delta), \ (\Delta) \in \{\Delta_M^{\text{diag}}\},$$

where $f^{\Delta_1, \cdots, \Delta_k} = f^{(\Delta)}$ are complex numbers such that $f^{(\Delta)} = 0$ except for a finite number of cubes $(\Delta) \in \{\Delta_M\}$.

Let $L_2(\Omega) \equiv L_2(\Omega, \mathcal{F}_W)$ be the class of all complex-valued r.v.'s X measurable with respect to the σ-field $\mathcal{F}_W := \{W(\Delta), \Delta \in \mathcal{J}_1 \cup \mathcal{J}_2 \cup \cdots\}$, such that $E|X|^2 < \infty$.

For a simple function $f \in S_M(\mathbb{R}^k)$, the multiple Wiener–Itô integral $I_k(f)$ is defined by

$$I_k(f) := \sum_{(\Delta) \in \{\Delta_M\}} f^{\Delta_1, \cdots, \Delta_k} W(\Delta_1) \cdots W(\Delta_k). \qquad (14.3.3)$$

For $k = 0$, $S_n(\mathbb{R}^0) = L_2(\mathbb{R}^0) = \mathbb{C}$ denotes constants. Also $I_0(c) = c$ for $c \in \mathbb{C}$. For simple functions, the following properties of Wiener–Itô integral are immediate: for any $k, m = 1, 2, \cdots$ and any $f, g \in S_M(\mathbb{R}^k)$,

$$I_k(f) = I_k(\text{sym} \, f), \qquad\qquad (14.3.4)$$
$$EI_k(f) = 0, \qquad\qquad (14.3.5)$$
$$E\{I_k(f)\overline{I_m(g)}\} = k!\langle f, \text{sym} \, g\rangle, \quad k = m, \qquad (14.3.6)$$
$$= 0, \qquad\qquad k \neq m,$$

where $\langle f, g\rangle = \int_{\mathbb{R}^k} f(x)\overline{g(x)} \, d^k x$ is the scalar product in $L_2(\mathbb{R}^k)$.

Property (14.3.6) implies that for $f \in S_M(\mathbb{R}^k)$, $k, M \geq 1$,

$$E|I_k(f)|^2 \leq k!\|f\|^2, \qquad\qquad (14.3.7)$$
$$= k!\|f\|^2, \quad \text{if } f \text{ is symmetric.}$$

The integral of (14.3.3) is a bounded linear operator from $\bigcup_{M=1}^\infty S_M(\mathbb{R}^k)$ to $L_2(\Omega)$. Since $\bigcup_{M=1}^\infty S_M(\mathbb{R}^k)$ is dense in $L_2(\mathbb{R}^k)$, then for any $f \in L_2(\mathbb{R}^k)$, there exists a sequence of simple functions $f_n \in S_{M_n}(\mathbb{R}^k)$, $n \geq 1$, such that

$\|f - f_n\| \to 0$ as $n \to \infty$. Therefore, I_k extends to a bounded linear operator $I_k : L_2(\mathbb{R}^k) \mapsto L_2(\Omega)$ in the usual way, by defining

$$I_k(f) := \text{l.i.m.}_{n \to \infty} I_k(f_n),$$

where l.i.m. stands for limit in mean square. It is clear that this $I_k(f)$ satisfies (14.3.4) to (14.3.7). Since $E|I_k(f)|^2 < \infty$, L_2-convergence implies $I_k(f_n) \to_D I_k(f)$.

Observe that the first-order Wiener–Itô integral $I_1(f) \sim \mathcal{N}(0, \|f\|^2)$ has a Gaussian distribution.

The following proposition provides a criterion for the convergence of multiple Wiener–Itô integrals:

$$I_{k,n}(f_n) = \int_{\mathbb{R}^k} f_n(x_1, \cdots, x_k) W_n(dx_1) \cdots W_n(dx_k), \qquad (14.3.8)$$

where the Gaussian random measures $W_n(dx), n = 1, 2, \cdots$, may vary with n, but for each n, the measure $W_n(dx)$ has the same distributional properties as that of the Gaussian measure $W(dx)$ given by (14.3.1). Typically, such Gaussian measures W_n arise as a result of scaling a given measure W, that is, $W_n(dx) = n^{-1/2} W(ndx)$.

Proposition 14.3.1. *Let $I_k(f)$ and $I_{k,n}(f_n)$ be multiple Wiener–Itô integrals, (14.3.2) and (14.3.8), where $f, f_n \in L_2(\mathbb{R}^k)$, and W, W_n, $n \geq 1$, are identically distributed Gaussian random measures. Then*

$$\|f_n - f\| \to 0, \qquad (14.3.9)$$

implies $I_{k,n}(f_n) \to_D I_k(f)$.

Proof. By (14.3.9) and (14.3.7),

$$E|I_{k,n}(f_n) - I_{k,n}(f)|^2 \leq k! \|f_n - f\|^2 \to 0.$$

Hence, $I_{k,n}(f_n) = I_{k,n}(f) + o_p(1) =_D I_k(f) + o_p(1) \to_D I_k(f)$.

Discrete integrals. Consider an "off-diagonal" k-tuple form

$$Q_k(h) = \sum_{j_1, \cdots, j_k \in \mathbb{Z}} h(j_1, \cdots, j_k) \zeta_{j_1} \cdots \zeta_{j_k} \qquad (14.3.10)$$

of the i.i.d. r.v.'s $\{\zeta_j\} \sim \text{IID}(0, 1)$ and real square summable weights $h(j_1, \cdots, j_k)$ vanishing on the diagonals of $\mathbb{Z}^k$:

$$h(j_1, \cdots, j_k) = 0 \quad \text{if } j_i = j_m \text{ for some } i \neq m, \qquad (14.3.11)$$

$$\sum_{j_1, \cdots, j_k \in \mathbb{Z}} h^2(j_1, \cdots, j_k) < \infty.$$

Note that (14.3.11) ensures that $Q_k(h)$ is well defined, $EQ_k(h) = 0$ and $EQ_k^2(h) < \infty$. Such a sum could be viewed as a *discrete multiple integral.*

The following proposition provides a criterion for the convergence in distribution of $Q_k(h_n)$ to a multiple Wiener–Itô integral.

Proposition 14.3.2. *Let $Q_k(h_n)$, $n \geq 1$, be as in (14.3.10) and (14.3.11). Suppose the weights $h_n(j_1, \cdots, j_k)$ are such that for a real-valued $f \in L_2(\mathbb{R}^k)$ the functions*

$$\widetilde{h}_n(x_1, \cdots, x_k) := n^{k/2} h_n([x_1 n], \cdots, [x_k n]), \ x_1, \cdots, x_k \in \mathbb{R}, \quad (14.3.12)$$

satisfy

$$\|\widetilde{h}_n - f\| \to 0. \quad (14.3.13)$$

Then $Q_k(h_n) \to_D I_k(f)$.

Proof. For a function $f_\epsilon \in S_M(\mathbb{R}^k)$, $M \geq 1$, indexed by $\epsilon > 0$, define its "discretization"

$$h_{\epsilon,n}(j_1, \cdots, j_k) := n^{-k/2} f_\epsilon(\frac{j_1}{n}, \cdots, \frac{j_k}{n}), \quad j_1, \cdots, j_k \in \mathbb{Z}.$$

Note that $h_{\epsilon,n}(j_1, \cdots, j_k)$ satisfies (14.3.11). We show below that $\forall \epsilon > 0$, there exists $f_\epsilon \in S_M(\mathbb{R}^k)$, $M \geq 1$, such that, as $n \to \infty$,

$$\text{Var}(Q_k(h_n) - Q_k(h_{\epsilon,n})) \leq \epsilon, \quad (14.3.14)$$

$$Q_k(h_{\epsilon,n}) \to_D I_k(f_\epsilon), \quad (14.3.15)$$

$$\text{Var}(I_k(f_\epsilon) - I_k(f)) \leq \epsilon. \quad (14.3.16)$$

Because of Lemma 4.2.1, these results prove (14.3.13).

Note that

$$\text{Var}(Q_k(h)) \leq k! \sum_{j_1, \cdots, j_k \in \mathbb{Z}} h^2(j_1, \cdots, j_k)$$

$$= k! \int_{\mathbb{R}^k} n^k h^2([x_1 n], \cdots, [x_k n]) d^k x = k! \|\widetilde{h}_n\|^2,$$

which implies

$$\text{Var}(Q_k(h_n) - Q_k(h_{\epsilon,n})) \leq k! \|\widetilde{h}_n - \widetilde{h}_{\epsilon,n}\|^2.$$

To prove (14.3.14) it remains to find a simple function f_ϵ such that

$$k! \|\widetilde{h}_n - \widetilde{h}_{\epsilon,n}\|^2 \leq \epsilon, \quad n \to \infty. \quad (14.3.17)$$

By assumption (14.3.13), for some $n_0 \geq 1$,

$$\|\tilde{h}_n - \tilde{h}_{n_0}\|^2 \leq 2\|\tilde{h}_n - f\|^2 + 2\|f - \tilde{h}_{n_0}\|^2 \tag{14.3.18}$$
$$\leq \epsilon/(3k!), \quad \forall n \geq n_0.$$

Next, given n_0 and $\epsilon > 0$, there exists $f_\epsilon \in S_M(\mathbb{R}^k)$, $M \geq 1$, such that

$$\|\tilde{h}_{n_0} - f_\epsilon\|^2 \leq \epsilon/(3k!). \tag{14.3.19}$$

The function $\tilde{h}_{\epsilon,n}$ derived from $h_{\epsilon,n}$ by applying rule (14.3.12), satisfies

$$\|f_\epsilon - \tilde{h}_{\epsilon,n}\|^2 = \int_{\mathbb{R}^k} \left(f_\epsilon(x_1, \cdots, x_k) - f_\epsilon(\frac{[x_1 n]}{n}, \cdots, \frac{[x_k n]}{n}) \right)^2 d^k x$$
$$\to 0, \qquad n \to \infty.$$

Hence, there exists $\tilde{n}_0 \geq 1$ such that

$$\|f_\epsilon - \tilde{h}_{\epsilon,n}\|^2 \leq \epsilon/(3k!), \quad \forall n \geq \tilde{n}_0.$$

From the above relations, for all $n \geq n_0 \vee \tilde{n}_0$ we find

$$\|\tilde{h}_n - \tilde{h}_{\epsilon,n}\|^2 \leq 3\|\tilde{h}_n - \tilde{h}_{n_0}\|^2 + 3\|\tilde{h}_{n_0} - f_\epsilon\|^2 + 3\|f_\epsilon - \tilde{h}_{\epsilon,n}\|^2 \leq \epsilon/k!,$$

which proves (14.3.17). From (14.3.18) and (14.3.19) we also find

$$\mathrm{Var}(I_k(f_\epsilon) - I_k(f)) \leq k!\|f_\epsilon - f\|^2$$
$$\leq 2k!\|f_\epsilon - \tilde{h}_{n_0}\|^2 + 2k!\|\tilde{h}_{n_0} - f\|^2 \leq \epsilon,$$

thereby completing the proof of (14.3.16).

To show (14.3.15), note that

$$Q_k(h_{\epsilon,n}) = \sum_{(\Delta) \in \{\Delta_M\}} f_\epsilon^{\Delta_1, \cdots, \Delta_k}$$
$$\times n^{-k/2} \sum_{j_1, \cdots, j_k} \zeta_{j_1} \cdots \zeta_{j_k} I\left(\frac{j_1}{n} \in \Delta_1, \cdots, \frac{j_k}{n} \in \Delta_k \right)$$
$$= \sum_{(\Delta) \in \{\Delta_M\}} f_\epsilon^{\Delta_1, \cdots, \Delta_k} W_n(\Delta_1) \cdots W_n(\Delta_k),$$
$$W_n(\Delta_i) := n^{-1/2} \sum_{j:j/n \in \Delta_i} \zeta_j, \quad i = 1, \cdots, k.$$

Since $\{\zeta_j\} \sim \mathrm{IID}(0,1)$ and the intervals Δ_i are disjoint, so $W_n(\Delta_i)$, $i \in \mathbb{Z}$, are independent r.v.'s, $EW_n(\Delta_i) \equiv 0$, and

$$EW_n^2(\Delta_i) = n^{-1} \sum_{j \in \mathbb{Z}} I(\tfrac{j}{n} \in \Delta_i) \to \int_{\Delta_i} dx, \quad n \to \infty, \forall i \in \mathbb{Z}.$$

Moreover, since the ζ_j's satisfy the CLT, for any $J \geq 1$, as $n \to \infty$,

$$(W_n(\Delta_{-J}), \cdots, W_n(\Delta_J)) \to_D (W(\Delta_{-J}), \cdots, W(\Delta_J)), \qquad (14.3.20)$$

where $W(dx)$ is the Gaussian random measure appearing in the definition of the Wiener–Itô integral. Recall that by definition, a simple function f_ϵ takes non-zero values on a finite number of cubes (Δ). Hence, with some large J, we can write

$$Q_k(h_{\epsilon,n}) = P(W_n(\Delta_{-J}), \cdots, W_n(\Delta_J)),$$

where P is a polynomial with the typical term $f_\epsilon^{\Delta_1, \cdots, \Delta_k} W_n(\Delta_1) \cdots W_n(\Delta_k)$, which has a finite variance. Thus, we can verify that (14.3.20) implies

$$Q_k(h_{\epsilon,n}) \to_D P(W(\Delta_{-J}), \cdots, W(\Delta_J)) \equiv I_k(f_\epsilon),$$

which proves (14.3.15). Proposition 14.3.2 is proved.

The proof of the above proposition easily extends to the joint convergence of discrete multiple integrals, as follows.

Proposition 14.3.3. *Let $Q_{k_p}(h_n^{(p)})$, $p = 1, \cdots r$, be $r \geq 1$ polynomial forms as in (14.3.10) with coefficients $h_n^{(p)}(j_1, \cdots, j_{k_p})$ satisfying (14.3.11) and such that for some real-valued functions $f^{(p)} \in L_2(\mathbb{R}^{k_p})$, the functions*

$$\widetilde{h}_n^{(p)}(x_1, \cdots, x_{k_p}) := n^{k_p/2} h_n^{(p)}([x_1 n], \cdots, [x_{k_p} n])$$

satisfy $\|\widetilde{h}_n^{(p)} - f^{(p)}\| \to 0$, $p = 1, \cdots, r$. Then

$$\left(Q_{k_1}(h_n^{(1)}), \cdots, Q_{k_r}(h_n^{(r)})\right) \to_D \left(I_{k_1}(f^{(1)}), \cdots, I_{k_r}(f^{(r)})\right).$$

Other properties of Wiener–Itô integrals. There is a close relation between Wiener–Itô integrals, Wick products, and Hermite polynomials. To discuss this relation, we first state the following lemma.

Lemma 14.3.1. *For any $k = 0, 1, \cdots$ and any $f \in L_2(\mathbb{R}^k)$ and a real function $g \in L_2(\mathbb{R})$,*

$$E\left[I_k(f) e^{-iI_1(g)}\right] \qquad (14.3.21)$$

$$= (-i)^k e^{-\|g\|^2/2} \int_{\mathbb{R}^k} f(x_1, \cdots, x_k) \prod_{i=1}^{k} g(x_i) d^k x.$$

Proof. It suffices to prove (14.3.21) for simple functions $f \in S_M(\mathbb{R}^k)$ and real $g \in S_M(\mathbb{R})$. Consider

$$E\big[I_k(f)e^{-\mathrm{i}I_1(g)}\big]$$
$$= \sum_{(\Delta)} f^{(\Delta)} E\big[W(\Delta_1)\cdots W(\Delta_k)e^{-\mathrm{i}\sum_\Delta g^\Delta W(\Delta)}\big]$$
$$= \sum_{(\Delta)} f^{(\Delta)} E\big[e^{-\mathrm{i}\sum' g^\Delta W(\Delta)}\big]$$
$$\times \prod_{p=1}^k E\big[W(\Delta_p)\exp\{-\mathrm{i}g^{\Delta_p}W(\Delta_p)\}\big],$$

where the sum $\sum'$ is over all $\Delta \neq \Delta_1,\cdots,\Delta_k$. But, because $|\Delta|$ is the Lebesgue measure of interval Δ,

$$E\Big[W(\Delta)e^{-\mathrm{i}g^\Delta W(\Delta)}\Big] = \mathrm{i}\partial E e^{-\mathrm{i}(g^\Delta + a)W(\Delta)}/\partial a\Big|_{a=0}$$
$$= \mathrm{i}\partial \exp\big\{-(1/2)(g^\Delta + a)^2|\Delta|\big\}/\partial a\Big|_{a=0}$$
$$= -\mathrm{i}g^\Delta|\Delta|\,e^{-(1/2)|g^\Delta|^2|\Delta|}$$
$$= -\mathrm{i}g^\Delta|\Delta|\,E e^{-\mathrm{i}g^\Delta W(\Delta)}.$$

Therefore, using the independence of the r.v.'s $W(\Delta)$ on disjoint intervals Δ, we find

$$E\Big[I_k(f)e^{-\mathrm{i}I_1(g)}\Big]$$
$$= (-\mathrm{i})^k \sum_{(\Delta)} f^{(\Delta)} E\Big[e^{-\mathrm{i}\sum' g^\Delta W(\Delta)}\Big] \prod_{p=1}^k g(\Delta_p)|\Delta_p|E e^{-\mathrm{i}g^{\Delta_p}W(\Delta_p)}$$
$$= (-\mathrm{i})^k E\Big[e^{-\mathrm{i}\sum_\Delta g^\Delta W(\Delta)}\Big] \sum_{(\Delta)} f^{(\Delta)} \prod_{p=1}^k g(\Delta_p)|\Delta_p|^k$$
$$= (-\mathrm{i})^k e^{-\|g\|^2/2} \int_{\mathbb{R}^k} f(x_1,\cdots,x_k) \prod_{i=1}^k g(x_i)d^k x.$$

Lemma 14.3.1 is proved.

Let $S^{\mathrm{real}}(\mathbb{R}) = S_1^{\mathrm{real}}(\mathbb{R}) \cup S_2^{\mathrm{real}}(\mathbb{R}) \cup \cdots$ be the class of all simple real-valued functions on $\mathbb{R}$. For a positive integer m, $g_i \in S^{\mathrm{real}}(\mathbb{R})$, $c_i \in \mathbb{C}$, $i = 1,\cdots,m$, let

$$\eta := \sum_{i=1}^m c_i e^{-\mathrm{i}I_1(g_i)}. \tag{14.3.22}$$

Lemma 14.3.2. *The class of all finite linear combinations of the form η of (14.3.22) is dense in $L_2(\Omega, \mathcal{F}_W)$, i.e., for any r.v. $X \in L_2(\Omega, \mathcal{F}_W)$ and any $\epsilon > 0$, there exists a r.v. η of the type (14.3.22), such that $E|X - \eta|^2 < \epsilon$.*

Proof. Introduce the σ-fields $\mathcal{F}_n := \sigma\{W(\Delta), \Delta \in \mathcal{J}_n\}$ and $\mathcal{F}_{\cup n} := \sigma\{W(\Delta), \Delta \in \mathcal{J}_1 \cup \cdots \cup \mathcal{J}_n\}$, $n \geq 1$. Observe that $\mathcal{F}_{\cup n}$ is monotone non-decreasing and $L_2(\Omega, \mathcal{F}_n) \subset L_2(\Omega, \mathcal{F}_{\cup n}) \subset L_2(\Omega, \mathcal{F}_W)$.

Given a r.v. $X \in L_2(\Omega, \mathcal{F}_W)$, the sequence $X_n := E[X|\mathcal{F}_{\cup n}]$ has the property $X_{n-1} = E[X_n|\mathcal{F}_{\cup(n-1)}]$. Hence, $\{X_n, \mathcal{F}_{\cup n}\}$ is a square integrable martingale and $X_n \to X$ a.s. and in $L_2(\Omega, \mathcal{F}_W)$. Therefore $\cup_{n=1}^{\infty} L_2(\Omega, \mathcal{F}_{\cup n})$ is dense in $L_2(\Omega, \mathcal{F}_W)$. Since $\mathcal{F}_{\cup n} \subset \mathcal{F}_N$ with $N = n!$, then $\cup_{n=1}^{\infty} L_2(\Omega, \mathcal{F}_n)$ is also dense in $L_2(\Omega, \mathcal{F}_W)$.

Hence, it suffices to show that the class of variables η (14.3.22) is dense in $L_2(\Omega, \mathcal{F}_n)$ for any $n \geq 1$. Define the σ-fields $\mathcal{F}_{n,J} := \sigma\{W(\Delta_j), \Delta_j := (j/n, (j+1)/n] \in \mathcal{J}_n, j = -J, -J+1, \cdots, J\}$, $J \geq 1$. Then $\cup_{J=1}^{\infty} L_2(\Omega, \mathcal{F}_{n,J})$ is dense in $L_2(\Omega, \mathcal{F}_n)$, and we can restrict the proof to $X \in L_2(\Omega, \mathcal{F}_{n,J})$, with fixed $J \geq 1$. Set $W_j := W(\Delta_j)$, $|j| \leq J$. Observe that the W_j's are independent Gaussian r.v.'s, with zero mean and variance $\sigma^2 := EW_j^2 = 1/n$. For notational simplicity, assume below that $\sigma^2 = 1$. Then the above X can be written as $X = f(W_{-J}, \cdots, W_J)$, for some $f \in L_2(\mathbb{R}^{2J+1}, d\mu^{2J+1})$, where $\mu(dx) = \phi(x)dx$ is the standard Gaussian measure. Since $L_2(\mathbb{R}^{2J+1}, d\mu^{2J+1})$ contains a dense subset

$$\left\{ f : f = \sum_{i=1}^{p} f_{i,-J}(x_1) \cdots f_{i,J}(x_J), \quad f_{ij} \in L_2(\mathbb{R}, d\mu), \ p \geq 1 \right\},$$

the above approximation problem reduces to the univariate case $J = 0$ or to $f \in L_2(\mathbb{R}, d\mu)$. Namely, it remains to show that the set of trigonometric polynomials

$$\eta(x) := \sum_{i=1}^{m} c_i e^{-ig_i x}, \ c_i \in \mathbb{C}, \ g_i \in \mathbb{R}, \ m = 1, 2, \cdots$$

is dense in $L_2(\mathbb{R}, d\mu)$, which follows from two well-known facts. First, any function $f \in L_2(\mathbb{R}, d\mu)$ can be approximated, in the $L_2(\mathbb{R}, d\mu)$-norm, by a continuous function $g(x), x \in \mathbb{R}$, with a compact support, and second, any continuous function g, in any interval $[-K, K]$ containing the support of g, can be uniformly approximated by a trigonometric polynomial (Fejér's theorem) that periodically extends to $|x| \geq K$. This completes the proof of Lemma 14.3.2.

The tensor product $f \otimes g \in L_2(\mathbb{R}^{k+m})$ of functions $f \in L_2(\mathbb{R}^k)$ and $g \in L_2(\mathbb{R}^m)$ is defined by

$$(f \otimes g)(x_1, \cdots, x_k, x_{k+1}, \cdots, x_{k+m})$$
$$= f(x_1, \cdots, x_k)g(x_{k+1}, \cdots, x_{k+m}).$$

Write $f^{\otimes k} = f \otimes \cdots \otimes f$ (k times).

The following proposition shows that if the integrand of a multiple Wiener–Itô integral is a tensor product of univariate functions, then the integral is a multivariate Hermite polynomial.

Proposition 14.3.4. *Let* $\psi_p \in L_2(\mathbb{R}), p = 1, \cdots, m$, *be a finite orthonormal set of real-valued functions:*

$$\langle \psi_p, \psi_q \rangle = \int_{\mathbb{R}} \psi_p(x)\psi_q(x)dx = 1, \quad p = q, \tag{14.3.23}$$
$$= 0, \quad p \neq q.$$

In addition, suppose $k_p \geq 1$, $p = 1, \cdots, m$, *is a collection of non-negative integers and set* $k := k_1 + \cdots + k_m$. *Then*

$$I_k \left(\psi_1^{\otimes k_1} \otimes \cdots \otimes \psi_m^{\otimes k_m} \right) = \prod_{p=1}^{m} H_{k_p}(I_1(\psi_p)). \tag{14.3.24}$$

In particular, for any real ψ *with* $\|\psi\|^2 = 1$,

$$I_k(\psi^{\otimes k}) = H_k(I_1(\psi)), \quad k = 1, 2, \cdots.$$

Proof. For notational simplicity we limit the proof to the case $m = 2$. Let $\xi_1 := I_{k_1+k_2} \left(\psi_1^{\otimes k_1} \otimes \psi_2^{\otimes k_2} \right)$ and $\xi_2 := H_{k_1}(I_1(\psi_1))H_{k_2}(I_1(\psi_2))$. Equality $\xi_1 = \xi_2$ in $L_2(\Omega)$ is equivalent to

$$E\xi_1\eta = E\xi_2\eta \tag{14.3.25}$$

for any η from a dense subset of $L_2(\Omega, \mathcal{F}_W)$, in particular, for any η as in (14.3.22). Hence, it suffices to verify (14.3.25) for any $\eta = e^{-iI_1(g)}$, $g \in S^{\text{real}}(\mathbb{R})$. Then, according to (14.3.21),

$$E\xi_1\eta = (-i)^{k_1+k_2} e^{-\|g\|^2/2} \langle \psi_1, g \rangle^{k_1} \langle \psi_2, g \rangle^{k_2}. \tag{14.3.26}$$

To evaluate the r.h.s. of (14.3.25), first observe that using the generating function of Hermite polynomials; see (2.4.3), we find

$$E\left[\prod_{i=1}^{2}\left(\sum_{k=0}^{\infty}\frac{u_i^k}{k!}H_k\left(I_1(\psi_i)\right)\right)e^{-iI_1(g)}\right] \qquad (14.3.27)$$

$$= Ee^{u_1I_1(\psi_1)+u_2I_1(\psi_2)-(u_1^2+u_2^2)/2-iI_1(g)}$$

$$= e^{\left\{-u_1^2-u_2^2+E[u_1I_1(\psi_1)+u_2I_1(\psi_2)-iI_1(g)]^2\right\}/2}$$

$$= e^{-iu_1\langle\psi_1,g\rangle-iu_2\langle\psi_2,g\rangle-\|g\|^2/2},$$

since $E[u_1I_1(\psi_1) + u_2I_1(\psi_2) - iI_1(g)]^2 = u_1^2\|\psi_1\|^2 + u_2^2\|\psi_2\|^2 + i^2\|g\|^2 - 2u_1u_2\langle\psi_1,\psi_2\rangle - 2i\langle\psi_1,g\rangle - 2i\langle\psi_2,g\rangle = u_1^2 + u_2^2 - \|g\|^2 - 2i\langle\psi_1,g\rangle - 2i\langle\psi_2,g\rangle$, see (14.3.23). The required equality (14.3.26) follows by taking partial derivatives $\frac{\partial^{k_1+k_2}}{\partial u_1^{k_1}\partial u_2^{k_2}}$ on both sides of (14.3.27) at $u_1 = u_2 = 0$, thereby completing the proof of Proposition 14.3.4.

Corollary 14.3.1. *For any real function $f \in L_2(\mathbb{R})$,*

$$I_k\left(f^{\otimes k}\right) = \|f\|^k H_k\left(\frac{I_1(f)}{\|f\|}\right), \quad k \geq 1.$$

The following corollary relates Wiener–Itô integrals to Wick products of Gaussian variables and also provides an explanation of the fact that in the physics literature, these notions are sometimes considered as synonymous.

Corollary 14.3.2. *For any $f_i \in L_2(\mathbb{R}), i = 1, \cdots, n,$*

$$I_n(f_1 \otimes \cdots \otimes f_n) = :I_1(f_1) \times \cdots \times I_1(f_n): . \qquad (14.3.28)$$

Proof. By the (multi)linearity property of both sides in (14.3.28), it suffices to prove (14.3.28) for f_i's taken from an orthonormal set of functions $\psi_p \in L_2(\mathbb{R}), p = 1, \cdots, m, m \leq n$, as in Proposition 14.3.4. For these f_i's, both sides of (14.3.28) are equal to the same product of Hermite polynomials, see (14.3.24), (14.1.2), and (14.1.1).

Corollary 14.3.3. *Any r.v. $X \in L_2(\Omega, \mathcal{F}_W)$ can be expanded as a series of multiple Wiener–Itô integrals: There exist $f_k \in L_2(\mathbb{R}^k), k \geq 0$, with $f_0 = I_0(f_0) = EX$, such that*

$$X = \sum_{k=0}^{\infty}\frac{1}{k!} I_k(f_k), \quad in \ L_2(\Omega, \mathcal{F}_W).$$

Moreover,

$$E|X|^2 = \sum_{k=0}^{\infty}\frac{1}{(k!)^2} E|I_k(f_k)|^2 = \sum_{k=0}^{\infty}\frac{1}{k!} \|\text{sym}f_k\|^2 < \infty. \qquad (14.3.29)$$

Proof. Because of Lemma 14.3.2, it suffices to show that any r.v. η in (14.3.22) can be expanded in a series of such integrals, in particular, that for any real function $g \in L_2(\mathbb{R})$,

$$e^{iI_1(g)} = e^{-\|g\|^2/2} \sum_{k=0}^{\infty} \frac{i^k}{k!} I_k(g^{\otimes k}), \qquad (14.3.30)$$

where the series converges in $L_2(\Omega, \mathcal{F}_W)$. Let $Z := I_1(g)/\|g\| \sim \mathcal{N}(0,1)$. By Corollary 14.3.1, the r.h.s. of (14.3.30) coincides with the Hermite expansion in (2.4.3) of the exponent e^{iZu}, $u = \|g\|$, which converges in $L_2(\mathbb{R}, \Phi)$. Relation (14.3.29) follows from the orthogonality property (14.3.6).

The next proposition gives formulas for a change of variables for multiple Wiener–Itô integrals. For $f \in L_2(\mathbb{R}^k)$, $\alpha \in \mathbb{R}$, and $\beta > 0$, let

$$(T_\alpha f)(x_1, \cdots, x_k) := f(x_1 + \alpha, \cdots, x_k + \alpha),$$
$$(R_\beta f)(x_1, \cdots, x_k) := f(\beta x_1, \cdots, \beta x_k).$$

Clearly, T_a and R_β are linear bounded operators in $L_2(\mathbb{R}^k)$.

Proposition 14.3.5. *For any $f \in L_2(\mathbb{R}^k)$, $a \in \mathbb{R}$, and $b > 0$,*

$$I_k(T_\alpha f) =_D I_k(f), \quad I_k(R_\beta f) =_D \beta^{-k/2} I_k(f). \qquad (14.3.31)$$

Proof. Since linear combinations of products such as $f = \prod_{p=1}^{k} I(a_p < x_p \le b_p)$ of indicators of disjoint intervals $(a_p, b_p] \cap (a_q, b_q] = \emptyset$, $p \ne q$, are dense in $L_2(\mathbb{R}^k)$, it suffices to prove the proposition for such f. For these indicator functions, $R_\beta(f) = \prod_{p=1}^{k} I(a_p/\beta < x_p \le b_p/\beta)$, $I_k(f) = \prod_{p=1}^{k} W((a_p, b_p])$, and $I_k(R_\beta f) = \prod_{p=1}^{k} W((a_p/\beta, b_p/\beta])$. Then the second equality in (14.3.31) follows from the scaling property of white noise:

$$(W((a_p, b_p]), \, p = 1, \cdots, k) =_D \left(\beta^{1/2} W((a_p/\beta, b_p/\beta]), \, p = 1, \cdots, k \right).$$

The first equality in (14.3.31) follows exactly in the same way from the translation invariance property:

$$(W((a_p, b_p]), \, p = 1, \cdots, k) =_D (W((a_p - \alpha, b_p - \alpha]), \, p = 1, \cdots, k).$$

Diagrams and moments of Wiener–Itô integrals. Multiple Wiener–Itô integrals have finite moments of arbitrary order. Moments and cumulants of Wiener–Itô integrals can be written as sums of integrals, which are closely related to the diagram formula of Theorem 14.1.1.

Let $I_{n_1}(f_1), \cdots, I_{n_k}(f_k)$, $f_i \in L_2(\mathbb{R}^{n_i})$, $i = 1, 2, \cdots, k$, be a collection of Wiener–Itô integrals of respective orders $n_1, \cdots, n_k$. Consider the tensor product

$$F := f_1 \otimes \cdots \otimes f_k \in L_2(\mathbb{R}^m), \qquad m := n_1 + \cdots + n_k.$$

It is convenient to write the arguments of the function F in the form of a table T having k rows $T_1, \cdots, T_k$, where

$$T_i = \{x_{i,1}, \cdots, x_{i,n_i}\}, \qquad i = 1, \cdots, k. \tag{14.3.32}$$

Recall the definitions of the classes P_T^c, Γ_T^c, and G_T of partitions, or diagrams, of T from (14.1.10) and Corollary 14.1.1.

Let $m = |T|$ be even. With each partition $\{V\} = (V_1, \cdots, V_{m/2}) \in P_T^c \cap G_T$, associate a function $F_{\{V\}}$ of $m/2$ real variables $y_1, \cdots, y_{m/2}$ defined by replacing each connected pair $V_j = (x_{i_1,j_1}, x_{i_2,j_2})$ of variables of F by a single variable y_j, $j = 1, \cdots, m/2$. For example, let $k = 3, n_1 = 2, n_2 = n_3 = 1$ and $\{V\} = (V_1, V_2), V_1 = (x_{1,1}, x_{2,1}), V_2 = (x_{1,2}, x_{3,1})$. Then

$$F_{\{V\}}(y_1, y_2) = f_1(y_1, y_2)f_2(y_1)f_3(y_2). \tag{14.3.33}$$

The connected variables cannot belong to the same row by the definition of P_T^c.

Proposition 14.3.6. *Let $f_i \in L_2(\mathbb{R}^{n_i})$, $i = 1, 2, \cdots, k$, and let T be the table having k rows T_i of (14.3.32).*

(i) *Suppose $m := |T|$ is even. Then*

$$E\{I_{n_1}(f_1) \times \cdots \times I_{n_k}(f_k)\} = \sum_{\{V\} \in P_T^c \cap G_T} \int_{\mathbb{R}^{m/2}} F_{\{V\}} d^{m/2}y, \tag{14.3.34}$$

$$\mathrm{Cum}(I_{n_1}(f_1), \cdots, I_{n_k}(f_k)) = \sum_{\{V\} \in \Gamma_T^c \cap G_T} \int_{\mathbb{R}^{m/2}} F_{\{V\}} d^{m/2}y. \tag{14.3.35}$$

Moreover, for any partition $\{V\} \in P_T^c \cap G_T$,

$$\int_{\mathbb{R}^{m/2}} |F_{\{V\}}(y_1, \cdots, y_{m/2})| d^{m/2}y \leq \prod_{i=1}^{k} \|f_i\| < \infty. \tag{14.3.36}$$

(ii) *Suppose $m := |T|$ is odd. Then*

$$E\{I_{n_1}(f_1) \times \cdots \times I_{n_k}(f_k)\} = \mathrm{Cum}(I_{n_1}(f_1), \cdots, I_{n_k}(f_k)) = 0.$$

Proof. (i) The proof of (14.3.36) is standard and follows by repeated application of the C-S inequality. For example, for $F_{\{V\}}$ in (14.3.33), we find

$$\int \int |F_{\{V\}}(y_1, y_2)| dy_1 dy_2$$
$$\leq \left(\int f_2^2(y_1) dy_1 \right)^{1/2} \int \left(\int f_1^2(y_1, y_2) dy_1 \right)^{1/2} |f_3(y_2)| dy_2$$
$$\leq \|f_2\| \|f_1\| \|f_3\|.$$

Next, we shall prove (14.3.34) and (14.3.35) for simple functions $f_i \in S_M(\mathbb{R}^{n_i})$, $i = 1, \cdots, k$. Since $I_n(f)$ is linear in f, it suffices to prove this for the indicators $f_i = I(\Delta_{i,1} \times \cdots \times \Delta_{i,n_i})$, $\Delta_{i,j} \in \mathcal{J}_M, \Delta_{i,j} \cap \Delta_{i,\ell} = \emptyset \ j \neq \ell, i = 1, \cdots, k$. For these f_i, $I_n(f_i) = \prod_{j=1}^{n_i} W(\Delta_{i,j}) =: W(\Delta_{i,1}) \times \cdots W(\Delta_{i,n_i})$: by independence of $W(\Delta_{i,j})$, $j = 1, \cdots, n_i$, see (14.1.2). Thus, the l.h.s.'s of (14.3.34) and (14.3.35) satisfy the diagram formulas in (14.1.14) and (14.1.15), respectively. To complete the proof, set $X_{i,j} := W(\Delta_{i,j})$, $(i,j) \in T$. For any partition $\{V\} = (V_1, \cdots, V_{m/2}) \in P_T^c \cap G_T$, $V_s = (\{i_s, j_s\}, \{i'_s, j'_s\})$, $s = 1, \cdots, m/2$, the contribution I_V in (14.1.11) equals 0 unless $\Delta_{i_s, j_s} = \Delta_{i'_s, j'_s}$ for each $s = 1, \cdots, m/2$, in which case $I_V = |\Delta|^{m/2}$, where $|\Delta| = |\Delta_{i,j}|$ is the Lebesgue measure. This proves the diagram formulas (14.3.34) and (14.3.35) for simple functions f_i. The extension of these formulas to a general $f_i \in L_2(\mathbb{R}^{n_i})$ follows by a standard limit argument using the bound in (14.3.36).

Claim (ii) follows from a similar argument as in (i) and the fact that $G_T = \emptyset$ when $|T| = m$ is odd.

Hermite processes. The Hermite process $\mathcal{H}_k = \{\mathcal{H}_k(t), t \geq 0\}$ of order $k = 1, 2, \cdots$ is defined as a multiple Wiener–Itô integral

$$\mathcal{H}_k(t) := a_{k,d} \int_{\mathbb{R}^k} \left\{ \int_0^t \prod_{i=1}^k (s - x_i)_+^{d-1} ds \right\} W(dx_1) \cdots W(dx_k), \quad (14.3.37)$$

where d is a real parameter and the constant $a_{k,d}$ is defined through the normalization condition $E\mathcal{H}_k^2(1) = 1$.

The process $\mathcal{H}_k$ appeared in the non-CLT of Corollary 4.7.2 as the limit of the partial-sum process of the non-linear transformations $h(X_j), j = 1, \cdots, n$, of a long memory stationary Gaussian process $\{X_j\}$. We now check that $\mathcal{H}_k$ is a self-similar process and has stationary increments, as is indirectly indicated by Theorem 3.4.1.

Proposition 14.3.7. *For each $k \geq 1$, the process $\mathcal{H}_k$ of (14.3.37)*

(i) *is well defined if and only if $\frac{1}{2} - \frac{1}{2k} < d < 1/2$;*

(ii) *is self-similar with parameter $H = 1 - k(\frac{1}{2} - d)$, has stationary increments, and is a.s. continuous;*

(iii) *is such that $\mathcal{H}_1$ is a fractional Brownian motion B_H with $H = \frac{1}{2} + d$.*

Proof. (i) Rewrite $\mathcal{H}_k(t) = I_k(f_t)$, where

$$f_t(x_1, \cdots, x_k) = a_{k,d} \int_0^t \prod_{i=1}^k (s - x_i)_+^{d-1} ds. \qquad (14.3.38)$$

Then

$$\|f_1\|^2 = 2a_{k,d}^2 \int_0^1 ds \int_0^{1-s} du \left(\int_0^\infty x^{d-1}(u+x)^{d-1} dx \right)^k$$

$$= 2a_{k,d}^2 B^k(d, 1 - 2d) \int_0^1 ds \int_0^{1-s} u^{(2d-1)k} du$$

$$= \frac{2a_{k,d}^2 B^k(d, 1 - 2d)}{(1 - k(1 - 2d))(2 - k(1 - 2d))},$$

since $\int_0^\infty x^{d-1}(1 + x)^{d-1} dx = B(d, 1 - 2d)$. The above calculation is legitimate if $0 < d < 1/2$ and $k(1 - 2d) < 1$, or $d \in (\frac{1}{2} - \frac{1}{2k}, \frac{1}{2})$, otherwise $\|f_1\|^2 = \infty$. Since f_1 is symmetric, from (14.3.6) we also obtain an expression for the normalizing constant in (14.3.37):

$$a_{k,d}^2 = \frac{(1 - k(1 - 2d))(2 - k(1 - 2d))}{2k! B^k(d, 1 - 2d)}. \qquad (14.3.39)$$

Proof of (ii). The self-similarity property with parameter H of $\mathcal{H}_k$ follows from Proposition 14.3.5 and the scaling property $f_{\beta t} = \beta^{1+k(d-1)} R_{1/\beta} f_t$ of f_t in (14.3.38). This and (14.3.31) gives $\{\mathcal{H}_k(\beta t)\} =_{fdd} \beta^H \{\mathcal{H}_k(t)\}$, $\forall \beta > 0$, with $H = 1 - k(\frac{1}{2} - d)$.

The fact that $\mathcal{H}_k$ has stationary increments follows from Proposition 14.3.5 and the property of increments $f_{t+\alpha} - f_\alpha = T_{-\alpha} f_t$. The a.s. continuity follows from this, (ii), and the tightness criterion in (4.4.5) with $\gamma = 1$, since for any $t, \tau \geq 0$, $E(\mathcal{H}_k(t+\tau) - \mathcal{H}_k(t))^2 = E\mathcal{H}_k^2(\tau) = \tau^{2d+1}$ with $2d + 1 > 1$.

Proof of (iii). This follows from (ii) and the fact that $\mathcal{H}_1$ is a Gaussian process.

Remark 14.3.1. Part (iii) of Proposition 14.3.7 provides a stochastic integral representation of fractional Brownian motion (fBm) $B_{d+1/2} = \mathcal{H}_1$, $0 < d < 1/2$. More generally, for $-1/2 < d < 1/2$, $d \neq 0$,

$$B_{d+1/2}(t) = \frac{a_{1,d}}{d} \int_{\mathbb{R}} \left((t-x)_+^d - (-x)_+^d \right) W(dx). \qquad (14.3.40)$$

Note that for any $0 < |d| < 1/2$,

$$\frac{d^2}{a_{1,d}^2} = \frac{d\, B(d, 1-2d)}{1+2d} = \int_{\mathbb{R}} \left((1-x)_+^d - (-x)_+^d \right)^2 dx.$$

For $d = 0$, representation (14.3.40) is

$$B_{1/2}(t) = W((0,t]) = \int_{(0,t]} W(dx).$$

Observe that $\lim_{d \to 0} d\, B(d, 1-2d)/(1+2d) = 1$.

Note: The above discussion has benefitted from Major (1981), where some additional properties of these integrals and their applications can be found. See also Dobrushin (1979), Dobrushin and Major (1979), and Taqqu (1979).

14.4 Limit of traces of Toeplitz matrices

In this section we will study the asymptotic behavior of the trace $\operatorname{tr}(B_1 R_1 B_2 R_2)$ of a product of four Toeplitz matrices

$$B_j := \left(b_{j,t-s} \right)_{t,s=1,\cdots,n}, \qquad R_j := \left(r_{j,t-s} \right)_{t,s=1,\cdots,n}, \qquad j = 1, 2,$$

with the entries

$$b_{j,t} = \int_{\Pi} e^{itx} h_j(x) dx, \qquad r_{j,t} = \int_{\Pi} e^{itx} f_j(x) dx, \quad j = 1, 2, \ t \in \mathbb{Z}. \quad (14.4.1)$$

The functions $h_j \in L_1(\Pi)$ and $f_j \in L_1(\Pi)$ are even and possibly complex-valued, i.e., $\overline{h_j(x)} = h_j(-x)$ and $\overline{f_j(x)} = f_j(-x)$, $\forall x \in \Pi$, so that the $n \times n$ matrices B_j, R_j, $j = 1, 2$, are real, although not necessary symmetric.

Consider the following two sets of assumptions.

$$f_1 \in L_{q_1}(\Pi), \ f_2 \in L_{q_2}(\Pi), \ h_1 \in L_{q_3}(\Pi), \ h_2 \in L_{q_4}(\Pi), \qquad (14.4.2)$$

for some $q_j \geq 1$, $j = 1, 2, 3, 4$, $1/q_1 + 1/q_2 + 1/q_3 + 1/q_4 \leq 1$.

$$|f_j(u)| \leq C|u|^{-\alpha_j}, \ |h_j(u)| \leq C|u|^{-\beta_j}, \ \forall u \in \Pi, \ j = 1, 2, \qquad (14.4.3)$$

for some $-1 < \alpha_j, \beta_j < 1$, $j = 1, 2$, such that

$$\alpha_1 \alpha_2 \geq 0, \ \beta_1 \beta_2 \geq 0, \ \alpha_1 + \alpha_2 + \beta_1 + \beta_2 < 1.$$

We are now ready to state the following:

Lemma 14.4.1. *Under either (14.4.2) or (14.4.3),*

$$n^{-1}\mathrm{tr}\big(B_1 R_1 B_2 R_2\big) \to (2\pi)^3 \int_\Pi \prod_{j=1}^2 (f_j(u)h_j(u))du. \qquad (14.4.4)$$

Remark 14.4.1. The above result is useful for studying the asymptotics of quadratic forms of stationary variables. For example, let $\{X_{1,t}, X_{2,t}, t \in \mathbb{Z}\}$ and $\{Y_{1,t}, Y_{2,t}, t \in \mathbb{Z}\}$ be two mutually orthogonal bivariate stationary processes with zero means, $EX_{i,t}Y_{j,s} \equiv 0, \forall\, i,j = 1,2,\, t,s \in \mathbb{Z}$, and cross-covariances

$$r_{ij,t-s}^X := \mathrm{Cov}(X_{i,s}, X_{j,t}), \quad r_{ij,t-s}^Y := \mathrm{Cov}(Y_{i,s}, Y_{j,t}), \quad i,j = 1,2.$$

Consider the quadratic forms

$$\mathcal{Q}_j := \sum_{t,s=1}^n b_{j,t-s} X_{j,t} Y_{j,s}, \quad j = 1,2,$$

with real coefficients $b_{j,t}, j = 1,2$. Then

$$\mathrm{Cov}(\mathcal{Q}_1, \mathcal{Q}_2) = \sum_{t_1,s_1,t_2,s_2=1}^n b_{1,t_1-s_1} r_{12,t_2-t_1}^X b_{2,t_2-s_2} r_{12,s_2-s_1}^Y \qquad (14.4.5)$$

$$= \mathrm{tr}\big(B_1 R_{12}^X B_2' R_{12}^Y\big),$$

where $B_2' := \big(b_{2,s-t}\big)_{t,s=1,\cdots,n}$ is the transposed matrix. Let f_1 and f_2 in (14.4.1) be the cross-spectral densities of $\{X_{1,t}\}, \{X_{2,t}\}$ and $\{Y_{1,t}\}, \{Y_{2,t}\}$, respectively. Note that they may be complex-valued. Then, the variance of the difference $\mathcal{Q}_1 - \mathcal{Q}_2$ can be written as

$$\mathrm{Var}(\mathcal{Q}_1 - \mathcal{Q}_2) \qquad\qquad\qquad\qquad\qquad\qquad (14.4.6)$$

$$= \mathrm{Var}(\mathcal{Q}_1) - 2\mathrm{Cov}(\mathcal{Q}_1, \mathcal{Q}_2) + \mathrm{Var}(\mathcal{Q}_2)$$

$$= \mathrm{tr}\big(B_1 R_{11}^X B_1' R_{11}^Y\big) - 2\mathrm{tr}\big(B_1 R_{12}^X B_2' R_{12}^Y\big) + \mathrm{tr}\big(B_2 R_{22}^X B_2' R_{22}^Y\big).$$

In particular, we may want to approximate $\mathcal{Q}_1$ by a "simpler" quadratic form $\mathcal{Q}_2$ whose asymptotic behavior is easier to obtain, and such that the variance on the l.h.s. of (14.4.6) is asymptotically negligible with respect to the variance of $\mathcal{Q}_1$. According to Lemma 14.4.1 above, such an approximation is possible provided all traces on the r.h.s. of (14.4.6) tend to the same limit in (14.4.4). This approach is actually used in the proofs of the CLTs for quadratic forms in Sections 6.3, 11.4, and 13.4.

Proof of Lemma 14.4.1. Recall the definitions of the signed measure μ_n on Π^4 and the bilinear form $\langle \cdot, \cdot \rangle_n$ on $L_2(\Pi^3)$ from (6.3.14) and (6.3.21). Let $\mu_n(F) := \int_{\Pi^4} F(u) d\mu_n(u)$, where

$$F(u) = F(u_1, \cdots, u_4) := f_1(u_1) f_2(u_2) h_1(u_3) h_2(u_4). \qquad (14.4.7)$$

Note that

$$\begin{aligned}
\mathrm{tr}(B_1 R_1 B_2 R_2) &= \sum_{k_1, k_2, k_3, k_4 \in \mathbb{Z}} b_{1, k_1 - k_2} r_{1, k_2 - k_3} b_{2, k_3 - k_4} r_{2, k_4 - k_1} \\
&= (2\pi)^2 n\, \mu_n(F). \qquad (14.4.8)
\end{aligned}$$

Suppose (14.4.2) holds. For $K > 0$, define the truncated functions

$$\begin{aligned}
f_j^K(u) &:= f_j(u) I(|f_j(u)| < K), \\
h_j^K(u) &:= h_j(u) I(|h_j(u)| < K), \quad j = 1, 2, \\
F^K(u_1, u_2, u_3, u_4) &:= f_1^K(u_1) f_2^K(u_2) h_1^K(u_3) h_2^K(u_4).
\end{aligned}$$

By (14.4.8) and the linearity of $\mu_n(\cdot)$,

$$(2\pi)^{-2} n^{-1} \mathrm{tr}(B_1 R_1 B_2 R_2) = \mu_n(F^K) + \rho_{n,K},$$

$$\rho_{n,K} \le C \sum_{j=1}^{4} \int_{\Pi^4} |F|\, I(B_{K,j})\, d|\mu_n| \; =: \; C \sum_{j=1}^{K} I_{j,K},$$

where

$$\begin{aligned}
B_{K,1} &:= \{u \in \Pi^4 : |f_1(u_1)| \ge K\}, & B_{K,2} &:= \{u \in \Pi^4 : |f_2(u_2)| \ge K\}, \\
B_{K,3} &:= \{u \in \Pi^4 : |h_1(u_3)| \ge K\}, & B_{K,4} &:= \{u \in \Pi^4 : |h_2(u_4)| \ge K\}.
\end{aligned}$$

By Lemma 6.3.3, as $K \to \infty$,

$$\begin{aligned}
\lim_{K \to \infty} \lim_n \mu_n(F^K) &= 2\pi \lim_{K \to \infty} \int_{\Pi} f_1^K(u) f_2^K(u) h_1^K(u) h_2^K(u) du \\
&= 2\pi \int_{\Pi} f_1(u) f_2(u) h_1(u) h_2(u) du.
\end{aligned}$$

Hence, (14.4.4) will follow from

$$\lim_{K \to \infty} \lim_{n \to \infty} \sup I_{j,K} = 0, \qquad j = 1, 2, 3, 4. \qquad (14.4.9)$$

We shall prove (14.4.9) for $j = 1$ only. The proofs for the other values of j are similar. By Hölder's inequality, with $\sum_{j=1}^{4} q_j^{-1} = 1$, $q_j > 1$,

$$\begin{aligned}
I_{1,K} &\le \left(\int_{\Pi^4} |f_1(u_1)|^{q_1} I(B_{K,1})\, d|\mu_n| \right)^{1/q_1} \left(\int_{\Pi^4} |f_2(u_2)|^{q_2}\, d|\mu_n| \right)^{1/q_2} \\
&\quad \times \left(\int_{\Pi^4} |h_1(u_3)|^{q_3}\, d|\mu_n| \right)^{1/q_3} \left(\int_{\Pi^4} |h_2(u_4)|^{q_4}\, d|\mu_n| \right)^{1/q_4}.
\end{aligned}$$

By Lemma 6.3.3, as $K \to \infty$,

$$I_{1,K} \leq C \Big(\int_{\Pi} |f_1(u_1)|^{q_1} I(B_{K,1}) \, du_1 \Big)^{1/q_1} \Big(\int_{\Pi^4} |f_2(u_2)|^{q_2} \, du_2 \Big)^{1/q_2}$$

$$\times \Big(\int_{\Pi^4} |h_1(u_3)|^{q_3} \, du_3 \Big)^{1/q_3} \Big(\int_{\Pi^4} |h_2(u_4)|^{q_4} \, du_4 \Big)^{1/q_4} \to 0.$$

This proves (14.4.9) and also completes the proof of (14.4.4) under (14.4.2).

Now suppose (14.4.3) holds. The case $\alpha_j \geq 0$ and $\beta_j \geq 0$, $j = 1, 2$, follows from the previous case, since here f_j and h_j, $j = 1, 2$, satisfy (14.4.2) with $q_1 < 1/\alpha_1$, $q_2 < 1/\alpha_2$, $q_3 < 1/\beta_1$, and $q_4 < 1/\beta_4$.

Now suppose $\text{sgn}(\alpha_j) \neq \text{sgn}(\beta_j)$, $j = 1, 2$. By symmetry, it suffices to consider the case $\alpha_1 \geq 0, \alpha_2 \geq 0$, $\beta_1 \leq 0$, and $\beta_2 \leq 0$ only. Again, it suffices to prove (14.4.9) for $j = 1$. As f_1 is bounded outside a neighborhood of the origin $u = 0$, there is a sequence $\epsilon = \epsilon(K) \to 0$, $K \to \infty$, such that

$$\{|f_1(u)| \geq K\} \subset \{|u| < \epsilon\}.$$

Set $W_1 := \{u \in \Pi^4 : |u_1| \leq \epsilon, |u_2| > 2\epsilon\}$ and $W_2 := \{u \in \Pi^4 : |u_1| \leq \epsilon, |u_2| \leq 2\epsilon\}$. Then

$$I_{1,K} \leq \sum_{i=1}^{2} \int_{\Pi^4} |F| I(W_i) \, d|\mu_n| \; =: \; J_1(\epsilon) + J_2(\epsilon).$$

It remains to show that

$$\lim_{\epsilon \to 0} \limsup_{n \to \infty} J_i(\epsilon) = 0, \quad i = 1, 2. \tag{14.4.10}$$

First, consider $J_1(\epsilon)$. Since $\beta_j \leq 0$ and $\alpha_j \geq 0$, then $|F(u)| I(W_1) \leq C|u_1|^{-\alpha_1} \epsilon^{-\alpha_2}$ and hence

$$J_1(\epsilon) \leq C\epsilon^{-\alpha_2} \int_{\Pi^4} |u_1|^{-\alpha_1} I(W_1) \, d|\mu_n|.$$

Let $\psi_n(u)$ be the same as in (6.3.18). Then, for $|u_1| \leq \epsilon$ and $2\epsilon \leq |u_2| \leq \pi$, $|\psi_n(u_1 - u_2)| \leq C/(n\epsilon)$, where $C > 0$ does not depend on n or ϵ. Using the same argument as for (6.3.19), it follows that for any $\eta \in (0, 1)$ and $\epsilon > 0$,

$$J_1(\epsilon) \leq C\epsilon^{-\alpha_2} n \int_{\Pi^2} |u_1|^{-\alpha_1} I(|u_1| \leq \epsilon, |u_2| \geq 2\epsilon) \, \psi_n^{2-\eta}(u_1 - u_2) du_1 du_2$$

$$\leq C\epsilon^{-\alpha_2} n(n\epsilon)^{-2+\eta} \int_{\Pi} |u_1|^{-\alpha_1} du_1 \; \to \; 0,$$

which proves (14.4.10) for $i = 1$.

Next, consider $J_2(\epsilon)$. Split $J_2(\epsilon) = \int_{\Pi^4} |F| I(W_2, |u_3| \leq |u_4|) d|\mu_n| + \int_{\Pi^4} |F| I(W_2, |u_3| > |u_4|) d|\mu_n| =: J_2^+(\epsilon) + J_2^-(\epsilon)$. It suffices to prove (14.4.10) for $J_2^+(\epsilon)$ only since $J_2^-(\epsilon)$ can be treated analogously.

From the definition of F and the assumptions in (14.4.3) we have the bound

$$0 \leq |F(u)| \leq C|u_1|^{-\alpha_1}|u_2|^{-\alpha_2}|u_3|^{-\beta_1}|u_4|^{-\beta_2}, \quad \forall u \in \Pi^4.$$

To show that $J_2^+(\epsilon)$ satisfies (14.4.10), without loss of generality assume that the exponents α_1 and α_2 are as large as possible so that

$$\alpha_i + \beta_i > 0, \quad i = 1, 2. \tag{14.4.11}$$

Indeed, if say $\alpha_1 + \beta_1 > 0$ and $\alpha_2 + \beta_2 \leq 0$ then $\alpha_1 + \alpha_2 + \beta_1 + \beta_2 \leq \alpha_1 + \beta_1 < 1$ since $\alpha_1 < 1$ and $\beta_1 \geq 0$ and hence $\alpha_1 + \tilde{\alpha}_2 + \beta_1 + \beta_2 < 1$ for $\tilde{\alpha}_2 := -\beta_2 + \tilde{\delta}$ and sufficiently small $\tilde{\delta} > 0$. Moreover, $\tilde{\alpha}_2 > \alpha_2$, $0 < \tilde{\alpha}_2 < 1$, and $\tilde{\alpha}_2 + \beta_2 > 0$ by definition. In other words, setting $\tilde{\alpha}_1 := \alpha_1$ and $\tilde{\beta}_i = \beta_i$, $i = 1, 2$, the new quadruple $\tilde{\alpha}_i, \tilde{\beta}_i$, $i = 1, 2$, will satisfy (14.4.11) and (14.4.3) and it will dominate the original quadruple α_i, β_i, $i = 1, 2$, in the sense that $|u_1|^{-\alpha_1}|u_2|^{-\alpha_2}|u_3|^{-\beta_1}|u_4|^{-\beta_2} \leq C|u_1|^{-\tilde{\alpha}_1}|u_2|^{-\tilde{\alpha}_2}|u_3|^{-\tilde{\beta}_1}|u_4|^{-\tilde{\beta}_2}, \forall u \in \Pi^4$.

So, we split the set $W_2^+ := W_2 \cap \{|u_3| \leq |u_4|\} \subset \Pi^4$ into four disjoint sets: $W_2^+ = B_{11} \cup B_{22} \cup B_{12} \cup B_{21}$, where

$$B_{11} := \left\{ u \in W_2^+ : |u_1| > |u_3|^{\nu_1}|u_4|^{\nu_2}/2, \ |u_2| > |u_3|^{\nu_1}|u_4|^{\nu_2}/2 \right\},$$

$$B_{22} := \left\{ u \in W_2^+ : |u_1| \leq |u_3|^{\nu_1}|u_4|^{\nu_2}/2, \ |u_2| \leq |u_3|^{\nu_1}|u_4|^{\nu_2}/2 \right\},$$

$$B_{12} := \left\{ u \in W_2^+ : |u_1| > |u_3|^{\nu_1}|u_4|^{\nu_2}/2, \ |u_2| \leq |u_3|^{\nu_1}|u_4|^{\nu_2}/2 \right\},$$

$$B_{21} := \left\{ u \in W_2^+ : |u_1| \leq |u_3|^{\nu_1}|u_4|^{\nu_2}/2, \ |u_2| > |u_3|^{\nu_1}|u_4|^{\nu_2}/2 \right\},$$

and where $\nu_i := \alpha_i/(\alpha_1 + \alpha_2) \in [0, 1]$, $i = 1, 2$. Accordingly, write

$$J_2^+(\epsilon) = \sum_{i,j=1}^{2} \int_{\Pi^4} |F(u)| I(B_{ij}) d|\mu_n| =: \sum_{i,j=1}^{2} T_{ij}(\epsilon). \tag{14.4.12}$$

Let

$$A := \alpha_1 + \alpha_2 + \beta_1 + \beta_2, \quad v(u, u_3, u_4) := n^{-1/2}|D_n(u - u_3)D_n(u - u_4)|.$$

Use the definition of μ_n in (6.3.14), (14.4.11), $\nu_i(\alpha_1 + \alpha_2) = \alpha_i$, $i = 1, 2$, $A < 1$, and Lemma 6.3.3(ii), to give

$$T_{11}(\epsilon) \tag{14.4.13}$$

$$\leq C \int_{|u_3| \leq |u_4| \leq 4\epsilon} \left(\int_{|u| \leq 2\epsilon} v(u, u_3, u_4) du \right)^2 \frac{du_3 du_4}{|u_3|^{\alpha_1 + \beta_1}|u_4|^{\alpha_2 + \beta_2}}$$

$$\leq C \int_{\Pi^2} I(|u_3| \leq 4\epsilon) \left(\int_{\Pi} v(u, u_3, u_4) du \right)^2 \frac{du_3 du_4}{|u_3|^A}$$

$$\leq C \int_{\Pi^4} |u_3|^{-A} I(|u_3| \leq 4\epsilon) d|\mu_n|$$

$$\leq C \int_{\Pi} |u_3|^{-A} I(|u_3| \leq 4\epsilon) du_3 = C\epsilon^{1-A} \to 0, \quad \epsilon \to 0.$$

Next, consider $T_{22}(\epsilon)$. Recall from (2.1.7) that $|D_n(u)| \leq \pi n^\kappa |u|^{\kappa-1}$ for $0 < \kappa < 1$ and $u \in \Pi$. Also, $|u_3| \leq |u_4| \leq \pi$ and $u \in B_{22}$ imply $|u_i| \leq |u_3|^{\nu_1} |u_4|^{\nu_2}/2 \leq |u_4|^{\nu_1+\nu_2}/2 = |u_4|/2$ and $|u_i - u_4| \geq |u_4|/2$, $i = 1, 2$. We shall use

$$|D_n(u_i - u_4)| \leq Cn^{\kappa_{i4}} |u_i - u_4|^{-1+\kappa_{i4}} \leq Cn^{\kappa_{i4}} |u_4|^{-1+\kappa_{i4}}, \quad (14.4.14)$$

$$|D_n(u_i - u_3)| \leq Cn^{\kappa_{i3}} |u_i - u_3|^{-1+\kappa_{i3}}, \quad i = 1, 2,$$

with suitably chosen $\kappa_{ij} \in (0,1)$ defined by (14.4.15) below. Recall that $A = \alpha_1 + \alpha_2 + \beta_1 + \beta_2 \in (0,1)$, $\alpha_i > 0$, $i = 1, 2$, by (14.4.11), and $\beta_1 + \beta_2 < 0$ (the case $\beta_1 + \beta_2 = 0$ implies $\beta_1 = \beta_2 = 0$ and is covered by (14.4.2)). Next, define

$$\kappa_{13} := \alpha_1 + \beta_1 - \delta, \quad \kappa_{23} := \alpha_2 + \beta_2 - \delta, \quad (14.4.15)$$

$$\kappa_{i4} := 2\delta, \quad i = 1, 2,$$

where $\delta > 0$ is sufficiently small. The above definitions together with (14.4.3) and (14.4.11) imply

$$\kappa_{ij} \in (0,1), \quad A < \kappa_{13} + \kappa_{14} + \kappa_{23} + \kappa_{34} = A + 2\delta < 1, \quad (14.4.16)$$

$$\alpha_1 > \kappa_{13}, \quad \alpha_2 > \kappa_{23},$$

where the last two inequalities follow because $\beta_i \leq 0, i = 1, 2$. We find

$$\int_\Pi I(B_{22}) |u_1|^{-\alpha_1} v(u_1, u_3, u_4) du_1$$

$$\leq Cn^{\kappa_{13}+\kappa_{14}-1/2} |u_4|^{-1+\kappa_{14}} \int_{\mathbb{R}} |u_1|^{-\alpha_1} |u_1 - u_3|^{-1+\kappa_{13}} du_1$$

$$\leq Cn^{\kappa_{13}+\kappa_{14}-1/2} |u_4|^{-1+\kappa_{14}} |u_3|^{-\alpha_1+\kappa_{13}},$$

where we used (6.3.12) and the inequality $\alpha_1 > \kappa_{13}$ in (14.4.16). Similarly,

$$\int_\Pi I(B_{22}) |u_2|^{-\alpha_2} v(u_2, u_3, u_4) du_2 \quad (14.4.17)$$

$$\leq Cn^{\kappa_{23}+\kappa_{24}-1/2} |u_4|^{-1+\kappa_{24}} |u_3|^{-\alpha_2+\kappa_{23}}.$$

Hence,

$$T_{22}(\epsilon) \leq Cn^{-1+(\kappa_{13}+\kappa_{23}+\kappa_{14}+\kappa_{24})} U, \quad (14.4.18)$$

where

$$U := \int_0^\pi u_3^{-\alpha_1-\alpha_2-\beta_1+\kappa_{13}+\kappa_{23}} \, du_3 \int_{u_3}^\pi u_4^{-2-\beta_2+\kappa_{14}+\kappa_{24}} du_4$$

$$= \int_0^\pi u_3^{-1-A+\kappa_{13}+\kappa_{23}+\kappa_{14}+\kappa_{24}} \, du_3 < \infty,$$

since $\kappa_{13} + \kappa_{23} + \kappa_{14} + \kappa_{24} > A$ from (14.4.16). In the above proof, we also used the fact that $-2 - \beta_2 + \kappa_{14} + \kappa_{24} < -1$, which follows from $-\beta_2 < 1$ and $\kappa_{14} + \kappa_{24} = 2\delta$ being small enough. From (14.4.18) and (14.4.16) we find $T_{22}(\epsilon) = o(1)$.

Next, we shall show that $T_{12}(\epsilon) \to 0$ as $\epsilon \to 0$. Here, to bound the inner integral with respect to u_2 we use the bounds in (14.4.14) but with different exponents, namely

$$\tilde{\kappa}_{23} := (A \wedge \alpha_2) - 2\delta, \quad \tilde{\kappa}_{24} := \delta, \tag{14.4.19}$$

where $\delta > 0$ is small enough (recall that $A \wedge \alpha_2 > 0$, because of (14.4.11) and $\beta_2 \leq 0$). The choice in (14.4.19) implies

$$\tilde{\kappa}_{23}, \tilde{\kappa}_{24} \in (0,1), \quad A > \tilde{\kappa}_{23} + \tilde{\kappa}_{24}, \quad \alpha_2 > \tilde{\kappa}_{23}, \tag{14.4.20}$$

$$\alpha_1 + \alpha_2 + \beta_1 < 1 + \tilde{\kappa}_{23}.$$

To check the last inequality, assume first that $\alpha_2 \leq A$. Then, $\tilde{\kappa}_{23} = \alpha_2 - 2\delta$ and $1 + \tilde{\kappa}_{23} = 1 + \alpha_2 - 2\delta > \alpha_1 + \alpha_2 + \beta_1$ hold because $\beta_1 \leq 0$ and $\alpha_1 < 1$. Next, let $\alpha_2 > A$, then $\tilde{\kappa}_{23} = A - 2\delta$ and $1 + \tilde{\kappa}_{23} = 1 + \alpha_1 + \alpha_2 + \beta_1 + \beta_2 - 2\delta > \alpha_1 + \alpha_2 + \beta_1$ is equivalent to $1 + \beta_2 - 2\delta > 0$, which holds because $\beta_2 > -1$.

The inequality $\alpha_2 > \tilde{\kappa}_{23}$, and arguing as in (14.4.17), give

$$\int_\Pi I(B_{12})|u_2|^{-\alpha_2} v(u_2, u_3, u_4) du_2 \leq C n^{\tilde{\kappa}_{23}+\tilde{\kappa}_{24}-1/2}|u_4|^{-1+\tilde{\kappa}_{24}}|u_3|^{-\alpha_2+\tilde{\kappa}_{23}}.$$

Using (6.3.17) and (6.3.12) as for (6.3.18), for any $\rho > 0$ we can find $C > 0$ such that for any $x, y \in \Pi$,

$$\int_\Pi |D_n(z - x)D_n(z - y)| dz \leq \frac{Cn}{(1 + n|x - y|)^{1-\rho}}. \tag{14.4.21}$$

Using the above bounds together with $\nu_1 + \nu_2 = 1$ and the definition of the set B_{12}, we obtain

$$T_{12}(\epsilon) \tag{14.4.22}$$

$$\leq C n^{\tilde{\kappa}_{23}+\tilde{\kappa}_{24}-1} \int_{|u_3|<|u_4|<\pi} du_3 \, du_4 \, |u_3|^{-\alpha_1\nu_1-\alpha_2-\beta_1+\tilde{\kappa}_{23}}$$

$$\times |u_4|^{-1-\alpha_1\nu_2-\beta_2+\tilde{\kappa}_{24}} \int_\Pi |D_n(u_1 - u_3)D_n(u_1 - u_4)| du_1$$

$$\leq Cn^{\tilde{\kappa}_{23}+\tilde{\kappa}_{24}} \int_{0<u_3<u_4<\infty} du_3\, du_4\, |u_3|^{-\alpha_1\nu_1-\alpha_2-\beta_1+\tilde{\kappa}_{23}}$$
$$\times |u_4|^{-1-\alpha_1\nu_2-\beta_2+\tilde{\kappa}_{24}}(1+n|u_3-u_4|)^{-1+\rho}$$
$$= Cn^{A-1}V,$$

where the integral

$$V := \int_0^\infty \frac{dy}{y^{A-\tilde{\kappa}_{23}-\tilde{\kappa}_{24}}} \int_0^1 \frac{dz}{z^{\alpha_1\nu_1+\alpha_2+\beta_1-\tilde{\kappa}_{23}}(1+y(1-z))^{1-\rho}} < \infty.$$

To check this claim, let $V(y) := \int_0^1 \{\cdots\} dz$ be the inner integral in the above V so that $V = \int_0^\infty y^{-A+\tilde{\kappa}_{23}+\tilde{\kappa}_{24}} V(y) dy$. Since $\lambda := \alpha_1\nu_1 + \alpha_2 + \beta_1 - \tilde{\kappa}_{23} \leq \alpha_1 + \alpha_2 + \beta_1 - \tilde{\kappa}_{23} < 1$, by the last inequality in (14.4.20), so $V(y)$ converges and is bounded for $y > 0$. In fact,

$$V(y) \leq C(1+y)^{-1+\rho}, \quad \forall y > 0.$$

To see this, split, for $y > 2$, $V(y) = \int_0^{1/2} \cdots + \int_{1/2}^1 \cdots =: V_1(y) + V_2(y)$ and note that as $y \to \infty$,

$$V_1(y) \leq (1+(y/2))^{-1+\rho} \int_0^{1/2} z^{-\lambda} dz = O(y^{-1+\rho}),$$

$$V_2(y) \leq C \int_{1/2}^1 (1+y(1-z))^{-1+\rho} dz$$
$$= C \int_0^{1/2} (1+yx)^{-1+\rho} dx = O(y^{-1+\rho}).$$

Therefore,

$$V \leq C \int_0^\infty \frac{dy}{y^{A-\tilde{\kappa}_{23}-\tilde{\kappa}_{24}}(1+y)^{1-\rho}} < \infty,$$

where the last inequality follows from $0 < A - \tilde{\kappa}_{23} - \tilde{\kappa}_{24} < 1$, see (14.4.20), by choosing $\rho > 0$ small enough. This proves $\lim_{n\to\infty} T_{12}(\epsilon) = 0$ and the proof of $\lim_{n\to\infty} T_{21}(\epsilon) = 0$ is analogous. This also ends the proof of (14.4.10) and that of Lemma 14.4.1.

The crucial condition for the limit of $\text{tr}(B_1 R_1 B_2 R_2)$ in Lemma 14.4.1 is $A := \alpha_1 + \alpha_2 + \beta_1 + \beta_2 < 1$. From the proof of this lemma, we can obtain a robust asymptotics of the above trace in the case when the condition $A < 1$ is violated, as seen in the following corollary.

Corollary 14.4.1. *Assume all the conditions of (14.4.3) hold except that $A = \alpha_1 + \alpha_2 + \beta_1 + \beta_2 \geq 1$. Then, for any $A' > A$,*

$$\text{tr}(B_1 R_1 B_2 R_2) = O(n^{A'}).$$

Proof. Without loss of generality, assume $A > 1$. It suffices to prove

$$\tilde{T}_{ij} := \int_{\Pi^4} |F(u)| I(\tilde{B}_{ij}) d|\mu_n| = O(n^{A'-1}), \quad i, j = 1, 2, \quad (14.4.23)$$

where F is defined by (14.4.7) and the sets $\tilde{B}_{ij}$ are similar to the B_{ij} in (14.4.12), with the only difference that W_2 in the definition of B_{ij} is now replaced by Π^4. As for (14.4.13) and using (14.4.21) with $\rho > 0$ small enough, we find

$$\tilde{T}_{11} \leq Cn \int_{\Pi^2} \frac{du_3 du_4}{|u_3|^{\alpha_1+\beta_1} |u_4|^{\alpha_2+\beta_2} (1 + n|u_3 - u_4|)^{2-2\rho}}$$

$$\leq Cn^{A-1} \int_{\mathbb{R}^2} \frac{dx dy}{|x|^{\alpha_1+\beta_1} |y|^{\alpha_2+\beta_2} (1 + |x - y|)^{2-2\rho}}$$

$$\leq Cn^{A-1} \int_{\mathbb{R}} \frac{dx}{|x|^{\alpha_1+\beta_1} (1 + |x|)^{\alpha_2+\beta_2}},$$

where we also used $0 < \alpha_i + \beta_i < 1$, $i = 1, 2$, and Lemma 6.3.2. Since the last integral converges, this proves (14.4.23) for $i = j = 1$, with $A' = A$.

The argument for $\tilde{T}_{22}$ is similar to that of $T_{22}(\epsilon)$ in the proof of Lemma 14.4.1 and gives the bound

$$\tilde{T}_{22} \leq Cn^{-1+(\kappa_{13}+\kappa_{23}+\kappa_{14}+\kappa_{24})} U,$$

where κ_{ij} and U are the same as in (14.4.18) and satisfy the conditions in (14.4.16) with the exception of $\mathcal{K} := \kappa_{13}+\kappa_{23}+\kappa_{14}+\kappa_{24} < 1$. In particular, we have again that $U < \infty$ since $\mathcal{K} > A$. Since $\mathcal{K}$ can be chosen arbitrarily close to A, this proves (14.4.23) for $i = j = 2$. Finally, the estimation of $\tilde{T}_{12}$ coincides with that of $T_{12}(\epsilon)$ in (14.4.22) and leads to a similar bound $\tilde{T}_{12} \leq Cn^{A-1} V$. Note that the inequalities $0 < A - \tilde{\kappa}_{23} - \tilde{\kappa}_{24} < 1$ used in the proof of $V < \infty$ are again satisfied for $\tilde{\kappa}_{23} = \alpha_2 - 2\delta$ and $\tilde{\kappa}_{24} = \delta$ as in (14.4.19) since $A > 1$, $\alpha_2 < 1$, and $A - \tilde{\kappa}_{23} - \tilde{\kappa}_{24} = \alpha_1 + \beta_1 + \beta_2 + \delta < 1$ because $\alpha_1 < 1$ and $\beta_1 + \beta_2 \leq 0$. This proves (14.4.23), and also completes the proof of the corollary.

Chapter 15

Bibliography

Abadir, K.M., Distaso, W., and Giraitis, L. (2007). Nonstationarity-extended local Whittle estimation. *J. of Econometrics*, **141**, 1353–1384.

Abadir, K.M., Distaso, W., and Giraitis, L. (2009). Two estimators of the long-run variance: beyond short memory. *J. of Econometrics*, **150**, 56–70.

Abadir, K.M., Distaso, W., and Giraitis, L. (2011). An I(d) model with trend and cycles. *J. of Econometrics*, **163**, 186–199.

Abramowitz, M. and Stegun, I. (1972). *Handbook of Mathematical Functions with Formulas, Graphs, and Mathematical Tables.* Dover, New York.

Adenstedt, R.K. (1974). On large sample estimation for the mean of stationary random sequence. *Ann. Statist.*, **2**, 1095–1107.

An, H.-Z. and Cheng, B. (1991). A Kolmogorov-Smirnov type statistic with application to test for nonlinearity in time series. *Int. Statist. Rev.*, **59**, 287–307.

Andel, J. (1986). Long memory time series models. *Kybernetika*, **22**, 105–123.

Andrews, D.W.K. (1993). Tests for parameter instability and structural change with unknown change point. *Econometrica*, **61**, 821–856.

Andrews, D.W.K. and Monahan, J.C. (1992). An improved heteroscedasticity and autocorrelation consistent covariance matrix estimator. *Econometrica*, **60**, 953–966.

Andrews, D.W.K. and Sun, Y. (2004). Adaptive polynomial Whittle estimation of long-range dependence. *Econometrica*, **72**, 569–614.

Arcones, M.A. (1994). Limit theorems for nonlinear functions of a stationary Gaussian sequence of vectors. *Ann. Probab.*, **22**, 2242–2274.

Arteche, J. (2004). Gaussian semiparametric estimation in long memory in stochastic volatility and signal plus noise models. *J. of Econometrics*, **119**, 131–154.

Avram, F. and Taqqu, M.S. (1987). Generalized powers of strongly dependent random variables. *Ann. Probab.*, **15**, 767–775.

Baillie, R.T. (1996). Long memory processes and fractional integration in econometrics. *J. of Econometrics*, **73**, 5–59.

Baillie, R.T. and Bollerslev, T. (1994). The long memory of the forward premium. *J. of Intern. Money Fin.*, **13**, 309–324.

Baillie, R.T. and Bollerslev, T. (2000). The forward premium anomaly is not as bad as you think. *J. of Intern. Money Fin.*, **19**, 471–488.

Baillie, R.T. and Kapetanios, G. (2010). Estimation and inference for impulse response weights from strongly persistent processes. Preprint, Queen Mary, University of London.

Bardet, J.-M. and Surgailis, D. (2011). Measuring the roughness of random paths by increment ratios. *Bernoulli*, **17**, 749–780.

Bates, D.M. and Watts, D.G. (1988). *Nonlinear Analysis and its Applications.* John Wiley & Sons, New York.

Beran, J. (1992a). Statistical methods for data with long-range dependence (with discussions). *Statist. Sci.*, **7**, 404–427.

Beran, J. (1992b). M-estimators of location for data with slowly decaying parameters. *J. Amer. Statist. Assoc.*, **86**, 704–708.

Beran, J. (1994). *Statistics for Long-Memory Processes.* Chapman & Hall, New York.

Beran, J., Schutzner, M., and Ghosh, S. (2010). From short to long memory: aggregation and estimation. *Comput. Statist. Data Anal.*, **54**, 2432–2442.

Bhansali, R., Giraitis, L., and Kokoszka, P. (2007a). Convergence of quadratic forms with nonvanishing diagonal. *Statist. Prob. Lett.*, **77**, 726–734.

Bhansali, R., Giraitis, L., and Kokoszka, P. (2007b). Approximations and limit theory for quadratic forms of linear processes. *Stoch. Proc. Appl.*, **117**, 71–95.

Bhattacharya, R.N., Gupta, V.K., and Waymire, E. (1983). The Hurst effect under trends. *J. Appl. Prob.*, **20**, 649–662.

Billingsley, P. (1968). *Convergence of Probability Measures.* John Wiley & Sons, New York.

Bingham, N., Goldie, C., and Teugels, J. (1987). *Regular Variation.* Cambridge University Press, Cambridge.

Bollerslev, T. (1986). Generalized autoregressive conditional heteroskedasticity. *J. of Econometrics*, **31**, 307–327.

Booth, G.G., Kaen, R.R., and Koveos, P.E. (1982). R/S analysis of foreign exchange rates under two international money regimes. *J. Monetary Econ.*, **10**, 407–415.

Box, G.E.P. and Jenkins, G.M. (1970). *Times Series Analysis. Forecasting and Control.* Holden-Day, San Francisco.

Breidt, F.J., Crato, N., and de Lima, P. (1998). On the detection and estimation of long memory in stochastic volatility. *J. of Econometrics*, **83**, 325–348.

Breuer, P. and Major, P. (1983). Central limit theorems for non-linear functionals of Gaussian fields. *J. Multiv. Anal.*, **13**, 425–441.

Brillinger, D.R. (1981). *Time Series: Data Analysis and Theory*. Holden-Day, San Francisco.

Brockwell, P.J. and Davis, R.A. (1991). *Time Series: Theory and Methods (2nd ed.)*. Springer Series in Statistics. Springer-Verlag, New York.

Buckley, M.J. (1991). Detecting a smooth signal: optimality of cusum based procedures. *Biometrika*, **78**, 253–262.

Chen, W.W. and Deo, R. (2006). Estimation of misspecified long memory models. *J. of Econometrics*, **134**, 257–281.

Cheung, Y.W. (1993). Long memory in foreign-exchange rates. *J. Bus. Econ. Statist.*, **11**, 93–101.

Chung, C.F. (1996). Estimating a generalized long memory process. *J. of Econometrics*, **73**, 237–259.

Craigmile, P.F. (2003). Simulating a class of stationary Gaussian processes using the Davies–Harte algorithm, with application to long memory processes. *J. Time Series Anal.*, **24**, 505–511.

Crámer, H. and Kadelka, D. (1986). On weak convergence of integral functionals of stochastic processes with applications to processes taking paths in L_p^E. *Stoch. Proc. Appl.*, **21**, 305–317.

Crato, N. and de Lima, P.J.F. (1994). Long-range dependence in the conditional variance of stock returns. *Econ. Lett.*, **45**, 281–285.

Csörgő, S. and Mielniczuk, J. (1995a). Nonparametric regression under long-range dependent normal errors, *Ann. Statist.*, **23**, 1000–1014.

Csörgő, S. and Mielniczuk, J. (1995b). Distant long-range dependent sums and regression estimation. *Stoch. Proc. Appl.*, **59**, 143–155.

Csörgő, S. and Mielniczuk, J. (1996). The empirical process of a short-range dependent stationary sequence under Gaussian subordination. *Probab. Th. Rel. Fields*, **104**, 15–25.

Csörgő, S. and Mielniczuk, J. (2000). The smoothing dichotomy in random design regression with long-memory errors based on moving averages. *Statist. Sinica*, **10**, 771–787.

Dacunha-Castelle, V. and Fermin, L. (2006). Disaggregation of long memory processes on C^∞ class. *Elect. Comm. in Probab.*, **11**, 35–44.

Dahlhaus, R. (1989). Efficient parameter estimation for self-similar processes. *Ann. Statist.*, **17**, 1749–1766.

Dahlhaus, R. (1995). Efficient location and regression estimation for long range dependent regression models. *Ann. Statist.*, **23**, 1029–1047.

Dalla, V., Giraitis, L., and Hidalgo, J. (2006). Consistent estimation of the memory parameter for nonlinear time series. *J. Time Series Anal.*, **27**, 211–251.

Davies, R.B. and Harte, D.S. (1987). Tests for Hurst effect. *Biometrika*, **74**, 95–102.

Davis, R.A. and Mikosch, T. (1999). The maximum of the periodogram of a non-Gaussian sequence. *Ann. Probab.*, **27**, 522–536.

Davis, R.A. and Mikosch, T. (2009). Extreme value theory for GARCH processes. In: Andersen, T.G., Davis, R.A., Kreiss, J.-P. and Mikosch, T. (eds.), *Handbook of Financial Time Series*, 187–200. Springer-Verlag, Berlin.

Davydov, Y.A. (1970). The invariance principle for stationary processes. *Theor. Probab. Appl.*, **15**, 487–498.

Dedecker, J. and Prieur, C. (2007). An empirical central limit theorem for dependent sequences. *Stoch. Proc. Appl.*, **117**, 121–142.

Dedecker, J., Doukhan, P., Lang, G., Léon, J.R., Louhichi, S., and Prieur, C. (2007). *Weak Dependence: Models, Theory and Applications*. Lecture Notes in Statistics 190. Springer-Verlag, New York.

Dehling, H. and Taqqu, M.S. (1989). The empirical process of some long range dependent sequences with an application to U-statistics. *Ann. Statist.*, **17**, 1767–1783.

Dehling, H., Mikosch, T., and Sørensen, M., eds. (2002). *Empirical Process Techniques for Dependent Data*. Birkhäuser, Boston.

Deo, R.S. and Hurvich, C.M. (1998). Linear trend with fractionally integrated errors. *J. Time Series Anal.*, **19**, 379–397.

Deo, R. and Hurvich, C. (2001). On the log periodogram regression estimator of the memory parameter in long memory stochastic volatility models. *Econometric Theory*, **17**, 686–710.

Ding, Z. and Granger, C.W.J. (1996). Modeling volatility persistence of speculative returns: a new approach. *J. of Econometrics*, **73**, 185–215.

Dobrushin, R.L. (1979). Gaussian and their subordinated self-similar random generalized fields. *Ann. Probab.*, **7**, 1–28.

Dobrushin, R. L. and Major, P. (1979). Non-central limit theorems for non-linear functionals of Gaussian fields. *Z. Wahr. verw. Gebiete*, **50**, 27–52.

Doukhan, P., Oppenheim, G., and Taqqu, M.S., eds. (2003). *Theory and Applications of Long-Range Dependence*. Birkhäuser, Boston.

Durbin, J. (1973). *Distribution Theory for Tests Based on the Sample Distribution Function*. CBMS Regional Conference Series in Applied Mathematics 9. SIAM, Philadelphia.

Embrechts, P. and Maejima, M. (2002). *Selfsimilar Processes*. Princeton University Press, Princeton, NJ.

Embrechts, P., Klüppelberg, C., and Mikosch, T. (1997). *Modelling Extremal Events. For Insurance and Finance.* Springer-Verlag, Berlin.

Engel, C. (1996). The forward discount anomaly and the risk premium: a survey of recent evidence. *J. Empirical Finance*, **3**, 123–192.

Engle, R.F. (1982). Autoregressive conditional heteroscedasticity with estimates of the variance of United Kingdom inflation. *Econometrica*, **50**, 987–1008.

Fan, J. and Yao, Q. (1998). Efficient estimation of conditional variance functions in stochastic regression. *Biometrika*, **85**, 645–660.

Feinsilver, P.J. (1978). *Special Functions, Probability Semigroups, and Hamiltonian Flows.* Lecture Notes in Mathematics 696. Springer-Verlag, New York.

Fox, R. and Taqqu, M.S. (1986). Large sample properties of parameter estimates for strongly dependent stationary Gaussian time series. *Ann. Statist.*, **14**, 517–532.

Fox, R. and Taqqu, M.S. (1987). Central limit theorems for quadratic forms in random variables having long-range dependence. *Probab. Th. Rel. Fields*, **74**, 213–240.

Gastwirth, J.L. and Rubin, H. (1975). The behavior of robust estimators on dependent data. *Ann. Statist.*, **3**, 1070–1100.

Geweke, J. and Porter-Hudak, S. (1983). The estimation and application of long memory time series models. *J. Time Series Anal.*, **4**, 221–238.

Giraitis, L. (1985). Central limit theorem for functionals of a linear process. *Lithuanian Math. J.*, **25**, 25–35.

Giraitis, L. and Koul, H.L. (1997). Estimation of the dependence parameter in linear regression with long memory errors. *Stoch. Proc. Appl.*, **71**, 207–224.

Giraitis, L. and Leipus, R. (1995). A generalized fractionally differencing approach in long memory modeling. *Lithuanian Math. J.*, **35**, 65–81.

Giraitis, L. and Surgailis, D. (1985). CLT and other limit theorems for functionals of Gaussian process. *Z. Wahr. verw. Gebiete*, **70**, 191–212.

Giraitis, L. and Surgailis, D. (1986). Multivariate Appell polynomials and the central limit theorem. In: Eberlein, E. and Taqqu, M.S., (eds.), *Dependence in Probability and Statistics*, 21–71. Birkhäuser, Boston.

Giraitis, L. and Surgailis, D. (1989). Limit theorems for polynomials of a linear process with long range dependence. *Lithuanian Math. J.*, **29**, 290–311.

Giraitis, L. and Surgailis, D. (1990). A central limit theorem for quadratic forms in strongly dependent linear variables and its application to asymptotic normality of Whittle's estimate. *Probab. Th. Rel. Fields*, **86**, 87–104.

Giraitis, L. and Surgailis, D. (1999). Central limit theorem for the empirical process. *J. Statist. Plann. Inference*, **80**, 81–93.

Giraitis, L. and Surgailis, D. (2002). ARCH-type bilinear models with double long memory. *Stoch. Proc. Appl.*, **100**, 275–300.

Giraitis, L. and Taqqu, M.S. (1999). Whittle estimator for finite-variance non-Gaussian time series with long-memory. *Ann. Statist.*, **27**, 178–203.

Giraitis, L., Hidalgo, J., and Robinson, P.M. (2001). Gaussian estimation of parametric spectral density with unknown pole. *Ann. Statist.*, **29**, 987–1023.

Giraitis, L., Kokoszka, P., and Leipus, R. (2000). Stationary ARCH models: dependence structure and central limit theorems. *Econometric Theory*, **16**, 3–22.

Giraitis, L., Koul, H.L., and Surgailis, D. (1996). Asymptotic normality of regression estimators with long memory errors. *Statist. Probab. Lett.*, **29**, 317–335.

Giraitis, L., Leipus, R., and Philippe, A. (2006). A test for stationarity versus trends and unit roots for a wide class of dependent errors. *Econometric Theory*, **22**, 989–1029.

Giraitis, L., Leipus, R., and Surgailis, D. (2009). ARCH(∞) models and long memory properties. In: Andersen, T.G., Davis, R.A., Kreiss, J.-P., and Mikosch, T. (eds.), *Handbook of Financial Time Series*, 71–84. Springer-Verlag, Berlin.

Giraitis, L., Robinson, P.M., and Samarov, A. (1997). Rate optimal semiparametric estimation of the memory parameter of the Gaussian time series with long range dependence. *J. Time Series Anal.*, **18**, 49–60.

Giraitis, L., Robinson, P.M., and Samarov, A. (2000). Adaptive semiparametric estimation of the memory parameter. *J. Multiv. Anal.*, **72**, 183–207.

Giraitis, L., Robinson, P.M., and Surgailis, D. (2000). A model for long memory conditional heteroscedasticity. *Ann. Appl. Probab.*, **10**, 1002–1024.

Giraitis, L., Kokoszka, P., Leipus, R., and Teyssière, G. (2003). Rescaled variance and related tests for long memory in volatility and levels. *J. of Econometrics*, **112**, 265–294.

Giraitis, L., Leipus, R., Robinson, P.M., and Surgailis, D. (2004). LARCH, leverage and long memory. *J. Financial Econometrics*, **2**, 177–210.

Goetzmann, W.N. (1993). Patterns in three centuries of stock market prices. *J. of Business*, **66**, 249–270.

Gradshteyn, I.S. and Ryzhik, I.M. (2000). *Table of Integrals, Series and Products*. Academic Press, New York.

Granger, C.W.J. (1980). Long memory relationships and the aggregation of dynamic models. *J. of Econometrics*, **14**, 227–238.

Granger, C.W.J. and Joyeux, R. (1980). An introduction to long-memory time series models and fractional differencing. *J. Time Series Anal.*, **1**, 15–29.

Gray, H.L., Zhang, N.-F., and Woodward, W.A. (1989). On generalized fractional processes. *J. Time Series Anal.*, **10**, 233–257.

Gray, H.L., Zhang, N.-F., and Woodward, W.A. (1994). On generalized fractional processes – a correction. *J. Time Series Anal.*, **15**, 561–562.

Guo, H. and Koul, H.L. (2007). Nonparametric regression with heteroscedastic long memory errors. *J. Statist. Plann. Inference*, **137**, 379–404.

Guo, H. and Koul, H.L. (2008). Asymptotic inference in some heteroscedastic regression models with long memory design and errors. *Ann. Statist.*, **36**, 458–487.

Guttorp, P. and Lockhart, R.A. (1988). On the asymptotic distribution of quadratic forms in uniform order statistics. *Ann. Statist.*, **16**, 433–449.

Hajék, J. (1969). *Nonparametric Statistics*. Holden-Day, San Francisco.

Hall, P. and Hart, J. (1990). Nonparametric regression with long-range dependence. *Stoch. Proc. Appl.*, **36**, 339–351.

Hall, P. and Heyde, C.C. (1980). *Martingale Limit Theory and Applications*. Academic Press, New York.

Hall, P., Jing, B.-Y., and Lahiri, S.N. (1998). On the sampling window method for long-range dependent data. *Statist. Sinica*, **8**, 1189–1204.

Hall, P., Koul, H.L., and Turlach, B.A. (1997). Note on convergence rates of semiparametric estimators of dependence index. *Ann. Statist.*, **25**, 1725–1739.

Hannan, E.J. (1957). The variance of the mean of a stationary process. *J. Royal Statist. Soc. B*, **19**, 282–285.

Hannan, E.J. (1973). The asymptotic theory of linear time series models. *J. Appl. Probab.*, **10**, 130–145.

Hart, J. (1997). *Nonparametric Smoothing and Lack-of-fit Tests*. Springer-Verlag, New York.

Harvey, A.C. (1998). Long memory in stochastic volatility. In: Knight, J. and Satchell, S. (eds.), *Forecasting Volatility in the Financial Markets*, 307–320. Butterworth & Heinemann, Oxford.

Harvey, A.C., Ruiz, E., and Shephard, N. (1994). Multivariate stochastic variance models. *Rev. Econ. Studies*, **61**, 247–264.

Hassler, U. (1994). (Mis)specification of long memory in seasonal time series. *J. Time Series Anal.*, **15**, 19–30.

Haubrich, J.G. and Lo, A.W. (2001). The sources and nature of long-term dependence in the business cycle. *Econometric Rev.* (Fed. Reserve Bank Cleveland, Q II), **37**, 15–30.

Henry, M. and Robinson P.M. (1996). Bandwidth choice in Gaussian semiparametric estimation of long range dependence. In: Robinson, P. and Rosenblatt, M. (eds.), *Athens Conference on Applied Probability and*

Time Series Analysis, II, In Memory of E.J. Hannan. Lecture Notes in Statistics 115, 220–232. Springer-Verlag, New York.

Hidalgo, J. (2005). Semiparametric estimation for stationary processes whose spectra have an unknown pole. *Ann. Statist.*, **33**, 1843–1889.

Hidalgo, F.J. and Robinson, P.M. (2002). Adapting to unknown disturbance autocorrelation in regression with long memory. *Econometrica*, **70**, 1545–1581.

Ho, H.-C. and Hsing, T. (1996). On the asymptotic expansion of the empirical process of long memory moving averages. *Ann. Statist.*, **24**, 992–1024.

Hosking, J.R.M. (1981). Fractional differencing. *Biometrika*, **68**, 165–176.

Huber, P.J. (1981). *Robust Statistics*. Wiley Series in Probability and Mathematical Statistics. John Wiley & Sons, New York.

Hurst, H.E. (1951). Long-term store capacity of reservoirs. *Trans. Amer. Math. Eng.*, **116**, 770–799.

Hurst, H.E. (1956). Methods of using long-term storage in reservoirs. *Proceedings of the Institution of Civil Engineering*, Part I, 519–590.

Hurvich, C.M. and Beltrao, K.I. (1994). Automatic semiparametric estimation of the memory parameter of a long-memory time series. *J. Time Series Anal.*, **15**, 285–302.

Hurvich, C.M. and Brodsky, J. (2001). Broadband semiparametric estimation of the memory parameter of a long-memory time series using fractional exponential model. *J. Time Series Anal.*, **22**, 221–249.

Hurvich, C.M. and Chen, W.W. (2000). An efficient taper for potentially overdifferenced long-memory time series. *J. Time Series Anal.*, **21**, 155–180.

Hurvich, C.M. and Deo, R. (1998). Plug-in selection of the number of frequencies in regression estimates of the memory parameter of a long memory time series. *J. Time Series Anal.*, **20**, 331–341.

Hurvich, C.M. and Soulier, P. (2009). Stochastic volatility models with long memory. In: Andersen, T.G., Davis, R.A., Kreiss, J.-P., and Mikosch, T. (eds.), *Handbook of Financial Time Series*, 345–354. Springer-Verlag, Berlin.

Hurvich, C.M., Deo, R., and Brodsky, J. (1998). The mean squared error of Geweke and Porter-Hudak's estimator of a long memory time series. *J. Time Series Anal.*, **19**, 19–46.

Hurvich, C.M., Moulines, E., and Soulier, P. (2005). Estimation of long memory in volatility. *Econometrica*, **73**, 1283–1328.

Ibragimov, I.A. and Linnik, Yu.V. (1971). *Independent and Stationary Sequences of Random Variables*. Wolters-Noordhoff, Groningen.

Ivanov, A.V. and Leonenko, N.N. (2004). Asymptotic theory of nonlinear regression with long-range dependence. *Math. Methods of Statist.*, **13**, 153–178.

Ivanov, A.V. and Leonenko, N.N. (2008). Semiparametric analysis of long-range dependence in nonlinear regression. *J. Statist. Plann. Inference*, **138**, 1733–1753.

Jaeckel, L.A. (1972). Estimating regression coefficients by minimizing the dispersion of the residuals. *Ann. Statist.*, **43**, 1449–1458.

Jennrich, R.I. (1969). Asymptotic properties of non-linear least squares estimators. *Ann. Statist.*, **40**, 633–643.

Jowett, G.H. (1955). The comparison of means of sets of observations from sections of independent stochastic series. *J. Royal Statist. Soc.* B, **17**, 208–227.

Khmaladze, E.V. and Koul, H.L. (2004). Martingale transforms goodness-of-fit tests in regression models. *Ann. Statist.*, **32**, 995–1034.

Kiefer, J. (1959). *K*-sample analogues of the Kolmogorov–Smirnov and Cramér–v. Mises tests. *Ann. Math. Statist.*, **30**, 420–447.

Kirman, A. and Teyssière, G. (2002). Microeconomic models for long-memory in the volatility of financial time series. *Studies in Nonlinear Dynamics and Econometrics*, **5**, 281–302.

Koul, H.L. (1977). Behavior of robust estimators in the regression model with dependent errors. *Ann. Statist.*, **5**, 681–699.

Koul, H.L. (1992a). *Weighted Empiricals and Linear Models*. IMS, Hayward, CA.

Koul, H.L. (1992b). M-estimators in linear models with long range dependence. *Statist. Probab. Lett.*, **14**, 153–164.

Koul, H.L. (1996). Asymptotics of M-estimators in non-linear regression with long-range dependent errors. In: Robinson, P. and Rosenblatt, M. (eds.), *Athens Conference on Applied Probability and Time Series Analysis, II, In Memory of E.J. Hannan*. Lecture Notes in Statistics 115, 272–290. Springer-Verlag, New York.

Koul, H.L. (2002). *Weighted Empirical Processes in Dynamic Nonlinear Models (2nd ed.)*. Lecture Notes in Statistics 166. Springer-Verlag, New York.

Koul, H.L. (2006). Model diagnostics via martingale transforms: a brief review. In: Fan, J. and Koul, H.L. (eds.), *Frontiers in Statistics*, 183–206. Imperial College Press, London.

Koul, H.L. and Baillie, R.T. (2003). Asymptotics of M-estimators in non-linear regression with long memory designs. *Statist. Probab. Lett.*, **61**, 237–252.

Koul, H.L. and Mukherjee, K. (1993). Asymptotics of R-, MD- and LAD-estimators in linear regression models with long range dependent errors. *Probab. Th. Rel. Fields*, **95**, 525–553.

Koul, H.L. and Stute, W. (1999). Nonparametric model checks in time series. *Ann. Statist.*, **27**, 204–237.

Koul, H.L. and Surgailis, D. (1997). Asymptotic expansion of M estimators with long-memory errors. *Ann. Statist.*, **25**, 818–850.

Koul, H.L. and Surgailis, D. (2000a). Second-order behavior of M estimators in linear regression with long-memory errors. *J. Statist. Plann. Inference*, **91**, 399–412.

Koul, H.L. and Surgailis, D. (2000b). Asymptotic normality of the Whittle estimator in linear regression models with long memory errors. In Proc. of 19th Rencontres Franco-Belges de Statisticiens (Marseille, 1998). *Statist. Inference Stoch. Process.*, **3**, 129–147.

Koul, H.L. and Surgailis, D. (2001). Asymptotics of empirical processes of long memory moving averages with infinite variance. *Stoch. Proc. Appl.*, **91**, 309–336.

Koul, H.L. and Surgailis, D. (2002). Asymptotic expansion of the empirical process of long memory moving averages. In: Dehling, H., Mikosch, T., and Sørensen, M. (eds.), *Empirical Process Techniques for Dependent Data*, 213–239. Birkhäuser, Boston.

Koul, H.L. and Surgailis, D. (2008). Testing of a sub-hypothesis in linear regression models with long memory covariates and errors. *Applications of Mathematics*, **53**, 235–248.

Koul, H.L. and Surgailis, D. (2009). Testing of a sub-hypothesis in linear regression models with long memory errors and deterministic design. *J. Statist. Plann. Inference*, **139**, 2715–2730.

Koul, H.L. and Surgailis, D. (2010). Goodness-of-fit testing under long memory. *J. Statist. Plann. Inference*, **140** (the Emanuel Parzen festschrift), 3742–3753.

Koul, H.L., Baillie, R.T., and Surgailis, D. (2004). Regression model fitting with a long memory covariance process. *Economic Theory*, **20**, 485–512.

Kuiper, N.H. (1960). Tests concerning random points on a circle. *Proc. Koninkl. Nederl. Akad. Van Wettenschappen, Ser. A.*, **63**, 38–47.

Künsch, H. (1987). Statistical aspects of self-similar processes. In: Prohorov, Yu.A. and Sazonov. V.V. (eds.), *Proceedings of the 1st World Congress of the Bernoulli Society*, **1**, 67–74. Science Press, Utrecht.

Kwiatkowski, D., Phillips, P.C.B., Schmidt, P., and Shin, Y. (1992). Testing the null hypothesis of stationarity against the alternative of a unit root: how sure are we that economic time series have a unit root? *J. of Econometrics*, **54**, 159–178.

Lahiri, S.N. (2003). A necessary and sufficient condition for asymptotic independence of discrete Fourier transforms under short- and long-range dependence. *Ann. Statist.*, **31**, 613–641.

Lamperti, J.W. (1962). Semi-stable stochastic processes. *Trans. Amer. Math. Soc.*, **104**, 62–78.

Lavancier, F., Philippe, A., and Surgailis, D. (2010). A two-sample test for comparison of long memory parameters. *J. Multiv. Anal.*, **101**, 2118–2136.

Lawrance, A.J. and Kottegoda, N.T. (1977). Stochastic modeling of river flow time series. *J. Royal Statist. Soc. A*, **140**, Part 1, 1–30.

Lee, H.S. and Amsler, C. (1997). Consistency of the *KPSS* unit root test against fractionally integrated alternative. *Econ. Lett.*, **55**, 151–160.

Lehmann, E.L. (1966). Some concepts of dependence. *Ann. Math. Statist.*, **37**, 1137–1153.

Lehmann, E.L. (1983). *Theory of Point Estimation.* John Wiley & Sons, New York.

Lehmann, E.L. (1986). *Testing Statistical Hypotheses (2nd ed.).* John Wiley & Sons, New York.

Leipus, R., Oppenheim, G., Philippe, A., and Viano, M.-C. (2006). Orthogonal series density estimation in a disaggregation scheme, *J. Statist. Plann. Inference*, **136**, 2547–2571.

Leonov, V.P. and Shiryaev, A.N. (1959). On a method of calculation of semi-invariants. *Theory Probab. Appl.*, **4**, 319–329.

Li, L. (2003). On Koul's minimum distance estimators in the regression models with long memory moving averages. *Stoch. Proc. Appl.*, **105**, 257–269.

Li, W.K. (2004). *Diagnostic Checks in Time Series.* Chapman & Hall/CRC, Boca Raton, FL.

Li, W., Yu, C., Carriquiry, A., and Kliemann, W. (2011). The asymptotic behavior of the *R/S* statistic for fractional Brownian motion. *Statist. Prob. Lett.*, **81**, 83–91.

Lindner, A.M. (2009). Stationarity, mixing, distributional properties of GARCH(p, q)-processes. In: Andersen, T.G., Davis, R.A., Kreiss, J.-P., and Mikosch, T. (eds.), *Handbook of Financial Time Series*, 43–69. Springer-Verlag, Berlin.

Lo, A. (1991). Long term memory in stock market prices. *Econometrica*, **59**, 1279–1313.

Lobato, I. and Robinson, P.M. (1996). Average periodogram estimation of long memory. *J. of Econometrics*, **73**, 303–324.

Lobato, I. and Robinson, P.M. (1998). A nonparametric test for I(0). *Rev. Econ. Studies*, **68**, 475–495.

Loève, M. (1968). *Probability Theory (3rd ed.).* D. Van Nostrand Company, Princeton, NJ.

Major, P. (1981). *Multiple Wiener–Itô Integrals.* Lecture Notes in Mathematics 849. Springer-Verlag, New York.

Mandelbrot, B.B. and Taqqu, M.S. (1979). Robust *R/S* of long run serial correlation. In Proc. of the 42nd Session of the I.S.I., *Bullet. of the I.S.I.*, **48**(2), 69–104.

Mandelbrot, B.B. and van Ness, J.W. (1968). Fractional Brownian motions, fractional noise and applications. *SIAM Rev.*, **10**, 422–437.

Mandelbrot, B.B. and Wallis, J.R. (1969a). Some long-run properties of geophysical records. *Water Resources Research*, **5**, 321–340.

Mandelbrot, B.B. and Wallis, J.R. (1969b). Robustness of rescaled range R/S in the measurement of noncyclic long-run statistical dependence. *Water Resources Research*, **5**, 967–988.

Masry, E. and Mielniczuk, J. (1999). Local linear regression estimation for time series with long-range dependence. *Stoch. Proc. Appl.*, **82**, 173–193.

Maynard, A. and Phillips, P.C.B. (2001). Rethinking an old empirical puzzle: econometric evidence on the forward discount anomaly. *J. Appl. Econometrics*, **16**, 671–798.

Merlevède, F., Peligrad, M., and Utev, S. (2006). Recent advances in invariance principles for stationary sequences. *Probability Surveys*, **3**, 1–36.

Mikosch, T., Resnick, S., Rootzén, H., and Stegeman, A. (2002). Is network traffic approximated by stable Lévy motion or fractional Brownian motion? *Ann. Appl. Probab.*, **12**, 23–68.

Móricz, F. (2006). Absolutely convergent Fourier series and function classes. *J. Math. Anal. Appl.*, **324**, 1168–1177.

Moulines, E. and Soulier, P. (1999). Broad band log-periodogram regression of time series with long range dependence. *Ann. Statist.*, **27**, 1415–1439.

Moulines, E. and Soulier, P. (2003). Semiparametric spectral estimation for fractional processes. In: Doukhan, P., Oppenheim, G., and Taqqu, M.S. (eds.), *Theory and Applications of Long-Range Dependence*, 251–302. Birkhäuser, Boston.

Mukherjee, K. (1994). Minimum distance estimation in linear models with long-range dependent errors. *Statist. Probab. Lett.*, **21**, 347–355.

Mukherjee, K. (1999). The asymptotic behaviour of a class of L-estimators under long-range dependence. *Canad. J. Statist.*, **27**, 345–360.

Natanson, I.P. (1964). *Theory of Functions of a Real Variable*, 1. Ungar Publishing Co., New York.

Nelson, D.B. (1991). Conditional heteroskedasticity in asset returns: a new approach. *Econometrica*, **59**, 347–370.

Newey, W.K. and West, K.D. (1987). A simple, positive semi-definite, heteroscedasticity and autocorrelation consistent covariance matrix. *Econometrica*, **55**, 703–708.

Nielsen, M.Ø. (2005). Semiparametric estimation in time-series regression with long-range dependence. *J. Time Series Anal.*, **26**, 279–304.

Nourdin, I., Pecati, G., and Podolskij, M. (2011). Quantitative Breuer-Major Theorems. *Stoch. Proc. Appl.*, **121**, 793–812.

Nualart, D. and Peccati, G. (2005). Central limit theorems for sequences of multiple stochastic integrals. *Ann. Probab.*, **33**, 177–193.

Oppenheim, G. and Viano, M.-C. (2004). Aggregation of random parameters Ornstein–Uhlenbeck or AR processes: some convergence results. *J. Time Series Anal.*, **25**, 335–350.

Palma, W. (2007). *Long-Memory Time Series*. John Wiley & Sons, New York.

Palma, W. and Chan, N.H. (2005). Efficient estimation of seasonal long-range dependent processes. *J. Time Series Anal.*, **26**, 863–892.

Park, K. and Willinger, W. (2000). Self-similar network traffic: an overview. In: Park, K. and Willinger, W. (eds.), *Self-Similar Network Traffic and Performance Evaluations*, 1–38, John Wiley & Sons, New York.

Peccati, G. and Taqqu, M.S. (2011). *Wiener Chaos: Moments, Cumulants and Diagrams*. Bocconi & Springer Series, Vol. 1. Springer-Verlag, New York.

Peligrad, M. and Utev, S. (2006). Central limit theorem for stationary linear processes. *Ann. Probab.*, **34**, 1608–1622.

Phillips, P.C.B. (2007). Unit root log periodogram regression. *J. of Econometrics*, **138**, 104–124.

Phillips, P.C.B. and Solo, V. (1992). Asymptotics for linear processes. *Ann. Statist.*, **20**, 971–1001.

Phillips, P.C.B., Sun, Y. and Jin, S. (2007). Long run variance estimation and robust regression testing using sharp origin kernels with no truncation. *J. Statist. Plann. Inference*, **137**, 985–1023.

Ploberger, W., Krämer, W., and Kontrus, K. (1989). A new test for structural stability in the linear regression model. *J. of Econometrics*, **40**, 307–318.

Pollard, D. (1984). *Convergence of Stochastic Processes*. Springer Series in Statistics. Springer-Verlag, New York.

Porter-Hudak, S. (1990). An application of seasonal fractionally differenced model to the monetary aggregates. *J. Amer. Statist. Assoc.*, **85**, 338–344.

Puplinskaitė, D. and Surgailis, D. (2010). Aggregation of random-coefficient AR(1) process with infinite variance and idiosyncratic innovations. *Adv. Appl. Probab.*, **42**, 509–527.

Rao, C.R. (1973). *Linear Statistical Inference and its Applications*. John Wiley & Sons, New York.

Reisen, V., Rodrigues, A., and Palma, W. (2006). Estimation of seasonal fractionally integrated processes. *Comput. Statist. Data Anal.*, **50**, 568–582.

Resnick, S. (1992). *Adventures in Stochastic Processes*. Birkhäuser, Boston.

Ripple, W. (1957). The capacity of storage reservoirs for water supply. *Proc. Inst. Civ. Eng.*, **71**, 270–278.

Robinson, P.M. (1978). Statistical inference for a random coefficient autoregressive model. *Scand. J. Statist.*, **5**, 163–168.

Robinson, P.M. (1991). Testing for strong serial correlation and dynamic conditional heteroskedasticity in multiple regression. *J. of Econometrics*, **47**, 67–84.

Robinson, P.M. (1994). Semiparametric analysis of long-memory time series. *Ann. Statist.*, **22**, 515–539.

Robinson, P.M. (1995a). Log-periodogram regression of time series with long range dependence. *Ann. Statist.*, **23**, 1048–1072.

Robinson, P.M. (1995b). Gaussian semiparametric estimation of long range dependence. *Ann. Statist.*, **23**, 1630–1661.

Robinson, P.M. (1997). Large-sample inference for nonparametric regression with dependent errors. *Ann. Statist.*, **25**, 2054–2083.

Robinson, P.M. (2001). The memory of stochastic volatility models. *J. of Econometrics*, **101**, 195–218.

Robinson, P.M. (2003). *Long-memory Time Series.* Advanced Texts in Econometrics. Oxford University Press, Oxford.

Robinson, P.M. (2005a). Efficiency improvements in inference on stationary and nonstationary fractional time series. *Ann. Statist.*, **33**, 1800–1842.

Robinson, P.M. (2005b). Robust covariance matrix estimation: HAC estimates with long memory/antipersistence correction. *Econometric Theory*, **21**, 171–180.

Robinson, P.M. and Henry, M. (1999). Long and short memory conditional heteroscedasticity in estimating the memory parameter of levels. *Econometric Theory*, **15**, 299–336.

Robinson, P.M. and Hidalgo, F.J. (1997). Time series regression with long-range dependence. *Ann. Statist.*, **25**, 77–104.

Robinson, P.M. and Marinucci, D. (2001). Narrow-band analysis of nonstationary processes. *Ann. Statist.*, **29**, 947–986.

Rosenblatt, M. (1952). Limit theorems associated with variants of the von Mises statistic. *Ann. Math. Statist.*, **23**, 617–623.

Rosenblatt, M. (1961). Independence and dependence. In: *Proceedings of the Fourth Berkeley Symp. Math. Statist. Probab.*, 411–443. University of California Press, Berkeley.

Rozanov, Yu.A. (1967). *Stationary Random Processes.* Translated from the Russian by A. Feinstein. Holden-Day, San Francisco.

Samorodnitsky, G. (2007). *Long Range Dependence. Foundations and Trends in Stochastic Systems*, **1** (3). (Now Publishers, Inc., Hanover, MA).

Samorodnitsky, G. and Taqqu, M.S. (1994). *Stable Non-Gaussian Random Processes. Stochastic Models with Infinite Variance.* Chapman & Hall, New York.

Sansone, G. (1959). *Orthogonal Functions.* Interscience Publishers, New York.

Seber, G.A.F. and Wild, C.J. (1989). *Nonlinear Regression.* J. Wiley and Sons, New York.

Shimotsu, K. and Phillips, P.C.B. (2005). Exact local Whittle estimation of fractional integration. *Ann. Statist.,* **33**, 1890–1933.

Shin, Y. and Schmidt, P. (1992). The *KPSS* stationarity test as a unit root test. *Econ. Lett.,* **38**, 387–505.

Soulier, P. (2001). Moment bounds and central limit theorem for functions of Gaussian vectors. *Statist. Prob. Lett.,* **54**, 193–203.

Sowell, F. (1992). Maximum likelihood estimation of stationary univariate fractionally integrated time series models. *J. of Econometrics,* **53**, 165–188.

Stout, W. (1974). *Almost Sure Convergence.* Academic Press, New York.

Stute, W. (1997). Nonparametric model checks for regression. *Ann. Statist.,* **25**, 613–641.

Stute, W., Thies, S., and Zhu, L. (1998). Model checks for regression: an innovation process approach. *Ann. Statist.,* **26**, 1916–1934.

Stute, W. and Zhu, L. (2002). Model checks for generalized linear models. *Scand. J. Statist.,* **29**, 535–545.

Sun, Y. and Phillips, P.C.B. (2003). Nonlinear log-periodogram regression for perturbed fractional processes. *J. of Econometrics,* **115**, 355–389.

Surgailis, D. (1982). Zones of attraction of self-similar multiple integrals. *Lithuanian Math. J.,* **22**, 327–340.

Surgailis, D. (2000). Long-range dependence and Appell rank. *Ann. Probab.,* **28**, 478–497.

Surgailis, D. (2003). Non-CLTs: U-statistics, multinomial formula and approximations of multiple Itô–Wiener integrals. In: Doukhan, P., Oppenheim, G., and Taqqu, M.S. (eds.), *Theory and Applications of Long-range Dependence,* 129–142. Birkhäuser, Boston.

Surgailis, D. (2004). Stable limits of sums of bounded functions of long-memory moving averages with finite variance. *Bernoulli,* **10**, 327–355.

Surgailis, D. (2008). A quadratic ARCH(∞) model with long memory and Lévy stable behavior of squares. *Adv. Appl. Probab.,* **40**, 1198–1222.

Surgailis, D. and Viano, M.-C. (2002). Long memory properties and covariance structure of the *EGARCH* model. *ESAIM: Probability and Statistics,* **6**, 311–329.

Surgailis, D., Teyssière, G., and Vaičiulis, M. (2008). The increment ratio statistic. *J. Multiv. Anal.,* **99**, 510–541.

Szegö, G. (1975). *Orthogonal Polynomials.* American Math. Society, New York.

Taniguchi, M. and Kakizawa, Y. (2000). *Asymptotic Theory of Statistical Inference for Time Series.* Springer-Verlag, New York.

Taqqu, M.S. (1975). Weak convergence to fractional Brownian motion and to the Rosenblatt process. *Z. Wahr. verw. Gebiete*, **31**, 287–302.

Taqqu, M.S. (1977). Law of the iterated logarithm for sums of non-linear functions of Gaussian variables that exhibit a long range dependence. *Z. Wahr. verw. Gebiete*, **40**, 203–238.

Taqqu, M.S. (1979). Convergence of integrated processes of arbitrary Hermite rank. *Z. Wahr. verw. Gebiete*, **50**, 53–83.

Taqqu, M.S. (2003). Fractional Brownian motion and long-range dependence. In: Doukhan, P., Oppenheim, G., and Taqqu, M.S. (eds.), *Theory and Applications of Long-range Dependence*, 5–38. Birkhäuser, Boston.

Taqqu, M.S. and Teverovsky, V. (1997). Robustness of Whittle-type estimators for time series with long-range dependence. Heavy tails and highly volatile phenomena. *Comm. Statist. Stochastic Models*, **13**, 723–757.

Taylor, S.J. (1994). Modelling stochastic volatility: a review and a comparative study. *Mathematical Finance*, **4**, 183–204.

Terrin, N. and Taqqu, M.S. (1990). A noncentral limit theorem for quadratic forms of Gaussian stationary sequences. *J. Theor. Probab.*, **3**, 449–475.

Teverovsky, V. and Taqqu, M.S. (1997). Testing for long range dependence in the presence of shifting means or a slowly declining trend, using a variance-type estimator. *J. Time Series Anal.*, **18**, 279–304.

Teverovsky, V., Taqqu, M.S., and Willinger, W. (1998). Stock market prices and long-range dependence. *Finance & Stochastics*, **3**, 1–13.

Teyssière, G. and Kirman, A., eds. (2007). *Long Memory in Economics*. Springer-Verlag, Berlin.

Titchmarsh, E.C. (1986). *Introduction to the Theory of Fourier Integrals (3rd ed.)*. Chelsea Publishing Company, New York.

Tsay, R.S. (2002). *Analysis of Financial Time Series*. Wiley Series in Probability and Statistics. John Wiley & Sons, New York.

Tsay, W.J. (1998). On the power of Durbin–Watson statistic against fractionally integrated processes. *Econometric Reviews*, **17**, 361–386.

Velasco, C. (1999a). Gaussian semiparametric estimation of nonstationary time series. *J. Time Series Anal.*, **20**, 87–127.

Velasco, C. (1999b). Non-stationary log-periodogram regression. *J. of Econometrics*, **91**, 325–371.

Velasco, C. and Robinson, P.M. (2000). Whittle pseudo-maximum likelihood estimates for non-stationary time series. *J. Amer. Statist. Assoc.*, **95**, 1229–1243.

von Bahr, B. and Esséen, C.G. (1965). Inequalities for the rth absolute moment of a sum of random variables, $1 \le r \le 2$. *Ann. Math. Statist*, **36**, 299–303.

von Neumann, J. (1941). Distribution of the ratio of the mean square successive difference to the variance. *Ann. Math. Statist.*, **12**, 367–395.

Watson, G. S. (1961). Goodness-of-fit tests on a circle. *Biometrika*, **48**, 109–114.

White, H. (1980). A heteroscedasticity-consistent covariance matrix estimator and a direct test for heteroscedasticity. *Econometrica*, **48**, 817–838.

Whittle, P. (1953). Estimation and information in stationary time series. *Ark. Mat.*, **2**, 423–443.

Willinger, W., Paxson, V., Reidi, R., and Taqqu, M.S. (2003). Long-range dependence and data network traffic. In: Doukhan, P., Oppenheim, G., and Taqqu, M.S. (eds.), *Theory and Applications of Long-range Dependence*, 373–408. Birkhäuser, Boston.

Wold, H. (1938). *A Study in the Analysis of Stationary Time Series*. Diss. Stockholm, Uppsala.

Woodward, W.A., Cheng, Q.C., and Gray, H.L. (1998). A k-factor GARMA long-memory model. *J. Time Series Anal.*, **19**, 485–504.

Wu, W.B. (2003). Empirical processes of long-memory sequences. *Bernoulli*, **9**, 809–831.

Wu, W.B. (2007). M-estimation of linear models with dependent errors. *Ann. Statist.*, **35**, 495–521.

Wu, W.B. and Shao, X. (2007). A limit theorem for quadratic forms and its applications. *Econometric Theory*, **23**, 930–951.

Wu, W.B. and Woodroofe, M. (2004). Martingale approximations for sums of stationary processes. *Ann. Probab.*, **32**, 1674–1690.

Yajima, Y. (1988). On estimation of a regression model with long-memory stationary errors. *Ann. Statist.*, **16**, 791–807.

Yajima, Y. (1991). Asymptotic properties of the LSE in a regression model with long-memory stationary errors. *Ann. Statist.*, **19**, 158–177.

Zaffaroni, P. (2004). Contemporaneous aggregation of linear dynamic models in large economies. *J. of Econometrics.* **120**, 75–102.

Zygmund, A. (2002). *Trigonometric Series, Vols I, II, (3rd ed.)*. Cambridge University Press, Cambridge.

Author Index

Abadir, K.M., 263, 275, 278, 283
Abramowitz, M., 503
Adenstedt, R.K., 274
Amsler, C., 304
An, H.-Z., 485
Andel, J., 185
Andrews, D.W.K., 273, 283, 304
Arcones, M.A., 100
Arteche, J., 258, 264
Avram, F., 29, 106, 110, 337, 338

Bailey, N., vii
Baillie, R.T., vi, 264, 394, 412, 446,
    463, 482
Bardet, J.-M, 304
Bates, D.M., 394
Beltrao, K.I., 264
Beran, J., vi, 201, 224, 375
Bhansali, R., 142, 173
Bhattacharya, R.N., 304
Billingsley, P., 74–76, 306
Bingham, N., 18, 22
Bollerslev, T., 53, 394, 446, 482
Booth, G.G., 382, 483
Box, G.E.P., v, 175
Breidt, F.J., 256
Breuer, P., 100
Brillinger, D.R., 62, 525

Brockwell, P.J., vi, 15, 17, 131, 157,
    192, 206
Brodsky, J., 263, 264
Buckley, M.J., 448

Carriquiry, A., 224
Chan, N.H., 191
Chen, W.W., 263
Cheng, B., 485
Cheng, Q.C., 191
Cheung, Y.W., 304, 383, 446, 483
Chung, C.F., 189, 191
Craigmile, P.F., 191
Cramér, H., 270
Crato, N., 256, 304
Csörgő, S., 445, 486

Dacunha-Castelle, V., 201
Dahlhaus, R., vii, 221, 274
Dalla, V., vii, 263, 264
Davies, R.B., 191, 258
Davis, R.A., vi, 15, 17, 54, 131, 157,
    192, 206
Davydov, Y.A., 88
de Lima, P.J.F., 256, 304
Dedecker, J., 34
Dehling, H., vi, 4, 307, 308, 328
Deo, R., 264, 275

Ding, Z., vi
Distaso, W., 263, 275, 278
Dobrushin, R.L., vii, 106, 330
Doukhan, P., vi, 34
Durbin, J., 509, 510

Embrechts, P., 52, 54
Engel, C., 482
Engle, R.F., 53
Esséen, C.G., 30

Fan, J., 445
Feinsilver, P.J., 527
Fermin, L., 201
Fox, R., 161, 173, 221

Gastwirth, J.L., 357
Geweke, J., 225, 263
Ghosh, S., 201
Giraitis, L., 29, 54, 62, 100, 106, 142,
    173, 191, 221, 263, 275, 278, 288,
    296, 298, 304, 328, 375, 394, 527
Goetzmann, W.N., 304
Goldie, C., 18, 22
Gradshteyn, I.S., 42, 176, 178
Granger, C.W.J., v, vi, 174, 180, 197
Gray, H.L., 185, 191
Guo, H., 429, 445, 465, 479, 482,
    484, 486, 518
Gupta, V.K., 304
Guttorp, P., 142

Hajék, J., 373
Hall, P., 29, 30, 263, 274, 429
Hannan, E.J., 157, 206, 221, 283,
    394
Hart, J., 429, 447
Harte, D.S., 191, 258
Harvey, A.C., 256, 264
Hassler, U., 191
Haubrich, J.G., 201

Henry, M., 264, 273, 304
Heyde, C.C., 29, 30
Hidalgo, J., 191, 263, 389
Ho, H.-C., 106, 328, 350
Hosking, J.R.M., v, 175, 180, 185
Hsing, T., 106, 328, 350
Huber, P.J., 357, 395
Hurst, H.E., 222, 223, 304
Hurvich, C.M., 54, 258, 263, 264,
    275

Ibragimov, I.A., 34, 63, 88
Ivanov, A.V., 275

Jaeckel, L., 357
Jenkins, G.M., v, 175
Jennrich, R.I., 394
Jin, S., 283
Jing, B.-Y., 274
Jowett, G.H., 283
Joyeux, R., v, 175, 180

Kadelka, D., 270
Kaen, R.R., 383
Kakizawa, Y., vi
Kapetanios, G., 264
Khmaladze, E.V., 454, 485
Kiefer, J., 291
Kirman, A., vi, 304
Kliemann, W., 224
Klüppelberg, C., 54
Kokoszka, P., 53, 142, 173, 288
Kontrus, K., 304
Kottegoda, N.T., v
Koul, A., vii
Koul, H.L., 76, 106, 263, 306, 313,
    328, 337, 350, 351, 357–359, 361,
    369, 374, 375, 378, 382, 394, 412,
    429, 445, 453–455, 463, 465, 479,
    482, 484–486, 505, 513, 518
Koul, S., vii

Koveos, P.E., 383
Krämer, W., 304
Kuiper, N.H., 289, 290
Künsch, H., 263
Kwiatkowski, D., 298, 304

Léon, J.R., 34
Lahiri, S.N., 131, 274
Lamperti, J.W., 51, 77
Lang, G., 34
Lavancier, F., 304
Lawrance, A.J., v
Lee, H.S., 304
Lehmann, E.L., 305, 357, 373
Leipus, R., vii, 53, 54, 191, 201, 288, 296
Leonenko, N.N., 275
Leonov, V.P., 62, 525
Li, L., 376
Li, W., 224
Lindner, A.M., 53
Linnik, Yu.V., 34, 63, 88
Lo, A., 201, 288, 304, 383, 483
Lobato, I., 263, 264, 304
Lockhart, R.A., 142
Loève, M., 9
Louhichi, S., 34

Maejima, M., 52
Major, P., 100, 106, 330
Mandelbrot, B.B., v, 223, 224, 304
Marinucci, D., 131
Masry, E., 486
Maynard, A., 394, 446, 482
Merlevède, F., 88
Mielniczuk, J., 445, 486
Mikosch, T., vi, 54, 131
Monahan, J.C., 283
Móricz, F., 128
Moulines, E., 258, 263
Mukherjee, K., 313, 328, 374–376

Natanson, I.P., 29, 32
Nelson, D.B., 54, 256
Newey, W.K., 283
Nielsen, M.Ø., 390
Nourdin, I., 100
Nualart, D., 173

Oppenheim, G., vi, 201

Palma, W., vi, 191
Park, K., 54
Paxson, V., vi
Peccati, G., 62, 100, 173
Peligrad, M., 88
Philippe, A., 201, 296, 304
Phillips, P.C.B., vii, 88, 131, 263, 264, 283, 394, 446, 482
Ploberger, W., 304
Podolskij, M., 100
Pollard, D., 74, 75
Porter-Hudak, S., 191, 225, 263
Prieur, C., 34
Puplinskaitė, D., 54, 201

Rao, C.R., 395, 488
Reidi, R., vi
Reisen, V., 191
Resnick, S., 54, 454
Ripple, W., 222
Robinson, P.M., vi, vii, 53, 54, 122, 131, 157, 191, 201, 221, 226, 256, 263, 264, 271, 273, 275, 283, 304, 389, 445
Rodrigues, A., 191
Rootzén, H., 54
Rosenblatt, M., 106, 291, 518
Rozanov, Yu.A., 29, 32
Rubin, H., 357
Ruiz, E., 264
Ryzhik, I.M., 42, 176, 178

Samarov, A., 264
Samorodnitsky, G., vi, 45, 52, 54
Sansone, G., 29
Schmidt, P., 304
Schutzner, M., 201
Seber, G.A.F., 394, 395
Shao, X., 173
Shephard, N., 264
Shimotsu, K., 263
Shin, Y., 304
Shiryaev, A.N., 62, 525
Solo, V., 88
Sørensen, M., vi
Soulier, P., 54, 258, 263, 473, 474
Sowell, F., 185
Stegeman, A., 54
Stegun, I., 503
Stout, W., 32, 179, 199
Stute, W., 453, 455, 485, 486
Sun, Y., 264, 273, 283
Surgailienė, R., vii
Surgailis, D., 29, 54, 62, 100, 106,
    110, 173, 201, 221, 256, 275, 304,
    328, 337, 339, 350, 378, 382, 394,
    412, 463, 505, 518, 527
Szegö, G., 186

Taniguchi, M., vi
Taqqu, M.S., vi, vii, 4, 29, 45, 52,
    54, 62, 88, 106, 110, 161, 173, 203,
    221, 224, 304, 307, 308, 328, 337,
    338, 518
Taylor, S.J., 264
Terrin, N., 173
Teugels, J., 18, 22
Teverovsky, V., 203, 304
Teyssière, G., vi, 304
Teyssière, G., 288

Thies, S., 453
Titchmarsh, E.C., 29
Tsay, R.S., 506
Tsay, W.J., 304
Turlach, B., 263

Utev, S., 88

Vaičiulis, M., 304
van Ness, J.W., v
Velasco, C., 221, 263
Viano, M.-C., 201, 256, 275
von Bahr, B., 30
von Neumann, J., 447

Wallis, J.R., v, 223
Watson, G.S., 289, 292
Watts, D.G., 394
Waymire, E., 304
West, K.D., 283
White, H., 283
Whittle, P., 205
Wild, C.J., 394, 395
Willinger, W., vi, 54, 304
Wold, H., 38
Woodroofe, M., 88
Woodward, W.A., 185, 191
Wu, W.B., 88, 173, 350, 351

Yajima, Y., 274
Yao, Q., 445
Yu, C., 224

Zaffaroni, P., 201
Zhang, N.-F., 185
Zhu, L., 453, 485
Zygmund, A., 7, 9, 22

Subject Index

Aggregation, 197, 200
Aggregation procedure, 200
α-mixing, 33
Appell polynomials, 27, 337
Appell rank, 28
Appell rank of ψ, 339
ARCH(∞) model, 52
ARCH(p) model, 53
ARFIMA$(0, d, 0)$ process, 175
ARFIMA(p, d, q) process, 180
ARMA(p, q) process, 16
Assumption FDD, 267
Asymptotic expansion, 328
Asymptotic normality of Whittle estimator, 215
Asymptotic uniform linearity of M scores, 361
Asymptotic uniform linearity of R scores, 369
Asymptotically first-order equivalent statistics, 353
Autocorrelation function, 10
Autocovariance function, 9

Bandwidth, 229
Bartlett approximation, 126
Bochner's theorem, 10
Brownian motion, 49

Cholesky decomposition, 493
CLT for martingale-difference arrays, 29
CLT for LS estimator, 359
CLT for sums of a linear process, 62
Consistency of local Whittle estimator, 230
Consistency of Whittle estimator, 210
Covariance stationary process, 9
Cramér–von Mises-type test, 506
Cramér–Wold device, 61
Cross validation estimator, 470
Cumulant, 56
Cumulant criterion for CLT, 61
Cumulants of Wiener–Itô integrals, 541

Davies–Harte algorithm, 191
Diagram formulas, 519
Dirichlet kernel, 8
Disaggregation, 201
Discrete Fourier transform, 111
Discrete multiple integral, 534
Dispersion, 358
Double Wiener–Itô integral, 410
Durbin–Watson statistic, 304

EGARCH model, 256
Empirical process, 305
Ergodic theorem, 32
Estimators of the long-run variance, 275

Forward premium, 446
Fractional Brownian motion, 49
Fractional Gaussian noise, 50
Frequency domain, 34

GARCH(p, q) model, 53
GARMA models, 185
GARMA($0, d, 0$) process, 186
GARMA(p, d, q) model, 185
Gaussian process, 9
Gaussian subordinate process, 313
Geweke–Porter-Hudak estimator, 226

H-rank, 25
HAC estimator, 275
Hermite polynomials, 22
Hermite process, 106, 543
Hermite rank, 24
Hermite rank of a family of functions, 309
Heteroscedastic non-parametric regression, 415
Hodges–Lehmann estimator, 354
Huber M estimator, 354
Huber score, 356
Hurst exponent, 223

I(d) process, 36
Innovations, 38

Joint cumulants, 57

Kernel-type estimators, 415
Kolmogorov formula, 205

Kolmogorov statistic, 289
Kolmogorov test, 506
KPSS statistic, 288

Lack-of-fit test, 447
LAD estimator, 356, 395
Lamperti theorem, 51
LARCH process, 53
Least-squares estimator, 355
Leonov–Shiryaev formulas, 59
Lindeberg–Feller condition, 133
Linear filter, 13
Linear process, 38
Linear regression model, 353
Linear trend, 265
Linear trend model, 266
Lipschitz class Λ_α, 8
Lobato–Robinson test, 285, 304
Local Whittle estimator, 228
Log-periodogram estimator, 225
Long memory, 1, 34
Long-run variance, 44
Lusin's theorem, 32

M estimators, 355
m-dependent, 33
MAC estimator, 277
Marked residual empirical process, 447
Martingale, 13
Martingale differences, 12
Martingale-difference errors, 449
Memory parameter, 36
Method of moments for CLT, 56
Mixture density, 198
Modified R/S test, 288
Moments of Wiener–Itô integrals, 541
Moving-average process, 38
Multivariate Appell polynomial, 520

Negative memory, 34

Non-central limit theorem, 55, 101, 329

Non-linear regression model, 395

Normal score estimator, 357

Omnibus test, 506

Optimal bandwidth m_{opt}, 244

Parametric regression model with random design, 352

Parseval's identity, 8

Partial-sum process, 50

Periodogram, 15, 111

Polynomial form, 107

R estimator, 356

R/S estimation method, 222

R/S test, 287

Regularly varying function, 18

Residual empirical process, 512

Rosenthal's inequality, 30

Score function, 356

Self-similar process, 46

Semiparametric process, 202

Short memory, 34

Signal plus noise process, 245

Slowly varying function, 18

Slutsky's theorem, 62

Spectral density, 11

Spectral distribution function, 10

Spectral representation, 10

Spot returns, 446

Stochastic volatility process, 54

Strictly stationary process, 9

Subordinated process, 88, 89

Testing a subhypothesis, 488

Testing for breaks, 293

Testing for long memory, 285

Tightness criterion, 75

Time domain, 34

Transfer function, 13

Translation invariance of periodogram, 133

Uniform consistency rate, 416, 425

Uniform design, 415

Uniform reduction principle, 4, 307

Upper bound for variances, 82

V/S test, 295

van der Waard estimator, 357

Volatility models, 255

von Neumann test, 448

Watson statistic, 289

Weak convergence in uniform metric, 74

Weighted periodograms, 132

Weighted residual empirical process, 305

Weighted sums of ε_j^2, 87

Weighted sums of m-dependent process, 80

White noise, 12

Whittle estimators, 3, 205, 273, 382

Wick products, 519

Wilcoxon estimator, 357

Wold decomposition, 38

Zygmund slowly varying function at 0, 21

Zygmund slowly varying function at ∞, 19